W9-AWJ-886

Precalculus Mathematics
A Functions Approach

Precalculus Mathematics
A Functions Approach

Floyd F. Helton

University of the Pacific

Margaret L. Lial

American River College

Scott, Foresman and Company
Glenview, Illinois

Dallas, Tex. Oakland, N.J. Palo Alto, Cal. Tucker, Ga. London, England

Cover Illustration As the smaller circle, with diameter one-half that of the larger circle, rolls along the internal circumference of the larger circle, its diameter envelops the four-pointed curve shown in white, called an astroid. This curve is related to the cycloid described on p. 64, as its method of generation suggests. The same curve would be traced out by a point on the circumference of a circle with diameter one-fourth that of the larger circle, as the smaller circle rolled along the inner circumference of the larger circle.

Section 8.6 on Linear Programming was adapted from pp. 505–509 of *Algebra and Trigonometry,* 2e, by Margaret L. Lial and Charles D. Miller. Copyright © 1980 by Scott, Foresman and Company.

Library of Congress Cataloging in Publication Data

Helton, Floyd Franklin,
Precalculus mathematics.

Includes index.
1. Mathematics—1961– . 2. Functions.
I. Lial, Margaret L. II. Title.
QA39.2.H4 1983 515 82-20506
ISBN 0-673-15507-2

Printed in the United States of America.
23456-RRC-8786858483

Preface

Precalculus classes usually include students from various disciplines. This text was developed in response to the needs of these diverse groups of students. Topics included in the text were selected to provide a solid foundation for those going on to calculus as well as to satisfy the needs of students taking the course as part of the general curriculum.

Key Features

- Examples are often annotated to help prepare students for the exercises. (For example, see pp. 4 and 94.)
- Numerous realistic applications appear throughout the text in discussions, examples, and exercise sets. (See pp. 149–51 and 211–15.)
- Important concepts are introduced and illustrated by examples. (See pp. 34–35, 57–58.)
- Each section is followed by an extensive exercise set. (For example, see pp. 161–63.)
- Each chapter ends with a comprehensive set of review exercises (see pp. 70–71, 133–35) and a set of miscellaneous exercises (see pp. 170–72).
- More than 3000 exercises throughout the text give students ample opportunity to master text material.
- Calculator use is integrated into the text. (For example, see pp. 75 and 141.)
- More than 600 detailed figures, most of them two-color, illustrate important concepts and serve as problem-solving tools. (See pp. 67 and 114–15.)
- Historical notes appear throughout the text. (For example, see pp. 69 and 276.)
- Extensive tables, with instructions for their use (including interpolation), are given in the appendix.

Organization

The central theme of the text is that of the function—its application and role in mathematics. The basic properties of real numbers, intervals, and inequalities are discussed in Chapter 1 along with an introduction to the coordinate plane. The general concept of a function is more thoroughly developed in Chapter 2. The significant features of the elementary functions, first discussed here, are specifically related to the different classes of functions in the following chapters.

This approach permits the use of general function concepts as tools both for problem solving and for the study of any particular class of functions and their related graphs. For example, the idea of the inverse of a function is developed in Chapter 2 and then used in the discussion of exponential functions. Similarly, the notion of a periodic function is introduced here and then related to the circular and trigonometric functions in Chapters 5 and 6. This repeated use of the properties of functions, as each new class is introduced, provides continual reinforcement of the essential ideas of the function.

Separate chapters on algebraic functions (Chapter 3), exponential and logarithmic functions (Chapter 4), circular functions (Chapter 5), and trigonometric functions (Chapter 6) are included. Chapter 7 gives a thorough development of all essential topics of analytic geometry in the plane. Finally, there are individual chapters on Systems of Equations and Inequalities (Chapter 8), Complex Numbers (Chapter 9), and Sequences and Series (Chapter 10).

Applications

In order to stimulate student interest and illustrate the role of mathematics in everyday life, a larger number and greater variety of simple applications are included in this text than in most precalculus texts. This is particularly true with respect to the quadratic functions (maximum-minimum), the exponential functions (law of growth and decay), the circular functions (periodic phenomena), and the relationship of the conic sections to applications involving natural phenomena.

Algebra Review

A well integrated review of elementary algebra is provided. Some basic ideas and tools are presented in Chapter 1. Additional algebraic concepts such as factoring, solving equations and inequalities, and operations with algebraic expressions, are then incorporated at various points as part of the general development.

Pedagogical Aids

- Mathematical terms are in boldface when they are first used and defined.
- Key definitions and results are identified.
- Important formulas and statements are printed in color.
- Explanatory comments (in color) clarify steps in examples and procedures.
- Chapter review exercises provide a representative sample of exercises to work as preparation for chapter examinations.
- Answers to all odd-numbered exercises are at the back of the text.

Instructor's Guide

The Instructor's Guide to the text contains suggestions for its use, answers to all the even-numbered exercises, detailed solutions to selected exercises, and hints for working many others. The guide also contains an extensive bank of carefully graded test questions. These test items are keyed to chapter sections for use in preparing chapter tests, final examinations, etc. Additionally, there are short quizzes, arranged by topic, for quick testing on significant topics or on a series of related topics.

Acknowledgments

The authors owe a debt of gratitude to the many students who endured temporary and inadequate text materials as these ideas were being developed and class-tested.

Thanks are also due to the reviewers of the text for their careful reading of the initial draft and for their suggestions as the project developed. They include: Professors Donald F. Devine, Western Illinois University; Gregory M. Dotseth, University of Northern Iowa; Joe W. Fitzpatrick, University of Texas at El Paso; Louis Hoelzle, Bucks County Community College; Harvey B. Keynes, University of Minnesota, Twin Cities; Russell C. Magnusun, Bellevue Community College; Larry Riemersma, Hillsborough Community College; Philip W. Schaefer, University of Tennessee; Betty Travis, University of Texas at San Antonio; and Jack Wadhams, Golden West College. The authors are particularly grateful to Professor Albert G. Fadell, State University of New York at Buffalo, who read several versions of the manuscript and made many useful suggestions for improvement.

Finally, the staff of Scott, Foresman and Company has been most helpful. Special thanks go to Margaret Prullage and Bill Poole for their genuine interest and careful work with the endless details of preparation for publication.

Floyd F. Helton
Margaret L. Lial

Contents

6 The Trigonometric Functions and Applications 233

7 Coordinate Geometry in the Plane 280

8 Systems of Equations and Inequalities 345

9 Complex Numbers 391

10 Sequences and Series 421

Appendixes

Precalculus Mathematics
A Functions Approach

1

The Real Number System

We are beginning a study of the elementary functions of mathematics. A good understanding of these functions is essential for the study of statistics, economics and business administration, the sciences (both biological and physical), engineering, and, of course, mathematics itself.

The notion of a function occurs everywhere in the applications of mathematics. We are all familiar with *graphs,* which are pictures of functions. The one in Figure 1.1 shows the growth of population in the U.S. in this century. The graph shows that population depends upon the year and increases steadily with time. However, there is more to the idea of function than a dependence relationship. In the next chapter the concept will be made more precise as we study functions in greater detail.

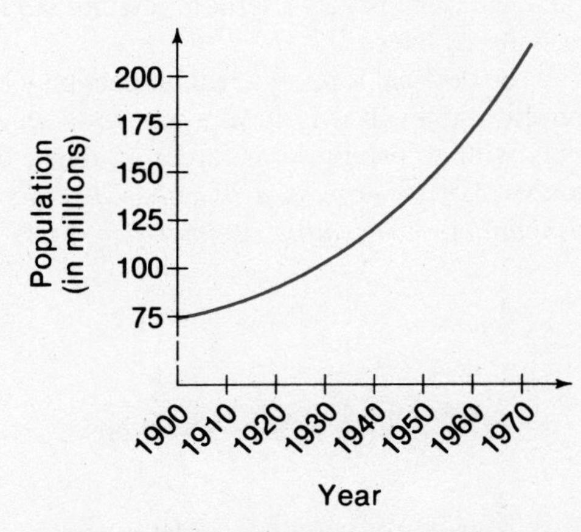

Figure 1.1 Picture of a function

1.1 The Real Numbers

The numbers we ordinarily use are those which can be written in decimal form. They are called **real numbers,** and we will describe some of their important features. To do this it is helpful to represent them geometrically on a line.

In Figure 1.2, two points O and U and a direction on a line have been chosen arbitrarily. We call O the **origin** and identify it with the number 0. Point U, which corresponds to the number 1, determines segment OU of unit length. The direction from O to U (indicated by the arrow at the end of the line), is **positive** and the opposite direction **negative.** Next, the unit length is used to mark off points in both directions corresponding to the positive and negative integers. Similarly, we can identify each of the fractions (positive and negative) with a point on the line by first dividing unit segments into two, three, four, or more, parts, as shown in Figure 1.2.

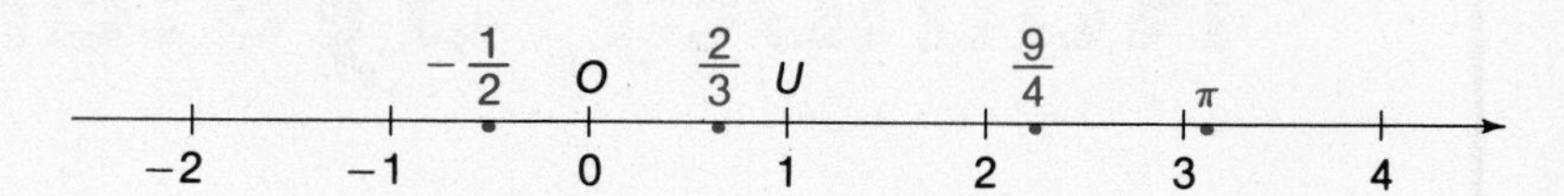

Figure 1.2 The number line

Every integer and every fraction can be expressed as the ratio of two integers, p/q, where q is not zero. Such numbers are called **rational numbers.** However, the rational numbers do not use up all the points of a line. So we let a number be assigned to every point of the line, and those which cannot be expressed as ratios are called **irrational numbers.** Thus, there is a one-to-one pairing of real numbers with points of a line.

When numbers have been assigned to the points of a line, it is called a **number line** or **coordinate line.** Each real number is called the **coordinate** of the point to which it belongs. In practice, the terms *number* and *point* are used interchangeably. Thus, the phrases "the point 3" and "a point x" may refer either to the numbers 3 and x or to the points to which they correspond. This identification of real numbers with points on a line is very useful in relating the ideas of algebra to those of geometry, as we will see later.

The decimal form of a real number may terminate (that is, have a finite number of nonzero digits), it may have a repeating sequence of digits, or it may continue infinitely with no pattern in the order of digits. It is an important fact that *every rational number can be expressed as either a terminating or repeating decimal, and every terminating or repeating decimal represents a rational number*. For example,

$$\frac{3}{4} = 3 \div 4 = 0.75 \qquad \text{or} \qquad 0.75000000\ldots,$$

$$\frac{2}{3} = 2 \div 3 = 0.666\ldots,$$

$$\frac{4}{11} = 4 \div 11 = 0.363636\ldots.$$

As the examples show, a fraction can be expressed as a terminating or repeating decimal by dividing the numerator by the denominator.

The fact that a repeating decimal represents a fraction is illustrated as follows for 0.919191 . . . . Let

$$N = 0.919191 \ldots.$$

Multiply each side of the equality by 100.

$$100N = 91.919191 \ldots$$

Subtract the first line from the second, and then divide.

$$99N = 91.000 \ldots$$

$$N = \frac{91}{99}$$

From these examples it can be seen that *the unending, nonrepeating decimals are irrational numbers and,* conversely, *irrational numbers are unending nonrepeating decimals.* Familiar examples are $\pi = 3.141592654 \ldots$ and $\sqrt{2} = 1.414213562 \ldots$. However, every irrational number (nonrepeating decimal) can be approximated as closely as we wish by a rational number (terminating or repeating decimal). This is what we do when we use the rational number 3.14 (or 3.142, or 3.1416) for π or 1.4 (or 1.41, or 1.414) for $\sqrt{2}$. The electronic calculator does this automatically. For example, the above decimal for π, 3.141592654, was displayed by a hand calculator that gives results to ten places.

Since the real numbers completely fill up the entire number line, we say that the real number system is **complete.** We use this property whenever we assume that there is a number to measure the length of every line segment. This is one of the most important properties of the real numbers. The relationships among the different kinds of real numbers mentioned above are shown in Figure 1.3 below.

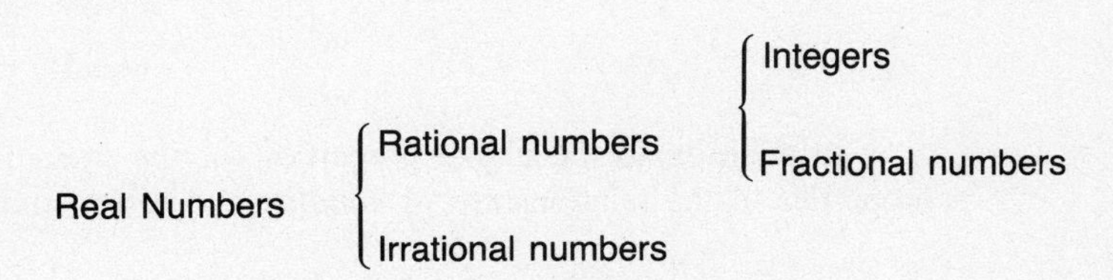

Figure 1.3 The Real Number System

The rules for operating with real numbers are familiar to you. The most important ones are listed here for review and reference.

Properties of the Operations for Real Numbers

If a, b, and c are any real numbers, then:

Closure properties $a + b$ and $a \cdot b$ are real numbers.

Commutative properties	$a + b = b + a$ and $a \cdot b = b \cdot a$.
Associative properties	$(a + b) + c = a + (b + c)$ and $(a \cdot b) \cdot c = a \cdot (b \cdot c)$.
Distributive property	$a \cdot (b + c) = a \cdot b + a \cdot c$ and $(b + c) \cdot a = b \cdot a + c \cdot a$.
Identity properties	$a + 0 = 0 + a = a$ and $a \cdot 1 = 1 \cdot a = a$.
Inverse properties	For each real number a there is a real number, called the **negative** of a and denoted by $-a$, such that $a + (-a) = 0$. For each real number a, except 0, there is a real number, called the **reciprocal** of a and denoted $1/a$, such that $a \cdot (1/a) = 1$.

We agree to write $a(bc)$ for $a \cdot (b \cdot c)$, $a(b + c)$ for $a \cdot (b + c)$, $a - b$ for $a + (-b)$, and a/b for $a \cdot (1/b)$. The following examples illustrate these properties.

EXAMPLE 1 Factor $6x - 3y$.

SOLUTION

$$\begin{aligned} 6x - 3y &= 3(2x) + 3(-y) \\ &= 3\big(2x + (-y)\big) && \text{distributive property} \\ &= 3(2x - y). && \text{the agreement above} \end{aligned}$$

●

EXAMPLE 2 Find the product $(x - y)(x + y)$.

SOLUTION

$$\begin{aligned} (x - y)(x + y) &= (x - y)x + (x - y)y && \text{distributive property} \\ &= (x^2 - xy) + (xy - y^2) && \text{distributive property (again)} \\ &= x^2 + (-xy + xy) - y^2 && \text{associative property for addition} \\ &= x^2 + 0 - y^2 && \text{inverse property for addition} \\ &= x^2 - y^2. && \text{identity property for addition} \end{aligned}$$

●

In addition to the above properties for the operations of algebra we also use properties of the relationships of *equality* and *order* among real numbers.

DEFINITION 1.1 If a and b denote the same number, then we say that ***a* is equal to *b*** and write $\boldsymbol{a = b}$.

The most important properties of the equals relation are the following.

Properties of Equality

Reflexive	$a = a$.
Symmetric	If $a = b$, then $b = a$.

Transitive	If $a = b$ and $b = c$, then $a = c$.
Equal-additions	If $a = b$ and $c = d$, then $a + c = b + d$.
Equal-multiplications	If $a = b$ and $c = d$, then $a \cdot c = b \cdot d$.

These properties are basic to solving equations. We give a simple example.

EXAMPLE 3 Solve $2x - 6 = 0$ for x.

SOLUTION

$$2x - 6 = 0$$

$$(2x - 6) + 6 = 0 + 6 \quad \text{equal-additions}$$

$$2x + (-6 + 6) = 6 \quad \text{associative property and identity for addition}$$

$$2x + 0 = 6 \quad \text{inverse for addition}$$

$$2x = 6 \quad \text{identity for addition}$$

$$\frac{1}{2}(2x) = \frac{1}{2}(6) \quad \text{equal-multiplications}$$

$$x = 3. \quad \bullet$$

Of course, we ordinarily abbreviate all this, but only because we are familiar with these rules.

Order Associating numbers with points on a number line automatically determines an order relation among the real numbers. This is spelled out in the following *geometric* definition of order.

DEFINITION 1.2 Suppose a real number a corresponds to point A and a real number b corresponds to point B. Then, when A is to the left of B (or equivalently, B is to the right of A), we say that ***a* is less than *b*** (or equivalently, ***b* is greater than *a***) and write $\boldsymbol{a < b}$ (or $\boldsymbol{b > a}$).*

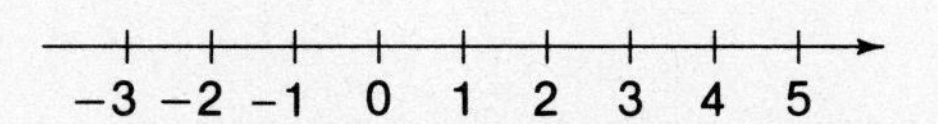

Figure 1.4 Order on the number line

Thus, as shown in Figure 1.4, $2 < 5$, $-3 < -1$, $2 > 0$, $1 > -2$, and so on.

An equivalent *algebraic* definition of order is the following.

*The symbols < and > were first used about 1600.

DEFINITION 1.3 The real number a is less than the real number b, written $a < b$, if and only if there is a positive real number x such that $a + x = b$ or, equivalently, if $b - a$ is a positive number.

For example, $1 < 4$ since $4 - 1 = 3$ is positive,

$-3 < -1$ since $-1 - (-3) = 2$ is positive.

If we wish to indicate that x is either 0 or a positive number, we write $x \geq 0$, read "x is positive or zero," or "x is greater than or equal to 0," or "x is nonnegative." Similar remarks apply to the symbol $x \leq 0$.

From the above definition for $a < b$ we can establish the fundamental algebraic properties of this order relation. Similar properties hold for $\leq$, $>$, and $\geq$.

Properties of Order Relations

Transitive property	If $a < b$ and $b < c$, then $a < c$.
Equal-additions or **Equal-subtractions**	If $a < b$, then $a + c < b + c$ and $a - c < b - c$.
Equal-multiplications or **Equal-divisions**	If $a < b$ and $c > 0$, then $ac < bc$ and $\dfrac{a}{c} < \dfrac{b}{c}$. If $a < b$ and $c < 0$, then $ac > bc$ and $\dfrac{a}{c} > \dfrac{b}{c}$.

As an illustration of the transitive property,

$$2 < 5 \text{ and } 5 < 9, \text{ so } 2 < 9,$$

$$-4 < -2 \text{ and } -2 < -1, \text{ so } -4 < -1.$$

The equal-additions property says that we may add (or subtract) the same number to both members of an inequality without changing the direction of the inequality. For example,

$$\text{if } x - 2 < 5, \text{ then } (x - 2) + 2 < 5 + 2, \quad x < 7;$$

$$\text{if } x + 1 > 3, \text{ then } (x + 1) - 1 > 3 - 1, \quad x > 2.$$

The first equal-multiplications property says that both members of an inequality can be multiplied or divided by the same *positive* number without changing the direction of the inequality. For example,

$$\text{if } \frac{1}{2}x < 1, \text{ then } 2\left(\frac{1}{2}x\right) < 2(1), \quad x < 2;$$

$$\text{if } 2x > 6, \text{ then } \frac{2x}{2} > \frac{6}{2}, \quad x > 3.$$

The second equal-multiplications property says that if both members of an inequality are multiplied or divided by the same *negative* number, the direction of the inequality must be reversed. For example,

$$\text{if} \quad -\frac{1}{2}x < 3, \qquad\qquad \text{if} \quad -2x > 1,$$

$$\text{then} \quad -2\left(-\frac{1}{2}x\right) > -2(3), \qquad\qquad \text{then} \quad \frac{-2x}{-2} < \frac{1}{-2},$$

$$x > -6; \qquad\qquad x < -\frac{1}{2}.$$

Except for multiplication or division by a negative number, these properties are similar to the corresponding properties for equalities.

The following example shows how one of these properties of inequalities is proved. Proofs of the other properties are left to the exercises.

EXAMPLE 4 Prove that if $a < b$, then $a + c < b + c$.

SOLUTION If $a < b$, then, by Definition 1.3, there is a positive number x such that

$$a + x = b.$$

By the equal-additions property,

$$(a + x) + c = b + c.$$

The commutative and associative properties for addition can be used to rewrite the left side as follows:

$$(a + c) + x = b + c.$$

Then, by Definition 1.3,

$$a + c < b + c. \qquad \bullet$$

The following examples illustrate the use of these properties to solve inequalities.

EXAMPLE 5 Solve the inequality $2x - 1 < 5$ for x.

SOLUTION First, write down the given inequality.

$$2x - 1 < 5$$

Then, $\quad (2x - 1) + 1 < 5 + 1 \quad$ equal-additions

$$2x < 6$$

and $\quad x < 3. \quad$ equal-divisions $\bullet$

EXAMPLE 6 Solve the inequality $-3x + 2 < x - 1$ for x.

SOLUTION

$$-3x + 2 < x - 1$$

$$-3x < x - 3 \quad \text{equal-subtractions}$$

$$-4x < -3 \quad \text{equal-subtractions}$$

$$x > \frac{3}{4} \quad \text{equal-divisions}$$

In order to deal more fully with inequalities, we need the notions of set, number interval, and absolute value. These will be discussed in the next two sections.

EXERCISES 1.1

1. Locate approximately on a number line: 1.3, 22/7, $2\frac{3}{4}$, $-(-3)$, $-\frac{5}{2}$, 1.3333 . . . , 0.60000.

2. Which of the following are rational and which irrational? How do you know? $\frac{2}{3}$, -3.5, $\sqrt{3}$, 0.123123123 . . . , 1.23456789101112 . . . , 3.444444 . . . .

Show that each of the following numbers may be represented by a terminating or repeating decimal.

3. $\frac{6}{25}$ **4.** $\frac{5}{8}$ **5.** $\frac{2}{11}$ **6.** $\frac{1}{7}$ **7.** $\frac{3}{10}$

8. $\frac{1}{3}$ **9.** $\frac{1}{9}$ **10.** $\frac{7}{16}$ **11.** $\frac{5}{11}$ **12.** $\frac{4}{33}$

Show that the following repeating decimals are equivalent to the indicated fractions.

13. $0.666\ldots = \frac{2}{3}$

14. $0.222\ldots = \frac{2}{9}$

15. $0.272727\ldots = \frac{3}{11}$

16. $0.324324324\ldots = \frac{12}{37}$

17. $0.142857142857\ldots = \frac{1}{7}$

18. $0.999\ldots = 1$

19. In each of the following nonterminating decimals, enough digits are given to show a pattern. How do you know that they are not rational numbers?
(a) 0.246810 . . . (b) 0.1020406 . . . (c) 0.112233 . . . .

Place the proper symbol ($<$, $=$, or $>$) between the numbers of each pair.

20. 2, 5 **21.** -2, -5 **22.** $\sqrt{2}$, 1 **23.** 6, -1

24. 22/7, π **25.** 3/4, 9/12 **26.** -2, 0 **27.** $\sqrt{3}$, 1.7

28. 2/3, 12/18 **29.** 0.6, 5/8

Each of the following statements holds because of one of the following properties: (a) commutative, (b) associative, (c) distributive, (d) inverse, (e) identity. State which one in each case.

30. $a + (b + c) = (a + b) + c$

31. $a + (b + c) = (c + b) + a$

32. $a + (b + c) = a + (c + b)$

33. $x(y + z) = xy + xz$

34. $(u + v)w = uw + vw$

35. $(x + y) + z = x + (y + z)$

36. $u + (v + w) = v + (w + u)$

37. $4x + 2y = 2(2x + y)$

38. $-x^2 + xy = -x(x - y)$

39. $5 + (-5) = 0$

40. $-4 + 4 = 0$

41. $x + 0 = x$

42. $0 - y = -y$

43. $3\left(\frac{1}{3}\right) = 1$

44. $\left(\frac{1}{x}\right)x = 1$

Use the properties of addition and multiplication to show that each of the following is a true statement. Give any intermediate steps.

45. $x + (y + z) = (z + y) + x$

46. $(x + y) + z = x + (z + y)$

47. $u(v + w) = uv + uw$

48. $(x + y)z = xz + yz$

49. $a \cdot (b - c) = a \cdot b - a \cdot c$

50. $-u(v - w) = -uv + uw$

51. $2 + xy - xy = 2$

52. $3 - [(-7) + 7] = 3$

53. $x\left(1 + \frac{1}{x}\right) = x + 1$

54. $\frac{1}{u}(u - u^2) = 1 - u$

55. $3x - 6y = 3(x - 2y)$

56. $x^2 - 2ax = x(x - 2a)$

Solve the following inequalities for x.

57. $x + 3 < 4$

58. $x - 1 > 2$

59. $x - 2 < 5$

60. $x + 1 > 3$

61. $x + 2 < x - 3$

62. $2x - 3 < 5$

63. $2 < 2x + 4$

64. $4 - 3x < 3 - 4x$

65. $3x - 8 < 2x + 8$

66. $5x - 9 > 3x + 16$

Using the pattern of Example 4, prove the following.

67. If $a < b$, then $a - c < b - c$.

68. If $a < b$ and $c > 0$, then $ac < bc$.

69. If $a < b$ and $c > 0$, then $\frac{a}{c} < \frac{b}{c}$.

70. If $a < b$ and $c < 0$, then $ac > bc$.

71. If $a < b$ and $c < 0$, then $\frac{a}{c} > \frac{b}{c}$.

1.2 Sets, Intervals, and Inequalities

Sets A collection of objects is often treated as a single entity. Some examples are: a class of mathematics students, the collection of all integers, a flock of birds, and so on. The general term for any collection is **set,** and any individual object in the collection is called a **member** (or **element**) of the set. A capital letter is commonly used to name a set and lower-case letters to name its members.

A set may also be identified by a descriptive statement. Thus, we may refer to "the set R of rational numbers." Alternatively, a set may be defined by listing the members enclosed in **set braces,** { }. For example, the set of vowels is

$$\{a, e, i, o, u\}.$$

A set may also be described by using a variable. For example, the set of numbers x such that x is greater than 1 is written

$$\{x \mid x > 1\}.$$

This is called **set-builder notation.**

The equals sign may be used between two different notations for the same set. For example,

the set of positive even integers less than 10

$$= \{2, 4, 6, 8\} = \{x \mid x \text{ an integer, } x \text{ even, and } 0 < x < 10\}.$$

In this book we are primarily concerned with sets of numbers, such as the set I of integers, the set R of rational numbers, and the set $\mathbb{R}$ of real numbers. To say that 3/5 is a member of the set of rational numbers we write (in symbols)

$$\frac{3}{5} \in R \qquad \text{or} \qquad \frac{3}{5} \in \{x \mid x \text{ rational}\}.$$

Similarly, $\sqrt{2} \in \mathbb{R}$. To denote that a given object is *not* a member of a given set, we use the symbol $\notin$. For example,

$$\frac{3}{5} \notin I \qquad \text{and} \qquad \sqrt{2} \notin R.$$

A set may be finite or it may be infinite. The set of positive odd integers less than 7, $\{1, 3, 5\}$, has a finite number of members while the set of all positive odd integers is an infinite set. It is also useful to have the notion of a set with no members, the **empty set** or **null set,** denoted by $\emptyset$. For example, the set of numbers which are both positive and negative is the empty set.

If every member of a set A is also a member of set B, we say that A is a **subset** of B and write $A \subseteq B$. Thus, the set of integers is a subset of the set of rational numbers, and the set of vowels is a subset of the set of letters of the alphabet. Symbolically, we may write

$$I \subseteq R \qquad \text{or} \qquad \{a, e, i, o, u\} \subseteq \{a, b, c, \ldots, x, y, z\}.$$

Given any two sets A and B, the set made up of all members of either A or B is called the **union** of A and B, written $A \cup B$. For example, the union of the set of odd integers and the set of even integers is the set of all integers. The union of the set of vowels and the set of consonants is the set of letters of the alphabet.

From elementary algebra we know that some equations are satisfied by a single number and others by two or more numbers. In the last section we found that an inequality may be satisfied by an infinite set of numbers. In any case we may use the terms **solution** or **solution set** for the collection of all numbers satisfying a given equation or inequality. For example, if $2x = 6$, then $x = 3$; the solution is the single number 3 and the solution set is $\{3\}$. If $2x < 6$, then $x < 3$ and the solution set consists of all real numbers less than 3, that is, the set $\{x \mid x < 3\}$. The ideas and language of sets will be useful in the work to follow.

We are often concerned with only a subset of the real numbers (or part of the number line). In Figure 1.5 several such subsets, called **intervals,** are shown.

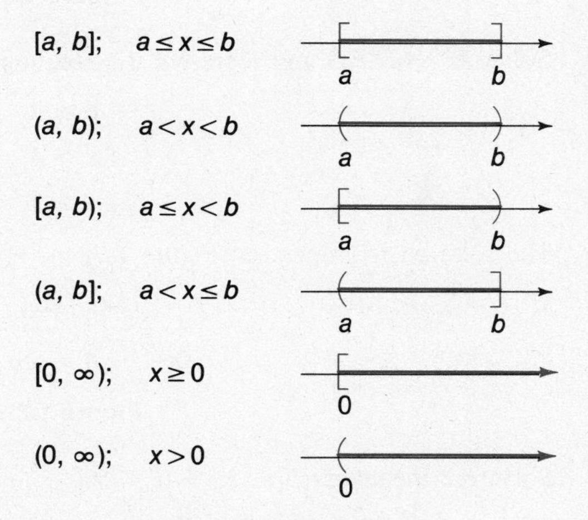

Figure 1.5 Intervals

A square bracket at the endpoint of an interval indicates that the point is included, whereas a parenthesis shows that the point is to be excluded. If both endpoints are included, the interval is called a **closed interval,** denoted $[a, b]$ or $a \le x \le b$, and read "the closed interval from a to b" or "x greater than or equal to a and less than or equal to b." If neither endpoint is included, the interval is called an **open interval,** denoted (a, b) or $a < x < b$ and read "the open interval from a to b" or "x is greater than a but less than b." If only one endpoint is included, the interval is called **half-open** or **half-closed** and written as either $(a, b]$ or $[a, b)$, depending on whether the right or left endpoint is included. The positive real numbers, $\{x \mid x > 0\}$, and the nonnegative real numbers, $\{x \mid x \ge 0\}$, are represented by infinite intervals on the number line and symbolized $(0, \infty)$ and $[0, \infty)$ respectively. The above notation using parentheses or brackets is called **interval notation.**

The expression $a < x < b$ is a compact way of writing three inequalities: $a < x$, $x < b$, and (by the transitive property) $a < b$. The expression $a < x > b$ is ambiguous and should be avoided.

EXAMPLE 1 Solve $2x - 1 < 3$ and show the solution graphically as an interval on the number line.

SOLUTION

$$2x - 1 < 3$$
$$2x < 4 \quad \text{Why?}$$
$$x < 2 \quad \text{Why?}$$

The solution is the infinite open interval shown in Figure 1.6. •

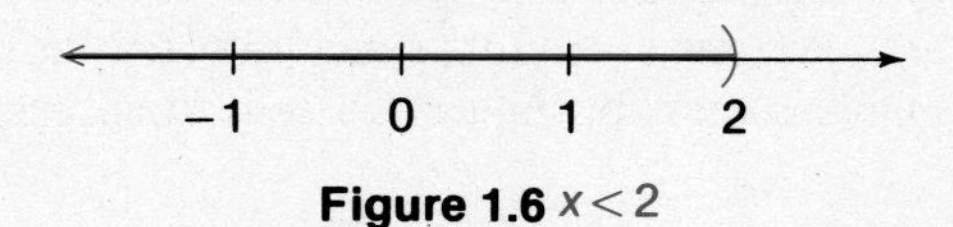

Figure 1.6 $x < 2$

EXAMPLE 2 Solve $2x - 1 \geq 3$ and represent the solution graphically.

SOLUTION

$$2x - 1 \geq 3$$
$$2x \geq 4 \quad \text{Why?}$$
$$x \geq 2 \quad \text{Why?}$$

The solution is graphed in Figure 1.7. •

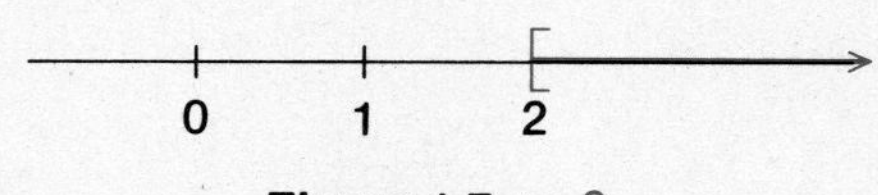

Figure 1.7 $x \geq 2$

EXAMPLE 3 Solve the inequality $5 < 2x + 1 < 9$.

SOLUTION As noted earlier, this represents the inequality $5 < 2x + 1$ and the inequality $2x + 1 < 9$. We must solve both these inequalities.

$$5 < 2x + 1 \qquad 2x + 1 < 9$$
$$2x > 4 \qquad 2x < 8$$
$$x > 2 \qquad x < 4$$

For both inequalities, $2 < x$ and $x < 4$, to be satisfied, $2 < x < 4$. That is, x is between 2 and 4, and the solution set is the open interval (2, 4). See Figure 1.8.

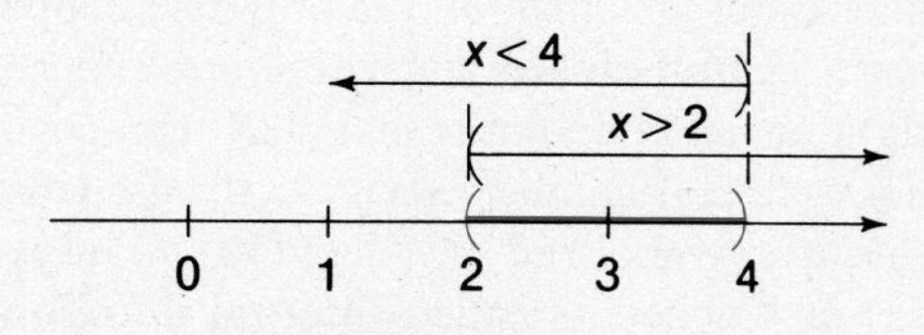

Figure 1.8 $2 < x < 4$

A simpler way to solve this kind of inequality is to combine the two parts, as shown below.

$5 < 2x + 1 < 9$	
$4 < 2x < 8$	subtracting 1 from each member
$2 < x < 4$	dividing each member by 2 ●

The methods used to solve linear inequalities can be extended to inequalities involving products and quotients. For instance, if $p(x)$ and $q(x)$ are expressions involving x, then, given the **product inequality** $p(x) \cdot q(x) > 0$, either $p(x)$ and $q(x)$ are both positive or $p(x)$ and $q(x)$ are both negative. On the other hand, given $p(x) \cdot q(x) < 0$, then either $p(x)$ is positive and $q(x)$ negative, or $p(x)$ is negative and $q(x)$ positive. Similar remarks apply to the fractional inequalities $p(x)/q(x) > 0$ and $p(x)/q(x) < 0$. The following examples illustrate these ideas.

EXAMPLE 4 Solve the inequality $(x - 1)(x + 3) > 0$.

SOLUTION The product is positive, so either both factors are positive or both are negative. The product is zero when $x = 1$ and when $x = -3$. These numbers separate the number line into three intervals:

$$x < -3, \qquad -3 < x < 1, \qquad \text{and} \qquad x > 1.$$

To analyze the situation, determine the sign of the factors in each interval and enter the results in a sign table, as shown below. In this example, if $x < -3$, then $x - 1$ is negative, so there is a minus sign under $x < -3$ and opposite $x - 1$ in the table. Similarly, if $-3 < x < 1$, then $x + 3$ is positive, and we put a plus sign in the table under $-3 < x < 1$ and opposite $x + 3$. The other signs in the table are determined in the same way.

	$x < -3$	$-3 < x < 1$	$x > 1$
$x - 1$	−	−	+
$x + 3$	−	+	+
	↑		↑

From the table we can see that both factors are negative when $x < -3$ and both are positive when $x > 1$. So the product is positive when x is outside the interval $-3 \leq x \leq 1$, that is, when $x > 1$ or when $x < -3$. (See Figure 1.9.) ●

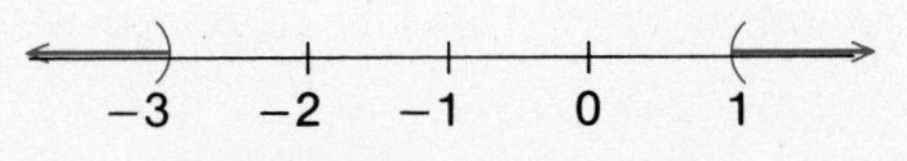

Figure 1.9 $x < -3$ or $x > 1$

EXAMPLE 5 Solve the inequality $(x - 2)(x + 2) < 0$.

SOLUTION Since the product is negative, one factor must be positive and the other negative. The product is zero when $x = -2$ and when $x = 2$, so we treat separately the intervals $x < -2$, $-2 < x < 2$, and $x > 2$. The results are shown in the table.

	$x < -2$	$-2 < x < 2$	$x > 2$
$x - 2$	$-$	$-$	$+$
$x + 2$	$-$	$+$	$+$
		↑	

The table shows that we must have $-2 < x < 2$. See Figure 1.10. •

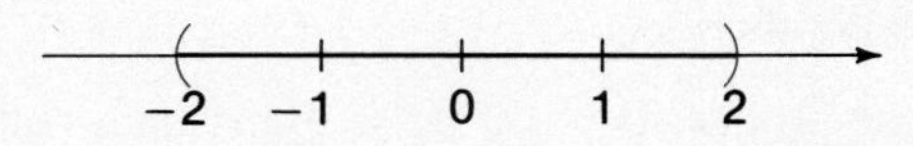

Figure 1.10 $-2<x<2$

In a similar way, we can show that if $x^2 - a^2 < 0$, then $-a < x < a$, and if $x^2 - a^2 > 0$, then $x < -a$ or $x > a$. The method of the last two examples will be used in Section 3.3 to solve quadratic inequalities.

EXAMPLE 6 Solve $\dfrac{2x - 1}{x - 2} < 1$.

SOLUTION We first rewrite this as an equivalent inequality with 0 on one side.

$$\frac{2x - 1}{x - 2} < 1$$

$$\frac{2x - 1}{x - 2} - 1 < 0$$

$$\frac{(2x - 1) - (x - 2)}{x - 2} < 0$$

$$\frac{x + 1}{x - 2} < 0$$

Since the quotient is negative, either the numerator is positive and the denominator is negative, or the denominator is positive and the numerator is negative. The fraction is zero when $x = -1$ and is undefined when $x = 2$. Thus, we consider the intervals

$$x < -1, \qquad -1 < x < 2, \qquad \text{and} \qquad x > 2,$$

and make a sign table.

	$x < -1$	$-1 < x < 2$	$x > 2$
$x + 1$	$-$	$+$	$+$
$x - 2$	$-$	$-$	$+$
		↑	

From the table we see that the signs of the numerator and denominator are different only when $-1 < x < 2$. Thus, the inequality is satisfied on the open interval $(-1, 2)$. See Figure 1.11.

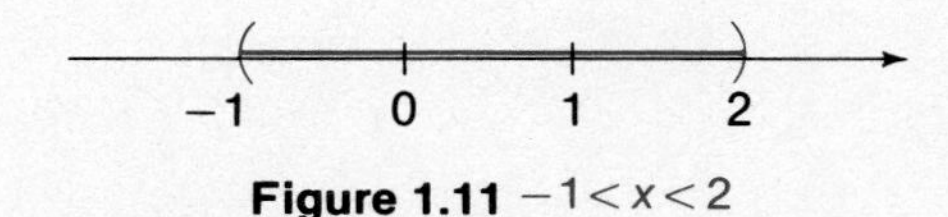

Figure 1.11 $-1<x<2$

Alternate method: We could also solve this inequality without using the sign table, as follows.

Case 1: $x + 1 > 0$ and $x - 2 < 0$

$$x > -1 \qquad x < 2$$

$$-1 < x < 2$$

Case 2: $x + 1 < 0$ and $x - 2 > 0$

$$x < -1 \qquad x > 2$$

Impossible

Since the second case cannot occur, $-1 < x < 2$ is the solution. •

EXAMPLE 7 Solve $\dfrac{2}{x+1} > \dfrac{1}{x}$.

SOLUTION

$$\frac{2}{x+1} > \frac{1}{x}$$

$$\frac{2}{x+1} - \frac{1}{x} > 0$$

$$\frac{2x - (x+1)}{x(x+1)} > 0$$

$$\frac{x-1}{x(x+1)} > 0$$

For this algebraic fraction to be positive, either all three factors in the numerator and denominator are positive or two of them are negative and the other positive. The fraction is zero when $x = 1$ and is undefined when $x = 0$ or $x = -1$. These three numbers -1, 0, and 1 divide the number line into four intervals:

$$x < -1, \qquad -1 < x < 0, \qquad 0 < x < 1, \qquad \text{and} \qquad x > 1.$$

We construct a sign table for the situation.

	$x < -1$	$-1 < x < 0$	$0 < x < 1$	$x > 1$
x	$-$	$-$	$+$	$+$
$x + 1$	$-$	$+$	$+$	$+$
$x - 1$	$-$	$-$	$-$	$+$
		↑		↑

From the table it can be seen that the inequality is satisfied if $-1 < x < 0$ or if $x > 1$ (the word *or* is used to indicate the union of two sets). Thus, the solution set consists of the two intervals $(-1, 0)$ and $(1, \infty)$. This is written in set notation as

$$\{x \mid -1 < x < 0\} \cup \{x \mid x > 1\},$$

or in interval notation as

$$(-1, 0) \cup (1, \infty).$$

See Figure 1.12. ●

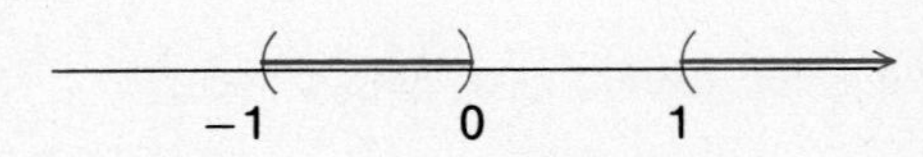

Figure 1.12 $-1 < x < 0$ or $x > 1$

The general procedure in such examples is to rewrite the inequality so it involves a single fractional expression and then analyze the cases for the numerator and denominator. A table like the one shown above is the easiest way to determine the intervals that satisfy the inequality.

EXERCISES 1.2

Identify each of the following sets by listing the members in set braces.

1. The set of U.S. states bordering Mexico
2. The set of odd integers between 4 and 10

3. The set of positive integers less than 2

4. The set of months with 30 days

Give a verbal description of each of the following sets.

5. $\{x, y, z\}$

6. $\{2, 4, 6, 8\}$

7. $\{1, 2, 3, 4, 5, 6, 7, 8, 9, 10\}$

8. {Matthew, Mark, Luke, John}

For the set $A = \{1, 2, 3, 4, 5\}$, give the subset of A whose members are:

9. Numbers less than 4;

10. Even numbers;

11. Multiples of 2;

12. Greater than 5.

Write the solution set of each of the following.

13. $3x = -12$

14. $2x = 5$

15. $4x \leq 8$

16. $5x > 11$

Rewrite each of the following using inequality symbols and set notation, then graph the interval on a number line.

17. x is less than 2

18. x is greater than 3

19. x is between -2 and 3

20. x is less than 1 and greater than -3

21. x is less than 2 but not less than -1

22. x is greater than 0 but less than 4

23. x is negative or zero

24. x is positive

25. x is nonnegative

Graph each of the following inequalities on a number line and give the interval notation for it.

26. $1 < x < 2$

27. $-2 \leq x \leq 1$

28. $-3 \leq x \leq 3$

29. $0 < x < 3$

30. $-1 \leq x < 0$

31. $1 < x \leq 3$

32. $\{x \mid x \leq 2\}$

33. $\{x \mid x \geq 1\}$

34. $\{x \mid x > 0\}$

35. $\{x \mid x < -2\}$

36. $\{x \mid x < 0\} \cup \{x \mid x > 2\}$

37. $\{x \mid x < 1\} \cup \{x \mid x > 3\}$

Solve each of the following inequalities. State which properties of inequalities are used and illustrate the solution with a graph.

38. $2x - 3 < 7$

39. $x + 5 < 2x - 1$

40. $1 < 2x - 3 < 5$

41. $4 < -2x + 1 \leq 7$

42. $\dfrac{1}{x} < 1$

43. $\dfrac{2}{x - 1} > 1$

44. $\dfrac{1}{x + 1} > 2$

45. $\dfrac{1}{x - 2} < \dfrac{1}{x}$

1.3 Absolute Value

A positive number, such as 3, appears on the number line to the right of 0, and a negative number, say -3, is to the left of 0. The symbol x, however, may denote a positive number, a negative number, or zero. So the symbol "$-x$" does not necessarily denote a negative number. It simply indicates the number which is symmetrically opposite x from 0 on the number line. If x is positive, then $-x$ is negative, but if x is negative, $-x$ is positive. See Figure 1.13.

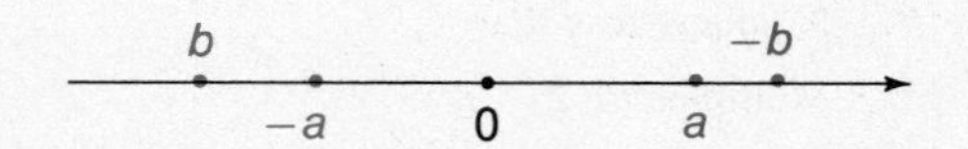

Figure 1.13 Opposites on the coordinate line

It follows from the above that if x is a number to the right of 0, its distance from 0 is the positive number x. However, if x is to the left of 0, its distance from 0 is the positive number $-x$. In either case, we call the distance of x from the origin the *absolute value* of x, symbolized by $|x|$. It is defined algebraically as follows.

DEFINITION 1.4

$$\begin{aligned} |x| &= x, && \text{if } x > 0; \\ |x| &= 0, && \text{if } x = 0; \\ |x| &= -x, && \text{if } x < 0. \end{aligned}$$

According to the definition, $|x|$ is x or $-x$, *whichever is nonnegative*. In any case, $|x| \geq 0$. For example,

$$|2| = 2 \quad \text{and} \quad |-2| = 2,$$

so that

$$|-2| = |2|.$$

Geometrically, 2 is the distance of both 2 and -2 from 0.

EXAMPLE 1 Give the absolute value of (a) $-\sqrt{2}$, (b) $3 - \sqrt{3}$, (c) $x - 4$.

SOLUTION (a) $|-\sqrt{2}| = \sqrt{2}$

(b) $3 - \sqrt{3} > 0$ because $3 > \sqrt{3}$. Therefore, $|3 - \sqrt{3}| = 3 - \sqrt{3}$.

(c) $|x - 4| = x - 4$ if $x \geq 4$, and $|x - 4| = -(x - 4) = 4 - x$ if $x \leq 4$, because in the latter instance $4 - x \geq 0$. ●

EXAMPLE 2 Solve for x: (a) $|x| = 5$; (b) $|x - a| = b$.

SOLUTION (a) If $|x| = 5$, then from the definition,

$$x = 5 \quad \text{or} \quad -x = 5$$
$$x = -5.$$

That is, $x = \pm 5$.

(b) If $|x - a| = b$, then

$$\begin{aligned} x - a &= b & \text{or} \quad -(x - a) &= b \\ x &= a + b & x - a &= -b \\ & & x &= a - b. \end{aligned}$$

Thus, the solution is $x = a \pm b$. ●

The distance between x and 0 is $|x|$, or $|x - 0|$. Similarly, $|a - b|$ is the distance between a and b on the number line. If a is to the right of b, then $a > b$ and this distance is $a - b$. If a is to the left of b, then this distance is $b - a$, or $-(a - b)$. See Figure 1.14.

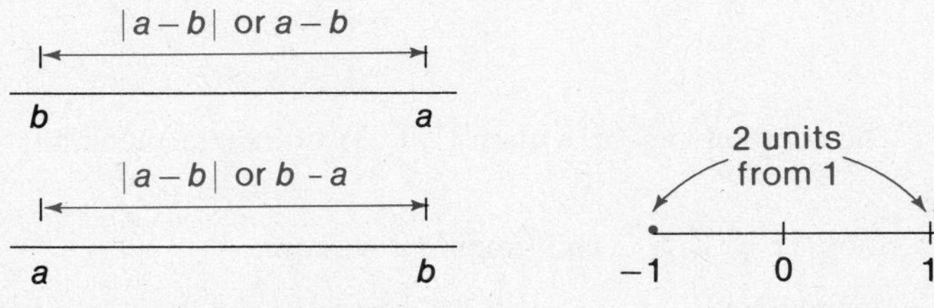

Figure 1.14 Distance between a and b

Figure 1.15 $|x - 1| = 2$

EXAMPLE 3 Solve $|x - 1| = 2$ and graph the solution.

SOLUTION Geometrically, $|x - 1| = 2$ means that x is 2 units from 1 on the number line in Figure 1.15. Thus, $x = -1$ or $x = 3$.

Algebraically, $|x - 1| = 2$ means that $x - 1 = 2$, or $x - 1 = -2$.

$$\begin{aligned} &\text{If} \quad x - 1 = 2, & &\text{if} \quad x - 1 = -2, \\ &\text{then} \quad x = 3; & &\text{then} \quad x = -1. \end{aligned}$$

The solution consists of the two numbers -1 and 3. ●

If $a > 0$, then $|x| < a$ means that x is less than a units from 0 on the number line. Algebraically, as seen in Figure 1.16,

$$|x| < a \text{ means that } -a < x < a.$$

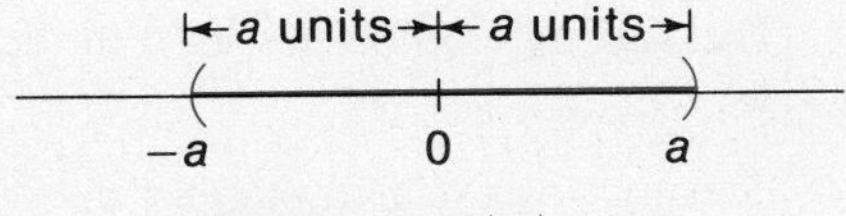

Figure 1.16 $|x| < a$

In the same way,

$$|x - c| < a \text{ means that } -a < x - c < a,$$

that is,

$$c - a < x < c + a.$$

In Figure 1.17, the geometric interpretation of $|x - c| < a$ is that x is less than a units from c.

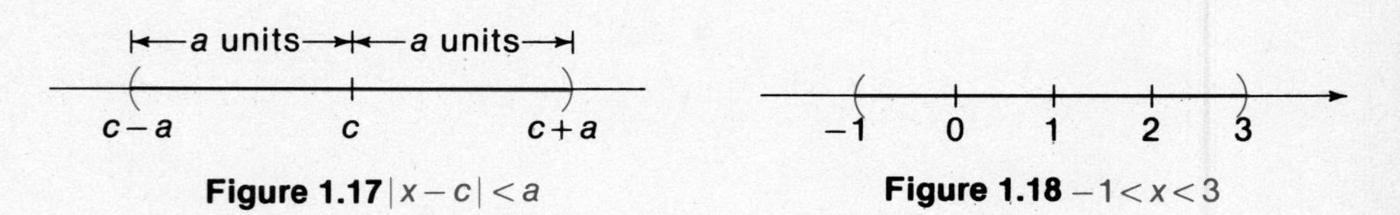

Figure 1.17 $|x - c| < a$

Figure 1.18 $-1 < x < 3$

EXAMPLE 4 Solve $|x - 1| < 2$ and graph the solution.

SOLUTION Geometrically, if $|x - 1| < 2$, then x is less than 2 units from 1. That is, x is between -1 and 3. See Figure 1.18.

Algebraically, $|x - 1| < 2$ means that

$$-2 < x - 1 < 2,$$

$$-1 < x < 3.$$

The solution can be written $(-1, 3)$ in interval notation. ●

EXAMPLE 5 Solve $|x - 1| > 2$ and graph the solution.

SOLUTION Geometrically, if $|x - 1| > 2$, then x is more than 2 units from 1. That is, x could be to the right of 3, $x > 3$, or to the left of -1, $x < -1$. So $x > 3$ or $x < -1$. See Figure 1.19.

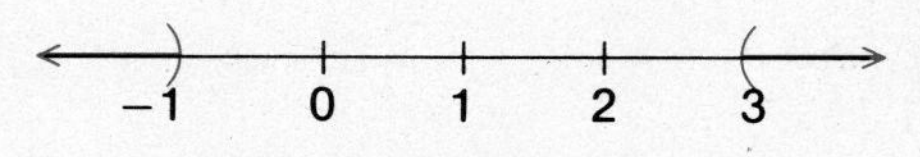

Figure 1.19 $x < -1$ or $x > 3$

Algebraically, $|x - 1| > 2$ means that

$$x - 1 > 2 \quad \text{or} \quad -(x - 1) > 2,$$

$$x > 3 \qquad\qquad x - 1 < -2,$$

$$x < -1.$$

The solution can be written in interval notation as $(-\infty, -1) \cup (3, \infty)$, or in set notation as $\{x \mid x < -1\} \cup \{x \mid x > 3\}$. ●

The algebraic properties used in the preceding examples may be summarized as follows:

For any positive real number a,

$|x| = a$ is equivalent to $x = a$ or $x = -a$.

$|x - a| = b$ is equivalent to $x - b = a$ or $x - b = -a$.

$|x| < a$ is equivalent to $-a < x < a$.

$|x - b| < a$ is equivalent to $-a < x - b < a$.

$|x| > a$ is equivalent to $x > a$ or $x < -a$.

$|x - b| > a$ is equivalent to $x - b > a$ or $x - b < -a$.

As the following example shows, there are several ways to describe a set on the number line. At times one of these may be more useful while at other times another may be preferred.

EXAMPLE 6 Given the interval described by $-1 < x < 3$, (a) show it on a number line, (b) describe it verbally, (c) write it in absolute-value notation, and (d) write it in interval notation.

SOLUTION (a)

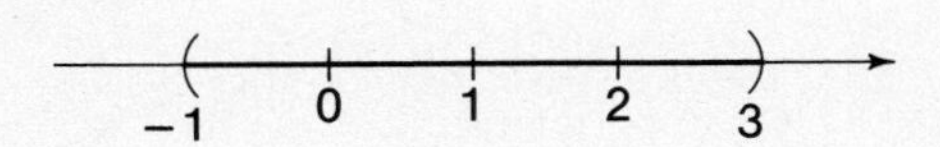

(b) As shown in the graph of part (a), x is greater than -1 and less than 3.

(c) We can write $|x - 1| < 2$, since 1 is two units from -1 and 3 in the graph of part (a).

(d) This is the open interval $(-1, 3)$. ●

The following examples illustrate the solution of inequalities involving absolute value of fractional expressions.

EXAMPLE 7 Solve the inequality $\left|\frac{1}{x}\right| < 2$.

SOLUTION By definition, if $x > 0$, then

$$\left|\frac{1}{x}\right| = \frac{1}{x}.$$

In this case, the given inequality is equivalent to

$$\frac{1}{x} < 2.$$

$$1 < 2x \qquad \text{multiplying both sides by } x, \text{ which is } > 0$$

$$\frac{1}{2} < x. \qquad \text{dividing both sides by 2}$$

If $x < 0$, then

$$\left|\frac{1}{x}\right| = -\frac{1}{x},$$

and the given inequality is equivalent to

$$-\frac{1}{x} < 2,$$

$$\frac{-1}{x} < 2,$$

$$-1 > 2x, \qquad \text{multiplying both sides by } x \text{, which is} < 0$$

$$-\frac{1}{2} > x. \qquad \text{dividing both sides by 2}$$

The solution is the union of two intervals: $(-\infty, -1/2) \cup (1/2, \infty)$. ●

In general,

$$\text{if } \left|\frac{1}{x}\right| < c, \qquad \text{then } x < -\frac{1}{c} \quad \text{or} \quad x > \frac{1}{c}.$$

In the same way we can show that

$$\text{if } \left|\frac{1}{x}\right| > c, \qquad \text{then } -\frac{1}{c} < x < \frac{1}{c}.$$

See exercise 29.

EXAMPLE 8 Solve the inequality $\left|\dfrac{1}{x-1}\right| < 3$.

SOLUTION From above,

$$x - 1 < -\frac{1}{3} \qquad \text{or} \qquad x - 1 > \frac{1}{3},$$

$$x < \frac{2}{3} \qquad \text{or} \qquad x > \frac{4}{3}.$$

So the solution is $(-\infty, 2/3) \cup (4/3, \infty)$. ●

The preceding sections briefly covered some of the important ideas about real numbers that are essential for later use. Our treatment has depended rather heavily upon geometric intuition, whereas the concept of a real number is actually quite sophisticated and rather elusive. Mathematicians from the time of the ancient Greeks struggled to formulate a sufficiently precise definition, but success came only around 1870 with the work of two German mathematicians, Cantor and Dedekind. However, a discussion of their work is beyond the scope of this text.

EXERCISES 1.3

Give the value of each of the following.

1. $|4 - 1|$ **2.** $|1 - 4|$ **3.** $|3|$ **4.** $|-1|$

5. $|0|$ **6.** $|3 - 3|$ **7.** $|x - y|$ **8.** $|x - 1|$

9. $|x + 2|$ **10.** $|2x - 1|$

Solve the following equations and inequalities. Graph the solution in each case.

11. $|x - 3| = 2$

12. $|2x - 4| = 1$

13. $|3x - 2| = 0$

14. $|x - 5| = -3$

15. $|x - 3| < 2$

16. $|x - 3| > 2$

17. $|x + 1| > 2$

18. $|2 - x| < 1$

19. $\left|\frac{1}{x}\right| \leq 5$

20. $\left|\frac{-1}{x - 3}\right| < 1$

21. $\left|\frac{1}{x + 1}\right| < 2$

22. $\left|\frac{1}{x - 2}\right| < 1$

23. $\left|\frac{1}{3x - 1}\right| > 1$

24. $\left|\frac{1}{2x - 1}\right| < 3$

25. $\left|\frac{x}{x - 1}\right| < 2$

26. $\left|\frac{x}{1 - x}\right| > 3$

27. $|x^2 - 1| \geq 0$

28. $|x^2 - a^2| > 0$

29. Show that if $\left|\frac{1}{x}\right| > c$, then $-\frac{1}{c} < x < \frac{1}{c}$.

Example 6 illustrates four ways of describing intervals. Each of the following intervals is described in one of these four ways. Describe it in each of the other three.

30. $-2 \leq x \leq 1$

31. $0 < x < 2$

32. x is between -1 and 2

33. x is between 2 and 5, including 2 and 5

34. x is less than 3 but greater than or equal to 2

35. x is less than 2 but not less than -1

36. $|x - 3| \leq 1$

37. $|x + 2| > 3$

1.4 The Coordinate Plane

The useful idea of a number line can be extended to the plane. One advantage of the number line is that negative numbers play roles just like the positive numbers. At one time people had only a vague concept of negative numbers which were sometimes referred to as "false" or "fictitious" numbers. Individuals had everyday experience with counting numbers and fractional numbers, but not with negative numbers like those that arose in connection with equations such as $x + 1 = 0$. The number line thus provides an important visual and conceptual model not only for the negative but for the positive numbers as well.

In the 17th century the French mathematician-philosopher Descartes extended the idea of graphing numbers on the number line to graphing pairs of numbers in the plane. The ideas he introduced unified algebra and geometry and were the beginnings of **analytic geometry** or **coordinate geometry.** Because of the work of Descartes the terms *Cartesian coordinates* and *Cartesian plane* are often used. Analytic geometry will be studied in some detail in later chapters, but a few of its elementary ideas will be needed sooner and will be discussed in this section.

Cartesian Coordinates Let two coordinate lines OX and OY be perpendicular to each other and let the point O be the origin on both of them. (See Figure 1.20.) We call these lines **coordinate axes** (the ***x*-axis** and the ***y*-axis**). The coordinate axes separate the plane into four **quadrants:** I, II, III, and IV, as indicated in Figure 1.20. Let P be *any* point in the plane of the axes and drop perpendiculars PA and PB to the x-axis and y-axis, respectively. Each of these perpendiculars intersects an axis at a point on a number line and so identifies a real number. We call these numbers a and b, respectively, and assign to the point P the **ordered pair** of numbers (a, b). These numbers are called **coordinates** of P, the first one being the **abscissa,** or ***x*-coordinate,** and the second one the **ordinate,** or ***y*-coordinate** of the point P. (We use the term *ordered pair* because, in general, the pair (a, b) identifies a different point than does the pair (b, a). (See Figure 1.21.)

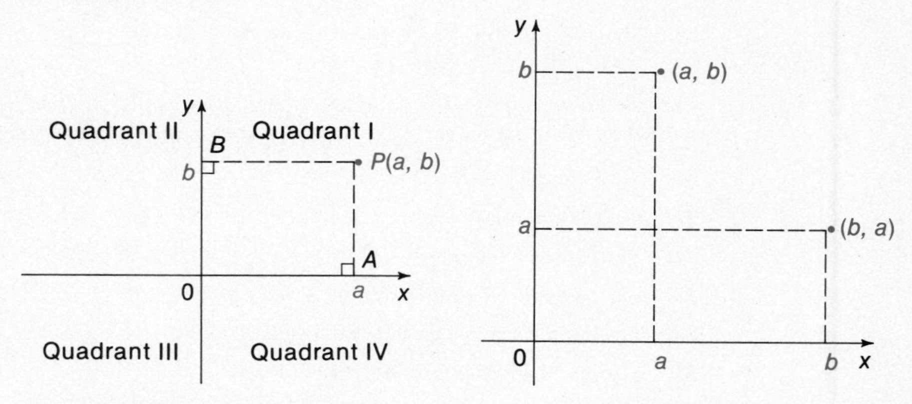

Figure 1.20 Cartesian plane

Figure 1.21

The coordinates of a given point in the plane depend upon where the axes have been placed and upon the scale chosen for these axes. The important observation, however, is that, once the axes have been placed,

each point in the plane corresponds to exactly one ordered pair of real numbers
and
each ordered pair of real numbers corresponds to exactly one point in the plane.

This is a consequence of the completeness of the set of real numbers on the number line. We now call the plane the **coordinate plane** and refer to the point P as either (a, b) or $P(a, b)$. In the same way, we could extend the idea of coordinates to three-dimensional space.

It is conventional, as was done above, to locate the coordinate axes horizontally and vertically, with positive directions to the right and upward. This is not necessary and other orientations are sometimes used. Also, for most of our work the same scale will be used on both axes, a convention that in certain instances may not always be possible or desirable. In Figure 1.1, for example, the horizontal axis had time measured in decades, while the vertical axis displayed population numbers in millions.

Symmetry Symmetry with respect to a point and symmetry with respect to a line are familiar ideas from elementary geometry. They are illustrated in Figure 1.22. If the lengths of segments AP and PB on the line AB are the same, then points A and B are **symmetric with respect to the point *P*.** See Figure 1.22a. If line l is the perpendicular bisector of the line segment PQ, then points P and Q are **symmetric with respect to the line *l*.** See Figure 1.22b.

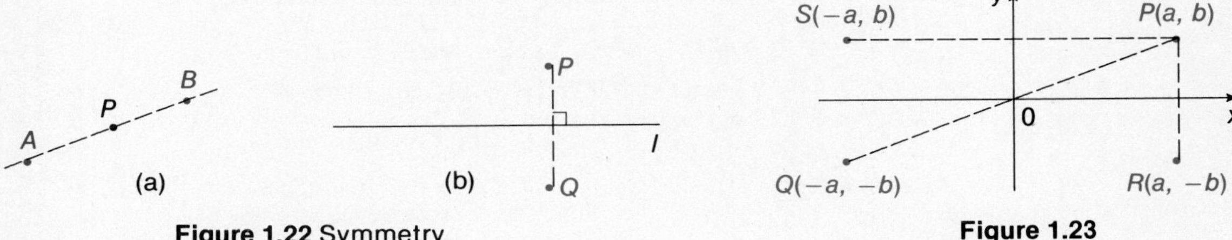

Figure 1.22 Symmetry

Figure 1.23

In Figure 1.23, the same kinds of symmetry are represented by coordinates. Thus,

$P(a, b)$ and $Q(-a, -b)$ are symmetric with respect to the origin;
$P(a, b)$ and $R(a, -b)$ are symmetric with respect to the x-axis;
$P(a, b)$ and $S(-a, b)$ are symmetric with respect to the y-axis.

In Figure 1.24, it can be proved that the points $P(a, b)$ and $T(b, a)$ are symmetric with respect to the 45-degree line bisecting the first and third quadrants. (See the exercises.)

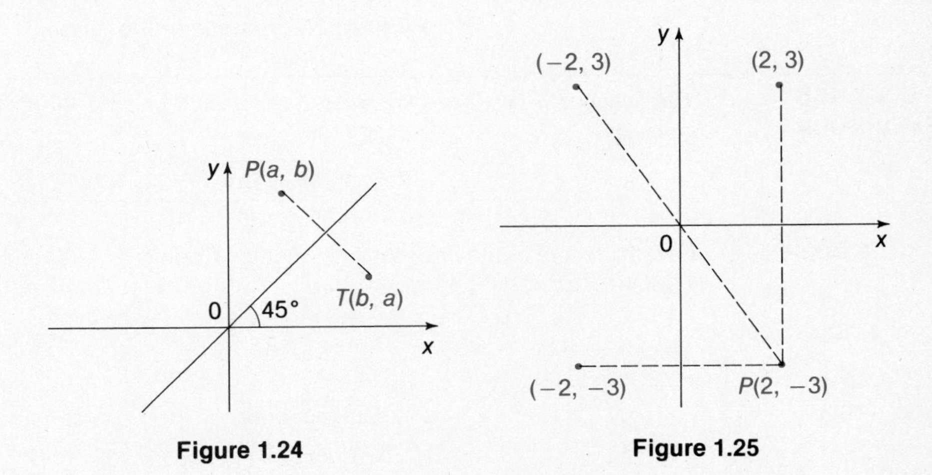

Figure 1.24

Figure 1.25

EXAMPLE 1 In Figure 1.25, find the point symmetric to the point $P(2, -3)$ with respect to: (a) the x-axis; (b) the y-axis; (c) the origin.

SOLUTION (a) (2, 3) is symmetric to P with respect to the x-axis.

(b) $(-2, -3)$ is symmetric to P with respect to the y-axis.

(c) $(-2, 3)$ is symmetric to P with respect to the origin. •

Distance We first consider horizontal and vertical line segments in the coordinate plane. In Figure 1.26a, let perpendiculars be dropped to the x-axis from the endpoints of the horizontal line segment PQ. If p and q are the real number coordinates of these points on the x-axis, then the length of PQ (often written $\overline{PQ}$) is $|p - q|$. Similarly, $|s - t|$ is the length of the vertical segment ST.

Now let $P_1(x_1, y_1)$ and $P_2(x_2, y_2)$ be *any* two points in the coordinate plane and d the length of the segment P_1P_2. (See Figure 1.26b.) Then, the preceding discussion, together with the Pythagorean theorem of elementary geometry, gives

$$\begin{aligned} d^2 = \overline{P_1P_2}^2 &= \overline{P_1Q}^2 + \overline{QP_2}^2 \\ &= |x_1 - x_2|^2 + |y_1 - y_2|^2 \\ &= (x_1 - x_2)^2 + (y_1 - y_2)^2. \end{aligned}$$

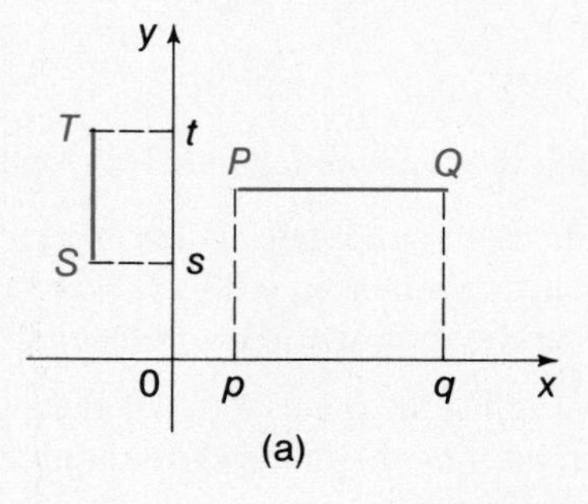

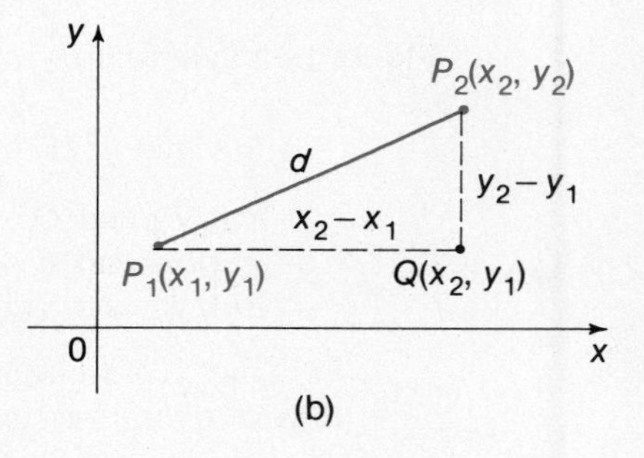

Figure 1.26 Distance in the plane

DISTANCE FORMULA The distance d between two points (x_1, y_1) and (x_2, y_2) in the coordinate plane is given by

$$d = \sqrt{(x_1 - x_2)^2 + (y_1 - y_2)^2}.$$

EXAMPLE 2 Find the length of the line segment joining the points $(-4, 3)$ and $(2, -1)$. (See Figure 1.27.)

SOLUTION

$$\begin{aligned} d &= \sqrt{(x_1 - x_2)^2 + (y_1 - y_2)^2} \\ &= \sqrt{(-4 - 2)^2 + (3 + 1)^2} \\ &= \sqrt{(-6)^2 + 4^2} \\ &= \sqrt{52} = \sqrt{4(13)} = 2\sqrt{13}. \end{aligned}$$ •

In using the distance formula, it does not matter which of the points is called (x_1, y_1) and which (x_2, y_2). (Why?) Also, it is generally simpler to calculate d^2 first because then the radical does not have to be carried along with the calculations. This is illustrated in the next example.

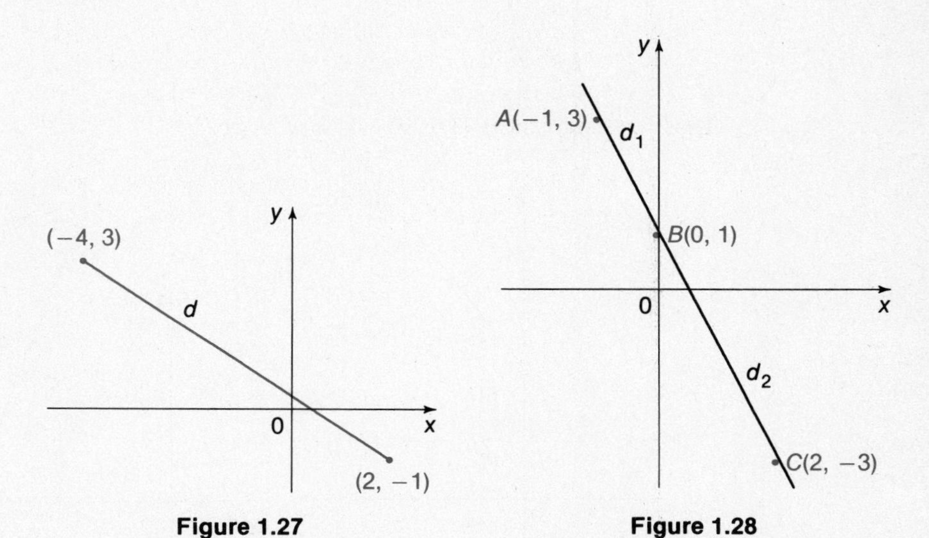

Figure 1.27

Figure 1.28

EXAMPLE 3 Use the distance formula to show that the points $A(-1, 3)$, $B(0, 1)$, and $C(2, -3)$ are **collinear** (lie on the same straight line). See Figure 1.28.

SOLUTION If we let $d_1 = \overline{AB}$, $d_2 = \overline{BC}$, and $d = \overline{AC}$, we need to show that $d = d_1 + d_2$.

$$d_1^2 = (0 + 1)^2 + (1 - 3)^2 = 1 + 4 = 5$$

$$d_2^2 = (2 - 0)^2 + (-3 - 1)^2 = 4 + 16 = 20$$

$$d^2 = (2 + 1)^2 + (-3 - 3)^2 = 9 + 36 = 45$$

So, $d_1 = \sqrt{5}$, $d_2 = \sqrt{20} = 2\sqrt{5}$, and $d = \sqrt{45} = 3\sqrt{5}$, and hence $d = d_1 + d_2$. Thus, the points A, B, and C are on the same straight line. ●

The point on a line segment midway between the endpoints of the segment is called the **midpoint** of the segment. The next two examples illustrate this idea.

EXAMPLE 4 Find the coordinates of the midpoint of the line segment which joins the points $P(-1, 1)$ and $Q(2, 5)$. See Figure 1.29 on the next page.

SOLUTION Let $M(x, y)$ designate the midpoint. Then, in the figure, triangles PAM and MBQ are congruent. Since corresponding sides of congruent triangles have equal lengths,

$$\begin{aligned} \overline{PA} &= \overline{MB} \\ x - (-1) &= 2 - x \\ x + 1 &= 2 - x \\ 2x &= 1 \\ x &= \frac{1}{2} \end{aligned} \qquad \text{and} \qquad \begin{aligned} \overline{AM} &= \overline{BQ} \\ y - 1 &= 5 - y \\ 2y &= 6 \\ y &= 3. \end{aligned}$$

Thus, the midpoint is (1/2, 3). ●

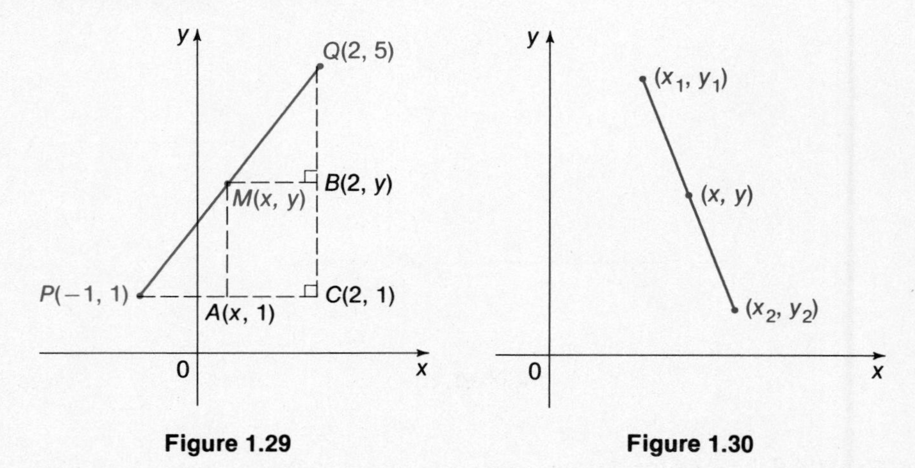

Figure 1.29

Figure 1.30

The same procedure is used with Figure 1.30 to get the following result.

MIDPOINT FORMULA

The midpoint of the line segment joining (x_1, y_1) and (x_2, y_2) has coordinates

$$\left(\frac{1}{2}(x_1 + x_2), \frac{1}{2}(y_1 + y_2)\right).$$

Equation of a Curve In Figure 1.31, the 45-degree line bisects the first and third quadrants. From elementary geometry we know that $y = x$ for every point (x, y) on this line. Conversely, for points not on this line, $y \neq x$. Thus, the equation $y = x$ identifies this line uniquely. We call it an *equation of the line*. Similarly, $y = 0$ is the equation of the x-axis, $x = 0$ is the equation of the y-axis, and $y = c$ (c a constant) is the equation of any line parallel to the x-axis. Equations of lines will be treated in considerable detail in Chapter 7.

The distance formula is used to obtain the equation of a circle. Suppose, for example, the radius of a circle is r and its center is at the origin O. Then, from Figure 1.32 and the distance formula,

$$\begin{aligned} (x - 0)^2 + (y - 0)^2 &= r^2, \\ x^2 + y^2 &= r^2. \end{aligned}$$

Since the distance from the origin to any point *not* on the circle is *not* r, this equation identifies the circle uniquely.

We cannot actually list the coordinates of all the points of a curve. However, an equation for a curve provides us (often quite simply) with an *algebraic description* for all points of the curve. Deriving equations of curves and using them to study the curves algebraically is the essential purpose of the subject called ''analytic geometry'' (Chapter 7).

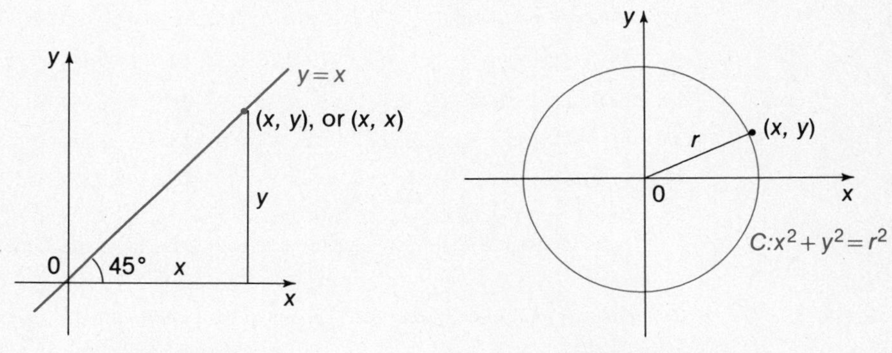

Figure 1.31 **Figure 1.32**

EXERCISES 1.4

Always draw figures to illustrate and help explain a geometric situation. Be careful to indicate coordinates of significant points.

1. Suppose that a line through $(2, -3)$ is perpendicular to the y-axis. What is true about the ordinates (y-values) of all points on the line?
2. Draw a figure and give the vertices of (a) a square of side 2 with sides parallel to the coordinate axes and center at the origin, (b) an equilateral triangle of side 2 with one side on the x-axis and the third vertex on the y-axis.
3. Identify the point symmetric to $(-2, 3)$ with respect to (a) the origin, (b) the x-axis, (c) the y-axis, (d) the line whose equation is $y = x$. Then repeat the exercise for the point $(-3, -2)$.
4. The origin is the midpoint of a line segment with $(3, 2)$ as one endpoint. Determine the other endpoint.
5. Find the remaining vertex of a rectangle with vertices $(3, 4)$, $(-5, 4)$, and $(3, -1)$.

Find the distance between the following points.

6. $(1, 7)$ and $(-3, -5)$
7. $(2, -8)$ and $(-3, 4)$
8. $(5, 4)$ and $(-2, 3)$
9. $(-2, 5)$ and $(-3, -1)$
10. $(5, 0)$ and $(0, 7)$
11. $(0, 8)$ and (x, y)
12. (u, v) and (c, d)
13. (a, b) and $(a - b, a + b)$

Use the distance formula to determine whether the following points are collinear. Draw a figure. Note that we cannot decide from the figure alone if the points are collinear.

14. $(-8, -2)$, $(4, 6)$, $(1, 4)$

15. $(2, -1)$, $(-1, -3)$, $(3, -1/3)$

16. $(0, 5/2)$, $(1, 5)$, $(-2, -1)$

17. $(-1, 3)$, $(4, 2)$, $(2, 12/5)$

The following sets of points are vertices of triangles in the coordinate plane. Use the distance formula to determine whether each triangle is ***isosceles*** *(two sides equal),* ***equilateral*** *(all sides equal),* ***right*** *(one right angle), or none of these. Draw a figure in each case and guess the answer before verifying it.* [*Hint:* In a right triangle with sides a, b, and c, $a^2 + b^2 = c^2$.]

18. $(6, -5)$, $(1, 5)$, $(-2, -1)$

19. $(-2, 8)$, $(-1, 1)$, $(3, 3)$

20. $(-4, 1)$, $(-2, 4)$, $(6, 2)$

21. $(-2, 0)$, $(8, 0)$, $(3, 5\sqrt{3})$

22. $(3, -1)$, $(3, -3)$, $(7, -2)$

23. $(-2, 4)$, $(6, 2)$, $(5, -2)$

24. $(-2, 3)$, $(-6, -3)$, $(1, 1)$

25. $(1, 1)$, $(-1, 3)$, $(4, 4)$

26. Show that the quadrilateral with vertices $(4, 1)$, $(-3, 2)$, $(-2, -3)$, and $(5, -4)$ is a parallelogram by showing that its opposite sides have the same length.

Calculate the midpoint of the line segment whose endpoints are as follows.

27. $(1, 5)$ and $(-3, -1)$

28. $(-1, 1)$ and $(3, 3)$

29. $(-2, 3)$ and $(6, 2)$

30. $(4, 1)$ and $(-3, 2)$

31. $(-1, -2)$ and $(5, -4)$

32. $(-2, -3)$ and $(3, 1)$

33. As suggested in this section, show that points (a, b) and (b, a) are symmetric with respect to the line whose equation is $y = x$. [*Hint:* Using Figure 1.24, denote by (c, c) the point M of intersection of the line (where $y = x$) with segment PT. Then show that OM is the perpendicular bisector of PT. That is, using the distance formula, show that $TM = MP$ and $OT = OP$.]

34. Use the procedure outlined in Example 4 to derive the midpoint formula. Then use this formula to check the results of Example 4.

35. Explain why, in the distance formula, it does not matter which is considered the first point and which the second.

36. The point $(x, 1)$ is $2\sqrt{5}$ units from the point $(2, 5)$. Determine the value of x.

37. Find the point on the y-axis equidistant from points $(1, 3)$ and $(-1, -1)$. [*Hint:* Call the point $(0, y)$.]

Find the equation that x and y must satisfy in order that point (x, y) be on the circle in each of the following.

38. Center $(0, 5)$ and radius 3

39. Center $(-1, 0)$ and radius 2

40. Center $(3, 4)$ and radius 5

41. Center $(2, 3)$ and radius 4

42. Use the distance formula and the definition of a circle to derive the equation of the circle of radius r having its center at (a, b).

Chapter 1 Review Exercises

Solve each of the following equations or inequalities and graph the solution set.

1. $|x + 2| = 1$

2. $|x + 1| = 2$

3. $|x - 6| = 2$

4. $|2 - x| = 3$

5. $|2x - 1| = 1$

6. $|3 - 2x| = 2$

7. $-2x < 4$

8. $4x \leq 2x - 3$

9. $x + 3 > 5x + 7$

10. $3x + 1 > 7$

11. $|x - 3| < 2$

12. $|x + 1| < 2$

13. $|2 - x| > 3$

14. $|2x - 1| > 3$

15. $x + 3 < 2x + 7 < x + 12$

16. $1 < 2x - 3 \leq 4$

17. $2 > 2x + 4 > -4$

18. $x \leq 2x + 1 < 3x - 6$

19. $\dfrac{1}{x} > -1$

20. $\dfrac{1}{x + 1} < 2$

21. $\dfrac{1}{x - 1} < \dfrac{2}{x}$

22. $\dfrac{3}{x - 2} > \dfrac{1}{x}$

If point (a, b) is in the second quadrant, in which quadrant is each of the following points? Sketch a figure.

23. $(a, -b)$

24. $(-a, -b)$

25. $(-b, -a)$

26. $(-a, b)$

27. (b, a)

28. $(b, -a)$

Sketch a figure and then use the distance formula to determine which of the following sets of points are collinear:

29. $(-3, 5)$, $(2, 3)$, $(8, 1)$

30. $(2, 3)$, $(3, 5)$, $(1, 1)$

31. $(0, 3)$, $(1, 7)$, $(-1, 0)$

Let each of the following sets of points be the vertices of a triangle. Sketch a figure and use the distance formula to determine the kind of triangle in each case: equilateral, isosceles, right, or none of these.

32. $(0, 6)$, $(9, -6)$, $(-3, 0)$

33. $(1, -3)$, $(5, 0)$, $(2, 4)$

34. $(-2, 1)$, $(2, 4)$, $(3, -3/2)$

35. $(-2, 4)$, $(-4, 1)$, $(6, 2)$

36. Show that the triangle with vertices $(1, 8)$, $(3, 2)$, and $(9, 4)$ is an isosceles right triangle.

By calculating lengths, show the following.

37. The points $(-4, 1)$, $(2, 4)$, $(5, -2)$, and $(-1, -5)$ are vertices of a square.

38. The points $(-2, 4)$, $(-4, 1)$, $(6, 2)$, and $(4, -1)$ are vertices of **a parallelogram** (opposite sides the same length).

39. A **median** of a triangle is a line segment joining a vertex of the triangle to the midpoint of the opposite side. (a) Find the midpoints of the sides of the triangle whose vertices are $(-2, -1)$, $(2, 7)$, and $(4, 3)$. (b) Calculate the lengths of the medians of the triangle.

40. Points on the perpendicular bisector of a line segment are **equidistant** (the same distance) from the endpoints of the segment. Use this fact to show that $(-3/2, 7/2)$ is on the perpendicular bisector of the segment joining $(1, 1)$ and $(2, 3)$.

Find c such that:

41. $(1, c)$ is $\sqrt{5}$ units from $(3, 3)$;

42. $(c, 2)$ is 5 units from $(6, 6)$.

Find a point on the x-axis which is equidistant from the given pair of points. [*Hint:* Label the point $(x, 0)$.]

43. $(-1, -1)$ and $(3, 5)$

44. $(-3, 2)$ and $(5, 6)$

Find a point on the y-axis equidistant from the given pair of points. [*Hint:* Label the point $(0, y)$.]

45. $(-1, -3)$ and $(-7, 1)$

46. $(3, 4)$ and $(-2, 6)$

Write the equation of each of the following lines.

47. Parallel to the x-axis and 2 units below it

48. Parallel to the y-axis and 1 unit to the left of it

49. Bisecting the second and fourth quadrants

Use the distance formula to write the equation of each of the following circles.

50. Center $(1, 1)$ and radius $\sqrt{2}$

51. Center $(2, -1)$ and radius $\sqrt{5}$

52. Center $(-2, -3)$ and radius 4

53. Center $(-1, 2)$ and radius 3

Chapter 1 Miscellaneous Exercises

Prove.

1. $|-x| = |x|$.

2. $|x| = \sqrt{x^2}$, that is, $|x|^2 = x^2$.
(This is a useful alternative definition of absolute value.)

Prove each of the following theorems from the definition of absolute value. Simplify the proof by considering the separate cases: (a) $x > 0$, $y > 0$, (b) $x > 0$, $y < 0$, (c) $x < 0$, $y > 0$, (d) $x < 0$, $y < 0$.

3. $|xy| = |x| \cdot |y|$, that is, the absolute value of a product is equal to the product of the absolute values of its factors.

4. $\left|\dfrac{x}{y}\right| = \dfrac{|x|}{|y|}$, that is, the absolute value of a quotient is the quotient of the absolute values of numerator and denominator.

5. $|x + y| \leq |x| + |y|$ (the **triangle inequality**), that is, the absolute value of a sum is not greater than the sum of the absolute values of the addends. [*Hint:* Show that if x and y have the same sign or if either is zero, then equality holds, whereas if they differ in sign, inequality holds. For cases (b) and (c), consider the cases $x + y > 0$ and $x + y < 0$ separately.]

6. $|x - y| \geq |x| - |y|$.

7. $\big||x| - |y|\big| \leq |x - y|$.

8. Use the pattern of Example 4 to find the coordinates of the points M and N which divide the line segment joining $(-1, -1)$ and $(2, 5)$ into three equal parts. There will be three congruent triangles in this situation.

9. In the figure, suppose that points P and Q divide the segment P_1P_2 into thirds. Use similar triangles, as in the midpoint formulas earlier, to show that

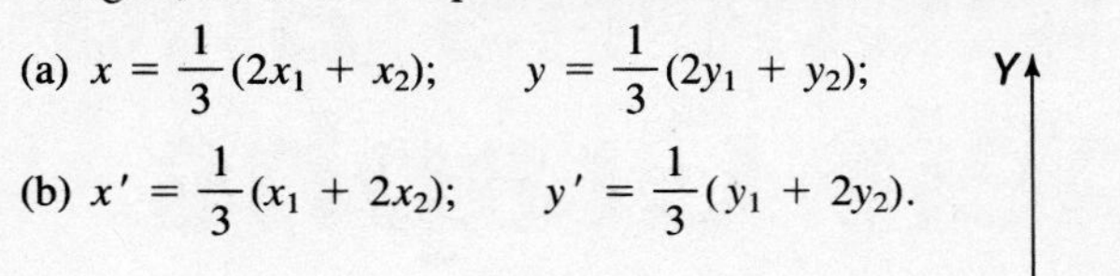

(a) $x = \frac{1}{3}(2x_1 + x_2)$; $\quad y = \frac{1}{3}(2y_1 + y_2)$;

(b) $x' = \frac{1}{3}(x_1 + 2x_2)$; $\quad y' = \frac{1}{3}(y_1 + 2y_2)$.

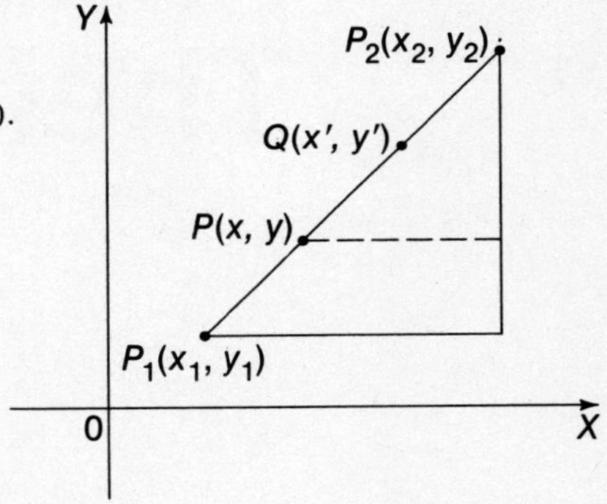

2

Functions and Their Graphs

Relationships between quantities may be observed all around us: the price of corn and the amount planted are related, as are the distance covered by a falling body and the time it falls, the Fahrenheit temperature and the corresponding Celsius reading, and so on. Some of these relationships are the special sort called *functions*. It is standard procedure in the sciences, social and behavioral as well as physical, to seek a mathematical function to describe a situation and then to use mathematical techniques in order to solve the real-world problems for which the function provides a *mathematical model*.

2.1 The Function Concept

To introduce the notion of a function, several simple examples of functions are given. The common features of these examples are brought out in the formal definition which follows them.

EXAMPLE 1 The formula $y = x^2$ assigns exactly one real number y to each real number x. Some (x, y) values satisfying $y = x^2$ are shown in the table at the top of the next page. ●

x	$y = x^2$
-2	4
0	0
3/4	9/16
1	1
2	4
3	9

EXAMPLE 2 The table below shows the populations of several small towns in California. To each town there corresponds exactly one number, its population. ●

Town	Population
Alpine	1900
Boron	2500
Colfax	800
Diablo	500
Escalon	2400

EXAMPLE 3 The graph in Figure 2.1 shows that for each x in the interval $[a, b]$ there corresponds exactly one y in the interval $[c, d]$. A graph like this might be recorded by some scientific instrument, such as a seismograph or cardiograph. ●

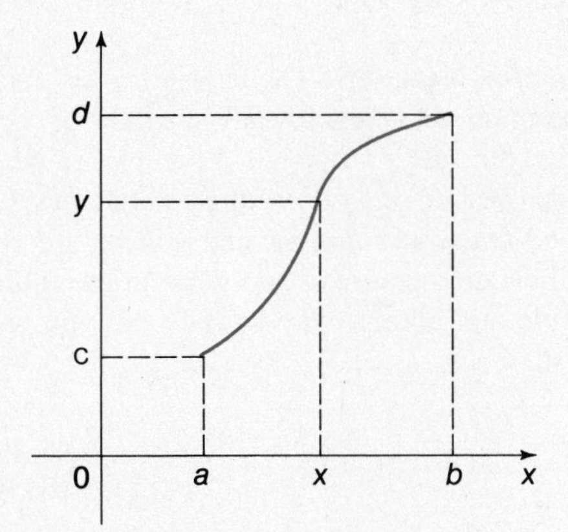

Figure 2.1

EXAMPLE 4 To each point on the coordinate line there corresponds exactly one real number. ●

In each of the preceding examples, there are *two sets* of objects (not necessarily numbers) and a *correspondence* or *rule* which assigns to each member of the first set *exactly one* member of the second set. Such a correspondence is called a *function* of the first set into the second set.

DEFINITION 2.1 A **function** is a correspondence between two sets X and Y such that to each member of the first set, X, there corresponds exactly one member of the second set, Y. The first set, X, is the **domain** of the function; a member of the second set assigned by the function rule is called the **image** (or **function value**) of the corresponding first member; and the set, Y, of all images of first members is called the **range** of the function.

Thus, the basic ideas relating to functions are a domain, a correspondence or rule, and a range. In the four examples above, the respective domains are the set of all real numbers x, a set of five specific towns, the real number interval $[a, b]$, and the set of points on a line. The corresponding ranges are, respectively, the set of all nonnegative numbers ($y \geq 0$), a set of five population numbers (listed), the real number interval $[c, d]$, and the set of all real numbers.

The notion of function is quite specific and yet very inclusive. The types of elements of the domain and range are unrestricted. In the examples, they included sets of numbers, towns, and points. The range may be identical with the domain. This is the case with the function defined by $y = 2x$, for example, where both the domain and the range are the set of all real numbers. The rule of correspondence may be given in an unlimited variety of ways, provided only that it assigns exactly one member of the range to each member of the domain. In the examples above, the correspondences were given by a formula, a table, a graph, and a verbal statement.

Function Notation The letter f (or any letter, for that matter) may be used to refer to a function. If x denotes any member of the domain of f, $f(x)$ denotes the image of x under the rule f. Thus, in Example 1, the equation $y = f(x)$ or $f(x) = x^2$, read "y equals f of x" or "f of x equals x squared," describes the function. This terminology is called *function notation,* and will be a common part of the language of much of the mathematics you read and write in the future. Among other advantages, it provides a simple and precise way to state the rule for finding function values. For example, if

$$f(x) = 2x + 1,$$

then

$$f(1) = 2(1) + 1 = 3$$

$$f(0) = 2(0) + 1 = 1$$

$$f\left(-\frac{1}{4}\right) = 2\left(-\frac{1}{4}\right) + 1 = \frac{1}{2}$$

$$f(u) = 2u + 1$$

$$f(x + h) = 2(x + h) + 1$$

and so on.

The familiar formulas of geometry and science define functions. For example, the equations

$$A = f(r) = \pi r^2,\ r > 0, \qquad \text{or} \qquad A(r) = \pi r^2,\ r > 0,$$

describe the area of a circle as a function of its radius. The geometric nature of the situation here requires that to each positive value of r there is a corresponding positive value of A. So both the domain and the range are the set of positive real numbers.

In the same way, the equations

$$s = f(t) = 16t^2,\ t > 0, \qquad \text{or} \qquad s(t) = 16t^2,\ t > 0,$$

describe the distance a free-falling body drops as a function of elapsed time. Also,

$$F = f(C) = 32 + \frac{9}{5}C$$

gives the Fahrenheit temperature F as a function of the Celsius (centigrade) temperature C.

Since a function is determined by a domain and a rule of correspondence, any letter may be used to denote it. Similarly, the letter x used for any member of the domain may be replaced by any other letter. Exactly the same function is identified by $f(x) = x^2$, $g(x) = x^2$, $s(t) = t^2$, $A(s) = s^2$, and so on, since there is the same domain (set of real numbers) and the same rule of correspondence (square the member of the domain). The letter x, t, or s is called the **independent variable,** or **argument,** of the function. In $y = f(x)$, y is called the **dependent variable.** We often use letters suggested by the situation. For example, the area of a circle as a function of the radius was given above by $A(r) = \pi r^2$, and the Fahrenheit temperature as a function of the Celsius temperature by $f(C) = 32 + (9/5)C$.

Historically, the idea of function first became important in the seventeenth century when scientists began serious study of variable quantities. The dependent variable was called a "function of" the independent variable, and it was thought of as being defined by some formula. The concept of function, however, is broader than that of a dependence, and also *need not involve a formula,* as illustrated in the examples above.

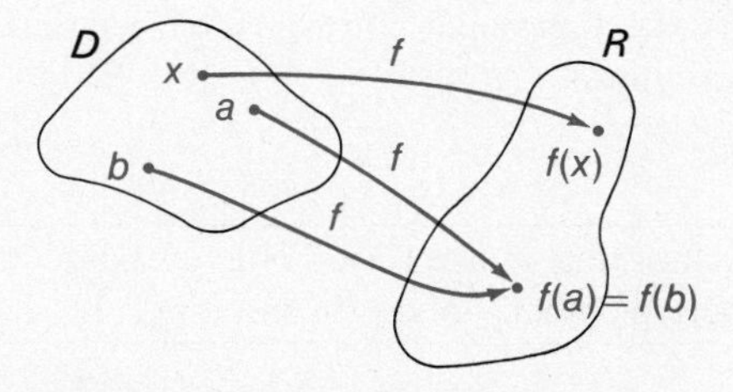

Figure 2.2 Function as a mapping

A useful way of describing a function is by a *mapping,* as shown in Figure 2.2. Each element x of the domain D is "mapped" to the corresponding $f(x)$ in the range R.

The correspondence, or mapping, is represented by an arrow from x to $f(x)$. For instance, the mapping of Example 1 can be indicated by

$$x \rightarrow f(x), \qquad x \xrightarrow{f} y, \qquad \text{or} \qquad x \xrightarrow{f} x^2.$$

Observe that the definition of a function allows a given member of the range to be the image of more than one member of the domain. For example, different persons may have the *same* last name, different books the *same* number of pages, and in Figure 2.2, a and b the *same* image, that is, $f(a) = f(b)$.

Sometimes it is helpful to think of a function as a machine which acts upon any member x of the domain to produce the function value $f(x)$. (See Figure 2.3.) A hand calculator is such a "function machine": if you enter a number and then press the x^2 key, for example, the machine displays the square of the number. Similarly, it acts as a "square root machine": the function $f(x) = \sqrt{x}$. These are functions of a single variable. In later chapters we will consider both of these and many more particular functions of a real variable in some detail.

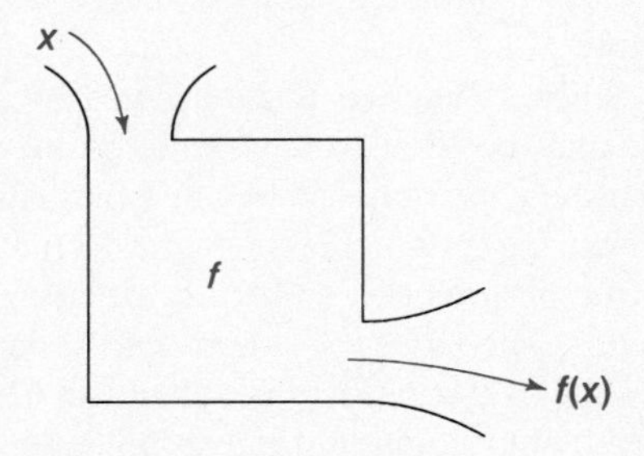

Figure 2.3 A function machine

The calculator also deals with functions of two variables. For example, it may assign to the pair x and y the value $x + y$, the value $x - y$, the value $x \cdot y$, or the value $x \div y$. Each of these is a function value for the pair x and y, and is denoted $f(x, y)$.

Regardless of the manner of describing a function, the correspondence establishes a set of ordered pairs. In Example 1 these pairs may be denoted by (x, y), (x, x^2), or $(x, f(x))$. Sometimes we say the function *is* the set of ordered pairs, since they identify the functional correspondence precisely. Thus, we have the following alternate definition.

DEFINITION 2.2 A **function** is a set S of ordered pairs (x, y) such that for each x in the domain X, there is exactly one y in the range Y.

The set of ordered pairs $\{(1, 2), (-3, 0), (7, 4)\}$ is a function, but the set $\{(1, 2), (-3, 0), (7, 4), (1, -5)\}$ is not since it pairs both 2 and -5 with 1. The tables of Examples 1 and 2 are functions because they pair exactly one member of the second column with each member of the first column.

Rate of Change A fundamental concept of calculus is that of rate of change of a function. The following example introduces this idea and at the same time illustrates how to use function notation.

EXAMPLE 5 Let $f(x) = 2x$. Find $f(5)$, $f(1)$, $f(5) - f(1)$, and $\dfrac{f(5) - f(1)}{5 - 1}$.

SOLUTION

$$f(5) = 2(5) = 10$$
$$f(1) = 2(1) = 2$$
$$f(5) - f(1) = 10 - 2 = 8$$

This difference is the change in the value of f as x changes from 1 to 5. The quotient

$$\frac{f(5) - f(1)}{5 - 1} = \frac{8}{4} = 2$$

gives the **average change** in the value of f as x changes from 1 to 5. See Figure 2.4a, where one can see that $8/4 = 2$ measures the steepness (slope) of the line. ●

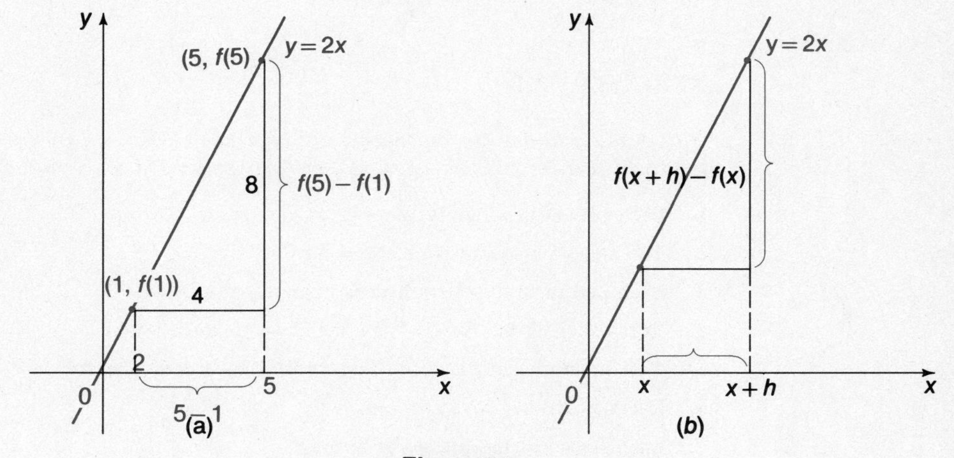

Figure 2.4

This same procedure can be used to calculate a *formula* for the average change in the value of f for any x. For example, if $f(x) = 2x$, then as the value of x changes from x to $x + h$, we have

$$f(x) = 2x,$$
$$f(x + h) = 2(x + h) = 2x + 2h,$$
$$f(x + h) - f(x) = (2x + 2h) - 2x = 2h,$$
$$\frac{f(x + h) - f(x)}{h} = \frac{2h}{h} = 2.$$

That is, f changes by 2 units for each unit of change in x. In this example, the average change in f is a constant, that is, f changes at a constant rate. (See Figure 2.4b.) The next example shows that this is not generally the case.

EXAMPLE 6 Find the average change in f for any x if $f(x) = x^2$.

SOLUTION

$$f(x) = x^2$$
$$f(x + h) = (x + h)^2$$
$$= x^2 + 2hx + h^2$$
$$f(x + h) - f(x) = x^2 + 2hx + h^2 - x^2$$
$$= 2hx + h^2$$
$$\frac{f(x + h) - f(x)}{h} = \frac{2hx + h^2}{h}$$
$$= \frac{\cancel{h}(2x + h)}{\cancel{h}}$$
$$= 2x + h$$

Here the average change in the value of f depends not only upon h, the change in the value of x, but also upon the value of x. That is, the average change is a function of x. ●

EXERCISES 2.1

Which of the following correspondences is a function? If not, tell why not. For each function, identify the domain and the range. (Note: The domain and range are sets.*)*

1. Each person in a family has a given first name.

2. The area of a square is determined by its side.

3. Each person in this class had two parents.

4. Every shelf in the library has a certain number of books.

5. The fingerprint of the index finger identifies a particular person.

6. Every house has several rooms.

7. Each President has a state of birth.

8. Each point on a map corresponds to a particular point on the earth.

9. The formula $x^2 + y^2 = 1$ assigns two values of y to each value of x in the interval $-1 < x < 1$.

10. n is the number of letters in the word name of the integer x.

Write a formula for the function described by each of the following: With each first number x there is associated a second number y which is

11. One-half the first number;

12. Two more than the first number;

13. π times the square of the first number;

14. The cube of the first number;

15. The positive square root of the first number.

16. If $f(x) = 3x - 1$, calculate $f(-1)$, $f(0)$, $f(2)$, $f(y)$, $f(x + u)$.

17. If $g(x) = \sqrt{x - 1}$, calculate $g(1)$, $g(2)$, $g(4)$, $g(u)$, $g(x + 1)$.

18. If $h(x) = |x - 1|$, calculate $h(3)$, $h(-3)$, $h(0)$, $h(u)$.

The specific heat of a liquid is a function of its temperature. A partial table of laboratory data is shown below.

t	5.7	7.1	9.2	11.4	13.4	15.7	17.9	19.2	22.0
$H(t)$	0.40	0.40	0.40	0.41	0.41	0.41	0.41	0.42	0.42

Call this function H and from the table give each of the following.

19. $H(7.1)$

20. $H(13.4)$

21. $H(17.9)$

22. $H(5.7)$

23. $H(22.0)$

24. $H(9.2)$

Write the formula which describes each of the following correspondences.

25. The area A of a square as a function of the length s of a side

26. The perimeter p of a square as a function of the length s of a side

27. The area A of an equilateral triangle as a function of the length s of a side

28. The area A of a triangle of height 3 as a function of the length b of the base

29. One side a of a right triangle with hypotenuse 7 as a function of the other side b

30. The formula $f(a + b) = f(a) + f(b)$ is generally not true. Show that it is true for $f(x) = 3x$ but not for $f(x) = 3x + 2$.

31. The formula $f(ab) = f(a) \cdot f(b)$ is generally not true. Show that it is true for $f(x) = x$ but not true for $f(x) = 2x$.

For each of the following functions, calculate (a) $f(x + h)$, *(b)* $f(x + h) - f(x)$, *and (c)* $\dfrac{f(x + h) - f(x)}{h}$. *(See Examples 5 and 6.)*

32. $f(x) = 3$

33. $f(x) = x$

34. $f(x) = -2x + 1$

35. $f(x) = -2x^2$

36. $f(x) = 1/x$

37. $f(x) = \sqrt{x}$

The formula $F(C) = 32 + \frac{9}{5}C$ *expresses the Fahrenheit temperature reading F as a function of the Celsius reading C. Calculate the following.*

38. $F(0)$

39. $F(100)$

40. $F(10)$

41. $F(50)$

42. $F(-20)$

43. $F(25)$

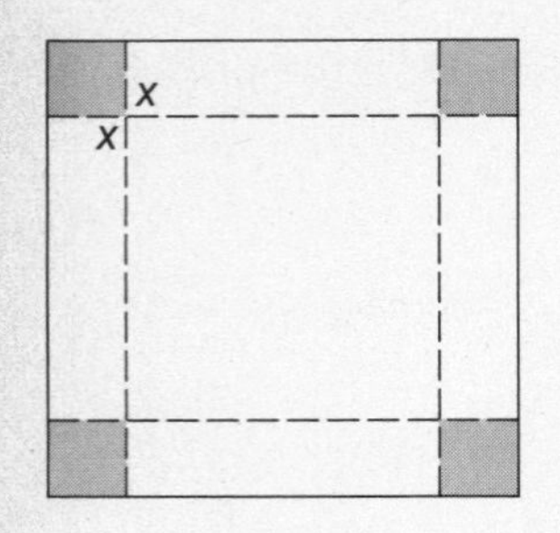

44. A metal rectangular container without a top is made by cutting out small square corners from a 12-inch square sheet of metal and then turning up the sides, as shown in the figure. If x represents the length of a side of the square pieces removed, write the formula which expresses the volume V of the resulting container as a function of x. What is the domain of this function? Explain.

45. The formula for the area A of a square of side s is $A = a(s) = s^2$ and for the perimeter is $P = p(s) = 4s$. Solve for s as a function of P and substitute in the formula $A = s^2$ to obtain a formula for the area of a square as a function of its perimeter.

46. The area A of a circle of radius r is $A = a(r) = \pi r^2$ and its circumference is $C = c(r) = 2\pi r$. Solve for r as a function of C and substitute in the formula $A = \pi r^2$ to obtain the area of a circle as a function of its circumference.

47. If the side of a square of side x is increased by 2 units, find a formula which expresses the amount the area is increased as a function of x. That is, find $a(x + 2) - a(x)$.

48. Two circles have the same center. The inner circle has radius 3 and the outer circle radius $3 + x$. Find the area of the ring-shaped region between the circles as a function of x. That is, find $a(3 + x) - a(3)$.

2.2 Real Functions

The definition of function puts no restriction on the kinds of objects in either the domain or the range. In particular, the domain and range need not be sets of numbers. However, our interest will be almost exclusively with functions for which *both* the domain and range are subsets of the real numbers. These are **real-valued functions of a real variable,** or **real functions.** Also, although the function rule need not be given by a formula or equation, most of the functions we work with will be of this type. Several important functions are described in the following examples.

EXAMPLE 1 Let $f(x) = 3$. If the domain is taken to be the set of real numbers, f assigns exactly one real number (actually, the same number, 3) to each real number x of the domain. It is therefore a function. Its range is the set with the single member 3. In general,

$f(x) = c$ defines a **constant function.**

The function $f(x) = 0$ is a particular case, called the **zero function.** •

Observe that in Example 1 $f(x)$ does not "depend" upon the value of x. It is the same for all x. This illustrates the important fact that the concept of function is *not necessarily a relationship of dependence, but one of correspondence.*

EXAMPLE 2 Let $f(x) = x$. The domain may be taken as the set of all real numbers. Then the image of any number is the number itself, and the range is the same as the domain.

$f(x) = x$ defines the **identity function.** •

EXAMPLE 3 Let $f(x) = x^2$. The domain is the set of all real numbers and the range is the set of all nonnegative real numbers. (Why?)

$$f(x) = x^2 \quad \text{defines the } \textbf{squaring function.}$$ •

EXAMPLE 4 Let $f(x) = \sqrt{x}$, $x \geq 0$. In order for $\sqrt{x}$ to be a real number, x must be a nonnegative number. Also, every nonnegative number is a square root. So the domain and range are the same, the set of nonnegative real numbers.

$$f(x) = \sqrt{x} \quad \text{defines the } \textbf{square-root function.}$$ •

EXAMPLE 5 Let $f(x) = 1/x$, $x \neq 0$. Since division by zero is undefined, $x \neq 0$ and thus the domain is the set of all nonzero real numbers. Also, every nonzero real number is the reciprocal of some nonzero real number. (Why?) Thus, the domain and range are the same sets.

$$f(x) = \frac{1}{x} \quad \text{defines the } \textbf{reciprocal function.}$$ •

EXAMPLE 6 Let $f(x) = |x| = \begin{cases} x, & \text{if } x \geq 0, \\ -x, & \text{if } x < 0. \end{cases}$

The domain is the set of all real numbers. Since $|x| \geq 0$, the range is the set of all nonnegative real numbers.

$$f(x) = |x| \quad \text{defines the } \textbf{absolute value function.}$$ •

EXAMPLE 7 Let $f(x) = [\![x]\!]$. The symbol $[\![x]\!]$ means "the greatest integer less than or equal to x." For example,

$$[\![5]\!] = 5, \qquad [\![8.3]\!] = 8, \qquad [\![\sqrt{2}]\!] = 1, \qquad [\![\pi]\!] = 3, \qquad [\![-1.4]\!] = -2,$$

and so on. In general,

$$\begin{aligned} &\text{if} \quad -2 \leq x < -1, \quad \text{then } [\![x]\!] = -2; \\ &\text{if} \quad -1 \leq x < 0, \quad \text{then } [\![x]\!] = -1; \\ &\text{if} \quad 0 \leq x < 1, \quad \text{then } [\![x]\!] = 0; \\ &\text{if} \quad 1 \leq x < 2, \quad \text{then } [\![x]\!] = 1, \end{aligned}$$

and so on. That is, to evaluate $[\![x]\!]$, locate x in an interval with successive integers as endpoints and select the left endpoint as the value $[\![x]\!]$.

$$f(x) = [\![x]\!] \quad \text{defines the } \textbf{greatest integer function.}$$ •

Its domain is the set of all real numbers x and its range is the set of all integers. (Why?)

As an illustration, the greatest integer function [\![age]\!] gives a person's age as of the last birthday. Thus, for someone 18 years, 3 months old, [\![age]\!] = 18. Similarly, it describes what the merchant probably expects you to think when he advertises a price of \$7.99, because $[\![7.99]\!] = 7$.

The function in each of the preceding examples is a real function, since in each case both the domain and the range are sets of real numbers. Example 6 shows that there may be a different way of describing a function for different parts of its domain. Another such example is given by

$$f(x) = \begin{cases} -1, & \text{if } x < -1; \\ x, & \text{if } -1 \le x < 1; \\ 2, & \text{if } x \ge 1. \end{cases}$$

Functions like these occur frequently in mathematics and its applications. For instance, consider the ''car-rental'' function: A certain car rents for 35 dollars for each 24-hour day, plus \$9.75 for each additional hour. If x is the number of hours the car is used in the partial day and C the cost in dollars for the partial day, then C is a function of x such that

$$C(x) = 9.75, \quad 0 < x \le 1;$$
$$C(x) = 19.50, \quad 1 < x \le 2;$$
$$C(x) = 29.25, \quad 2 < x \le 3;$$

and so on.

Natural Domain The domain of a real-valued function is sometimes restricted by algebraic considerations. In particular, *the square root of a negative number and division by zero* must be avoided.

EXAMPLE 8 For which real numbers are the following functions defined? (a) $f(x) = \sqrt{1 - x}$, (b) $f(x) = \sqrt{x^2 - 1}$, (c) $f(x) = 1/(x + 1)$.

SOLUTION (a) If $f(x) = \sqrt{1 - x}$, then $1 - x \ge 0$ or $x \le 1$ in order for $f(x)$ to be a real number.

(b) If $f(x) = \sqrt{x^2 - 1}$ is to be real, then $x^2 \ge 1$, that is, either $x \ge 1$ or $x \le -1$. See Example 4, Section 1.2.

(c) If $f(x) = 1/(x + 1)$, then $x \ne -1$. ●

The above example shows that the function rule is used only where it makes sense. Specifically, unless some particular restriction is given, we assume that the domain is the one given by the following definition.

DEFINITION 2.3 The **natural domain (domain of definition)** of a real function is the largest set of real numbers for which the function rule assigns a real number as image.

A function is determined by its domain and its rule of correspondence. If either the domain or rule is changed, a different function is defined.

DEFINITION 2.4 Given f and g, $\boldsymbol{f = g}$ on some domain if and only if $f(x) = g(x)$ for every x in that domain.

(Although we say that f and g are equal functions, there is actually only one function with two different names.)

EXAMPLE 9 Suppose that $f(x) = 2x + 3$ and $g(x) = 2(x + 3/2)$. Each of these formulas assigns the same value to $f(x)$ and $g(x)$ for any given x. For instance, $f(1) = 2(1) + 3 = 5$ and $g(1) = 2(1 + 3/2) = 2 + 3 = 5$. So $f = g$ on the set of real numbers. ●

EXAMPLE 10 Let $f(x) = x/x$ and $g(x) = 1$. The domain of f is the set of all nonzero real numbers, while that for g is the set of all real numbers. Since their domains are different, f and g are different functions. (Their ranges, of course, are the same: the set with the single number 1.) ●

EXAMPLE 11 The expression $(x^2 - 1)/(x - 1)$ defines a function for all real numbers except 1. For $x = 1$, it becomes the meaningless expression 0/0. For $x \neq 1$, it has the same value as $x + 1$. So, if we define functions f and g by

$$f(x) = \frac{x^2 - 1}{x - 1}, \quad x \neq 1 \text{ (for this to be defined)},$$

and

$$g(x) = x + 1, \quad x \neq 1,$$

then $f = g$, since they both have the same value for each value of $x \neq 1$. If we had not put the specific restriction that $x \neq 1$ on the function g, it would be defined for all real x and so would not be the same as f. ●

Relations Not every correspondence between the members of two sets is a function. For instance, to each state in the U.S. correspond two senators and to each person ordinarily correspond two or more names. For these correspondences we use the term *relation,* not function.

DEFINITION 2.5 A **relation** is a correspondence between two sets that associates with each member of the first set *one or more* members of the second set.

A function is a particular case of a relation for which each member of the domain has exactly one image. The following examples make this distinction clear.

EXAMPLE 12 The formula $y^2 = x$ defines a correspondence $x \to y$ for all $x > 0$. However, since $y = \pm\sqrt{x}$, the correspondence assigns two y-images to each positive x, namely $\sqrt{x}$ and $-\sqrt{x}$. The correspondence is a relation but not a function. ●

EXAMPLE 13 In the correspondence defined by the set of ordered pairs $\{(1, 2), (3, 5), (-1, 0), (1, 6)\}$ the number 1 has two different images, 2 and 6. So this correspondence is a relation but not a function. ●

The formula $y^2 = x$ of Example 12 above defines y **implicitly** in terms of x. On the other hand, the formula $y = x^2$ defines y **explicitly** in terms of x. In the first case, a value of y is implied by the formula and, in the second, it is given directly in terms of x.

EXAMPLE 14 The equation $2x - y = 1$ defines y implicitly as a function of x, and $y = 2x - 1$ defines y explicitly as a function of x. ●

EXAMPLE 15 The equation $x^2 + y^2 = 1$ defines y implicitly in terms of x. We can solve it for y and obtain $y = \pm\sqrt{1 - x^2}$, which defines y explicitly in terms of x. The correspondence is a relation, not a function. ●

The same terms (domain, range, argument, image, and so on) are used with relations as with functions. Relations which are not also functions will be of particular concern when we consider inverses of functions later (Section 2.5).

EXERCISES 2.2

Identify the domain and the range for each of the following functions. Explain your answer in each case.

1. The identity function

2. The squaring function

3. The square root function

4. The reciprocal function

5. The absolute value function

6. The greatest integer function

What is the natural domain of each of the following functions?

7. $f(x) = x^2 - 1$

8. $f(x) = x + \dfrac{1}{x}$

9. $f(x) = \sqrt{x + 1}$

10. $f(x) = \sqrt{1 - x^2}$

11. $f(x) = \dfrac{1}{x - 2}$

12. $f(x) = \dfrac{1}{2x + 3}$

13. $f(x) = \dfrac{1}{x^2 + 1}$

14. $f(x) = \dfrac{1}{2x - 1}$

15. $f(x) = \dfrac{1}{x^2 - 1}$

16. $f(x) = \dfrac{x + 1}{3 - x}$

17. $f(x) = \dfrac{1}{\sqrt{x - 1}}$

18. $f(x) = \dfrac{1}{\sqrt{x^2 - 1}}$

19. $f(x) = \dfrac{1}{\sqrt{x^2 + 1}}$

20. $f(x) = \dfrac{1}{\sqrt{x + 1}}$

Each of the following equations expresses a relation between x and y. By solving for y in terms of x, determine whether or not y is a function of x. Explain your answer.

21. $x + y = 1$

22. $x - 2y = 4$

23. $x^2 - y^2 = 1$

24. $2x^2 + 3y^2 = 6$

25. $x^2y = 1$

26. $xy^2 = 1$

27. $xy - 2y = 1$

28. $xy = 1$

29. $y^2 = 4 - x$

30. $y^2 + x - 1 = 0$

31. $xy = 1 - y$

32. $xy - 1 = x$

33. $x \geq |y|$

34. $|x| \leq y$

35. $\dfrac{y - 1}{y + 2} = x$

36. $\dfrac{y - 2}{x} = y$

37. $\dfrac{x - 1}{x + 2} = y$

38. $\dfrac{x - 2}{y} = x$

39. The formula $A(x) = x^2$ for the area of a square of side x does not define the squaring function of Example 3. Why not?

40. Suppose that first-class postage is 20 cents for each ounce or part of an ounce. If $p(x)$ is the amount of postage on x ounces of mail, show that $p(x) = [\![x]\!] + 1$ or, equivalently, $p(x) = [\![x + 1]\!]$. Then, as in Example 7, write the formulas for $p(x)$ for each unit interval from 0 to 5.

2.3 Graphs of Functions

Many functions can be represented in the coordinate plane. If $y = f(x)$ defines a real function, then both x and y are real numbers, and the function f determines a set of ordered pairs, (x, y), of real numbers.

DEFINITION 2.6 If f is a real function, then the set of points $(x, f(x))$ is called the **graph** of f.

The graphs of several simple functions are used in the following examples to illustrate how the graph of a function exhibits significant characteristics of the function.

EXAMPLE 1 Graph the function defined by the set of ordered pairs $\{(-1, 2), (2, 1), (4, 3)\}$.

SOLUTION The graph is simply the set of points shown in Figure 2.5. ●

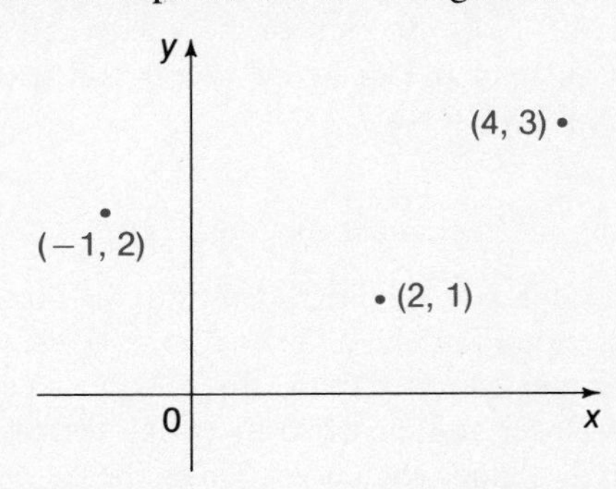

Figure 2.5

EXAMPLE 2 Graph the identity function $f(x) = x$.

SOLUTION The corresponding pairs are represented by (x, x), in which the $f(x)$-value is the same as the x-value. These points all lie on the line bisecting the first and third quadrants in the coordinate plane. See Figure 2.6. ●

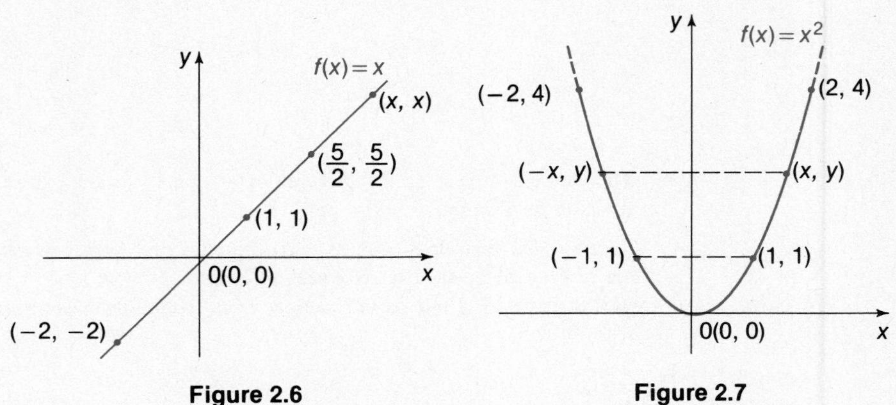

Figure 2.6 **Figure 2.7**

EXAMPLE 3 Graph the squaring function $f(x) = x^2$.

SOLUTION Since $f(x) \geq 0$, no part of the graph lies below the x-axis. Also, since $(-x)^2 = x^2$, for each point $(x, f(x))$ on the graph, the point $(-x, f(-x))$ symmetric to it with respect to the y-axis is also on the graph. Plot several points and use this information to complete the graph as in Figure 2.7. ●

In drawing an unbroken curve for these graphs we are using the completeness of the real number line. That is, since there is a real number for each point of the x-axis, there is a point on the curve corresponding to each point on the real axis for which the function is defined.

EXAMPLE 4 Graph the square root function $f(x) = \sqrt{x}$.

SOLUTION Since $f(x) \geq 0$, no part of the graph is below the x-axis. And, since $x \geq 0$ (Why?), none of it is to the left of the y-axis. Also, $f(0) = 0$ and the values of $f(x)$ increase as x increases. Plot a few points and use these facts to sketch the graph shown in Figure 2.8. ●

EXAMPLE 5 Graph the reciprocal function $f(x) = 1/x$.

SOLUTION The value $x = 0$ is excluded from the domain, so the graph does not cross the y-axis, as seen in Figure 2.9. Since $f(x) > 0$ whenever $x > 0$ and $f(x) < 0$ whenever $x < 0$, the graph is located in the first and third quadrants only. In the first quadrant, the curve gets closer and closer to the x-axis without reaching it as one moves in the positive direction along the curve. This is because as x increases without bound through posi-

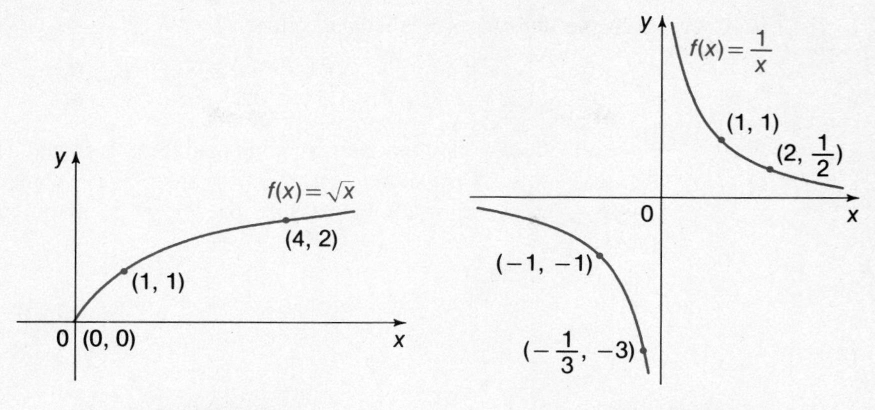

Figure 2.8 **Figure 2.9**

tive values, $f(x)$ decreases toward zero without ever becoming zero. And, as x stays positive but gets closer and closer to zero, $f(x)$ increases without bound. We write this symbolically as $f(x) \to 0$ as $x \to \infty$, and $f(x) \to \infty$ as $x \to 0$ (from the positive side). This is read "$f(x)$ approaches 0 as x approaches ∞, and $f(x)$ approaches ∞ as x approaches 0." The same kind of analysis applies to the third quadrant. Note that the two branches of the curve are symmetric with respect to the origin. ●

A straight line is called an **asymptote** of a curve if the distance of the curve from the line approaches (but does not reach) zero. Example 5 shows that the coordinate axes are asymptotes of the graph of the reciprocal function. We will meet asymptotes later with graphs of other functions.

EXAMPLE 6 Graph the constant function $f(x) = 3$.

SOLUTION All pairs $(x, f(x))$ are of the form $(x, 3)$, so the graph is a line parallel to the x-axis and 3 units above it. See Figure 2.10. ●

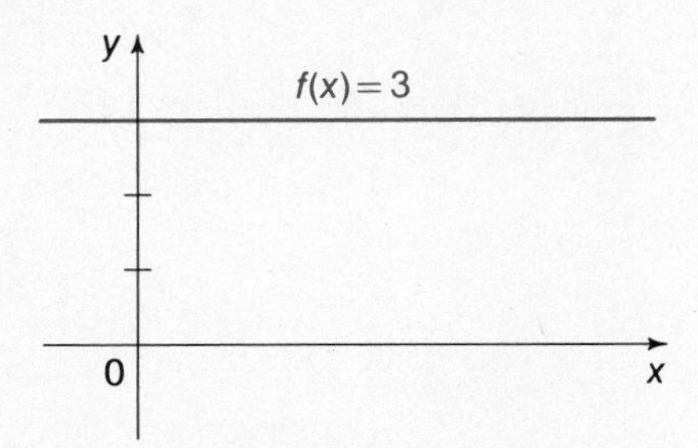

Figure 2.10

EXAMPLE 7 Graph the absolute value function $f(x) = |x|$.

SOLUTION From the definition of absolute value,

$$f(x) = \begin{cases} x, & \text{if } x \geq 0, \\ -x, & \text{if } x < 0. \end{cases}$$

Since all values of this function are nonnegative, $y \geq 0$, and none of the graph is below the x-axis. Also, since either $f(x) = x$ or $f(x) = -x$, we get the graph in Figure 2.11. Observe that $|-x| = |x|$ means that the graph is symmetric with respect to the y-axis. ●

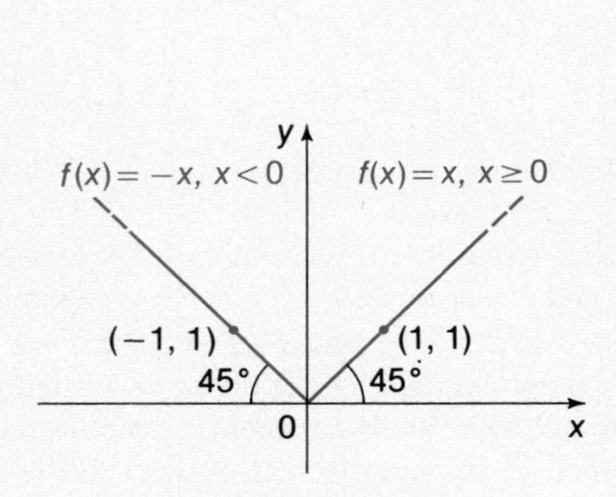

Figure 2.11

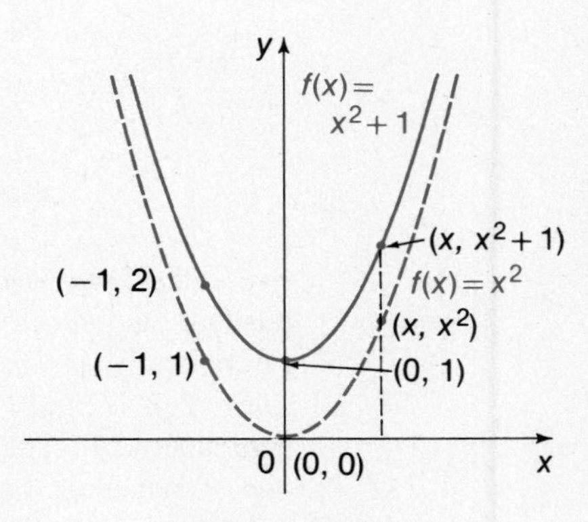

Figure 2.12

EXAMPLE 8 Graph $f(x) = x^2 + 1$.

SOLUTION Here the value that f assigns to each x is just 1 more than the corresponding value of the squaring function of Example 3. Thus, the graph is obtained by shifting the graph of the squaring function *one unit upward*. See Figure 2.12. ●

EXAMPLE 9 Graph the function

$$f(x) = \begin{cases} -1, & \text{if } x < 0; \\ 0, & \text{if } x = 0; \\ 1, & \text{if } x > 0. \end{cases}$$

SOLUTION The graph is shown in Figure 2.13. ●

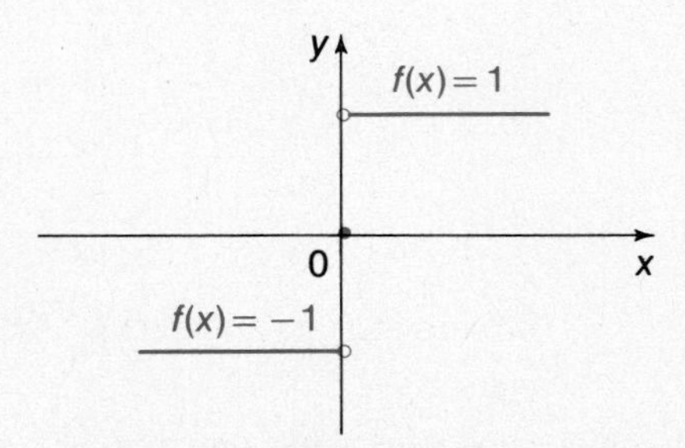

Figure 2.13

The graph of a function f is often referred to as the *graph of the equation* $y = f(x)$. For example, the curve of Figure 2.12 may be considered either the graph of the function $f(x) = x^2 + 1$ or the graph of the equation $y = x^2 + 1$.

In the graphing done in this section, little attention was paid to plotting many individual points. It is much more efficient first to *analyze the function* (or equation) and from this get the major features of the graph. A graph often shows the most important characteristics of a function more clearly than any other means. It provides a view of the "whole" function not given by individual function values. Analytic geometry (Chapter 7) and calculus provide much more powerful methods of studying both functions and graphs than are available here.

Relations are graphed in the same way as functions. The only significant difference occurs when the relation is not also a function. Consider, for example, the relation defined by $y^2 = x$. Here the correspondence assigns *two* values, $\sqrt{x}$ and $-\sqrt{x}$, to each value of x. The graph is symmetric with respect to the x-axis, as shown in Figure 2.14. This graph has the same size and shape as the graph of $y = x^2$ (Figure 2.7), but the roles of x and y are interchanged.

From the fact that a function assigns exactly one y to each x, it follows that a vertical line crosses the graph of a function in at most *one* point. If the relation is not a function, a vertical line may cross the graph in more than one point, as shown in Figure 2.14. This will be an important consideration in treating the inverse of a function in Section 2.5.

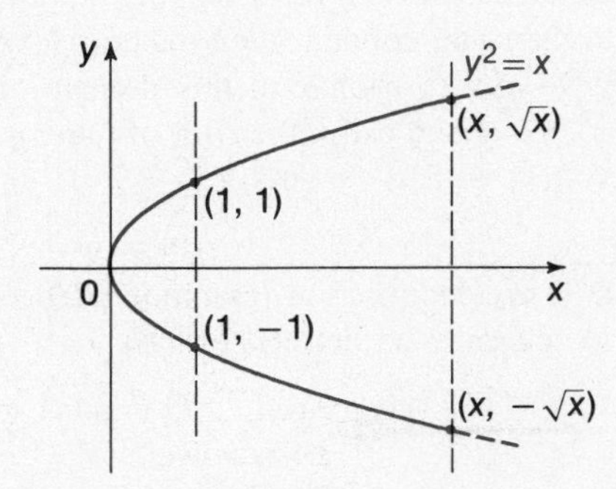

Figure 2.14

EXERCISES 2.3

Graph the function defined by the following sets.

1. $\{(1, 2), (2, 1), (3, 0), (-1, 3)\}$
2. $\{(2, -1), (3, 1), (0, 2), (1, 0)\}$

Graph the functions defined by $y = f(x)$.

3. $f(x) = -2$
4. $f(x) = -x$. [*Hint:* Each y-value is the negative of the corresponding y-value for the identity function.]

5. $f(x) = x + 1$. [*Hint:* See Example 8.]

6. $f(x) = -2x$

7. $f(x) = \frac{1}{2}x$

8. $f(x) = x - 1$

9. $f(x) = -x^2$

10. $f(x) = x^2 - 1$

11. $f(x) = |x|$

12. $f(x) = \begin{cases} 1, & \text{if } x < 0 \\ -1, & \text{if } x \geq 0 \end{cases}$

13. $f(x) = \begin{cases} 1, & \text{if } x \geq 0 \\ -x, & \text{if } x < 0 \end{cases}$

14. $f(x) = \begin{cases} 0, & \text{if } x \leq 0 \\ x, & \text{if } x > 0 \end{cases}$

15. $f(x) = \begin{cases} x^2, & \text{if } x < 0 \\ 0, & \text{if } x \geq 0 \end{cases}$

16. Boyle's Law relating pressure and volume of a confined gas under constant temperature is $pv = c$, where c is a constant. Write p as a function of v and graph it for the case $c = 3$. How is this graph related to that of the reciprocal function?

17. Graph the greatest integer function (Example 7, Section 2.2)

18. Graph the "car-rental" function described following Example 8, Section 2.2.

2.4 Construction of Functions from Other Functions

The algebraic operations on numbers may be extended in a natural way to real functions because of the rules for real numbers. For example, if f and g are two functions with the same domain, then the correspondence $x \to f(x) + g(x)$ assigns the number $f(x) + g(x)$ to each x in this domain. If we let $h(x) = f(x) + g(x)$, we can write $h = f + g$ and call h the **sum** of f and g. As an example, if $f(x) = x^2$ and $g(x) = x$, then $h(x) = f(x) + g(x) = x^2 + x$.

DEFINITION 2.7 If D is the intersection (common part) of the domains of f and g, the following functions may be defined on D:

Sum: $(f + g)(x) = f(x) + g(x)$
Difference: $(f - g)(x) = f(x) - g(x)$
Product: $(f \cdot g)(x) = f(x) \cdot g(x)$
Quotient: $(f/g)(x) = f(x)/g(x),\ g(x) \neq 0$
Scalar multiple: $(cf)(x) = c \cdot f(x)$

EXAMPLE 1 If $f(x) = x$ and $g(x) = \sqrt{1 - x^2}$, find $(f + g)(x)$ and give its domain.

SOLUTION Let $(f + g)(x) = h(x)$. Then

$$\begin{aligned} h(x) &= f(x) + g(x) \\ &= x + \sqrt{1 - x^2}. \end{aligned}$$

The domain of f is the set of all reals, and the domain of g is the set of reals in the interval $[-1, 1]$. So the domain of h is $[-1, 1]$, the common part of the domains of f and g. ●

EXAMPLE 2 Write the difference, product, and quotient of the functions f and g of Example 1 and identify the domain of each.

SOLUTION

$$(f - g)(x) = x - \sqrt{1 - x^2}$$

The domain is the closed interval $[-1, 1]$.

$$(fg)(x) = x\sqrt{1 - x^2}$$

The domain is also $[-1, 1]$.

$$\left(\frac{f}{g}\right)(x) = \frac{x}{\sqrt{1 - x^2}}$$

Here the domain is the open interval $(-1, 1)$, since we must have $x^2 < 1$. ●

EXAMPLE 3 Given $f(x) = x^2 - 1$ and $g(x) = x - 1$, find $\frac{f}{g}$ and give its domain.

SOLUTION Both f and g are defined for all real numbers x. If we let $h = f/g$, then

$$h(x) = \frac{x^2 - 1}{x - 1}$$

which is defined everywhere except for $x = 1$, since $g(1) = 0$. ●

A **scalar multiple function** $c \cdot f(x)$ is the product of a constant function and a variable function. The area formula $A(r) = \pi r^2$ is such a function: the constant is π and f is the squaring function with the restricted domain $\{x \mid x > 0\}$. Another example is **Boyle's Law,** $pv = c$, relating pressure and volume of a confined gas. This can be written as

$$p(v) = \frac{c}{v} = c\left(\frac{1}{v}\right),$$

which is the product of a constant function and the reciprocal function. This function can also be considered the quotient of the constant function $f(v) = c$ and the identity function $g(v) = v$.

Composition of Functions Another important way in which two functions may be used to construct a new function is illustrated in Figure 2.15. In (a), the mappings $f{:}x \to y$ and $g{:}y \to z$ combine to give the mapping $h{:}x \to z$. Of course, the range of f must be a subset of the domain of g. (Why?) In function notation, the above composite correspondence is expressed by $h(x) = g(y) = g[f(x)]$, read "g of f of x." The new function h is often called a *function of a function* or a *composite function* and is denoted by $g \circ f$. Figure 2.15b shows this situation represented by a function machine.

A simple illustration of a composite function is provided by the relationship yards → feet → inches. If x = the number of yards, y = the number of feet, and z = the number of inches, then $z = 12y$, $y = 3x$, and so $z = 12y = 12(3x) = 36x$.

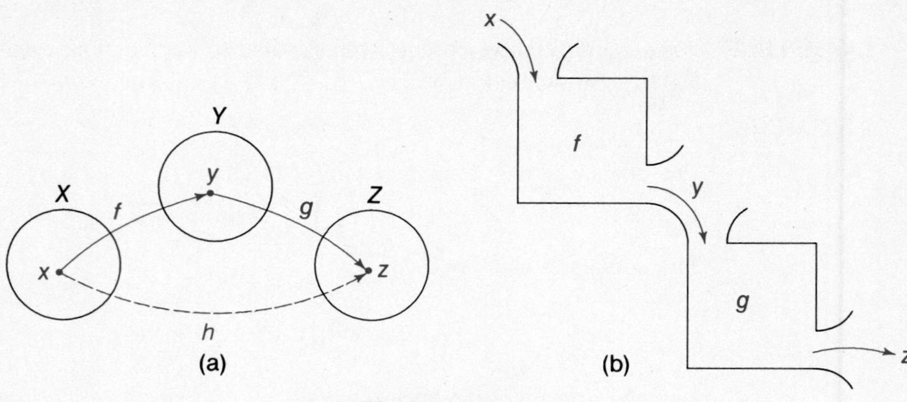

Figure 2.15

DEFINITION 2.8 If f is a function from set X into set Y and g is a function from set Y into set Z, then the **composite function $g \circ f$** is given by

$$(g \circ f)(x) = g[f(x)].$$

Here the domain of $g \circ f$ is restricted to those members x of the domain of f for which $g[f(x)]$ has meaning. The composite function $f \circ g$ is similarly defined and similar restrictions apply. In the notation $g[f(x)]$ it is often convenient to refer to f as the "inner function" and g as the "outer function."

EXAMPLE 4 Given $f(x) = x^2$ and $g(x) = x + 1$, find (a) $g \circ f$ and (b) $f \circ g$.

SOLUTION The domain and range of both functions are the set of all real numbers.

(a) $$\begin{aligned}(g \circ f)(x) &= g[f(x)] \\ &= f(x) + 1 \\ &= x^2 + 1.\end{aligned}$$

Note that we constructed $g \circ f$ by substituting the inner function $f(x)$ for x in the outer function $g(x)$. The domain of $g \circ f$ is the set of all real numbers, and the range is the set of all real numbers greater than or equal to 1.

(b) $$\begin{aligned}(f \circ g)(x) &= f[g(x)] \\ &= [g(x)]^2 \\ &= (x + 1)^2.\end{aligned}$$

In this case we substituted $g(x)$ for x in $f(x)$. The domain here is also the set of all real numbers. The range is the set of all non-negative real numbers. Compare this with the next example. ●

EXAMPLE 5 Given $f(x) = \dfrac{1}{x}$ and $g(x) = |x|$, find (a) $f \circ g$ and (b) $g \circ f$.

SOLUTION The domain and range of f are both the set of all nonzero real numbers. The domain of g is the set of all real numbers and its range is the set of all nonnegative real numbers.

(a) $$(f \circ g)(x) = f[g(x)] = \frac{1}{g(x)} = \frac{1}{|x|}.$$

In order to construct $f[g(x)]$, we must limit the domain to those x in the domain of g for which f has meaning, that is, $x \neq 0$.

(b) $$(g \circ f)(x) = g[f(x)] = |f(x)| = \left|\frac{1}{x}\right|.$$

Here $g \circ f$ is defined for all the domain of f, $x \neq 0$. ●

In Example 5, $f \circ g = g \circ f$, since they both have the same domain and same rule. This is not always the case, as Example 4 illustrates. In particular, if $f[g(x)] = g[f(x)] = x$ for some domain, the functions f and g have a special relationship which will be discussed in the next section.

Not only is $f \circ g$ not generally equal to $g \circ f$, but note also $f \circ g$ is not the same as $f \cdot g$, that is, *the composite of two functions is not the same as their product*. For instance, in Example 4 above, $f[g(x)] = (x + 1)^2$ but $(fg)(x) = x^2(x + 1)$, which is not equal to $(x + 1)^2$.

It may not be possible to construct both $f \circ g$ and $g \circ f$. As an example, let $f(x) = \sqrt{x - 1}$ and $g(x) = -x^2$. The domain of f is $\{x \mid x \geq 1\}$ whose members are all positive, but $g(x)$ has no positive values, so that $f[g(x)]$ is not defined. Formally, we would have $f[g(x)] = \sqrt{-x^2 - 1}$, which is not a real number. Check that $g[f(x)] = 1 - x$, for $x \geq 1$.

EXAMPLE 6 Suppose a stone is thrown into a pool of water, setting up circular waves moving outward at the rate of 10 feet per minute. Use composite function language to find the area enclosed by a circular wave as a function of the time elapsed.

SOLUTION The area as a function of the radius r is given by $A(r) = \pi r^2$, and the radius as a function of time is given by $r(t) = 10t$, where t is the time elapsed (in minutes). Then, the area as a function of time is given by the composite of these two functions:

$$a(t) = A[r(t)] = \pi(r(t))^2 = \pi(10t)^2 = 100\pi t^2.$$ ●

It is often convenient to think of a function as the composition of two simpler functions. As an example, suppose that $h(x) = \sqrt{x - 1}$. If we let $f(x) = \sqrt{x}$ and $g(x) = x - 1$, then

$$f(g(x)) = \sqrt{g(x)} = \sqrt{x - 1} = h(x).$$

That is, we may think of h as the composition of f with g. Observe that the domain of g is the set of all reals and the domain of f is the set of nonnegative reals, while the domain of h is the set of reals greater than or equal to 1. The domain of h is thus only a subset of the common part of the domains of f and g.

The significance of the construction of functions as described in this section is that it permits us to treat more complicated functions by dealing with their simpler component functions.

EXERCISES 2.4

Write the (a) sum, (b) difference, (c) product, and (d) quotient of each of the following pairs of functions, and give the domain of each function and constructed function.

1. $f(x) = x$ and $g(x) = \sqrt{x}$

2. $f(x) = x^2$ and $g(x) = x - 1$

3. $f(x) = \dfrac{1}{x}$ and $g(x) = x$

4. $f(x) = x^2 - 1$ and $g(x) = x - 1$

5. $f(x) = x$ and $g(x) = x^2$

6. $f(x) = x^2$ and $g(x) = x^3$

7. $f(x) = x^3$ and $g(x) = 4$

8. $f(x) = 1 + x$ and $g(x) = x$

9. $f(x) = 3x$ and $g(x) = x + 2$

10. $f(x) = x + 1$ and $g(x) = 1$

For each of the following pairs of functions, calculate both $f \circ g$ and $g \circ f$. In each case indicate whether or not $f \circ g = g \circ f$. Give the domains of f, g, $f \circ g$, and $g \circ f$.

11. $f(x) = x^2$ and $g(x) = \dfrac{1}{x}$

12. $f(x) = x + 2$ and $g(x) = 3$

13. $f(x) = \sqrt{x}$ and $g(x) = x^2, x \geq 0$

14. $f(x) = 3x - 1$ and $g(x) = \dfrac{1}{3}(x + 1)$

15. $f(x) = x^4$ and $g(x) = \sqrt{x}, x \geq 0$

16. $f(x) = \sqrt{x^2 + 1}$ and $g(x) = x$

17. $f(x) = x$ and $g(x) = x^2$

18. $f(x) = c$ and $g(x) = x$

19. $f(x) = \dfrac{x - 1}{x + 1}$ and $g(x) = \dfrac{1}{x}$

20. $f(x) = x + 1$ and $g(x) = \dfrac{x + 1}{x - 1}$

Show that $f \circ g = g \circ f$ for each of the following.

21. $f(x) = x + 1$ and $g(x) = x - 1$

22. $f(x) = 2x + 1$ and $g(x) = \dfrac{1}{2}(x - 1)$

23. $f(x) = \sqrt{x}$ and $g(x) = x^2, x \geq 0$

24. $f(x) = \dfrac{1}{2}(x + 3)$ and $g(x) = 2x - 3$

25. $f(x) = \dfrac{2}{x}$ and $g(x) = \dfrac{2}{x}$

26. $f(x) = \dfrac{3}{x - 2}$ and $g(x) = \dfrac{2x + 3}{x}$

27. $f(x) = \sqrt{x + 1}$ and $g(x) = x^2 - 1,\ x \geq 0$

28. $f(x) = 3x$ and $g(x) = x/3$

29. $f(x) = \sqrt[3]{x}$ and $g(x) = x^3$

30. $f(x) = \dfrac{1}{2}(x^2 - 1)$ and $g(x) = \sqrt{2x + 1}$

2.5 Inverse of a Function

A function f is a correspondence *from* each x in its domain *to* its image $f(x)$ in the range. An important concept is obtained by considering the reverse of the function correspondence. Figure 2.16 exhibits this idea in general by means of a mapping. Note that the domain and range sets of f have been interchanged to obtain those of g. We now consider some particular examples.

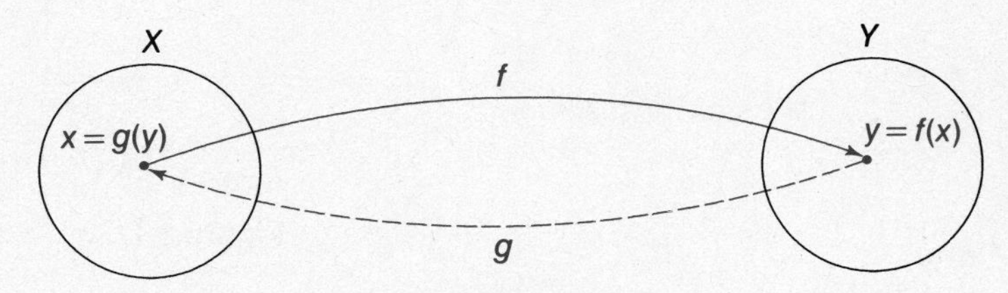

Figure 2.16

EXAMPLE 1 Let f and g be two functions defined by the following tables.

x	1	2	3	4
$f(x)$	3	-1	2	7

x	3	-1	2	7
$g(x)$	1	2	3	4

The ordered pairs of f are (1, 3), (2, -1), (3, 2), and (4, 7). The numbers in each pair of g, (3, 1), (-1, 2), (2, 3), and (7, 4), are the same but their order is reversed from those of f. For instance $f(1) = 3$ and $g(3) = 1$, so that $g[f(1)] = g(3) = 1$ and $f[g(3)] = f(1) = 3$. ●

EXAMPLE 2 Let $y = f(x) = 2x + 1$. Here the image of x is obtained by doubling x and then adding 1, which gives, in turn,

$$x, \quad 2x, \quad 2x + 1.$$

To reverse the rule of correspondence we must first *subtract* 1 from y and then *halve* the result, getting

$$y, \quad y - 1, \quad \frac{1}{2}(y - 1).$$

Thus we obtain $x = \frac{1}{2}(y - 1) = g(y)$. ●

In the above examples, two functions are so related that the directions of the correspondences are reversed and the domain and range interchanged.

DEFINITION 2.9 Suppose f is a function with domain X and range Y, and g is a function with domain Y and range X. Also suppose that $g[f(x)] = x$ for every x in X and $f[g(y)] = y$ for every y in Y. Then g is the **inverse function** of f, written $g = f^{-1}$, and f is the inverse function of g, $f = g^{-1}$. (See Figure 2.17.)

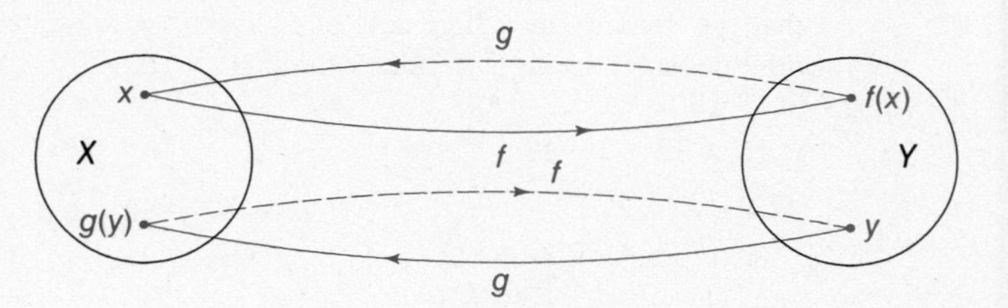

Figure 2.17

The notation f^{-1} is read "f inverse." (The symbol "f^{-1}" should not be confused with an exponent. Here f^{-1} does not mean $1/f$.) The functions f and g in Examples 1 and 2 above are inverse functions. The definition means that

$$f^{-1}[f(x)] = x \text{ for every } x \text{ in its domain } X, \text{ and}$$
$$f[f^{-1}(y)] = y \text{ for every } y \text{ in its domain } Y.$$

Each of the functions f and f^{-1} "undoes" what the other "does" to x under its rule. The relationship is a mutual one, that is f and g are *inverses of each other*. Using the language of inverse functions, the illustration in Example 1 can be written $f[f^{-1}(1)] = 1$ and $g[g^{-1}(3)] = 3$.

Given a function f, the method used to find its inverse depends upon the nature of f. If f is defined by a simple algebraic equation, such as $f(x) = 2x + 1$ in Example 2 above, there are several easy methods.

Method (1). The first method uses the definition directly: for every x,

$$f[f^{-1}(x)] = x, \qquad \text{by definition 2.9}$$
$$2[f^{-1}(x)] + 1 = x. \qquad \text{substituting } f^{-1}(x) \text{ for } x \text{ in } f(x) = 2x + 1$$

Solve this equation for $f^{-1}(x)$.

$$2[f^{-1}(x)] = x - 1$$

$$f^{-1}(x) = \frac{1}{2}(x - 1)$$

Method (2). The image y of the number x is obtained by applying the rule $y = f(x)$ to x. Now to reverse this rule, we wish to start with y and obtain the x of which y is the image. This is done by solving for x in terms of y to get $x = f^{-1}(y)$.

$$y = 2x + 1$$

$$-2x = 1 - y$$

$$x = \frac{1}{2}(y - 1)$$

The final equation defines x explicitly in terms of y: $f^{-1}(y) = (1/2)(y - 1)$. Now, in order to use x for the independent variable, we write

$$y = f^{-1}(x) = \frac{1}{2}(x - 1).$$

Method (3). As in Example 2, the steps used in constructing $f(x)$ from x may be reversed. Thus, since $f(x)$ was obtained from x in the following stages:

$$x, \quad 2x, \quad 2x + 1,$$

if we reverse the steps we get

$$y, \quad y - 1, \quad \frac{1}{2}(y - 1).$$

This gives $x = f^{-1}(y) = (1/2)\ (y - 1)$ or, with x as the independent variable,

$$f^{-1}(x) = \frac{1}{2}(x - 1).$$

We now apply the definition of inverse function to check that $f(x) = 2x + 1$ and $f^{-1}(x) = (1/2)(x - 1)$ are in fact inverses.

$$\begin{aligned} f[f^{-1}(x)] &= 2f^{-1}(x) + 1 \\ &= 2\left[\frac{1}{2}(x - 1)\right] + 1 \\ &= x. \end{aligned}$$

$$\begin{aligned} f^{-1}[f(x)] &= \frac{1}{2}[f(x) - 1] \\ &= \frac{1}{2}[(2x + 1) - 1] \\ &= x. \end{aligned}$$

Thus, $f[f^{-1}(x)] = f^{-1}[f(x)] = x$. The reader should check the result of Example 3 in the same way.

EXAMPLE 3 Given $f(x) = -\frac{1}{2}x + 3$, calculate the formula for $f^{-1}(x)$ in each of the three ways just illustrated.

Method (1). $f[f^{-1}(x)] = x$ by definition

$$-\frac{1}{2}[f^{-1}(x)] + 3 = x \qquad \text{substituting } f^{-1}(x) \text{ for } x \text{ in } f(x)$$

$$-\frac{1}{2}f^{-1}(x) = x - 3$$

$$f^{-1}(x) = -2(x - 3) \qquad \text{solving for } f^{-1}(x)$$

Method (2). Let $y = -\frac{1}{2}x + 3$. Then

$$-\frac{1}{2}x = y - 3,$$

$$x = -2(y - 3) = f^{-1}(y), \qquad \text{solving for } x$$

$$f^{-1}(x) = -2(x - 3). \qquad \text{interchanging } x \text{ and } y$$

Method (3). $f(x)$ is obtained from x in the stages:

$$x, \quad -\frac{1}{2}x, \quad -\frac{1}{2}x + 3.$$

Then, reversing these steps from last to first, we get: y, $y - 3$, $-2(y - 3)$, so that $f^{-1}(y) = -2(y - 3)$, and

$$f^{-1}(x) = -2(x - 3). \qquad \bullet$$

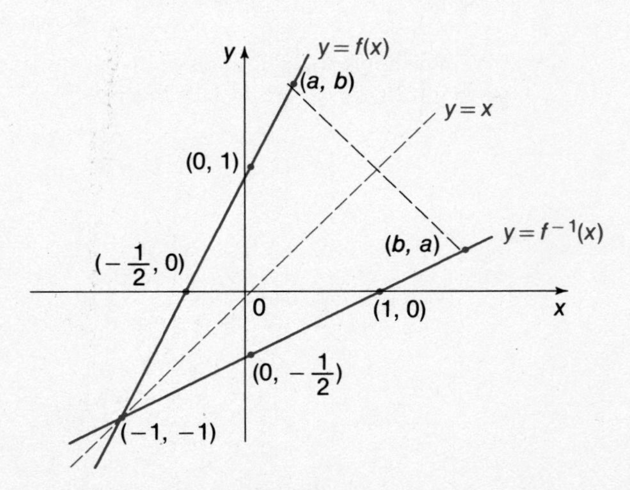

Figure 2.18

Graphs of Inverse Functions There is an important geometric relationship between the graphs of a function and its inverse. From the meaning of inverse function, if the pair (x, y) belongs to f, the pair (y, x) belongs to f^{-1}. For example, the pair $(3, 7)$ belongs to $y = 2x + 1$ while the pair $(7, 3)$ belongs to the inverse $y = (1/2)(x - 1)$. In Chapter 1 we noted that the points (a, b) and (b, a) are symmetric with respect to the line $y = x$. As a consequence, the graph of f^{-1} is a geometric reflection in this line of the graph of f. That is, if the coordinate plane were folded along this line, the graphs of f and f^{-1} would coincide. Figure 2.18 shows this for the graphs of $f(x) = 2x + 1$ and $f^{-1}(x) = (1/2)(x - 1)$.

Since $f(x) = x$ implies $f^{-1}(x) = x$, *the identity function is its own inverse*. Similarly, if $f(x) = 1/x$, then $f^{-1}(x) = 1/x$. So *the reciprocal function is its own inverse* also. See Figure 2.19 and 2.20 and note the symmetry about the line $y = x$.

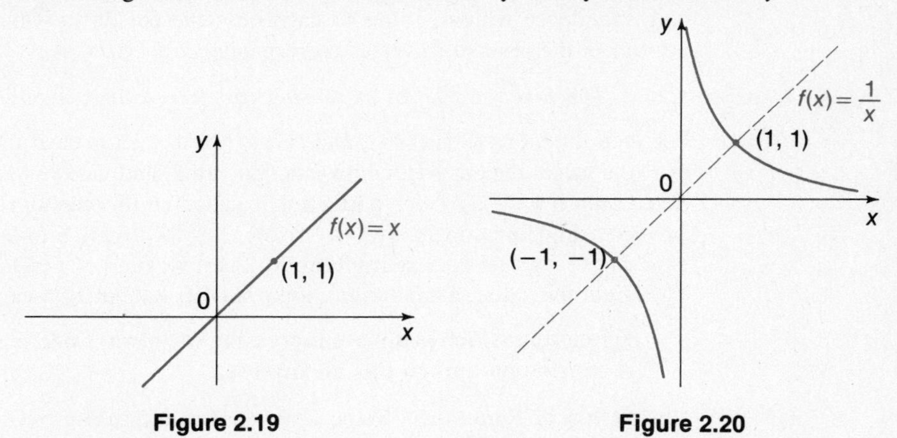

Figure 2.19

Figure 2.20

Existence of Inverses Not every function has an inverse function. For example, the squaring function, $y = f(x) = x^2$, assigns exactly one real number y to each real number x, but the reverse correspondence, $x = y^2$ or $y = \pm\sqrt{x}$, assigns to each positive x both the positive and negative square roots of x. (See Figure 2.21.) So the reverse correspondence is *not* a function. The reader should check that the inverse correspondence of the absolute value function $f(x) = |x|$ is not a function.

If a function satisfies one of two simple criteria it has an inverse:

(1) *The function f is a one-to-one correspondence.*

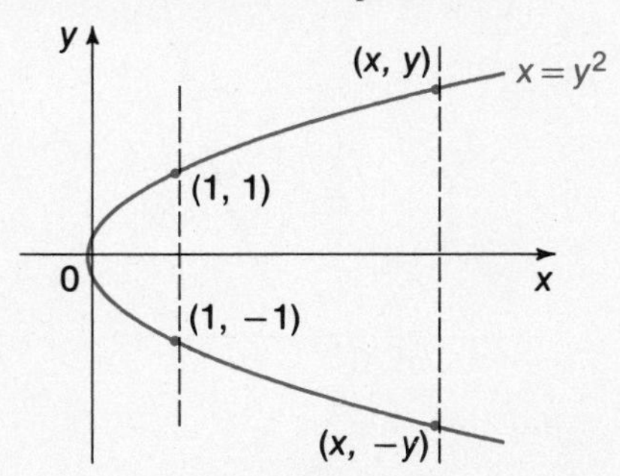

Figure 2.21

For example, to each person there corresponds exactly one nose. Not only does each person have one nose, but each nose belongs to just one person. The correspondence between the set of persons and set of noses is an example of a *one-to-one correspondence*.

Similarly, the formula $y = 2x + 1$ assigns one number y to each number x and, conversely, each y corresponds to only one x. So this is also a one-to-one correspondence or function. On the other hand, the formula $y = x^2$ assigns only one y to each x, but each positive y is assigned to two different x's. This correspondence is *not* one-to-one.

A one-to-one correspondence has an inverse.

For example, let the correspondence described above, person → nose, be called f. The correspondence which assigns to each nose the person to which it belongs, nose → person, is the reverse (inverse) correspondence f^{-1}.

(2) *The function f is an increasing (or decreasing) function.*

For such functions as $f(x) = x$ and $f(x) = 2x + 1$, it is clear that the larger the value of x, the larger the corresponding function value, and vice versa; that is, $f(x_1) > f(x_2)$ if and only if $x_1 > x_2$. Such a function is called an **increasing function.** The graph of an increasing function is always "rising" to the right. Similarly, the function $f(x) = 1/x$ $(x > 0)$, is a **decreasing function,** that is, $f(x_1) < f(x_2)$ if and only if $x_1 > x_2$. The larger the value of the variable x the smaller the function value. (See Figure 2.20.)

A function which is always increasing (or always decreasing) over its domain is one-to-one and so has an inverse.

Restriction of Functions Some functions are increasing over only part of their domains while decreasing over other parts, as for example $f(x) = x^2$, which is increasing for $x \geq 0$ and decreasing for $x \leq 0$. (See Figure 2.7.) However, if we let $g(x) = x^2$, $x \geq 0$ and $h(x) = x^2$, $x \leq 0$ (see Figure 2.22a), then $g^{-1}(x) = \sqrt{x}$ is the inverse of g and $h^{-1}(x) = -\sqrt{x}$ is the inverse of h. (See Figure 2.22b.) Thus, even if a function is not one-to-one over its entire domain, it may be so over some subset of the domain. It then has an inverse over this restricted domain of the function. We use this fact later in treating the inverses of the circular functions (Section 5.11).

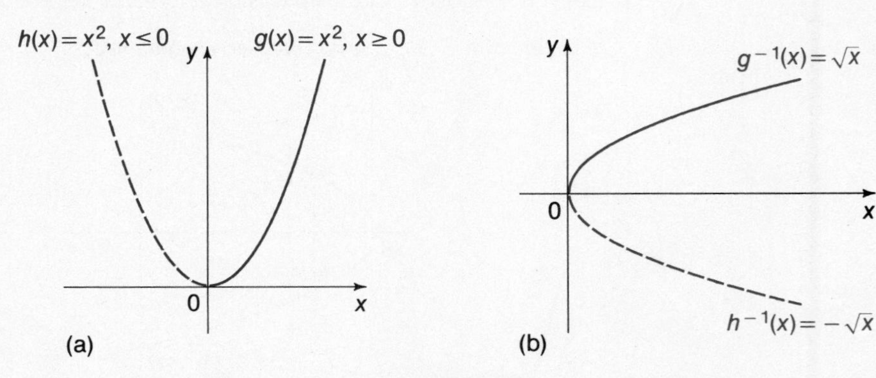

Figure 2.22

EXERCISES 2.5

Prove that f and g are inverse functions by showing that $f[g(x)] = x$ and $g[f(x)] = x$, for all real values of x.

1. $f(x) = x + 1$ and $g(x) = x - 1$
2. $f(x) = 1 - x$ and $g(x) = 1 - x$
3. $f(x) = 3x - 2$ and $g(x) = \frac{1}{3}(x + 2)$
4. $f(x) = 2x - 3$ and $g(x) = \frac{1}{2}(x + 3)$
5. $f(x) = x^2, x \geq 0,$ and $g(x) = \sqrt{x}, x \geq 0$
6. $f(x) = \sqrt{x + 1}$ and $g(x) = x^2 - 1$
7. $f(x) = \frac{2}{x}$ and $g(x) = \frac{1}{2}(1 - y)$
8. $f(x) = 1 - 2x$ and $g(x) = \frac{1}{2}(1 - x)$
9. $f(x) = \frac{3}{x - 2}$ and $g(x) = \frac{2x + 3}{x}$
10. $f(x) = \sqrt{2x + 1}$ and $g(x) = \frac{1}{2}(x^2 - 1)$

For each of the following functions f, calculate f^{-1} by the three methods illustrated in this section. Then check by calculating $f[f^{-1}(x)]$ and $f^{-1}[f(x)]$.

11. $f(x) = 1 - x$
12. $f(x) = x^3$
13. $f(x) = 2 - 3x$
14. $f(x) = x^3 + 1$
15. $f(x) = \sqrt{1 - 4x^2}$
16. $f(x) = 2 + x$
17. $f(x) = \frac{3}{x}$
18. $f(x) = 3 + 2x$
19. $f(x) = x^2 - 1$
20. $f(x) = \frac{x}{x - 1}$
21. From the relation between Fahrenheit and Celsius temperatures, $F(C) = 32 + \frac{9}{5}C$, express the Celsius reading C as a function of the Fahrenheit reading F. Then show that the functions F and C are inverses.
22. From the formula $A = P + Prt$ for the amount of money at the end of t years if P dollars are invested at r percent interest, express t as a function of amount A. Then show that the functions A and t are inverses.

Show that each of the following functions is its own inverse.

23. The identity function, $f(x) = x$.
24. The reciprocal function, $f(x) = \frac{1}{x}$.

2.6 Periodic Functions

All about us are phenomena which are repeating or periodic in nature. Familiar examples are the alternation of day and night, the ebb and flow of the tides, the motion of the moon about the earth, the oscillation of a pendulum, the up-and-down motion of a weight on a spring, and biological rhythms such as heartbeat and breathing. To study such phenomena mathematically requires functions whose values vary in a periodic fashion. The circular functions (Chapter 5) are significant in this respect.

EXAMPLE 1 Imagine a wheel having radius 1 rolling in a straight line on a flat surface. A point on the rim of the wheel traces out a curve called a **cycloid.** Figure 2.23 illustrates that each complete revolution of the wheel will trace out an arc congruent to that traced out by any other complete revolution. Thus in the given case the point will touch the line along which the wheel is rolling at intervals of 2π (the circumference of the wheel). Also, the given point on the rim will reach a given height repeatedly at intervals of 2π. Thus, if $y = f(x)$ is the equation of the curve, then $f(x) = f(x + 2\pi) = f(x + 4\pi)$, and so on. We say that the function f is periodic and has period 2π. ●

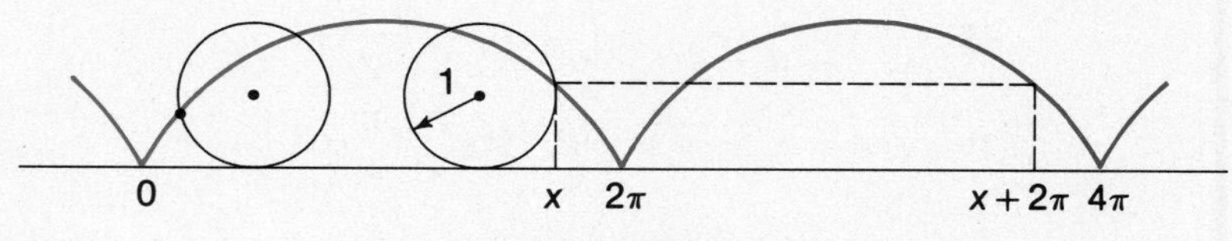

Figure 2.23

DEFINITION 2.10 Let f be a real function and suppose that for some fixed real number p, $f(x + p) = f(x)$ for every x in the domain of f. Then f is **periodic,** and the smallest such positive value of p is called the **period** of f.

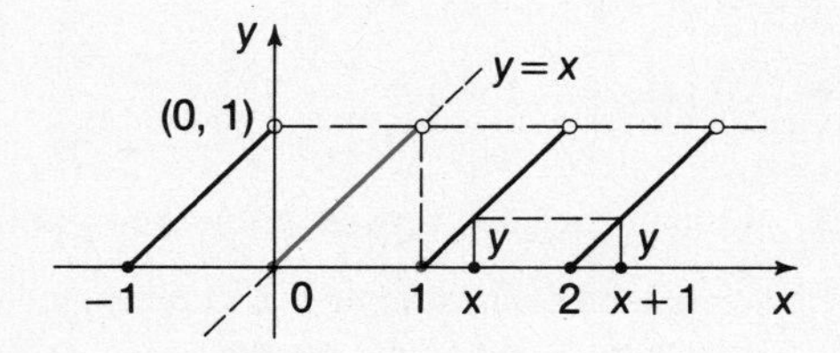

Figure 2.24

EXAMPLE 2 The function f whose graph is shown in Figure 2.24 is periodic, that is, it repeats cyclically. From $x = 0$ to $x = 1$, the graph has equation $y = x$. Also, $f(x + 1) = f(x)$ and no number smaller than 1 has this property. Thus, the period is 1 and the periodic function f can be defined as follows:

$$f(x) = \begin{cases} x, & \text{for } 0 \leq x < 1; \\ f(x \pm 1), & \text{for all } x. \end{cases}$$ ●

The special significance of a periodic function is that *if its values for a complete period are known,* all its values can be calculated. For example, using the values of f from Example 2 in the interval $0 \le x < 1$, we can calculate $f(x)$ for all x in the interval $1 \le x < 2$ as follows:

$$f(1) = f(0 + 1) = f(0) = 0,$$

$$f\left(\frac{5}{4}\right) = f\left(\frac{1}{4} + 1\right) = f\left(\frac{1}{4}\right) = \frac{1}{4},$$

$$f\left(\frac{3}{2}\right) = f\left(\frac{1}{2} + 1\right) = f\left(\frac{1}{2}\right) = \frac{1}{2},$$

$$f(2) = f(1 + 1) = f(1) = 0,$$

and so on. Using function values for this interval, we can calculate those for $2 \le x < 3, \ldots, n \le x < n + 1$.

In the same way, for the interval $-1 \le x < 0$,

$$f(-1) = f(-1 + 1) = f(0) = 0,$$

$$f\left(-\frac{1}{3}\right) = f\left(-\frac{1}{3} + 1\right) = f\left(\frac{2}{3}\right) = \frac{2}{3},$$

and so on. Note that we used the fact that the period is 1 to write $f(x) = f(x + 1)$, where $x + 1$ is in an interval where the function values are already known.

EXERCISES 2.6

Using the results of Example 2 above, calculate the following values.

1. $f\left(\frac{9}{4}\right)$ **2.** $f\left(\frac{5}{2}\right)$ **3.** $f\left(\frac{8}{3}\right)$ **4.** $f(3)$

5. $f(-1)$ **6.** $f\left(-\frac{1}{2}\right)$ **7.** $f(5)$ **8.** $f(-3)$

Exercises 9–12 refer to the figure below. The graph consists of a set of congruent semicircles of radius 1. *The equation of a circle centered at the origin with radius* 1 *is* $x^2 + y^2 = 1$, *so the equation of the middle semicircle in the figure is* $y = \sqrt{1 - x^2}$. Thus, the function is defined as follows:

$$f(x) = \begin{cases} \sqrt{1 - x^2}, & -1 \le x < 1; \\ f(x + 2), & \text{for all } x. \end{cases}$$

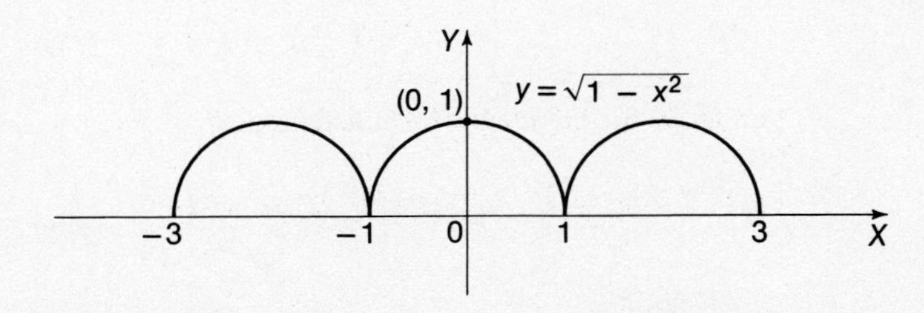

9. What is the period for this function?

10. Indicate this on the figure as was done on the figures with Examples 1 and 2.

11. What is the domain and the range of this function?

12. Calculate $f(-1)$, $f(0)$, $f\left(\frac{1}{2}\right)$, and $f(2)$.

Exercises 13–15 refer to the figure below.

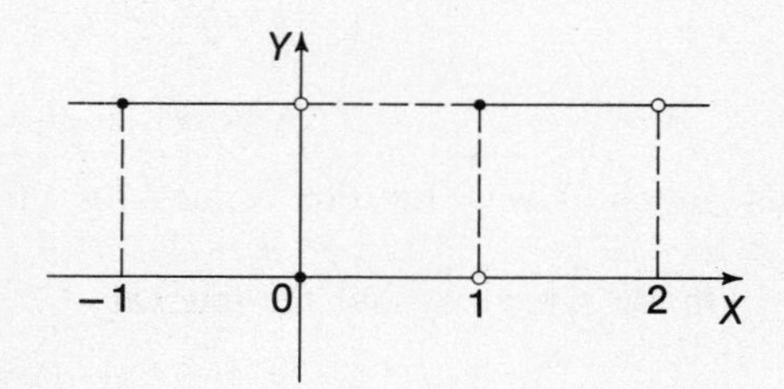

13. Verify that the function graphed in the figure is given by

$$f(x) = \begin{cases} 1, \text{ for } -1 \le x < 0; \\ 0, \text{ for } 0 \le x < 1; \\ f(x+2) = f(x) \text{ for all } x. \end{cases}$$

14. What is its domain? Its range? Its period?

15. Calculate $f\left(-\frac{1}{2}\right), f\left(\frac{1}{4}\right), f\left(\frac{1}{2}\right), f\left(\frac{3}{2}\right)$.

For any function with period p it can be shown that $f(x + np) = f(x)$ for any integer n. Use the definition of periodicity to prove the following particular cases.

16. $f(x + 2p) = f(x)$. [*Hint:* $f(x + 2p) = f((x + p) + p) = f(x + p)$.]

17. $f(x + 3p) = f(x)$

18. $f(x - p) = f(x)$.

2.7 Odd and Even Functions. Zeros of Functions

Odd and Even Functions Another useful feature of certain functions is that of being *odd* or *even*. For example, for the identity function $f(x) = x$,

$$f(-x) = -x = -f(x).$$

Similarly, for the reciprocal function, $f(x) = \dfrac{1}{x}$,

$$f(-x) = -\frac{1}{x} = -f(x).$$

That is, replacing the variable x by its negative changes the sign on the corresponding function value. We call such a function an *odd function*. Geometrically, *the graph of an odd function is symmetric with respect to the origin*. See Figures 2.25 and 2.26.

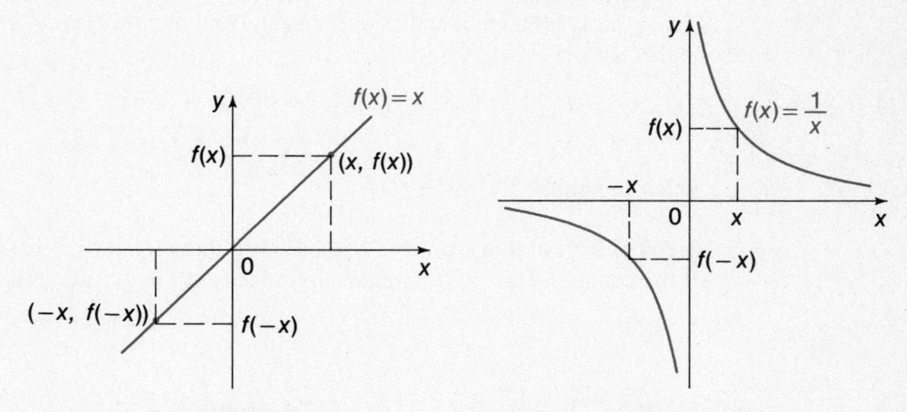

Figure 2.25

Figure 2.26

On the other hand, if $f(x) = x^2$, then

$$f(-x) = (-x)^2 = x^2 = f(x).$$

Also, for $f(x) = |x|$,

$$f(-x) = |-x| = |x| = f(x).$$

In each of these cases, if x is replaced by its negative, the corresponding function value is unchanged. Such functions are called *even functions*. The *graphs of even functions are symmetric with respect to the y-axis*. See Figures 2.27 and 2.28.

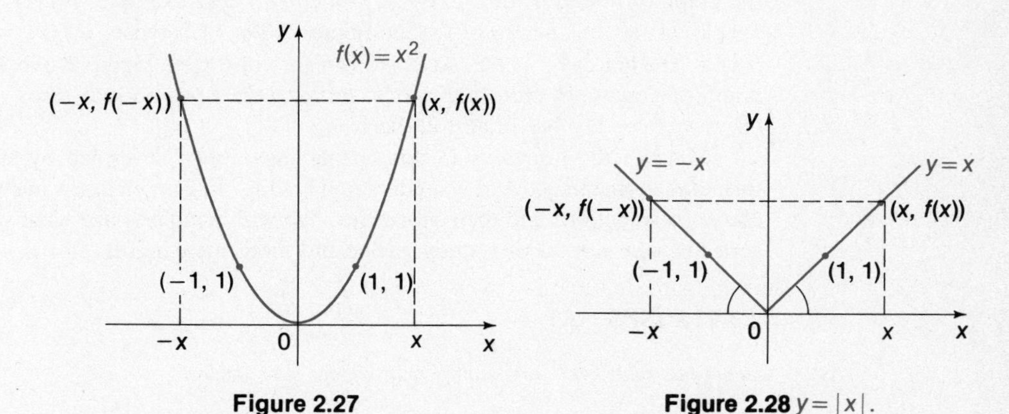

Figure 2.27

Figure 2.28 $y = |x|$.

DEFINITION 2.11 If $f(-x) = -f(x)$ for all x in the domain of f, then f is an **odd function.**

If $f(-x) = f(x)$ for all x in the domain of f, then f is an **even function.**

The terms *odd* and *even* probably come from the fact that odd powers (such as x, x^3, x^5) of negative numbers are negative and even powers (such as x^2, x^4, x^6) of negative numbers are positive. For example, if $f(x) = 3x^4 + x^2 - 1$, then f is an even function, and if $f(x) = x^3 - 2x$, then f is an odd function. The zero function, $f(x) = 0$, is both even and odd. (Why?) The function $f(x) = x^3 + x^2$, however, is neither even nor odd, since

$$\begin{aligned} f(-x) &= (-x)^3 + (-x)^2 \\ &= -x^3 + x^2, \end{aligned}$$

which is neither $f(x)$ or $-f(x)$.

Zeros of a Function If r is a number such that $f(r) = 0$, r is a **zero** of f and a **root** of the equation $f(x) = 0$. Graphically, a zero of f is the x-coordinate of a point where

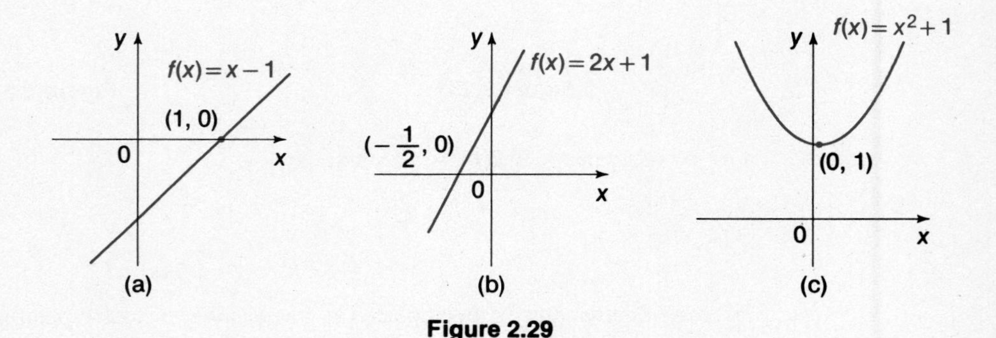

Figure 2.29

the graph of f crosses the x-axis (an x-intercept). For example, if $f(x) = x - 1$, then $f(1) = 0$ and 1 is a zero of f. (See Figure 2.29a.) Likewise, if $f(x) = 2x + 1$, then $f(x) = 0$ when $x = -1/2$, so $-1/2$ is a zero of f. (See Figure 2.29b.) Of course, the graph of f may not cross the x-axis; in that case f has no real zeros. An example is $f(x) = x^2 + 1$. (See Figure 2.29c.)

Most of the functions in this chapter have been described by simple algebraic formulas so that they could be understood readily. In later chapters, we will study other classes of functions and their properties. We will be able to use what we have learned here to make this further study easier and more meaningful.

EXERCISES 2.7

Identify as odd, even, or neither, and explain your answer.

1. $f(x) = x^3 - x$

2. $f(x) = x^2 + 1$

3. $f(x) = \sqrt{x}$

4. $f(x) = x + x^2$

5. $f(x) = x^2 - x + 1$

6. $f(x) = \sqrt{1 + x^2}$

7. $f(x) = x^4 - x^2$

8. $f(x) = x + \dfrac{1}{x}$

9. $f(x) = 3$

10. $f(x) = x^2 - x$

11. $f(x) = x^5 - x^3$

12. $f(x) = \sqrt{x + x^3}$

By inspection, identify the zeros (if any) of each of the following functions.

13. $f(x) = 1 + x$ **14.** $f(x) = 2x$ **15.** $f(x) = x^2 - 3$

16. $f(x) = 3x^2 + 2$ **17.** $f(x) = \sqrt{x - 1}$ **18.** $f(x) = (x + 1)(x - 2)$

19. $f(x) = 2x - 1$ **20.** $f(x) = x^2 - x$ **21.** $f(x) = 1 - x^2$

22. $f(x) = \frac{1}{x}$ **23.** $f(x) = |x|$ **24.** $f(x) = (x - a)(x - b)$

2.8 Function—a Bit of History

The idea of *function* is the central theme of our study in this course. It is one of the most important concepts of mathematics and permeates all of it as a unifying principle. As with any idea, the function concept was at first only roughly perceived. The term itself was introduced by Descartes about 1630 to refer to expressions such as x^n.

Later in the same century Leibniz (one of the inventors of calculus) used the term to denote any quantity connected with a curve, such as a point on the curve. The notion of a function as any expression involving constants and literal numbers (as it is often seen to be in elementary algebra) was presented by the great Swiss mathematician Euler in a book published in 1748. *(Introductio in analysin infinitorum.)* There he stated that "a function of a variable quantity is any analytic expression made from the variable quantity and from numbers and constant quantities." Thus, expressions such as $ax^2 + bx + c$, $(9/5)C + 32$, and so on, would be functions according to Euler.

The above view was held until about 1800. As explained in this chapter, the concept of a function is much more inclusive than Euler's description implies. For one thing, not all functions can be described by formulas. For example, the number of primes smaller than the positive integer n is a function of n, but there is no known formula for this function. A function was first thought of as a *dependence* relationship, but its use is seen in a broader context as a *correspondence*.

The modern interpretation of function as a rule or a correspondence was formulated by Dirichlet at the beginning of the last century. The symbol $f(x)$ is due to Euler who, together with Leibniz, did much to simplify notation. Jean Bernouilli somewhat earlier (1702) had experimented with notations for functions. One of them was ϕx for our $f(x)$. Interestingly enough, the idea of picturing (graphing) the way a quantity varies was used as early as 1361 (Oresme).

Functions are classified into groups according to their characteristics. Certain functions are called *elementary functions*. The classes of elementary functions studied in this course are related as shown below.

- Elementary Functions
  - algebraic
    - polynomial
    - rational
    - irrational
  - transcendental (nonalgebraic)
    - exponential
    - logarithmic
    - circular (trigonometric)
    - inverse circular

Algebraic functions will be discussed in Chapter 3, exponential and logarithmic functions in Chapter 4, circular functions in Chapter 5, and trigonometric functions in Chapter 6.

Chapter 2 Review Exercises

Use the definition of a function to explain why each of the following either does (or does not) describe a function. If a function is indicated, identify its domain and range.

1. y is a number such that $y = x^2 + 2x$.

2. y is a number such that $y^2 + 1 = x$.

3. Each biological species has a specific number of chromosomes.

4. Counting determines a number for each collection of objects.

5. Every house has several rooms.

6. The surface area of a sphere is given by the formula $S = 4\pi r^2$.

7. A salesman is assigned several states for his territory.

8. Each country is composed of its citizens.

9. Each person standing in the sunlight has a shadow.

10. Each state has its capital city.

Find the average change in the value of each function f corresponding to a change, h, in the value of x.

11. $f(x) = 3x - 1$ **12.** $f(x) = x^2 + 1$ **13.** $f(x) = x^2 + 2x$

14. $f(x) = \sqrt{2x}$ **15.** $f(x) = x^3$

What is the set of admissible values of x (natural domain of f) in each of the following?

16. $f(x) = \sqrt{x - 1}$ **17.** $f(x) = \sqrt{x^2 - 1}$

18. $f(x) = \dfrac{1}{x^2 + 1}$ **19.** $f(x) = \dfrac{1}{x - 1}$

20. $f(x) = \dfrac{1}{\sqrt{1 - x^2}}$ **21.** $f(x) = \dfrac{1}{x(x - 1)}$

22. $f(x) = \dfrac{\sqrt{x + 1}}{x}$ **23.** $f(x) = \dfrac{1}{2x - 1}$

Which of the following relations between x and y define y as a function of x. Why or why not?

24. $x + 3y = 1$ **25.** $x^2 + 2y = 1$

26. $2x + y^2 = 1$ **27.** $|x| + |y| = 1$

For each of the following, find f[g(x)] and g[f(x)]. Identify those cases in which f and g are inverses and explain why.

28. $f(x) = x^2$ and $g(x) = \dfrac{1}{x^2 - 1}$ **29.** $f(x) = \dfrac{1}{x}$ and $g(x) = \sqrt{x}$

30. $f(x) = 2 - 3x$ and $g(x) = \frac{1}{3}(2 - x)$ **31.** $f(x) = \frac{1}{x - 1}$ and $g(x) = \frac{x + 1}{x}$

32. $f(x) = x + 1$ and $g(x) = x^2 - 2$

Graph each of the following.

33. The identity function

34. The squaring function

35. The square root function

36. The reciprocal function

37. The absolute value function

38. The greatest integer function

Graph each of the following and identify the domain and the range in each case.

39. $f(x) = \begin{cases} -1, & 0 \le x \le 2 \\ 2, & 2 < x \le 4 \end{cases}$

40. $f(x) = \begin{cases} x, & \text{if } x \le 1 \\ 2, & \text{if } 1 < x \le 2 \end{cases}$

41. $f(x) = \begin{cases} x, & x < 0 \\ x + 1, & x \ge 0 \end{cases}$

42. $f(x) = |x - 1|$

Find the inverse of each of the following functions (if it exists) by the three methods of the text examples. If f^{-1} does not exist, explain how you know. Check each f^{-1}.

43. $f(x) = 3x - 1$

44. $f(x) = \sqrt{1 + x}$

45. $f(x) = \frac{x - 5}{2x + 7}$

46. $f(x) = \frac{2x + 3}{x - 1}$

For each of the following, calculate $f^{-1}(x)$ and graph both $y = f(x)$ and $y = f^{-1}(x)$ on the same set of axes.

47. $f(x) = 2x$

48. $f(x) = x + 1$

49. $f(x) = 2x - 1$

50. Given the periodic function defined by $f(x) = \begin{cases} 1, & 0 \le x < 1 \\ -1, & 1 \le x < 2 \\ f(x + 2), \text{ all } x, \end{cases}$

state the period of f and write the following values: (a) $f\left(\frac{1}{3}\right)$, (b) $f\left(\frac{3}{2}\right)$, (c) $f\left(-\frac{1}{2}\right)$, (d) $f(-1)$, (e) $f\left(\frac{5}{4}\right)$.

Identify each of the following functions as even, odd, or neither. Justify your answer.

51. $f(x) = 4$

52. $f(x) = 2x$

53. $f(x) = x^3 - 3x$

54. $f(x) = x^4 + x^2$

55. $f(x) = \sqrt{x^2 - 4}$

56. $f(x) = \frac{x^3 + x}{x^2}$

In your own words, express clearly what is meant by each of the following concepts, and give an illustration of each.

57. A function

58. Domain and range of a function

59. Graph of a function

60. Inverse of a function

61. Odd and even functions

62. Periodic function

63. Equal functions

64. Composite function

Chapter 2 Miscellaneous Exercises

For each of the functions defined by the graphs below:

(a) identify the domain and range;

(b) find the value at $x = -1, 0, \frac{1}{2}$, *and* 3;

(c) write a formula which defines the function.

1.

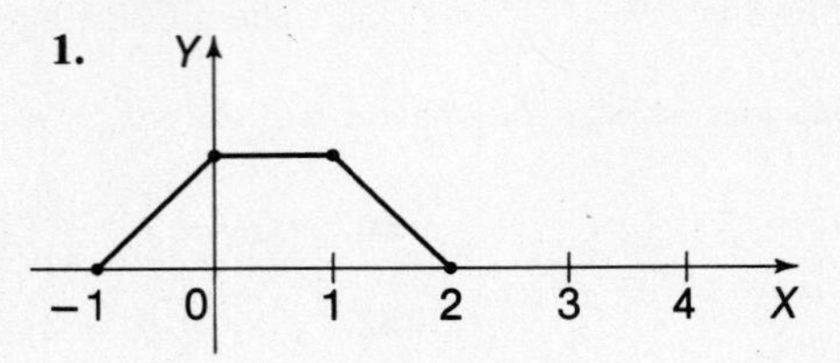

2.

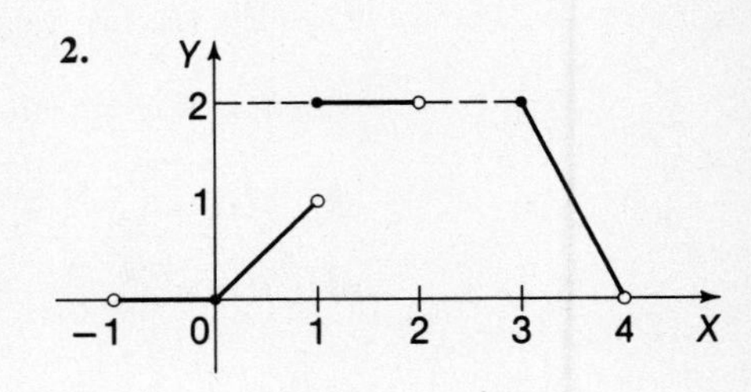

3. The law of gravity defines the force of attraction F of the earth for a mass of one unit as a function of its distance x from the center of the earth by the formulas

$$F = f(x) = \begin{cases} \left(\frac{c}{r^3}\right)x, & 0 \le x \le r; \\ \frac{c}{x^2}, & x > r; \end{cases}$$

where c is constant and r is the radius of the earth. Use these formulas to determine the force F on a unit mass (a) halfway from the surface to the center of the earth, (b) at the center of the earth, and (c) at a distance beyond the surface of the earth equal to 2 earth-radii. *Express results in functional notation.*

4. The following data were obtained from observations and then graphed. Do the points lie on a straight line? Prove your answer.

x	1.4	4.5	12.2
y	−2.0	2.4	13.5

5. Graph the function f if $f(x) = \begin{cases} -1, \text{ if } x < -1 \\ x, \text{ if } -1 \le x \le 1 \\ 1, \text{ if } x > 1 \end{cases}$

6. Show that if both f and g are even functions, then so also are (a) $f + g$, (b) $f - g$, (c) $f \cdot g$, and (d) f/g.

7. Show that an even function does not have an inverse.

8. Show that (a) the product of two even functions is even, (b) the product of two odd functions is even, and (c) the product of an even and an odd function is odd.

3

Algebraic Functions

Much of the material on polynomials in the first part of this chapter will be familiar from elementary algebra. Here, however, it will be presented in the context of the function concept developed in Chapter 2.

3.1 Polynomials and Polynomial Functions

Expressions such as

$$x^2 + 3x - 2, \qquad \frac{(x + 1)^2}{x^3 + 2x}, \qquad \text{and} \qquad \sqrt{x^2 - 1}$$

are called **algebraic expressions.** Each of them is constructed from the real variable x and real numbers by means of the algebraic operations of addition, subtraction, multiplication, division, and finding roots. In the first expression above, only the operations of addition and multiplication are used (subtraction is really addition of the negative). This expression is called a **polynomial.** The second expression also involves the operation of division and is called a **rational algebraic expression.** (It is called a *rational* expression because it is the ratio of two polynomials.) The relationships of these separate types of algebraic expressions are shown in Figure 3.1.

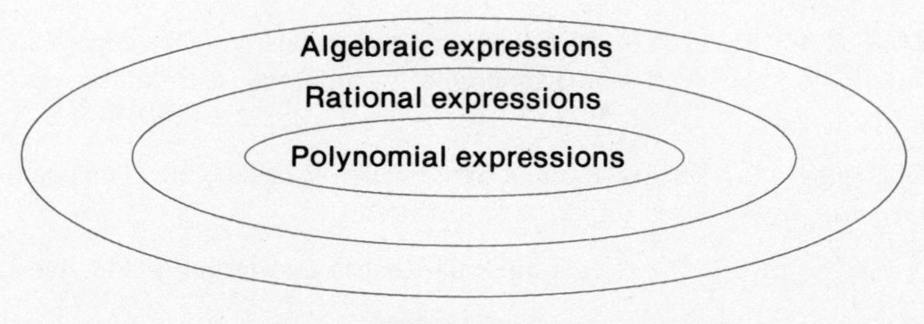

Figure 3.1

Polynomial Functions Any algebraic expression can be used to define an **algebraic function.** For instance, the rule

$$f(x) = x^3 + 3x - 2$$

defines f as a *polynomial function* of x.

Starting with the identity function $f_1(x) = x$ and multiplication of functions, we build up the squaring function $f_2(x) = [f_1(x)]^2 = x^2$, the cubing function $f_3(x) = x^3$, and, in general, the **power function** $f_n(x) = x^n$. Next, multiplication by a constant a produces $g_1(x) = af_1(x) = ax$, $g_2(x) = ax^2$, and so on. Finally, by addition, we obtain the family of **polynomial functions** described by the following equations.

$p_0(x) = a_0$	constant function
$p_1(x) = a_1x + a_0$	linear function
$p_2(x) = a_2x^2 + a_1x + a_0$	quadratic function
$p_3(x) = a_3x^3 + a_2x^2 + a_1x + a_0$	cubic function
$p_4(x) = a_4x^4 + a_3x^3 + a_2x^2 + a_1x + a_0$	quartic function
$\vdots$	

In general,

$p(x) = a_nx^n + a_{n-1}x^{n-1} + \cdots + a_1x + a_0$. *n*th degree polynomial function

The constants a_0, a_1, a_2, and so on, assume *only* real values, and only real values are assigned to the variable x. Then the polynomial or polynomial function (the two terms are used interchangeably) is real-valued and we call it a **real polynomial.**

Let us review some of the polynomial terminology learned in elementary algebra. The **degree** of a polynomial (and of the corresponding polynomial function) is the highest power of the variable in the polynomial. The polynomial function $p_0(x) = ax^0$ of degree zero is the constant function $f(x) = a$ (recall that $x^0 = 1$) introduced in the preceding chapter. In particular, if a is zero, that is, if $f(x) = 0$ for all x, f is called the **zero function** and does not have a degree.

The constants a_0, a_1, a_2, and so on, are called **coefficients,** and each of the expressions of the form a_kx^k is called a **term** of the polynomial. The term containing the highest power of x is called the **leading term** and its coefficient is the **leading coefficient.** A polynomial with a single term is called a **monomial,** one with two terms is a **binomial,** and one with three terms is a **trinomial.**

EXAMPLE 1 For each of the following polynomials, give the degree and the leading coefficient, and decide if it is a monomial, a binomial, a trinomial, or none of these.
(a) $3x^2 + 5x + 7$ (b) $x^3 - 1$ (c) $2 - x + 5x^2 + 3x^3 - x^4$

SOLUTION (a) The polynomial $3x^2 + 5x + 7$ has degree 2 and leading coefficient 3. Since it has three terms, it is a trinomial.

(b) $x^3 - 1$ is a binomial (it has two terms) of degree 3, with leading coefficient 1.

(c) The leading coefficient in $2 - x + 5x^2 + 3x^3 - x^4$ is -1, since $-x^4$ is the term of highest power, and the degree is 4. ●

Two polynomials are **equal** if and only if they have identical coefficients for every power of the variable. For example,

$$ax^3 + bx^2 + cx + d = x^2 - 2x + 5$$

if and only if $a = 0$, $b = 1$, $c = -2$, and $d = 5$.

Polynomial functions are relatively easy to evaluate, as shown in the following example. For this reason, one of the important uses of polynomials is to approximate more complicated functions. We will see how this is done in later chapters as we study different kinds of functions.

EXAMPLE 2 Evaluate the polynomial function $f(x) = 2x^3 + 4x + 5$ at 0, -1, and 2.

SOLUTION Replace x, in turn, by 0, -1, and 2 in the expression for $f(x)$.

$$f(0) = 2(0)^3 + 4(0) + 5 = 5$$
$$f(-1) = 2(-1)^3 + 4(-1) + 5 = -1$$
$$f(2) = 2(2)^3 + 4(2) + 5 = 29 \quad ●$$

Calculation of Polynomial Function Values With a little practice the calculation of polynomial function values on a small hand calculator (without a memory) can be quite rapid. Consider the polynomial function $p(x) = 2x^3 + 3x^2 - x + 4$. Notice that it can be thought of as being constructed in the following stages.

2	write the first coefficient
$2x$	multiply by x
$2x + 3$	add the next coefficient
$2x^2 + 3x$	multiply by x
$2x^2 + 3x - 1$	add the next coefficient
$2x^3 + 3x^2 - x$	multiply by x
$2x^3 + 3x^2 - x + 4$	add the last coefficient

Briefly, $2x^3 + 3x^2 - x + 4 = [(2x + 3)x - 1]x + 4$. The following procedure shows how to use this to calculate $p(5)$ on a calculator.

enter 2	2	$(= 2)$
multiply by 5	$2(5)$	$(= 10)$
add 3	$2(5) + 3$	$(= 13)$
multiply by 5	$2(5^2) + 3(5)$	$(= 65)$
add -1	$2(5^2) + 3(5) - 1$	$(= 64)$

$$\begin{array}{lll} \text{multiply by 5} & 2(5^3) + 3(5^2) - 1(5) & (= 320) \\ \text{add 4} & 2(5^3) + 3(5^2) - 1(5) + 4 & (= 324) \end{array}$$

It is no more difficult to calculate $p(1.23)$ by this method than to calculate $p(5)$.

Polynomial Algebra The rest of this section gives a brief review of the arithmetic operations on polynomials.

EXAMPLE 3 Given the polynomial functions $p(x) = 2x^3 + 3x + 7$ and $q(x) = x^2 + x - 2$, find (a) $p(x) + q(x)$, (b) $p(x) - q(x)$, (c) $q(x) \cdot p(x)$, (d) $p(x)/q(x)$.

SOLUTION (a) The sum, $p(x) + q(x)$, is found by adding coefficients of like powers.

$$\begin{array}{rcl} p(x) &=& 2x^3 + 0x^2 + 3x + 7 \\ q(x) &=& + 1x^2 + 1x - 2 \\ \hline p(x) + q(x) &=& 2x^3 + x^2 + 4x + 5 \end{array}$$

(b) To find the difference, $p(x) - q(x)$, subtract coefficients.

$$\begin{array}{rcl} p(x) &=& 2x^3 + 0x^2 + 3x + 7 \\ q(x) &=& + 1x^2 + 1x - 2 \\ \hline p(x) - q(x) &=& 2x^3 - x^2 + 2x + 9 \end{array}$$

(c) The product, $q(x) \cdot p(x)$, is found by multiplying $p(x)$ by each term of $q(x)$, then adding.

$$\begin{array}{rcll} p(x) &=& 2x^3 + 3x + 7 & \\ q(x) &=& x^2 + x - 2 & \\ \hline & & 2x^5 + 3x^3 + 7x^2 & = x^2 \cdot p(x) \\ & & 2x^4 + 3x^2 + 7x & = x \cdot p(x) \\ & & -4x^3 - 6x - 14 & = -2p(x) \\ \hline q(x) \cdot p(x) &=& 2x^5 + 2x^4 - x^3 + 10x^2 + x - 14 & = (x^2 + x - 2)p(x) \end{array}$$

(d) The quotient, $p(x)/q(x)$, may be found by long division.

$$\begin{array}{rll} & 2x - 2 & \text{quotient} \\ x^2 + x - 2 \,\big) & 2x^3 + 0x^2 + 3x + 7 & \\ & 2x^3 + 2x^2 - 4x & = 2x(x^2 + x - 2) \\ \hline & -2x^2 + 7x + 7 & \\ & -2x^2 - 2x + 4 & = -2(x^2 + x - 2) \\ \hline & 9x + 3 & \text{remainder} \end{array}$$

$$\frac{p(x)}{q(x)} = \frac{2x^3 + 3x + 7}{x^2 + x - 2} = 2x - 2 + \frac{9x + 3}{x^2 + x - 2}.$$

Addition and multiplication may be done more directly by using the properties of operations with real numbers. For example,

$$\begin{aligned} p(x) + q(x) &= (2x^3 + 3x + 7) + (x^2 + x - 2) \\ &= (2 + 0)x^3 + (0 + 1)x^2 + (3 + 1)x + (7 - 2) \\ &= 2x^3 + x^2 + 4x + 5. \end{aligned}$$

That is, the commutative and distributive properties are used to add coefficients of like powers. Each pair of coefficients in parentheses in the example above corresponds to a pair in a column in the arrangement shown in Example 3(a). Similarly,

$$\begin{aligned} p(x) \cdot q(x) &= (2x^3 + 3x + 7)(x^2 + x - 2) \\ &= (2x^3 + 3x + 7)x^2 + (2x^3 + 3x + 7)x + (2x^3 + 3x + 7)(-2) \\ &= (2x^5 + 3x^3 + 7x^2) + (2x^4 + 3x^2 + 7x) + (-4x^3 - 6x - 14) \\ &= 2x^5 + 2x^4 - x^3 + 10x^2 + x - 14. \end{aligned}$$

That is, the distributive property is used successively and then coefficients of like powers are added. Observe that each expression in parentheses in the next to last line above corresponds to a line in Example 3(c).

As shown in Example 3(a)–(c), the sum, difference, and product of two polynomials is a polynomial. The quotient of two polynomials is not necessarily a polynomial; it may be written either as a rational expression or as the sum of a polynomial and a rational expression, as in 3(d). The degree of the sum of two polynomials is the same as that of the polynomial of greater degree, and the degree of the product of two polynomials is the sum of their degrees.

Factoring An important operation on polynomials is **factoring:** expressing a polynomial as a product of polynomials. Facility in factoring depends upon knowing certain standard products. The most important ones are listed here for review.

1. $a(x + c) = ax + ac$ — distributive property
2. $(x + a)(x + b) = x^2 + (a + b)x + ab$ — special trinomial
3. $(ax + b)(cx + d) = acx^2 + (ad + bc)x + bd$ — general trinomial
4. $(x + a)^2 = x^2 + 2ax + a^2$ — trinomial squares
5. $(x + a)(x - a) = x^2 - a^2$ — difference of square
6. $(x + a)(x^2 - ax + a^2) = x^3 + a^3$ — sum of cubes
7. $(x - a)(x^2 + ax + a^2) = x^3 - a^3$ — difference of cubes

These products may all be checked by multiplication. Their use in factoring is illustrated by the following examples.

EXAMPLE 4 Factor the following polynomials by using the distributive property. (This process is often called *removing a common factor*.)

SOLUTION (a) $ax + ay = a(x + y)$

(b) $x^2 - 2x = x(x - 2)$

(c) $3x^2 + 6x - 9 = 3(x^2 + 2x - 3)$ •

The next two examples illustrate the use of the formulas to factor a special trinomial (coefficient of x^2 is 1) and a general trinomial, (2 and 3) in the above list.

EXAMPLE 5 (a) $x^2 + 5x + 6 = (x + 2)(x + 3)$

(b) $a^2 - a - 2 = (a - 2)(a + 1)$

(c) $x^2 + 2x - 8 = (x + 4)(x - 2)$ •

EXAMPLE 6 (a) $2u^2 - 5u - 3 = (2u + 1)(u - 3)$

(b) $4x^2 - 5x - 6 = (4x + 3)(x - 2)$

(c) $6y^2 + 13y - 5 = (2y + 5)(3y - 1)$ •

In Example 7, the expressions to be factored are trinomial squares. As shown by Product 4 in the preceding list, they are squares of binomials.

EXAMPLE 7 (a) $x^2 + 2x + 1 = (x + 1)^2$

(b) $a^2 - 6a + 9 = (a - 3)^2$

(c) $4z^2 + 4z + 1 = (2z)^2 + 2(2z) + 1 = (2z + 1)^2$ •

Each of the expressions to be factored in Example 8 is a difference of squares (Product 5).

EXAMPLE 8 (a) $x^2 - 1 = (x + 1)(x - 1)$

(b) $y^4 - 9 = (y^2)^2 - (3)^2 = (y^2 + 3)(y^2 - 3)$

(c) $4b^2 - 25 = (2b)^2 - (5)^2 = (2b + 5)(2b - 5)$ •

The expressions in the final two examples are factored using the formulas for the sum and the difference of cubes (Products 6 and 7).

EXAMPLE 9 (a) $x^3 + 8 = x^3 + 2^3 = (x + 2)(x^2 - 2x + 4)$

(b) $8z^3 + 27 = (2z)^3 + 3^3 = (2z + 3)(4z^2 - 6z + 9)$

(c) $64a^3 + 1 = (4a)^3 + 1^3 = (4a + 1)(16a^2 - 4a + 1)$ •

EXAMPLE 10 (a) $a^3 - 8 = a^3 - 2^3 = (a - 2)(a^2 + 2a + 4)$

(b) $1 - 8x^3 = 1^3 - (2x)^3 = (1 - 2x)(1 + 2x + 4x^2)$

(c) $64z^3 - 27 = (4z)^3 - 3^3 = (4z - 3)(16z^2 + 12z + 9)$ •

Factoring will be used throughout the rest of this book.

EXERCISES 3.1

Which of the following algebraic expressions are polynomials or rational expressions?

1. $x + 4$

2. $\dfrac{3}{x^2} + \dfrac{2}{x} + 1$

3. $\sqrt{3x + 5}$

4. $\dfrac{4}{5}x^3 + x - \sqrt{2}$

5. $\dfrac{2x}{x^3 + 3}$

6. $1 + \dfrac{3}{\sqrt{x}}$

7. $\sqrt[4]{(x + 1)^3}$

8. $2 + \dfrac{1}{x} + x^2$

9. $5x^2 + \dfrac{1}{2}x - 3$

10. $x + \dfrac{1}{\sqrt{x}}$

11. $1 - x^2 + 2x$

12. $1 + \sqrt{x}$

13. $\dfrac{x^2 + 1}{x^3 + x - 2}$

14. $\sqrt[5]{x^2 - 1}$

Find the sum, difference, product, and quotient of each of the following pairs of polynomials. Write each quotient as the sum of a polynomial and a rational expression.

15. $2x^2 + 5x - 8$ and $3x^2 - 6x + 1$

16. $x^3 + 4x^2 - x + 2$ and $5x^2 + 2x - 3$

17. $2x^4 + x^3 - 3x - 5$ and $x^2 + x + 1$

18. $2x^3 + x + 1$ and $3x - 2$

By successive multiplications, find each of the following powers of the binomial $x + y$.

19. $(x + y)^3$

20. $(x + y)^4$

21. $(x + y)^5$

22. $(x + y)^6$

Factor each of the following polynomials.

23. $20x - 5$

24. $a^2x + ay + 2az$

25. $x^2 - 2x - 15$

26. $3x^2 - 5x - 2$

27. $x^2 - 10x + 25$

28. $9 + 12x + 4x^2$

29. $16x^2 - 49$

30. $\dfrac{x^2}{9} - \dfrac{1}{4}$

31. $8 - 27x^3$

32. $10x^2 + 29x - 21$

33. $3x^2 + 5x - 12$

34. $10x^2 + 7x - 12$

35. $16x^2 + 24x + 9$

36. $3x + 6y$

37. $2x + 4y - 6z$

38. $x^2 - 8x + 12$

39. $3x^2 + 10x - 8$

40. $x^2 + 4x + 4$

41. $9x^2 + 6x + 1$

42. $9x^2 - 16$

43. $25x^2 - y^2$

44. $8x^3 + 1$

45. $12x^3 + 21x^2 + 15x$

46. $14x^2 - 41x + 15$

47. $12x^2 + 7x - 12$

48. $\frac{1}{4}x^2 + x + 1$

Use the method described in this section to calculate the indicated polynomial function values with a hand calculator.

49. $f(7);\ f(x) = 3x^2 + 5x - 2$

50. $f(4);\ f(x) = 5x^3 + 12x^2 - 36x - 16$

51. $f(-2);\ f(x) = 6x^4 - 5x^3 + 2x$

52. $f(-3);\ f(x) = 2x^5 + 3x^4 - x^3 - 4x^2 + x + 5$

53. $f(0.23);\ f(x) = x^4 - 10x^2 - 7$

54. $f(2.1);\ f(x) = x^3 - 5x^2 + 3x - 8$

55. $f(5.6);\ f(x) = 3x^3 + 2x^2 - 12x - 8$

56. $f(12.5);\ f(x) = x^6 + 3x^5 - 2$

3.2 Linear Functions and Their Graphs

The simplest polynomial function (other than a constant function) is the polynomial function of the first degree. This function, the **linear function,** is represented by

$$f(x) = ax + b,$$

where a and b are constants (real numbers) and $a \neq 0$. Examples are

$$f(x) = x, \qquad f(x) = x + 2, \qquad f(x) = 3x - 1.$$

Many familiar formulas represent linear functions. One of these is the temperature conversion formula, $F(C) = (9/5)C + 32$. Another is $A(t) = Prt + P$, the formula for the amount A accumulated in t years from investing P dollars at r percent annual interest. The domain of these two functions may be limited by the nature of the situations. However, the domain of the general linear function, $f(x) = ax + b$, is the set of all real numbers, and so is its range. For, if $y = ax + b$, then $x = (y - b)/a$, so that for every y there is an x such that $y = f(x)$. This implies that y may assume any real value.

If we equate a linear function to zero, we obtain a **linear equation** $ax + b = 0$. Its solution is the single number $-b/a$. This number is called the **root** of the equation and the **zero** of the function.

EXAMPLE 1 Find the root of the equation $3x + 6 = 0$ and the zero of the function f, where $f(x) = 3x + 6$.

SOLUTION If $3x + 6 = 0$, then $3x = -6$, and $x = -2$ is the root of the equation and the zero of f. So the intersection of the graph of f with the x-axis is $(-2, 0)$. •

The term *linear* comes from the following fundamental property of linear functions.

THEOREM 3.1 The graph of a linear function is an oblique (neither horizontal nor vertical) straight line and, conversely, every oblique straight line is the graph of a linear function.

The proof of this theorem will not be given here. In Section 7.1 graphs of straight lines will be discussed in some detail, and this theorem will be proved there.

Observe that every vertical line has an equation of the form $x = c$ (c a constant), and conversely, only vertical lines have equations of this form. This equation does not define a function. (Why?)

The equation of a horizontal line is $y = k$ (k a constant). This equation defines a constant function but not a linear function, since $a = 0$ in the general definition.

Graphs of Linear Functions From elementary geometry we know that two points determine a straight line. So, knowing that the graph of a linear function is a straight line, we need only two of its points to determine the graph. The simplest points to calculate are the *intercepts:* the **x-intercept** is the first coordinate of the point at which the line crosses the x-axis, and the **y-intercept** is the second coordinate of the point at which the line crosses the y-axis. The x-intercept is found by solving $f(x) = 0$ for x; the y-intercept by letting $x = 0$ in $f(x)$. (That is, the y-intercept is $f(0)$.)

EXAMPLE 2 Graph $f(x) = 2x + 1$.

SOLUTION We first find the intercepts. If $x = 0$, then $f(0) = 1$, and if $f(x) = 0$, then $x = -1/2$. So the intercept points are $(-1/2, 0)$ and $(0, 1)$. The graph is shown in Figure 3.2. ●

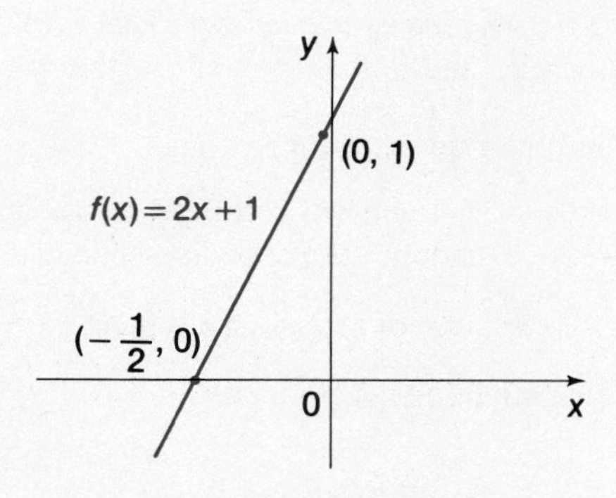

Figure 3.2

Any two points may be used to graph a linear function, as the next example illustrates.

EXAMPLE 3 Graph $f(x) = 3x - 2$ on the interval $-1 \le x \le 2$.

SOLUTION Here the domain is restricted to the closed interval $[-1, 2]$. If we let $y = f(x) = 3x - 2$, then when $x = -1$, $y = -5$, and when $x = 2$, $y = 4$. The graph is the *line segment* shown in Figure 3.3. ●

EXAMPLE 4 Graph the identity function $f(x) = x$.

SOLUTION This is the particular linear function $f(x) = ax + b$ for which $a = 1$ and $b = 0$. Its graph is the line bisecting the first and third quadrants of the coordinate plane. Since this graph has only one intercept, we need some other point satisfying the equation, such as $(-1, -1)$ or $(2, 2)$. See Figure 3.4 ●

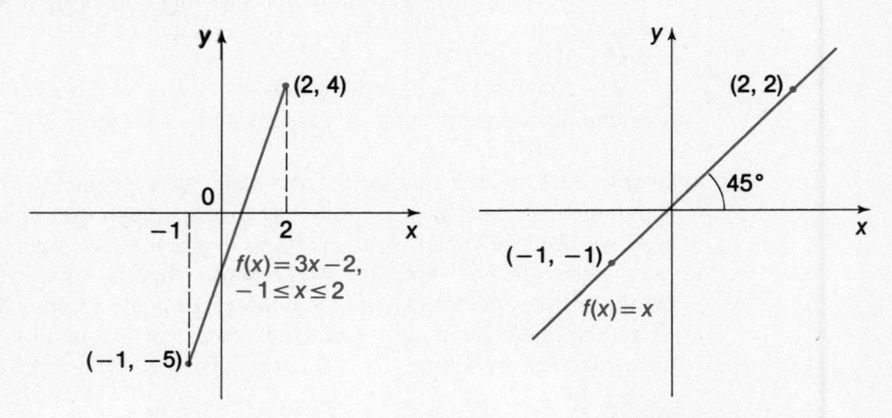

Figure 3.3

Figure 3.4

Rate of Change and Slope In Section 2.1 the idea of *rate of change* of a function was introduced. This concept is closely related both to the graph of a function and to many of the applications of functions. It is an important fact that *the rate of change of any linear function is a constant*. This is illustrated for a particular function in the next example, and is then proved for the general case in Example 6.

EXAMPLE 5 Find the rate of change of $f(x) = 2x - 3$.

SOLUTION Let $y = f(x)$ and let x change by an amount h from x to $x + h$. Then y becomes $f(x + h)$ and the corresponding change in y is $f(x + h) - f(x)$. The average change in y for each unit of change in x is then $\dfrac{f(x + h) - f(x)}{h}$. The following steps show these calculations for $f(x) = 2x - 3$.

$$\begin{aligned} f(x) &= 2x - 3 \\ f(x + h) &= 2(x + h) - 3 = 2x + 2h - 3 \\ f(x + h) - f(x) &= 2h \\ \frac{f(x + h) - f(x)}{h} &= 2 \end{aligned}$$

Thus, the *average* rate of change in the value of $f(x)$ for each unit of change in x is 2. See Figure 3.5. ●

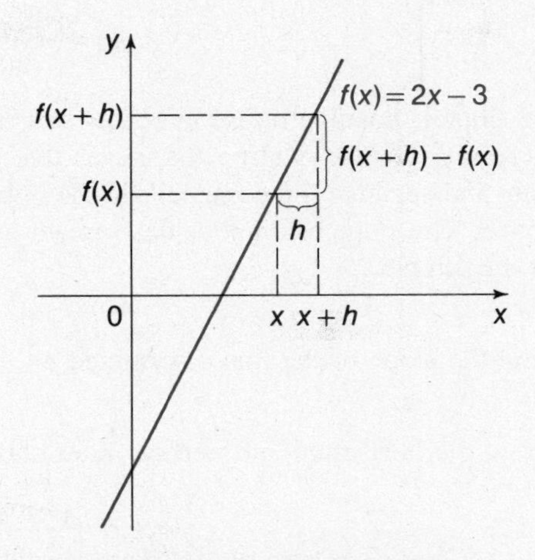

Figure 3.5 Rate of change of a linear function

EXAMPLE 6 Let $f(x) = ax + b$ and find the rate of change of $f(x)$.

SOLUTION

$$f(x) = ax + b$$
$$f(x + h) = a(x + h) + b$$
$$= ax + ah + b$$
$$f(x + h) - f(x) = ah$$
$$\frac{f(x + h) - f(x)}{h} = a \qquad \text{constant}$$

The general case shows that the rate of change of a linear function is given by a, the coefficient of x. Note that Example 5 illustrates this fact. ●

Observe that the rate of change can be described as follows:

$$\frac{f(x + h) - f(x)}{h} = \frac{\text{change in value of } y}{\text{change in value of } x}.$$

Geometrically, this is the ratio of "rise" of the line which is the graph of f to the "run" of the line. It measures the *steepness* of the line. In Section 7.1 we will discuss lines in detail and there this ratio, a, defines the **slope** of the line $y = ax + b$.

Also observe that if $x = 0$, then $y = b$, so that b is the y-intercept of the line $y = ax + b$. In Examples 2–4 above, we have

(2) $y = 2x + 1$; slope 2, y-intercept 1

(3) $y = 3x - 2$; slope 3, y-intercept -2

(4) $y = x$; slope 1, y-intercept 0.

Since

$$\text{slope} = \frac{\text{change in } f(x)}{\text{increase in } x},$$

the slope is positive if $f(x)$ increases as x increases and negative if $f(x)$ decreases as x increases. Geometrically, this means that the slope is positive if the line rises to the right and negative if the line falls to the right. Check this with Figures 3.2, 3.3, and 3.4 above. The slope of a line is determined by any two of its points, as illustrated in the next example.

EXAMPLE 7 Find the slope of the line determined by the two points $(-3, 1)$ and $(1, 4)$. See Figure 3.6.

SOLUTION Draw the horizontal and vertical lines shown in the figure. Then

$$\begin{aligned}\text{slope} &= \frac{\text{change in } y}{\text{change in } x}\\ &= \frac{4 - 1}{1 + 3} = \frac{3}{4}.\end{aligned}$$ ●

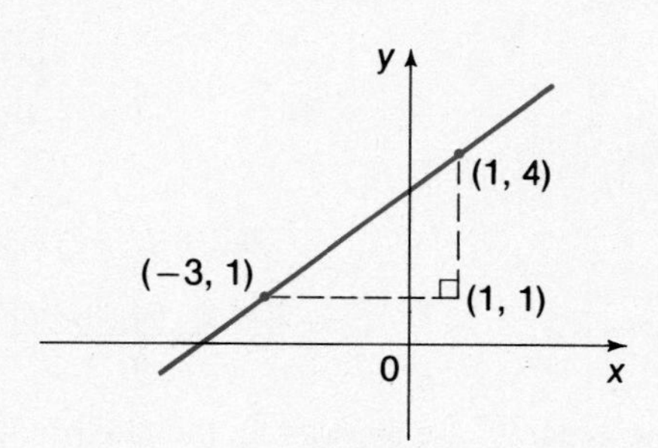

Figure 3.6

We will note instances of slope in the rest of this section. In Section 7.1 there is a detailed treatment of straight lines, and the ideas discussed here will be considered more fully there. For example, there are several useful forms in which the equation of a line may be written. In that context, equations (2)–(4) above are called the slope-intercept form of the equation of the given lines.

Applications of Linear Functions Much of the usefulness of mathematical functions is that they can be *mathematical models* of "real-life" situations. Models are usually only approximations to the real world situation, but in the physical sciences they are frequently very good approximations, limited only by the precision of the physical measurements used. For example, the formula

$$v(t) = 32t$$

describes very closely the velocity (in feet per second) of a body which has fallen freely for t seconds. The formula

$$p(v) = \frac{c}{v},$$

called Boyle's Law and usually written $pv = c$, is a good approximation of the relation between pressure and volume of a confined gas. On the other hand, the geometric formula

$$C(r) = 2\pi r$$

states the exact relation of the circumference of a circle to its radius.

The functions $v(t)$ and $C(r)$ above are instances of **linear function models.** Such functions are frequently very useful in the behavioral sciences—business, economics, education, psychology, sociology. The reason is that the rate of change of the function values is constant. For $v(t)$ above this rate is 32 and for $C(r)$ it is 2π.

The following examples of linear function models from economics relate price, supply, and demand. Usually, as demand for an item increases, so does the price. As the supply increases, the price usually decreases. That is, price is a function of the demand and also a function of the supply.

EXAMPLE 8 Suppose that price and supply of some product are related by the function P given by

$$p = P(x) = 40 - \frac{2}{3}x,$$

where p is the price (in dollars) and x is the supply (number of units). (a) Find the price when there are 30 units of supply. (b) Find the supply for which the corresponding price is 25 dollars per unit. (c) Graph $P(x)$.

SOLUTION (a) If $x = 30$, then

$$\begin{aligned} p &= 40 - \frac{2}{3}(30) \\ &= 40 - 20 \\ &= 20 \text{ (dollars).} \end{aligned}$$

(b) If $p = 25$, then

$$\begin{aligned} 25 &= 40 - \frac{2}{3}x \\ \frac{2}{3}x &= 40 - 25 = 15 \\ 2x &= 45 \\ x &= 22.5 \text{ (about 22) units.} \end{aligned}$$

(c) Since x must be a whole number, the domain of P is the nonnegative integers. Also, $p > 0$, because a negative price would be meaningless in this situation. If $x = 0$, $p = 40$, and if $p = 0$, $x = 60$, so the intercept points are (0, 40) and (60, 0). These two points determine the graph. In Figure 3.7a, a continuous line has been drawn, but only the points corresponding to integral values of x have meaning for the given function. ●

EXAMPLE 9 For the product in Example 8, suppose the demand (x) and the price (p) are related by the function Q given by

$$p = Q(x) = \frac{2x}{3}.$$

Graph the function $Q(x)$ on the same coordinate plane with the price-supply function p of Example 8, and find the point where the graphs intersect.

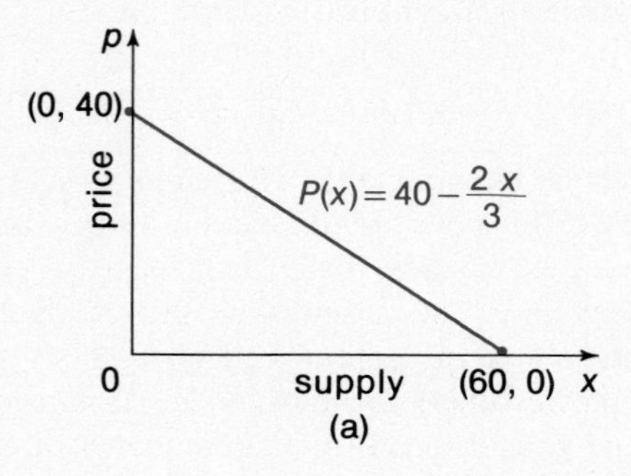

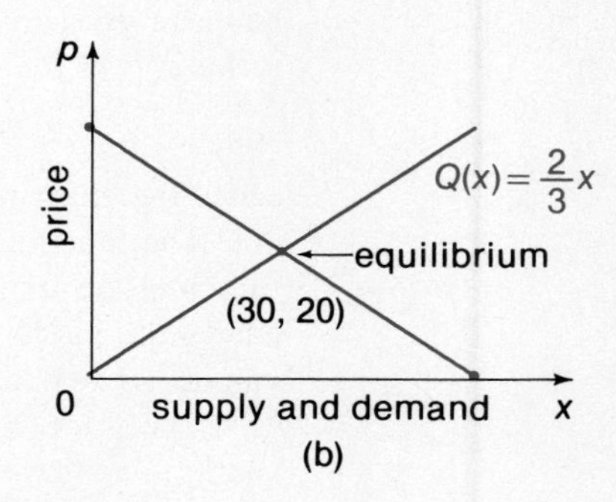

Figure 3.7

SOLUTION See Figure 3.7b for the graph. The price at which supply equals demand is called the **equilibrium price.** The coordinates of this point on the graph are found by solving the equations of supply and demand simultaneously.

$$P(x) = Q(x)$$

$$40 - \frac{2}{3}x = \frac{2}{3}x$$

$$\frac{4}{3}x = 40$$

$$4x = 120$$

$$x = 30$$

When $x = 30$, $P(x) = 40 - \frac{2}{3}(30) = 40 - 20 = 20$, and so the equilibrium price is 20 dollars and it occurs when the demand is 30. The corresponding point of intersection of the graphs (30, 20) is shown in Figure 3.7b. ●

Notice that the slope of the supply curve, $-2/3$, is the change (decrease) in the price per unit of increase in the supply. Similarly, the slope of the demand curve, $2/3$, is the change (increase) in the price per unit of increase in demand.

EXERCISES 3.2

Graph each of the following linear functions by first determining the intercepts on the coordinate axes.

1. $f(x) = x - 2$

2. $f(x) = 3x - 4$

3. $f(x) = -x + 3$

4. $f(x) = \frac{1}{3}x - 1$

5. $f(x) = 2x - 3, \quad 0 \leq x \leq 2$. [Use endpoints.]

6. $f(x) = -2x + 1, \quad -2 \leq x \leq 1$. [Use endpoints.]

For what value of x does the given $f(x)$ have the indicated value in each of the following?

7. $f(x) = 2x + 1; \quad 3$

8. $f(x) = -5x + 2; \quad 4$

9. $f(x) = 3x - 4; \quad 11$

10. $f(x) = -2x + 1; \quad -2$

11. $f(x) = 4x - 2; \quad 5$

12. $f(x) = 2x - 3; \quad -19$

Sketch the graph and calculate the slope of the straight line determined by each of the following pairs of points.

13. $(2, 1)$ and $(3, 4)$

14. $(1, 4)$ and $(-1, 0)$

15. $(-2, 3)$ and $(1, -1)$

16. $(-3, 0)$ and $(0, -2)$

Find the slope and the y-intercept of the graph of each of the following linear functions.

17. $f(x) = 3x - 4$

18. $f(x) = -2x + 1$

19. $f(x) = -x$

20. $f(x) = 2$

21. Graph the function $F(C) = 32 + \frac{9}{5}C$, relating Fahrenheit and Celsius temperatures. What is the physical interpretation of the intercept on each coordinate axis?

22. The pressure p of water is proportional to its depth d. That is, $p = kd$, where k is some constant. If measurements give $p = 1.02$ when $d = 0.8$, calculate the **constant of proportionality** k and then graph $p = f(d)$.

23. In a certain manufacturing process, there is a fixed daily **start-up cost** of b dollars and a cost of a dollars per unit produced. So if x units are produced per day, the total cost is given by $C(x) = ax + b$, that is, daily costs are a linear function of the number of units produced. Graph the function for the case $a = 5$ and $b = 20$.

24. The velocity of a body thrown upward is given by $v(t) = 120 - 32t$. Graph $v(t)$. The v-intercept is called the **initial velocity** (when $t = 0$) of the body. What is the initial velocity in this case?

25. Suppose car rental charges are 20 dollars per day, plus 25 cents for each mile driven. Let x be the number of miles driven in a single day of use and $C(x)$ the total cost for the day. Write the cost function $C(x)$ and graph.

26. If $C(x) = 2.7x + 100$ represents the cost of producing x items, find (a) the start-up cost (see Exercise 23); (b) the cost per item; (c) the cost of producing 15 items.

27. Suppose a taxi charges 2 dollars plus 60 cents per mile. (a) Write the cost function. (b) Calculate the cost for 5 miles. (c) Graph the function.

As in Examples 8 and 9, graph the following supply and demand curves on the same set of axes and calculate the equilibrium price and the equilibrium demand.

28. Supply: $p(x) = 38 - \frac{2x}{3}$; demand: $p(x) = \frac{3x}{5}$

29. Supply: $p(x) = 69 - \frac{3x}{4}$; demand: $p(x) = \frac{2x}{5}$

3.3 Quadratic Functions

A polynomial function represented by

$$f(x) = ax^2 + bx + c,$$

where a, b, and c are constants and $a \neq 0$, is called a **quadratic function.** We assume $a \neq 0$ because otherwise the formula for $f(x)$ reduces to $f(x) = bx + c$, which is a linear function. However, either b or c or both, may be zero. Some examples of quadratic functions are:

$$f(x) = x^2 - x - 2, \qquad g(x) = 3x^2 + 2x, \qquad h(x) = 4x^2 + 1, \qquad k(x) = x^2.$$

The domain of a quadratic function is the set of all reals. Later in this section we will learn how to determine the range.

In the last section we showed that a linear function has a constant rate of change. The rate of change of a quadratic function is a linear function. An illustration of this was given in Example 6, Section 2.1. The general case is shown as follows.

$$\begin{aligned}
f(x) &= ax^2 + bx + c \\
f(x + h) &= a(x + h)^2 + b(x + h) + c \\
&= a(x^2 + 2hx + h^2) + b(x + h) + c \\
&= ax^2 + 2ahx + ah^2 + bx + bh + c \\
f(x + h) - f(x) &= (ax^2 + 2ahx + ah^2 + bx + bh + c) - (ax^2 + bx + c) \\
&= 2ahx + ah^2 + bh \\
&= h(2ax + ah + b) \\
\frac{f(x + h) - f(x)}{h} &= \frac{\not{h}(2ax + ah + b)}{\not{h}} \\
&= 2ax + (ah + b)
\end{aligned}$$

This is a linear function, since a, h, and b are all constants.

Many formulas from geometry and from science are quadratic functions. One example is the formula for the area of a circle, $A(r) = \pi r^2$, which expresses the area A as a quadratic function of the radius r. Some other important formulas are discussed in the following example which concerns a falling body under the influence of gravity.

EXAMPLE 1 (a) If a body falls freely (is dropped from rest), its distance s (in feet) from the starting point is given by

$$s(t) = \frac{1}{2}gt^2,$$

where g (about 32 in this context) is the gravitational constant and t is the time in seconds. See Figure 3.8a.

(b) If the body is given an initial velocity v_0, the corresponding formula is

$$s(t) = \frac{1}{2}gt^2 + v_0t.$$

Here v_0 is taken as negative for upward motion. See Figure 3.8b.

(c) If the starting point for the body is a vertical distance s_0 below the level from which distance is measured, then its distance below this level at time t is

$$s(t) = \frac{1}{2}gt^2 + v_0t + s_0.$$

See Figure 3.8c. ●

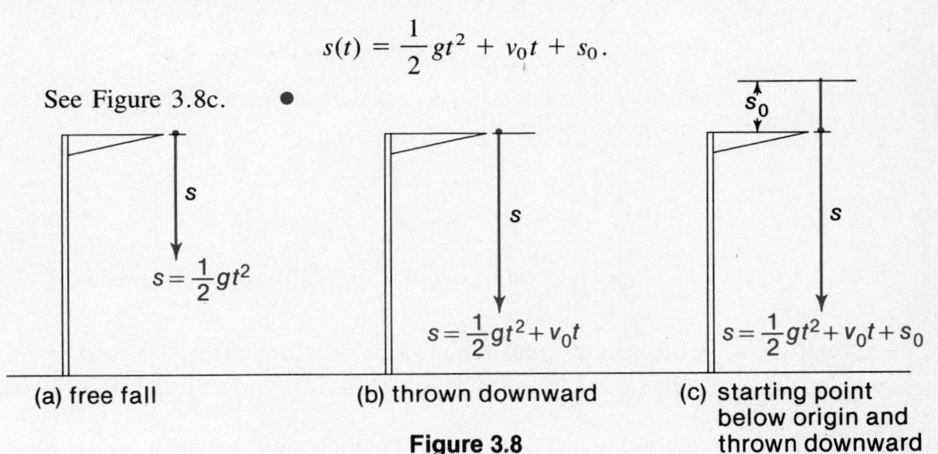

Figure 3.8

Completing the Square Many of the uses of quadratic functions (including graphing) depend upon a procedure called *completing the square*. To see how to do it, we begin with the formula for the square of a binomial,

$$\begin{aligned}(x + a)^2 &= x^2 + 2ax + a^2 \\ &= x^2 + (2a)x + \left[\frac{1}{2}(2a)\right]^2.\end{aligned}$$

(See Section 3.1 on factoring.) Note that on the right side of the above equation, the constant term a^2 is *the square of one-half the coefficient of* x in the quadratic expression on the left side. Thus, to "complete the square" for the quadratic $x^2 + 6x$ we write

$$\begin{aligned}x^2 + 6x + \square &= x^2 + 6x + \left(\frac{6}{2}\right)^2 \\ &= x^2 + 2(3)x + 3^2 \\ &= (x + 3)^2.\end{aligned}$$

To carry out the work above, we ask: "What should be put in the box in order to get the square of a linear polynomial?" The answer is "$(6/2)^2$," and the linear polynomial is $x + 3$.

Similarly, for $x^2 - 4x$,

$$\begin{aligned} x^2 - 4x + \square &= x^2 - 4x + (-2)^2 \\ &= x^2 - 4x + 4 \\ &= (x - 2)^2. \end{aligned}$$

The next example illustrates the process of **completing the square** on a binomial, that is, writing the binomial in the form $(x + a)^2 + c$.

EXAMPLE 2 Complete the square on $f(x) = x^2 + 8x$.

SOLUTION

$$\begin{aligned} f(x) &= x^2 + 8x \\ &= x^2 + 8x + (4)^2 - (4)^2 \\ &= x^2 + 8x + 16 - 16 \end{aligned}$$

Here we have *added* 16 in order to *complete* the square and *subtracted* 16 to "balance" the given expression. Then

$$\begin{aligned} f(x) &= (x^2 + 8x + 16) - 16 \\ &= (x + 4)^2 - 16. \end{aligned}$$

●

The following examples further illustrate the process.

EXAMPLE 3 Complete the square on (a) $x^2 + 12x$, (b) $x^2 - 10x$, (c) $x^2 + 3x$.

SOLUTION

(a)
$$\begin{aligned} x^2 + 12x &= (x^2 + 12x + \square) - \square \\ &= (x^2 + 12x + 36) - 36 \\ &= (x + 6)^2 - 36 \end{aligned}$$

(b)
$$\begin{aligned} x^2 - 10x &= (x^2 - 10x + \square) - \square \\ &= (x^2 - 10x + 25) - 25 \\ &= (x - 5)^2 - 25 \end{aligned}$$

(c)
$$\begin{aligned} x^2 + 3x &= (x^2 + 3x + \square) - \square \\ &= \left(x^2 + 3x + \frac{9}{4}\right) - \frac{9}{4} \\ &= \left(x + \frac{3}{2}\right)^2 - \frac{9}{4} \end{aligned}$$

●

The above examples illustrate the process of completing the square on quadratics of the form $x^2 + bx$. This method is extended, in the following examples, to quadratics of the form $ax^2 + bx$ and $ax^2 + bx + c$.

EXAMPLE 4 Complete the square for $2x^2 + 4x$.

SOLUTION First factor out 2, the coefficient of x^2, to obtain the form $x^2 + bx$. Then complete the square on the expression in parentheses.

$$\begin{aligned} 2x^2 + 4x &= 2(x^2 + 2x) \\ &= 2(x^2 + 2x + 1 - 1) \\ &= 2(x^2 + 2x + 1) - 2 \\ &= 2(x + 1)^2 - 2 \end{aligned}$$ ●

EXAMPLE 5 Complete the square on $4x^2 - 16x + 6$.

SOLUTION

$$\begin{aligned} 4x^2 - 16x + 6 &= 4(x^2 - 4x) + 6 \\ &= 4[(x^2 - 4x + 4) - 4] + 6 \\ &= 4(x^2 - 4x + 4) - 16 + 6 \\ &= 4(x - 2)^2 - 10 \end{aligned}$$ ●

The process of completing the square may be used to find the *greatest* (or *least*) value of a quadratic function. To say that k is the **greatest** value of $f(x)$ means that $f(x) \le k$ for all values of x in the domain and $f(x) = k$ for at least one value of x. Similarly, if $f(x) \ge c$ on some domain and $f(x) = c$ for at least one value of x in that domain, then c is the **least** value of $f(x)$ there. The terms **maximum** and **minimum** are also used instead of greatest and least.

EXAMPLE 6 Find the greatest (or least) value of the function $f(x) = x^2 + 8x$.

SOLUTION First, write $f(x)$ in the equivalent form shown in Example 2.

$$\begin{aligned} f(x) &= x^2 + 8x \\ &= (x + 4)^2 - 16 \\ &= -16 + (x + 4)^2 \\ f(x) &= -16 + (\text{a nonnegative expression}) \end{aligned}$$

When $f(x)$ is written as the sum of a constant and a nonnegative expression, as in the last two lines above, it is easy to see that the constant must be the *least* value of $f(x)$. For, when $x = -4$, $f(x) = -16$, and when x is any other number, $(x + 4)^2$ is positive and therefore $f(x) > -16$. Thus, $f(x) \ge -16$ for all x, that is, the least (or minimum) value of $f(x)$ is -16. Since $f(x) \ge -16$, the range of f is the set of reals greater than or equal to -16. ●

EXAMPLE 7 Find the greatest (or least) value of $f(x) = -3x^2 + 4x + 1$.

SOLUTION First, factor out the coefficient of x^2 from the first two terms to obtain an expression of the form $x^2 + 2ax$. Then, complete the square on the expression in parentheses.

$$\begin{aligned} f(x) &= -3x^2 + 4x + 1 \\ &= -3\left(x^2 - \frac{4}{3}x\right) + 1 \\ &= -3\left[x^2 - \frac{4}{3}x + \left(-\frac{2}{3}\right)^2\right] + 1 + 3\left(-\frac{2}{3}\right)^2 \end{aligned}$$

Here we have, in effect, inserted -3 times the term $(-2/3)^2$ in the bracket, so that we must add $3(-2/3)^2$ to balance the equation.

$$\begin{aligned} f(x) &= -3\left(x - \frac{2}{3}\right)^2 + \frac{7}{3} \\ &= \frac{7}{3} - 3\left(x - \frac{2}{3}\right)^2 \\ &= \frac{7}{3} - \text{(a nonnegative expression)} \end{aligned}$$

When $f(x)$ is written as the difference of a constant and a nonnegative expression, an argument similar to the one used in Example 6 shows that the constant must be the *greatest* value of f, that is,

$$f(x) \leq \frac{7}{3}.$$

Thus, $f(x)$ is at most 7/3 and, in fact, $f(x) = 7/3$ when $x = 2/3$. So the greatest (or maximum) value of $f(x)$ is 7/3. The result here also shows that the range of f is the set of real numbers less than or equal to 7/3. ●

EXAMPLE 8 Express the general quadratic function in terms of the square of a linear function by the method of completing the square.

SOLUTION We begin by writing the equation for the general quadratic function.

$$\begin{aligned} f(x) &= ax^2 + bx + c \\ &= a\left(x^2 + \frac{b}{a}x\right) + c && \text{factoring out the coefficient of } x^2 \\ &= a\left[x^2 + \frac{b}{a}x + \left(\frac{b}{2a}\right)^2\right] + c - \frac{b^2}{4a} && \text{adding and subtracting } a\left(\frac{b}{2a}\right)^2 \\ &= a\left(x + \frac{b}{2a}\right)^2 + \frac{4ac - b^2}{4a} \end{aligned}$$

An argument similar to those used in Examples 6 and 7 shows that the greatest (or least) value of the general quadratic is

$$\frac{4ac - b^2}{4a} \quad \text{when} \quad x = -\frac{b}{2a}.$$

It is a least value if $a > 0$ and a greatest value if $a < 0$. (Why?) ●

EXAMPLE 9 Suppose that 1000 feet of fencing is available for enclosing a rectangular region. What are the dimensions of the largest (in area) rectangle which can be so enclosed?

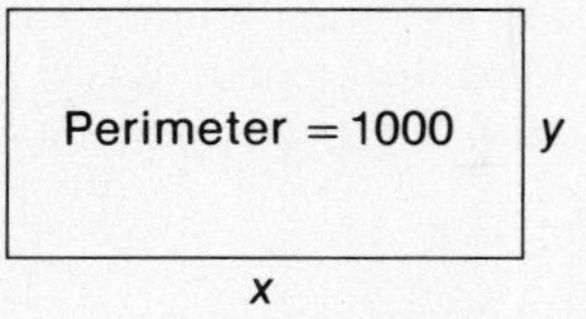

Perimeter = 1000 y

x

Figure 3.9

SOLUTION Let x and y be the dimensions of the rectangle, as shown in Figure 3.9. Then

$$\text{area} = A = xy \qquad \text{and} \qquad \text{perimeter} = 2x + 2y = 1000.$$

The solution to the resulting system of equations,

$$A = xy$$
$$1000 = 2x + 2y$$

may be found by solving the second equation for y and substituting in the first equation.

$$x + y = 500$$
$$y = 500 - x$$
$$A = xy$$
$$= x(500 - x)$$

Hence, the area A is a quadratic function of the side x and may be written

$$A(x) = 500x - x^2, \quad 0 < x < 500.$$

The problem thus reduces to finding the greatest value of a quadratic, which may be done by completing the square.

$$A(x) = 500x - x^2$$
$$= -(x^2 - 500x + 250^2) + 250^2$$
$$= -(x - 250)^2 + 250^2$$

Since $A(x) = 250^2 - (x - 250)^2$, $A(x) \leq 250^2$. That is, the greatest value of the area A is 250^2. This value is obtained when $x = 250$. Then $y = 500 - x = 500 - 250 = 250$.

Since x and y are equal, the rectangle of greatest area is a square of side 250. It can be shown that a square would have resulted *whatever* the length of fencing given. (See Exercise 47 at the end of this section.) ●

Note the restriction imposed on the domain of A in this example. As a mathematical function, A has as its domain the set of all reals. However, because A represents area, x represents a dimension and so must be positive and less than half the perimeter, that is, $0 < x < 500$.

Quadratic Equations If the general quadratic polynomial is set equal to zero, the quadratic equation $ax^2 + bx + c = 0$ results. It will be shown that it has exactly two roots. Quadratic equations whose roots are real numbers will be discussed here and those whose roots are not real numbers will be treated in Chapter 9.

In some cases the solution can be found by factoring the quadratic into its linear factors. When the product of two factors is equal to zero, either one or both of them must be zero. Thus, the roots may be found by solving two linear equations.

EXAMPLE 10 Solve the equation $2x^2 + 5x - 3 = 0$.

SOLUTION

$$2x^2 + 5x - 3 = 0$$

$$(2x - 1)(x + 3) = 0 \qquad \text{factoring}$$

$$\left.\begin{aligned} 2x - 1 = 0, &\quad \text{when } x = \frac{1}{2} \\ x + 3 = 0, &\quad \text{when } x = -3 \end{aligned}\right\} \qquad \text{equating linear factors to zero and solving in both cases}$$

Check that 1/2 and -3 are roots by substituting them for x in the original equation.

This equation can also be solved by completing the square.

$$2x^2 + 5x - 3 = 0$$

$$2x^2 + 5x = 3 \qquad \text{adding 3 to both sides}$$

$$x^2 + \frac{5}{2}x = \frac{3}{2} \qquad \text{dividing both sides by 2}$$

Now complete the square for the left side and "balance" the equation by adding the same number to both sides.

$$x^2 + \frac{5}{2}x + \left(\frac{5}{4}\right)^2 = \frac{3}{2} + \left(\frac{5}{4}\right)^2 \qquad \text{completing square for left side and balancing}$$

$$\left(x + \frac{5}{4}\right)^2 = \frac{3}{2} + \frac{25}{16} = \frac{49}{16}$$

$$x + \frac{5}{4} = \pm\frac{7}{4} \qquad \text{taking the square root of both sides}$$

$$x = \frac{1}{2} \quad \text{or} \quad x = -3 \qquad \text{solving for } x \text{ in two cases}$$ ●

This same procedure may be used to solve the general quadratic equation and thus find a formula for the roots. From Example 8,

$$ax^2 + bx + c = a\left(x + \frac{b}{2a}\right)^2 + \frac{4ac - b^2}{4a},$$

so the general quadratic equation $ax^2 + bx + c = 0$ can be written as

$$a\left(x + \frac{b}{2a}\right)^2 + \frac{4ac - b^2}{4a} = 0$$

$$a\left(x + \frac{b}{2a}\right)^2 = \frac{b^2 - 4ac}{4a} \qquad \text{subtracting } (4ac - b^2)/4a \text{ from both sides}$$

$$\left(x + \frac{b}{2a}\right)^2 = \frac{b^2 - 4ac}{4a^2} \qquad \text{dividing both sides by } a$$

$$x + \frac{b}{2a} = \frac{\pm\sqrt{b^2 - 4ac}}{2a} \qquad \text{taking the square root of both sides}$$

$$x = \frac{-b \pm \sqrt{b^2 - 4ac}}{2a} \qquad \text{solving for } x$$

This is the **quadratic formula** of elementary algebra. It demonstrates the fact that the quadratic equation has two roots.

$$\boldsymbol{x_1 = \frac{-b + \sqrt{b^2 - 4ac}}{2a}} \quad \text{and} \quad \boldsymbol{x_2 = \frac{-b - \sqrt{b^2 - 4ac}}{2a}}.$$

The expression $b^2 - 4ac$ in this formula is called the **discriminant** of the quadratic function (equation). The roots of the equation are real and distinct if the discriminant is positive, real and identical if it is zero, and not real if it is negative.

EXAMPLE 11 Solve by the quadratic formula: (a) $3x^2 - 2x - 4 = 0$, (b) $x^2 + 2x + 5 = 0$.

SOLUTION (a) Here $a = 3$, $b = -2$, and $c = -4$ are substituted in the formula.

$$x = \frac{2 \pm \sqrt{(-2)^2 - 4(3)(-4)}}{2(3)}$$

$$= \frac{2 \pm \sqrt{52}}{6} = \frac{2 \pm \sqrt{4(13)}}{6}$$

$$= \frac{2 \pm 2\sqrt{13}}{6} = \frac{1}{3} \pm \frac{1}{3}\sqrt{13}$$

The roots are $x_1 = \frac{1}{3} + \frac{1}{3}\sqrt{13}$ and $x_2 = \frac{1}{3} - \frac{1}{3}\sqrt{13}$.

(b) Here $a = 1$, $b = 2$, and $c = 5$.

$$x = \frac{-2 \pm \sqrt{2^2 - 4(1)(5)}}{2(1)}$$

$$= \frac{-2 \pm \sqrt{-16}}{2} \qquad \bullet$$

Because -16 occurs under the radical, the roots are not real numbers. For now, the answer will be left in the above form. In Section 9.1 we will discuss the case when the discriminant is negative and the roots "imaginary."

An interesting illustration of the developments of this section is provided by the following examples.

EXAMPLE 12 Suppose a small plant manufactures radios. From experience it has been found that the cost per day for producing x units is given by the cost function $C(x) = x^2 + 3x + 4$. Suppose that the radios can be sold at $75 - 2x$ dollars per unit. Find the maximum profit.

SOLUTION The total received from the sale of x radios is the product of the number of units sold and the price per unit, or $x(75 - 2x)$. Since profit = receipts − cost,

$$\begin{aligned} \text{profit } P(x) &= x(75 - 2x) - (x^2 + 3x + 4) \\ &= -3x^2 + 72x - 4. \end{aligned}$$

Now complete the square to find the maximum profit.

$$\begin{aligned} P(x) &= -3(x^2 - 24x + 144) - 4 + 432 \\ &= -3(x - \mathbf{12})^2 + \mathbf{428}. \end{aligned}$$

Since $P(x)$ is greatest when $x = 12$, the profit is maximum for a production of 12 radios per day and this profit is \$428. ●

EXAMPLE 13 Find the **break-even point,** the value of x for which receipts equal costs, for the cost function of Example 12.

SOLUTION The break-even point occurs when receipts = cost, that is, when

$$\begin{aligned} x(75 - 2x) &= x^2 + 3x + 4. \\ 75x - 2x^2 &= x^2 + 3x + 4 \\ -3x^2 + 72x - 4 &= 0 \\ x &= \frac{-72 \pm \sqrt{(72)^2 - 4(-3)(-4)}}{2(-3)} \\ &= \frac{-72 \pm \sqrt{5136}}{-6} \\ &= \frac{-72 \pm 71.666}{-6} \\ &= 23.94 \text{ or } 0.06, \text{ to two decimal places.} \end{aligned}$$

So $x = 24$ units per day produces a revenue about the same as the cost. The other solution, $x = 0.06$, has no meaning for the given problem, since x must be a counting number. Such meaningless roots occur frequently when we use mathematical models for "real-world" situations. ●

Quadratic Inequalities In Section 1.2, Example 5, we solved the quadratic inequality $x^2 - 4 < 0$ by using the factored form and a sign-table. This method can be used to solve any quadratic inequality for which the quadratic polynomial can be factored.

EXAMPLE 14 Solve the inequality $x^2 - x - 6 > 0$.

SOLUTION We begin by factoring the polynomial.

$$x^2 - x - 6 > 0$$
$$(x + 2)(x - 3) > 0$$

Since the product is positive, either both factors are positive or both factors are negative. The product is zero when $x = -2$ and $x = 3$, so we consider separately the intervals $x < -2$, $-2 < x < 3$, and $x > 3$. The sign table is shown below.

	$x < -2$	$-2 < x < 3$	$x > 3$
$x - 3$	−	−	+
$x + 2$	−	+	+
	↑		↑

From the sign table, if $x < -2$ or if $x > 3$ the product is positive. That is, the inequality is satisfied when x is outside the interval $-2 \leq x \leq 3$. ●

EXAMPLE 15 Solve the inequality $2x^2 - x - 1 < 0$.

SOLUTION We first factor the polynomial.

$$2x^2 - x - 1 < 0$$
$$(2x + 1)(x - 1) < 0$$

Since the product is negative, one of the factors must be positive and the other negative. The product is zero when $x = -1/2$ and when $x = 1$, so we consider separately the intervals $x < -1/2$, $-1/2 < x < 1$, and $x > 1$.

	$x < -\frac{1}{2}$	$-\frac{1}{2} < x < 1$	$x > 1$
$2x + 1$	−	+	+
$x - 1$	−	−	+
		↑	

From the sign table above it is seen that the solution is $-1/2 < x < 1$. ●

EXERCISES 3.3

Complete the square on each of the following quadratics.

1. $x^2 - 2x$ **2.** $x^2 + 4x$ **3.** $x^2 - 5x$ **4.** $3x^2 + 10x + 6$

5. $2x^2 - 10x + 11$ **6.** $x^2 - 8x - 4$ **7.** $2x^2 + 4x - 1$ **8.** $3x^2 + 5x$

9. $5x^2 - 15x + 8$ **10.** $2x^2 + x + 5$

Find the greatest or least value of each of the following quadratic functions, as in Examples 6 and 7. Give the range of each function.

11. $f(x) = x^2 + 2x$

12. $f(x) = x^2 - 4x - 1$

13. $f(x) = -x^2 - 4x - 3$

14. $f(x) = 3x^2 + 10x + 6$

15. $f(x) = 2x^2 + x + 5$

16. $f(x) = 3x^2 + 5x + 2$

17. $f(x) = x(1 - x)$

18. $f(x) = -2x(x + 3)$

19. $f(x) = 2x^2 - 10x + 11$

20. $f(x) = 5x^2 - 15x + 8$

Solve each of the following quadratic equations by factoring (where possible), or by using the quadratic formula.

21. $x^2 - 2x - 8 = 0$

22. $x^2 - 5x - 6 = 0$

23. $x^2 + x = 12$

24. $x^2 - 8x - 4 = 0$

25. $5x^2 - 15x + 8 = 0$

26. $x^2 + 6x - 1 = 0$

27. $2x^2 + 3x - 3 = 0$

28. $3x^2 + 10x + 6 = 0$

29. $2x^2 + 4x - 1 = 0$

30. $x^2 - 2x - 8 = 0$

Solve the following quadratic inequalities.

31. $x^2 + x - 12 > 0$

32. $x^2 - 5x + 6 > 0$

33. $x^2 - 2x - 3 < 0$

34. $x^2 + 7x + 12 < 0$

35. $4x^2 - 11x + 6 > 0$

36. $2x^2 + 9x + 9 < 0$

In each case find the interval of values of x for which $f(x) \leq 0$ and for which $f(x) \geq 0$.

37. $f(x) = x^2 - 9$

38. $f(x) = 2x^2 - 10$

39. $f(x) = x^2 - 2x - 8$

40. $f(x) = 6x^2 + 7x - 5$

41. $f(x) = 8x^2 + 5x - 3$

42. $f(x) = 12x^2 + 11x - 15$

43. The formula for the height above the ground in feet of an object t seconds after it is thrown vertically upward at an initial speed v_0 is given by $h(t) = v_0t - 16t^2$. Find the greatest height reached if (a) $v_0 = 32$, (b) $v_0 = 48$, (c) $v_0 = 16$.

44. The height in feet of a projectile above the ground t seconds after being fired is given by $h(t) = 500\,t - 16t^2$. (a) What is the greatest height it reaches and after how much time? (b) At what time does it reach the ground again?

45. The distance s (in feet) a body falls in t seconds if it has initial velocity 64 feet per second is given by the formula $s(t) = 16t^2 + 64t$. Find the time required for the body to fall 1520 feet.

46. Suppose that in Example 9 of this section one side of the rectangular region is along an existing fence, so that only three sides require fencing. Show that the rectangle of maximum area has the same width as before but is twice as long.

47. Repeat Example 9 of this section for a given perimeter p, and thus show that the largest rectangle is a square of side $\frac{1}{4}p$.

48. A rectangular plot of ground is 6 rods longer than it is wide and its area is 27 square rods. Find its dimensions. $\left(\text{A } \textbf{rod} \text{ is } 16\frac{1}{2} \text{ feet}\right)$.

49. Two numbers differ by 1 and the sum of their squares is 41. Find the numbers.

50. (a) Find two numbers whose sum is 4 and which have the greatest product. (b) Repeat for sum 6. (c) Repeat for sum s. What general result do you observe?

3.4 Graphs of Quadratic Functions

The graph of a quadratic function is called a **parabola.** The parabola will be treated in more detail as one of the conic curves in Section 7.3. Only a brief treatment of the subject is given here.

EXAMPLE 1 Graph the squaring function, $f(x) = x^2$.

SOLUTION This was first discussed in Chapter 2. Since $f(0) = 0$, the origin (the point with coordinates (0, 0)) is on the graph. Also, since $f(x) = x^2 \geq 0$, no part of the graph is below the x-axis. The lowest (or highest) point on a parabola is called its **vertex.** Thus, the vertex of $f(x) = x^2$ is the origin. This function is even, that is, $f(-x) = f(x)$, so its graph is symmetric with respect to the y-axis (Section 2.7). For this reason, the y-axis is called an **axis of symmetry** of the curve. From the analysis just given, we know that the graph must have the general appearance shown in Figure 3.10. By calculating several points, such as (1, 1) and (2, 4), we can complete the sketch of the graph. ●

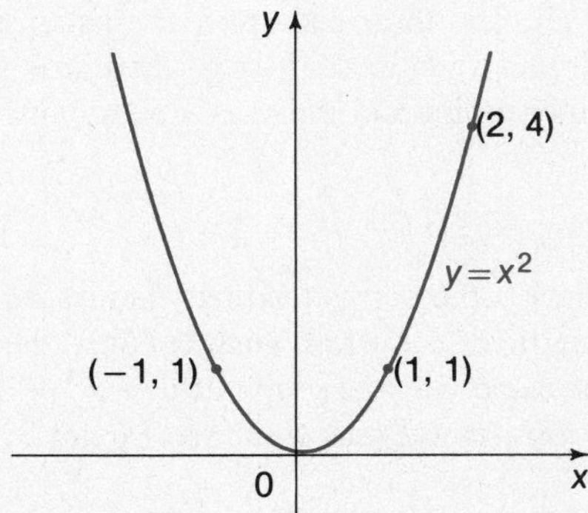

Figure 3.10

EXAMPLE 2 Graph $f(x) = 2x^2$.

SOLUTION If we let $y = f(x) = 2x^2$, then for each x the corresponding y is just twice that in $y = x^2$. For instance, given $y = x^2$, when $x = 1$, $y = 1$, but for $y = 2x^2$, when $x = 1$, $y = 2$, and so on. This curve also has a vertex at the origin and lies on or above the x-axis. Like the graph of $y = x^2$, it is symmetric with respect to the y-axis. The graph is shown in Figure 3.11, where the curve $y = x^2$ is also shown for reference. The curve $y = x^2$ has been "stretched" upward to produce the curve $y = 2x^2$. ●

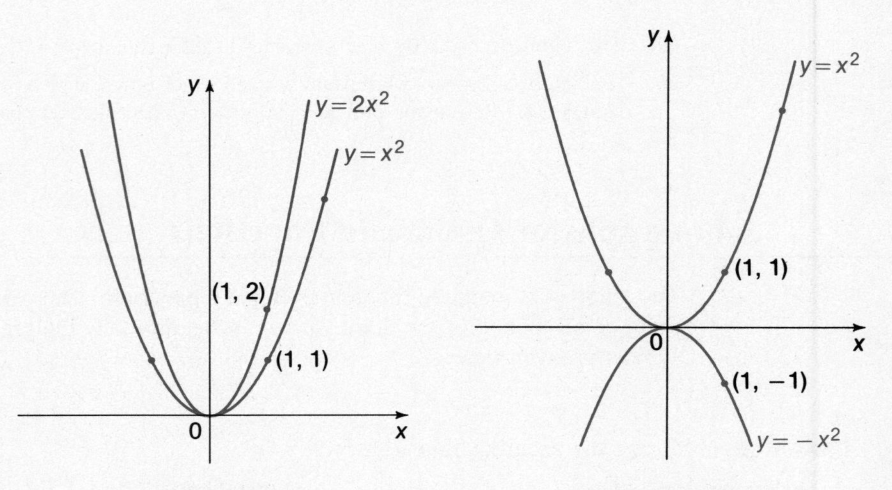

Figure 3.11 **Figure 3.12**

EXAMPLE 3 Graph $f(x) = -x^2$.

SOLUTION Here, if $y = -x^2$, each y is just the negative of the corresponding y for the curve $y = x^2$. So the graph of $y = -x^2$ is the geometric reflection in the x-axis of the graph of $y = x^2$. See Figure 3.12. ●

The last three examples are instances of graphs of the general family of "pure" quadratics, $y = f(x) = ax^2$, where a is a real constant. They all share the origin as a common point and the y-axis as a common axis of symmetry.

EXAMPLE 4 Graph $y = f(x) = x^2 + 2$.

SOLUTION For each x the corresponding y is just 2 *more* than the corresponding y in $y = x^2$. The geometric effect is to **translate** (shift) the graph of $y = x^2$ vertically upward two units. In the same way the graph of $y = x^2 - 1$ is obtained by shifting that of $y = x^2$ vertically downward one unit. See Figure 3.13. ●

The graph in each example up to now has been a member of the family given by $f(x) = ax^2 + c$, where a and c are constants. Observe that the graph opens upward if $a > 0$ and downward if $a < 0$. Also, c is the y-intercept, since $y = c$ when $x = 0$.

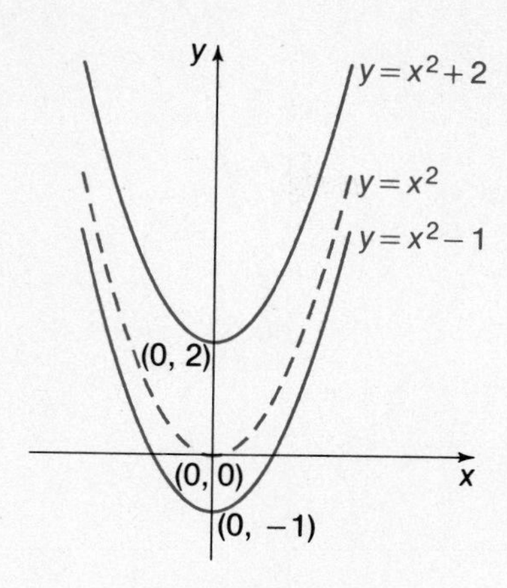

Figure 3.13

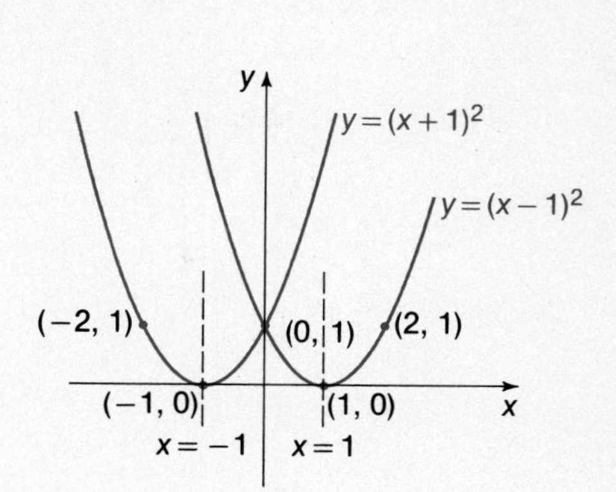

Figure 3.14

EXAMPLE 5 Graph $y = f(x) = (x + 1)^2$.

SOLUTION Since $(x + 1)^2 \geq 0$, the least value of y is 0 and it occurs when $x = -1$. So the vertex of the graph is $(-1, 0)$. The graph is symmetric to the line $x = -1$, since

$$f(-1 \pm a) = (-1 \pm a + 1)^2 = (\pm a)^2 = a^2,$$

for any a. Thus, we get

$$f(-2) = f(0) = 1,$$
$$f(-3) = f(1) = 4,$$

and so on. With this information we can sketch the curve, as shown in Figure 3.14. The effect of the quantity "$x + 1$" is to *translate* the curve $y = x^2$ one unit *to the left*. The graph of $y = (x - 1)^2$ is similarly obtained by translating the graph of $y = x^2$ one unit *to the right*. See Figure 3.14. ●

EXAMPLE 6 Graph $f(x) = x^2 + 2x + 3$.

SOLUTION Completing the square, we have

$$f(x) = (x + 1)^2 + 2.$$

From the preceding examples we see that the graph is obtained from that of $y = x^2$ by translating it one unit *to the left* and two units *upward*. The vertex is the point $(-1, 2)$, and the line $x = -1$ is the axis of symmetry. See Figure 3.15, top of next page. ●

Parabolas will be studied again in more detail in Chapter 7. Here, we note that careful study of the equation of a curve reveals its significant features. The graph can then be sketched without plotting many individual points.

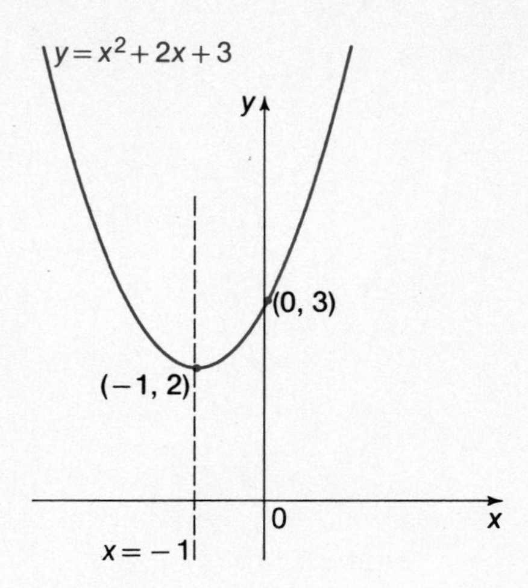

Figure 3.15

The Discriminant The real roots of the equation $f(x) = 0$ are the real zeros of the function f and the x-intercepts of the graph of f. For a quadratic function, $f(x) = 0$ means that $ax^2 + bx + c = 0$ and $x = (-b \pm \sqrt{b^2 - 4ac})/2a$. Thus, for a quadratic function with discriminant $b^2 - 4ac$, there are three possibilities:

1. If the discriminant is negative, there are no x-intercepts.
2. If the discriminant is 0, there is one x-intercept.
3. If the discriminant is positive, there are two distinct x-intercepts.

If the first possibility occurs, all the values of f have the same sign (they are all positive or all negative), so the graph of f is either entirely above or entirely below the x-axis. This occurred in Example 6, where

$$f(x) = x^2 + 2x + 3 \qquad \text{and} \qquad b^2 - 4ac = 2^2 - 4 \cdot 1 \cdot 3 < 0.$$

In Figure 3.15 the entire graph of $y = x^2 + 2x + 3$ is *above* the x-axis.

In the second case, when the discriminant is zero, the graph must be tangent to the x-axis. In Example 5,

$$f(x) = (x + 1)^2 = x^2 + 2x + 1 \qquad \text{and} \qquad b^2 - 4ac = 2^2 - 4 \cdot 1 \cdot 1 = 0.$$

Figure 3.14 shows that the graph of $y = (x + 1)^2$ is tangent to the x-axis.

In Figure 3.13 (lowest graph),

$$f(x) = x^2 - 1 \qquad \text{and} \qquad b^2 - 4ac = 0^2 - 4 \cdot 1 \cdot -1 = 4 > 0.$$

The graph of $y = x^2 - 1$ crosses the x-axis at two points.

EXERCISES 3.4

Without actually graphing, find the coordinates of the vertex and the equation of the axis of symmetry of each of the following curves. You may need to complete the square first.

1. $y = f(x) = -3x^2$ **2.** $y = f(x) = x^2 - 3$ **3.** $y = f(x) = x^2 - 4x$

4. $y = f(x) = x^2 + 2x$ **5.** $y = f(x) = x^2 - 2x + 1$ **6.** $y = f(x) = x^2 + 4x + 4$

7. $y = f(x) = 2x^2 + 3x$ **8.** $y = f(x) = 2x^2 - 4x + 5$

Graph each of the following quadratic functions.

9. $f(x) = 3x^2$ **10.** $f(x) = x^2 + 3$ **11.** $f(x) = 2x^2 - 1$

12. $f(x) = 2x^2 + 1$ **13.** $f(x) = -x^2 - 1$ **14.** $f(x) = -x^2 + 2$

15. $f(x) = x^2 + 4x$ **16.** $f(x) = -x^2 + 2x$ **17.** $f(x) = x^2 - x - 2$

18. $f(x) = x^2 - 2x + 1$ **19.** $f(x) = x^2 + x + 1$ **20.** $f(x) = 2x^2 + 4x - 5$

21. $f(x) = 3x^2 - 6x + 1$ **22.** $f(x) = -2x^2 + 3x - 2$

Use the graph of each of the following quadratic functions to determine for what values of x the function is (a) *positive or negative,* (b) increasing or decreasing.

23. $f(x) = x^2 + 2$ **24.** $f(x) = -x^2 + 1$ **25.** $f(x) = -x^2 + 3x$

26. $f(x) = x^2 - 2x$ **27.** $f(x) = 2x^2 - 4x + 1$ **28.** $f(x) = -3x^2 + 6x + 2$

29. For what value of c is the lowest point of the graph of $y = x^2 - 4x + c$ on the x-axis? Explain.

30. Suppose that the lowest point on the graph of $y = ax^2 + bx + c$ is its y-intercept. Show that $b = 0$ and the intercept is $(0, c)$.

3.5 Polynomials of Higher Degree

Synthetic Division In dealing with polynomials of higher degree, the division of a polynomial by a linear polynomial of the form $x - c$ is quite important. There is a shortcut method for doing this division, called *synthetic division,* which we will now describe. We first write the usual step-by-step arrangement for calculating $(2x^3 - 5x^2 - 4x + 7) \div (x - 3)$ on the left below.

$$
\begin{array}{r|l}
 & 2x^2 + x - 1 \quad \text{quotient} \\
\hline
x - 3 & 2x^3 - 5x^2 - 4x + 7 \\
 & 2x^3 - 6x^2 \\
\hline
 & \quad x^2 - 4x \\
 & \quad x^2 - 3x \\
\hline
 & \qquad -x + 7 \\
 & \qquad -x + 3 \\
\hline
 & \qquad\quad 4 \quad \text{remainder}
\end{array}
$$

$$
\begin{array}{rr|rrrr}
1 & -3 & 2 & -5 & -4 & +7 \\
\hline
 & & \textcircled{2} & -6 & & \\
\hline
 & & & 1 & (-4) & \\
 & & & \textcircled{1} & -3 & \\
\hline
 & & & & -1 & (+7) \\
 & & & & \textcircled{-1} & +3 \\
\hline
 & & & & & 4
\end{array}
$$

On the right we have written the "skeleton": just the numerical coefficients in the same relative positions as on the left. Note that each of the circled numbers is a repetition of the one directly above it and is also the same as the corresponding coefficient in the quotient above. The number following each circled number is -3 times the circled number. The numbers in parentheses (including the sign) are repetitions of the coefficients of the dividend directly above. Now, if we eliminate the coefficient of x in the divisor because it is always 1, the circled repetitions, and the numbers in parentheses, we can write this in the compact arrangement shown at the left below. Here the first line shows just the coefficients of the dividend, together with the constant term of the divisor. The middle line shows the numbers that followed the circled ones above, in the *same* column positions. The last line shows the circled numbers, together with the remainder. This line is the result of subtracting, from the top line, -3 times successive entries in the last line. These are the coefficients of the quotient and the remainder.

$$\begin{array}{lrrrrrl} \text{Divisor} \longrightarrow & -3 \,| & 2 & -5 & -4 & +7 & \longleftarrow \text{Dividend} \\ & & & -6 & -3 & +3 & \\ \hline \text{Quotient} \longrightarrow & & 2 & +1 & -1 & +4 & \longleftarrow \text{Remainder} \end{array} \qquad \begin{array}{rrrrr} 3\,| & 2 & -5 & -4 & +7 \\ & & 6 & +3 & -3 \\ \hline & 2 & +1 & -1 & +4 \end{array}$$

A slight modification of this scheme is shown on the right. Here 3 has been used as the divisor instead of -3 and the numbers of the third line are found by successively *adding* 3 times the numbers in the last line to those in the first line. Since adding 3 times these numbers is equivalent to subtracting -3 times them, the numbers in the last line are the same as before—namely, the coefficients of the quotient and remainder. The advantage is that addition is ordinarily easier than subtraction. When division is arranged in this last form (at the right above), it is called **synthetic division.**

EXAMPLE 1 Use synthetic division to divide $(3x^4 - x^3 + 2x^2 + 4x - 5)$ by $(x + 2)$.

SOLUTION Since the divisor is $x + 2 = x - (-2)$, we use the divisor -2. First, write down the coefficients of the dividend (first line). Next, bring down the leading coefficient, 3, multiply it by -2, and then add the product -6 to -1 to get -7. Then multiply -7 by -2 and add the product, 14, to 2, and so on. These calculations may be performed in one continuous operation with a hand calculator.

$$\begin{array}{lrrrrrrl} \text{Divisor} \longrightarrow & -2\,| & 3 & -1 & +2 & +4 & -5 & \longleftarrow \text{Dividend} \\ & & & -6 & +14 & -32 & +56 & \\ \hline \text{Quotient} \longrightarrow & & 3 & -7 & +16 & -28 & +51 & \longleftarrow \text{Remainder} \end{array}$$

Thus, the quotient is $3x^3 - 7x^2 + 16x - 28$ and the remainder is 51. The quotient will always be of degree one less than the dividend. This result can be written

$$(3x^4 - x^3 + 2x^2 + 4x - 5) \div (x + 2) = 3x^3 - 7x^2 + 16x - 28 + \frac{51}{x + 2}.$$

It can be checked by showing that

$$(x + 2)(3x^3 - 7x^2 + 16x - 28) + 51 = 3x^4 - x^3 + 2x^2 + 4x - 5. \quad \bullet$$

It is essential that a "0" fill the place of any missing power. The following illustrates this.

EXAMPLE 2 Use synthetic division to calculate $(x^3 + 2x + 1) \div (x - 1)$.

SOLUTION Since the x^2-term is missing, we use 0 for its coefficient.

$$\begin{array}{r|rrrr} 1 & 1 & +0 & +2 & +1 \\ & & 1 & +1 & +3 \\ \hline & 1 & +1 & +3 & +4 \end{array}$$

So $(x^3 + 2x + 1) \div (x - 1) = x^2 + x + 3 + \dfrac{4}{x - 1}$. ●

As we saw in Section 3.1 and in the preceding examples, the quotient of two polynomials may not be a polynomial. It may be written as the sum of a polynomial and a rational fraction. For example,

$$\frac{x^3 + 1}{x} = x^2 + \frac{1}{x}.$$

This general fact is expressed in the following theorem.

THEOREM 3.2 **Division Algorithm* for Polynomials** If $f(x)$ and $h(x)$ are polynomials, with $h(x) \neq 0$, then there are unique polynomials $q(x)$ and $r(x)$ such that

$$\boldsymbol{f(x) = h(x) \cdot q(x) + r(x),}$$

where either $r(x) = 0$ or the degree of $r(x)$ is less than that of $h(x)$.

Observe that this statement accounts for the usual procedure for checking division: multiply the divisor by the quotient and add the remainder to obtain the dividend. We will not prove this theorem here but will use it in what follows.

EXAMPLE 3 Find $q(x)$ and $r(x)$ in the division algorithm for (a) $f(x) = x^3 + 1$ and $h(x) = x$, (b) $f(x) = 2x^3 + 3x + 7$ and $h(x) = x^2 + x - 2$.

SOLUTION (a) We saw above that $\dfrac{x^3 + 1}{x} = x^2 + \dfrac{1}{x}$.

Here, $q(x) = x^2$ and $r(x) = 1$. Written as in the division algorithm

$$x^3 + 1 = x \cdot x^2 + 1.$$

(b) In Example 3(d), section 3.1, we found

$$(2x^3 + 3x + 7) \div (x^2 + x - 2) = 2x - 2 + \frac{9x + 3}{x^2 - x - 2}.$$

Here, $q(x) = 2x - 2$ and $r(x) = 9x + 3$. Rewritten,

$$2x^3 + 3x + 7 = (x^2 + x - 2)(2x - 2) + 9x + 3.$$ ●

---

*The term **algorithm** refers to any specific procedure for solving some particular type of problem. For example, the arithmetic processes of finding products and quotients are algorithms.

Two powerful consequences of the division algorithm theorem will now be given. If the divisor $h(x)$ is $x - c$, the remainder is either 0 or its degree is less than 1, that is, its degree is 0. Thus, $r(x)$ is either 0 or some constant k, and

$$f(x) = (x - c)q(x) + k.$$

Then $f(c) = (c - c)q(c) + k = k$, where k may be 0. This result is stated in the following theorem.

THEOREM 3.3 **Remainder Theorem** If a polynomial $f(x)$ is divided by $x - c$, then the remainder is $f(c)$.

EXAMPLE 4 Let $f(x) = 2x^3 + x - 7$ and find $f(4)$.

SOLUTION We could find $f(4)$ by substituting 4 for x in $f(x)$:

$$f(4) = 2(4)^3 + 4 - 7 = 125.$$

However, let us divide $f(x)$ by $x - 4$, using synthetic division.

$$\begin{array}{r|rrrr} 4 & 2 & 0 & +1 & -7 \\ & & 8 & +32 & +132 \\ \hline & 2 & +8 & +33 & +125 \end{array}$$

As expected, the remainder is 125. Thus, by the remainder theorem, we can use synthetic division to obtain $f(4) = 125$. ●

EXAMPLE 5 Given $f(x) = 3x^3 + 7x^2 + x - 4$, calculate $f(1)$ and $f(-4/3)$.

SOLUTION Use synthetic division to divide $f(x)$ by $x - 1$ and by $x + \frac{4}{3}$ respectively.

$$\begin{array}{r|rrrr} 1 & 3 & +7 & +1 & -4 \\ & & +3 & +10 & +11 \\ \hline & 3 & +10 & +11 & +7 \end{array} \qquad \begin{array}{r|rrrr} -\frac{4}{3} & 3 & +7 & +1 & -4 \\ & & -4 & -4 & +4 \\ \hline & 3 & +3 & -3 & +0 \end{array}$$

The remainders are, respectively, 7 and 0, so that $f(1) = 7$ and $f(-4/3) = 0$, by the remainder theorem. Note that in the latter case,

$$\begin{aligned} f(x) &= \left(x + \frac{4}{3}\right)(3x^2 + 3x - 3) + 0 \\ &= \left(x + \frac{4}{3}\right)(3x^2 + 3x - 3), \end{aligned}$$

that is, $x + 4/3$ is a factor of $f(x)$. ●

The example illustrates the following corollary* to the remainder theorem.

---

* A corollary is a secondary theorem which follows from some theorem with very little proof.

THEOREM 3.4 **Factor Theorem** A polynomial has $x - c$ as a factor if an only if $f(c) = 0$.

Proof: From the remainder theorem, the remainder after division by $x - c$ is $f(c)$. Then $f(x) = (x - c)q(x)$ if and only if $f(c) = 0$. But this is equivalent to saying that $x - c$ is a factor of $f(x)$.

EXAMPLE 6 Factor $f(x) = 2x^2 - 5x - 3$ and use the results to illustrate the factor theorem.

SOLUTION

$$\begin{aligned} f(x) &= (2x + 1)(x - 3) \\ &= 2\left(x + \frac{1}{2}\right)(x - 3) \end{aligned}$$

Thus, both $x + 1/2$ and $x - 3$ are factors of $f(x)$. By synthetic division or substitution, we may verify that $f(-1/2) = 0$ and $f(3) = 0$. ●

In section 3.1 we factored quadratic polynomials whose factors had rational coefficients. Now we can use the factor theorem together with the quadratic formula to factor *any* quadratic polynomial.

EXAMPLE 7 Factor $f(x) = x^2 + 2x - 2$.

SOLUTION Solving the quadratic equation $x^2 + 2x - 2 = 0$ gives the roots $-1 + \sqrt{3}$ and $-1 - \sqrt{3}$. By the factor theorem,

$$\begin{aligned} f(x) &= (x - (-1 + \sqrt{3}))\,(x - (-1 - \sqrt{3})) \\ &= (x + 1 - \sqrt{3})(x + 1 + \sqrt{3}). \end{aligned}$$ ●

EXAMPLE 8 Use the factor theorem to find a polynomial function f whose zeros are -2, 1, and 3.

SOLUTION Since $f(-2) = f(1) = f(3) = 0$, by the factor theorem $x + 2$, $x - 1$, and $x - 3$ are factors of f. So

$$\begin{aligned} f(x) &= (x + 2)(x - 1)(x - 3) \\ &= x^3 - 2x^2 - 5x + 6. \end{aligned}$$ ●

EXERCISES 3.5

Use synthetic division to calculate the quotient and remainder in each of the following.

1. $(x^2 - 5x + 10) \div (x - 1)$

2. $(x^3 - 2x^2 + x - 3) \div (x + 3)$

3. $(2x^3 + x^2 - 8x - 4) \div (x - 2)$

4. $(2x^4 - 3x^2 + 4) \div (x + 2/3)$

5. $(3x^4 + 14x^3 - 4x^2 - 11x - 2) \div (x + 1)$

Use synthetic division and the remainder theorem to calculate the indicated function values.

6. If $f(x) = x^3 - 2x^2 + 3x + 4$, find $f(3)$, $f(-2)$, $f(0)$.

7. If $f(x) = 2x^4 + 6x^3 + x^2 + 3x - 6$, find $f(-3)$, $f(1)$, $f(2)$.

8. If $f(x) = -x^3 - 2x + 4$, find $f(1.3)$, $f(-2.1)$.

9. If $f(x) = 2x^3 - 13x^2 + 27x - 18$, find $f(3/2)$, $f(1/2)$.

10. If $f(x) = x^5 + 1$, find $f(1)$, $f(-1)$.

11. If $f(x) = x^4 - 16$, find $f(1)$, $f(2)$, $f(3)$.

Use synthetic division and the factor theorem to determine whether the given linear polynomial $x + c$ is a factor of the indicated polynomial. If it is, give the corresponding factorization.

12. $x - 2$; $3x^3 + 10x^2 + 9x + 2$

13. $x + 1$; $2x^3 - 5x^2 - x + 6$

14. $x + 3$; $2x^4 + 6x^3 + x^2 + 3x - 6$

15. $x - 1$; $8x^3 - 10x^2 - x + 3$

16. $x + 2$; $x^4 + 16$

17. $x + 2$; $x^4 - 16$

Use the factor theorem and the quadratic formula to factor each of the following quadratic polynomials.

18. $f(x) = x^2 + 6x - 1$

19. $f(x) = x^2 - 6x + 7$

20. $f(x) = x^2 - 2x - 5$

21. $f(x) = 4x^2 - 12x - 9$

22. $f(x) = 3x^2 + 5x + 1$

23. $f(x) = 2x^2 + 2x - 1$

Use the factor theorem to determine a polynomial function f which has the given numbers as its zeros.

24. 2, −1, and 1

25. $\frac{1}{2}$, 0, and 1

26. $\frac{1}{2}$, 0, 1, and −2

27. $\sqrt{2}$ and $-\sqrt{2}$

28. $\frac{1}{2}$, 3, and −2

29. 1, 1, −2, and −2.

3.6 Zeros of a Polynomial Function

We have seen that a linear polynomial function has a single zero and a quadratic polynomial function has at most two zeros. It is a fact that a polynomial of degree n has at most n distinct real zeros.

THEOREM 3.5 If $p(x)$ is a real polynomial of degree n, then $p(x)$ has at most n distinct real zeros.

Proof: We show that $p(x)$ can have *no more than n* distinct zeros by showing that the assumption that it has more leads to a contradiction. Suppose that it had $n + 1$ distinct zeros: $x_1, x_2, x_3, \ldots x_n$, and k. Then

$$p(x_1) = p(x_2) = p(x_3) = \cdots = p(x_n) = p(k) = 0.$$

By the factor theorem, $(x - x_1)$, $(x - x_2)$, . . . , $(x - x_n)$ are all factors of $p(x)$. Since the product has degree n, we must have

$$p(x) = a(x - x_1)(x - x_2)(x - x_3) \ldots (x - x_n).$$

Since k is a zero of p, $p(k) = 0$, and

$$0 = p(k) = (k - x_1)(k - x_2)(k - x_3) \ldots (k - x_n).$$

But k is distinct from each of the x_i's so that none of the factors on the right-hand side of the above equation can be zero. This is a contradiction, because the product of two or more numbers cannot be zero unless at least one of them is zero. Hence $p(x)$ *cannot* have more than n distinct zeros.

In Example 11, Section 3.3, we saw that the roots of a quadratic equation are not necessarily real numbers. In Chapter 9 we will see that the introduction of complex numbers allows a quadratic equation to have exactly two roots (not necessarily distinct), real or complex. It is a fact that no new kinds of numbers are needed to provide for roots of polynomial equations of *any* degree. This follows from the *fundamental theorem of algebra,* first proved in 1799 by the German mathematician Gauss.

The zeros of a polynomial may not be distinct. For example,

$$p(x) = x^2 - 2x + 1 = (x - 1)(x - 1).$$

In such a case, $p(x)$ has 1 as a **double zero.** In general, if $x - c$ occurs m times as a factor of a polynomial $p(x)$, we say that c is a **zero of multiplicity m** of $p(x)$.

Rational Zeros We have seen that the zeros of a quadratic polynomial may be calculated from the quadratic formula. There are similar formulas for calculating the zeros of cubic and quartic (third and fourth degree) polynomials. The derivation of these is complicated and not appropriate here. There is no general formula for the zeros of polynomials of degree $n \geq 5$, however. In fact, after many years of trying to find general methods for solving for zeros of polynomials of higher degrees, it was proved in the early nineteenth century that the zeros of a polynomial of degree five or more cannot be expressed in terms of the coefficients of the polynomial.

There is a procedure for identifying all the *rational* zeros of a polynomial function having *integral* coefficients. The theorem used is called the *theorem on rational zeros.*

THEOREM 3.6 Suppose that $\boldsymbol{f(x) = a_n x^n + a_{n-1}x^{n-1} + \cdots + a_1 x + a_0,}$

where the a_k's are integers. Suppose f has a rational zero, p/q, with $q > 0$ and p/q in lowest terms. Then p is a factor of the constant term a_0 and q is a factor of the leading coefficient a_n.

Proof: Since p/q is a zero of f, then

$$f\left(\frac{p}{q}\right) = a_n\left(\frac{p}{q}\right)^n + a_{n-1}\left(\frac{p}{q}\right)^{n-1} + \cdots + a_1\left(\frac{p}{q}\right) + a_0 = 0.$$

Multiplying both sides by q^n gives

$$a_n p^n + a_{n-1}p^{n-1}q + \cdots + a_1 p q^{n-1} + a_0 q^n = 0.$$

Then

$$a_n p^n + a_{n-1}p^{n-1}q + \cdots + a_1 p q^{n-1} = -a_0 q^n,$$

$$p(a_n p^{n-1} + a_{n-1}p^{n-2}q + \cdots + a_1 q^{n-1}) = -a_0 q^n.$$

All the constants, a_k, p, q, and n, are integers, so both sides of the last equation are integers. Then, since p is a factor of the left side, it must be a factor of the right side. The fraction p/q is in lowest terms, that is, the only factors common to p and q are 1 and -1. Hence p is not a factor of q^n and so must be a factor of a_0, as we wished to prove.

To show that q is a factor of a_n, write

$$a_n p^n + a_{n-1}p^{n-1}q + \cdots + a_0 q^n = 0,$$
$$a_{n-1}p^{n-1}q + a_{n-2}p^{n-2}q^2 + \cdots + a_0 q^n = -a_n p^n,$$
$$q(a_{n-1}p^{n-1} + a_{n-2}p^{n-2}q + \cdots + a_0 q^{n-1}) = -a_n p^n.$$

Now we can use the same kind of argument to show that q is a factor of a_n as we did above to show that p is a factor of a_0. For example, the quadratic equation $2x^2 - x - 3 = 0$ has the roots $3/2$ and $-1 = -1/1$. The numerators, 3 and -1, are factors of the constant term -3 and the denominators, 2 and 1, are factors of the leading coefficient 2.

This theorem does *not* say that a given polynomial has a rational zero. It only tells us the form of any rational zeros which may exist.

EXAMPLE 1 Find any rational zeros of $f(x) = x^4 - 5x^3 + 5x^2 + 5x - 6$.

SOLUTION Since the leading coefficient is 1, the only possible zeros of f are divisors of 6, that is, $\pm 1, \pm 2, \pm 3, \pm 6$. First we test $x = 1$, using synthetic division.

$$\begin{array}{r|rrrrr} 1 & 1 & -5 & +5 & +5 & -6 \\ & & +1 & -4 & +1 & +6 \\ \hline & 1 & -4 & +1 & +6 & +0 \end{array}$$

So $x = 1$ is a zero of f and, by the factor theorem, $x - 1$ is a factor. The coefficients of the other factor appear in the last line of the synthetic division. So

$$f(x) = (x - 1)(x^3 - 4x^2 + x + 6).$$

All other rational zeros of f must be rational zeros of $x^3 - 4x^2 + x + 6$. We test $x = -1$, using synthetic division.

$$\begin{array}{r|rrrr} -1 & 1 & -4 & +1 & +6 \\ & & -1 & +5 & -6 \\ \hline & 1 & -5 & +6 & +0 \end{array}$$

Thus, $x = -1$ is a zero of f and $x + 1$ is a factor. Then

$$f(x) = (x - 1)(x + 1)(x^2 - 5x + 6).$$

The remaining zeros of f must be zeros of $x^2 - 5x + 6$ and so roots of the equation $x^2 - 5x + 6 = 0$. If

$$x^2 - 5x + 6 = 0,$$
$$(x - 2)(x - 3) = 0,$$
$$x = 2 \text{ or } x = 3.$$

So the zeros of f are 1, -1, 2, and 3. Incidentally, we have also solved the fourth degree equation $x^4 - 5x^3 + 5x^2 + 5x - 6 = 0$. Its roots are the above zeros of f. ●

EXAMPLE 2 Find any rational zeros of $f(x) = 2x^3 - 5x^2 + 1$.

SOLUTION The only possible rational zeros of f are $\pm 1/2$ and ± 1. We first test $x = 1/2$.

$$\begin{array}{r|rrrr} 1/2 & 2 & -5 & +0 & +1 \\ & & +1 & -2 & -1 \\ \hline & 2 & -4 & -2 & +0 \end{array}$$

Thus, $x = 1/2$ is a zero of f, $x - 1/2$ is a factor and

$$f(x) = \left(x - \frac{1}{2}\right)(2x^2 - 4x - 2).$$

The remaining zeros must be zeros of $2x^2 - 4x - 2$ and so roots of

$$2x^2 - 4x - 2 = 0, \text{ or } x^2 - 2x - 1 = 0.$$

This cannot be solved by factoring and so we use the quadratic formula.

$$\begin{aligned} x &= \frac{2 \pm \sqrt{4 - 4(1)(-1)}}{2} \\ &= \frac{2 \pm \sqrt{8}}{2} \\ &= \frac{2 \pm 2\sqrt{2}}{2} \\ &= 1 \pm \sqrt{2}. \end{aligned}$$

Thus, the only rational zero of f is $x = 1/2$, and there are two irrational zeros. ●

EXAMPLE 3 Determine whether the equation $x^4 - 2x^3 + 3x^2 - 2x - 3 = 0$ has any rational roots.

SOLUTION If $p(x) = x^4 - 2x^3 + 3x^2 - 2x - 3$, the only possible rational zeros of $p(x)$ are ± 1 and ± 3. Synthetic division shows that none of these is actually a zero of $p(x)$, so $p(x) = 0$ has no rational roots. (See Exercise 15, this section.) ●

The above examples illustrate the fact that an equation may have all rational roots, some rational roots, or no rational roots.

EXAMPLE 4 Factor $f(x) = 2x^3 - 5x^2 - 4x + 3$.

SOLUTION As in the two previous examples, we can show that the rational zeros of $f(x)$ are -1, 3, and 1/2. Then $x + 1$, $x - 3$, and $x - 1/2$ are factors of $f(x)$. Their product has degree three, so we can write, for some constant c,

$$2x^3 - 5x^2 - 4x + 3 = c(x + 1)(x - 3)\left(x - \frac{1}{2}\right)$$

$$= \frac{1}{2}c(x + 1)(x - 3)(2x - 1)$$

$$= \frac{c}{2}(2x^3 - 5x^2 - 4x + 3).$$

Since these polynomials are equal, their corresponding coefficients are equal, and thus $c/2 = 1$ and $c = 2$. Finally,

$$f(x) = (x + 1)(x - 3)(2x - 1). \quad \bullet$$

EXAMPLE 5 Factor the following polynomials and identify any multiple zeros: (a) $f(x) = x^3 - 5x^2 + 3x + 9$; (b) $x^5 + x^4 - 5x^3 - x^2 + 8x - 4$.

SOLUTION (a) Given $f(x) = x^3 - 5x^2 + 3x + 9$, the only possible rational zeros are factors of 9: ± 1, ± 3, ± 9. First we test $x = -1$.

$$\begin{array}{r|rrrr} -1 & 1 & -5 & +3 & +9 \\ & & -1 & +6 & -9 \\ \hline & 1 & -6 & +9 & +0 \end{array}$$

Thus, $x = -1$ is a zero of f, and

$$f(x) = (x + 1)(x^2 - 6x + 9)$$
$$= (x + 1)(x - 3)(x - 3).$$

Here $x - 3$ occurs twice as a factor and 3, the corresponding zero of f, has multiplicity 2.

(b) In the same way, check that

$$f(x) = x^5 + x^4 - 5x^3 - x^2 + 8x - 4$$
$$= (x + 2)(x + 2)(x - 1)(x - 1)(x - 1)$$

so that f has -2 as a zero of multiplicity 2 and 1 as a zero of multiplicity 3. •

The theorem on rational zeros can be applied to any polynomial with *rational* coefficients.

EXAMPLE 6 (a) Factor $f(x) = x^3 - 9x^2/2 + 7x/4 + 1$. (b) Solve the equation $f(x) = 0$.

SOLUTION (a) $f(x) = x^3 - \dfrac{9}{2}x^2 + \dfrac{7}{4}x + 1 = \dfrac{1}{4}(4x^3 - 18x^2 + 7x + 4).$

The zeros of f are the zeros of the polynomial $4x^3 - 18x^2 + 7x + 4$ (Why?), and the possibilities are ± 1, $\pm 1/4$, $\pm 1/2$, ± 2, and ± 4. Using synthetic division we find that $x = 4$ is the only rational zero and that

$$f(x) = \frac{1}{4}(x - 4)(4x^2 - 2x - 1).$$

(b) Let $f(x) = \frac{1}{4}(x - 4)(4x^2 - 2x - 1) = 0$. The solutions of this equation are $x = 4$ and the roots of the equation $4x^2 - 2x - 1 = 0$. We solve this equation by the quadratic formula.

$$\begin{aligned} x &= \frac{2 \pm \sqrt{(-2)^2 - 4(4)(-1)}}{2(4)} \\ &= \frac{2 \pm \sqrt{20}}{8} \\ &= \frac{2 \pm 2\sqrt{5}}{8} \\ &= \frac{1}{4} \pm \frac{1}{4}\sqrt{5}. \end{aligned}$$

So the roots of $f(x) = 0$ are 4, $\frac{1}{4} + \frac{1}{4}\sqrt{5}$, and $\frac{1}{4} - \frac{1}{4}\sqrt{5}$. •

EXERCISES 3.6

For each of the following functions (a) *determine the rational zeros, and* (b) *use the results of part* (a) *and the factor theorem to factor the polynomial.*

1. $f(x) = x^4 + x^3 - 2x - 4$

2. $f(x) = x^5 + x^3 - 2x^2 - 2$

3. $f(x) = x^3 + 3x^2 - 4x - 12$

4. $f(x) = 3x^3 - 7x^2 - 3x + 2$

5. $f(x) = 4x^4 - 7x^2 - 5x - 1$

6. $f(x) = 4x^4 - 4x^3 + 13x^2 + 12x + 3$

7. $f(x) = x^3 - 4x^2 + 4x$

8. $f(x) = 4x^4 + 8x^3 - 11x^2 + 3x$

9. $f(x) = x^3 - 6x^2 + 12x - 8$

10. $f(x) = 9x^4 + 6x^3 - 2x^2 - 16x + 8$

Solve the following equations.

11. $f(x) = x^3 - 3x^2 - \frac{7}{2}x - 2 = 0$ [*Hint:* Multiply both sides by 2 to obtain integral coefficients.]

12. $f(x) = x^3 - \frac{17}{2}x^2 + \frac{9}{2}x - 4 = 0$

13. $\frac{1}{4}x^4 - \frac{1}{2}x^3 + \frac{1}{4}x^2 - \frac{1}{2}x = 0$ [*Hint:* Multiply both sides by 4.]

14. $\frac{1}{4}x^4 - x^3 + \frac{1}{4}x^2 - \frac{3}{2}x + 9 = 0$

15. Complete the details of Example 3 of this section by showing that none of the proposed numbers is a zero of p.

16. Show that $\sqrt{2}$ is not a rational number by considering $p(x) = x^2 - 2$.

3.7 Graphing Polynomial Functions

We now know that the graph of a constant function is a horizontal line, the graph of a linear function is an oblique straight line, and the graph of a quadratic function is a parabola with a vertical axis of symmetry. The construction of graphs of higher degree polynomials in general can be done efficiently only with the help of calculus. A few relatively simple cases are discussed in the following examples. Synthetic division and the theorems of the last section will prove helpful.

EXAMPLE 1 Graph $f(x) = x^3$.

SOLUTION First, note that $f(0) = 0$, so that the graph passes through the origin. Also, $f(-x) = -f(x)$, that is, f is an odd function, so its graph is symmetric with respect to the origin. If $x > 0$, then $f(x) > 0$ and if $x < 0$, then $f(x) < 0$, so that the graph lies only in the first and third quadrants. Finally, since $x_2{}^3 > x_1{}^3$ whenever $x_2 > x_1$ (check this), f is an increasing function and its graph rises continually to the right (Section 2.5).

Using the information thus extracted from the equation, we can sketch the graph, as shown in Figure 3.16. Some points have been plotted to set the scale. ●

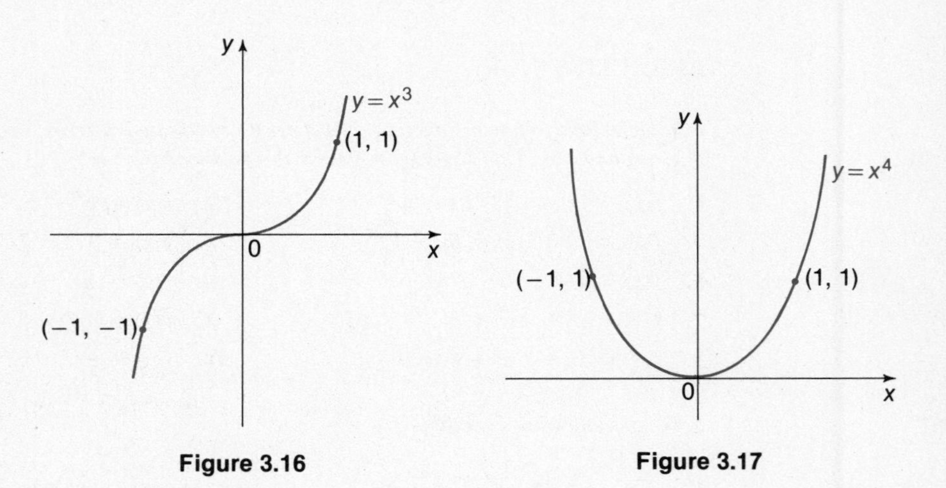

Figure 3.16 **Figure 3.17**

EXAMPLE 2 Graph $y = f(x) = x^4$.

SOLUTION Here, $f(0) = 0$, so the graph passes through the origin. Since $f(-x) = x^4 = f(x)$, f is an even function and its graph is symmetric with respect to the y-axis. Also, $f(x) \geq 0$ for all x, so that no part of the graph is below the x-axis. Finally, f is a decreasing function for $x < 0$ and an increasing function for $x > 0$.

A sketch of the graph, based on this information, is shown in Figure 3.17. Although this curve has the general appearance of a parabola, the only points it has in common with the parabola $y = x^2$ are the origin and the points (1, 1) and (−1, 1). ●

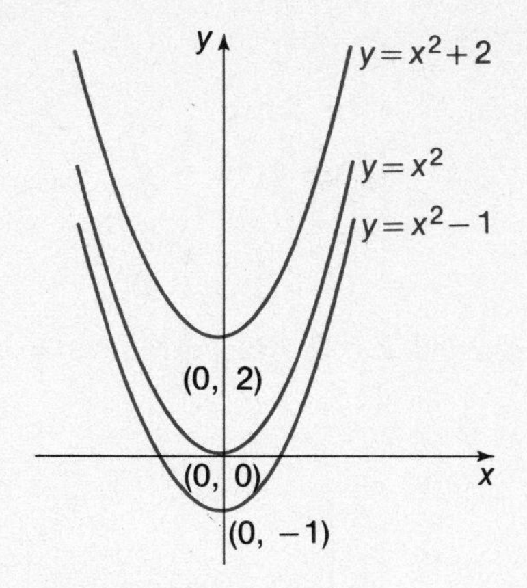

Figure 3.18

Figure 3.19

In Section 3.4 we used the idea of translation in graphing quadratic polynomials, as illustrated in Figures 3.18 and 3.19. Figure 3.18 shows the graph of some equations of the form $y = x^2 + c$, which are obtained from the graph of $y = x^2$ by a vertical translation of c units (upward if $c > 0$ and downward if $c < 0$). Note that c is the y-intercept. Similarly, in Figure 3.19 the graphs of some equations of the form $y = (x + c)^2$ are obtained by a horizontal translation of the graph of $y = x^2$. The translation is to the left if $c > 0$ and to the right if $c < 0$.

Exactly the same procedures give the graphs of functions such as $y = (x + c)^3$ and $y = x^4 + c$ from those of $y = x^3$ and $y = x^4$. In Figures 3.20 and 3.21 these procedures have been used to sketch the graphs of $y = (x - 1)^3$ and $y = x^4 + 1$.

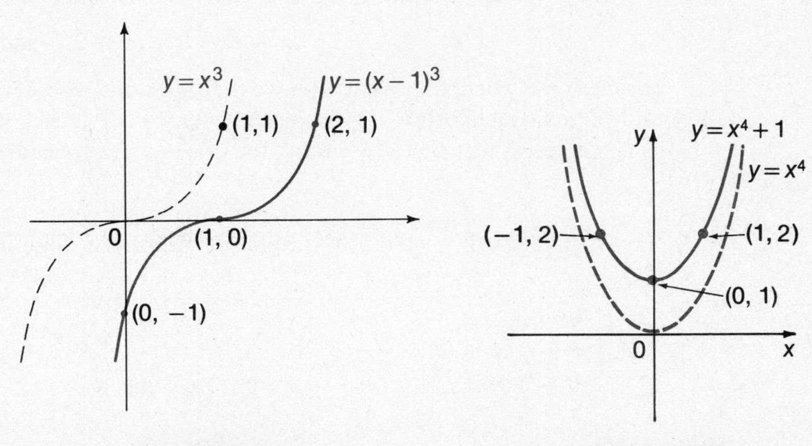

Figure 3.20

Figure 3.21

EXAMPLE 3 Graph $f(x) = x^3 - 4x$.

SOLUTION First, we factor the polynomial $x^3 - 4x$. Since

$$\begin{aligned} f(x) &= x^3 - 4x \\ &= x(x^2 - 4) \\ &= x(x + 2)(x - 2), \end{aligned}$$

$f(x) = 0$ for $x = 0$, $x = -2$, and $x = 2$. Hence the curve intersects the x-axis at $(-2, 0)$, $(0, 0)$, and $(2, 0)$.

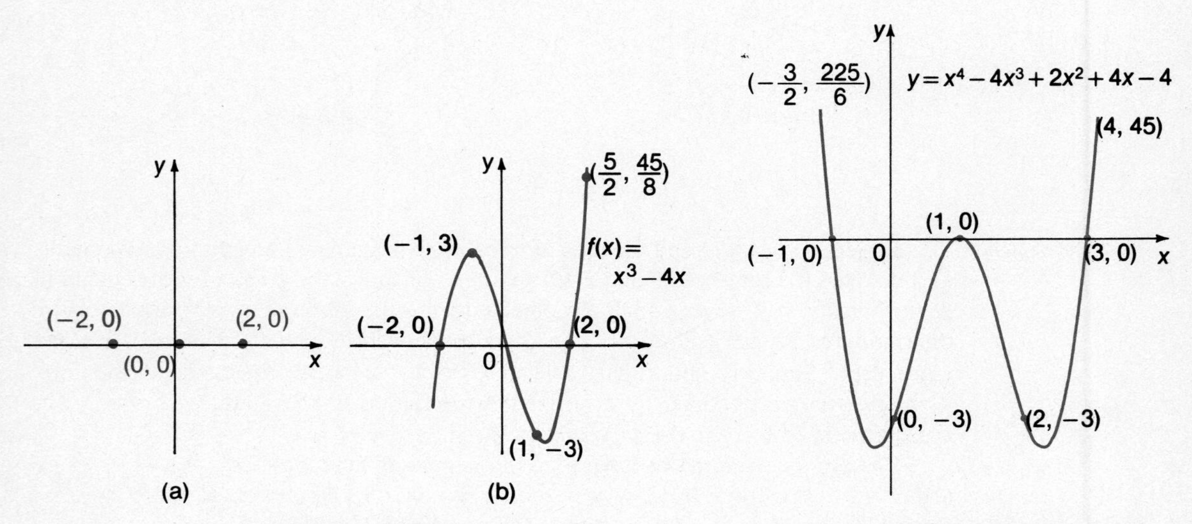

Figure 3.22

Figure 3.23

Now we analyze the graph separately for the intervals

$$x < -2, \quad -2 < x < 0, \quad 0 < x < 2, \quad \text{and} \quad x > 2$$

to determine where the graph is above the x-axis and where it is below. The results are shown in the table below. For instance, when $x < -2$, each of the three factors of $f(x)$ is negative, so that $f(x) < 0$. Thus, the graph of $f(x)$ is below the x-axis for $x < -2$.

Interval	Sign of x	Sign of $x - 2$	Sign of $x + 2$	Sign of $f(x)$	Location of graph
$x < -2$	−	−	−	−	below x-axis
$-2 < x < 0$	−	−	+	+	above x-axis
$0 < x < 2$	+	−	+	−	below x-axis
$x > 2$	+	+	+	+	above x-axis

To locate the curve more accurately, we calculate the coordinates of several points, at least one in each of the intervals of the table. For example, when $x = -1$, $y = 3$, when $x = 1$, $y = -3$, and when $x = 5/2$, $y = 45/8$. In this particular example, we may take advantage of symmetry. Since $f(-x) = -x^3 + 4x = -f(x)$, f is an odd function and its graph is symmetric with respect to the origin. The final graph is shown in Figure 3.22. ●

Without calculus we cannot locate the high or low points on the graph. However, synthetic division and the remainder theorem may be used to find the coordinates of as many points as desired.

EXAMPLE 4 Graph $f(x) = x^4 - 4x^3 + 2x^2 + 4x - 3$.

SOLUTION Using the theorem on rational zeros and synthetic division, we find that the rational zeros of f are -1, 1, 1, and 3. (1 is a double zero.) Hence the curve intersects the x-axis at $(-1, 0)$, $(1, 0)$, and $(3, 0)$. By inspection, we see that when $x = 0$, $f(x) = -3$.

Now, using the factor theorem, we can write $f(x) = (x + 1)(x - 1)^2(x - 3)$ and then determine whether the graph is above or below the x-axis in each of the intervals, as in the last example. The table shows the results.

Interval	Sign of $x + 1$	Sign of $(x - 1)^2$	Sign of $x - 3$	Sign of $f(x)$	Location of graph
$x < -1$	$-$	$+$	$-$	$+$	above x-axis
$-1 < x < 1$	$+$	$+$	$-$	$-$	below x-axis
$1 < x < 3$	$+$	$+$	$-$	$-$	below x-axis
$x > 3$	$+$	$+$	$+$	$+$	above x-axis

A sketch of the graph is shown in Figure 3.23. Several points, those for which $x = -3/2$, $x = 0$, $x = 2$, and $x = 4$, have been plotted for more careful placement of the graph. Check that for $x > 3$, as x increases, all four factors of $f(x)$ increase without bound. Likewise, for $x < -1$, as x decreases each of the four factors decreases without bound, so that their product increases without bound. Thus, the curve rises both for $x < -1$ and $x > 3$, as indicated in the figure. Observe that, since $(x - 1)^2$ has the same sign on both sides of $x = 1$, the curve touches the x-axis at $x = 1$ without crossing it. This situation occurs with every double zero of a function. ●

EXAMPLE 5 Graph $f(x) = x^3 + 3x + 2$.

SOLUTION The only possible rational zeros of $f(x)$ are ± 1 and ± 2. Testing by means of synthetic division shows that none of these is a zero of $f(x)$, so that the graph has no rational x-intercepts. Since $f(x) = 2$ when $x = 0$, the y-intercept is 2.

Since both x^3 and $3x + 2$ increase with x for all x, f is the sum of increasing functions and hence is an increasing function over the whole number line.

The graph of a polynomial function is an unbroken curve, so it must cross the x-axis between the values of x for which $f(x)$ has opposite signs. In this case, since $f(-1) = -2$ and $f(0) = 2$, there must be an intercept between -1 and 0. We can get a more exact value for this intercept by taking a value of x between -1 and 0, say -0.5, and calculating $f(-0.5) = 0.375$. (Use a hand calculator with either synthetic division or direct computation.) Since $f(-1)$ is negative and $f(-0.5)$ is positive, the intercept is between -1 and -0.5. (See Figure 3.24.) This step-by-step procedure can be repeated until any degree of precision needed is obtained for the value of the x-intercept. Finally, we calculate additional points, such as $(1, 6)$ and $(-1, -2)$ to complete a satisfactory graph, as shown in Figure 3.24. ●

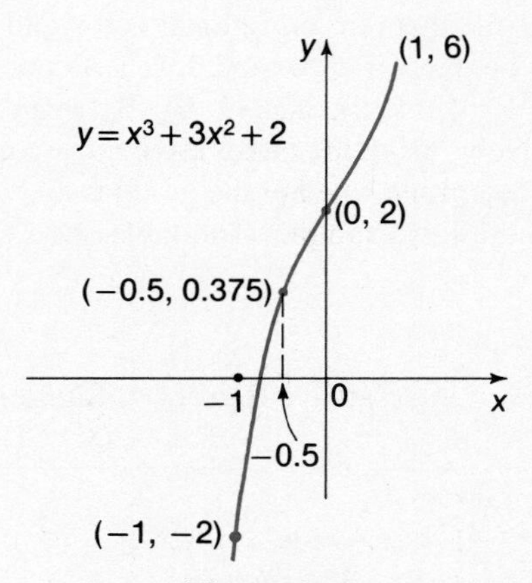

Figure 3.24

These last three examples illustrate how polynomial functions may be graphed by algebraic methods. There are still many refinements which have not been considered here. The methods of calculus make the analysis of polynomial functions so simple and so precise that most of such graphing is left until the study of calculus.

Several further observations may be made here, based upon the examples given:

(1) The fact that the graph of a polynomial is a smooth, unbroken curve is assumed.

(2) Analyzing the polynomial by means of a sign table, calculating any rational zeros, determining intervals of increasing or decreasing function values, and so on, eliminates the need for calculating many individual points on the curve.

(3) Typical graphs of a linear, a quadratic, a cubic, and a quartic polynomial function are shown in Figure 3.25. Observe the pattern of no "humps" on the graph in (a), one in (b), two in (c), and three in (d). It can be shown that, in general, the graph of a polynomial of degree n has *at most* $n - 1$ "humps". In the figures associated with Examples 1, 2, and 5, the graphs all have fewer than the maximum number of "humps."

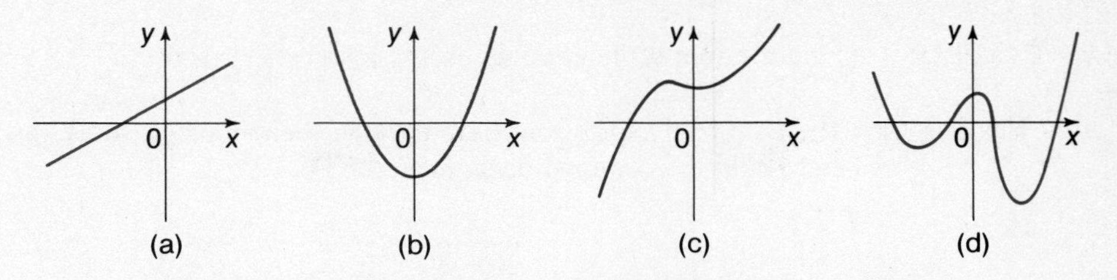

Figure 3.25 Graphs of polynomials of degrees 1, 2, 3, and 4

EXERCISES 3.7

Graph the following polynomials.

1. $f(x) = -x^3$

2. $f(x) = -x^4$

3. $f(x) = x^4 + 1$

4. $f(x) = x^3 - 1$

5. $f(x) = x^5$

6. $f(x) = x^6$

7. $f(x) = (x + 1)^3$

8. $f(x) = (x - 1)^4$

9. $f(x) = (x+1)^4$

10. $f(x) = (x - 2)^3$

11. $f(x) = x(x - 1)$

12. $f(x) = x(x + 2)$

13. $f(x) = (x + 1)(x - 2)$

14. $f(x) = (x - 1)(x + 3)$

15. $f(x) = x^3 - 2x^2$

16. $f(x) = x - x^3$

17. $f(x) = x^4 - 16x^2$

18. $f(x) = x^2 - x^4$

19. $f(x) = x^3 - x^2 + x - 1$

20. $f(x) = x^3 + x^2 + 2x + 2$

21. $f(x) = x^3 + x + 1$

22. $f(x) = 1 - x - x^3$

23. $f(x) = x^3 - 2x - 7$

24. $f(x) = x^3 + 3x - 5$

25. $f(x) = x^4 - 3x^2 + 5$

26. $f(x) = x^4 + 2x^2 - 5$

27. $f(x) = x^4 + 4x^3 + 2x^2 - 12x - 15$

28. $f(x) = x^4 - 5x^3 + 6x^2 - x - 1$

3.8 Rational Expressions and Partial Fractions

Rational Expressions The construction of a polynomial involves only the algebraic operations of addition and multiplication. If division is involved we get a much larger class of algebraic expressions, those which may be written as quotients of polynomials, called **rational expressions.** Examples are

$$\frac{1}{x}, \qquad \frac{x}{x-1}, \qquad \frac{x+2}{x^2+3x}, \qquad \frac{x^3-x+1}{2x-5}.$$

Polynomials behave like integers and rational expressions behave like rational numbers. The rules for operating with rational numbers are used to perform the operations on rational expressions.

EXAMPLE 1 Find the sum of the rational expressions $\frac{x}{x+1}$ and $\frac{2}{x+1}$.

SOLUTION These rational expressions have the same denominator, so we just add the numerators and retain the common denominator.

$$\frac{x}{x+1} + \frac{2}{x+1} = \frac{x+2}{x+1} \qquad \bullet$$

If the denominators are not the same, the fractions are rewritten as equivalent fractions with a common denominator. The next example illustrates this.

EXAMPLE 2 Find the sum of the rational expressions:
(a) $\frac{1}{x+1}$ and $\frac{2}{x-1}$; (b) $\frac{x}{x^2-x-2}$ and $\frac{2}{x+1}$.

SOLUTION (a)
$$\begin{aligned} \frac{x}{x+1} + \frac{2}{x-1} &= \frac{x(x-1)}{(x+1)(x-1)} + \frac{2(x+1)}{(x+1)(x-1)} \\ &= \frac{x^2-x}{x^2-1} + \frac{2x+2}{x^2-1} \\ &= \frac{(x^2-x)+(2x+2)}{x^2-1} \\ &= \frac{x^2+x+2}{x^2-1} \end{aligned}$$

(b)
$$\begin{aligned} \frac{x}{x^2-x-2} + \frac{2}{x+1} &= \frac{x}{(x-2)(x+1)} + \frac{2}{x+1} \\ &= \frac{x}{(x-2)(x+1)} + \frac{2(x-2)}{(x-2)(x+1)} \\ &= \frac{x+2(x-2)}{(x-2)(x+1)} \\ &= \frac{3x-4}{(x-2)(x+1)} \qquad \bullet \end{aligned}$$

As with arithmetic fractions, the product of rational expressions is obtained by taking the products of numerators and denominators separately.

EXAMPLE 3 Find the following products:

(a) $\frac{x}{x+1} \cdot \frac{x^2-3}{x+2}$; (b) $\frac{x^3}{x+1} \cdot \frac{x^2+x}{x^2}$.

SOLUTION (a) $\frac{x}{x+1} \cdot \frac{x^2-3}{x+2} = \frac{x(x^2-3)}{(x+1)(x+2)}$

$$= \frac{x^3 - 3x}{x^2 + 3x + 2}$$

(b) $\frac{x^3}{x+1} \cdot \frac{x^2+x}{x^2} = \frac{x^3[x\cancel{(x+1)}]}{\cancel{(x+1)}x^2} = x^2$

We may reduce a numerical fraction such as 12/15 to lowest terms, 4/5, by dividing numerator and denominator by the common factor 3. Similarly, we reduce a rational expression to lowest terms by dividing numerator and denominator by their common factors. In Example 3(b), the common factors are x^2 and $x + 1$. This requires, of course, that we first factor where possible. ●

Division of rational algebraic expressions is performed by multiplying the first expression (the dividend) by the reciprocal of the second expression (the divisor). The reciprocal is formed by inverting the rational expression.

EXAMPLE 4 Divide $\frac{x}{x+1}$ by $\frac{x+2}{x^2-x-2}$.

SOLUTION

$$\frac{x}{x+1} \div \frac{x+2}{x^2-x-2} = \frac{x}{x+1} \cdot \frac{x^2-x-2}{x+2}$$

$$= \frac{x[(x-2)\cancel{(x+1)}]}{\cancel{(x+1)}(x+2)}$$

$$= \frac{x^2-2x}{x+2}.$$

As in Example 3(b), the result is factored and reduced to lowest terms. ●

Observe that the sum, difference, product, and quotient of two rational expressions is a rational expression. That is, the rational expressions are **closed** (have the closure property) under the rational algebraic operations.

A rational expression whose numerator is a polynomial of smaller degree than that of the denominator is called *proper*. If the degree of the numerator is at least as great as that of the denominator, the rational expression is called *improper*.

From the statement of the division algorithm (Section 3.5),

$$f(x) = h(x) \cdot q(x) + r(x),$$

with $r(x)$ of degree 0 or less than that of $h(x)$, we can write

$$\frac{f(x)}{h(x)} = q(x) + \frac{r(x)}{h(x)}.$$

That is, any improper rational expression may be written as the sum of a polynomial and a proper rational expression. The following illustrate this.

$$\frac{x^2 + 1}{x} = x + \frac{1}{x}$$

$$\frac{x^5 + 3x^2 + 1}{x^2 - 2} = x^3 + 2x + 3 + \frac{4x + 7}{x^2 - 2}$$

$$\frac{2x^3 + 3x}{x^3 - 1} = 2 + \frac{3x + 2}{x^3 - 1}$$

In these examples the rational expression has been written as the sum of a polynomial and a *proper* rational expression. This form is sometimes very useful in working with rational expressions.

Partial Fractions We have seen how to write the sum of several rational expressions as a single rational expression. This process may be reversed and a rational expression broken up into the sum of several simpler fractional expressions, called **partial fractions.** Since an improper rational expression can be written as the sum of a polynomial and a proper rational expression, we need to consider only proper rational expressions. The procedure is explained in the following examples.

EXAMPLE 5 Write $\frac{3x + 1}{x^2 - 1}$ as a sum of fractions in lowest terms.

SOLUTION Since

$$x^2 - 1 = (x + 1)(x - 1),$$

the possible denominators of the partial fractions are $x + 1$ and $x - 1$. To find the numerators, let

$$\frac{3x + 1}{x^2 - 1} = \frac{A}{x + 1} + \frac{B}{x - 1},$$

where A and B are constants to be determined. Multiply both members of the above equation by the common denominator, $x^2 - 1$, to obtain

$$\begin{aligned} 3x + 1 &= A(x - 1) + B(x + 1) \\ &= Ax - A + Bx + B \\ 3x + 1 &= (A + B)x + (-A + B). \end{aligned}$$

Since the polynomial on the right is equal to the polynomial on the left, corresponding coefficients must be equal. That is,

$$A + B = 3,$$

and

$$-A + B = 1.$$

Adding these two equations gives

$$2B = 4,$$

$$B = 2.$$

Substituting 2 for B in $A + B = 3$ gives

$$A + 2 = 3,$$
$$A = 1.$$

The solution* is $A = 1$, $B = 2$, and thus

$$\frac{3x + 1}{x^2 - 1} = \frac{1}{x + 1} + \frac{2}{x - 1}.$$

This can be checked by adding the rational expressions. ●

EXAMPLE 6 Write $\dfrac{x^3 - 3x - 1}{x^2 - x - 2}$ as a sum of fractions.

SOLUTION First, since this is not a proper fraction, we use the division algorithm to express the given fraction as follows.

$$\frac{x^3 - 3x - 1}{x^2 - x - 2} = x + 1 + \frac{1}{x^2 - x - 2}$$

Now, let

$$\frac{1}{x^2 - x - 2} = \frac{1}{(x + 1)(x - 2)}$$
$$= \frac{A}{x + 1} + \frac{B}{x - 2}.$$

Multiply through by $x^2 - x - 2$.

$$1 = A(x - 2) + B(x + 1)$$
$$0x + 1 = (A + B)x + (-2A + B)$$

So $A + B = 0$ and $-2A + B = 1$. Solving this system of equations, we find $A = -1/3$ and $B = 1/3$.

$$\frac{1}{x^2 - x - 2} = \frac{-\frac{1}{3}}{x + 1} + \frac{\frac{1}{3}}{x - 2} = \frac{-1}{3(x + 1)} + \frac{1}{3(x - 2)}$$

Finally,

$$\frac{x^3 - 3x - 1}{x^2 - x - 2} = x + 1 - \frac{1}{3(x + 1)} + \frac{1}{3(x - 2)}.$$ ●

The same procedure may be used for any rational expression whose denominator can be factored into distinct *linear* factors. In fact, calculus makes use of this procedure and, with some modifications, extends it to any rational expression whose denominator

*The solution of systems of equations is reviewed in some detail in Chapter 8.

can be factored into linear and quadratic factors (distinct or not) or powers of linear and quadratic factors.

EXERCISES 3.8

First factor and then reduce to lowest terms, as illustrated in Examples 3 and 4.

1. $\dfrac{x^2 + xy}{xy + xz}$

2. $\dfrac{x^2 - y^2}{x^2 + xy}$

3. $\dfrac{x^2 - xy - 2y^2}{x^2 + 2xy + y^2}$

4. $\dfrac{x^2 + 3x}{x^2 + x - 6}$

5. $\dfrac{2x^2 + 7x + 5}{x^2 - x - 2}$

6. $\dfrac{6x^2 - 7x - 3}{12x^2 - 16x - 3}$

Perform the indicated operations and reduce the result to lowest terms. Leave denominators in factored form.

7. $\dfrac{1}{x + 1} + \dfrac{1}{x - 1}$

8. $\dfrac{2x}{x + 2} + \dfrac{x}{x - 3}$

9. $\dfrac{x - 1}{x + 1} - \dfrac{x + 1}{x - 1}$

10. $\dfrac{x}{x^2 - 1} - \dfrac{x + 1}{x - 1}$

11. $\dfrac{6x}{x - y} \cdot \dfrac{x^2 - y^2}{3}$

12. $\dfrac{x^2 - 4}{x + 3} \cdot \dfrac{2x + 6}{3x - 6}$

13. $\dfrac{x + 1}{1 - x} \cdot \dfrac{1 - x^2}{1 + 2x + x^2}$

14. $\dfrac{x^3}{x - 1} \div \dfrac{x^2}{x^2 - x}$

15. $\dfrac{2x - 1}{15x^3} \div \dfrac{6x - 3}{5x}$

16. $\dfrac{(x - y)^2}{x + y} \div \dfrac{2(x - y)}{3(x + y)}$

Express as a sum of partial fractions, or as the sum of a polynomial and partial fractions.

17. $\dfrac{1}{x(x + 1)}$

18. $\dfrac{2}{(x - 1)(x - 2)}$

19. $\dfrac{x}{(x + 2)(x - 3)}$

20. $\dfrac{x + 1}{(x - 2)(x + 3)}$

21. $\dfrac{x - 1}{x^2 + 6x + 8}$

22. $\dfrac{1}{x^2 + x - 30}$

23. $\dfrac{5x - 10}{x^2 - x - 6}$

24. $\dfrac{x}{2x^2 - x - 1}$

25. $\dfrac{x^3 + 2}{x^2 + 4x}$

26. $\dfrac{x^2 + 4x}{x + 8}$

27. $\dfrac{x^3 + 2x^2 - x}{x + 2}$

28. $\dfrac{2x^3 - x}{x^2 + 2x - 3}$

29. $\dfrac{x^2}{x^2 - 1}$

30. $\dfrac{x^4}{16x - x^3}$

3.9 Rational Functions and Their Graphs

The rational algebraic expressions are used to define the class of functions called *rational functions*.

DEFINITION 3.1 A **rational function** $f(x)$ is the quotient of two polynomial functions:

$$f(x) = \frac{p(x)}{q(x)}, \quad q(x) \neq 0.$$

The domain of a rational function is the set of all x for which $q(x) \neq 0$. The polynomial functions are those rational functions for which $q(x) = 1$.

A rational function may not always be written as a single fraction. For example,

$$f(x) = 1 + \frac{2}{x} = \frac{x + 2}{x},$$

and so $f(x)$ is a rational function. Similarly,

$$g(x) = \frac{1}{x} - \frac{x}{x + 1} = \frac{(x + 1) - x^2}{x(x + 1)} = \frac{-x^2 + x + 1}{x^2 + x}$$

is a rational function.

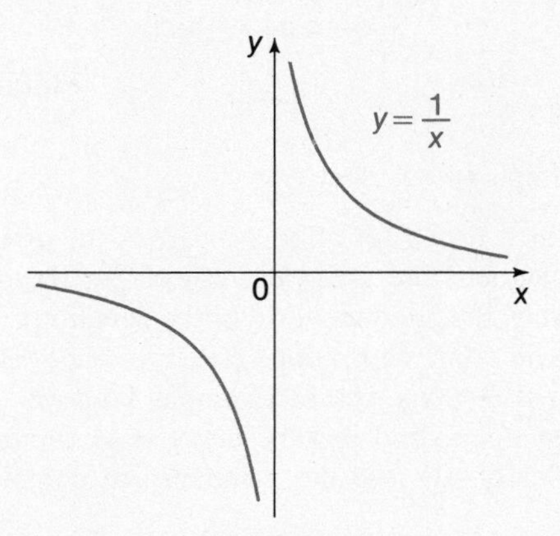

Figure 3.26

Graphing Rational Functions Rational functions which are not polynomials have denominators with degree > 0. A simple illustration of the graph of a rational function is given by the reciprocal function, described in detail in Example 5, Section 2.3. The graph is repeated in Figure 3.26. Observe that as x approaches 0 from the positive side, written $x \to 0^+$, y increases without bound, written $y \to \infty$. Also, as x approaches 0 through negative values, written $x \to 0^-$, y decreases without bound, written $y \to -\infty$. Thus, as x becomes smaller the curve comes closer to the y-axis. We call the y-axis a **vertical asymptote** to the curve and we say that the curve approaches the y-axis **asymptotically.** Similarly, the x-axis is a **horizontal asymptote,** as shown in the figure.

Figure 3.27a illustrates a vertical and horizontal asymptote and 3.27b an **oblique** (neither vertical nor horizontal) **asymptote** to a curve. The idea of asymptotes is essential in discussing graphs of rational functions which are not polynomials.

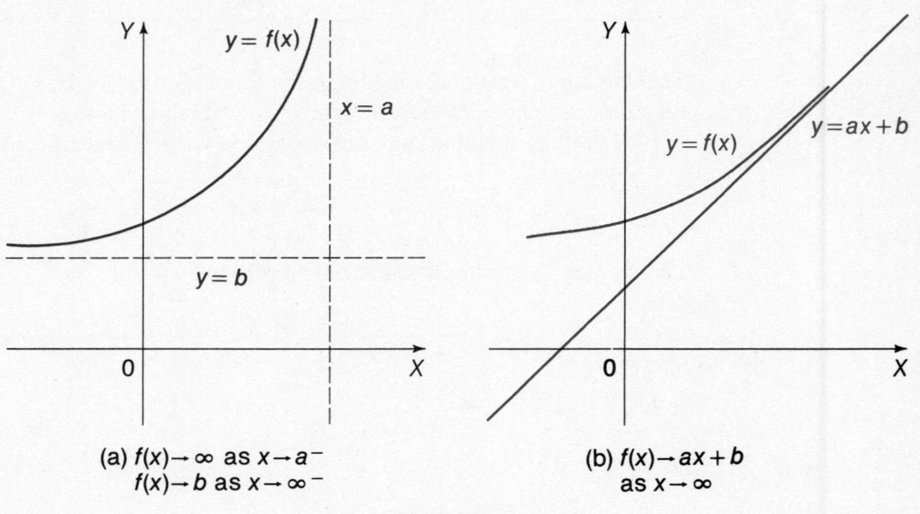

Figure 3.27

EXAMPLE 1 Graph $f(x) = 1/x^2$.

SOLUTION Since $y > 0$ for all x, no part of the graph is below the x-axis. The third and fourth quadrants are *excluded regions* for the graph. Since $x \neq 0$, there is no y-intercept. Also, the alternate form of the equation, $x = \pm 1/\sqrt{y}$, tells us that $y \neq 0$ and so there is no x-intercept. Since $f(-x) = f(x)$, f is an even function and its graph is symmetric to the y-axis. As in Example 1 above, we can show that the positive y-axis is an asymptote and that the curve is asymptotic to both the positive and negative x-axis. Finally, we use the symmetry to complete the graph shown in Figure 3.28. ●

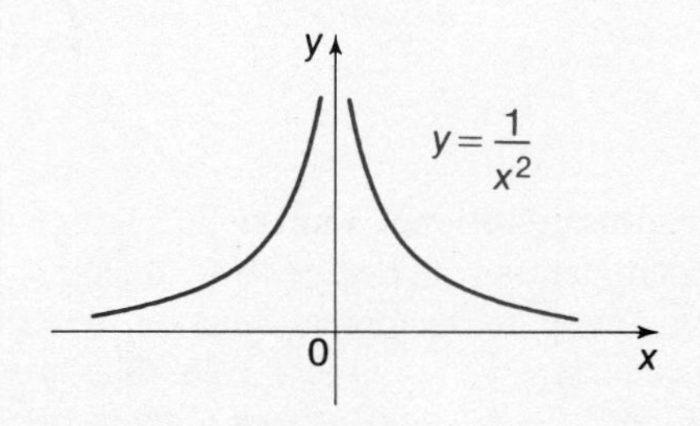

Figure 3.28

EXAMPLE 2 Graph $f(x) = \dfrac{1}{x - 1}$.

Recall the idea of translation of a graph. (See Example 5, Section 3.4.) We have the same situation here. The graph of $y = f(x) = 1/(x - 1)$ is obtained by translating the graph of $y = 1/x$ one unit to the right. The vertical asymptote is also translated to become the vertical line $x = 1$. The resulting graph is shown in Figure 3.29. ●

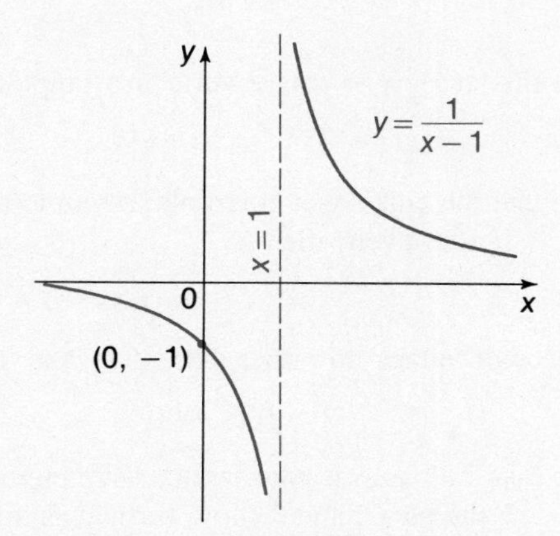

Figure 3.29

EXAMPLE 3 Graph $f(x) = \dfrac{1}{(x + 1)^2}$.

SOLUTION This graph is obtained by translating the graph of $f(x) = 1/x^2$ (Figure 3.28) one unit to the left. The vertical asymptote of $y = 1/x^2$ is translated also, so that the vertical asymptote is $x = -1$. See Figure 3.30. ●

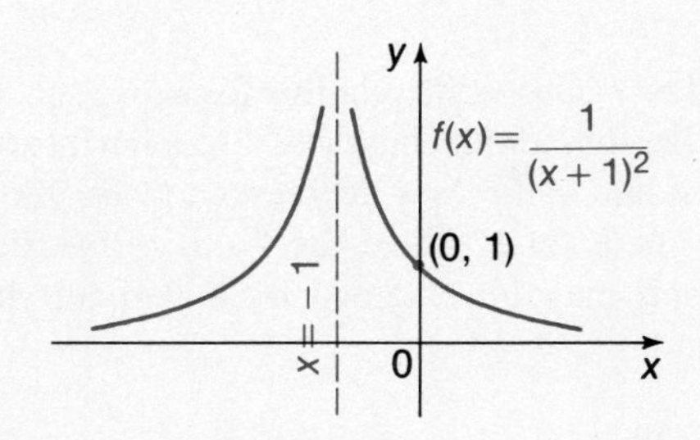

Figure 3.30

Each of the preceding examples is an instance of a rational function of the form

$$f(x) = \frac{1}{(x - a)^n},$$

where n is a positive integer. First, consider the case when n is odd as in $f(x) = 1/x$ and $f(x) = 1/(x - 1)$. When x approaches a through values greater than a, $f(x)$ increases positively without bound. This is written

$$f(x) \to \infty \quad \text{as} \quad x \to a^{+}.$$

When x approaches a through values less than a, then $f(x)$ decreases without bound, and we write

$$f(x) \to -\infty \quad \text{as} \quad x \to a^{-}.$$

In this case, $x = a$ is a vertical asymptote. Similarly,

$$f(x) \to 0^{+} \quad \text{as} \quad x \to \infty \qquad \text{and} \qquad f(x) \to 0^{-} \quad \text{as} \quad x \to -\infty,$$

so that the x-axis is a horizontal asymptote. See Figures 3.26 and 3.29 as illustrations.

If n is even, then

$$f(x) \to \infty \quad \text{as} \quad x \to a$$

through values on either side of a, and so $x = a$ is a vertical asymptote. Also,

$$f(x) \to 0^{+} \quad \text{as} \quad x \to \infty \quad \text{or} \quad x \to -\infty,$$

so that the x-axis is a horizontal asymptote. See Figures 3.28 and 3.30 for illustrations.

Examples 2 and 3 show vertical asymptotes other than the y-axis. The next example illustrates horizontal asymptotes not the x-axis. This situation occurs when the numerator and denominator are both linear polynomials.

EXAMPLE 4 Graph $f(x) = \dfrac{2x + 3}{x}$.

SOLUTION By division

$$f(x) = \frac{2x + 3}{x}$$
$$= 2 + \frac{3}{x}.$$

When $f(x)$ is written in this form, we see that $f(x) \to 2$ as x increases without bound in either direction. Thus, $y = 2$ is a horizontal asymptote. Also note that $3/x$ gets larger in absolute value as x decreases toward zero from either side, so $f(x)$ also increases in absolute value. Thus, the y-axis is a vertical asymptote. (See Figure 3.31.) This same graph may be obtained by a vertical shift of the graph of $y = 3/x$ two units upward. ●

An asymptote which is neither vertical nor horizontal is called an *oblique asymptote*. If the numerator of a rational expression has degree one more than the denominator, the graph has an oblique asymptote. This is illustrated in the next example.

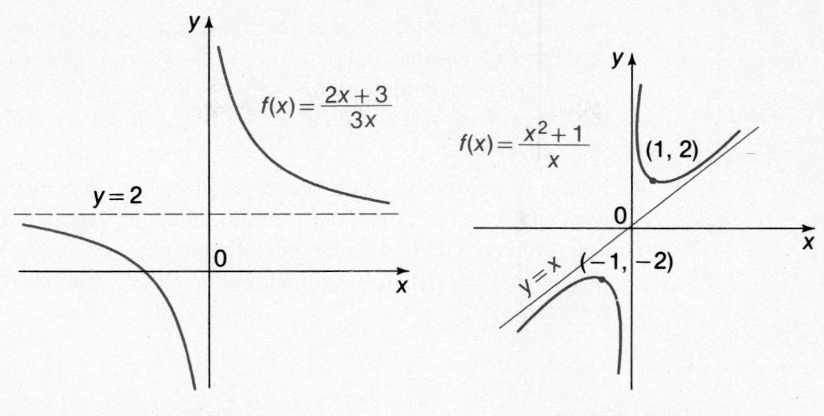

Figure 3.31 **Figure 3.32**

EXAMPLE 5 Graph $f(x) = \dfrac{x^2 + 1}{x}$.

SOLUTION Use division to write
$$f(x) = \frac{x^2 + 1}{x} = x + \frac{1}{x},$$
and then investigate for intercepts, symmetry, and asymptotes. First, since $f(0)$ is undefined, x cannot be 0 and there is no y-intercept. By inspection, y cannot be 0 so that there is no x-intercept. (Let $y = 0$ and try to solve for x.) If $x > 0$, then $y > 0$, and if $x < 0$, then $y < 0$, so the graph lies entirely in the first and third quadrants. Also, $f(-x) = -f(x)$, so that f is an odd function and its graph is symmetric with respect to the origin.

Now consider the portion of the graph in the first quadrant. From $y = x + 1/x$, we see that for $x > 0$, $y > x$, that is, the curve is *above* the line $y = x$ in the first quadrant. Similarly, if $x < 0$, then $y < x$, and the curve is *below* the line $y = x$ in the third quadrant.

Finally, from $y = x + 1/x$, we have $y \to x$ as $x \to \infty$ or $x \to -\infty$, that is, the curve approaches the line $y = x$ asymptotically. Since $y \to \infty$ as $x \to 0^+$ and $y \to -\infty$ as $x \to 0^-$, the y-axis is a vertical asymptote. See Figure 3.32. ●

The discussion has been limited to relatively simple examples because calculus is needed for efficient treatment of graphs of rational functions. The following steps are helpful in determining the graphs of such functions.

1. Find any intercepts, symmetries, and translations.
2. Find any vertical asymptotes by setting the denominator equal to 0. If the denominator is zero for $x = a$ but the numerator is not, then $x = a$ is a vertical asymptote.

3. Find any horizontal asymptotes by investigating how y varies as x increases numerically without bound. If $y \to c$, where c is some constant, then $y = c$ is a horizontal asymptote. This will always occur when the numerator and denominator are both linear functions, and c is found by dividing the numerator by the denominator.
4. Find any oblique asymptotes. This situation always occurs when the numerator is of degree one more than the denominator. Division of the numerator by the denominator yields the sum of a linear function and a proper rational expression. Then let x increase numerically without bound to find the equation of the asymptote.

EXERCISES 3.9

Graph the following rational functions. For each function, identify the domain, the intercepts, and the asymptotes of the graph.

1. $f(x) = -\dfrac{1}{x}$ **2.** $f(x) = \dfrac{1}{x^3}$ **3.** $f(x) = \dfrac{1}{x^4}$

4. $f(x) = \dfrac{2}{x-4}$ **5.** $f(x) = \dfrac{4}{x+1}$ **6.** $f(x) = \dfrac{-3}{x+2}$

7. $f(x) = \dfrac{1}{(x+1)^3}$ **8.** $f(x) = \dfrac{1}{(x-1)^2}$ **9.** $f(x) = \dfrac{2}{(x-2)^2}$

10. $f(x) = \dfrac{1}{(2x+1)^2}$ **11.** $f(x) = 2 + \dfrac{1}{x}$ **12.** $f(x) = 1 - \dfrac{1}{x^2}$

13. $f(x) = \dfrac{x^2-1}{x}$ **14.** $f(x) = \dfrac{x^2-1}{x+1}$ **15.** $f(x) = \dfrac{x}{x-1}$

16. $f(x) = \dfrac{2x}{1+x}$ **17.** $f(x) = \dfrac{x+2}{2x-1}$ **18.** $f(x) = \dfrac{x-1}{3x+4}$

3.10 Irrational Algebraic Functions

A function such as

$$f(x) = \sqrt{x+1} \qquad \text{or} \qquad g(x) = x^2 + \sqrt{x}$$

involves root extraction and is called an **irrational algebraic function.** Domains of such functions tend to be more restricted than those for polynomial and rational functions. The domain must not only exclude zeros of denominators but also those values of the variable x which require taking an even root of a negative number.

EXAMPLE 1 Find the domain of (a) $f(x) = \dfrac{1}{x^2 - 1}$; (b) $g(x) = \dfrac{1}{\sqrt{x^2 - 1}}$.

SOLUTION (a) The domain of

$$f(x) = \frac{1}{x^2 - 1} = \frac{1}{(x + 1)(x - 1)}$$

excludes 1 and -1, the zeros of the denominator. The domain is all real numbers except 1 and -1.

(b) The domain of

$$g(x) = \frac{1}{\sqrt{x^2 - 1}}$$

must exclude not only 1 and -1 but also all x such that $x^2 - 1 < 0$. Since the solution of this inequality is $-1 < x < 1$, the entire closed interval $[-1, 1]$ is excluded from the domain of x. The domain may be written $(-\infty, -1) \cup (1, \infty)$. ●

EXAMPLE 2 Graph $y = \sqrt{1 - x^2}$.

SOLUTION Since $1 - x^2$ must be ≥ 0 in order for the square root to be a real number, the domain is the closed interval $[-1, 1]$, while the range is the interval $[0, 1]$. If we square both members of $y = \sqrt{1 - x^2}$, we obtain $x^2 + y^2 = 1$, the equation of the unit circle centered at the origin. Since the radical indicates the positive root, $y = \sqrt{1 - x^2}$ is the equation of the upper semicircle ($y \geq 0$) of this circle. See Figure 3.33. ●

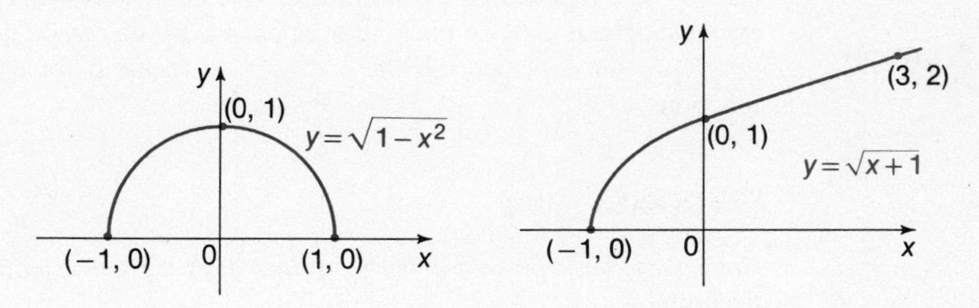

Figure 3.33 **Figure 3.34**

EXAMPLE 3 Graph $y = \sqrt{x + 1}$.

SOLUTION Since $x + 1$ must be ≥ 0, the domain is the set of all reals greater than or equal to -1. The range is the set of nonnegative reals. The curve intersects the axes at $(-1, 0)$ and $(0, 1)$, and the function values increase as x increases from -1. Additional points, such as $(3, 2)$, are found to help sketch the graph as shown in Figure 3.34. ●

Notice that this curve could have been obtained directly by shifting the graph of the square-root function $y = \sqrt{x}$, discussed in section 2.2, one unit to the left. (See Figure 2.8.)

EXAMPLE 4 Graph $y = \dfrac{1}{\sqrt{x-1}}$.

SOLUTION First of all, $y > 0$. For y to be real, x must be >1. So the graph must lie in the quadrant above the x-axis and to the right of the vertical line $x = 1$. As $x \to 1$ from values >1, $x - 1 \to 0$, so that y increases without bound. Also, as $x \to \infty$, $y \to 0$. Hence the graph is asymptotic to the positive portion of the line $x = 1$ and to the positive x-axis, as shown in Figure 3.35. ●

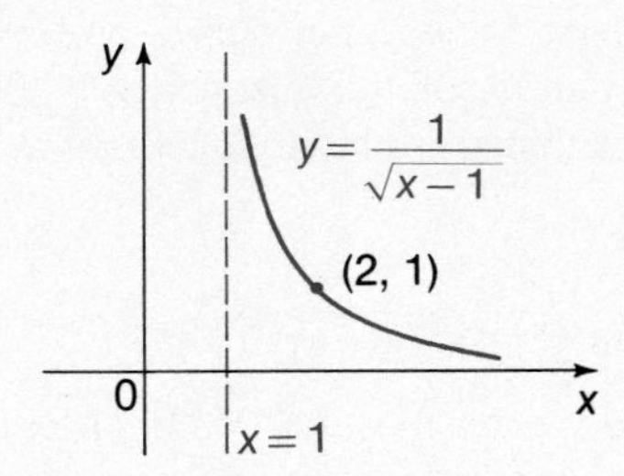

Figure 3.35

The wide variety of possibilities for irrational functions makes it impossible to classify and study them by families as was done with other algebraic functions. The examples given indicate the kind of approach which can be taken when graphing them with only the tools of algebra. Calculus is required for a thorough study of these functions.

EXERCISES 3.10

Graph the following irrational functions. Identify the domain of the function and the intercepts of the graph.

1. $f(x) = 2 - \sqrt{x}$ [*Hint:* Write $y = -\sqrt{x} + 2$ and use translation.]
2. $f(x) = 1 + \sqrt{x}$ [*Hint:* Use translation.]
3. $f(x) = \sqrt{x - 1}$
4. $f(x) = \sqrt{x + 2}$
5. $f(x) = \sqrt{1 - x}$
6. $f(x) = \sqrt{1 + x}$
7. $f(x) = \dfrac{1}{\sqrt{x}}$
8. $f(x) = \dfrac{\sqrt{x} + 1}{\sqrt{x}}$
9. $f(x) = \sqrt{x^2 - 1}$ [*Hint:* Note that $f(x) < x$ and that $f(x)$ approaches x as x increases without bound.]
10. $f(x) = \sqrt{x^2 + 1}$

11. $f(x) = \sqrt{x^2 - 4}$

12. $f(x) = \sqrt{x^2 + 4}$

13. $f(x) = \dfrac{1}{\sqrt{x + 1}}$

14. $f(x) = \dfrac{1}{\sqrt{1 - x}}$

15. $f(x) = \dfrac{1}{\sqrt{x^2 + 4}}$

16. $f(x) = \dfrac{1}{\sqrt{x + 2}}$

Chapter 3 Review Exercises

Identify the type of function in each case.

1. $f(x) = -2x + 3$

2. $f(x) = \dfrac{x^2 - 3x}{x + 1}$

3. $f(x) = (x - 2)^2$

4. $f(x) = \sqrt{4 - x^2}$

5. $f(x) = \dfrac{1}{x} + x$

6. $f(x) = x^3 - 2x$

Factor the following.

7. $x^2 - 3x - 4$

8. $x^2 - 8x + 15$

9. $-x^2 + 7x - 12$

10. $-x^2 - 3x - 2$

11. $3x^2 + 10x - 8$

12. $2x^2 - 5x - 3$

Find the least (or greatest) value of each of the following.

13. $f(x) = x^2 + 4x$

14. $f(x) = -x^2 + 2x$

15. $f(x) = 2x^2 - 4x + 5$

16. $f(x) = 3x^2 + 6x + 1$

17. $f(x) = 2x^2 + 3x + 2$

18. $f(x) = -2x^2 - 5x - 4$

Solve the following quadratic equations and inequalities.

19. $x^2 - 2x - 3 = 0$

20. $2x^2 + 3x - 2 = 0$

21. $2x^2 + 4x - 1 = 0$

22. $4x^2 + 12x - 7 = 0$

23. $x^2 - 2x + 5 = 0$

24. $2x^2 - 6x + 7 = 0$

25. $x^2 - 3x + 2 < 0$

26. $x^2 - x - 6 \leq 0$

27. $2x^2 + 3x - 2 \geq 0$

28. $-x^2 + 4x - 3 > 0$

Use synthetic division to calculate the quotient and remainder in each of the following.

29. $(x^3 - 2x^2 + x - 3) \div (x - 2)$

30. $(x^3 + 2x^2 - 12) \div (x + 3)$

31. $(2x^4 - 3x^2 + 4) \div \left(x + \dfrac{3}{2}\right)$

32. $(x^4 - 10x^2 - 7) \div \left(x - \dfrac{2}{3}\right)$

Find the indicated function values by using synthetic division and the remainder theorem. Check by calculator.

33. $f(3)$ and $f(-2)$ if $f(x) = x^3 - 2x^2 + 3x + 4$

34. $f\left(\dfrac{1}{3}\right)$ and $f\left(-\dfrac{1}{2}\right)$ if $f(x) = 6x^4 - 10x^3 - 4x^2 + 3x + 1$

35. $f(-1)$ and $f(2)$ if $f(x) = 2x^3 - 3x^2 - 4$

36. $f(5)$ and $f(2)$ if $f(x) = x^4 - 10x^2 - 7$

Determine the rational zeros (if any) of the following polynomials. Use the factor theorem to factor any for which there is at least one rational zero.

37. $f(x) = x^3 - 2x^2 - 5x + 6$

38. $f(x) = x^3 - 4x^2 + x + 6$

39. $f(x) = 4x^4 - 7x^2 - 5x - 1$

40. $f(x) = 6x^4 + 2x^3 + 7x^2 + x + 2$

Graph each of the following polynomials.

41. $f(x) = 3x - 4$

42. $f(x) = -2x + 3$

43. $f(x) = 3x^2$

44. $f(x) = x^2 - 1$

45. $f(x) = x^2 - 2x$

46. $f(x) = 2x^2 - 6x$

47. $f(x) = x^2 + 4x + 2$

48. $f(x) = x^2 + x - 1$

49. $f(x) = (x - 2)^2$

50. $f(x) = (x + 3)^2$

51. $f(x) = x^3 + 2$

52. $f(x) = x^4 - 2$

53. $f(x) = (x - 1)^4$

54. $f(x) = (x + 2)^3$

55. $f(x) = x^3 + x$

56. $f(x) = x^4 - 8x$

57. $f(x) = x^4 - 5x^2 + 4$

58. $f(x) = x^3 - 2x^2 - 2x + 4$

Express each of the following as a sum of partial fractions.

59. $\dfrac{1}{(x + 1)(x - 2)}$

60. $\dfrac{x - 1}{(x + 2)(x - 3)}$

61. $\dfrac{2x + 1}{x^2 - 4}$

62. $\dfrac{3x + 1}{x^2 - x - 12}$

63. $\dfrac{x^2 - 2}{x + 1}$

64. $\dfrac{x^3}{2x^2 - 3x - 2}$

Graph each of the following functions.

65. $f(x) = 1 + \dfrac{1}{x}$

66. $f(x) = \dfrac{x^2 - 2x}{x}$

67. $f(x) = \dfrac{1}{x^2 - 1}$

68. $f(x) = \dfrac{1}{x^2 + 1}$

69. $f(x) = \sqrt{x - 1}$

70. $f(x) = \sqrt{4 - x^2}$

71. Suppose that a price-supply relation is given by $p(x) = 50 - \dfrac{5}{4}x$ and the corresponding price-demand relation by $p(x) = \dfrac{4}{5}x$. Graph the supply and demand curves and determine the equilibrium price and the equilibrium demand.

72. Suppose the cost function for producing x items is $C(x) = 15x + 60$ and the revenue function is $R(x) = 20x$. What is the break-even point?

73. Find two numbers differing by 6 which have the least product.

74. A projectile is thrown straight up from the ground with an initial velocity of 96 feet per second. Its height after t seconds is given by $s(t) = 96t - 16t^2$, where s is measured in feet and t in seconds. (a) What is the maximum height reached by the projectile? (b) How long does it take to reach this height? (c) How much time is required to go up and then return to the ground? How do you know?

75. Determine the least distance from the point (4, 1) to the line whose equation is $y = x$.

[*Hint:* Sketch the figure and label any point on the line (x, x). Let $f(x)$ equal the square of the distance between $(4, 1)$ and (x, x) and then calculate the least value of this squared distance.]

76. A dollmaker finds that handmade dolls can be sold for 15 dollars each. A model for the cost function is $C(x) = 2x^2 - x + 20$, where x is the number of dolls produced each day. Find (a) the number that should be produced for greatest profit and (b) the "break-even" point.

77. Suppose that 1500 yards of fencing are used to enclose a rectangular yard and to put one cross fence parallel to a side, partitioning the yard into two rectangular regions. Determine the dimensions of the yard of largest combined area under these conditions. (The figure is essential in seeing how to set up the problem).

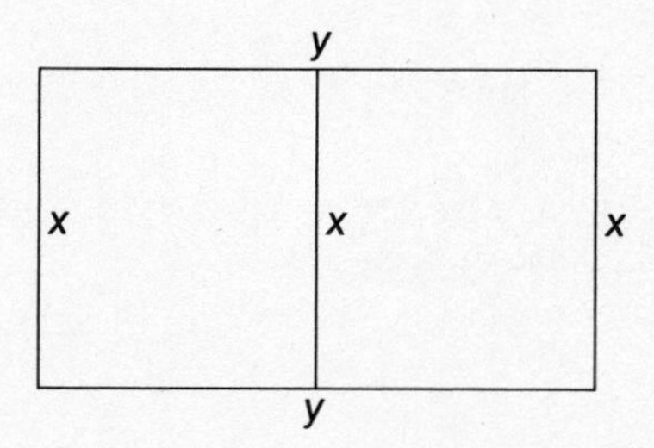

78. The power delivered by a 30-volt generator with resistance 2 ohms is given by the formula $p(x) = 30x - 2x^2$, where x is the measure of the current in amperes. For what value of x (current) is the power $p(x)$ greatest? What is the value of this greatest power (in watts)?

79. In the figure, a rectangle is inscribed in the region in the first quadrant enclosed by the line $x + y = 1$. The area of such a rectangle is $A = xy = x(1 - x) = x - x^2$. Determine the dimensions of the largest such rectangle. [*Hint:* Find x for which $A(x) = x - x^2$ is greatest.]

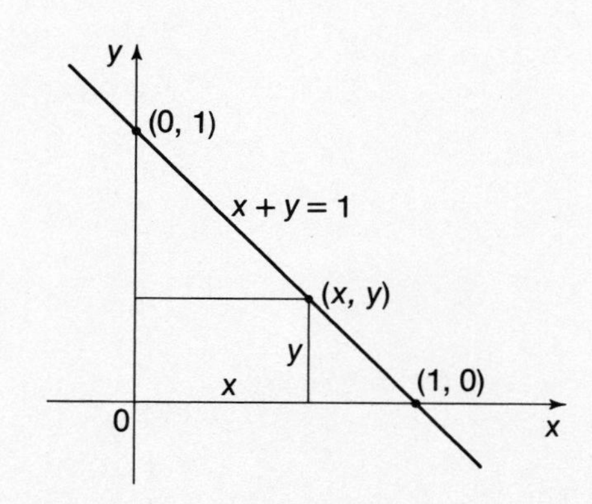

Chapter 3 Miscellaneous Exercises

1. Show that the composite function formed by the composition of two linear polynomial functions is a linear polynomial function. [*Hint:* Let $f(x) = ax + b$ and $g(x) = cx + d$. Then calculate $f \circ g$ and $g \circ f$ and observe that each is of the form $p(x) = Ax + B$.]

2. The quadratic formula gives the roots of $ax^2 + bx + c = 0$ as $x_1 = (-b + \sqrt{D})/2a$ and $x_2 = (-b - \sqrt{D})/2a$, where D is the discriminant $b^2 - 4ac$. Show that the sum of the roots $x_1 + x_2 = -b/a$ and the product of the roots $x_1x_2 = c/a$. [For example, the sum of the roots of $2x^2 - 3x + 1$ is 3/2 and their product is 1/2.]

Use the result of Exercise 2 to calculate the sum and the product of the roots of each of the following quadratic equations. Check by actually finding the roots and then adding and multiplying them.

3. $x^2 - 5x + 6 = 0$

4. $2x^2 - 7x + 3 = 0$

5. $3x^2 + 11x = 4$

6. $x^2 + 2x + 4 = 0$

7. $4x^2 = 5 - 19x$

8. $2x^2 - 5x = 5x$

9. Explain why $f(x) = ax^2 + bx + c$ has a least but not a greatest value if $a > 0$. [See Example 8, Section 3.3.]

4

Exponential and Logarithmic Functions

The functions studied in the preceding chapter are algebraic, that is, their defining formulas are constructed from a variable and the real numbers using only the algebraic operations (addition, subtraction, multiplication, division, and root extraction). In this and the next two chapters two important classes of *nonalgebraic*, or **transcendental** *functions* will be presented. The discussion begins with the exponential and the logarithmic functions, which are closely related to the exponents and logarithms of algebra. These functions have extensive applications to population growth, chemical reactions, radioactive decay, financial investments, and psychological learning theory, to name but a few.

4.1 Exponents

It is customary to abbreviate the product $a \cdot b$ by ab, the product $a \cdot a$ by a^2, the product $a \cdot a \cdot a$ by a^3, and so on. However, until about 1600 it was common to write $aaaa$, for example, for the product $a \cdot a \cdot a \cdot a$ which we abbreviate a^4. It was Descartes who introduced this shorter notation, called *exponential notation*.

DEFINITION 4.1 If a is a real number and n a positive integer, then the product $a \cdot a \cdot a \ . \ . \ . \ a$, where a occurs as a factor n times, is written in **exponential notation** as a^n. More concisely,

$$a^n = \underbrace{a \cdot a \cdot a \ . \ . \ . \ a.}_{n \text{ factors}}$$

The expression a^n is called an **exponential,** a is the **base,** and n is the **exponent.** The exponent n must be a positive integer because n is the number of factors of a.

The rules for exponents are summarized in the following theorem.

THEOREM 4.1 **Laws of Exponents** If a and b are real numbers and p and q are positive integers, then

1. $a^p a^q = a^{p+q}$;
2. $\dfrac{a^p}{a^q} = a^{p-q}$, if $p > q$, $a \neq 0$;

   $\dfrac{a^p}{a^q} = \dfrac{1}{a^{q-p}}$, if $p < q$, $a \neq 0$;
3. $(a^p)^q = a^{pq}$;
4. $(ab)^p = a^p b^p$;
5. $\left(\dfrac{a}{b}\right)^p = \dfrac{a^p}{b^p}$, $b \neq 0$.

EXAMPLE 1 Use the laws of exponents to simplify the following expressions: (a) x^3x^4, (b) $\dfrac{y^5}{y^3}$, (c) $\dfrac{(r+s)^3}{(r+s)^5}$, (d) $(x^2)^5$, (e) $(3x)^2$, (f) $\left(\dfrac{2}{x}\right)^3$.

SOLUTION (a) $x^3x^4 = x^{3+4} = x^7$

(b) $\dfrac{y^5}{y^3} = y^{5-3} = y^2$

(c) $\dfrac{(r+s)^3}{(r+s)^5} = \dfrac{1}{(r+s)^{5-3}} = \dfrac{1}{(r+s)^2}$

(d) $(x^2)^5 = x^{2\cdot 5} = x^{10}$

(e) $(3x)^2 = 3^2x^2 = 9x^2$

(f) $\left(\dfrac{2}{x}\right)^3 = \dfrac{2^3}{x^3} = \dfrac{8}{x^3}$ ●

The proofs of the laws of exponents follow directly from the definition. For example, since $4^2 = 4 \cdot 4$ and $4^3 = 4 \cdot 4 \cdot 4$,

$$\begin{aligned} 4^2 \cdot 4^3 &= (4 \cdot 4)(4 \cdot 4 \cdot 4) && \text{definition of exponent} \\ &= (4 \cdot 4 \cdot 4 \cdot 4 \cdot 4) && \text{associativity of multiplication} \\ &= 4^5 && \text{definition of exponent} \\ &= 4^{2+3}. \end{aligned}$$

The corresponding general law may be shown in the same way:

$$a^p a^q = \underbrace{(a \cdot a \cdot a \; . \; . \; . \; a)}_{p \text{ factors}} \cdot \underbrace{(a \cdot a \cdot a \; . \; . \; . \; a)}_{q \text{ factors}}$$

$$= \underbrace{a \cdot a \cdot a \; . \; . \; . \; a}_{p + q \text{ factors}}$$

$$= a^{p+q}.$$

The other laws are similar and are left as exercises.

The preceding discussion applies only to positive integral exponents. Before extending it to rational exponents, the following definition from algebra is needed.

$$\sqrt[n]{b} = a \quad \textbf{if and only if} \quad a^n = b$$

In the above, either a and b are both nonnegative real numbers and n is a positive integer, or a and b are both negative and n is an odd positive integer. (If b is negative and n is even, then complex numbers are needed to define $\sqrt[n]{b}$. (See Chapter 9.) The number $\sqrt[n]{b}$ is called the **principal *n*th root** of b. For example,

$\sqrt[4]{16} = 2$ because $2^4 = 16$, and 2 is the principal 4th root of 16.

$\sqrt[3]{-8} = -2$ because $(-2)^3 = -8$, and -2 is the principal 3rd root of -8.

DEFINITION 4.2

1. $a^0 = 1, \quad a \neq 0$
2. $a^{-p} = \dfrac{1}{a^p}, \quad a \neq 0$
3. $a^{1/q} = \sqrt[q]{a}$
4. $a^{p/q} = (a^{1/q})^p = (\sqrt[q]{a})^p$ and $a^{p/q} = (a^p)^{1/q} = \sqrt[q]{a^p}$
5. $a^{-p/q} = \dfrac{1}{a^{p/q}}, \quad a \neq 0$

(Negative and rational exponents, defined above, were introduced into mathematics around 1650.)

EXAMPLE 2 (a) $(3x)^0 = 1; \qquad 17^0 = 1$

(b) $2^{-3} = \dfrac{1}{2^3} = \dfrac{1}{8}; \qquad x^{-1} = \dfrac{1}{x}, \quad x \neq 0$

(c) $8^{2/3} = (\sqrt[3]{8})^2 = 2^2 = 4; \qquad 8^{2/3} = \sqrt[3]{8^2} = \sqrt[3]{64} = 4$

(d) $8^{-2/3} = \dfrac{1}{8^{2/3}} = \dfrac{1}{4}$ ●

Rational exponents have been defined so that they satisfy the same laws as the positive integral exponents. For example, if $a \neq 0$, $a^p/a^p = 1$ and, if the second law of exponents is to hold, $a^p/a^p = a^{p-p} = a^0$. Thus, for consistency a^0 must be 1 for every real number $a \neq 0$. Similar remarks apply to the other definitions and to the other laws of exponents. Details are left as exercises.

Now that a^n has been defined for all rational values of n so that the laws of exponents hold, it is natural to want to give suitable meaning to irrational exponents. For example, what meaning could be given to 2^π and $5^{\sqrt{2}}$? We saw in Chapter 1 that every irrational number has a nonterminating decimal representation and can be approximated as closely as we wish by a rational number. For example,

$$\pi = 3.141592654\ldots,$$

and 3.14, 3.141, 3.1415, 3.14159, and so on, are increasingly better approximations to π. It is reasonable to assume that $2^{3.14}$, $2^{3.141}$, $2^{3.1415}$, $2^{3.14159}$, and so on, are getting closer and closer to 2^π. This is illustrated in the table, where the successive entries agree to more and more decimal places.

x	2^x
3	8.00000
3.1	8.57419
3.14	8.81524
3.141	8.82135
3.1415	8.82441
3.14159	8.82496
3.141592	8.82497

In higher-level mathematics courses, this idea is used to define 2^π and other expressions having irrational exponents, and it can be shown that such exponents obey the laws of exponents introduced earlier. Therefore, in what follows, whenever a variable exponent appears we assume that it may be any real number, rational or irrational.

Scientific Notation The laws of exponents can be used to simplify calculations involving those very large or very small numbers that often occur in scientific work. Such numbers are first written in **scientific notation,** that is, in the form $c \times 10^k$, where $1 \leq c < 10$ and k is the appropriate exponent. For instance,

$$\begin{aligned} 12{,}300 &= 1230 \times 10 && \text{or } 1230 \times 10^1 \\ &= 123 \times 100 && \text{or } 123 \times 10^2 \\ &= 12.3 \times 1000 && \text{or } 12.3 \times 10^3 \\ &= 1.23 \times 10{,}000 && \text{or } 1.23 \times 10^4. \end{aligned}$$

To obtain the factor c so that it is between 1 and 10, we must move the decimal point to the left four places. This is then balanced by multiplying by 10^4. So in scientific notation, $12{,}300 = 1.23 \times 10^4$.

Similarly, to write 0.00123 in scientific notation we move the decimal point three places to the right to obtain $c = 1.23$ and then divide by 10^3 (that is, multiply by 10^{-3}). Thus, in scientific notation, $0.00123 = 1.23 \times 10^{-3}$.

EXAMPLE 3 Write each of the following numbers in scientific notation.
(a) Light travels 9,460,000,000,000 kilometers in one year;
(b) a molecule of oxygen weighs 0.000000000000000000000053 grams.

SOLUTION (a) $9{,}460{,}000{,}000{,}000 = 9.46 \times 10^{12}$
$\uparrow$

(b) $0.000000000000000000000053 = 5.3 \times 10^{-23}$
$\uparrow$

The arrow marks the location of the decimal point for scientific notation, and the number of places is then counted off for the exponent on 10. ●

EXAMPLE 4 Use scientific notation to calculate (a) the product of 0.00013 and 2100, (b) $\dfrac{92{,}900{,}000{,}000 \times 0.000000034}{0.000065}$.

SOLUTION (a)
$$\begin{aligned} 0.00013 \times 2100 &= (1.3 \times 10^{-4}) \times (2.1 \times 10^3) \\ &= (1.3 \times 2.1) \times (10^{-4} \times 10^3) \\ &= 2.73 \times 10^{-1} \\ &= 0.273. \end{aligned}$$

(b)
$$\begin{aligned} \frac{92{,}900{,}000{,}000 \times 0.000000034}{0.000065} &= \frac{(9.29 \times 10^{10}) \times (3.4 \times 10^{-8})}{6.5 \times 10^{-5}} \\ &= \frac{9.29 \times 3.4}{6.5} \times \frac{10^{10} \times 10^{-8}}{10^{-5}} \\ &= 48.6 \times 10^7 \\ &= 486{,}000{,}000. \end{aligned}$$ ●

If the result of a calculation on a scientific hand calculator exceeds the display capacity of the calculator, the result is given in scientific notation. For example, if the display capacity is ten decimal places, the result 231,000,000,000,000 would appear as shown below, meaning 2.31×10^{14}.

2.31	14

EXERCISES 4.1

Use the laws of exponents to simplify.

1. $(2x^3y^2)(4x^{-2}y^{-3})$

2. $(2x^{-2})^{-3}$

3. $x^4y^3/x^{-3}y^{-3}$

4. $a^3 \cdot a^0 \cdot a^{-2}$

5. $10^7 10^{-4}/10^3$

6. $(xy^3)^{-2}/(x^2y)^{-3}$

7. $(xy)^5/xy^2$

8. $(a^{-3}/a)^{-2}$

9. $(3^{\sqrt{2}})^{\sqrt{2}}$

10. $(2^{\pi})^{1/\pi}$

11. $2^{\frac{1}{2}}2^{\frac{1}{4}}$

12. $(a^2b^4)^{\frac{1}{2}}$

13. $(x^2/y^6)^{\frac{1}{2}}$

14. $(8x^3)^{\frac{2}{3}}$

15. $x^{\frac{1}{3}}x^{\frac{1}{2}}$

16. $\dfrac{x^{\frac{3}{4}}}{x^{\frac{2}{3}}}$

Write in radical form and simplify, where appropriate.

17. $9^{\frac{1}{2}}$

18. $27^{\frac{1}{3}}$

19. $81^{\frac{1}{4}}$

20. $(0.04)^{\frac{1}{2}}$

21. $\left(\dfrac{1}{32}\right)^{\frac{1}{5}}$

22. $4^{\frac{3}{2}}$

23. $16^{\frac{3}{4}}$

24. $125^{\frac{2}{3}}$

25. $(-8)^{\frac{2}{3}}$

26. $(0.001)^{\frac{2}{3}}$

27. $32^{\frac{3}{5}}$

28. $8^{-\frac{2}{3}}$

29. $4^{-\frac{1}{2}}$

30. $16^{-\frac{3}{4}}$

31. $9^{-\frac{3}{2}}$

32. $64^{-\frac{5}{4}}$

Write as powers of 10.

33. 0.1

34. 100

35. 0.01

36. 0.001

37. 10

38. 1

39. 1000

40. 0.0001

Write in ordinary decimal notation.

41. 10^4

42. 10^{-4}

43. 10^{-7}

44. 2.76×10^7

45. 3.67×10^{-6}

46. 6.27×10^6

47. 5.74×10^5

48. 4.66×10^0

49. 8.42×10^{-4}

Express in scientific notation.

50. 5280

51. 0.01745

52. 300

53. 4.6

54. 345,000

55. 0.001

56. 0.427

57. 78

58. 0.00062

Calculate, using scientific notation.

59. The number of centimeters in 200 yards, if 1 inch equals approximately 2.54 cm.

60. The number of cubic centimeters in 100 cubic meters.

61. The distance to the sun in meters, if the sun is 93 million miles from the earth and 1 mile equals approximately 1.609 kilometers.

62. A gram is 0.002205 pound. If the mass of the earth is 5.97×10^{27} grams, what is its mass in pounds?

63. An angstrom is a unit of length equal to 10^{-8} centimeters. Red light has a wavelength of 8000 angstroms. If 1 cm. = 0.3937 in., what is the wavelength of red light in inches?

64. The **parsec** and **astronomical unit** are units of astronomical distance. The parsec is 3.08×10^{13} kilometers and the astronomical unit is 1.495×10^8 kilometers. How many astronomical units are there in 1.5 parsecs?

65. The law of exponents $a^pa^q = a^{p+q}$ was proved in this section. Use a similar procedure to prove the remaining laws. Assume p and q are positive integers.
(a) $\dfrac{a^p}{a^q} = a^{p-q}$, if $p > q$, (b) $\dfrac{a^p}{a^q} = a^{q-p}$, $p < q$, (c) $(a^p)^q = a^{pq}$, (d) $(ab)^p = a^pb^p$, and (e) $\left(\dfrac{a}{b}\right)^p = \dfrac{a^p}{b^p}$, if $b \neq 0$.

66. Show that the definition $a^{-p} = \dfrac{1}{a^p}$ satisfies the five laws of exponents.

67. Show that the definition $a^{p/q} = (\sqrt[q]{a})^p = \sqrt[q]{a^p}$ satisfies the five laws of exponents.

4.2 Exponential Functions

From now on we assume that the exponential a^x has been defined for real numbers a and x with $a > 0$, and that exponential expressions satisfy the five laws of exponents of Theorem 4.1. Then, for any given $a > 0$, $f(x) = a^x$ defines a real function of x, called an **exponential function with base *a*.**

EXAMPLE 1 Sketch the graphs of the exponential functions $f(x) = 2^x$ and $f(x) = 3^x$ on the same coordinate system.

SOLUTION In the second line of the table below are listed several function values for $f(x) = 2^x$. (These are readily calculated with a hand calculator.) If we plot the corresponding points and connect them with a smooth curve we get the graph in Figure 4.1. The graph rises steadily to the right, and becomes steeper as values of x get larger. It crosses the y-axis at (0, 1) and is asymptotic to the negative x-axis.

x	-3	-2	-1	0	3/2	2	7/3	3
2^x	1/8 0.13	1/4 0.25	1/2 0.50	1 1.00	$2\sqrt{2}$ 2.83	4 4.00	$4\sqrt[3]{2}$ 5.04	8 8.00
3^x	1/27 0.04	1/9 0.11	1/3 0.33	1 1.00	$3\sqrt{3}$ 5.20	9 9.00	$9\sqrt[3]{3}$ 12.98	27 27.00

In the same table are also tabulated several values of $f(x) = 3^x$. For $x > 0$, these values are greater than the corresponding values of 2^x because $3 > 2$ and so $3^x > 2^x$. For $x < 0$, the values of 3^x are smaller than the corresponding values for 2^x. This is shown in Figure 4.1 where the graph of $y = 3^x$ is above that of $y = 2^x$ for $x > 0$, and below it for $x < 0$. ●

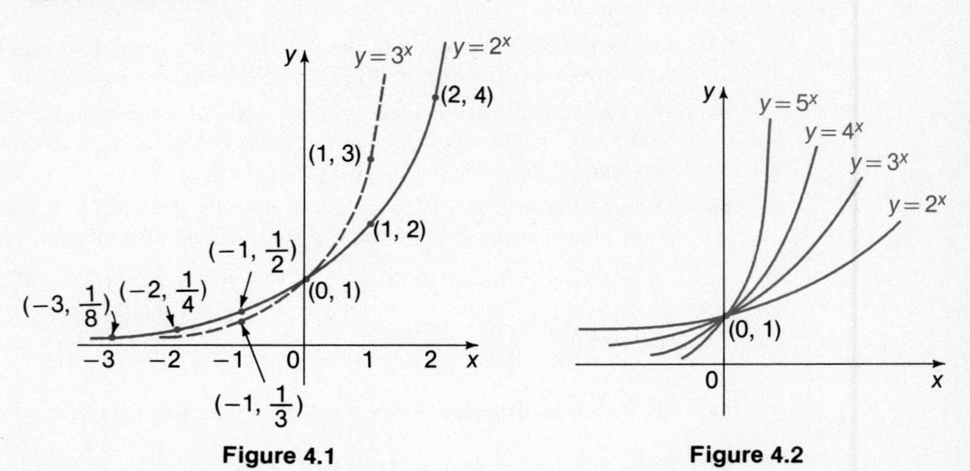

Figure 4.1

Figure 4.2

In the same way, we can construct graphs for $y = 4^x$, $y = 5^x$, and so on, to obtain a "family" of *exponential curves* with equation $y = a^x$, $a > 1$. As shown in Figure 4.2, they share the following characteristics:

1. rising to the right
2. crossing the y-axis at $(0, 1)$
3. asymptotic to the negative x-axis.

EXAMPLE 2 Sketch the graph of the exponential function $f(x) = \left(\frac{1}{2}\right)^x$.

SOLUTION This is an example of an exponential function of the form $f(x) = a^x$ with $0 < a < 1$. For $a = 1/2$,

$$f(x) = \left(\frac{1}{2}\right)^x = \frac{1}{2^x} = 2^{-x}.$$

Several values of this function are shown in the table, along with corresponding values of the function $g(x) = 2^x$.

x	-2	-1	0	1	2
2^{-x}	4	2	1	$\frac{1}{2}$	$\frac{1}{4}$
2^x	$\frac{1}{4}$	$\frac{1}{2}$	1	2	4

From the ordered pairs shown in the table above, and from Figure 4.3, it is easy to see that the graph of this function is just a reflection in the y-axis of the graph of $g(x) = 2^x$. For example, $f(1) = 1/2 = g(-1)$, $f(-1) = 2 = g(1)$, and so on. ●

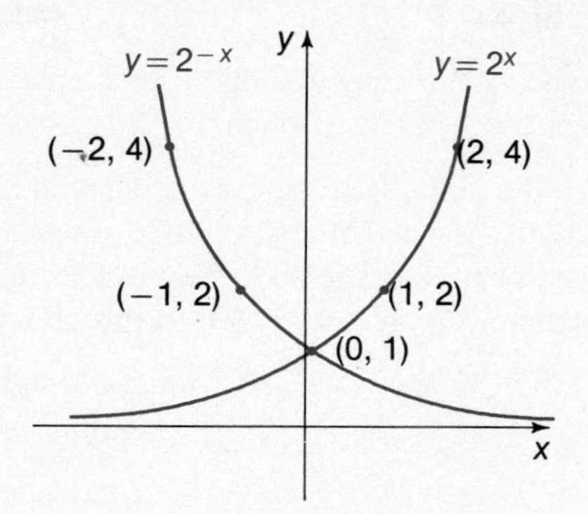

Figure 4.3

In summary, every member of the family of exponential functions $f(x) = a^x$ has as domain the set of all reals x and as range the set of positive reals y (if $a \neq 1$). The function is increasing for $a > 1$, constant for $a = 1$, and decreasing for $0 < a < 1$. Each graph intersects the y-axis at (0, 1), and is asymptotic to the positive x-axis if $0 < a < 1$ and to the negative x-axis if $a > 1$. (See Figure 4.4.) The function also has an inverse since for $a \neq 1$ it is either always increasing (if $a > 1$) or always decreasing (if $0 < a < 1$). (See Section 2.5.)

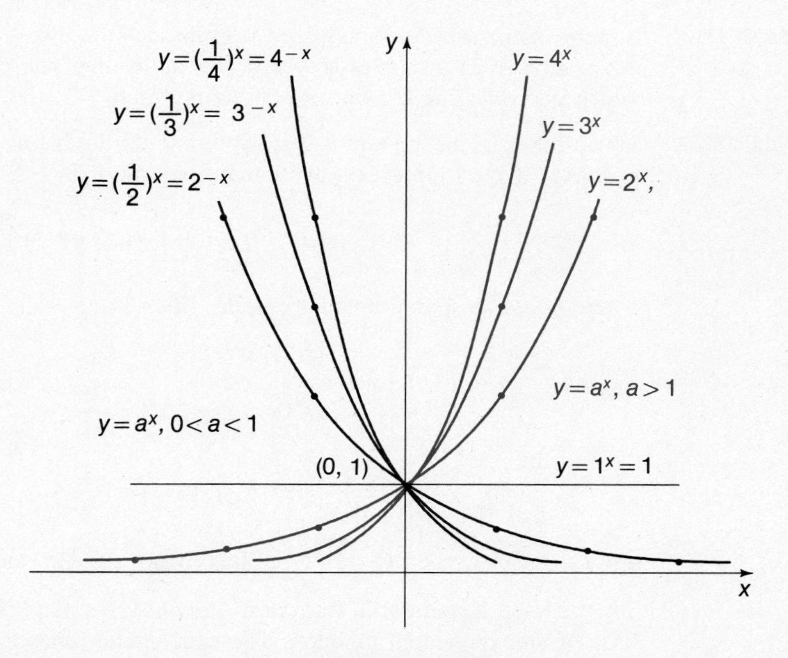

Figure 4.4

The property of always increasing or always decreasing makes these functions important in describing a wide variety of situations mathematically, as is shown in the following examples.

EXAMPLE 3 Suppose that the population of a bacterial culture doubles each hour. Write an equation which describes its growth.

SOLUTION Since the number of bacteria doubles in an hour, if it is 1000 at the beginning, it will be 2000 at the end of 1 hour, 4000 at the end of 2 hours, 8000 at the end of 3 hours, and so on. That is, at the end of each hour the number of bacteria has been multiplied by a factor of 2. If we let $N(t)$ represent the population at the end of t hours, then

$$N(0) = 1000 = 1000(1) = 1000(2^0),$$
$$N(1) = 2000 = 1000(2) = 1000(2^1),$$
$$N(2) = 4000 = 1000(4) = 1000(2^2),$$
$$N(3) = 8000 = 1000(8) = 1000(2^3),$$

and so on. Thus the growth above is described by the exponential function

$$N(t) = 1000(2^t). \qquad \bullet$$

In a situation like Example 3, we say that the population is **increasing exponentially.** We call $N(0)$ the **initial population** and often write N_0 for $N(0)$. Then for *any* population doubling each successive unit of time, $N(t) = N_0(2^t)$.

EXAMPLE 4 Suppose some substance dissolves at a rate such that two-thirds of the substance present at any given instant dissolves in the succeeding unit of time. Write an equation which describes the amount present at any time.

SOLUTION If two-thirds of the amount is dissolved, one-third remains. Then, as in the preceding example, if $Q(t)$ represents the amount present at time t, we can write

$$Q(t) = Q_0\left(\frac{1}{3}\right)^t = Q_0(3^{-t}),$$

where Q_0 is the initial amount present. Then

$$Q(0) = Q_0(3^0) = Q_0,$$
$$Q(1) = Q_0(3^{-1}) = \frac{1}{3}Q_0,$$
$$Q(2) = Q_0(3^{-2}) = \frac{1}{9}Q_0,$$

and so on. Here the quantity Q is **decreasing exponentially.** $\bullet$

The Natural Exponential Function Any positive number except 1 can be used as a base of an exponential function. The exponential function for which the base is the irrational number 2.71828 . . . , denoted by e, is of particular importance. The function $f(x) = e^x$ is called **the (natural) exponential function.** Its graph lies between the curves of $y = 2^x$ and $y = 3^x$. See Figure 4.5.

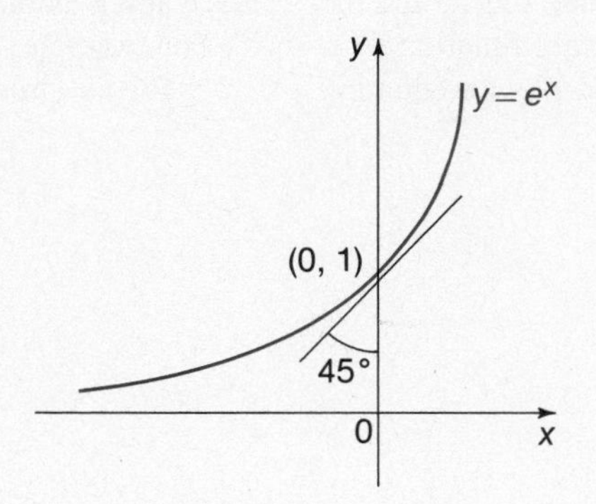

Figure 4.5

Tables of values of e^x and e^{-x} are given in the Appendix, and they may also be obtained directly with a hand calculator. The number e is a fundamental number which arises naturally in mathematics. In addition to its occurrence in both theoretical and applied mathematics it is utilized in the estimation of rates of natural growth and decay, appears in various formulas used in engineering and science, and is employed in statistical calculations involved with insurance, finance, and economics. How its value is calculated is explained in calculus. It was denoted by the letter e by the Swiss mathematician Euler about 200 years ago. (Euler was also responsible for the use of the symbol π for the ratio of the circumference of a circle to its diameter and for the functional notation $f(x)$.)

The next section describes and illustrates a wide variety of applications of the exponential function. Before studying them we need to show that *any* exponential function can be expressed in terms of the natural exponential function. For instance, in Example 3 we had $N(t) = 1000(2^t)$. Let us suppose that

$$2^t = e^{kt},$$

where k is a constant to be determined. Then

$$(2)^t = (e^k)^t,$$
$$2 = e^k.$$

From Table A in the Appendix, $e^k = 2$ for $k \approx 0.69$ (the symbol "$\approx$" means "approximately equal to"). Thus, we may write

$$N(t) \approx 1000e^{0.69t}$$

for the exponential function of Example 3. Similarly, in Example 4, where $Q(t) = Q_0(1/3)^t \approx Q_0(0.33)^t$, we find from the tables or a calculator that $e^{-1.1} \approx 0.33$.

Then the function Q may be written

$$Q(t) \approx Q_0 e^{-1.1t}.$$

In general, $a^t = e^{kt}$ if $e^k = a$. We can use the tabular values for e^x and e^{-x} to obtain this k, and then replace any general exponential function $f(t) = a^t$ by the exponential function $f(t) = e^{kt}$. For example, each of the exponential curves of Figure 4.4 also has an equation $y = e^{kx}$ for an appropriate choice of k. See Figure 4.6.

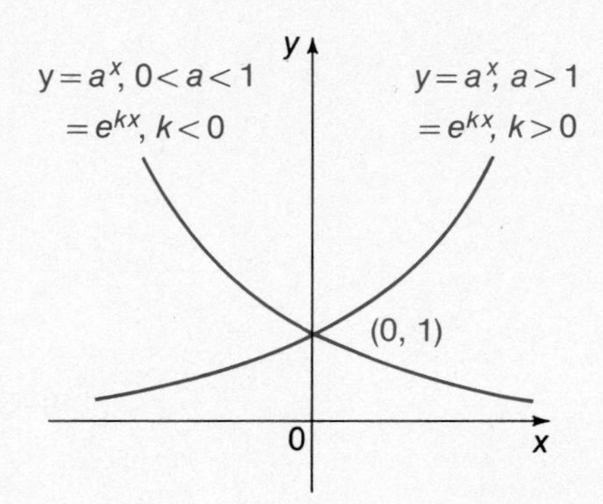

Figure 4.6

EXERCISES 4.2

Use the table of exponential function values or a calculator to evaluate the following.

1. e^2 **2.** e^{-1} **3.** $e^{0.5}$ **4.** $e^{3.4}$ **5.** $e^{6.2}$ **6.** $e^{-1.5}$

7. e^{-4} **8.** $e^{-2.3}$ **9.** $e^{1.6}$ **10.** $e^{-0.3}$ **11.** e^5 **12.** e^{-3}

From the table of function values, find t if given the following.

13. $e^t = 3$

14. $e^t = 1$

15. $e^t = 2.718$

16. $e^t = 9.974$

17. $e^t = 5.474$

18. $e^{-t} = 0.0183$

19. $e^{-t} = 0.1653$

20. $e^t = 99.484$

21. $e^{-t} = 0.0067$

22. $e^{-t} = 0.7408$

Graph each set of functions on the same coordinate plane and identify several points on each graph.

23. $y = 1^x, y = 2^x, y = 3^x, y = 4^x$

24. $y = 3^x, y = 3^{-x}$

25. $y = 2^x, y = 2^x + 1, y = 2^{x+1}$ [*Hint:* $2^{x+1} = 2^x 2^1 = 2 \cdot 2^x$.]

26. $y = 3^{-x}, y = 3^{-x} + 2, y = 3^{-x} - 1$

27. $y = e^x, y = e^x - 1, y = 2e^x$

28. Because $f(t) = 2^{t+1} = 2^t 2^1 = 2(2^t)$, the function values of $f(t) = 2^{t+1}$ are just double the corresponding values of $g(t) = 2^t$. What is the corresponding relation to $g(t)$ of $h(t) = 2^{t+2}$? Of $k(t) = 2^{t+3}$?

Suppose that water flows into a tank at a variable rate such that the quantity of water is doubled in each successive minute. If the tank fills in 10 minutes, how long did it take to fill the tank

29. Half full; **30.** One-fourth full; **31.** One-eighth full?

32. Write an exponential function describing this situation.

33. What part of the tank was full initially ($t = 0$)?

From the tables determine k so that $e^k = a$ for each of the following.

34. $a = 4$ **35.** $a = 5$ **36.** $a = -2$ **37.** $a = -3$

4.3 Law of Exponential Growth

Examples 3 and 4 of the previous section indicate the special significance of exponential functions. An exponential function is either

1. increasing for all real values of its argument and this increase is more rapid the greater the value of the argument, or
2. decreasing for all real values of its argument and this decrease is less rapid the greater the value of the argument.

These characteristics are shown by the increasing (or decreasing) steepness of the exponential curves. Because of this special property, exponential functions can be used to describe "growth" and "decay" situations which commonly occur in natural circumstances like those described in the last section. Such situations are described by functions of the form $q(t) = q_0e^{kt}$. Here $q(t)$ represents the quantity of the given substance at any time t, q_0 is the initial quantity $(q(0) = q_0)$, and k is a constant of proportionality called the **rate,** which is determined by observation or experiment. The formula

$$\mathbf{q(t) = q_0e^{kt}}$$

is called the **law of exponential growth,** or simply the **law of growth.** If $k < 0$, then q decreases with increasing time t and the formula is called the **law of negative growth,** or **law of decay.**

EXAMPLE 1 The population of the U.S. in 1970 was 203 million, or 2.03×10^8, and in 1980 it was 227 million, or 2.27×10^8. What was the annual growth rate during the 1970s? (Assume that the law of exponential growth applies.)

SOLUTION Write the law of growth in this case as

$$N(t) = N_0e^{kt},$$

where $N(t)$ is the population at time t. If we take 1970 as time $t = 0$, then 1980 corresponds to $t = 10$. Then $N_0 = N(0) = 2.03 \times 10^8$ and $N(10) = 2.27 \times 10^8$. We substitute these values in the above equation.

$$2.27 \times 10^8 = (2.03 \times 10^8)e^{k(10)}$$

$$e^{10k} = \frac{2.27}{2.03} \approx 1.1182$$

$$10k \approx 0.112 \qquad \text{from Table A}$$

$$k \approx 0.0112 = 1.12\%$$

The annual growth rate during the 1970s was approximately 1.12% (about 1 1/8%). ●

Radioactive Decay Any radioactive substance, such as radium or uranium, continuously emits particles. Hence the amount present continuously decreases (is a decreasing function of time). This process is called **radioactive decay** and it is described mathematically by the law of negative exponential growth.

EXAMPLE 2 Suppose a given radioactive substance loses one-fourth of its mass in one year. How much time will it take to lose one-half of its mass?

SOLUTION Let q_0 denote the initial amount. At the end of 1 year, $(1/4)q_0$ has been lost, so that $q(1) = (3/4)q_0$.

$$q(t) = q_0 e^{kt}$$

$$\frac{3}{4}q_0 = q_0 e^{k \cdot 1}$$

$$e^k = \frac{3}{4}$$

Substituting 3/4 for e^k in the law of decay gives the equation for this particular situation.

$$q(t) = q_0\left(\frac{3}{4}\right)^t$$

Note that $(3/4)^t$ decreases as time t increases.

Alternatively, if we keep e as a base and solve for k, we have

$$e^k = \frac{3}{4} = 0.75,$$

$$k \approx -0.29, \ q(t) = q_0 e^{-0.29t}. \qquad \text{from Table A or a calculator}$$

At the time T when one-half of the original substance has decayed, $q(T) = (1/2)q_0$, so that

$$\frac{1}{2}q_0 = q_0 e^{-0.29T},$$

$$e^{-0.29T} = \frac{1}{2} = 0.5,$$

$$-0.29T \approx 0.69, \qquad \text{from Table A or a calculator}$$

$$T \approx 2.4 \text{ (years)}.$$ ●

The time it takes for one-half of a radioactive substance to decay is called the **half-life** of the substance. One-half of the remaining substance would decay in the next half-life period. That is, in Example 2 one-fourth of the *original* substance would decay in the *next* 2.4 years. Thus, in 4.8 years three-fourths of the original sample would have decayed. This is illustrated in Figure 4.7.

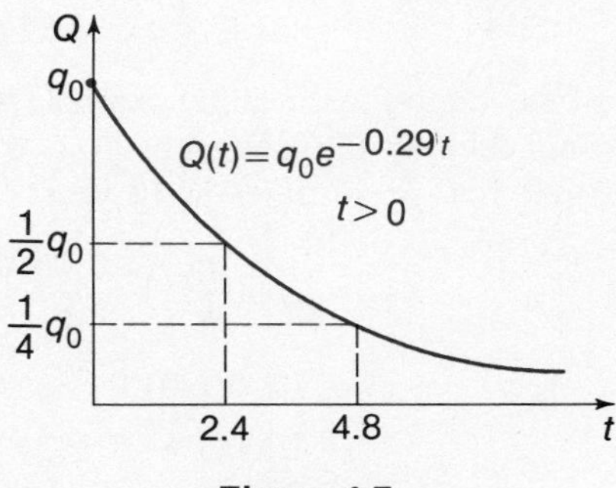

Figure 4.7

Note that in the law of growth (or decay) there are four quantities: q, q_0, k, and t. If any three of them are known, the remaining one can be calculated.

Compound Interest If \$100 is invested at 8% simple interest, it yields 8% of \$100, or 0.08(\$100) = \$8 interest per year. The **accumulated amount,** principal *and* interest, at the end of one year is then

$$A_1 = 100 + 0.08(100) = 100(1 + 0.08) = 100(1.08).$$

If this accumulated amount is left at 8% interest for another year, the total amount at the end of the second year is

$$\begin{aligned} A_2 &= A_1 + 0.08A_1 \\ &= A_1(1 + 0.08) = A_1(1.08) \\ &= [100(1.08)](1.08) \\ &= 100(1.08)^2. \end{aligned}$$

Similarly, at the end of three years, the total amount is

$$A_3 = 100(1.08)^3.$$

We say that the interest has been **compounded annually.**

The above example illustrates the general formula

$$\boldsymbol{A(n) = A_n = P(1 + r)^n}$$

for the amount A_n accumulated in n years if a principal amount P is invested at a rate of interest r compounded annually. This basic formula is used for loans, investments, insurance, and so on.

EXAMPLE 3 Suppose \$1000 is invested at 6% interest compounded annually. Find the accumulated amount at the end of 10 years.

SOLUTION Substitute $n = 10$, $P = 1000$, and $r = 0.06$ in the formula.

$$\begin{aligned} A_{10} &= 1000(1.06)^{10} \\ &\approx 1000(1.790848) && \text{using a calculator} \\ &\approx 1790.85, \quad \text{or } \$1790.85 \end{aligned}$$

If the interest in the above example were compounded quarterly (every 3 months), we could think of the interest as (1/4) 6%, or 1 1/2% per quarter, compounded *four* times per year or 40 times in 10 years. Then we would have

$$\begin{aligned} A_{40} &= 1000\left(1 + \frac{0.06}{4}\right)^{10(4)} \\ &= 1000(1.015)^{40} \\ &\approx 1000(1.814018) && \text{using a calculator} \\ &\approx 1814.02, \quad \text{or } \$1814.02. \end{aligned}$$

The general formula illustrated above is

$$\boldsymbol{A = P\left(1 + \frac{r}{k}\right)^{nk}},$$

where k is the number of times per year interest is compounded.

At one time interest was compounded only annually, semiannually, or quarterly. One may, of course, consider interest compounded as frequently as one wishes. Today it is common for interest to be compounded daily. Letting t represent the number of years, we get the following formulas for amounts accumulated in t years if \$1 is invested at an interest rate r compounded as indicated.

Annually ($k = 1$): $A = (1 + r)^t$

Semiannually ($k = 2$): $A = \left(1 + \frac{r}{2}\right)^{2t}$

Quarterly ($k = 4$): $A = \left(1 + \frac{r}{4}\right)^{4t}$

Monthly ($k = 12$): $A = \left(1 + \frac{r}{12}\right)^{12t}$

Daily ($k = 365$): $A = \left(1 + \frac{r}{365}\right)^{365t}$

k times per year: $A = \left(1 + \frac{r}{k}\right)^{kt}$

Now imagine that interest is compounded more and more frequently, that is, let k increase without limit. In this case we say that interest is **compounded continuously.** Let us rewrite the general formula as follows.

$$A = \left(1 + \frac{r}{k}\right)^{kt} = \left[\left(1 + \frac{r}{k}\right)^{k/r}\right]^{rt}$$

$$= \left[\left(1 + \frac{1}{x}\right)^{x}\right]^{rt} \qquad \text{letting } x = \frac{k}{r}$$

It can be shown that as x in this formula takes on arbitrarily large values, the expression $(1 + 1/x)^x$ comes closer and closer to e. Thus, as k takes on arbitrarily large values, $(1 + r/k)^{k/r}$ approaches e, and the formula for the amount accumulated from principal P at interest compounded continuously is

$$\boldsymbol{A = Pe^{rt}}.$$

This is an example of the exponential law of growth and is called the **compound-interest law.** It is fundamental in many investment matters such as insurance, annuities, and perpetuities.

EXAMPLE 4 What is the amount accumulated in 5 years if \$100 is invested at 8% compounded continuously?

SOLUTION Use $P = 100$, $r = 0.08$, and $t = 5$ in the compound-interest law.

$$\begin{aligned} A &= Pe^{rt}. \\ A &= 100e^{0.08(5)} \\ &= 100e^{0.4} \\ &\approx 100(1.49182), && \text{using a calculator} \\ &\approx 149.18, \quad \text{or } \$149.18. && \bullet \end{aligned}$$

EXAMPLE 5 How much time is required for a principal amount invested at 8% to be doubled?

SOLUTION When the principal has doubled, $A = 2P$. Let us call this time T. Substituting in the formula $A = Pe^{rt}$, we have

$$2P = Pe^{rT},$$

$$e^{rT} = 2.$$

Substituting 0.08 for r gives

$$e^{0.08T} = 2.$$

Since $e^{0.69} = 2$, from Table A,

$$0.08T = 0.69,$$

$$T = \frac{0.69}{0.08} = 8.6.$$

Thus, the time required to double money invested at 8% compounded continuously is about 8.6 years. •

Carbon Dating When cosmic rays from outer space strike nitrogen in the upper atmosphere, they produce a radioactive form of carbon called carbon-14 or ^{14}C. The ^{14}C decays, but since it is also being continuously produced, the percentage of ^{14}C relative to the stable carbon present remains constant. Living plants and animals absorb carbon into their cells in these same proportions. When the cells die this absorption stops, but the radioactive decay continues. This reduces the percentage of ^{14}C present. If this percentage is determined, it can be used with the known half-life of ^{14}C, about 5600 years, to calculate the approximate time the cells have been dead. This procedure is called **carbon dating.** It is used extensively by geologists and archeologists to estimate ages of fossils and artifacts.

EXAMPLE 6 Suppose a certain fossil contains 60% of "normal" carbon-14. How old is it?

SOLUTION First, the known half-life of ^{14}C is used to determine the rate k of decay.

$$q(t) = q_0 e^{-kt}$$

$$\frac{1}{2} q_0 = q_0 e^{-k(5600)}$$

$$e^{-5600k} = 0.5$$

$$-5600k \approx -0.7 \qquad \text{using the tables}$$

$$k \approx 0.000125$$

Thus, the rate of decay of ^{14}C is 0.0125%. For ^{14}C, the specific law of decay is

$$q(t) \approx q_0 e^{-0.000125t}.$$

To determine the value of t for which $q(t) = 0.6q_0$, substitute in the above equation.

$$0.6q_0 = q_0 e^{-0.000125t}$$

$$e^{-0.000125t} = 0.6$$

$$-0.000125t \approx -0.5 \qquad \text{using the tables}$$

$$t \approx 4000$$

Thus, the age of the fossil is about 4000 years. ●

In Section 3.2 we saw that a linear function has a *constant rate of change,* corresponding to the uniform steepness of the graph. Then in Section 3.3, it was shown that a quadratic function has a *linear rate of change*. Here we have found that the exponential function describes a special kind of change of a variable quantity, called **exponential growth.** We often express the fact that something is growing at this rate by saying that it is increasing *exponentially*.

EXERCISES 4.3

1. Suppose that the population N of bacteria in a culture at the end of t hours is given by $N(t) = 1000e^{0.64t}$. Calculate the number present (a) initially, and (b) at the end of 2 hours. (c) How much time is required to double the initial number?

2. If an exponential rate of population growth of 5% is assumed,

$$N(t) = N_0 e^{0.05t}.$$

How long will it take a population of 100 to grow to 300?

3. If a certain population grows exponentially at a 5% rate, what initial population would grow to 10,000 in 20 years?

4. A certain bacterial population is initially 10,000 and after 10 days is 25,000. Assume that the growth is exponential, and calculate (a) the rate of growth, and (b) the population at the end of 20 days.

5. The population of the world in 1980 has been estimated as 4.4×10^9, and the rate of growth as 2% per year. Assuming this rate continued, what would be the population in (a) 1990, (b) 2000? (c) If the same rate held prior to 1980, what was the world population in 1970?

6. Cobalt-56 is a radioactive substance with a 1% rate of decay. (a) Calculate its half-life. (b) How much of an original sample of 5 grams would remain at the end of 100 years?

7. Calculate the half-life of a radioactive substance whose rate of decay k is (a) 0.07, (b) 0.0014, (c) 0.003.

8. Radium decays according to the exponential law $q(t) = q_0 e^{-0.0004t}$, where t is measured in years. (a) What is the half-life of radium? (b) To what amount would an initial quantity of 100 grams of radium be reduced after 5000 years?

9. Atmospheric pressure decreases exponentially with increases in elevation, that is, $p(h) = p_0 e^{-kh}$, $k > 0$. The sea-level atmospheric pressure is 14.7 pounds per square inch and the pressure at 10,000 feet is 11.4 pounds per square inch. What is the atmospheric pressure at the top of Mt. Whitney (15,000 feet)?

10. If $1000 is invested with interest compounded continuously and yields $2000 in 10 years, what is the rate of interest?

11. If $1000 is invested at 4.75% interest compounded continuously, how much will have accumulated at the end of (a) 1 year, (b) 5 years, (c) 10 years?

12. If $1000 is invested with interest compounded continuously, how much is accumulated in one year if the rate of interest is (a) 5%, (b) 6%, (c) 7.5%?

13. If $1000 is invested at 8% interest, what is the amount at the end of one year if it is compounded (a) monthly, (b) quarterly, (c) semiannually?

14. A certain savings institution advertises that savings at 7.75% compounded continuously yields the equivalent of 8% (approximately) simple interest. Confirm this by calculating the amount accumulated in one year by investing $100 at 7.75% compounded continuously.

15. How long will it take to double an amount invested if it is compounded continuously at (a) 6.75%, (b) 7.5%, (c) 8%?

16. How much must be deposited at 10% interest compounded quarterly to accumulate $1000 at the end of one year?

4.4 Logarithmic Functions

One of the significant properties of an exponential function $f(x) = a^x$ $(a > 0,\ a \neq 1)$ is that it is either always increasing or always decreasing on its entire domain. Hence *it has an inverse* (recall Section 2.5). The inverse function of an exponential function is called a *logarithmic function*.

DEFINITION 4.3 If $f(x) = a^x$ $(a > 0,\ a \neq 1)$, then the inverse of f is the **logarithmic function** $f^{-1}(x) = \log_a x$, read "the logarithm of x to the base a." That is,

$$\mathbf{y = \log_a x \quad \text{if and only if} \quad x = a^y.}$$

EXAMPLE 1 The following table lists several pairs of equivalent statements of the form $x = a^y$ and $y = \log_a x$.

Exponential	Logarithmic
$2^3 = 8$	$\log_2 8 = 3$
$10^2 = 100$	$\log_{10} 100 = 2$
$a^0 = 1$	$\log_a 1 = 0$
$3^{-2} = \dfrac{1}{9}$	$\log_3 \left(\dfrac{1}{9}\right) = -2$
$y = e^4$	$\log_e y = 4$
$a^1 = a$	$\log_a a = 1$

The statements $\log_a 1 = 0$ and $\log_a a = 1$ are particularly important. ●

The graphs of a function and its inverse are reflections of each other in the line $y = x$. This fact has been used to graph $y = \log_a x$ in Figure 4.8.

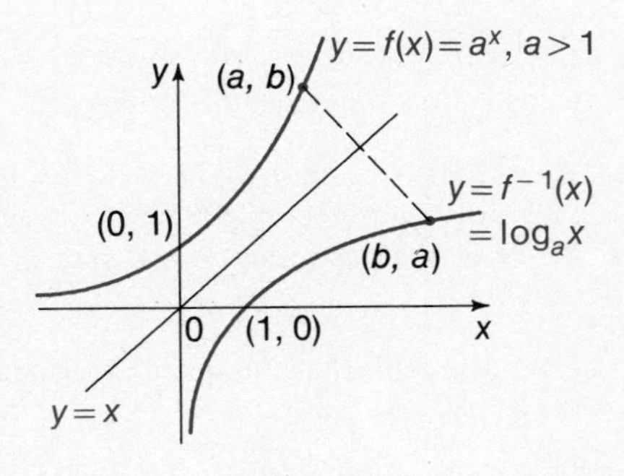

Figure 4.8

The graphs illustrate the interchange in the roles of x and y in the functions f and f^{-1}. The graph of $y = a^x$ crosses the y-axis at (0, 1) and is asymptotic to the negative x-axis, while the graph of $y = \log_a x$ crosses the x-axis at (1, 0) and is asymptotic to the negative y-axis. The domain of the exponential function is the set of all reals and its range is the set of positive reals. Its inverse, the logarithmic function, has as its domain the set of positive reals and as range the set of all reals. Both functions are increasing if $a > 1$ (as in Figure 4.8) and decreasing if $0 < a < 1$.

The values of a logarithmic function are called **logarithms.** For example,

$$\text{since } 10^2 = 100, \quad \text{then } \log_{10} 100 = 2.$$

We say that 2 is the logarithm of 100 to the base 10. Note that *a logarithm is an exponent*. Since the domain of a logarithmic function is the set of all positive real numbers, every positive number has a unique logarithm for a given base. Also, as Figure 4.8 shows,

$$\begin{aligned} \log_a x &< 0, \quad \text{if } 0 < x < 1, \\ \log_a x &= 0, \quad \text{if } x = 1, \\ \log_a x &> 0, \quad \text{if } x > 1. \end{aligned}$$

Laws of Logarithms If the laws of exponents (Section 4.1) are restated in the language of logarithms, we get the following laws of logarithms.

THEOREM 4.2 **Laws of Logarithms**

1. $\log_a xy = \log_a x + \log_a y$
2. $\log_a \dfrac{x}{y} = \log_a x - \log_a y$
3. $\log_a x^p = p \log_a x$

In words, these laws are stated as:

1. The logarithm of a product is the *sum* of the logarithms of the factors.
2. The logarithm of a quotient is the logarithm of the numerator *minus* the logarithm of the denominator.
3. The logarithm of a power is the *product* of the exponent and the logarithm of the base of the power.

The first law of logarithms may be proved as follows: Let $u = \log_a x$ and $v = \log_a y$, so that $a^u = x$ and $a^v = y$. By the first law of exponents,

$$xy = a^u a^v = a^{u+v}.$$

Changing to logarithmic form and substituting for u and v gives

$$\begin{aligned} \log_a xy &= u + v \\ &= \log_a x + \log_a y. \end{aligned}$$

The proofs of the other laws of logarithms are similar and are left as exercises.

The laws of exponents hold for any positive base $a \neq 1$. Therefore, the laws and properties of logarithms given above are independent of the particular base a. For this reason, in situations where no particular base is referred to we write simply "$\log x$." (The notation $\log x$ is sometimes used for base 10 logarithms.)

EXAMPLE 2 Use the laws of logarithms to write the following in terms of $\log 2$ and $\log 3$: (a) $\log 6$; (b) $\log 1.5$; (c) $\log 16$; (d) $\log \frac{1}{3}$.

SOLUTION (a) $\log 6 = \log(2)(3) = \log 2 + \log 3$

(b) $\log 1.5 = \log \frac{3}{2} = \log 3 - \log 2$

(c) $\log 16 = \log 2^4 = 4 \log 2$

(d) $\log \frac{1}{3} = \log 1 - \log 3 = 0 - \log 3 = -\log 3$ •

We would not normally use logarithms for such simple calculations as in the above example. But, they illustrate the fact that logarithms make multiplication, division, and raising to powers easier by replacing them by the simpler operations of addition, subtraction, and multiplication, respectively.

A very important consequence of the second law of logarithms, illustrated in Example 2(d) above, is that

$$\begin{aligned} \log \frac{1}{x} &= \log 1 - \log x \\ &= 0 - \log x \\ &= -\log x. \end{aligned}$$

That is, *the logarithm of the reciprocal of a number is the negative of the logarithm of the number*.

Another important fact is that $\log(x + y) \neq \log x + \log y$. That is, the logarithm of the sum of two numbers is not the sum of the logarithms of the numbers. *There are no laws for logarithms of sums and differences.*

Logarithms were originally developed to aid in calculations (see the Appendix). This use has become less important because of the widespread use of the hand-held calculator. In calculus, however, it is often helpful to use the laws of logarithms to rewrite the logarithm of an expression in a different form.

EXAMPLE 3 Use the laws of logarithms to rewrite the following expressions: (a) $\log \sqrt{xy^3}$; (b) $\log(\sqrt[3]{x}/yz^2)$.

SOLUTION (a) $\log \sqrt{xy^3} = \log(xy^3)^{1/2}$

$$= \frac{1}{2} \log xy^3$$

$$= \frac{1}{2}(\log x + \log y^3)$$

$$= \frac{1}{2}(\log x + 3 \log y)$$

$$= \frac{1}{2} \log x + \frac{3}{2} \log y$$

(b) $\log \dfrac{\sqrt[3]{x}}{yz^2} = \log \sqrt[3]{x} - \log(yz^2)$

$$= \log x^{1/3} - (\log y + \log z^2)$$

$$= \frac{1}{3} \log x - \log y - 2 \log z \qquad \bullet$$

EXAMPLE 4 The formula for the period of oscillation of a simple pendulum is $T = 2\pi\sqrt{g/L}$. Find an expression for $\log T$.

SOLUTION

$$\log T = \log 2\pi\left(\frac{g}{L}\right)^{1/2}$$

$$= \log 2\pi + \frac{1}{2} \log\left(\frac{g}{L}\right)$$

$$= \log 2\pi + \frac{1}{2}(\log g - \log L)$$

$$= \log 2 + \log \pi + \frac{1}{2} \log g - \frac{1}{2} \log L \qquad \bullet$$

These examples also show how the laws of logarithms are used in calculations.

Since $f(x) = e^x$ is a particularly important function, so also is its inverse, $f^{-1}(x) = \log_e x$. It is called **the (natural) logarithm function,** and its values are called **natural logarithms.** The notation $\ln x$ is often used for $\log_e x$, that is,

$$y = \ln x \quad \text{if and only if} \quad x = e^y.$$

Since our number system is based on 10, logarithms to the base 10, or **common logarithms,** are particularly suited for aiding in carrying out numerical calculations. Logarithms to the base e are also helpful. They not only provide useful simplifications for many mathematical formulas but, additionally, arise quite naturally in applications such as the estimation of rates of growth and decay. For these reasons they are called **natural logarithms.**

EXAMPLE 5 Graph $y = \ln x^2$.

SOLUTION Because $\ln x^2 = 2 \ln x$, the equation $y = \ln x^2$ may be graphed by simply doubling the corresponding ordinates (y-values) of $y = \ln x$. (These values may be found in Table C in the Appendix.) The table below shows several sets of corresponding values and the graph is shown in Figure 4.9. ●

x	0	0.5	1	2	3	4	5
$\ln x$	–	−0.7	0	0.7	1.1	1.4	1.6
$\ln x^2$	–	−1.4	0	1.4	2.2	2.8	3.2

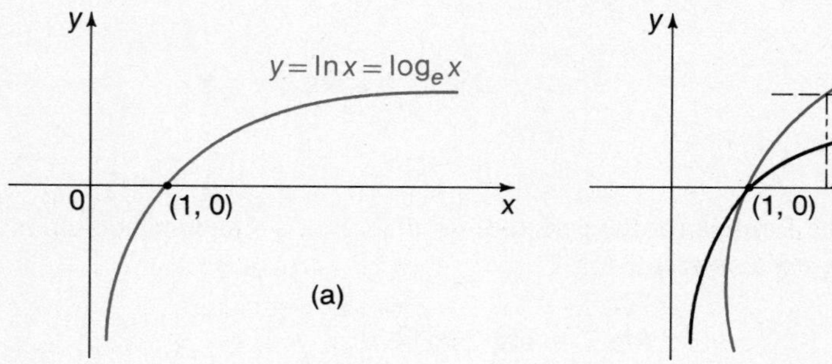

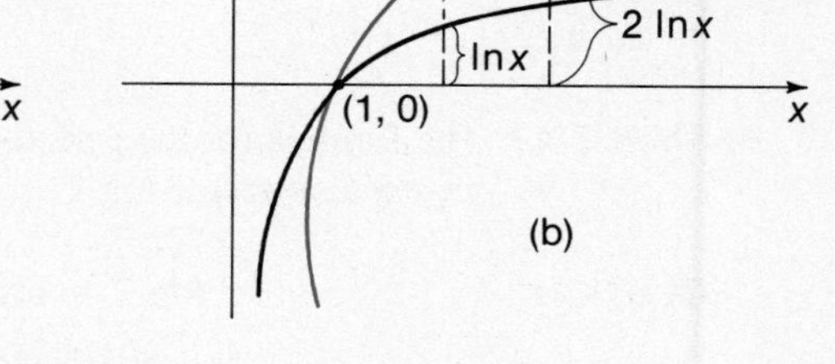

Figure 4.9

EXAMPLE 6 In Example 1, Section 4.3, we found that

$$e^{10k} = 1.1182.$$

Since $x = y$ implies that $\ln x = \ln y$, we have

$$\ln(e^{10k}) = \ln 1.1182.$$

Then by the third law of logarithms,

$$\begin{aligned} 10k \cdot \ln e &= \ln 1.1182, \\ 10k &= \ln 1.1182, && \text{since } \ln e = 1 \\ &\approx 0.112, && \text{from Table C} \\ k &\approx 0.0112. && \text{the same result as before} \end{aligned}$$

●

EXAMPLE 7 If \$1000 invested at interest compounded continuously accumulates to \$2000 in 10 years, what is the rate of interest?

SOLUTION

$$\begin{aligned} A(t) &= Pe^{rt} \\ &= 1000\, e^{rt} \end{aligned}$$

When $t = 10$, $A = 2000$, so that

$$2000 = 1000\, e^{r(10)},$$
$$e^{10r} = 2,$$
$$10r = \ln 2 \approx 0.6931, \qquad \text{taking natural logarithms of both sides}$$
$$r = 0.069 \quad \text{or } 6.9\%. \qquad \bullet$$

Several important and useful formulas involving the exponential and logarithmic functions follow from the definition of inverse, $f(f^{-1}(x)) = x$ and $f^{-1}(f(y)) = y$. If $f(x) = a^x$, then $f^{-1}(x) = \log_a x$, and from $f(f^{-1}(x)) = x$ we get

$$\boldsymbol{a^{\log_a x} = x.}$$

Similarly, from $f^{-1}(f(y)) = y$, we get

$$\boldsymbol{\log_a a^y = y.}$$

In particular, for base e logarithms,

$$\boldsymbol{e^{\ln x} = x} \qquad \text{and} \qquad \boldsymbol{\ln e^y = y.}$$

By the 17th century, numerical calculations, particularly those concerned with astronomy, had become increasingly involved and time consuming. In response, logarithms were invented to decrease the amount of time and labor spent in calculating products, quotients, and powers. As a result scientists, engineers, and all those concerned with mathematics made extensive use of logarithmic tables. The invention of the slide rule enabled those interested in simplifying rather complex calculations to make use of the laws of logarithms mechanically. Until the advent of the electronic hand calculator, every science and engineering student had a slide rule and made frequent use of it. Today, the direct use of logarithms in calculations is quite limited. The logarithmic functions themselves and their properties are more significant now.

Tables of logarithms are provided in the Appendix, together with brief descriptions of the tables and examples of their use. This will serve those who have not studied logarithms, those who may want some review, and those who do not have ready access to a hand calculator.

EXERCISES 4.4

Write each of the following logarithmic statements in an equivalent exponential form:

1. $\log_2 16 = 4$

2. $\log_5 125 = 3$

3. $\log_8 4 = \frac{2}{3}$

4. $\log_{10} 1000 = 3$

5. $\log_e 1 = 0$

6. $\log_4 8 = \frac{3}{2}$

7. $\log_{10} 1 = 0$

8. $\log_{10} 0.01 = -2$

9. $\log_e e = 1$

10. $\ln x = 2$

Write each of the following exponential statements in the equivalent logarithmic form:

11. $2^5 = 32$

12. $5^{-2} = \frac{1}{25}$

13. $8^{1/3} = 2$

14. $10^2 = 100$

15. $4^{-1/2} = \frac{1}{2}$

16. $16^{1/2} = 4$

17. $4^3 = 64$

18. $2^{-3} = \frac{1}{8}$

19. $3^0 = 1$

20. $e^2 = x$

Write the given statement in the alternate form, as indicated.

	Exponential	*Logarithmic*
21.	$3^2 = 9$	________
22.	$16^{1/2} = 4$	________
23.	________	$\log_5 25 = 2$
24.	$10^1 = 10$	________
25.	$27^{2/3} = 9$	________
26.	$a^0 = 1$	________
27.	________	$\log_4 64 = 3$
28.	________	$\log_{10} 10 = 1$
29.	$e^{-1.5} = 0.223$	________
30.	________	$\log_3 w = z$
31.	$u = e^v$	________

Use the graph of $y = \ln x$ in Figure 4.9 to help graph each of the following.

32. $y = -\ln x$

33. $y = \ln(-x), \quad x < 0$

34. $y = \ln \frac{1}{x}$

35. $y = \ln |x|$

36. $y = 1 + \ln x$

37. $y = \ln x - 1$

38. $y = 1 - \ln x$

Use the laws of logarithms to change the form of each of the following, as in $\log \frac{x}{1 + x} = \log x - \log(1 + x)$.

39. $\log x^2\sqrt{1 + x}$

40. $\log \sqrt{\frac{1 + x}{1 - x}}$

41. $\log \frac{xy}{z}$

42. $\log \frac{x^2 z^3}{y^3}$

43. $\log \sqrt{xy}$

44. $\log \sqrt{x^2}$

45. $\log \sqrt{\frac{x^2 y}{z^3}}$

46. $\ln \sqrt{\frac{(x - 1)^2}{x^2 + 2}}$

47. $\log \sqrt{\frac{x}{y}}$

48. $\ln x^{1/2}y^{1/4}$

Find the common logarithm (base 10) of each of the following. [Hint: First write the corresponding exponential form.]

49. 10,000

50. 0.0001

51. $\sqrt{1000}$

52. $\sqrt{0.10}$

53. 0.1

54. 0.01

Complete each of the following by looking at the corresponding exponential form.

55. $\log_3 9 =$ ____

56. $\log_2 \sqrt{2} =$ ____

57. $\log_4 1 =$ ____

58. $\log_3\left(\frac{1}{9}\right) =$ ____

59. $\log_5 125 =$ ____

60. $\log_{17} 1 =$ ____

Given $\log_{10}2 \approx 0.301$, $\log_{10}3 \approx 0.477$, and $\log_{10}5 \approx 0.699$, use the laws of logarithms to calculate the logarithm of each of the following:

61. 6

62. 10

63. 4/5

64. 2/3

65. 12

66. 24

67. $\sqrt{125}$

68. $\sqrt[3]{60}$

69. An important formula in astronomy is $M = m + 5 \log_{10} r$, where M is the absolute brightness of a star, m is its apparent brightness, and r is its distance in parsecs. The apparent brightness m can be determined by observation. If 1 parsec $\approx$ 3.26 light-years, find (a) the absolute brightness of the brightest star, Sirius, whose distance is 8.7 light-years and whose apparent brightness is -1.6; (b) the distance of the nearest star, Proxima, whose apparent brightness is 0.1 and whose absolute brightness is 4.7.

Important applications of exponential and logarithmic functions occur in connection with human responses to stimuli. For example, the response of the human ear to change in sound intensity is expressed by $n = 10 \log_{10}(p_1/p_2)$, where n is the number of decibels of difference sensed by the ear between two sounds and p_1 and p_2 are the levels of power intensity (in watts) of the two sounds. What is the decibel difference between the ear's response to sound produced by amplifiers with the following powers?

70. 40 and 50

71. 45 and 60

72. 70 and 80

Prove the following laws of logarithms.

73. $\log_a \frac{x}{y} = \log_a x - \log_a y$

74. $\log_a x^p = p \log_a x$

75. Show that an exponential function $f(x) = a^x$ satisfies the statement $f(x + y) = f(x) \cdot f(y)$.

76. Show that a logarithmic function $f(x) = \log_a x$ satisfies the statement $f(xy) = f(x) + f(y)$.

4.5 Exponential and Logarithmic Equations

By definition, if $y = a^x$, then $\log_a y = x$. In the last section we made use of the fact that if $a^x = a^y$, then $x = y$. These are simple examples of exponential and logarithmic equations that occur frequently in mathematics. In the following examples we use the properties:

1. $a^x = a^y$ if and only if $x = y$; $\quad a \neq 1$
2. $\log_a u = \log_a v$ if and only if $u = v$.

These laws are equivalent ways of saying the same thing, in either the language of exponents or that of logarithms.

EXAMPLE 1 Solve the equation $2^{x-1} = 8$ for x.

SOLUTION

$$\begin{aligned} 2^{x-1} &= 8 \\ 2^{x-1} &= 2^3 \\ x - 1 &= 3 \qquad \text{by Property (1)} \\ x &= 4 \end{aligned}$$

●

EXAMPLE 2 Solve the equation $\log_3(x + 1) = \log_3(2x - 5)$.

SOLUTION

$$\begin{aligned} \log_3(x + 1) &= \log_3(2x - 5) \\ x + 1 &= 2x - 5 \qquad \text{by Property (2)} \\ x &= 6 \end{aligned}$$

Because the logarithmic function has a restricted domain, it is necessary to check this solution. Substitution of $x = 6$ in the given equation yields $\log_3 7 = \log_3 7$. ●

EXAMPLE 3 Solve $\log_3 x = 1/2$ for x.

SOLUTION By the definition of a logarithm, $\quad x = 3^{1/2} = \sqrt{3} \approx 1.732.$ ●

EXAMPLE 4 Solve $4^x = 5$ for x.

SOLUTION This equation could be solved by using logarithms to any convenient base. Since we have natural logarithm tables available, we will use base e logarithms.

$$\begin{aligned} 4^x &= 5 \\ \ln 4^x &= \ln 5 \qquad \text{by Property (2)} \\ x \ln 4 &= \ln 5 \qquad \text{Theorem 4.2c} \\ x &= \frac{\ln 5}{\ln 4} \\ &\approx \frac{1.609}{1.386} \approx 1.16 \qquad \text{Table C or a calculator} \end{aligned}$$

●

EXAMPLE 5 Solve $\log x^2 = 4 \log 2 - 3 \log \dfrac{1}{2}$.

SOLUTION By Theorem 4.2c,

$$\begin{aligned}
\log x^2 &= \log 2^4 - \log\left(\frac{1}{2}\right)^3 \\
&= \log \frac{2^4}{\left(\frac{1}{2}\right)^3} \\
&= \log \frac{16}{\frac{1}{8}} = \log(16)(8) \\
x^2 &= 16(8) \qquad \text{by Property (2),} \\
x &= \pm 4(2\sqrt{2}) = \pm 8\sqrt{2}.
\end{aligned}$$

Verify that both answers are solutions to the given equation. •

EXAMPLE 6 First-order chemical reactions (those for which the rate of reaction is directly proportional to the concentration of the reacting substance) are given by the law

$$-\ln C = kt + a,$$

where C is the concentration at time t and a is a constant related to the initial concentration. Show that the concentration satisfies the law of exponential decay.

SOLUTION We can write this in equivalent *exponential form* as follows.

$$\begin{aligned}
-\ln C &= kt + a \\
\ln \frac{1}{C} &= kt + a \qquad \text{since } -\ln C = \ln C^{-1} = \ln \frac{1}{C} \\
\frac{1}{C} &= e^{kt+a} \qquad \text{changing to exponential form} \\
C &= e^{-(kt+a)} \\
C &= e^{-kt}e^{-a}
\end{aligned}$$

Since C is a function of time t and e^{-a} is a constant, the last equation above may be written as

$$C(t) = C_0 e^{-kt}. \qquad \text{letting } C_0 = e^{-a}$$

Thus, the concentration C satisfies the law of exponential decay. •

Change of Base Any positive number except 1 may be used as the base of a system of logarithms. As has been mentioned, the commonly used bases are 10 and e. There are situations where it is necessary to change from one base to another. For example, since $1000 = 10^3$, $\log_{10} 1000 = 3$. To find $\ln 1000$, we write

$$\begin{aligned}
\ln 1000 &= \ln 10^3 \\
&= 3 \ln 10 \\
&\approx 3(2.303) = 6.909.
\end{aligned}$$

That is, ln 1000 is the product of ln 10 and $\log_{10}1000$. In general, if $x = \log_{10}N$, then

$$
\begin{aligned}
N &= 10^x, \\
\ln N &= \ln 10^x \\
&= x \ln 10 \\
&\approx 2.303x = 2.303 \log_{10}N.
\end{aligned}
$$

Thus, the natural logarithm of a number is obtained by multiplying the common logarithm of the number by ln 10 (approximately 2.303). This fact may be used to construct a table of natural logarithms from a table of common logarithms.

Similarly, suppose that x is the natural logarithm of a number N.

$$
\begin{aligned}
\ln N &= x \\
e^x &= N \\
\log_{10}e^x &= \log_{10}N \\
x \log_{10}e &= \log_{10}N \\
\log_{10}N &= x \log_{10}e \\
&\approx x(0.4343) \\
\log_{10}N &\approx (0.4343)\ln N
\end{aligned}
$$

So the common logarithm of any number is the product of its natural logarithm and $\log_{10}e$ (approximately 0.4343).

Notice that the natural logarithm of any number is larger (by a factor of 2.303, about 2 1/3) than the common logarithm. Conversely, the common logarithm is smaller than the natural logarithm (less than half as large). This is to be expected since $e < 10$ and must be raised to a higher power than 10 to yield N.

For the general case of changing base, let x and y be the logarithms of N to the respective bases a and b. That is,

$$
\begin{aligned}
x = \log_a N \quad &\text{and} \quad y = \log_b N, \\
a^x = N \quad &\text{and} \quad b^y = N,
\end{aligned}
$$

so that

$$a^x = b^y.$$

Take the logarithm of each side to the base a.

$$
\begin{aligned}
\log_a a^x &= \log_a b^y \\
x \log_a a &= y \log_a b \\
x &= y \log_a b \qquad \text{since } \log_a a = 1 \\
\log_a N &= (\log_b N) \log_a b
\end{aligned}
$$

Thus, we have the following theorem.

THEOREM 4.3

$$\boldsymbol{\log_a N = (\log_b N) \log_a b, \qquad \text{or} \qquad \log_b N = \frac{\log_a N}{\log_a b}.}$$

It is helpful to think of this as log N (new base) = log N (old base) divided by logarithm of new base relative to the old base.

EXAMPLE 7 Use base 10 logarithms to find (a) $\log_5 9$, (b) $\log_3 14$.

SOLUTION (a) $\log_5 9 = \dfrac{\log_{10} 9}{\log_{10} 5} \approx \dfrac{0.954}{0.699} \approx 1.365.$

(b) $\log_3 14 = \dfrac{\log_{10} 14}{\log_{10} 3} \approx \dfrac{1.146}{0.477} \approx 2.403.$ ●

EXAMPLE 8 Given $\log_{10} 3 \approx 0.477$, calculate ln 3.

SOLUTION Since $\ln 10 \approx 2.303$,

$$\begin{aligned} \ln 3 &\approx 2.303 \log_{10} 3 \\ &\approx 2.303(0.477) \\ &\approx 1.10 \end{aligned}$$ ●

EXAMPLE 9 Given $\ln 5 \approx 1.609$, calculate $\log_{10} 5$.

SOLUTION Since $\log_{10} e \approx 0.4343$,

$$\begin{aligned} \log_{10} 5 &\approx 0.4343 \ln 5 \\ &\approx 0.4343(1.609) \\ &\approx 0.699 \end{aligned}$$ ●

Approximating Polynomials for Exponential and Logarithmic Functions In Section 3.1 it was mentioned that polynomial functions are used to approximate more complicated functions. In calculus it is shown that better and better approximations of e^x are given by the successive polynomials,

$$\begin{aligned} &1, \\ &1 + x, \\ &1 + x + \frac{x^2}{2}, \\ &1 + x + \frac{x^2}{2} + \frac{x^3}{6}, \\ &1 + x + \frac{x^2}{2} + \frac{x^3}{6} + \frac{x^4}{24}, \end{aligned}$$

and so on. For instance, if $x = 1$, we obtain the following values approximating e from the above polynomials: 1.00, 2.00, 2.50, 2.67, 2.71. There are similar polynomial approximations for ln x. It is in this way that tables of logarithms and exponential function values are constructed. In Chapter 10 we will discuss the idea of extending these polynomials infinitely.

EXERCISES 4.5

Rewrite in exponential form and then solve for x.

1. $x = \log_{1/2} 4$ **2.** $\log_2 64 = x$ **3.** $\log_x 64 = 3$ **4.** $\log_4 x = 3$

5. $x = \log_{1/3} \frac{1}{9}$ **6.** $x = \log_9 27$ **7.** $\log_{1/2} x = -5$ **8.** $\log_x 10 = \frac{1}{2}$

9. $\log_x 0.1 = -1$ **10.** $x = \log_5 125$

Solve the following exponential and logarithmic equations.

11. $5^x = 10$ **12.** $3^{2x+1} = 5^x$ **13.** $\log_2 x^4 = 4$

14. $6^x = 24$ **15.** $3^{3x-1} = 2^{x+1}$ **16.** $\log_x 64 = 3$

17. $\log x = 2 \log 5 + 3 \log 4$ **18.** $\log x = 3 \log 5 - 2 \log 4$

19. $\log x^2 - \log \frac{2x}{5} = 2.64$ **20.** $\log 3x^2 - \log \frac{6x}{5} = 1.46$

21. $\log_2(x - 2) + \log_2 x = 3$ **22.** $\log_6(x + 9) + \log_6 x = 2$

Given the following common logarithms, calculate the corresponding natural logarithms.

23. $\log 4 = 0.602$ **24.** $\log 13 = 1.114$ **25.** $\log 250 = 2.398$

26. $\log 6 = 0.778$ **27.** $\log 22 = 1.342$ **28.** $\log 480 = 2.681$

Given the following natural logarithms, calculate the corresponding common logarithms.

29. $\ln 7 = 1.946$ **30.** $\ln 2 = 0.693$ **31.** $\ln 24 = 3.18$

32. $\ln 3 = 1.10$ **33.** $\ln 9 = 2.197$ **34.** $\ln 35 = 3.56$

Calculate each of the following.

35. $\log_5 10$ **36.** $\log_2 5$ **37.** $\log_4 6$ **38.** $\log_3 8$ **39.** $\log_6 12$ **40.** $\log_7 9$

41. Use the first four terms in the appropriate approximating polynomial to approximate (a) e^2; (b) e^{-1}.

42. Use the polynomial approximation $\ln(1 + x) = x - \frac{x^2}{2} + \frac{x^3}{3} - \frac{x^4}{4}$ to approximate

(a) ln 2; [*Hint:* Use $x = 1$ in the series.]
(b) ln 1.5.

Chapter 4 Review Exercises

Solve for x.

1. $2^x = 8$ **2.** $3^x = 12$ **3.** $3^{x^2+x} = 9$

4. $3^{x^2} = 81$ **5.** $2^{-4x} = 64$ **6.** $2^{3x} = 16$

7. $16^x = 8$ **8.** $10^{2x}(10^3) = 100{,}000$ **9.** $\log_2 x^4 = 4$

10. $x = \log_4 16$ **11.** $x = \log_{1/2} 2$ **12.** $e^{\ln x} = 2$

13. $\ln e^x = 1$ **14.** $\log_3(x + 1) = 2$

15. $\log_4 2x = \log_4 3 + \log_4 8$ **16.** $\log_3 x^2 = -2$

In each case, express x in terms of logarithms.

17. $A = P(1 + r)^x$ **18.** $A = Pe^{rx}$ **19.** $pv^x = c$ **20.** $q = q_0 e^{kx}$

Use the laws of logarithms to rewrite as sums or differences, without exponents.

21. $\log \dfrac{x^3y^2}{z^5}$ **22.** $\log x^3\sqrt{y}$ **23.** $\log z^2\sqrt{\dfrac{x}{y}}$ **24.** $\log \sqrt[3]{\dfrac{xy^2}{z}}$

25. $\log \sqrt{\dfrac{x^5}{x^2 - 1}}$

Use the laws of logarithms to write as single logarithms.

26. $\log x - \log z$ **27.** $\dfrac{1}{2}\log x + \dfrac{1}{2}\log y$

28. $3 \log x - 2 \log y$ **29.** $\log x + 3 \log y - 2 \log z$

30. $2 \log x + \dfrac{1}{2}\log y - 3 \log z$ **31.** $\log 3 + \log 4 - \log 1.5$

Graph on the same set of axes.

32. $y = 3^x$, $y = 3^{x+1}$, $y = 3^x + 1$ **33.** $y = \ln x$, $y = 1 - \ln x$, $y = \ln(x - 1)$

34. The bacterial count of a culture increases from 400 to 1000 in 5 hours. (a) Calculate the rate of growth and write the formula for the law of growth. (b) What is the count after 6 hours? [Assume the law of exponential growth applies.]

35. If an original sample of 10 grams of a radioactive substance having half-life 1000 years has decayed to 6 grams, how old is it?

36. Calculate the half-life T of the radioactive substance whose law of decay is given (t in years).
(a) Beryllium: $q(t) = q_0 e^{-(1.5\times10^{-7})t}$
(b) Carbon-14: $q(t) = q_0 e^{-(1.4\times10^{-4})t}$
(c) Strontium-90: $q(t) = q_0 e^{-(2.8\times10^{-2})t}$

37. Use the formula $A(t) = Pe^{rt}$ to calculate the time required to double an amount invested at interest compounded continuously if (a) $r = 4\%$, (b) $r = 6\%$, (c) $r = 7\%$, (d) $r = 10\%$.

38. How much must be invested at 7% compounded continuously to yield \$1000 in 5 years?

If $f(x) = e^{-x}$, $g(x) = \ln x$, and $h(x) = \dfrac{1}{x}$, *write (and simplify) each of the following:*

39. $f[g(x)]$ **40.** $g[f(x)]$ **41.** $f[h(x)]$ **42.** $h[f(x)]$ **43.** $g[h(x)]$ **44.** $h[g(x)]$

Chapter 4 Miscellaneous Exercises

1. If $f(x) = a^x$, show that $f(x + 1) = af(x)$, $f(x + 2) = a^2f(x)$, and in general, $f(x + k) = a^kf(x)$.

2. Suppose a ball is dropped from a height of 20 feet and repeatedly bounces back one-half the distance it has just dropped. Thus, if the original height is h and $f(n)$ represents the height reached in the nth bounce, then $f(1) = \frac{1}{2}h$, $f(2) = \frac{1}{2}\left(\frac{1}{2}h\right) = h\left(\frac{1}{2}\right)^2$, and so on, so that $f(n) = h\left(\frac{1}{2}\right)^n$. Use this formula to calculate the height reached on the tenth bounce.

3. The population of California in 1920 was 3,476,000 and in 1970 it was 19,950,000. (a) Assuming the exponential law of growth, what was the equivalent annual rate of growth for this half-century? (b) Assuming this same rate of growth, what will the population of California be in the year 2000?

The half-life of several radioactive substances is given below. For each of them determine the time required to reduce the mass to (a) one-fourth the initial mass and (b) one-fifth the initial mass. [Hint: In each case, write $q = q_0e^{kt}$ and use the half-life to determine the rate of decay k].

4. Carbon-14, 5600 years
5. Uranium-238, 4.51×10^9 years
6. Uranium-224, 3.64 days
7. Plutonium, 36 minutes

8. Suppose a tank contains 1000 gallons of sea water. The salinity (ratio of salt to water) is reduced by running fresh water into the tank at the rate of 10 gallons per minute and letting the mixture run out at the same rate. If the salty water is kept uniformly mixed, it can be shown that the salinity s after t minutes is given by $s(t) = s_0e^{-0.01t}$, where s_0 is the initial salinity of the sea water. (a) After one hour the salinity is what fraction of the original salinity? (b) How much time is required to reduce the salinity to one-half its initial value?

9. In a certain electric circuit, when a switch is opened the current strength i decreases exponentially according to the law $i(t) = i_0e^{(-R/L)t}$, where i_0, R, and L are constants depending upon the particular circuit. Suppose that $i_0 = 5$, $R = 1.2$, and $L = 1$, and calculate the current i at the following times. (a) $t = 0$, (b) $t = 1$, (c) $t = 10$.

10. Newton's **law of cooling** states that the rate of change of the temperature difference between an object and its surrounding medium is proportional to the temperature difference. In calculus it is shown that this fact is expressed by the exponential law of decay, $D(t) = D_0e^{-kt}$, $k > 0$, where t is the time in minutes and D is the temperature difference. Suppose that a hot piece of metal has a temperature of 120° and the surrounding air has a temperature of 80°. If the metal cools to 100° in 15 minutes, what is its temperature at the end of (a) 30 minutes? (b) 1 hour?

11. A certain savings institution advertises that money invested at 8% compounded daily is very nearly equivalent to a simple interest rate of 8 1/3%. Show that this is true by calculating the amount accumulated by investing one dollar for one year at 8% compounded daily.

12. In the compound interest laws $A_n = P(1 + r)^n$ and $A = Pe^{rt}$, the value of P is called the **present value** of the investment. It is the amount that must be invested now in order to accumulate to the amount A in n (or t) years. Calculate the present value of $1000 in 10 years if interest of 8% is compounded (a) annually; (b) continuously.

Graph each of the following. [*Hint: Recall translation of graphs, section 3.7.*]

13. $y = 1 + \ln x$ **14.** $y = \ln(x + 1)$ **15.** $y = \ln(x - 1)$

Graph each of the following. [*Hint: First use the laws of logarithms.*]

16. $y = \ln \sqrt{x}$ **17.** $y = \ln \dfrac{1}{x}$

Write an expression for $f[g(x)]$ for the following functions.

18. $f(x) = e^x$ and $g(x) = \dfrac{1}{2}x$ **19.** $f(x) = \ln x$ and $g(x) = x^2$

20. $f(x) = \ln x$ and $g(x) = e^x$ **21.** $f(x) = e^x$ and $g(x) = \ln x$

22. An important formula from chemistry is $pH = -\log_{10}[H+]$, where pH is the hydrogen potential of a solution and $[H+]$ is the value of the concentration of hydrogen ions in solution. Calculate pH if $[H+]$ is (a) 4.2×10^{-6}, (b) 0.6×10^{-7}.

It is often useful to write logarithmic versions of formulas. The following are some examples.

23. The area K of a triangle with sides of lengths a, b, and c, respectively, is $K = \sqrt{s(s - a)(s - b)(s - c)}$, where s is the semiperimeter, $s = \dfrac{1}{2}(a + b + c)$. Write a simple logarithmic formula for $\log K$.

24. The radius r of a circle inscribed in a triangle whose sides have lengths a, b, and c, respectively, is given by $r = \sqrt{\dfrac{1}{s}(s - a)(s - b)(s - c)}$. Write a simple logarithmic formula for $\log r$.

25. The ideal gas law is $pv = \dfrac{RT}{m}$, where p is pressure, v is volume, and R, T, and m are constants depending upon the particular gas. Write a simple logarithmic formula for $\log p$.

The following formulas involving temperatures and pressures appear in certain studies of the atmosphere:

(1) $$\frac{T}{T_0} = \left(\frac{p}{p_0}\right)^{(k-1)/k}$$

(2) $$\log p - \log p_0 = \frac{1}{lR}[\log(T_0 - lz) - \log T_0].$$

26. Write the first formula in logarithmic form and then use it to solve for k.

27. Write the second formula in exponential form and then use it to solve for p.

28. Solve the compound-interest law $A = P(1 + r)^t$ (a) for t; (b) for r.

29. A certain gas law is $pv^n = c$, where p is pressure, v is volume, and c is a constant. Solve for v.

30. The current i in a certain electric circuit is given by $i = \dfrac{E}{R}(1 - e^{-(R/L)t})$, where E is voltage, R is resistance, L is inductance, and t is time. Solve for t.

In calculating the time T required for money to double at compound interest, we found $e^{rT} = 2$. Solve for T as a function of r and use the formula to calculate the time T corresponding to the following interest rates.

31. $r = 5\%$ **32.** $r = 6\%$ **33.** $r = 7\%$ **34.** $r = 10\%$

Given the exponential law of decay $q(t) = q_0 e^{-kt}$, $k > 0$, show the following.

35. The half-life T is given by $T = (1/k)\ln 2$.

36. The law of decay may be written $q(t) = q_0(2)^{-t/T}$.

37. Derive the general compound interest formula $A_n = P(1 + r)^n$.

[*Hint:* $A_1 = P + Pr = P(1 + r)$.]

38. Use the approximating polynomials for e^x (Section 4.5) to approximate $\sqrt{e}$ by using $x = \frac{1}{2}$. Calculate at least five successive approximations.

39. If a switch in a certain kind of electrical circuit is opened, the current increases according to the law

$$i = \frac{E}{R}(1 - e^{-(R/L)t}),$$

where i is the current at time t, and E, R, and L are constants determined by the particular circuit. Take $E = 12$, $R = 3$, and $L = 1$. (a) What is the initial current value? (b) After how many seconds has the current increased to 3 units? (c) The current increases asymptotically toward 4 units. Explain.

In calculus it is found that a very important combination of the exponential functions is $f(x) = \frac{1}{2}(e^x - e^{-x})$. Use the tables to calculate the following.

40. $f(0)$

41. $f(1)$

42. $f(2.5)$

43. $f(-1)$

44. $f(-2)$

45. If $f(x) = \frac{1}{2}(e^x - e^{-x})$, calculate $f^{-1}(x)$. [*Hint:* Let $y = \frac{1}{2}(e^x - e^{-x})$, so that $2y = e^x - e^{-x}$, $2ye^x = e^{2x} - 1$, or $(e^x)^2 - 2y(e^x) - 1 = 0$. Now use the quadratic formula to solve for e^x and then for x.]

5

The Circular Functions

The exponential and logarithmic functions studied in the preceding chapter are related to the natural law of growth and similar phenomena. Another class of transcendental (nonalgebraic) functions, the *circular functions,* are considered in this chapter. Because these functions are periodic, they are used to study many phenomena which vary cyclically. Among them are sound waves, electric currents, oscillatory motions, vibrating strings, and various biological processes. They are called circular functions because of their relationship to the circle.

5.1 The Sine and Cosine Functions

In Section 1.1 a one-to-one correspondence between the real numbers and the points on a coordinate line was described. Now imagine the entire coordinate line ''wrapped'' around a circle of radius 1 centered at the origin of a coordinate plane (see Figure 5.1).

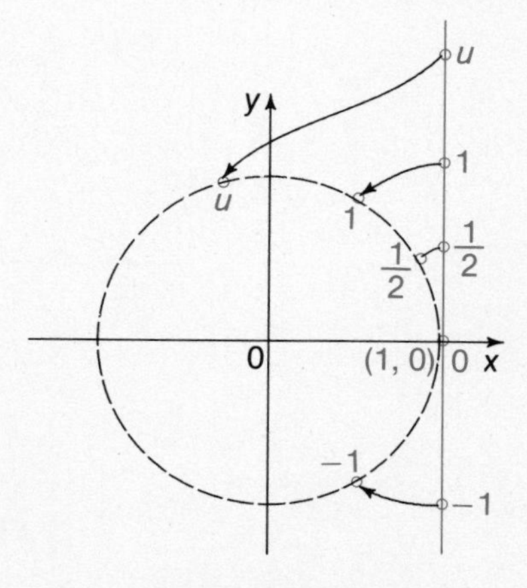

Figure 5.1

This circle is called the **standard unit circle** or, briefly, the **unit circle.** The origin of the coordinate line is placed at the point (1, 0) on the circle, and the positive part of the number line is wrapped around the circle in a counterclockwise direction. Just as the distance between two points on a line is the length of the segment between them, the distance between two points on a circle, measured along the circle, is the length of the arc between them, or the **arc length.** By means of the correspondence described above, a real number u on the number line is mapped to a point P with coordinates (x, y) on the unit circle. The distance from 0 to u then corresponds to the arc length from (1, 0) to (x, y). See Figure 5.2.

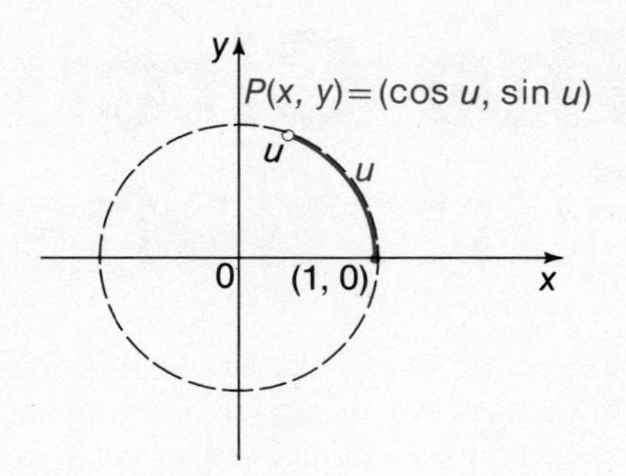

Figure 5.2

The above mapping establishes two important correspondences or functions: $u \to x$ and $u \to y$. The first correspondence is called the **cosine** function (abbreviated **cos**) and the second is the **sine** function (abbreviated **sin**).

DEFINITION 5.1 If u is the arc length from (1, 0) on the standard unit circle counterclockwise to the point $P(x, y)$, then

$$\cos u = x \qquad \text{and} \qquad \sin u = y.$$

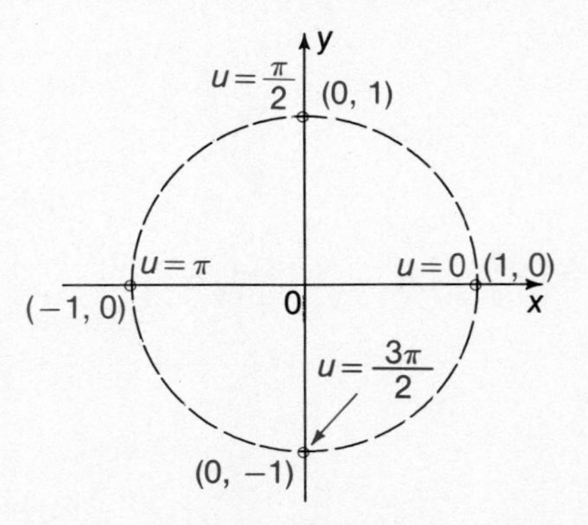

Figure 5.3

These functions are called **circular functions** and their relation to the unit circle is shown in Figure 5.2. They are real functions with domain the set of all real numbers u. Since the equation of the unit circle is $x^2 + y^2 = 1$ (Section 1.4), $|x| \leq 1$ and $|y| \leq 1$. Hence the range of each function is the closed interval $[-1, 1]$, that is, $-1 \leq \cos u \leq 1$ and $-1 \leq \sin u \leq 1$.

Many values of these functions are easily obtained as follows by means of elementary geometry. Cos u is the x-coordinate and sin u is the y-coordinate of the point of the unit circle corresponding to the arc length u. Since the arc length of the entire circle (the circumference) is 2π, the following correspondences may be seen from Figure 5.3.

u	$P(x, y)$	$\cos u$	$\sin u$
0	$(1, 0)$	1	0
$\frac{\pi}{2}$	$(0, 1)$	0	1
π	$(-1, 0)$	-1	0
$\frac{3\pi}{2}$	$(0, -1)$	0	-1
2π	$(1, 0)$	1	0

These are written $\cos 0 = 1$, $\cos \pi/2 = 0$, $\cos \pi = -1$, and so on, and $\sin 0 = 0$, $\sin \pi/2 = 1$, $\sin \pi = 0$, and so on. All these values for u are multiples of $\pi/2$. They are called **quadrantal values** of the variable u.

Circular function values for multiples of $\pi/6$ and $\pi/4$ are determined next. An arc that is one-fourth of a unit circle has length $\pi/2$, so the midpoint of the first-quadrant arc of the standard unit circle corresponds to $u = \pi/4$. (See Figure 5.4.) Since the x-coordinate and y-coordinate of this point are equal, x and y are found as follows.

$$\begin{aligned} x^2 + y^2 &= 1 \\ x^2 + x^2 &= 1 \\ 2x^2 &= 1 \\ x^2 &= \frac{1}{2} \\ x &= \frac{1}{\sqrt{2}} = \frac{\sqrt{2}}{2} \end{aligned}$$

Since $y = x$, $y = \sqrt{2}/2$, and thus

$$\cos \frac{\pi}{4} = x = \frac{\sqrt{2}}{2} \qquad \text{and} \qquad \sin \frac{\pi}{4} = y = \frac{\sqrt{2}}{2}.$$

By symmetry (See Figure 5.4),

$$\sin\frac{3\pi}{4} = \frac{\sqrt{2}}{2} \quad \text{and} \quad \cos\frac{3\pi}{4} = -\frac{\sqrt{2}}{2},$$

$$\sin\frac{5\pi}{4} = \cos\frac{5\pi}{4} = -\frac{\sqrt{2}}{2},$$

$$\sin\frac{7\pi}{4} = -\frac{\sqrt{2}}{2} \quad \text{and} \quad \cos\frac{7\pi}{4} = \frac{\sqrt{2}}{2}.$$

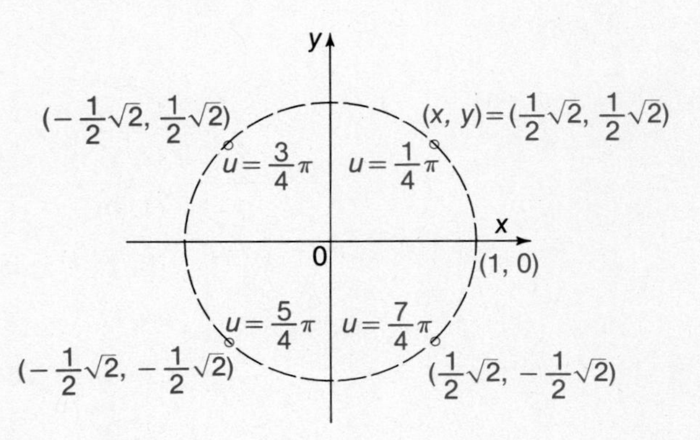

Figure 5.4

To see how to obtain circular function values for multiples of $\pi/6$, refer to Figure 5.5. The number $u = \pi/6$ is at a point one-third of the way along the first-quadrant arc from (1, 0) because $\pi/6 = (1/3)(\pi/2)$. Call this point $P(x, y)$. The point R on the circle corresponding to $-\pi/6$ has cartesian coordinates $(x, -y)$. (Why?) Now arc QP = arc PR, since arc $QP = \pi/2 - \pi/6 = \pi/3$ and arc $PR = \pi/6 + \pi/6 = \pi/3$. Then segment length QP = segment length PR, that is, by the distance formula,

$$(x - 0)^2 + (y - 1)^2 = (2y)^2$$

or

$$x^2 + y^2 - 2y + 1 = 4y^2.$$

Since (x, y) is on the standard unit circle, $x^2 + y^2 = 1$. Substituting 1 for $x^2 + y^2$ in the last equation gives

$$1 - 2y + 1 = 4y^2$$
$$4y^2 + 2y - 2 = 0$$
$$2y^2 + y - 1 = 0$$
$$(2y - 1)(y + 1) = 0$$
$$y = \frac{1}{2} \quad \text{or} \quad y = -1.$$

Since $y > 0$ (from Figure 5.5), take $y = 1/2$. Substituting this value of y into the equation $x^2 + y^2 = 1$ yields $x = \sqrt{3}/2$. Hence

$$\cos\frac{\pi}{6} = x = \frac{1}{2}\sqrt{3} \quad \text{and} \quad \sin\frac{\pi}{6} = y = \frac{1}{2}.$$

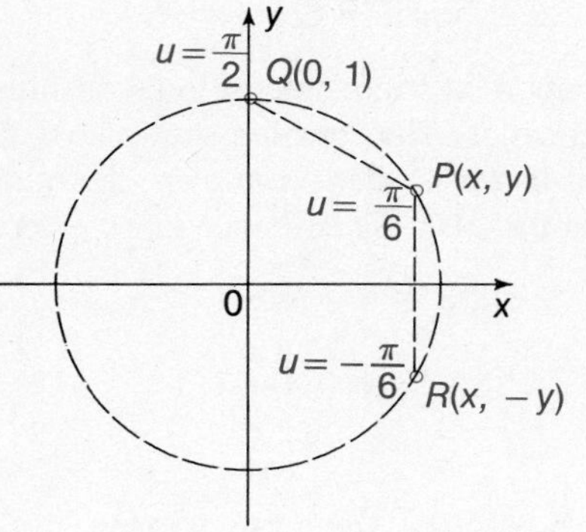

Figure 5.5

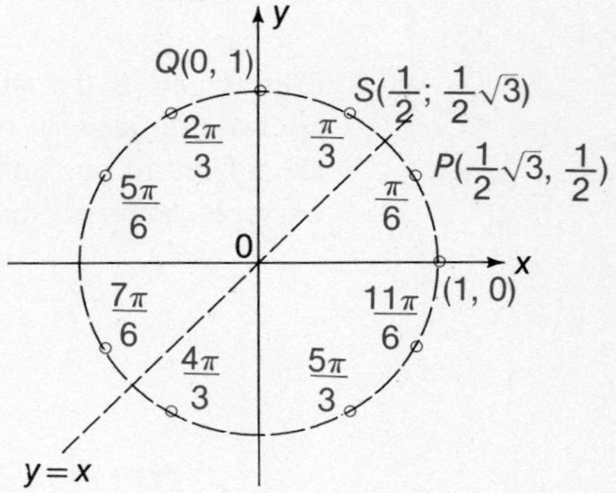

Figure 5.6

Now refer to Figure 5.6, where the point S corresponds to $\pi/3$. Then arc QS has length $\pi/2 - \pi/3 = \pi/6$, so the points P and S are symmetric with respect to the line $y = x$. Hence (Section 1.4) the coordinates of S are obtained by simply interchanging those of P, that is, S is the point $(1/2, \sqrt{3}/2)$. Then

$$\cos\frac{\pi}{3} = \frac{1}{2} \quad \text{and} \quad \sin\frac{\pi}{3} = \frac{\sqrt{3}}{2}.$$

By symmetry, the sine and cosine values for the other multiples of $\pi/6$, that is, $2\pi/3$, $5\pi/6$, $7\pi/6$, and so on, can be read from Figure 5.6.

The preceding results are summarized for the first quadrant in Figure 5.7 and the accompanying table. The corresponding function values for all multiples of $\pi/6$ and $\pi/4$ may then be read from the unit circle by symmetry.

u	$\sin u$	$\cos u$
0	0	1
$\frac{\pi}{6}$	$\frac{1}{2}$	$\frac{1}{2}\sqrt{3}$
$\frac{\pi}{4}$	$\frac{1}{2}\sqrt{2}$	$\frac{1}{2}\sqrt{2}$
$\frac{\pi}{3}$	$\frac{1}{2}\sqrt{3}$	$\frac{1}{2}$
$\frac{\pi}{2}$	1	0

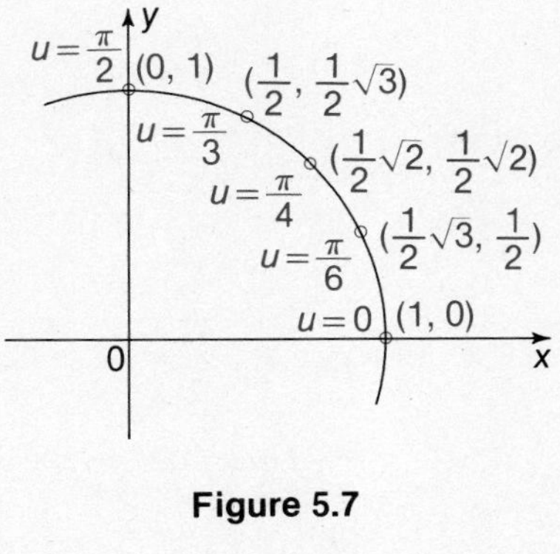

Figure 5.7

A Fundamental Identity If u is any real number then, $(\sin u)^2 + (\cos u)^2 = y^2 + x^2 = 1$. When $(\sin u)^2$ and $(\cos u)^2$ are written as $\sin^2 u$ and $\cos^2 u$, as is customary, we have the basic identity

$$\sin^2 u + \cos^2 u = 1.$$

This is one of the most important trigonometric relationships and appears in various contexts throughout mathematics. The equation shows how the sine and cosine values are related to one another. If one of them is known, the other can be calculated. For example, suppose that $\cos u = 4/5$. Then, since $\sin^2 u + \cos^2 u = 1$,

$$\begin{aligned} \sin^2 u &= 1 - \cos^2 u \\ &= 1 - \left(\frac{4}{5}\right)^2 \\ &= 1 - \frac{16}{25} = \frac{9}{25}, \\ \sin u &= \pm\frac{3}{5}. \end{aligned}$$

The positive sign is chosen if $0 < u < \pi$ and the negative sign if $\pi < u < 2\pi$. (Why?) The following formulas give each function in terms of (as a function of) the other. Of course, *the appropriate choice of sign must be made in each case.*

$$\sin u = \pm\sqrt{1 - \cos^2 u}$$
$$\cos u = \pm\sqrt{1 - \sin^2 u}$$

EXERCISES 5.1

1. Refer to Figure 5.7 and the accompanying table to draw a similar figure and construct the corresponding table for u in the (a) second quadrant; (b) third quadrant; (c) fourth quadrant.

Draw the standard unit circle, label it appropriately, and then read from it the following function values. Write them in their fractional-radical form, as in the text examples.

2. $\sin \frac{\pi}{6}$ **3.** $\cos \frac{1}{4}\pi$ **4.** $\sin \frac{\pi}{3}$ **5.** $\cos \frac{7\pi}{3}$

6. $\sin\left(-\frac{1}{2}\pi\right)$ **7.** $\cos\left(-\frac{1}{4}\pi\right)$ **8.** $\sin \frac{5\pi}{6}$ **9.** $\cos \frac{3\pi}{4}$

10. $\cos \frac{7\pi}{6}$ **11.** $\sin \frac{1}{4}\pi$ **12.** $\sin\left(-\frac{3\pi}{2}\right)$ **13.** $\cos \frac{1}{2}\pi$

14. $\sin \frac{7\pi}{2}$ **15.** $\cos \frac{11\pi}{3}$ **16.** $\cos(-\pi)$

Find the quadrant(s) containing u for the following circular function values of u.

17. $\sin u = \frac{1}{2}$ **18.** $\cos u = \frac{3}{4}$ **19.** $\sin u = -\frac{7}{8}$ **20.** $\cos u = -\frac{4}{5}$

21. $\sin u = \frac{1}{2}$ and $\cos u = \frac{1}{2}\sqrt{3}$

22. $\sin u = \frac{1}{2}\sqrt{3}$ and $\cos u = -\frac{1}{2}$

23. $\sin u = -\frac{3}{4}$ and $\cos u = -\frac{1}{4}\sqrt{7}$

24. $\sin u = -\frac{2}{3}$ and $\cos u = \frac{1}{3}\sqrt{5}$

Find the value of u if $0 \le u \le \pi$.

25. $\sin u = \frac{1}{2}$

26. $\cos u = -\frac{1}{2}$

27. $\cos u = \frac{1}{2}\sqrt{3}$

28. $\sin u = -\frac{1}{2}\sqrt{3}$

29. $\cos u = \frac{1}{2}\sqrt{2}$

30. $\cos u = -1$

31. $\sin u = 0$

32. $\sin u = 1$

33. $\cos u = 0$

34. $\sin u = -\frac{1}{2}\sqrt{2}$

Give two answers for each of the following.

35. What is $\sin u$ if $\cos u = 1/3$?

36. What is $\sin u$ if $\cos u = -3/4$?

37. What is $\cos u$ if $\sin u = 2/3$?

38. What is $\cos u$ if $\sin u = -3/5$?

5.2 Other Circular Functions

The sine and cosine functions are the basic circular functions. Other functions may be constructed from them in a variety of ways. (Recall Section 2.4.) Four quotients of these functions are given special names: **tangent, cotangent, secant,** and **cosecant.** They are abbreviated **tan, cot, sec,** and **csc,** and defined as follows.

DEFINITION 5.2

$$\tan u = \frac{\sin u}{\cos u} \qquad \sec u = \frac{1}{\cos u}, \quad \cos u \neq 0$$

$$\cot u = \frac{1}{\tan u} = \frac{\cos u}{\sin u} \qquad \csc u = \frac{1}{\sin u}, \quad \sin u \neq 0$$

Note that while the sine and cosine are defined for all real numbers, the tangent and secant functions are undefined for odd multiples of $\pi/2$ and the cotangent and cosecant are not defined for multiples of π.

EXAMPLE 1 Calculate (a) $\tan \pi/6$, (b) $\sec \pi/4$, (c) $\cot \pi$.

SOLUTION (a) $\tan \frac{\pi}{6} = \frac{\sin \pi/6}{\cos \pi/6} = \frac{\frac{1}{2}}{\frac{1}{2}\sqrt{3}} = \frac{1}{\sqrt{3}} = \frac{1}{3}\sqrt{3}$

(b) $\sec \dfrac{\pi}{4} = \dfrac{1}{\cos \pi/4} = \dfrac{1}{\frac{1}{2}\sqrt{2}} = \dfrac{2}{\sqrt{2}} = \sqrt{2}$

(c) $\cot \pi = \dfrac{\cos \pi}{\sin \pi} = \dfrac{-1}{0}$, which is meaningless, so $\cot \pi$ is undefined. ●

The six functions—sine, cosine, tangent, cotangent, secant, and cosecant—are referred to as the **elementary circular functions.** Each of them may be expressed directly in terms of the cartesian coordinates of the reference point (x, y) corresponding to u.

$$\sin u = y \qquad \cot u = \frac{x}{y}, \quad y \neq 0$$

$$\cos u = x \qquad \sec u = \frac{1}{x}, \quad x \neq 0$$

$$\tan u = \frac{y}{x}, \quad x \neq 0 \qquad \csc u = \frac{1}{y}, \quad y \neq 0$$

The last three definitions given at the beginning of this section may be written equivalently as

$$\sin u \cdot \csc u = 1,$$
$$\cos u \cdot \sec u = 1,$$
$$\tan u \cdot \cot u = 1.$$

The two functions of each pair are thus reciprocals of each other.

The term *circular* function comes from the way these functions are defined. Historically, they arose in a different context and because of it were called "trigonometric" functions. The reason for this will appear later (Section 6.2). It is now common practice to use the terms *circular function* and *trigonometric function* interchangeably.

We have been using the letter u for the argument of the circular functions. In that context, the letters x and y were reserved for cartesian coordinates on the unit circle. However, as we noted in Section 2.1, the symbol used to designate the argument of a function is immaterial. In what follows, we may use x, y, or any other letter and write $\sin x$, $\sin y$, and so on. The statement "$\sin^2 x + \cos^2 x = 1$" means the same as "$\sin^2 u + \cos^2 u = 1$", for example.

Signs of Function Values Sin u and cos u are coordinates of points on the unit circle and so they may be positive, negative, or zero. The zero function values occur only for quadrantal values of the variable u. By definition, $\sin u > 0$ in quadrants I and II and $\sin u < 0$ in quadrants III and IV. Similar remarks apply to the cosine function. These results are summarized in Figure 5.8. There are corresponding results for each of the other circular functions. Some of them are left to the exercises.

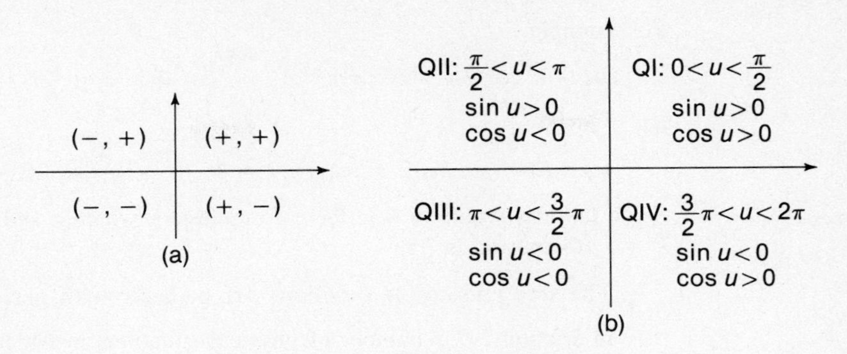

Figure 5.8

Odd and Even Functions From Figure 5.9 we see that $\sin(-u) = -\sin u$ and $\cos(-u) = \cos u$. For example, $\sin(-\pi/6) = -\sin \pi/6 = -1/2$ and $\cos(-\pi/6) = \cos \pi/6 = \sqrt{3}/2$.

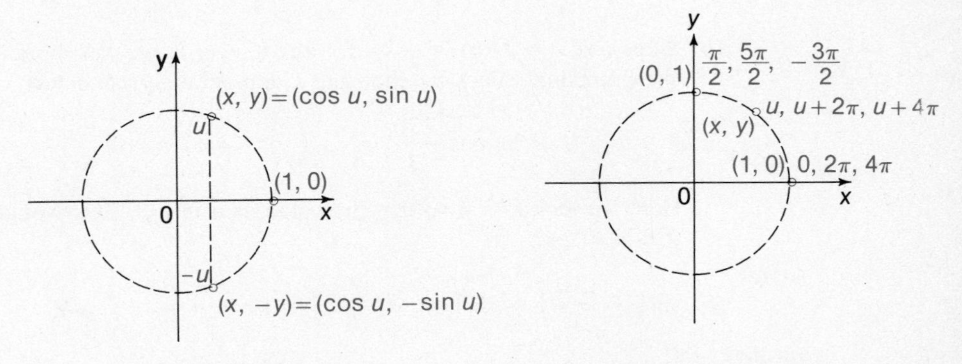

Figure 5.9 **Figure 5.10**

> **The sine is an odd function and the cosine is an even function.**

Using the definitions of the other circular functions, we can show that

> **the tangent, cotangent, and cosecant functions are odd and the secant function is even.**

Periodicity This is perhaps the most significant property of the circular functions. (If necessary, review the discussion of this concept in Section 2.6.) When the number line is "wrapped" around the unit circle infinitely often in each direction, the interval $0 \le u \le 2\pi$ is mapped exactly into the complete circle, and $u = 2\pi$ falls on the point (1, 0) along with $u = 0$. So also do $u = 4\pi, 6\pi, -2\pi, -4\pi$, and every even multiple of π. Similarly, $u = \pi/2, 5\pi/2, -3\pi/2, -7\pi/2$, all fall on the point (0, 1).

In general, for any real number u the numbers $u + 2k\pi$ $(k = \pm 1, \pm 2, \pm 3, \ldots)$ all correspond to the same point on the circle. (See Figure 5.10 above.)

For example,

$$\cos 0 = \cos 2\pi = \cos 4\pi = 1, \qquad \cos \pi = \cos(-\pi) = \cos(-3\pi) = -1,$$

and

$$\sin \pi/2 = \sin 5\pi/2 = \sin 9\pi/2 = 1, \qquad \sin(-3\pi/2) = \sin(-7\pi/2) = 1.$$

Since the least value of k in these formulas for which x and y repeat their values is $k = 1$, the period is 2π.

The sine and cosine functions are periodic with period 2π.

In Section 5.9 a number of physical situations in which this periodicity is extremely significant will be discussed. Here it will be used to obtain function values for $u > 2\pi$ or $u < 0$ in terms of their values for $0 < u < 2\pi$.

EXAMPLE 2 Find (a) $\sin 11\pi/4$; (b) $\cos 25\pi/3$; (c) $\sin(-13\pi/3)$.

SOLUTION (a) $\sin \dfrac{11\pi}{4} = \sin\left(\dfrac{3\pi + 8\pi}{4}\right) = \sin\left(\dfrac{3\pi}{4} + 2\pi\right) = \sin \dfrac{3\pi}{4} = \dfrac{1}{2}\sqrt{2}$

(b) Since $\cos(u + 2k\pi) = \cos u$ for any integer k, we can shorten the calculation of such function values by choosing k appropriately, as below.

$$\cos \frac{25\pi}{3} = \cos\left(\frac{\pi}{3} + 8\pi\right) = \cos \frac{\pi}{3} = \frac{1}{2}$$

Here we took $k = 4$ so that the value of u in $u + 2k\pi$ would be in the interval $0 \le u < 2\pi$.

(c) $\sin\left(-\dfrac{13\pi}{3}\right) = \sin\left(6\pi - \dfrac{13\pi}{3}\right) = \sin \dfrac{5\pi}{3} = -\dfrac{1}{2}\sqrt{3}.$

Note that the appropriate multiple of 2π, namely 6π, was added to $-13\pi/3$ so that the resulting argument would be in the interval $(0, 2\pi)$. ●

The periodicity of the sine and cosine functions can be used with the definitions of the other circular functions to show their periodicity. For example, we will show that the tangent function has period π. If u corresponds to (x, y) on the unit circle, then $\tan u = y/x$. Since $u + \pi$ is symmetric to u with respect to the origin,

$$\tan(u + \pi) = \frac{-y}{-x} = \frac{y}{x} = \tan u.$$

Since the other functions are reciprocals of the sine, cosine, and tangent, they have the same respective periods.

EXAMPLE 3 Find (a) $\tan 5\pi/4$; (b) $\cot 5\pi/2$; (c) $\sec 13\pi/3$; (d) $\csc 17\pi/6$.

SOLUTION (a) $\tan \dfrac{5\pi}{4} = \tan\left(\pi + \dfrac{\pi}{4}\right) = \tan \dfrac{\pi}{4} = 1$

(b) $\cot \frac{5\pi}{2} = \cot\left(2\pi + \frac{\pi}{2}\right) = \cot\left(\pi + \frac{\pi}{2}\right) = \cot \frac{\pi}{2} = 0$

(c) $\sec \frac{13\pi}{3} = \sec\left(4\pi + \frac{\pi}{3}\right) = \sec \frac{\pi}{3} = 2$

(d) $\csc \frac{17\pi}{6} = \csc\left(2\pi + \frac{5\pi}{6}\right) = \csc \frac{5\pi}{6} = 2.$ ●

In Section 5.1 the definitions of the sine and cosine functions were used to derive exact function values for special values of the argument. In this section these values and the definitions were used to evaluate the other circular functions for the same values of the argument. In the following sections formulas will be developed which may be used to determine other special circular function values. However, in general, circular function values cannot be calculated so directly (see Section 5.12), and thus are listed in tables for reference. Such tables, together with instructions for using them, are found in the Appendix.

EXERCISES 5.2

1. Copy and fill in the following table.

u	$\sin u$	$\cos u$	$\tan u$	$\cot u$	$\sec u$	$\csc u$
0						
$\frac{\pi}{6}$						
$\frac{\pi}{4}$						
$\frac{\pi}{3}$						
$\frac{\pi}{2}$						

Give the indicated function values (if defined).

2. $\tan \frac{\pi}{3}$ **3.** $\sec \frac{\pi}{6}$ **4.** $\cot 0$ **5.** $\tan \frac{9\pi}{4}$ **6.** $\csc \frac{3\pi}{2}$

7. $\cot \frac{\pi}{3}$ **8.** $\tan \frac{7\pi}{6}$ **9.** $\tan \frac{3\pi}{4}$ **10.** $\sec \frac{2\pi}{3}$

Give the value of u in the interval $0 \le u \le \pi$.

11. $\tan u = 1$ **12.** $\cot u = \sqrt{3}$ **13.** $\sec u = 2$

14. $\csc u = 1$ **15.** $\tan u = 0$ **16.** $\sec u = -\sqrt{2}$

17. $\cot u = -\sqrt{3}$ **18.** $\csc u = 2$

Name three real numbers to which the following apply and explain.

19. The tangent function is not defined.

20. The cotangent function is not defined.

21. The secant function is not defined.

22. The cosecant function is not defined.

Use periodicity, as in the text examples, to find the following.

23. $\sin \dfrac{17\pi}{6}$ **24.** $\sin \dfrac{8\pi}{3}$ **25.** $\cos \dfrac{9\pi}{4}$ **26.** $\cos \dfrac{11\pi}{4}$ **27.** $\sin \dfrac{25\pi}{6}$

28. $\cos \dfrac{16\pi}{3}$ **29.** $\tan \dfrac{7\pi}{3}$ **30.** $\tan \dfrac{7\pi}{4}$ **31.** $\cot \dfrac{13\pi}{4}$ **32.** $\cot \dfrac{13\pi}{6}$

33. Show that the secant function is even.

Show that each of the following is an odd function.

34. tangent **35.** cotangent **36.** cosecant

5.3 Fundamental Identities

The statement $\sin^2 u + \cos^2 u = 1$ is called an **identity** (identical equality); it is true for *all* real values of the argument u. From it several other important identities are derived.

$$\sin^2 u + \cos^2 u = 1$$

$$\frac{\sin^2 u}{\cos^2 u} + \frac{\cos^2 u}{\cos^2 u} = \frac{1}{\cos^2 u}, \qquad \cos u \neq 0$$

$$\mathbf{\tan^2 u + 1 = \sec^2 u}$$

The last statement is true for all values of u for which the tangent and secant functions are defined, that is, when u is not an odd multiple of $\pi/2$. Similarly,

$$\mathbf{\cot^2 u + 1 = \csc^2 u}$$

for all values of u except multiples of π.

We will refer to the following as the **fundamental identities** for the circular functions:

(1) $\tan x = \dfrac{\sin x}{\cos x}$ **(2)** $\cot x = \dfrac{\cos x}{\sin x}$

(3) $\sin x \csc x = 1$ **(4)** $\cos x \sec x = 1$

(5) $\tan x \cot x = 1$ **(6)** $\sin^2 x + \cos^2 x = 1$

(7) $\tan^2 x + 1 = \sec^2 x$ **(8)** $\cot^2 x + 1 = \csc^2 x$

These are the basic tools for developing an unlimited variety of other identities and should be memorized. Each of the last three identities in the above list has three equivalent forms:

(6) $\sin^2 x + \cos^2 x = 1,\quad \sin^2 x = 1 - \cos^2 x,\quad \cos^2 x = 1 - \sin^2 x$

(7) $\tan^2 x + 1 = \sec^2 x,\quad \tan^2 x = \sec^2 x - 1,\quad \sec^2 x - \tan^2 x = 1$

(8) $\cot^2 x + 1 = \csc^2 x,\quad \cot^2 x = \csc^2 x - 1,\quad \csc^2 x - \cot^2 x = 1.$

These identities are used when it is necessary to change an expression in the circular functions into an *equivalent* one that is more convenient for the purpose at hand. This is illustrated in the following examples. The tools used are the fundamental identities and the usual rules of algebra.

EXAMPLE 1 Show that $\cos u + \tan u \sin u = \sec u$.

SOLUTION Since the left side is more complicated, we rewrite it using the fact that all the circular functions may be expressed in terms of the sine and cosine.

$$\begin{aligned}
\cos u + \tan u \sin u &= \cos u + \frac{\sin u}{\cos u} \cdot \sin u && \text{by (1)}\\
&= \frac{\cos^2 u}{\cos u} + \frac{\sin^2 u}{\cos u}\\
&= \frac{\cos^2 u + \sin^2 u}{\cos u}\\
&= \frac{1}{\cos u} && \text{by (6)}\\
&= \sec u && \text{by (4).}
\end{aligned}$$

●

EXAMPLE 2 Prove the identity $\dfrac{\tan x}{1 + \tan^2 x} = \sin x \cos x$.

SOLUTION Operate on the left side of the identity.

$$\begin{aligned}
\frac{\tan x}{1 + \tan^2 x} &= \frac{\tan x}{\sec^2 x}\\
&= \tan x\left(\frac{1}{\sec^2 x}\right)\\
&= \frac{\sin x}{\cos x} \cdot \cos^2 x\\
&= \sin x \cos x.
\end{aligned}$$

●

Note the use of $\tan^2 x + 1 = \sec^2 x$ as a shortcut in Example 2. It would have been more complicated to rewrite all functions in terms of sines and cosines. This is often the case if squares of functions are involved. In such cases, look at fundamental identities (6) through (8).

The algebraic operations together with the fundamental trigonometric identities are used in proving identities. It is usually best to work on only one side, commonly the more complicated looking one. Look at the other side for clues as to what substitutions will likely get you to that side.

EXAMPLE 3 Prove the identity $\sin^2 x - \sin^4 x = \cos^2 x - \cos^4 x$.

SOLUTION Begin by factoring out $\sin^2 x$.

$$\begin{aligned} \sin^2 x - \sin^4 x &= \sin^2 x(1 - \sin^2 x) \\ &= (1 - \cos^2 x)\cos^2 x \\ &= \cos^2 x - \cos^4 x \end{aligned}$$ ●

EXERCISES 5.3

1. Show that $\cot^2 u + 1 = \csc^2 u$. [*Hint:* Start with $\cos^2 u + \sin^2 u = 1$.]

Use the fundamental identities to reduce the first expression to the simpler one at the right.

2. $(1 - \sin^2 u)(\sec u \tan u)$; $\sin u$

3. $\csc u \cos u$; $\cot u$

4. $\cos x + \sin x \tan x$; $\sec x$

5. $\cos^2 x(1 + \tan^2 x)$; 1

6. $\sin u + \cos u \cot u$; $\csc u$

7. $\tan t \csc^2 t - \cot t$; $\tan t$

8. $\dfrac{\sin x + \tan x}{1 + \cos x}$; $\tan x$

9. $\dfrac{\csc x - \sec x}{\sin x - \cos x}$; $-\sec x \csc x$

10. $\dfrac{\cos x}{1 + \sin x} + \dfrac{\cos x}{1 - \sin x}$; $2 \sec x$

Prove each of the following identities. In each case start with one side and reduce to the other side.

11. $\sin x \sec x = \tan x$

12. $\csc^2 x - \csc^2 x \cos^2 x = 1$

13. $\dfrac{\csc x}{\tan x + \cot x} = \cos x$

14. $\cot x \cos x + \sin x = \csc x$

15. $\dfrac{\tan x - 1}{1 - \cot x} = \tan x$

16. $\cot u + \tan u = \sec u \csc u$

17. $\sec x - \cos x = \sin x \tan x$

18. $\tan^2 u - \sin^2 u = \sin^2 u \tan^2 u$

19. $\cot^2 x + 1 = \sec^2 x \cot^2 x$

20. $(\tan t + \cot t)^2 = \sec^2 t + \csc^2 t$

21. $\tan^4 u + \tan^2 u = \sec^4 u - \sec^2 u$

22. $\cot^2 x - \cos^2 x = \cot^2 x \cos^2 x$

23. $\csc^4 x - \cot^4 x = \csc^2 x + \cot^2 x$

24. $\dfrac{\sec x}{\cos x} - \dfrac{\tan x}{\cot x} = 1$

25. $1 - 2\sin^2 x = 2\cos^2 x - 1$

26. $\csc x + \cot x = \dfrac{\sin x}{1 - \cos x}$

5.4 The Cosine Sum-Difference Identities

From the fundamental identities a number of other special identities such as formulas for $\sin(u + v)$, $\cos 2u$, and $\tan \frac{1}{2}u$ may be derived. First, note that, in general, $\sin(u + v) \neq \sin u + \sin v$, $\cos 2u \neq 2 \cos u$, and so on. For example,

$$\sin \frac{\pi}{6} + \sin \frac{\pi}{3} = \frac{1}{2} + \frac{1}{2}\sqrt{3} = \frac{1}{2}(1 + \sqrt{3}),$$

but

$$\sin\left(\frac{\pi}{6} + \frac{\pi}{3}\right) = \sin \frac{1}{2}\pi = 1,$$

which is *not* the same as $\frac{1}{2}(1 + \sqrt{3})$. Similarly,

$$2 \cos \frac{\pi}{3} = 2\left(\frac{1}{2}\right) = 1,$$

but

$$\cos 2\left(\frac{\pi}{3}\right) = \cos \frac{2\pi}{3} = -\frac{1}{2},$$

so that

$$\cos 2\left(\frac{\pi}{3}\right) \neq 2 \cos \frac{\pi}{3}.$$

The formula for $\cos(u - v)$ is derived directly from the definition of the cosine function. Then the other formulas are obtained from this derived formula.

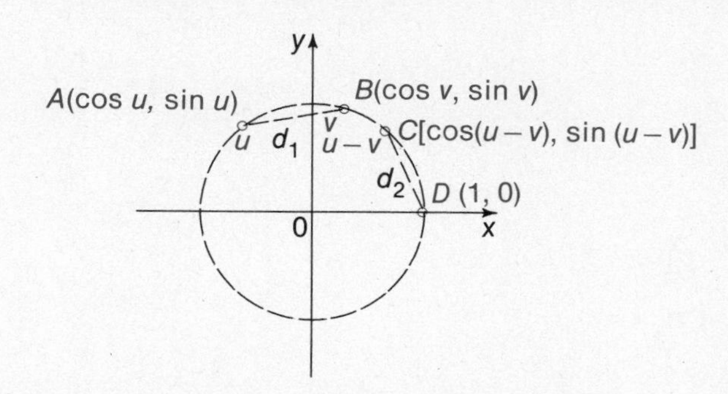

Figure 5.11

In Figure 5.11, points A and B on the unit circle correspond to the real numbers u and v, respectively. Then arc AB has length $u - v$. Now let C on the circle correspond to the real number $u - v$, that is, arc CD has the same length, $u - v$, as arc AB, but it starts at $(1, 0)$. The cartesian coordinates of A, B, C, and D are shown in the figure. Let

d_1 be the length of line segment AB and d_2 the length of line segment CD. Since arc AB has the same length as arc CD, $d_1 = d_2$. From the distance formula,

$$\begin{aligned} d_1^2 &= (\cos u - \cos v)^2 + (\sin u - \sin v)^2 \\ &= (\cos^2 u - 2\cos u \cos v + \cos^2 v) + (\sin^2 u - 2 \sin u \sin v + \sin^2 v) \\ &= (\cos^2 u + \sin^2 u) + (\cos^2 v + \sin^2 v) - 2(\cos u \cos v + \sin u \sin v) \\ &= 2 - 2(\cos u \cos v + \sin u \sin v). \end{aligned}$$

Similarly,

$$\begin{aligned} d_2^2 &= [\cos(u - v) - 1]^2 + [\sin(u - v) - 0]^2 \\ &= [\cos^2(u - v) - 2\cos(u - v) + 1] + \sin^2(u - v) \\ &= [\cos^2(u - v) + \sin^2(u - v)] + 1 - 2\cos(u - v) \\ &= 2 - 2\cos(u - v). \end{aligned}$$

Since $d_1 = d_2$, $d_1^2 = d_2^2$, and thus

$$2 - 2(\cos u \cos v + \sin u \sin v) = 2 - 2\cos(u - v),$$

or,

$$\mathbf{\cos(u - v) = \cos u \cos v + \sin u \sin v.} \qquad (1)$$

This is called the **cosine difference formula.** From it the other formulas of this section will be derived.

EXAMPLE 1 Calculate $\cos \pi/12$.

SOLUTION Since $\pi/12 = 4\pi/12 - 3\pi/12$, we have

$$\begin{aligned} \cos\frac{\pi}{12} &= \cos\left(\frac{4\pi}{12} - \frac{3\pi}{12}\right) \\ &= \cos\left(\frac{\pi}{3} - \frac{\pi}{4}\right) \\ &= \cos\frac{\pi}{3}\cos\frac{\pi}{4} + \sin\frac{\pi}{3}\sin\frac{\pi}{4} \\ &= \frac{1}{2}\cdot\frac{1}{2}\sqrt{2} + \frac{1}{2}\sqrt{3}\cdot\frac{1}{2}\sqrt{2} \\ &= \frac{1}{4}(\sqrt{2} + \sqrt{6}). \end{aligned}$$ ●

It should be noted that the cosine difference formula is an *identity*, that is, it is true for *any* choice of the real numbers u and v. This can be shown for u and v in quadrants other than those of Figure 5.11 by a proof similar to the one above.

In the cosine difference formula, let $u = 0$ and use the fact that $\sin 0 = 0$ and $\cos 0 = 1$. Then

$$\cos(0 - v) = \cos 0 \cos v + \sin 0 \sin v$$

becomes

$$\cos(-v) = \cos v. \tag{2}$$

This, of course, reaffirms that the cosine is an even function. (See Section 5.2.)
In the same way, letting $u = \pi/2$ in the cosine difference formula, we get

$$\cos\left(\frac{1}{2}\pi - v\right) = \sin v. \tag{3}$$

Letting $v = \pi/2 - u$ in formula (3) gives

$$\sin\left(\frac{1}{2}\pi - u\right) = \cos u. \tag{4}$$

EXAMPLE 2 Use the above formulas to calculate (a) $\cos(-\pi/4)$, (b) $\sin \pi/6$, (c) $\cos \pi/6$.

SOLUTION (a) $\cos\left(-\frac{1}{4}\pi\right) = \cos\frac{1}{4}\pi = \frac{1}{2}\sqrt{2}$ by (2)

(b) $\sin\frac{\pi}{6} = \cos\left(\frac{1}{2}\pi - \frac{\pi}{6}\right) = \cos\frac{\pi}{3} = \frac{1}{2}$ by (3)

(c) $\cos\frac{\pi}{6} = \sin\left(\frac{1}{2}\pi - \frac{\pi}{6}\right) = \sin\frac{\pi}{3} = \frac{\sqrt{3}}{2}$ by (4) ●

If v is replaced by $-v$ in formula (1), (which may be done since (1) is an identity),

$$\begin{aligned}\cos(u + v) &= \cos[u - (-v)]\\ &= \cos u \cos(-v) + \sin u \sin(-v)\\ &= \cos u \cos v - \sin u \sin v,\end{aligned}$$

since $\sin(-v) = -\sin v$ and $\cos(-v) = \cos v$. This gives the **cosine sum formula:**

$$\cos(u + v) = \cos u \cos v - \sin u \sin v.$$

EXAMPLE 3 Calculate $\cos 5\pi/12$.

SOLUTION

$$\begin{aligned}\cos\frac{5\pi}{12} &= \cos\left(\frac{2\pi}{12} + \frac{3\pi}{12}\right)\\ &= \cos\left(\frac{\pi}{6} + \frac{\pi}{4}\right)\\ &= \cos\frac{\pi}{6}\cos\frac{\pi}{4} - \sin\frac{\pi}{6}\sin\frac{\pi}{4}\\ &= \frac{1}{2}\sqrt{3}\cdot\frac{1}{2}\sqrt{2} - \frac{1}{2}\cdot\frac{1}{2}\sqrt{2}\\ &= \frac{1}{4}(\sqrt{6} - \sqrt{2})\end{aligned}$$ ●

EXERCISES 5.4

Use a sum or difference formula to calculate each of the following.

1. $\cos \frac{7\pi}{12}$ **2.** $\cos \frac{11\pi}{12}$ **3.** $\cos \frac{13\pi}{12}$

4. $\cos \frac{-17\pi}{12}$ **5.** $\cos \frac{25\pi}{12}$ **6.** $\sec \frac{\pi}{12}$

Use the appropriate sum or difference formula to prove each of the following identities.

7. $\cos(\pi - u) = -\cos u$ **8.** $\cos(\pi + u) = -\cos u$

9. $\cos\left(u + \frac{1}{2}\pi\right) = -\sin u$ **10.** $\cos\left(u - \frac{1}{2}\pi\right) = \sin u$

Prove the following identities.

11. $\cos(u + v) + \cos(u - v) = 2 \cos u \cos v$

12. $\cos(u + v) - \cos(u - v) = -2 \sin u \sin v$

13. $\cos(u + v) \cdot \cos(u - v) = \cos^2 u - \cos^2 v$

14. $\cos\left(u + \frac{\pi}{4}\right) = \frac{1}{2}\sqrt{2}(\cos u - \sin u)$

15. $\cos\left(u + \frac{3\pi}{2}\right) = -\sin u$

16. $\cos 2u = \cos(u + u) = \cos^2 u - \sin^2 u$

5.5 Sum-Difference Identities for Sine and Tangent

The formulas of the last section can be used as follows to derive a formula for $\sin(u + v)$.

$$\begin{aligned}
\sin(u + v) &= \cos\left[\frac{1}{2}\pi - (u + v)\right] && \text{by (3)} \\
&= \cos\left[\left(\frac{1}{2}\pi - u\right) - v\right] \\
&= \cos\left(\frac{1}{2}\pi - u\right) \cos v + \sin\left(\frac{1}{2}\pi - u\right) \sin v && \text{by (1)} \\
&= \sin u \cos v + \cos u \sin v && \text{by (3) and (4)}
\end{aligned}$$

Thus, we have the **sine sum formula:**

$$\boldsymbol{\sin(u + v) = \sin u \cos v + \cos u \sin v.}$$

Replacing v by $-v$ we obtain the **sine difference formula:**

$$\boldsymbol{\sin(u - v) = \sin u \cos v - \cos u \sin v}$$

EXAMPLE 1 Calculate (a) $\sin 7\pi/12$, (b) $\sin \pi/12$.

SOLUTION (a)
$$\sin\frac{7\pi}{12} = \sin\left(\frac{3\pi}{12} + \frac{4\pi}{12}\right)$$
$$= \sin\left(\frac{\pi}{4} + \frac{\pi}{3}\right)$$
$$= \sin\frac{\pi}{4}\cos\frac{\pi}{3} + \cos\frac{\pi}{4}\sin\frac{\pi}{3}$$
$$= \frac{1}{2}\sqrt{2}\cdot\frac{1}{2} + \frac{1}{2}\sqrt{2}\cdot\frac{1}{2}\sqrt{3}$$
$$= \frac{1}{4}(\sqrt{2} + \sqrt{6})$$

(b)
$$\sin\frac{\pi}{12} = \sin\left(\frac{4\pi}{12} - \frac{3\pi}{12}\right)$$
$$= \sin\left(\frac{\pi}{3} - \frac{\pi}{4}\right)$$
$$= \sin\frac{\pi}{3}\cos\frac{\pi}{4} - \cos\frac{\pi}{3}\sin\frac{\pi}{4}$$
$$= \frac{1}{2}\sqrt{3}\cdot\frac{1}{2}\sqrt{2} - \frac{1}{2}\cdot\frac{1}{2}\sqrt{2}$$
$$= \frac{1}{4}(\sqrt{6} - \sqrt{2}) \quad \bullet$$

The sine sum and sine difference formulas above may be used to get additional formulas as follows.

$$\sin(\pi - u) = \sin\pi\cos u - \cos\pi\sin u$$
$$= (0)\cos u - (-1)\sin u \quad = \sin u$$
$$\sin(2\pi - u) = \sin 2\pi\cos u - \cos 2\pi\sin u$$
$$= 0\cdot\cos u - 1\cdot\sin u \quad = -\sin u$$

The formulas just developed, together with others derived similarly, are called **reduction formulas.** We list them here together.

$$\sin(-u) = -\sin u \qquad \sin(2\pi - u) = -\sin u$$
$$\cos(-u) = \cos u \qquad \cos(2\pi - u) = \cos u$$
$$\sin(\pi - u) = \sin u \qquad \sin\left(\frac{1}{2}\pi + u\right) = \cos u$$
$$\cos(\pi - u) = -\cos u \qquad \cos\left(\frac{1}{2}\pi + u\right) = -\sin u$$
$$\sin(\pi + u) = -\sin u \qquad \sin\left(\frac{1}{2}\pi - u\right) = \cos u$$
$$\cos(\pi + u) = -\cos u \qquad \cos\left(\frac{1}{2}\pi - u\right) = \sin u$$

Sum and difference formulas for the tangent and cotangent functions are obtained from those for the sine and cosine functions. The **tangent sum formula** is derived as follows.

$$\tan(u + v) = \frac{\sin(u + v)}{\cos(u + v)}$$

$$= \frac{\sin u \cos v + \cos u \sin v}{\cos u \cos v - \sin u \sin v}$$

$$= \frac{\dfrac{\sin u \cancel{\cos v}}{\cos u \cancel{\cos v}} + \dfrac{\cancel{\cos u} \sin v}{\cancel{\cos u} \cos v}}{\dfrac{\cancel{\cos u} \cancel{\cos v}}{\cancel{\cos u} \cancel{\cos v}} - \dfrac{\sin u \sin v}{\cos u \cos v}}$$

Dividing each term of the fraction by $\cos u \cos v$

$$\boldsymbol{\tan(u + v) = \frac{\tan u + \tan v}{1 - \tan u \tan v}}$$

Replacing v by $-v$ in this formula gives the **tangent difference formula.**

$$\boldsymbol{\tan(u - v) = \frac{\tan u - \tan v}{1 + \tan u \tan v}}$$

EXAMPLE 2 Use the tangent sum formula to calculate $\tan \frac{5\pi}{12}$.

SOLUTION

$$\tan \frac{5\pi}{12} = \tan\left(\frac{2\pi}{12} + \frac{3\pi}{12}\right)$$

$$= \tan\left(\frac{\pi}{6} + \frac{\pi}{4}\right)$$

$$= \frac{\tan \frac{\pi}{6} + \tan \frac{\pi}{4}}{1 - \tan \frac{\pi}{6} \tan \frac{\pi}{4}}$$

$$= \frac{\frac{\sqrt{3}}{3} + 1}{1 - \frac{\sqrt{3}}{3}} = \frac{\sqrt{3} + 3}{3 - \sqrt{3}}$$

$$= \frac{\sqrt{3} + 3}{3 - \sqrt{3}} \cdot \frac{3 + \sqrt{3}}{3 + \sqrt{3}}$$

$$= \frac{(\sqrt{3} + 3)^2}{(3)^2 - (\sqrt{3})^2} = \frac{3 + 6\sqrt{3} + 9}{9 - 3}$$

$$= \frac{12 + 6\sqrt{3}}{6} = 2 + \sqrt{3}. \quad \bullet$$

The cotangent sum and difference formulas are derived in the same way as those for the tangent and are left as exercises.

EXERCISES 5.5

Use an appropriate sum or difference formula to calculate each of the following.

1. $\sin \frac{5\pi}{12}$

2. $\sin \frac{11\pi}{12}$

3. $\sin \frac{13\pi}{12}$

4. $\sin \frac{17\pi}{12}$

5. $\csc \frac{\pi}{12}$

6. $\tan \frac{5\pi}{12}$

7. $\tan \frac{7\pi}{12}$

8. $\tan \frac{11\pi}{12}$

9. $\cot \frac{5\pi}{12}$ $\left[\textit{Hint: } \cot \frac{5\pi}{12} = \frac{1}{\tan \frac{5\pi}{12}}\right]$

10. $\cot \frac{\pi}{12}$

Use the appropriate sum or difference formula to prove the following formulas.

11. $\sin(\pi + u) = -\sin u$

12. $\sin(\pi - u) = \sin u$

13. $\sin\left(u + \frac{1}{2}\pi\right) = \cos u$

14. $\sin\left(u - \frac{1}{2}\pi\right) = -\cos u$

15. $\tan(u + \pi) = \tan u$

16. $\tan(\pi - u) = -\tan u$

17. $\tan\left(u + \frac{1}{2}\pi\right) = -\cot u$

18. $\tan\left(u - \frac{1}{2}\pi\right) = -\cot u$

Use the suggestions made in the text to prove the following formulas.

19. Sine sum formula

20. Sine difference formula

21. Tangent difference formula

22. Cotangent sum formula

23. Cotangent difference formula

Prove the following identities.

24. $\sin(u + v) + \sin(u - v) = 2 \sin u \cos v$

25. $\sin(u + v) - \sin(u - v) = -2 \cos u \sin v$

26. $\sin(u + v) \cdot \sin(u - v) = \sin^2 u - \sin^2 v$

27. $\sin\left(\frac{1}{2}\pi - x\right) \cdot \sin\left(\frac{1}{2}\pi + x\right) = 1 - \sin^2 x$

28. $\sin\left(\frac{1}{4}\pi - x\right) = \cos\left(\frac{1}{4}\pi + x\right)$

29. $\frac{\sin(x - y)}{\cos x \cos y} = \tan x - \tan y$

30. $\frac{\sin(x - y)}{\sin(x + y)} = \frac{\tan x - \tan y}{\tan x + \tan y}$

31. $\frac{\sin(x + y)}{\sin(x - y)} = \frac{\cot x + \cot y}{1 + \cot x \cot y}$

5.6 Multiple-Value Identities

Now formulas can be developed for multiples of the argument u of a circular function, such as $\frac{1}{2}u$, $2u$, $3u$, and so on. These are called **half-value formulas, double-value formulas,** and so on. Similarly, when the argument u is used as the value of an angle, as will be done in Chapter 6, the formulas derived are commonly referred to as "half-angle formulas," "double-angle formulas," and so forth.

Double-Value Formulas If we let $v = u$ in the sine sum formula, $\sin(u + v) = \sin u \cos v + \cos u \sin v$, we obtain

$$\sin(u + u) = \sin u \cos u + \cos u \sin u,$$

or

$$\mathbf{\sin 2u = 2 \sin u \cos u.} \tag{1}$$

This is called the **sine double-value formula.**

Similarly,

$$\begin{aligned} \cos 2u &= \cos(u + u) \\ &= \cos u \cos u - \sin u \sin u \\ &= \cos^2 u - \sin^2 u. \end{aligned}$$

This identity can be expressed in two useful alternative forms.

$$\begin{aligned} \cos 2u &= \cos^2 u - \sin^2 u \\ &= (1 - \sin^2 u) - \sin^2 u \\ &= 1 - 2 \sin^2 u \end{aligned} \qquad \begin{aligned} \cos 2u &= \cos^2 u - \sin^2 u \\ &= \cos^2 u - (1 - \cos^2 u) \\ &= 2 \cos^2 u - 1 \end{aligned}$$

Thus, there are three forms of the **cosine double-value formula.**

$$\mathbf{\cos 2u = \cos^2 u - \sin^2 u} \tag{2a}$$

$$\mathbf{\cos 2u = 2 \cos^2 u - 1} \tag{2b}$$

$$\mathbf{\cos 2u = 1 - 2 \sin^2 u} \tag{2c}$$

Depending upon the particular situation, we may use formula (2a) involving both the sine and cosine, formula (2b) involving only the cosine, or formula (2c) involving only the sine. This will be illustrated presently.

To derive the tangent double-value formula we may either start with $\tan 2u = (\sin 2u)/(\cos 2u)$ and use (1) and (2) or begin with $\tan 2u = \tan(u + u)$ and then use the tangent-sum formula from the last section. In either case, the final result is

$$\mathbf{\tan 2u = \frac{2 \tan u}{1 - \tan^2 u}.} \tag{3}$$

EXAMPLE 1 Prove the identity $(\sin x - \cos x)^2 = 1 - \sin 2x$.

SOLUTION We work with the left side.

$$\begin{aligned}(\sin x - \cos x)^2 &= \sin^2 x - 2 \sin x \cos x + \cos^2 x \\ &= (\sin^2 x + \cos^2 x) - 2 \sin x \cos x \\ &= 1 - \sin 2x \qquad \bullet\end{aligned}$$

EXAMPLE 2 Prove the identity $\dfrac{1 - \tan^2 x}{1 + \tan^2 x} = \cos 2x$.

SOLUTION We begin by replacing $\tan x$ by $\sin x/\cos x$ and simplifying the resulting fraction.

$$\begin{aligned}\frac{1 - \tan^2 x}{1 + \tan^2 x} &= \frac{1 - \dfrac{\sin^2 x}{\cos^2 x}}{1 + \dfrac{\sin^2 x}{\cos^2 x}} \\ &= \frac{1 - \dfrac{\sin^2 x}{\cos^2 x}}{1 + \dfrac{\sin^2 x}{\cos^2 x}} \cdot \frac{\cos^2 x}{\cos^2 x} \\ &= \frac{\cos^2 x - \sin^2 x}{\cos^2 x + \sin^2 x} \\ &= \frac{\cos 2x}{1} = \cos 2x \qquad \bullet\end{aligned}$$

EXAMPLE 3 Express $\sin 3x$ in terms of $\sin x$.

SOLUTION

$$\begin{aligned}\sin 3x &= \sin(2x + x) \\ &= \sin 2x \cos x + \cos 2x \sin x, && \text{by the sine sum formula} \\ &= (2 \sin x \cos x)\cos x + (1 - 2 \sin^2 x)\sin x, && \text{by (1) and (2c)} \\ &= 2 \sin x \cos^2 x + (\sin x - 2 \sin^3 x) \\ &= 2 \sin x(1 - \sin^2 x) + (\sin x - 2 \sin^3 x) \\ &= 2 \sin x - 2 \sin^3 x + \sin x - 2 \sin^3 x \\ &= 3 \sin x - 4 \sin^3 x\end{aligned}$$

In the third line above, we used $\cos 2x = 1 - 2 \sin^2 x$ to reduce to terms in $\sin x$ only. •

Half-value Formulas If we replace u by $x/2$ and $2u$ by x in formula (2c), we get $\cos x = 1 - 2\sin^2(x/2)$. Then, solving for $\sin(x/2)$, we obtain the **sine half-value formula:**

$$\sin\frac{x}{2} = \pm\sqrt{\frac{1-\cos x}{2}}. \tag{4}$$

The positive sign must be used if $0 < x/2 < \pi$ and the negative if $\pi < x/2 < 2\pi$. Using (2c) enables us to solve directly for $\sin(x/2)$.

The **cosine half-value formula** is derived from (2b) in the same way.

$$\cos\frac{x}{2} = \pm\sqrt{\frac{1+\cos x}{2}}. \tag{5}$$

Here we use (2b) in order to solve for $\cos(x/2)$.

Next, using $\tan(x/2) = \sin(x/2)/\cos(x/2)$ with (4) and (5), we obtain the **tangent half-value formula:**

$$\tan\frac{x}{2} = \pm\sqrt{\frac{1-\cos x}{1+\cos x}}. \tag{6}$$

In each of these formulas (4)–(6), of course, we must choose the sign according to the quadrant of $x/2$.

EXAMPLE 4 Use the half-value formulas to calculate (a) $\sin \pi/12$, (b) $\tan 7\pi/12$.

SOLUTION (a)

$$\sin \pi/12 = \sin\frac{1}{2}\left(\frac{\pi}{6}\right) = \sqrt{\frac{1-\cos \pi/6}{2}}$$

$$= \sqrt{\frac{1-\sqrt{3}/2}{2}}$$

$$= \sqrt{\frac{2-\sqrt{3}}{4}}$$

$$= \frac{1}{2}\sqrt{2-\sqrt{3}}.$$

(b)

$$\tan\frac{7\pi}{12} = \tan\frac{1}{2}\left(\frac{7\pi}{6}\right)$$

$$= -\sqrt{\frac{1-\cos\frac{7\pi}{6}}{1+\cos\frac{7\pi}{6}}} = -\sqrt{\frac{1+\frac{\sqrt{3}}{2}}{1-\frac{\sqrt{3}}{2}}}$$

$$= -\sqrt{\frac{2+\sqrt{3}}{2-\sqrt{3}}} = -\sqrt{\frac{2+\sqrt{3}}{2-\sqrt{3}}\cdot\frac{2+\sqrt{3}}{2+\sqrt{3}}} \qquad \text{rationalizing the denominator}$$

$$= -\sqrt{\frac{4+4\sqrt{3}+3}{4-3}} = -\sqrt{7+4\sqrt{3}} \qquad \bullet$$

Formula (6) can be put in a simpler form by rationalizing the denominator as follows.

$$\begin{aligned}
\tan\frac{x}{2} &= \pm\sqrt{\frac{1-\cos x}{1+\cos x}} \\
&= \pm\sqrt{\frac{(1-\cos x)(1+\cos x)}{(1+\cos x)(1+\cos x)}} \\
&= \pm\sqrt{\frac{1-\cos^2 x}{(1+\cos x)^2}} \\
&= \pm\sqrt{\frac{\sin^2 x}{(1+\cos x)^2}} \\
&= \pm\left|\frac{\sin x}{1+\cos x}\right| && \text{See Exercise 2, Miscellaneous Exercises on Chapter 1} \\
&= \pm\frac{|\sin x|}{1+\cos x}, && \text{since } 1+\cos x > 0 \\
&= \frac{\sin x}{1+\cos x}
\end{aligned}$$

This last formula automatically gives the correct sign for $\tan \frac{1}{2}x$. For, $|\sin x| = \sin x$ if $\sin x \geq 0$, that is, if x is in quadrant I or II, so that $\frac{1}{2}x$ is in quadrant I. In this case, the formula gives a positive value for $\tan \frac{1}{2}x$. If $\sin x < 0$, $|\sin x| = -\sin x$. But then x is in quadrants III or IV, so that $\frac{1}{2}x$ is in quadrant II. Now the formula gives a negative value for $\tan \frac{1}{2}x$, as needed. Thus, we have the following alternative identity for $\tan x/2$.

$$\tan\frac{1}{2}x = \frac{\sin x}{1+\cos x} \tag{7}$$

Similarly, the radical expression for $\tan \frac{1}{2}x$ may be rationalized by multiplying both numerator and denominator of the radicand by $1-\cos x$ to obtain

$$\tan\frac{1}{2}x = \frac{1-\cos x}{\sin x}. \tag{8}$$

Details are left as an exercise.

All the formulas of this section are very useful for simplifying expressions in the circular functions and for solving equations involving the circular functions (next section).

EXAMPLE 5 Prove that $\cot x \sin 2x = 1 + \cos 2x$.

SOLUTION

$$\begin{aligned}
\cot x \sin 2x &= \frac{\cos x}{\sin x}\cdot 2\sin x\cos x, && \text{by (1)} \\
&= 2\cos^2 x \\
&= 1 + (2\cos^2 x - 1) \\
&= 1 + \cos 2x, && \text{by (2b)}
\end{aligned}$$

●

EXAMPLE 6 Prove that $\dfrac{2\cot x}{\csc^2 x - 2} = \tan 2x$.

SOLUTION Work on the left side.

$$\frac{2\cot x}{\csc^2 x - 2} = \frac{\dfrac{2\cos x}{\sin x}}{\dfrac{1}{\sin^2 x} - 2}$$

$$= \frac{2\sin x\cos x}{1 - 2\sin^2 x}$$

Multiplying numerator and denominator by $\sin^2 x$

$$= \frac{\sin 2x}{\cos 2x} = \tan 2x.$$

The following is an alternative solution.

$$\frac{2\cot x}{\csc^2 x - 2} = \frac{2\cot x}{(1 + \cot^2 x) - 2}$$

$$= \frac{2\left(\dfrac{1}{\tan x}\right)}{\dfrac{1}{\tan^2 x} - 1}$$

$$= \frac{2\tan x}{1 - \tan^2 x}$$

Multiplying numerator and denominator by $\tan^2 x$

$$= \tan 2x.$$

For convenience, the special identities developed in this chapter are listed here.

FUNDAMENTAL IDENTITIES

$$\tan x = \frac{\sin x}{\cos x} \qquad \cot x = \frac{\cos x}{\sin x}$$

$$\sin x \csc x = 1 \qquad \sin^2 x + \cos^2 x = 1$$

$$\cos x \sec x = 1 \qquad \tan^2 x + 1 = \sec^2 x$$

$$\tan x \cot x = 1 \qquad \cot^2 x + 1 = \csc^2 x$$

SUM FORMULAS

$$\sin(x + y) = \sin x \cos y + \cos x \sin y$$

$$\cos(x + y) = \cos x \cos y - \sin x \sin y$$

$$\tan(x + y) = \frac{\tan x + \tan y}{1 - \tan x \tan y}$$

DIFFERENCE FORMULAS

$$\sin(x - y) = \sin x \cos y - \cos x \sin y$$
$$\cos(x - y) = \cos x \cos y + \sin x \sin y$$
$$\tan(x - y) = \frac{\tan x - \tan y}{1 + \tan x \tan y}$$

DOUBLE-VALUE FORMULAS

$$\sin 2x = 2 \sin x \cos x \qquad \cos 2x = \cos^2 x - \sin^2 x$$
$$\tan 2x = \frac{2 \tan x}{1 - \tan^2 x} \qquad \cos 2x = 2\cos^2 x - 1$$
$$\cos 2x = 1 - 2\sin^2 x$$

HALF-VALUE FORMULAS

$$\sin \frac{1}{2}x = \pm\sqrt{\frac{1}{2}(1 - \cos x)} \qquad \tan \frac{1}{2}x = \frac{\sin x}{1 + \cos x}$$
$$\cos \frac{1}{2}x = \pm\sqrt{\frac{1}{2}(1 + \cos x)} \qquad \tan \frac{1}{2}x = \frac{1 - \cos x}{\sin x}$$
$$\tan \frac{1}{2}x = \pm\sqrt{\frac{1 - \cos x}{1 + \cos x}}$$

EXERCISES 5.6

Use the half-value formulas to calculate each of the following.

1. $\sin \frac{5\pi}{12}$ **2.** $\sin \frac{7\pi}{12}$ **3.** $\cos \frac{\pi}{12}$ **4.** $\cos \frac{11\pi}{12}$

5. $\sin \frac{\pi}{8}$ **6.** $\sin \frac{5\pi}{8}$ **7.** $\cos \frac{3\pi}{8}$ **8.** $\cos \frac{7\pi}{8}$

9. $\tan \frac{\pi}{8}$ **10.** $\tan \frac{\pi}{12}$ **11.** $\tan \frac{3\pi}{8}$ **12.** $\tan \frac{5\pi}{12}$

Derive the following formulas. [See Example 3.]

13. $\cos 3x = 4\cos^3 x - 3\cos x$ **14.** $\sin 4x = 2 \sin x \cos x - 4\sin^3 x \cos x$

15. $\cos 4x = 8\cos^4 x - 8\cos^2 x - 1$

Using the suggestions given in the text, derive the following formulas.

16. $\tan 2x = \frac{2\tan x}{1 - \tan^2 x}$, by two methods **17.** $\cos \frac{1}{2}x = \pm\sqrt{\frac{1}{2}(1 + \cos x)}$

18. $\tan \frac{1}{2}x = \frac{1 - \cos x}{\sin x}$

Prove the following identities.

19. $\cos^4 x - \sin^4 x = \cos 2x$ [*Hint:* Factor.] **20.** $\cos 2x + 1 = 2\cos^2 x$

21. $\tan u + \cot u = 2 \csc 2u$

22. $\cot x \sin 2x = 1 + \cos 2x$

23. $\dfrac{\sin x}{1 - \cos x} = \cot \dfrac{1}{2}x$

24. $\dfrac{\tan x}{1 + \tan^2 x} = \sin 2x$

25. $\csc u - \cot u = \tan \dfrac{1}{2}u$

26. $\tan\left(\dfrac{1}{2}x + \dfrac{1}{4}\pi\right) = \sec x + \tan x$

27. $\sin\left(\dfrac{\pi}{6} + x\right) = \cos\left(\dfrac{\pi}{3} - x\right)$

28. $\tan\left(u + \dfrac{1}{4}\pi\right) \cdot \tan\left(u - \dfrac{1}{4}\pi\right) = -1$

29. $\dfrac{1 + \cos 2x}{\sin 2x} = \cot x$

30. $\dfrac{\cot x - \tan x}{\cot x + \tan x} = \cos 2x$

31. $\dfrac{2 \cot x}{1 + \cot^2 x} = \sin 2x$

32. $\csc x - \cot x = \tan \dfrac{1}{2}x$

33. $\sin(x + y) \cdot \sin x + \cos(x + y) \cdot \cos x = \cos y$

34. $\cot \dfrac{1}{2}x - \tan \dfrac{1}{2}x = 2 \cot x$

35. $\cot x \cdot \tan(x + \pi) - \sin(\pi - x) \cdot \cos\left(\dfrac{1}{2}\pi - x\right) = \cos^2 x$

36. $\left(\sin \dfrac{1}{2}x + \cos \dfrac{1}{2}x\right)^2 = 1 + \sin x$

37. $1 + \tan x \tan 2x = \sec 2x$

38. $\tan x \sin 2x = 2 \sin^2 x$

39. $\dfrac{1 - \tan^2 x}{1 + \tan^2 x} = \cos 2x$

40. $\dfrac{\sin 2x}{\sin x} - \dfrac{\cos 2x}{\cos x} = \sec x$

41. $2 \cos^2 \dfrac{1}{2}x \tan x = \tan x + \sin x$

42. $\cot\left(u + \dfrac{1}{2}\pi\right) = \tan(u + \pi)$

5.7 Trigonometric Equations

One of the major uses of identities is solving trigonometric equations. Unlike an identity, which is true for all values of its argument, an equation is true for only certain values of the argument. Finding these values is called *solving the equation*. The process is illustrated in the following examples.

EXAMPLE 1 Solve $2 \cos x - 1 = 0$.

SOLUTION

$$2 \cos x - 1 = 0$$

$$2 \cos x = 1$$

$$\cos x = \frac{1}{2}$$

$$x = \frac{\pi}{3} \qquad \text{in the first quadrant}$$

$$x = \frac{5\pi}{3} \qquad \text{in the fourth quadrant}$$

●

The numbers which make the equation true are called **solutions** of the equation. Because the circular functions are periodic, a trigonometric equation has an infinite set of solutions. In the example above, since the period of the cosine is 2π, *all* solutions have the form

$$x = \frac{\pi}{3} + 2k\pi \quad \text{or} \quad x = \frac{5\pi}{3} + 2k\pi, \qquad k \text{ any integer.}$$

The particular solutions $\pi/3$ and $5\pi/3$ in the interval $0 \leq x < 2\pi$ are called the **basic solutions.** The complete set $\{\pi/3 + 2k\pi, 5\pi/3 + 2k\pi\}$ is called the **general solution.** Ordinarily only the basic solutions are given.

EXAMPLE 2 Solve $(\sin x - 1)(2 \cos x + 1) = 0$.

SOLUTION Setting each factor equal to zero, we obtain

$$\sin x - 1 = 0 \qquad \text{and} \qquad 2 \cos x + 1 = 0.$$

If $\sin x - 1 = 0$,

$$\sin x = 1,$$

$$x = \frac{\pi}{2},$$

for $0 \leq x < 2\pi$.

If $2 \cos x + 1 = 0$,

$$2 \cos x = -1,$$

$$\cos x = -\frac{1}{2}, \qquad x = \frac{2\pi}{3} \text{ or } \frac{4\pi}{3},$$

for $0 \leq x < 2\pi$.

Thus, the basic solutions are $\pi/2$, $2\pi/3$, and $4\pi/3$. The general solutions are $\pi/2 + 2k\pi$, $2\pi/3 + 2k\pi$, and $4\pi/3 + 2k\pi$. ●

EXAMPLE 3 Find the basic solutions for $\sin^2 x - \sin x - 2 = 0$.

SOLUTION We begin by factoring the left-hand side, which is quadratic in $\sin x$.

$$\sin^2 x - \sin x - 2 = 0$$

$$(\sin x - 2)(\sin x + 1) = 0$$

$$\sin x - 2 = 0 \qquad \text{and} \qquad \sin x + 1 = 0$$

If $\sin x - 2 = 0$, then $\sin x = 2$. But $-1 \leq \sin x \leq 1$, so this factor does not lead to a solution. If $\sin x + 1 = 0$, then

$$\sin x = -1,$$

$$x = \frac{3\pi}{2} \quad \text{for } 0 \leq x < 2\pi.$$

The only basic solution is $3\pi/2$. ●

The examples above were solved using algebraic methods. Some trigonometric equations require substitutions from the identities before they can be solved.

EXAMPLE 4 Find the basic solutions of $\sin^2 u + \cos u = 1$.

SOLUTION

$$\sin^2 u + \cos u = 1,$$
$$(1 - \cos^2 u) + \cos u = 1,$$
$$\cos^2 u - \cos u = 0,$$
$$\cos u(\cos u - 1) = 0.$$

If $\cos u = 0$,

$$u = \frac{1}{2}\pi \quad \text{or} \quad \frac{3\pi}{2},$$

for $0 \leq u < 2\pi$.

If $\cos u - 1 = 0$,

$$\cos u = 1,$$
$$u = 0,$$

for $0 \leq u < 2\pi$.

The basic solutions are 0, $\pi/2$, and $3\pi/2$. ●

EXAMPLE 5 Solve $\sec^2 x + \tan^2 x = 3$.

SOLUTION First, use an identity to get an equation with the tangent function only.

$$\sec^2 x + \tan^2 x = 3$$
$$(\tan^2 x + 1) + \tan^2 x = 3$$
$$2\tan^2 x = 2$$
$$\tan^2 x = 1$$
$$\tan x = \pm 1$$

If $\tan x = 1$,

$$x = \frac{\pi}{4} \quad \text{or} \quad \frac{5\pi}{4}, \quad 0 \leq x < 2\pi.$$

If $\tan x = -1$,

$$x = \frac{3\pi}{4} \quad \text{or} \quad \frac{7\pi}{4}, \quad 0 \leq x < 2\pi.$$

The basic solutions are $\pi/4$, $3\pi/4$, $5\pi/4$, and $7\pi/4$. Since the period of the tangent function is π, the general solutions are $\pi/4 + k\pi$ and $3\pi/4 + k\pi$. Note that all the basic solutions are included in the general solution. ●

EXAMPLE 6 Solve $\cos 2x = 1/2$.

SOLUTION In this case, we first solve for $2x$ and then determine all basic values of x.

$$\cos 2x = \frac{1}{2}$$
$$2x = \frac{\pi}{3} + 2k\pi \quad \text{or} \quad \frac{5\pi}{3} + 2k\pi$$
$$x = \frac{\pi}{6} + k\pi \quad \text{or} \quad \frac{5\pi}{6} + k\pi$$

The basic solutions are $\pi/6$, $5\pi/6$, $7\pi/6$, and $11\pi/6$, obtained by letting $k = 0$ or 1. ●

EXAMPLE 7 Solve $\sin^2 x - 6 \sin x + 4 = 0$

SOLUTION $(\sin x)^2 - 6(\sin x) + 4 = 0$. Then, by the quadratic formula,

$$\sin x = \frac{6 \pm \sqrt{36 - 4(4)}}{2} = \frac{6 \pm \sqrt{20}}{2}$$

$$= \frac{6 \pm 2\sqrt{5}}{2} = 3 \pm \sqrt{5}$$

$$\approx 3 \pm 2.236 \approx 5.236 \text{ or } 0.764.$$

Since $\sin x < 1$, 5.236 does not lead to a solution. If, however,

$$\sin x \approx 0.764$$

$$x \approx 0.869, \qquad \text{from Table D.}$$

Also, since $\sin (\pi - x) = \sin x$,

$$x = \pi - 0.869$$

$$\approx 2.273.$$

The basic solutions are (approximately) 0.869 and 2.273. ●

EXAMPLE 8 Solve $\tan^2 x + 2 \tan x = 1$.

SOLUTION The equation may be rewritten as

$$(\tan x)^2 + 2(\tan x) - 1 = 0.$$

Then, by the quadratic formula,

$$\tan x = \frac{-2 \pm \sqrt{4 + 4}}{2} = \frac{-2 \pm \sqrt{8}}{2}$$

$$= \frac{-2 \pm 2\sqrt{2}}{2}$$

$$= -1 \pm \sqrt{2}$$

$$\approx -1 \pm 1.414 \approx 0.414 \text{ or } -2.414.$$

If $\tan x \approx 0.414$, then from Table D in the Appendix,

$$x \approx 0.393 \quad \text{or} \quad 0.393 + \pi \qquad \text{by periodicity,}$$

$$\approx 0.393 \quad \text{or} \quad 3.535.$$

If $\tan x \approx -2.414$, then from Table D,

$$x \approx 1.964 \quad \text{or} \quad 1.964 + \pi$$

$$\approx 1.964 \quad \text{or} \quad 5.106.$$

Then, the basic solutions are (approximately) 0.393, 1.964, 3.535, and 5.106. ●

EXERCISES 5.7

Solve each of the following for x, $0 \leq x < 2\pi$.

1. $\cot x + \cos x = 0$
2. $\cos^2 x + \cos x = 2$
3. $\sin^2 x + 7 \sin x + 6 = 0$
4. $\sec^2 x - \tan x = 13$
5. $\cos 2x + \cos x = 0$
6. $\sin 2x + \sin x = 0$
7. $\tan 2x = \cot x$
8. $\sin 2x = \sin x$
9. $\sin 2x = \cos x$
10. $\cos x + \sin 2x = 0$
11. $2 \cos^2 x + \cos x = 1$
12. $\sec^2 x - 2 \tan x = 4$
13. $2 \sin^2 x + 2 \sin x - 1 = 0$
14. $\cos^2 x + \cos x + 1 = 0$
15. $\tan x = \cot x$
16. $2 \sin x + \tan x = 0$
17. $2 \cos^2 x + \sin x = 2$
18. $\cos 2x - \cos x + 1 = 0$
19. $\tan 2x = 3 \tan x$
20. $\cos 2x = \cos^2 x$
21. $\cos 2x + \sin x = 1$
22. $\sin^2 \frac{1}{2}x = \cos x$
23. $2 \sin^2 \frac{1}{2}x - \cos x = 1$
24. $3 \sin^2 x - \cos x = 2$
25. $\cos^2 x - 4 \sin x = 0$
26. $2 \tan^2 x - \tan x - 6 = 0$
27. $\sec^2 x + 2 \tan x = 4$

5.8 Graphs of the Sine and Cosine Functions

The most significant property of the circular functions—their periodicity—is shown by their graphs. In this section the graph of the sine function will be developed in detail by making use of this and other properties of the sine function. The graphs of the other circular functions may be obtained by similar procedures.

The Sine Curve Rather than plotting many individual points, we will develop the graph of the sine function by studying the function itself. Figure 5.12a shows the standard unit circle marked with special values of the circular coordinates.

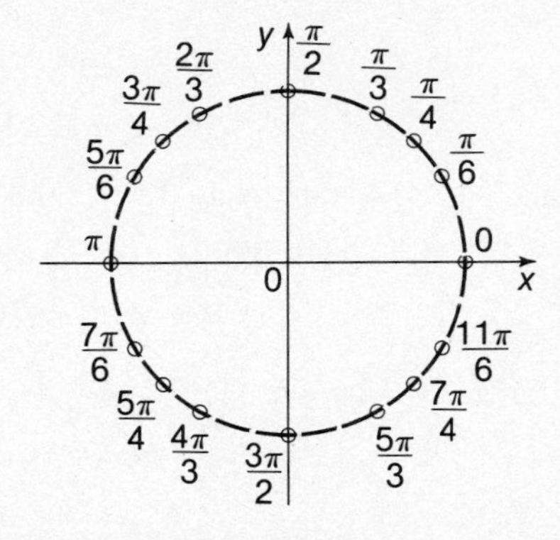

Figure 5.12 (a)

In Figure 5.12b the circle is shown "unwinding" along the x-axis. Each of the circular coordinates is mapped into the real number line (the x-axis) as shown. If x is the arc length from (1, 0) to the point P, then $\sin x$ is the ordinate (the y-value) of P. A horizontal line through a point such as $\pi/3$ on the circle and a vertical line through the same point on the x-axis intersect at the point $(\pi/3, \sin \pi/3)$ in the coordinate plane. If this procedure is repeated for each value of x we obtain the special points of the graph of the sine function. Some of them are shown in Figure 5.12b. The smooth curve

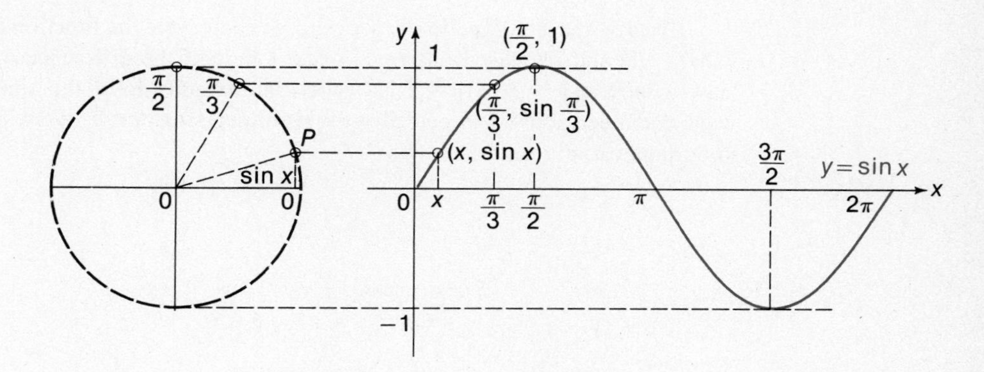

Figure 5.12 (b)

obtained by joining all these points is the **sine curve** or **sine wave.** Since the sine function is periodic with period 2π, the graph shown for the period $[0, 2\pi]$, called a **cycle,** is repeated for the intervals $[2\pi, 4\pi]$, $[4\pi, 6\pi]$, $[-2\pi, 0]$, and so on. The resulting graph of the sine function is shown in Figure 5.13.

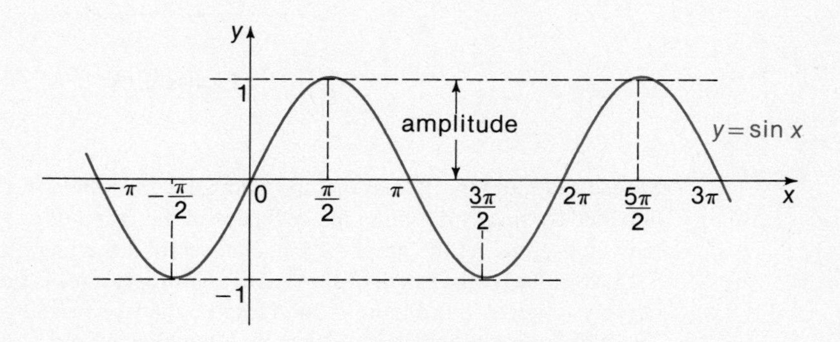

Figure 5.13

The graph in Figure 5.13 illustrates the following special properties of the sine function, previously found algebraically.

$-1 \leq \sin x \leq 1$	range $[-1, 1]$
$\sin(x + 2\pi) = \sin x$	period 2π
$\sin(-x) = -\sin x$	odd function
$\sin(\pi + x) = -\sin x$	reduction formula
$\sin(\pi - x) = \sin x$	reduction formula
$\sin x = 0$ if and only if $x = k\pi$, $k = 0, \pm 1, \pm 2, \ldots$	zeros at multiples of π

Because the sine function is periodic, in each cycle the function attains its greatest value $(+1)$ and its least value (-1) once. One-half the difference of the greatest and least values, $\frac{1}{2}[1 - (-1)] = 1$, is called the **amplitude** of the sine wave. The sine wave discussed above will be called the **standard sine curve.** We will now consider its important variations.

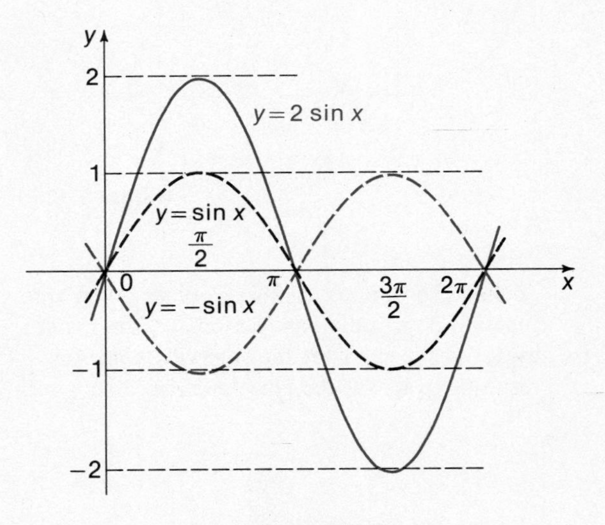

Figure 5.14

Sinusoidal Curves Recall that in Example 2, Section 3.4, the graph of $y = 2x^2$ was obtained by simply doubling the ordinates of the graph of $y = x^2$. The same procedure may be used to get the graph of $y = 2 \sin x$ from the graph of $y = \sin x$. (See Figure 5.14.) Similarly, the graph of $y = -\sin x$ is simply a reflection in the x-axis of the sine curve $y = \sin x$. See Figure 5.14. Observe that the period of each of these curves is 2π. The amplitude of $y = -\sin x$ is 1 but that of $y = 2 \sin x$ is 2. Similarly, the amplitude of $y = -3 \sin x$ is 3. In general, $y = a \sin x$ is the equation of a family of sine waves, each of which has period 2π and amplitude $|a|$.

Next we consider the graphs of the family whose general equation is $y = \sin kx$, where k is any real constant.

EXAMPLE 1 Graph $y = \sin 2x$.

SOLUTION Let $u = 2x$, so that $x = u/2$, and graph $y = \sin u$ as in Figure 5.15a. This is also the graph of $y = \sin 2x$. Both the x-scale and the u-scale are shown along the x-axis (each u-value is twice the corresponding x-value). If we remove the u-scale, we have Figure 5.15b. To make clear the relation of the graph of $y = \sin 2x$ to that of $y = \sin x$, we show them both in Figure 5.15c. ●

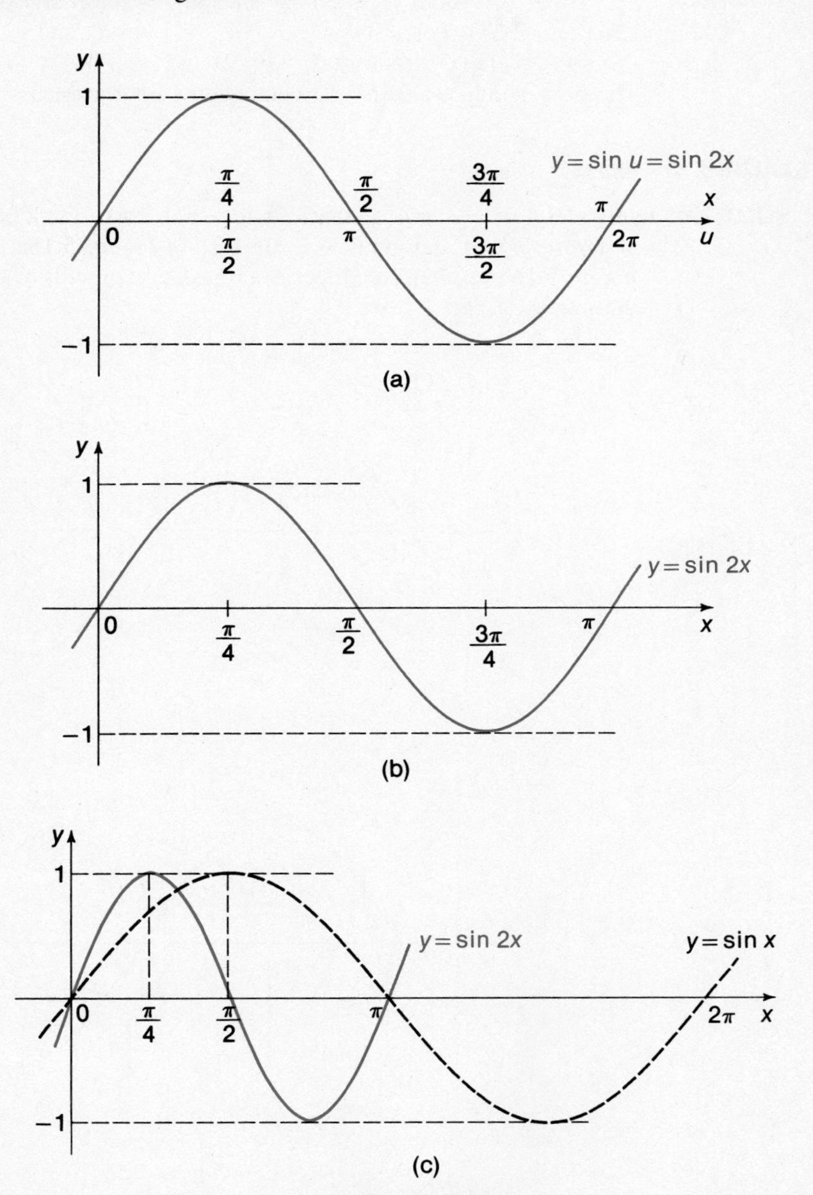

Figure 5.15

As seen in the above example, the amplitude of $y = \sin 2x$ is 1 and its period is π (just half that of $y = \sin x$). Similarly, $y = \sin 3x$ has amplitude 1 and period $(1/3)(2\pi)$, or $2\pi/3$. In general, any member of the family of curves $y = \sin kx$ has amplitude 1 and period $2\pi/|k|$. In Figure 5.15, the constant 2 "compressed" the standard sine curve by a factor of 2. In general, the positive constant k in the equation $y = \sin kx$ compresses the standard sine curve if $k > 1$ and stretches it if $k < 1$.

Note that both $y = \sin 2x$ and $y = 2 \sin x$ are compositions of functions (Section 2.4). For, if $y = f(u) = \sin u$ and $u = g(x) = 2x$, then $(f \circ g)(x) = f[g(x)] = \sin g(x) = \sin 2x$ and $(g \circ f)(u) = g[f(u)] = 2 \cdot f(u) = 2 \sin u$. The next example also uses a composition of functions.

EXAMPLE 2 Graph $y = \sin(x - \pi/4)$.

SOLUTION First, let $u = x - \pi/4$, so that $x = u + \pi/4$. Next, mark both an x-scale and a u-scale along the x-axis and graph $y = \sin u$ as in Figure 5.16a. If the u-scale is removed, Figure 5.16b results. The effect is a translation or shift of the standard sine curve $\pi/4$ units to the right. ●

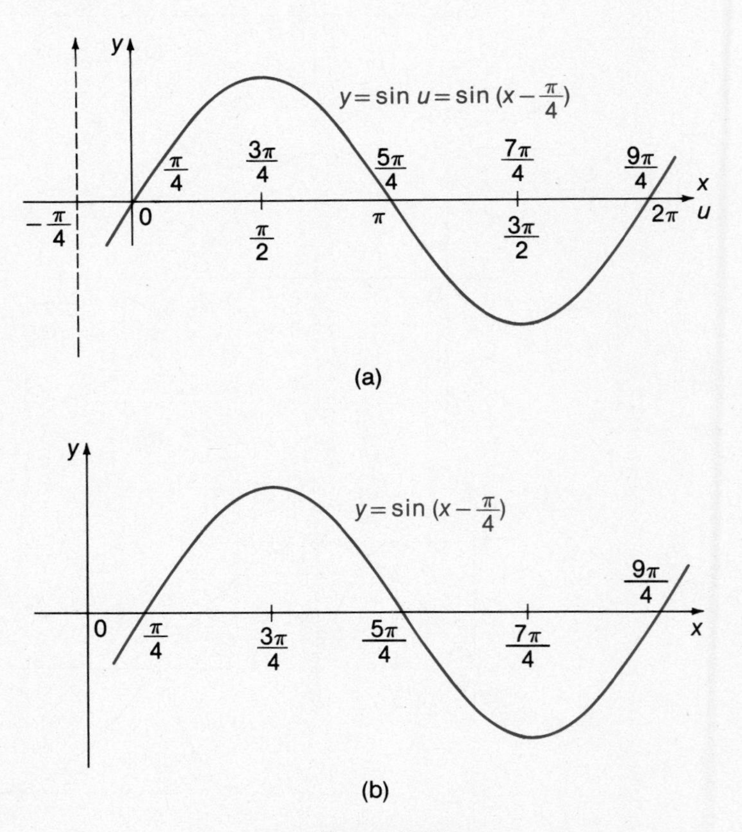

Figure 5.16

The graph of $y = \sin(x + \pi/4)$ is similarly obtained by shifting the standard sine curve $\pi/4$ units to the left. In general, a member of the family of curves $y = \sin(x - c)$ has the same shape as the standard sine wave but is shifted c units to the right if c is positive and $|c|$ units to the left if c is negative. (Refer again to Figure 3.14.)

If we consider successively the graphs of

$y = \sin x,$	standard sine curve
$y = a \sin x,$	vertical scale change
$y = a \sin kx,$	horizontal scale change
$y = a \sin k(x - c),$	horizontal translation

we build up by easy stages to the most general case, the function

$$f(x) = a \sin k(x - c).$$

This is called the **generalized sine function** and its graph is a **sinusoidal curve,** or **sinusoid.** All the curves discussed above are included in the general family of sinusoids. These are the patterns displayed on the screen of an oscilloscope. For the general sinusoid:

the **amplitude** is $|a|$,

the **period** is $2\pi/|k|$,

the **phase shift** is $|c|$.

See Figure 5.17. For the standard sine curve, $a = 1$, $k = 1$, and $c = 0$.

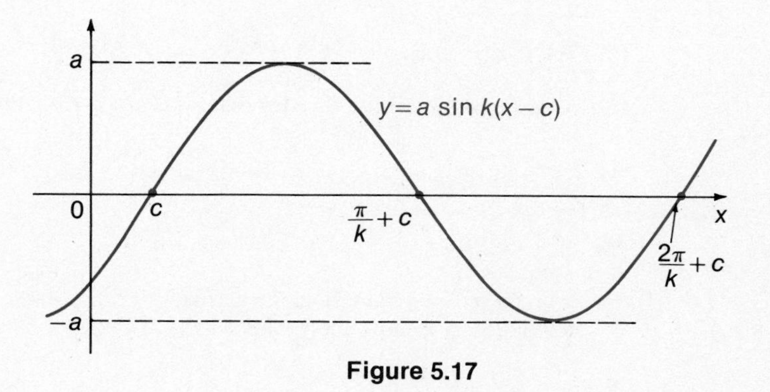

Figure 5.17

Many natural phenomena, such as sound waves, oscillations of a pendulum, and alternating electrical current, are periodic, with appropriate choice of the constants a, k, and c above. Of course, not all periodic functions are sinusoidal (see Section 2.6).

The Cosine Curve The preceding methods for graphing sine waves were given in some detail. The graphs of the other circular functions can be found by similar methods.

EXAMPLE 3 Graph $y = \cos x$.

SOLUTION If we use the fact that $\cos x = \sin(x + \frac{1}{2}\pi)$, the graph of $y = \cos x$ is simply that of $y = \sin(x + \frac{1}{2}\pi)$. This is a sine wave translated $\frac{1}{2}\pi$ units to the left relative to the standard sine curve, as shown in Figure 5.18.

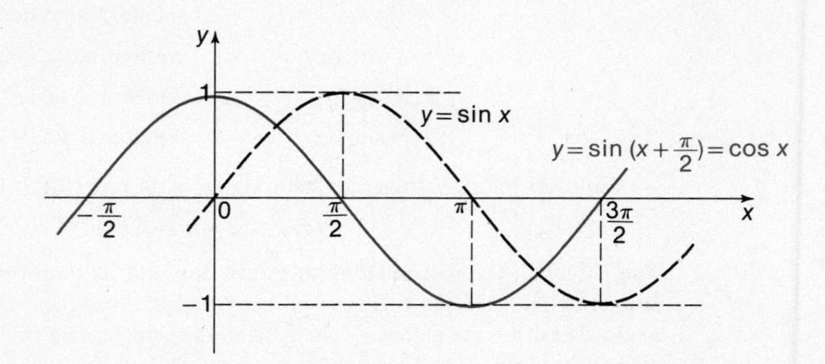

Figure 5.18

Example 3 shows that *the cosine curve is a sine wave*. For this reason, many applications use the sine and cosine curves interchangeably. An example is simple harmonic motion, which will be discussed in the next section.

EXERCISES 5.8

Graph each of the following. In each case, state the period, the amplitude, and the phase shift.

1. $y = -2 \sin x$

2. $y = 3 \sin x$

3. $y = \dfrac{1}{2} \sin x$

4. $y = -\dfrac{3}{2} \sin x$

5. $y = \sin 3x$

6. $y = \sin \dfrac{1}{2} x$

7. $y = \sin \pi x$ [*Hint:* Note the period and use it to identify the x-intercepts.]

8. $y = \sin 2\pi x$

9. $y = \sin\left(x + \dfrac{1}{2}\pi\right)$

10. $y = \sin\left(x - \dfrac{\pi}{6}\right)$

11. $y = \sin\left(x + \dfrac{\pi}{6}\right)$

12. $y = \sin\left(x + \dfrac{1}{4}\pi\right)$

13. $y = 2 \sin\left(x + \dfrac{\pi}{3}\right)$

14. $y = \sin(2x - \pi)$

15. $y = -\cos x$

16. $y = 2 \cos x$

17. $y = \cos 3x$

18. $y = 2 \cos 2x$

19. $y = \cos\left(x - \dfrac{1}{4}\pi\right)$

20. $y = -\cos\left(x + \dfrac{\pi}{3}\right)$

21. $y = 2\cos\left(x + \frac{1}{2}\pi\right)$

22. $y = \cos\left(x + \frac{\pi}{6}\right)$

Find the least and greatest values of the following.

23. $f(x) = 2 \sin x$

24. $f(x) = -\cos x$

25. $(x) = \sin\left(x + \frac{1}{4}\pi\right)$

26. $f(x) = \sin \frac{1}{2}x$

27. $f(x) = \frac{1}{2}\sin x$

28. $f(x) = -3 \cos 2x$

29. $f(x) = 1 + \sin x$

30. $f(x) = \cos x - 1$

5.9 Some Applications of Sine Wave Functions

Many periodic phenomena are described mathematically by sine wave functions. Some are given by (or at least approximated by) simple cases of these functions. Others require complicated combinations of them. We will begin with an example of simple oscillatory (back-and-forth) motion.

Simple Harmonic Motion Case I: In the simplest case of harmonic motion, as illustrated in Figure 5.19, *point $P(x, y)$ is considered to move around the unit circle counterclockwise at a uniform speed.* Let k be its **linear speed** (number of units P moves along the circle per unit of time). If we suppose that the point P is at $(1, 0)$ when $t = 0$, then the arc distance u is given by $u = kt$ after t units of time (distance equals rate times time). So $y = \sin u = \sin kt$. If a horizontal line through P intersects the y-axis at Q with coordinates $(0, y)$, then the formula $y = \sin kt$ describes the up-and-down motion of Q along the y-axis as a function of time t. This oscillatory motion is called **simple harmonic motion.**

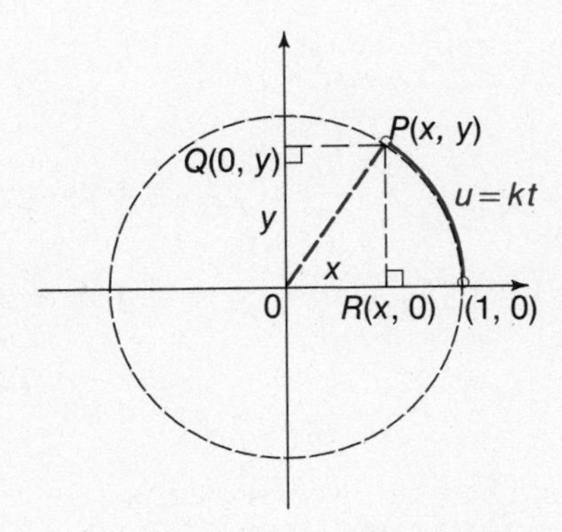

Figure 5.19

The amplitude of the motion is 1 and its period is $2\pi/k$. The moving points P and Q complete one cycle per period. The number of cycles per unit of time (called the

frequency) is the reciprocal of the period, $k/2\pi$. The back-and-forth motion of a point R along the x-axis is another instance of simple harmonic motion, with equation $x = \cos kt$.

The basic notion is extended in Cases II and III below to give a more general view of simple harmonic motion.

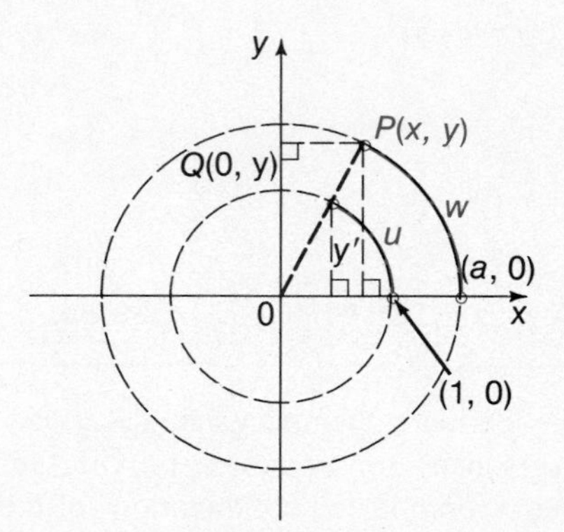

Figure 5.20

Case II: *The radius of the circle is allowed to be any number*. As in Figure 5.20, suppose that the outer circle has radius a and that P is at $(a, 0)$ when $t = 0$ and is moving counterclockwise at a uniform speed of k units per unit of time. After t units of time the arc length w is given by $w = kt$. Using geometric similarities in the figure, we have

$$\frac{w}{u} = \frac{a}{1} \quad \text{and} \quad \frac{y}{a} = \frac{y'}{1},$$

$$w = au \quad \text{and} \quad y = ay'.$$

From $y = ay'$, we get

$$\begin{aligned} y &= a \sin u \\ &= a \sin \frac{w}{a} && \text{since } w = au \\ &= a \sin \frac{kt}{a} && \text{since } w = kt \\ &= a \sin\left(\frac{k}{a}\right)t. \end{aligned}$$

Letting $b = k/a$, we write this as $y = a \sin bt$.

Case III: *The initial position of the point is anywhere on the circle*. Suppose that at time $t = 0$ the point P is at a distance w_o from $(a, 0)$, measured along the circle counterclockwise. Then t units of time later it will have moved a distance kt from w_o and so is along the circle at a distance $w = kt + w_o$ from $(a, 0)$. See Figure 5.21. Now we have

$$\begin{aligned} y &= a \sin \frac{w}{a} \qquad \text{as in Case I} \\ &= a \sin \frac{kt + w_o}{a} \\ &= a \sin\left(\frac{kt}{a} + \frac{w_o}{a}\right) \\ &= a \sin \frac{k}{a}\left(t + \frac{w_o}{k}\right). \end{aligned}$$

The final equation may be written as $y = a \sin b(t + c)$, where $b = k/a$ and $c = w_o/k$. Note that c = distance w_o/speed k = time for the particle to go from $(a, 0)$ to w_o along the circle.

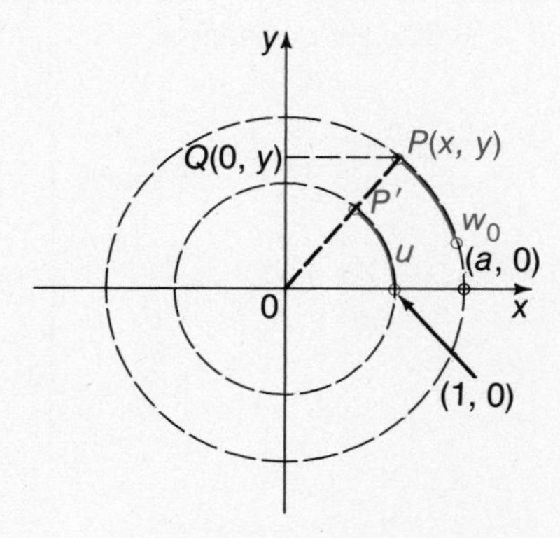

Figure 5.21

The three cases we have just described are then represented by the following formulas:
Case I: $y = \sin kt$, for motion on the unit circle, starting at the point $(1, 0)$ and moving k units per unit of time;
Case II: $y = a \sin bt$, where $b = k/a$, for motion on a circle of radius a, starting at the point $(a, 0)$ and moving k units per unit of time;
Case III: $y = a \sin b(t + c)$, where $b = k/a$ and $c = w_o/k$, for motion on a circle of radius a, starting w_o units from $(a, 0)$.

EXAMPLE 1 Write the equation of motion for a point moving at a linear speed of 5 around each of the following circles, and state the amplitude, period, and frequency.
(a) The unit circle, starting at the point $(1, 0)$

(b) A circle of radius 3, starting at the point (3, 0)

(c) A circle of radius 3, starting at the point $\left(\frac{3}{2}\sqrt{3}, \frac{3}{2}\right)$

SOLUTION (a) In this case, $y = \sin kt = \sin 5t$. So the amplitude is 1, the period is $2\pi/5$, and the frequency is $5/2\pi$.

(b) Here, $y = a \sin bt = 3 \sin(5/3)t$. The amplitude is 3, the period is $2\pi/(5/3) = 6\pi/5$, and the frequency is $5/6\pi$.

(c) In this case, $y = a \sin b(t + c) = 3 \sin (5/3)(t + \pi/30)$, since $w_o = \pi/6$ and $c = w_o/k = (\pi/6)/5 = \pi/30$. Hence the period, amplitude, and frequency are the same as in (b), but there is a phase shift of $\pi/30$. •

The sine-wave function $s(t) = a \sin b(t + c)$ defines the general case of simple harmonic motion. Its graph is a sine wave of amplitude a (the radius of the circle) and its period is $2\pi/b$, or $2\pi a/k$ (the time for P to go once around the circle). It is left as an exercise to show that this formula is equivalent to $s(t) = A \sin bt + B \cos bt$, where A and B are constants.

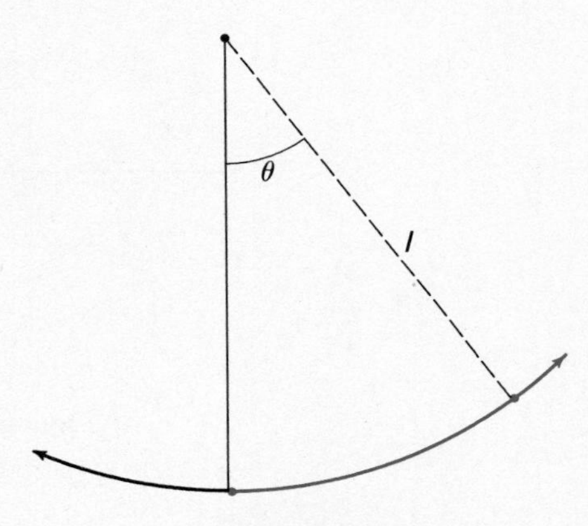

Figure 5.22

From the calculus it can be shown that for a pendulum, as shown in Figure 5.22, the angle θ is given by $\theta(t) = a \sin(\sqrt{g/l})t$, where a is the maximum angle of displacement of the pendulum arm from the vertical, g is the constant of gravitation (≈ 32), and l is the length of the pendulum arm. (Here, l is measured in feet and the angle θ in radians, where 1 radian $\approx 57°$ (see Section 6.1), and t is time in seconds. This formula is actually an approximation for small values of θ. The period of motion is $2\pi/\sqrt{32/l}$, or $\frac{1}{4}\pi\sqrt{2l}$.

EXAMPLE 2 Suppose a pendulum arm of length 2 feet is displaced 2 radians from the vertical and released. Write the formula for the motion of the pendulum and state the period and frequency.

SOLUTION In this case, $a = 2$ and $l = 2$.

$$\theta(t) = a \sin\left(\sqrt{\frac{g}{l}}\right)t$$
$$= 2 \sin\left(\sqrt{\frac{32}{2}}\right)t$$
$$= 2 \sin 4t$$

The period is $2\pi/4 = \pi/2 \approx 1.7$ (seconds) and the frequency is $2/\pi \approx 0.6$ (cycles per second). ●

Suppose a weight is placed on the end of a suspended spring and allowed to come to rest (see Figure 5.23). If it is then pulled down, stretching the spring, and released, (neglecting friction) it oscillates up and down periodically in simple harmonic motion. The equation for this motion is $s(t) = a \sin(\sqrt{k/m})t$, where a is the maximum vertical displacement from the position of rest, k is the spring constant, and m is the mass of the weight. Here the spring constant k is determined by the fact (Hooke's Law) that the force f required to stretch the spring a distance s is given by $f = ks$.

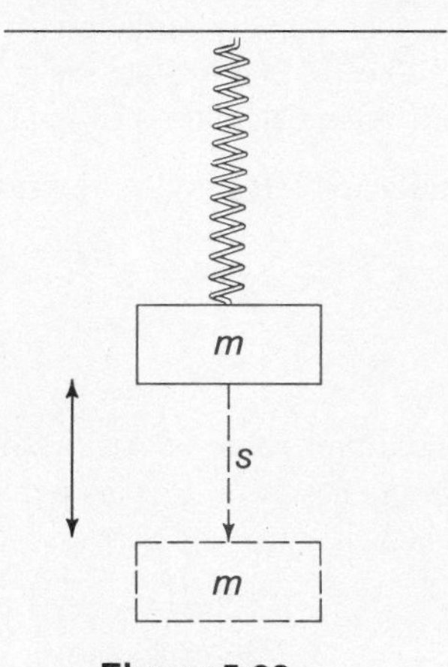

Figure 5.23

Mathematically, the formula for the motion of a weight on a spring is identical with that for the simple pendulum. Only the interpretations of the variables and constants are different. This is an example of the sort of situation that occurs quite frequently in mathematics.

The Rest of the Material in This Section is Optional.

Sound Waves Sound is produced by rapid alternations of air pressure at each point. For many simple musical sounds, if $p(t)$ represents the difference in pressure at time t from the pressure in the undisturbed air, $p(t) = a \sin kt$. The amplitude constant a and the period constant k depend upon the particular sound. Because $\sin kt \leq 1$, a represents the maximum pressure difference, and $k/2\pi$ is the frequency of the alternations of this difference. The greater the frequency, the higher the pitch of the sound. Its loudness is proportional to a^2. More complex sounds are composites of several simple sounds and are described mathematically by such expressions as $p(t) = a \sin k(t + c)$ or, equivalently, $p(t) = A \sin kt + B \cos kt$, or by much more complicated expressions in the sinusoidal functions.

Electric Currents An electric generator produces an alternating current of electricity by rotating a wire (or system of wires) between the poles of a magnet. The current flows alternately first in one and then in the opposite direction as the wire system turns through successive complete revolutions. The strength of the current is described mathematically by the sine wave function

$$I(t) = a \sin k(t - c),$$

where the constants a, k, and c depend upon the particular electric system. Thus, the current has maximum strength of a (amperes) and a frequency of $k/2\pi$ (cycles per second).

EXAMPLE 3 Suppose that an alternating current is described by the formula $I(t) = 5 \sin(120\pi t - 30\pi)$. State the maximum strength of the current, the frequency, and the period. Also calculate a time at which the current reaches its maximum strength.

SOLUTION The formula for $I(t)$ may be rewritten by factoring 120π out of the argument.

$$\begin{aligned} I(t) &= 5 \sin(120\pi t - 30\pi) \\ &= 5 \sin 120\pi\left(t - \frac{1}{4}\right) \end{aligned}$$

So the maximum value of I is 5 (amperes), the period is $2\pi/120\pi = 1/60$ (seconds), and the frequency is 60 (cycles per second). The greatest value of the current strength occurs when

$$\begin{aligned} 120\pi\left(t - \frac{1}{4}\right) &= \frac{\pi}{2}, \\ t - \frac{1}{4} &= \frac{1}{240} \\ t &= \frac{61}{240} \approx 0.25 \text{ (seconds).} \end{aligned}$$ ●

EXAMPLE 4 Write the equation describing an alternating 60-cycle current of maximum strength 20 amperes, if the maximum current occurs at time $t = 0$.

SOLUTION The general equation for alternating current is $I(t) = a \sin k(t - c)$.

Since frequency $= k/2\pi = 60$, $k = 120\pi$. Thus, $I(t) = 20 \sin 120\pi(t - c)$.

The maximum value of I occurs when $120\pi(t - c) = \pi/2$, so that $120\pi(0 - c) = \pi/2$, or $c = -1/240$. Then, finally,

$$I(t) = 20 \sin 120\pi\left(t + \frac{1}{240}\right). \qquad \bullet$$

Electromagnetic Waves Just as sound waves are produced by generating alternations in air pressure, electromagnetic waves are produced by exciting electrons into oscillatory motion. The perceptions of these waves may be as light, radio waves, x-rays, and so on, depending upon their wave lengths (or frequencies). For example, visible light to which the human eye is sensitive ranges in frequencies from 4×10^{14} cycles per second (red) to 7×10^{14} cycles per second (violet). In any case, wavelength $\times$ frequency $=$ total distance travelled per second $=$ speed of light $=$ 186,000 miles per second. Electromagnetic waves are described by the same form of sine wave function as electric current.

Damped Oscillatory Motion In the examples of the swinging pendulum and the stretched spring, we omitted the effect of friction, which causes the amplitude of the motion to diminish gradually until the pendulum or the weight comes to rest. We say that the motion has been ''damped'' by the force of friction. Most oscillatory motions are damped and the decrease in amplitude follows the pattern of exponential decay. In figure 5.24 the graphs of $s = \sin t$, $s = e^{-t}$, and $s = e^{-t} \sin t$ are shown. The latter represents a **damped oscillatory motion.** Observe that whenever $\sin t = 0$, $e^{-t} \sin t = 0$, so that the curve $s = e^{-t} \sin t$ has the same t-intercepts as $y = \sin t$. Also, whenever $\sin t = 1$, $s = e^{-t} \sin t = e^{-t}$ so that the curve $y = e^{-t} \sin t$ touches the exponential curve periodically but does not go above it.

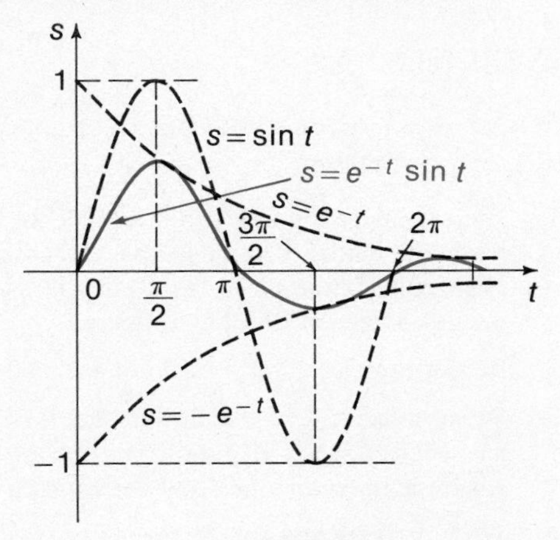

Figure 5.24

Shock absorbers are put on an automobile in order to damp oscillatory motion. Instead of the car oscillating up and down for a long while after hitting a bump or chuckhole, the oscillations are quickly damped out for a smoother ride. The general formula

$$s(t) = a\, e^{-kt} \sin b(t + c)$$

represents damped oscillatory motion, where the magnitudes of the constants a, b, c, and k determine amplitude, period, phase, and damping intensity, respectively.

Consider Figure 5.25, showing graphs of $s = e^t$, $s = \sin t$, and $s = e^t \sin t$. Here we have represented an up-and-down motion for which the amplitudes "build up" so that successive high points are higher and higher. This is a "forced" oscillatory motion caused by resonance in the system.

Most vibratory motions occurring naturally are damped motions. In calculus and in mechanics these motions are treated in detail.

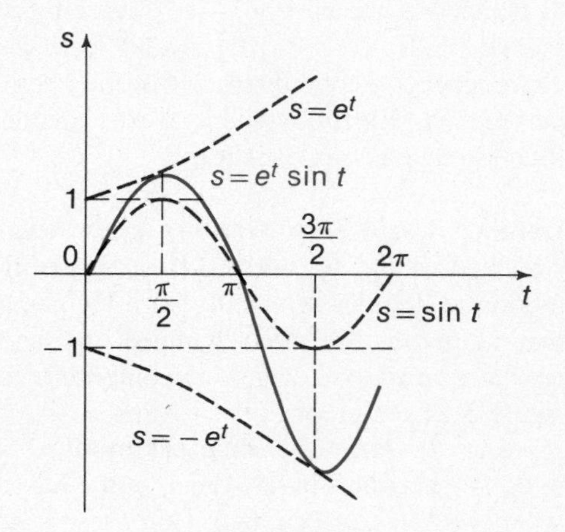

Figure 5.25

EXERCISES 5.9

If a hand calculator is not available, the use of the tables of logarithms will be helpful in some of the following exercises.

1. Write the equation and then determine the amplitude, period, and frequency of the simple harmonic motion determined by the uniform circular motion of a particle around a unit circle with linear speed (a) 2 units per second, (b) 3 units per second, (c) 4 units per second. Draw an illustrative figure.
2. Repeat exercise 1 for a circle of radius 2 units.
3. Suppose that in Exercise 2, the point P is initially (when $t = 0$) at the point $(1, \sqrt{3})$ and has a linear speed of 2 units per unit of time. Write the equation of the corresponding simple harmonic motion and determine its amplitude, period, and frequency.
4. What is the period and frequency of oscillation of a pendulum of length 1/2 foot?
5. How long should a pendulum be in order to have a period of one second?
6. Suppose that 4 pounds of force are required to stretch a spring 2 feet. (Since force $= ks$, $4 = k(2)$, so the spring constant is $k = 2$.) Let a mass of one unit be placed on the spring and allowed to come to rest. If the spring is then stretched 1/2 foot and released, what is the amplitude, period, and frequency of the resulting oscillatory motion?
7. The formula for the up-and-down motion of a weight on a spring is given by $s(t) = a \sin(\sqrt{k/m})t$. (a) Write the formulas for the period and frequency of the motion. (b) If the spring constant is $k = 4$, what mass m must be used to produce a period of 1 second?

The following exercises refer to the optional material in this section.

8. A certain generator produces alternating current whose strength is described by $I(t) = 10 \sin(120\pi t - 60\pi)$, where t is time in seconds and I is current in amperes. What is the maximum strength of the current, the frequency, period, and phase shift?

9. If a 60-cycle alternating current I has a maximum strength of 15 amperes and the current strength is 0 when $t = 0$, write the equation for $I(t)$.

10. (a) Expand $s(t) = a \sin b(t + c)$ to obtain the alternative form $s(t) = A \sin bt + B \cos bt$. (b) Write the formulas for A and B in terms of a, b, and c.

5.10 Graphs of Other Circular Functions

The graph of the tangent function may be developed by relating it to the standard unit circle, as was done for the sine function in Section 5.8. In Figure 5.26a, arc CB is of length x, so tht $\overline{OA} = \cos x$ and $\overline{AB} = \sin x$. Let CT be the tangent to the circle at $C(1, 0)$. Then, from similar triangles OCT and OAB,

$$\frac{\overline{CT}}{\overline{OC}} = \frac{\overline{AB}}{\overline{OA}} = \frac{\sin x}{\cos x} = \tan x.$$

But $\overline{OC} = 1$, so that $\overline{CT} = \tan x$. This fact is used in Figure 5.26b to determine points on the tangent curve.

As x approaches $\pi/2$, the value of $\tan x$ increases without bound, so that the graph becomes steeper and steeper and approaches the line $x = \pi/2$ asymptotically. (Recall Section 3.9.) The curve does not reach this line because $\tan \pi/2$ is undefined. We now have the portion of the curve shown in Figure 5.26b.

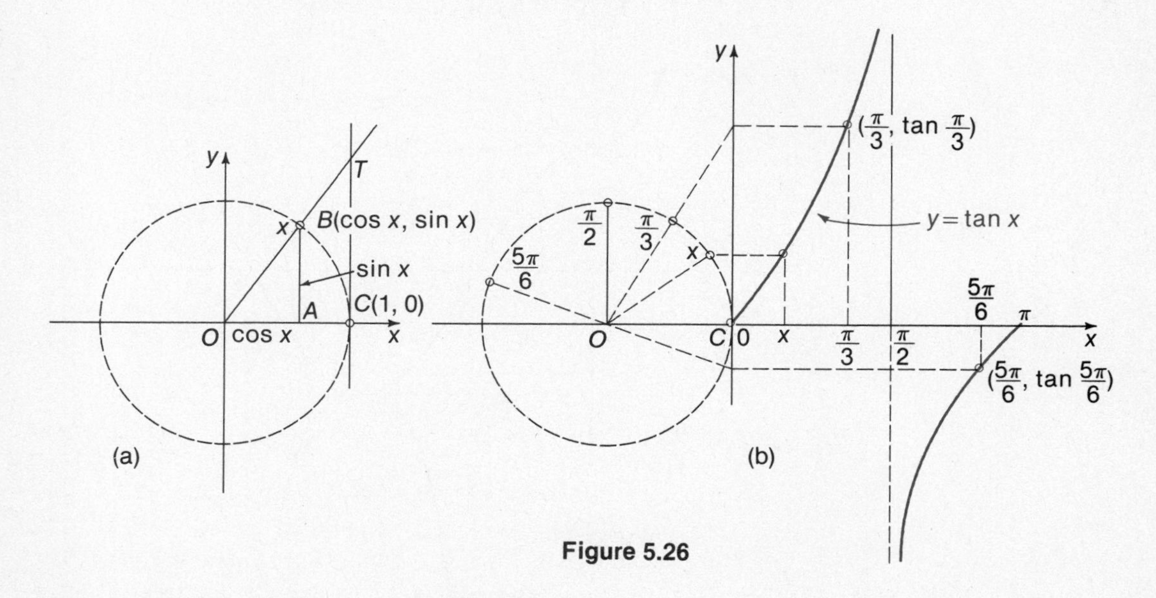

Figure 5.26

The portion of the tangent curve obtained above is now reproduced in Figure 5.27 as the part of the curve for the interval $0 \leq x < \pi/2$. We obtain the remainder of the tangent curve by using its properties. Since the tangent function is odd, that is, $\tan(-x) = -\tan x$, its graph is symmetric with respect to the origin. Thus, we have the portion of the curve in Figure 5.27 for the interval $-\pi/2 < x \leq 0$. Finally, the tangent function has period π, and so the cycle for the interval $-\pi/2 < x < \pi/2$ is repeated in the interval $\pi/2 < x < 3\pi/2$, and so on.

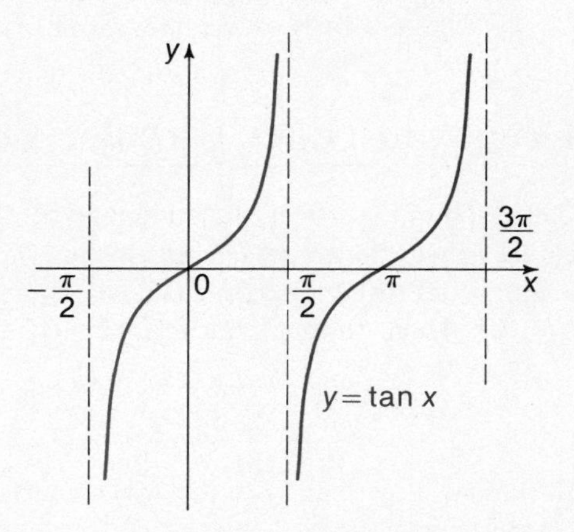

Figure 5.27

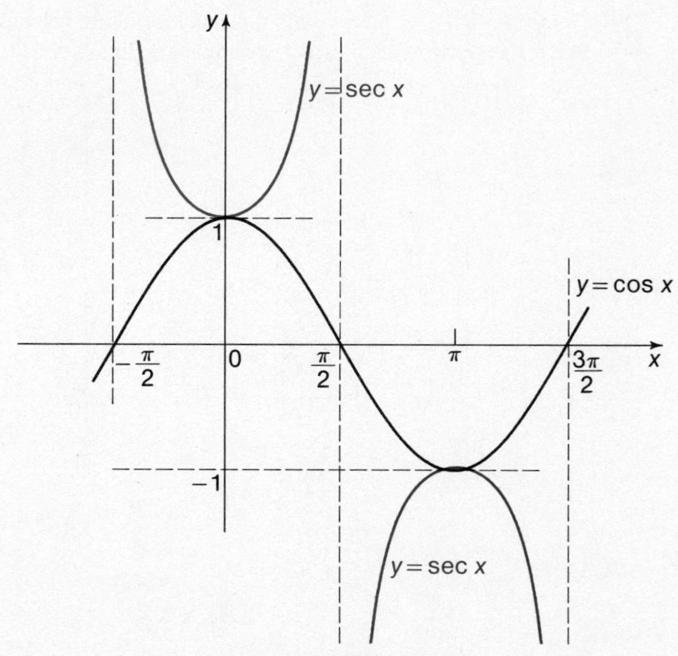

Figure 5.28

The graphs of the reciprocal functions—cotangent, secant, and cosecant—may be derived from those of the tangent, cosine, and sine already discussed. For example, to obtain the graph of $y = \sec x$, first sketch $y = \cos x$ (Figure 5.28), and then use the fact that $\sec x = 1/\cos x$. For example, when $\cos x = 0$, $\sec x$ is unbounded, when $\cos x = 1/2$, $\sec x = 2$, and when $\cos x = 1$, $\sec x = 1$. As shown in Figure 5.28, the curve is asymptotic to the lines $x = -\pi/2$, $x = \pi/2$, $x = 3\pi/2$, and so on. Since $|\cos x| \leq 1$, $|\sec x| \geq 1$, so that none of the curve lies between the lines $y = -1$ and $y = 1$. Because of periodicity, this cycle of the curve is repeated at intervals of 2π. The graphs of $y = \cot x$ and $y = \csc x$, shown in Figures 5.29 and 5.30, may be obtained in a similar fashion. Details are left to the exercises.

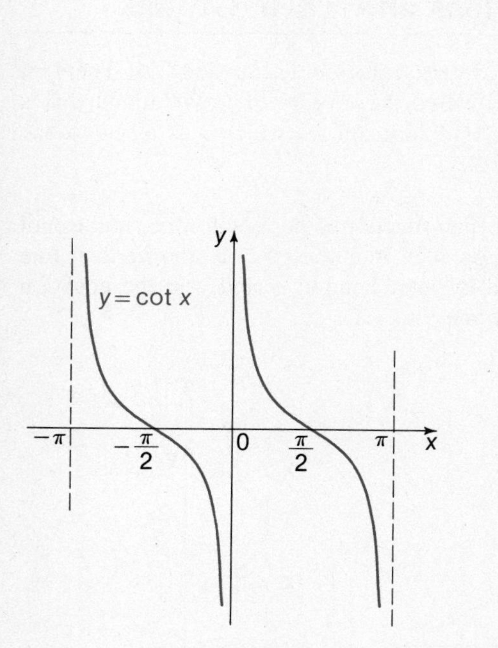

Figure 5.29

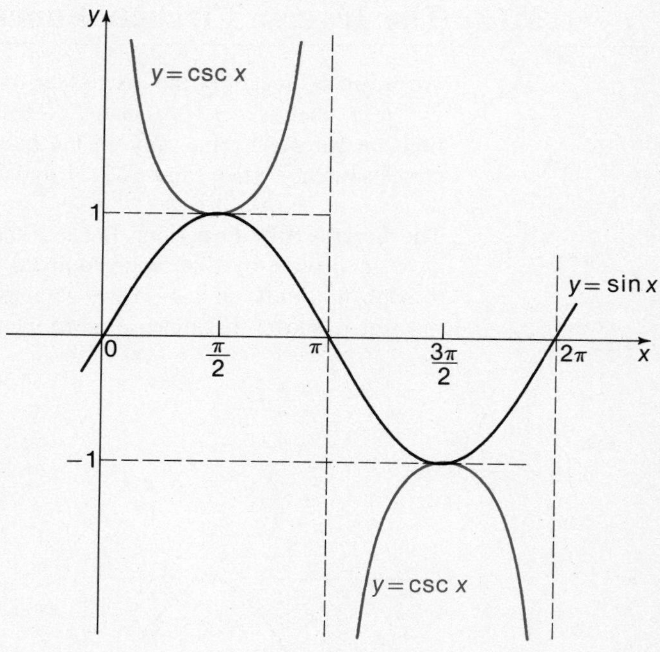

Figure 5.30

EXERCISES 5.10

Graph each of the following over one period. State the period.

1. $y = -\tan x$

2. $y = 2 \tan x$

3. $y = \frac{1}{3} \tan x$

4. $y = -\frac{1}{2} \tan x$

5. $y = \tan \frac{1}{2}x$

6. $y = \tan 2x$

7. $y = \tan\left(x - \frac{1}{4}\pi\right)$

8. $y = \tan\left(x + \frac{1}{2}\pi\right)$

9. $y = \tan(x + \pi)$

10. $y = \tan\left(x + \frac{1}{4}\pi\right)$

11. $y = -\cot x$

12. $y = \frac{1}{2} \cot x$

13. $y = -\cot 2x$ **14.** $y = \cot \frac{1}{3}x$ **15.** $y = \cot(x - \pi)$

16. $y = \cot\left(x - \frac{1}{2}\pi\right)$ **17.** $y = \cot\left(x + \frac{1}{2}\pi\right)$ **18.** $y = \cot\left(x + \frac{1}{4}\pi\right)$

Following the procedure of the text example for y = sec x, analyze each of the following functions and describe how their graphs, shown in Figures 5.29 and 5.30, are obtained.

19. cotangent **20.** cosecant

5.11 The Inverse Circular Functions and Their Graphs

We begin by recalling the discussion on inverse functions in Section 2.5. There, if $y = f(x)$, then $x = f^{-1}(y)$ and f^{-1} was called the *inverse* of f. We found that a function has an inverse only on the part of its domain for which it is a one-to-one correspondence. (See Figures 2.21 and 2.22.)

The Inverse Sine Function Because the sine function is not one-to-one, it does not have an inverse over its entire domain. As seen in Figure 5.31b, the vertical line through the graph of $x = \sin y$, obtained by interchanging x and y in the equation $y = \sin x$, shows that there is more than one y to each x.

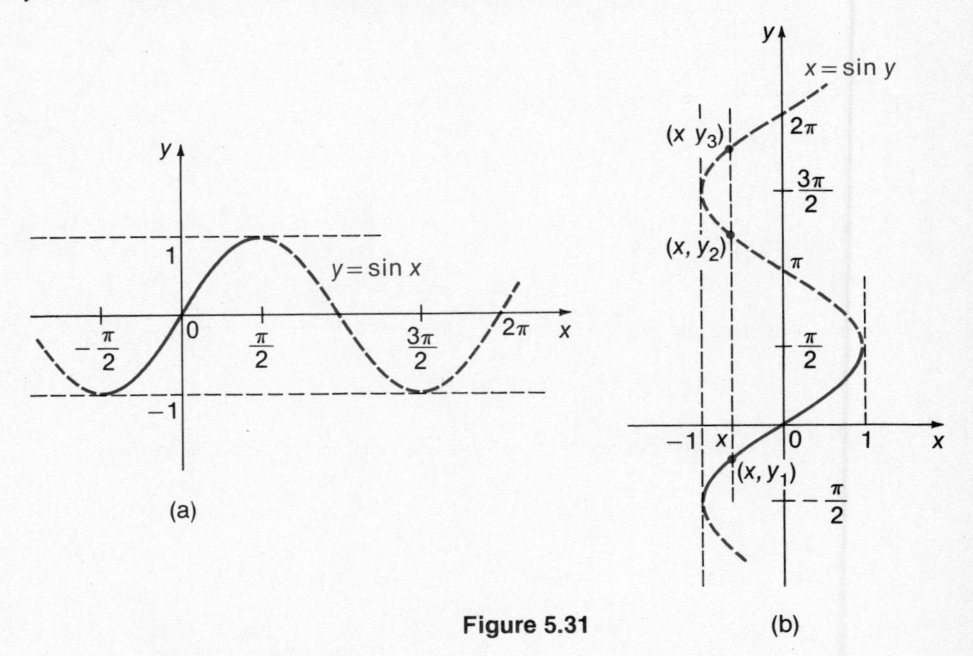

Figure 5.31

If the domain of $y = \sin x$ is restricted to the interval $[-\pi/2, \pi/2]$, as in Figure 5.31a, then for this interval the function $y = \sin x$ is one-to-one and so has an inverse on this domain. The result is shown in Figure 5.32.

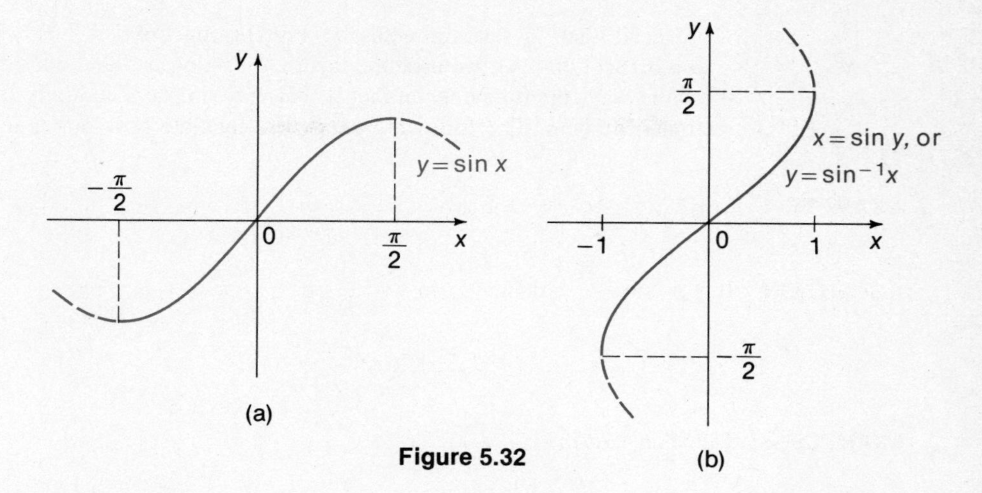

Figure 5.32

DEFINITION 5.3 The **inverse sine** function or **arcsine** function, denoted by $\sin^{-1}$ or arcsin (or Arcsin) is given by

$$y = \sin^{-1} x \quad \text{if and only if} \quad x = \sin y,$$

for $-1 \leq x \leq 1$ and $-\pi/2 \leq y \leq \pi/2$.

Thus, the inverse sine function has as its domain the closed interval $[-1, 1]$ and as range the closed interval $[-\pi/2, \pi/2]$. It is easy to see that a similar situation would occur if the domain of $y = \sin x$ were restricted to any interval on which the function is one-to-one, say $\pi/2 \leq x \leq 3\pi/2$.

In this definition, the expression $\sin^{-1}x$ is read "the number whose sine is x." The alternative expression *arcsin x* is read "arc sine x" and comes from the definition of sine. In Figure 5.33, $\sin u = b$, so "u is the arc whose sine is b," that is, $u = \arcsin b$.

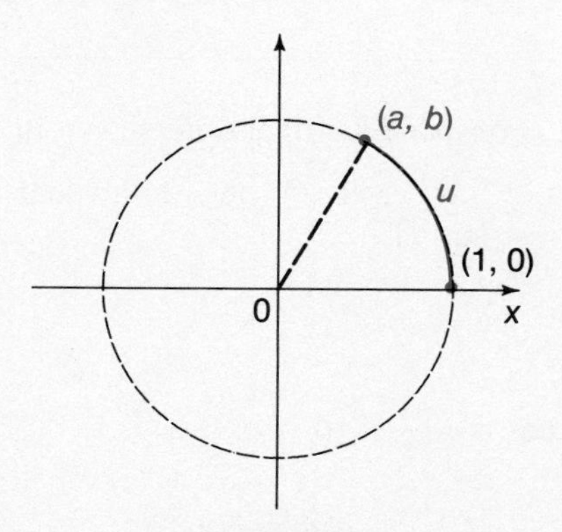

Figure 5.33

Recall that in dealing with the inverse functions $y = f(x) = a^x$ and $f^{-1}(x) = \log_a x$ (Section 4.4), we used the fact that $y = \log_a x$ if and only if $x = a^y$. Here, in the same way, the fundamental fact is that $y = \sin^{-1}x$ if and only if $x = \sin y$ for the restricted domain. The following examples illustrate how this is used.

EXAMPLE 1 Solve $y = \sin^{-1}\frac{1}{2}$ for y.

SOLUTION If $y = \sin^{-1}\frac{1}{2}$, then

$$\sin y = \frac{1}{2}, \quad -\frac{1}{2}\pi \le y \le \frac{1}{2}\pi,$$

$$y = \frac{\pi}{6}. \quad \bullet$$

EXAMPLE 2 Calculate $\arcsin(-\sqrt{2}/2)$.

SOLUTION If $y = \arcsin\left(-\frac{1}{2}\sqrt{2}\right)$, then

$$\sin y = -\frac{1}{2}\sqrt{2}, \quad -\frac{1}{2}\pi \le y \le \frac{1}{2}\pi$$

$$y = -\frac{1}{4}\pi \quad \bullet$$

EXAMPLE 3 Find $\sin^{-1}2$.

SOLUTION If $y = \sin^{-1}2$, then $\sin y = 2$, which is impossible. Hence $\sin^{-1}2$ is undefined. This is because 2 is outside the domain $[-1, 1]$ of the inverse sine function. $\bullet$

Other Inverse Circular Functions The same procedure is used to define the inverses of the other circular functions. The inverses of the sine, cosine, and tangent are the only ones of interest to us. Since the cosine function is one-to-one for the interval $0 \le x \le \pi$, and the tangent function is one-to-one on the interval $-\pi/2 < x < \pi/2$, the following definitions can be made.

DEFINITION 5.4 **(a) $y = \cos^{-1}x$ if and only if $x = \cos y$, where $-1 \le x \le 1$ and $0 \le y \le \pi$.**

(b) $y = \tan^{-1}x$ if and only if $x = \tan y$, where $-\infty < x < \infty$ and $-\pi/2 < y < \pi/2$.

The corresponding graphs are shown in Figure 5.34.

EXAMPLE 4 Find y in each of the following: (a) $y = \cos^{-1}1/2$, (b) $y = \arccos(-1)$, (c) $y = \tan^{-1}1$.

SOLUTION (a)

$$y = \cos^{-1}\frac{1}{2}$$

$$\cos y = \frac{1}{2}, \quad 0 \le y \le \pi$$

$$y = \frac{\pi}{3}$$

(b)

$$y = \arccos(-1)$$

$$\cos y = -1, \quad 0 \le y \le \pi$$

$$y = \pi$$

(c) $y = \tan^{-1} 1$

$$\tan y = 1, \quad -\frac{1}{2}\pi < y < \frac{1}{2}\pi$$

$$y = \frac{1}{4}\pi \quad \bullet$$

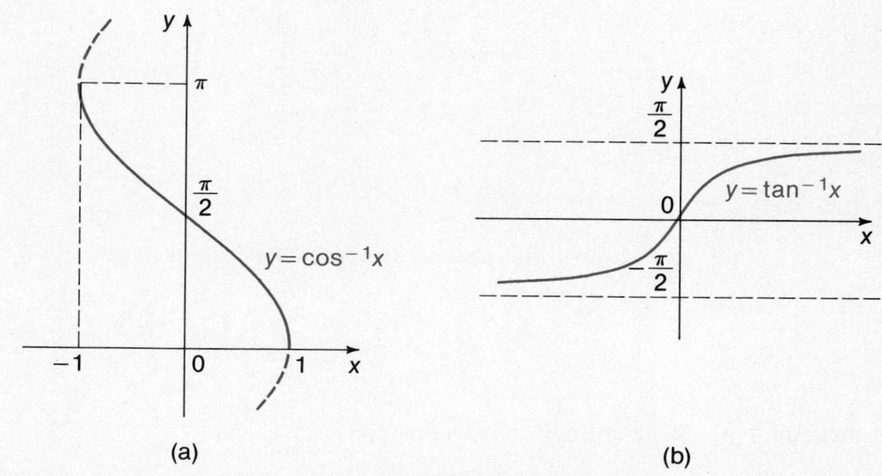

Figure 5.34

EXAMPLE 5 Solve the equation $\tan x - 3 = 0$, for $0 \le x < 2\pi$.

SOLUTION We may write

$$\tan x - 3 = 0,$$

$$\tan x = 3,$$

$$x = \arctan 3, \quad \text{or} \quad x = \pi + \arctan 3. \quad \bullet$$

The above example illustrates how the language of inverse trigonometric functions can be used in the solution of a trigonometric equation.

EXAMPLE 6 Solve $y = 3 \arccos \frac{1}{4}x$ for x.

SOLUTION First, solve for $\arccos \frac{1}{4}x$ by multiplying both sides by 1/3.

$$\frac{1}{3}y = \arccos \frac{1}{4}x$$

$$\cos \frac{1}{3}y = \frac{1}{4}x, \qquad \text{by definition of arccosine}$$

$$x = 4 \cos \frac{1}{3}y. \quad \bullet$$

EXAMPLE 7 Find $\sin(\cos^{-1}2/5)$.

SOLUTION Let $y = \cos^{-1}2/5$ and find $\sin y$.

$$\cos y = \frac{2}{5}, \quad 0 \le y \le \pi \qquad \text{by definition of } \cos^{-1}x$$

$$\begin{aligned}\sin^2 y &= 1 - \cos^2 y \\ &= 1 - \left(\frac{2}{5}\right)^2 \\ &= 1 - \frac{4}{25} = \frac{21}{25}. \\ \sin y &= \frac{1}{5}\sqrt{21}.\end{aligned}$$

The positive root is chosen because $0 \le y \le \pi$. Then

$$\sin\left(\cos^{-1}\frac{2}{5}\right) = \frac{1}{5}\sqrt{21}. \qquad \bullet$$

EXAMPLE 8 Write $\cos(\tan^{-1}x)$ in terms of x.

SOLUTION Let $y = \tan^{-1}x$. Then $x = \tan y$, and $-\pi/2 < y < \pi/2$. We use the identity $\sec^2 y = \tan^2 y + 1$ to find $\sec y$ and then find $\cos y$.

$$\begin{aligned}\tan^2 y + 1 &= x^2 + 1 \\ \sec^2 y &= x^2 + 1 \\ \sec y &= \sqrt{x^2 + 1}.\end{aligned}$$

The positive root is chosen because $\sec y > 0$ for $-\pi/2 < y < \pi/2$. Finally,

$$\begin{aligned}\cos y &= \frac{1}{\sec y} = \frac{1}{\sqrt{x^2 + 1}}, \\ \cos(\tan^{-1}x) &= \frac{1}{\sqrt{x^2 + 1}}. \qquad \bullet\end{aligned}$$

EXAMPLE 9 Graph $y = \arcsin \frac{1}{2}x$.

SOLUTION First, solve $y = \arcsin \frac{1}{2}x$ for x as follows.

$$\begin{aligned}\frac{x}{2} &= \sin y, \qquad -\frac{\pi}{2} \le y \le \frac{\pi}{2} \\ x &= 2 \sin y, \qquad -2 \le x \le 2\end{aligned}$$

Now, graph this by interchanging the values of x and y in the graph of $y = 2 \sin x$ (see Figure 5.14). The result is shown in Figure 5.35. $\bullet$

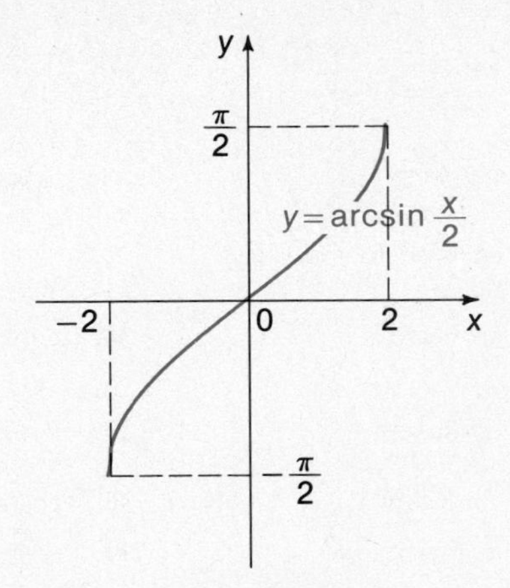

Figure 5.35

The key principles in solving problems like those in the preceding examples are: (1) for each statement involving an inverse function there is a corresponding statement involving the more familiar primary function, and (2) the domains of the inverse circular functions are restricted. Thus,

$$y = \arcsin x \quad \text{only if} \quad x = \sin y \quad \text{and} \quad -\frac{\pi}{2} \leq y \leq \frac{\pi}{2},$$

$$y = \arccos x \quad \text{only if} \quad x = \cos y \quad \text{and} \quad 0 \leq y \leq \pi,$$

$$y = \arctan x \quad \text{only if} \quad x = \tan y \quad \text{and} \quad -\frac{\pi}{2} < y < \frac{\pi}{2}.$$

EXERCISES 5.11

Find each of the following without using tables or a calculator.

1. $\arcsin \frac{1}{2}\sqrt{3}$ **2.** $\arcsin \frac{1}{2}\sqrt{2}$ **3.** $\sin^{-1} 1$ **4.** $\arccos\left(-\frac{1}{2}\sqrt{3}\right)$

5. $\arccos 0$ **6.** $\cos^{-1} \frac{1}{2}\sqrt{3}$ **7.** $\cos^{-1} \frac{1}{2}\sqrt{2}$ **8.** $\arctan \sqrt{3}$

9. $\arctan 0$ **10.** $\tan^{-1}\left(-\frac{1}{3}\sqrt{3}\right)$ **11.** $\arcsin\left(-\frac{1}{2}\right)$ **12.** $\arcsin\left(-\frac{1}{2}\sqrt{3}\right)$

13. $\sin^{-1} \sqrt{2}$ **14.** $\arccos(-1)$ **15.** $\arccos\left(-\frac{1}{2}\sqrt{2}\right)$ **16.** $\cos^{-1} 1$

17. $\cos^{-1}\left(\frac{3}{2}\right)$ **18.** $\arctan(-1)$ **19.** $\arctan\left(\frac{1}{3}\sqrt{3}\right)$ **20.** $\tan^{-1}(-\sqrt{3})$

Solve for x in terms of an inverse function.

21. $\sin x = \frac{1}{4}$ **22.** $3 \sin x + 1 = 0$ **23.** $y = \arccos 2x$ **24.** $y = 2 \arcsin \frac{1}{4}x$

25. $\cos x = -\dfrac{1}{3}$

26. $4 \cos x - 1 = 0$

27. $y = \arcsin 3x$

28. $y = -\arccos \dfrac{1}{2}x$

Find each of the following.

29. $\sin\left(\arccos \dfrac{1}{2}\right)$

30. $\cos\left(\sin^{-1} \dfrac{2}{3}\right)$

31. $\tan\left(\cos^{-1} \dfrac{3}{5}\right)$

32. $\cos(\arcsin 1)$

33. $\cos(\tan^{-1} 2)$

34. $\sin\left(\cos^{-1} \dfrac{3}{4}\right)$

35. $\sin(\arcsin x)$

36. $\cos(\arcsin x)$

37. $\sec(\arccos x)$

38. $\tan(\arccos x)$

39. $\cos(\arccos x)$

40. $\sin(\arccos x)$

41. $\cot(\arctan x)$

42. $\sec(\arctan x)$

43. $\sin(2 \arcsin x)$

44. $\cos(2 \arcsin x)$

45. $\sin(\arcsin 2x)$

46. $\cos(\arcsin 2x)$

Graph each of the following.

47. $y = 2 \arcsin x$

48. $y = \arcsin 2x$. [*Hint:* $2x = \sin y$]

49. $y = \arccos 3x$

50. $y = \dfrac{1}{2} \arcsin 3x$

51. $y = \dfrac{1}{2} \arccos x$

52. $y = \arcsin 3x$

53. $y = \dfrac{1}{4} \arccos \dfrac{1}{4}x$

54. $y = 2 \arccos 2x$

5.12 Approximating Circular Functions by Polynomials (Optional)

The following examples, in which factorial notation is used (with this notation, $2 \cdot 1$ is written as $2!$, $3 \cdot 2 \cdot 1$ as $3!$, and so on) give polynomial approximations for $\sin x$, $\cos x$, and $\arcsin x$.

$$\sin x \approx x - \frac{x^3}{3!} + \frac{x^5}{5!} - \frac{x^7}{7!},$$

$$\cos x \approx 1 - \frac{x^2}{2!} + \frac{x^4}{4!} - \frac{x^6}{6!},$$

$$\arcsin x \approx x + \frac{x^3}{6} + \frac{3x^5}{40}.$$

Notice that these formulas give exact values for $\sin 0$ and $\cos 0$, since $\sin 0 = 0$ and $\cos 0 = 1$.

EXAMPLE 1 Use the terms of the above polynomial for $\sin x$ to approximate $\sin 0.7$.

SOLUTION

$$\sin x \approx x - \frac{x^3}{3!} + \frac{x^5}{5!} - \frac{x^7}{7!}$$

$$\sin 0.7 \approx 0.7 \qquad \text{Using only the first term}$$

$$\sin 0.7 \approx 0.7 - \frac{(0.7)^3}{6} \qquad \text{Using two terms}$$

$$\approx 0.7 - 0.0572$$

$$\approx 0.6428$$

$$\sin 0.7 \approx 0.7 - \frac{(0.7)^3}{6} + \frac{(0.7)^5}{120} \qquad \text{Using three terms}$$

$$\approx 0.7 - 0.0572 + 0.0014$$

$$\approx 0.6442 \quad \bullet$$

This procedure may be continued by using higher and higher degree polynomials to obtain better and better approximations. On a hand calculator this calculation is relatively simple. In practice, the entries in tables of circular function values are obtained from these polynomials by means of a computer.

EXAMPLE 2 In calculus it is shown that the function $\arctan x$ is approximated by the polynomial $x - x^3/3 + x^5/5 - x^7/7$. Use this polynomial to find an approximate value for π.

SOLUTION

$$\arctan x \approx x - \frac{x^3}{3} + \frac{x^5}{5} - \frac{x^7}{7}$$

$$\arctan 1 \approx 1 - \frac{1}{3} + \frac{1}{5} - \frac{1}{7}$$

$$\frac{\pi}{4} \approx 1 - \frac{1}{3} + \frac{1}{5} - \frac{1}{7}$$

$$\pi \approx 4 - \frac{4}{3} + \frac{4}{5} - \frac{4}{7}$$

$$\pi \approx 4 - 1.3333 + 0.8000 - 0.5714$$

$$\pi \approx 2.8956.$$

Of course, more terms would be needed for a better approximation. •

In calculus and advanced mathematics, polynomials such as those above and in Section 4.5 are extended infinitely and play a fundamental role in the study of functions. A brief consideration of this idea appears in Section 10.3.

EXERCISES 5.12

Use the first three terms of the appropriate approximating polynomial to approximate each of the following. Compare results with those shown by a calculator.

1. sin 0 **2.** sin 0.1 **3.** sin 0.5 **4.** sin 1 **5.** cos 0 **6.** cos 0.5
7. cos 1 **8.** cos 1.5 **9.** arctan 0 **10.** arctan 0.1 **11.** arctan 0.5 **12.** arctan 1

Chapter 5 Review Exercises

Evaluate each of the following, either from memory or from the unit circle.

1. $\sin \frac{\pi}{4}$ **2.** $\cos\left(-\frac{\pi}{6}\right)$ **3.** $\sin \frac{7\pi}{3}$ **4.** $\cos \frac{2\pi}{3}$ **5.** $\tan \frac{5\pi}{4}$

6. $\tan\left(-\frac{\pi}{3}\right)$ **7.** $\tan \frac{8\pi}{3}$ **8.** $\sin\left(-\frac{3\pi}{4}\right)$ **9.** $\cos(-5\pi)$ **10.** $\cos \frac{17\pi}{3}$

Give the value(s) of u in fractional-radical form for the following exercises, if $0 \le u < 2\pi$.

11. $\sin u = \frac{1}{2}\sqrt{3}$ **12.** $\cos u = -1$ **13.** $\cos u = \frac{1}{2}\sqrt{3}$ **14.** $\sin u = -\frac{1}{2}$

15. $\sin u = 0$ **16.** $\tan u = \sqrt{3}$ **17.** $\tan u = -1$ **18.** $\cot u = -\sqrt{3}$

19. $\sin u = -\frac{1}{2}\sqrt{2}$ **20.** $\tan u = 0$

21. Given the value of one of the circular functions, calculate the other five function values. (a) $\sin u = 1/2$ and u in quadrant I, (b) $\tan u = -2/3$ and u in quadrant II, (c) $\sec u = -2$ and u in quadrant III.

Prove each of the following identities.

22. $\sec^2 x + \csc^2 x = \sec^2 x \csc^2 x$

23. $\cos(x + y)\cos(x - y) = \cos^2 x - \sin^2 y$

24. $\cot x - \tan x = 2 \cot 2x$

25. $(\sin x + \cos x)^2 = 1 + \sin 2x$

26. $4 \sin^2 \frac{1}{2}x \cos^2 \frac{1}{2}x = \sin^2 x$

27. $\csc x - \cot x = \tan \frac{1}{2}x$

28. $\dfrac{2 \tan x}{1 + \tan^2 x} = \sin 2x$

29. $\dfrac{1 - \tan^2 x}{1 + \tan^2 x} = \cos 2x$

30. $\dfrac{\tan x - 1}{1 - \cot x} = \tan x$

Solve for x, if $0 \le x < 2\pi$.

31. $2 \cos^2 x + 5 \sin x = 4$ **32.** $3 \sec^2 x + \tan x = 5$ **33.** $\cos 2x + \cos x = 0$

34. $\tan 2x = \cot x$ **35.** $\cos \frac{1}{2}x - \cos x = 1$ **36.** $\sin^2 x + 3 \sin x + 1 = 0$

Graph each of the following.

37. $y = 2\sin\left(x - \frac{\pi}{3}\right)$ **38.** $y = -\cos\left(x + \frac{\pi}{3}\right)$ **39.** $y = \tan\left(x + \frac{1}{4}\pi\right)$

40. $y = 1 + \sin x$ **41.** $y = \arcsin \frac{1}{2}x$ **42.** $y = 2 \arccos x$

Write from memory the indicated formulas for each of the functions: sine, cosine, and tangent.

43. The sum formulas

44. The difference formulas

45. The double-value formulas

46. The half-value formulas

47. Derive at least one of the formulas of each of the four sets of formulas in exercises 43–46 above.

Evaluate each of the following without the use of either tables or calculator.

48. $\arcsin \frac{1}{2}$ **49.** $\arccos\left(-\frac{1}{2}\right)$ **50.** $\arctan 1$

51. $\arccos \frac{1}{2}\sqrt{2}$ **52.** $\arcsin\left(-\frac{1}{2}\sqrt{3}\right)$ **53.** $\arctan(-1)$

In your own words, express clearly what is meant by:

54. $\sin u$, where u is any real number;

55. $\tan u$, where u is a real number;

56. the fact that the cosine function is even;

57. the fact that the cotangent function has period π.

58. For the cosine function, state the (a) domain, (b) range, (c) period, (d) amplitude.

59. Use the definition of the sine and cosine functions to prove that $\sin^2 u + \cos^2 u = 1$ for *all* real numbers.

Use the appropriate sum formula to prove the double-value formula for

60. The sine function

61. The cosine function

62. The tangent function

Given that $\sin 2 \approx 0.91$, *evaluate each of the following without the use of tables or calculator.*

63. $\sin(-2)$ **64.** $\sin(\pi - 2)$ **65.** $\sin(\pi + 2)$ **66.** $\sin\left(\frac{1}{2}\pi - 2\right)$

67. Write an equation of the simple harmonic motion of a point on the rim of a wheel of radius 0.5 meter, rotating such that the point is moving 0.8 meter per second. Then state the period, frequency, and amplitude of the motion of the point.

68. Calculate the period and the frequency of the oscillation of a pendulum 1.2 meters long.

69. Let a mass of 2 kilograms be attached to the end of a spring whose constant is $k = 1$. If the spring is stretched 1/4 meter, what is the period, frequency, and amplitude of the resulting simple harmonic motion?

Chapter 5 Miscellaneous Exercises

If $f(x) = e^x$, $g(x) = \ln x$, and $h(x) = \sin x$, write

1. $f[g(x)]$
2. $g[f(x)]$
3. $g[h(x)]$
4. $h[g(x)]$
5. $f[h(x)]$
6. $h[f(x)]$

Calculate:

7. $\ln \tan \frac{1}{4}\pi$
8. $e^{\sin \pi/6}$
9. $\sin \ln 1$
10. $\arctan e^0$

Prove the following identities. (They first appeared in an early mathematics book by Vieta, published in 1579.)

11. $\csc x + \cot x = \cot \frac{1}{2}x$
12. $\csc x - \cot x = \tan \frac{1}{2}x$
13. Distinguish clearly between the meanings of (a) $\sin^2 x$ and $\sin x^2$; (b) $\sin^{-1} x$, $\sin x^{-1}$, and $(\sin x)^{-1}$.

Use the meaning of periodicity and the fact that the sine and cosine functions have period 2π to show that each of the following functions also has period 2π.

14. $f(x) = \sin x - \cos x$
15. $f(x) = \sin x + \sin 2x$

Graph

16. $y = \sin^2 x$
17. $y = \sin |x|$
18. $y = \sin x \cos x$.
    [*Hint:* Use the double-value formula.]
19. $y = |\sin x|$
20. $y = x + \sin x$

Given the functions $f(x) = 3x$ and $g(x) = \cos x$.

21. Write $f[g(x)]$ and $g[f(x)]$.
22. Is $f[(g(x)] = g[f(x)]$ true?

Use the same functions f and g as above and state whether or not each of the following functions is periodic. If so, state the period.

23. $f(x)$
24. $g(x)$
25. $f[g(x)]$
26. $g[f(x)]$

6

The Trigonometric Functions and Applications

Chapter 5 was concerned with the circular functions, whose domains are sets of real numbers. Historically, these functions grew out of the use of certain ratios of the lengths of sides of a right triangle, called *trigonometric ratios* (Section 6.3). They have many applications in engineering, physics, astronomy, and other areas.

6.1 Angles and Angle Measure

An **angle** is the union of two **rays** (half-lines) having a common initial point. This common point is called the **vertex** of the angle, and the rays the **sides** of the angle. An angle may be thought of as resulting from the rotation of one of the rays about the vertex into the position of the other ray. The first ray is called the **initial side** and the other ray the **terminal side** of the angle. (See Figure 6.1.) The rotation may be in either

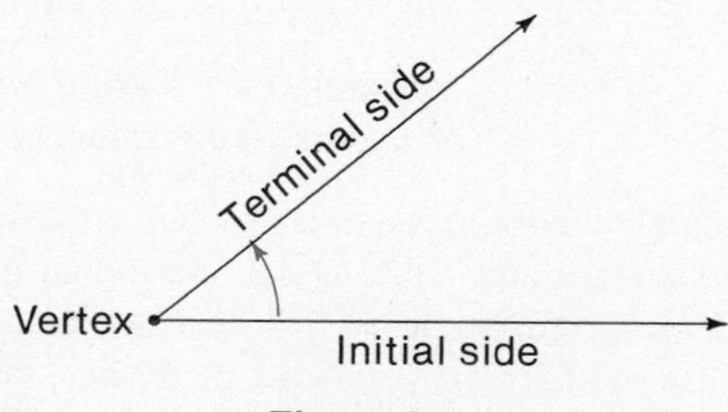

Figure 6.1

a clockwise or counterclockwise direction, and once the direction of rotation has been designated, the angle is called a **directed angle.** An angle is called **positive** if the rotation is counterclockwise, and **negative** if the rotation is clockwise. In Figure 6.2, the curved arrows indicate the direction of rotation. This figure also shows that an unlimited number of angles may have the same pair of sides. The angles of such a family are said to be **coterminal.**

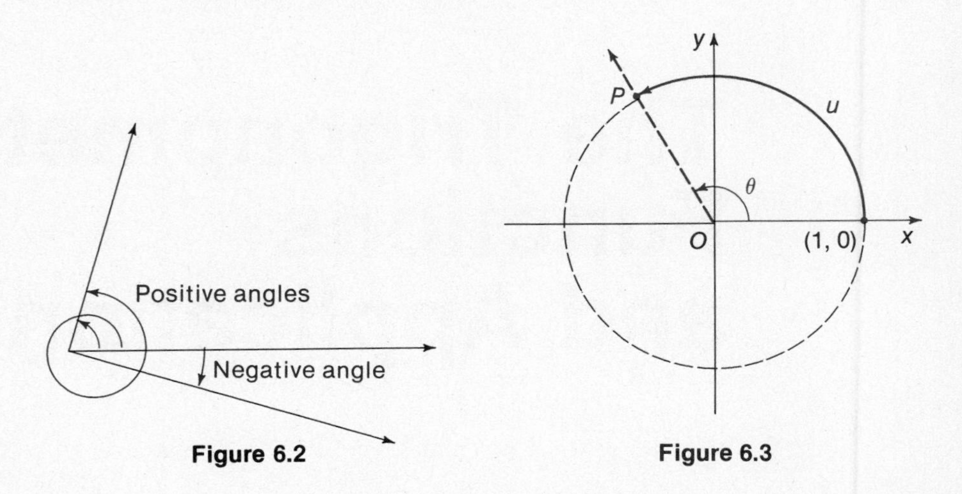

Figure 6.2 **Figure 6.3**

In Figure 6.3, the Greek letter θ (theta—Greek letters are often used to designate angles) denotes an angle having vertex at the origin O of the coordinate plane and initial side along the positive x-axis. Such an angle is in **standard position.** Any angle may be translated to standard position. If the terminal side of an angle in standard position falls along a coordinate axis, the angle is called a **quadrantal angle.** If it falls in, say, the second quadrant, as θ does in Figure 6.3, the angle is said to be a "second-quadrant angle."

Now suppose that angle θ is in standard position and that its terminal side intersects the unit circle in the point P whose circular coordinate is u. (See Section 5.1.) We can thus establish a one-to-one correspondence between the set of *directed angles* θ and the set of real numbers u. The number u is taken as the measure of the corresponding angle θ. Thus, for a right angle, the measure is $\pi/2$. The measure u is positive for a counterclockwise rotation and negative for a clockwise rotation. This accounts for our choice of the terms *positive* and *negative* directed angles.

The angle corresponding to $u = 1$ is the unit of angle measure, called a **radian.**

An angle of 1 radian with vertex at the center of a unit circle intercepts an arc of length 1.

An angle whose measure is 1 radian is shown in Figure 6.4. Since the circumference of the standard unit circle is 2π, the radian measure of an angle of one revolution is 2π and of a right angle is $\pi/2$. All quadrantal angles have radian measures which are multiples of $\pi/2$: $0, \pm\pi/2, \pm\pi, \pm 3\pi/2$, and so on. A first-quadrant angle has a radian measure greater than 0 and less than $\pi/2$, and so on.

In Section 1.1, the terms *number* and *point* were used interchangeably in relation to the number line. Similarly, from now on we will often say "angle u" for "the angle whose measure is u." Thus, in Figure 6.3, θ could designate either the angle or its

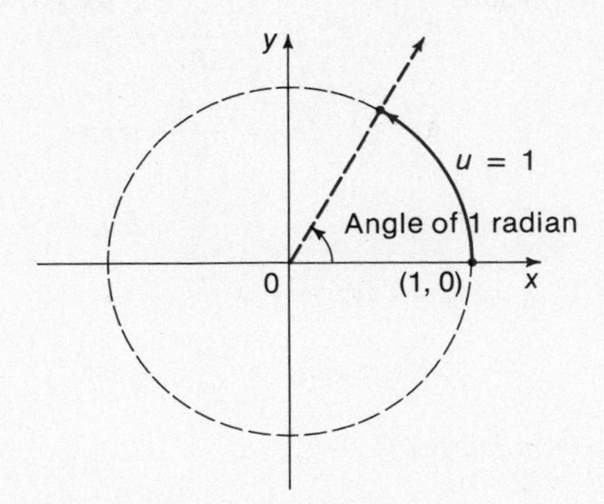

Figure 6.4

measure. We will write, for example, "a right angle equals $\pi/2$," for "a right angle has radian measure $\pi/2$." The context will make the meaning clear.

In elementary geometry the unit of angle measure is the **degree:** An angle of 1 degree is 1/360 of an angle of one revolution. This system originated with the Babylonians whose numeration was based on 60. As smaller units they took minutes and seconds, where 1 degree (1°) = 60 minutes ($60'$) and 1 minute = 60 seconds ($60''$). The degree-minute-second system of angle measure is the one used in geometry, surveying, navigation, and astronomy. However, in much of mathematics the system of radian measure is the natural one. We therefore need to be able to convert angle measure from one system of units to the other. The important relationships are listed below.

$$1 \text{ revolution} = 360^\circ = 2\pi \text{ radians } (\approx 6.283 \text{ radians})$$

$$\pi \text{ radians} = 180^\circ$$

$$1 \text{ radian} = \frac{180^\circ}{\pi}$$

$$1 \text{ degree} = \frac{\pi}{180} \text{ radians}$$

Only one of the above formulas,

$$\boldsymbol{\pi} \textbf{ radians} = \mathbf{180^\circ},$$

needs to be memorized. The others may be derived from it, as seen in the following example.

EXAMPLE 1 Convert (a) 75° to radian measure; (b) $\pi/3$ radians to degree measure; (c) 4 radians to degree measure; (d) a right angle to radian measure.*

SOLUTION (a) Since $180° = \pi$ radians,

$$1° = \frac{\pi}{180} \text{ radians}, \qquad \text{dividing by 180}$$

$$75° = \frac{75\pi}{180} \text{ radians}, \qquad \text{multiplying by 75}$$

$$= \frac{5\pi}{12} \text{ radians} \approx 1.3 \text{ radians.}$$

(b) π radians $= 180°$; therefore,

$$\frac{\pi}{3} \text{ radians} = \frac{180°}{3} = 60°$$

(c) π radians $= 180°$

$$1 \text{ radian} = \left(\frac{180}{\pi}\right)^{\circ}$$

$$4 \text{ radians} = 4\left(\frac{180}{\pi}\right)^{\circ}$$

$$= \left(\frac{720}{\pi}\right)^{\circ} \approx 229°$$

(d) 1 right angle $= 90°$; and,

$$90° = 90\left(\frac{\pi}{180}\right) \text{ radians} = \frac{\pi}{2} \text{ radians}$$ •

The use of radian measure often simplifies many problems involving angles. This is shown in the following derivations of two useful formulas. In Figure 6.5, we say that angle θ is **subtended** by the **arc length** (length of the arc) s and that the arc of length s on the circle is **intercepted** by the sides of the angle θ. The ratio of the radian measure of angle θ to the measure of an angle of 1 revolution (2π radians) is the same as the ratio of the arc length s to the circumference $2\pi r$ of the circle. That is,

$$\frac{\theta}{2\pi} = \frac{s}{2\pi r}.$$

Multiplying both sides by 2π gives

$$\boldsymbol{\theta = \frac{s}{r}} \qquad \text{or} \qquad \boldsymbol{s = r\theta}. \tag{1}$$

*On a scientific hand calculator, the conversions are made directly without separate calculations.

The first formula expresses the radian measure of a central angle as the ratio of the length of the intercepted arc to the radius of the circle. The second one says that the length of the intercepted arc is the product of the radius of the circle and the radian measure of the subtended angle. Observe that if $r = 1$, then $s = \theta$, so that *in a unit circle, the length of the intercepted arc is the measure of the central angle in radians.*

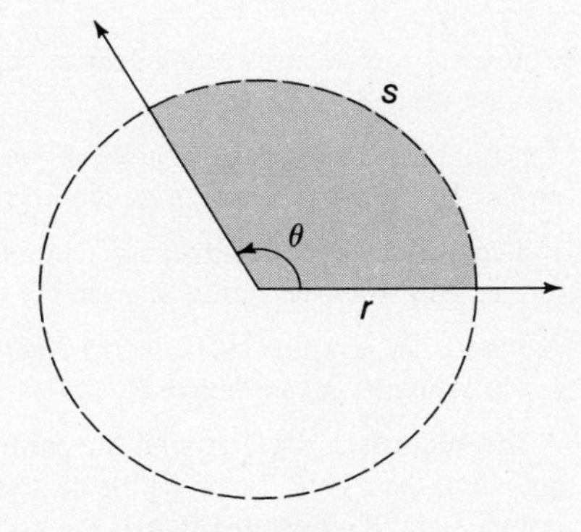

Figure 6.5

Now consider the area of the circular sector shaded in Figure 6.5. Here the ratio of the area of the sector to that of the circular region is the same as the ratio of the central angle θ to 2π (angle of one revolution). That is,

$$\frac{A \quad \text{(area of sector)}}{\text{Area of circular region}} = \frac{\theta \quad \text{(radians)}}{2\pi \quad \text{(radians)}},$$

$$\frac{A}{\pi r^2} = \frac{\theta}{2\pi},$$

$$A = \frac{1}{2} r^2 \theta. \tag{2}$$

EXAMPLE 2 Calculate (a) the angle θ subtended at the center of a circle of radius 8 inches by an arc of length 30 inches; (b) the area of the sector corresponding to this angle.

SOLUTION (a) By formula (1), $\theta = \dfrac{s}{r} = \dfrac{30}{8} = \dfrac{15}{4} = 3.75$ (radians).

(b) Substitute the value found in (a) for θ in formula (2).

$$A = \frac{1}{2} r^2 \theta = \frac{1}{2}(8)^2\left(\frac{15}{4}\right) = 120 \text{ (square inches).}$$ •

EXERCISES 6.1

1. Angles which are special fractional parts of one revolution are particularly important. Give both the degree measure and the radian measure of each angle in the following table.

Revolutions	1/12	1/8	1/6	1/4	1/3	3/8	5/12	1/2
Degree measure								
Radian measure (fractions of π)								

2. Convert 1 radian to its equivalent in degree measure (a) to the nearest minute, (b) to the nearest second.

3. An arc 15 inches in length subtends an angle of 3 radians. (a) What is the radius of the circle? (b) What is the area of the corresponding circular sector region?

4. If a central angle of 72° intercepts an arc of length 8π inches, (a) what is the radius of the circle, and (b) what is the area of the circular sector region?

5. A wheel has a radius of 12 inches. Through what angle does it turn in rolling 30 inches (a) in radians? (b) in degrees?

6. A **meridian** is a circle around the earth at its surface and passing through both poles, as shown in the figure. Let C be the center of the earth. Then the change in latitude between points A and B on the meridian is the angle θ. The radius of the earth is approximately 4000 miles. What distance s from A to B corresponds to a difference in latitude of 1 degree?

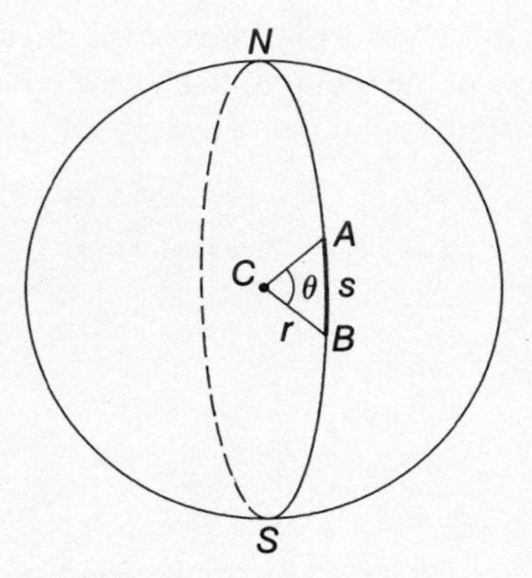

7. The diameter of the moon is about 2160 miles and it is approximately 240,000 miles from the earth. What is the measure, in degrees, of the angle θ subtended by the disk of the moon as seen from the earth? (See the figure. Since the arc length s is quite small compared to the distance r, we may use the chord length (2160) as a good approximation for s and the distance r' for r.)

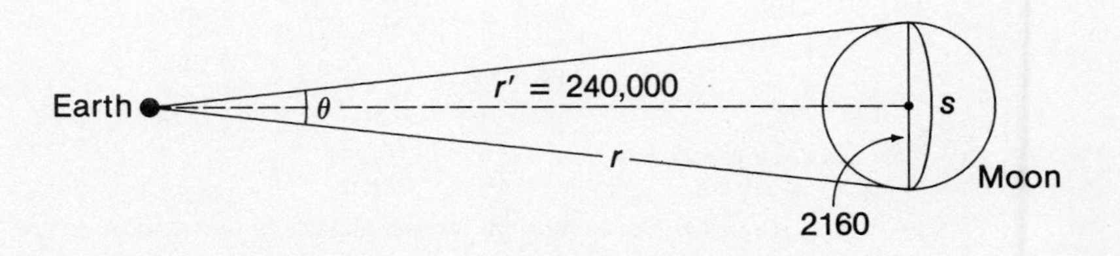

6.2 Trigonometric Functions of Angles

The primary circular functions, sine and cosine, were defined as correspondences between the set of all real numbers (the number line) and the set of either second or first coordinates of points on the unit circle, that is, the real numbers of the interval $[-1, 1]$. The domains of these functions are sets of real numbers. Similar correspondences between the set of all angles and the set of real numbers in the interval $[-1, 1]$ may be defined as follows.

DEFINITION 6.1 Let u be the radian measure of a directed angle θ. Then the **trigonometric functions sine** and **cosine** (denoted by **sin** and **cos**) of θ are defined by

$$\sin \theta = \sin u \qquad \text{and} \qquad \cos \theta = \cos u.$$

See Figure 6.6. For example, if θ has measure 30° ($\pi/6$ radians), then

$$\sin 30° = \sin \frac{\pi}{6} = \frac{1}{2}.$$

Since we agreed in the last section to use θ to denote both an angle and its measure, we can write this simply as $\sin 30° = 1/2$.

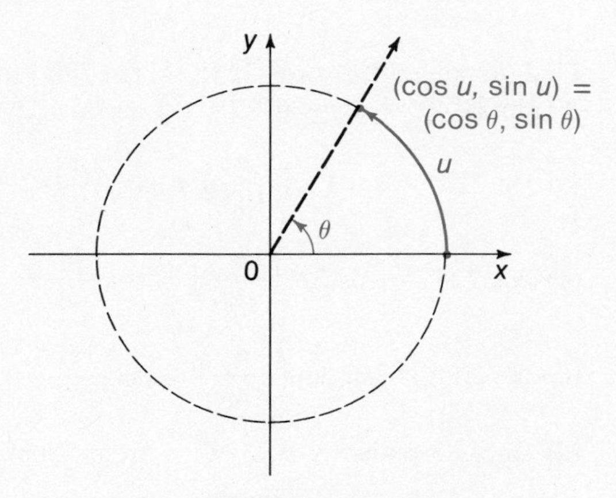

Figure 6.6

The domain of $\sin \theta$ is the set of directed angles θ. Thus, although the same names are used for these new functions, they are *different* from the circular functions because they have different domains.

The definitions of the trigonometric sine and cosine functions are patterned after those for the circular functions. There are similar definitions for the other trigonometric functions.

DEFINITION 6.2 For any angle θ, trigonometric functions **tangent (tan)**, **cotangent (cot)**, **secant (sec)**, and **cosecant (csc)** are defined as follows.

$$\tan\theta = \frac{\sin\theta}{\cos\theta} \qquad \sec\theta = \frac{1}{\cos\theta}$$

$$\cot\theta = \frac{1}{\tan\theta} = \frac{\cos\theta}{\sin\theta} \qquad \csc\theta = \frac{1}{\sin\theta}$$

For each of the properties of the circular functions there is a corresponding one for the trigonometric functions. Some examples are listed below.

$$\sin(\theta + 360°) = \sin(\theta + 2\pi) = \sin\theta \qquad \text{periodicity}$$

$$\sin^2\theta + \cos^2\theta = \sin^2 u + \cos^2 u = 1 \qquad \text{identity}$$

$$\cos(90° - \theta) = \cos\left(\frac{1}{2}\pi - u\right) = \sin u = \sin\theta \qquad \text{reduction formula}$$

$$\arcsin x = \theta \qquad \text{inverse functions}$$

Values of trigonometric functions of angles in radian measure are given in Table D, and values of these functions for angles in degree measure are in Table E. Explanations and examples of the use of these tables are given in the Appendix.

EXAMPLE 1 Find (a) sin 120°; (b) cos 225°; (c) tan 300°; (d) sin 0.83 radian; (e) tan 1.07 radians; (f) sin 21°40′; (g) cos 67°20′; (h) arcsin 1/2; (i) arctan 1.

SOLUTION

(a) $\sin 120° = \sin 120\left(\frac{\pi}{180}\right) = \sin\frac{2\pi}{3} = \frac{1}{2}\sqrt{3}$

(b) $\cos 225° = \cos 225\left(\frac{\pi}{180}\right) = \cos\frac{5\pi}{4} = -\frac{1}{2}\sqrt{2}$

(c) $\tan 300° = \tan 300\left(\frac{\pi}{180}\right) = \tan\frac{5\pi}{3} = -\sqrt{3}$

(d) $\sin 0.83$ radian ≈ 0.7379 from Table D

(e) $\tan 1.07$ radians ≈ 1.827 from Table D

(f) $\sin 21°40' \approx 0.3692$ from Table E

(g) $\cos 67°20' \approx 0.3854$ from Table E

(h) $\arcsin\frac{1}{2} = \frac{\pi}{6}$, or 30°

(i) $\arctan 1 = \frac{\pi}{4}$, or 45° ●

Exactly the same procedures used to prove identities and solve equations in the circular functions are used with trigonometric functions. However, the argument (θ in the examples below) denotes an angle rather than a real number.

EXAMPLE 2 Prove the identity $\csc \theta \cos \theta = \cot \theta$.

SOLUTION We will work on the left side.

$$\csc \theta \cos \theta = \left(\frac{1}{\sin \theta}\right)\cos \theta$$

$$= \frac{\cos \theta}{\sin \theta} = \cot \theta.$$

EXAMPLE 3 Solve $\cos^2\theta + \cos \theta = 0$ for θ, $0° \le \theta < 360°$.

SOLUTION The left member of the equation can be factored.

$$\cos^2\theta + \cos \theta = 0$$

$$\cos \theta(\cos \theta + 1) = 0$$

If $\cos \theta = 0$, then $\theta = 90°$ or $270°$, since $\pi/2 = 90°$ and $3\pi/2 = 270°$. If $\cos \theta + 1 = 0$, then $\cos \theta = -1$, and $\theta = 180°$, since π radians $= 180°$.

EXERCISES 6.2

1. Complete the following table without referring to the tables in the Appendix.

Degree measure	0	30	45	60	90	120	135	150	180
Radian measure (fractions of π)									
Sine									
Cosine									

Give the value of the following without using the tables.

2. sin 225° **3.** tan 240° **4.** csc 330° **5.** cos 300° **6.** cot 225°

7. cos 210° **8.** cot 300° **9.** sec 315° **10.** sin 240° **11.** tan 330°

Use the appropriate table from the Appendix to write each of the following values.

12. sin 0.68 radian

13. cos 0.14 radian

14. tan 0.71 radian

15. sin 1.12 radians

16. cos 0.97 radian

17. sin 43°10′

18. $\cos 15°20'$ **19.** $\tan 36°30'$ **20.** $\cot 51°40'$ **21.** $\sec 78°50'$

Prove the following identities.

22. $\cos\theta + \tan\theta \sin\theta = \sec\theta$

23. $\cot\theta \cos\theta + \sin\theta = \csc\theta$

24. $\tan^2\theta - \sin^2\theta = \sin^4\theta \sec^2\theta$

25. $(\sin\theta + \cos\theta)^2 = 1 + \sin 2\theta$

26. $\cos^4\theta - \sin^4\theta = \cos 2\theta$

27. $\csc\theta - \cot\theta = \tan\frac{1}{2}\theta$

28. $\cot\theta \sin 2\theta = 1 + \cos 2\theta$

29. $2 \sin 30° \cos 30° = \frac{1}{2}\sqrt{3}$

30. $\sin 17° \cos 13° + \cos 17° \sin 13° = \frac{1}{2}$

31. $\sin(30° + \theta) = \cos(60° - \theta)$

32. $\tan 15° = 2 - \sqrt{3}$

33. $\tan(\theta + 45°)\tan(\theta - 45°) = -1$

What is the degree measure of each angle θ less than 360° for which

34. $\sin\theta = -\frac{1}{2}\sqrt{2}$ **35.** $\cos\theta = 1$ **36.** $\tan\theta = -1$ **37.** $\cot\theta = \sqrt{3}$

38. $\csc\theta = \frac{2}{\sqrt{3}}$ **39.** $\sec\theta = 2$ **40.** $\sin\theta = -1$ **41.** $\cos\theta = \frac{1}{2}$

42. $\sin\theta = -\frac{1}{2}\sqrt{3}$ **43.** $\tan\theta = \frac{1}{\sqrt{3}}$

Solve for all angles θ, $0 \le \theta < 2\pi$.

44. $2\sin^2\theta + \sin\theta - 1 = 0$

45. $3\sin\theta - 2\cos^2\theta = 0$

46. $\tan^2\theta - 2\tan\theta = 3$

47. $3\sin^2\theta - 2 = \cos\theta$

48. $\sin 2\theta + \sin\theta = 0$

49. $\cos 2\theta + 3\sin\theta + 1 = 0$

50. $\sin 3\theta = 1$

51. $\cos 2\theta = \frac{1}{2}$

52. $\cos^2 2\theta - \cos 2\theta + 2 = 0$

53. $\sin^2 3\theta + \sin 3\theta = 2$

6.3 Trigonometric Ratios and Right-Angle Trigonometry

Trigonometric Ratios A triangle has six parts—three sides and three angles. For convenience, the sides are labeled a, b, and c and the angles respectively opposite them α, β, and γ (Greek letters alpha, beta, and gamma), as in Figure 6.7.

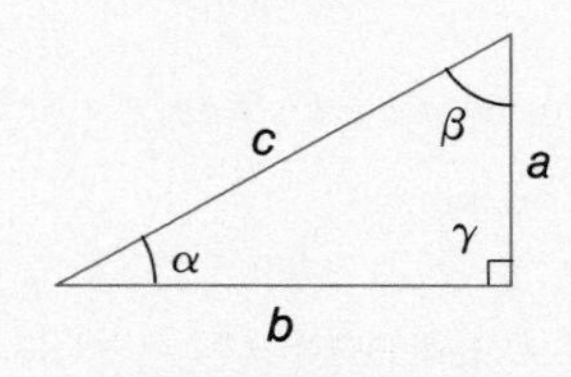

Figure 6.7

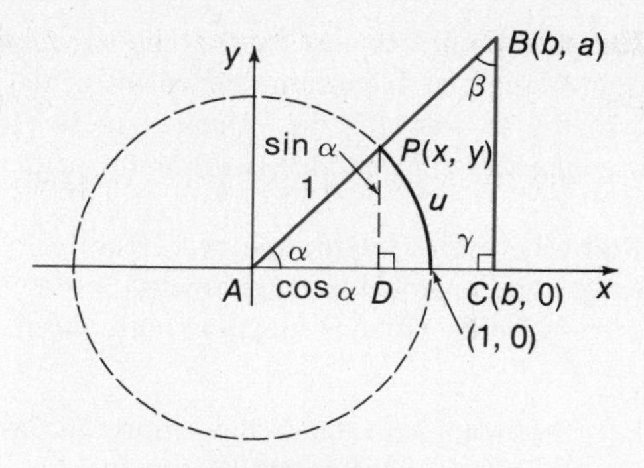

Figure 6.8

The trigonometric functions of the angles of a right triangle may be expressed as ratios of the lengths of the sides of the triangle. In Figure 6.8, angle α of triangle ACB has been placed in standard position, so $x = \cos u = \cos \alpha$ and $y = \sin u = \sin \alpha$. From similar triangles ACB and ADP, we have

$$\frac{a}{c} = \frac{\overline{CB}}{\overline{AB}} = \frac{\overline{DP}}{\overline{AP}} = \frac{\sin \alpha}{1} = \sin \alpha,$$

$$\frac{b}{c} = \frac{\overline{AC}}{\overline{AB}} = \frac{\overline{AD}}{\overline{AP}} = \frac{\cos \alpha}{1} = \cos \alpha.$$

From these,

$$\tan \alpha = \frac{\sin \alpha}{\cos \alpha} = \frac{\frac{a}{c}}{\frac{b}{c}} = \frac{a}{b}.$$

Rewriting these formulas, we get

$$\boldsymbol{\sin \alpha = \frac{a}{c} = \frac{\text{length of side opposite } \alpha}{\text{length of hypotenuse}}},$$

$$\boldsymbol{\cos \alpha = \frac{b}{c} = \frac{\text{length of side adjacent to } \alpha}{\text{length of hypotenuse}}},$$

$$\boldsymbol{\tan \alpha = \frac{a}{b} = \frac{\text{length of side opposite } \alpha}{\text{length of side adjacent to } \alpha}}.$$

The following are the corresponding formulas for the reciprocal functions.

$$\boldsymbol{\cot \alpha = \frac{b}{a} = \frac{\text{length of side adjacent to } \alpha}{\text{length of side opposite } \alpha}}$$

$$\boldsymbol{\sec \alpha = \frac{c}{b} = \frac{\text{length of hypotenuse}}{\text{length of side adjacent to } \alpha}}$$

$$\boldsymbol{\csc \alpha = \frac{c}{a} = \frac{\text{length of hypotenuse}}{\text{length of side opposite } \alpha}}$$

The preceding formulas express the trigonometric functions of the acute angles of a right triangle as **trigonometric ratios** of the lengths of the sides of the triangle. It is essential to note that the values of these ratios do *not* depend upon the size of the triangle. They are *functions of angles* of the triangle.

Right-Triangle Trigonometry Historically, the trigonometric ratios were used to solve certain problems in geometry, surveying, and navigation. Only later were the corresponding circular functions introduced.

EXAMPLE 1 Let θ be an angle in standard position and suppose than its terminal side passes through (3, 4). What are the trigonometric function values of θ?

SOLUTION In Figure 6.9, $(\overline{OP})^2 = 3^2 + 4^2 = 25$, so $\overline{OP} = 5$. Then,

$$\sin\theta = \frac{\text{opposite side}}{\text{hypotenuse}} = \frac{4}{5}, \qquad \csc\theta = \frac{\text{hypotenuse}}{\text{opposite side}} = \frac{5}{4}$$

$$\cos\theta = \frac{\text{adjacent side}}{\text{hypotenuse}} = \frac{3}{5}, \qquad \sec\theta = \frac{\text{hypotenuse}}{\text{adjacent side}} = \frac{5}{3}$$

$$\tan\theta = \frac{\text{opposite side}}{\text{adjacent side}} = \frac{4}{3}, \qquad \cot\theta = \frac{\text{adjacent side}}{\text{opposite side}} = \frac{3}{4}$$

●

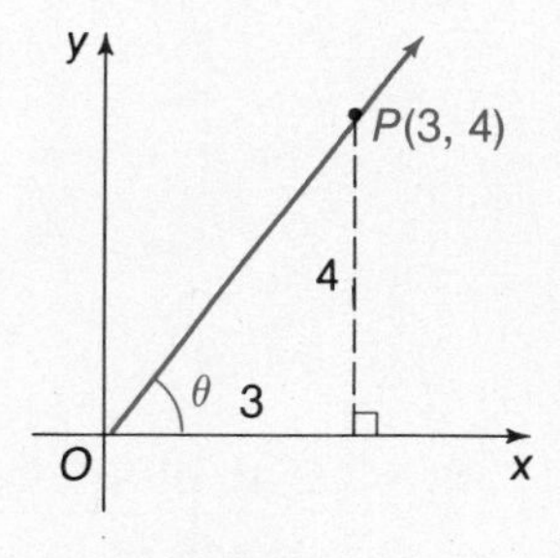

Figure 6.9

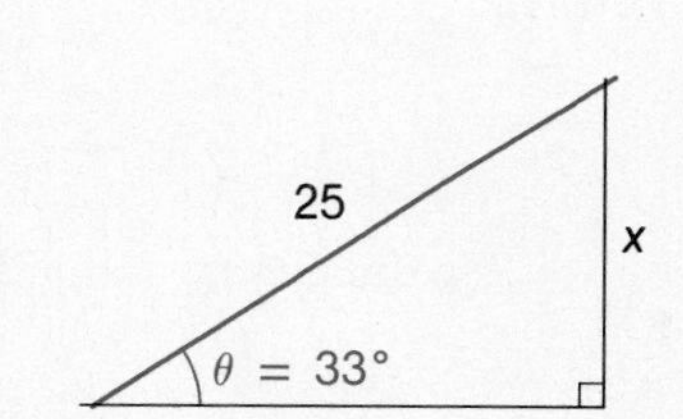

Figure 6.10

EXAMPLE 2 The right triangle in Figure 6.10 has a hypotenuse of 25 and an acute angle of 33°. Find the length of the side opposite the given angle.

SOLUTION To calculate the length of the side opposite the given angle, we use the sine ratio.

$$\sin 33^\circ = \frac{x}{25}$$

$$x = 25 \sin 33^\circ$$

$$\approx 25(0.545) \qquad \text{from the tables*}$$

$$\approx 13.6. \qquad ●$$

---

*Here, as in the examples to follow, calculations have been done with the aid of the tables. With a scientific hand calculator they can be done more directly. In that case the angle results will appear in degrees and decimal parts of a degree, rather than in degrees and minutes.

Example 2 is representative of the many situations in which the trigonometric ratios are used to calculate a distance which cannot be measured directly. We will give additional examples after considering the general situation.

In elementary geometry we learned that, given three parts of a triangle (at least one of them a side), we could construct the triangle, using only a straight edge and compass. The numerical analogue here is that, given the *measures* of three parts of a triangle (at least one of which is a side), we can *calculate* the measures of the remaining three parts. We call this procedure **solving a triangle.** Figure 6.11 shows why one of the parts must be a side. The similar (but *different*) triangles pictured there all have corresponding angles equal.

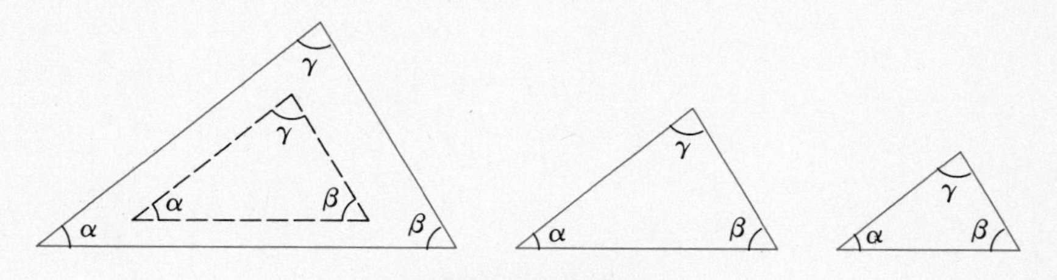

Similar triangles

Figure 6.11

The following relationships hold for the right triangle in Figure 6.12 below.

$$\frac{a}{c} = \sin\alpha = \cos\beta, \quad \text{or} \quad a = c\sin\alpha = c\cos\beta$$

$$\frac{b}{c} = \cos\alpha = \sin\beta, \quad \text{or} \quad b = c\cos\alpha = c\sin\beta$$

$$\frac{a}{b} = \tan\alpha = \cot\beta, \quad \text{or} \quad a = b\tan\alpha = b\cot\beta$$

$$\alpha + \beta = 90°$$

$$a^2 + b^2 = c^2$$

Inspection of these formulas shows that if the right angle and two sides or the right angle, one side, and another angle are given, the remaining three parts can be found.

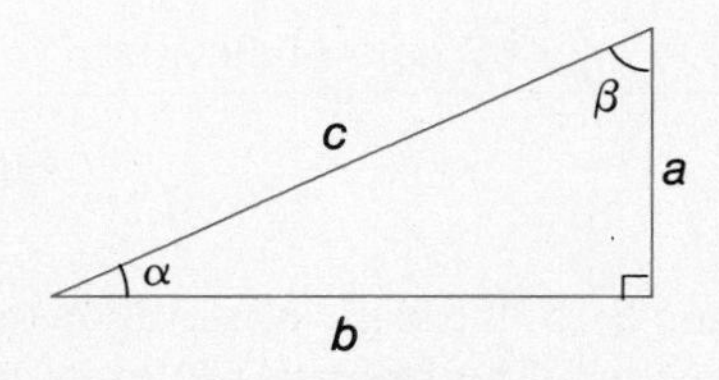

Figure 6.12

EXAMPLE 3 Given $\alpha = 35°$ and $c = 10$, solve the right triangle shown in Figure 6.13.

SOLUTION Since one of the acute angles is known, the other is found by subtraction.

$$\beta = 90° - \alpha = 90° - 35° = 55°$$

The opposite side a and the adjacent side b can be found from the trigonometric ratios.

$$\sin \alpha = \frac{a}{10} \qquad \cos \alpha = \frac{b}{10}$$

$$\begin{aligned} a &= 10 \sin \alpha & b &= 10 \cos \alpha \\ &= 10 \sin 35° & &= 10 \cos 35° \\ &\approx 10(0.5736) & &\approx 10(0.8192) \\ a &\approx 5.74 & b &\approx 8.19 \end{aligned}$$

●

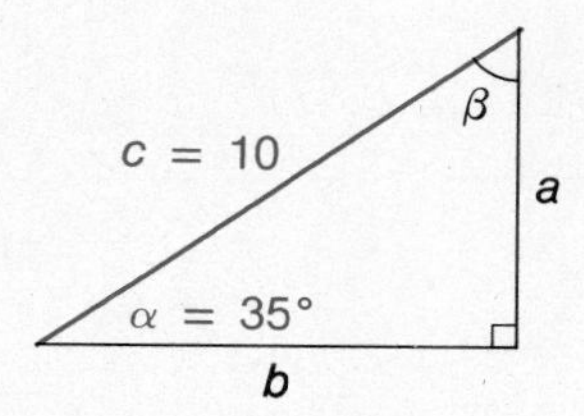

Figure 6.13

In applications of trigonometry such as the preceding example, we must often deal with approximate numbers. Guidelines for doing so, especially when a hand calculator is used, are given in Appendix 1. They will be followed in the applied examples in the remainder of this chapter and in succeeding chapters.

The **angle of elevation** of an object which is above the viewer is the angle between the horizontal and the line of sight from the viewer to the object. A similar definition applies to the **angle of depression** of an object below the viewer. See Figure 6.14.

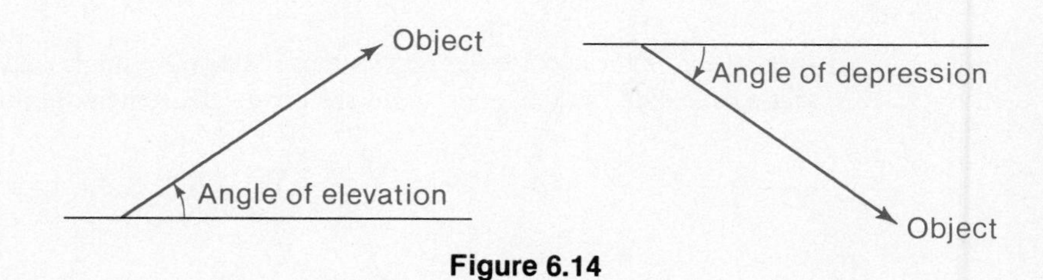

Figure 6.14

EXAMPLE 4 In Figure 6.15, the angle of elevation of the top of a tree is 60° as seen from a point 42 feet (measured horizontally) from its base. How high is the tree?

SOLUTION From the figure,

$$\tan \alpha = \frac{h}{42}$$
$$h = 42 \tan \alpha$$
$$= 42 \tan 60°$$
$$= 42\sqrt{3} \approx 42(1.73)$$
$$h \approx 72.7 \text{ (feet).} \quad \bullet$$

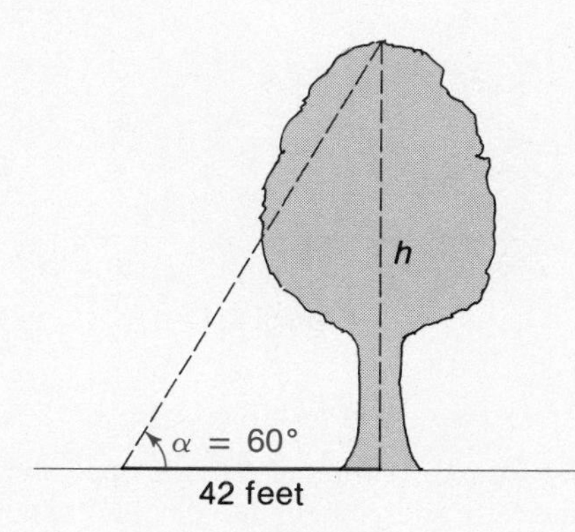

Figure 6.15

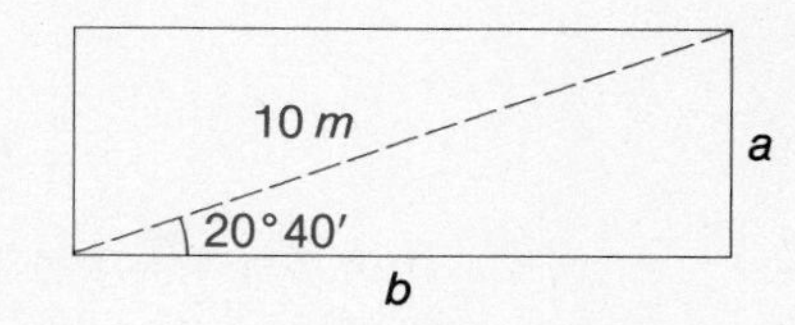

Figure 6.16

EXAMPLE 5 If the diagonal of a rectangle is 10 meters long and makes an angle of 20°40′ with the longer side, what are the dimensions of the rectangle?

SOLUTION From Figure 6.16,

$$\frac{a}{10} = \sin 20°40'$$
$$a = 10 \sin 20°40'$$
$$\approx 10(0.3529), \quad \text{from Table E}$$
$$a \approx 3.53$$

$$\frac{b}{10} = \cos 20°40'$$
$$b = 10 \cos 20°40'$$
$$\approx 10(0.9356), \quad \text{from Table E}$$
$$b \approx 9.36$$

Thus, the rectangle is 3.53 meters by 9.36 meters.

The results can be checked by the Pythagorean theorem.

$$a^2 + b^2 \approx (3.53)^2 + (9.36)^2$$
$$\approx 12.46 + 87.61$$
$$\approx 100$$

This is very nearly the value of $c^2 = 10^2 = 100$. •

A triangle which has no right angle is called an **oblique triangle.** The next example concerns solving an oblique triangle.

EXAMPLE 6 With the conventional notation, suppose that for a certain oblique triangle $\alpha = 31°$, $\beta = 57°$, and $a = 14$. Calculate the measures of the remaining parts of the triangle.

SOLUTION The remaining angle, γ, may be found by subtraction.

$$\gamma = 180° - (\alpha + \beta) = 180° - (31° + 57°) = 92°.$$

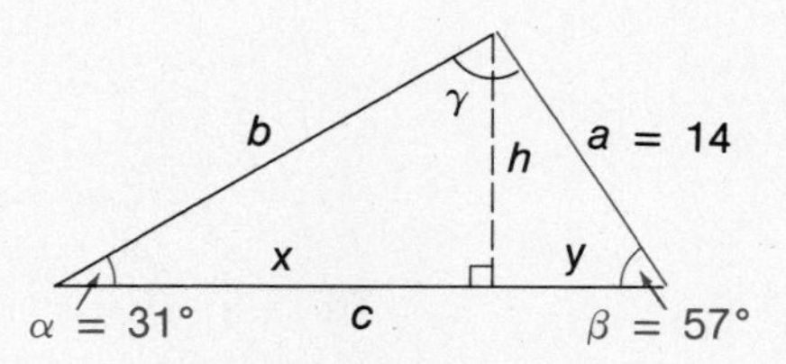

Figure 6.17

Now drop a perpendicular from the vertex of γ to the opposite side, as shown in Figure 6.17. Call its length h and let x and y be the lengths into which it divides the opposite side c.

$$\begin{aligned} \sin\beta &= \frac{h}{a} & \csc\alpha &= \frac{b}{h} \\ h &= a\sin\beta & b &= h\csc\alpha \\ &= 14\sin 57° & &\approx 11.7\csc 31° \\ &\approx 14(0.8367) & &\approx 11.7(1.942) \\ h &\approx 11.7 & b &\approx 22.7 \end{aligned}$$

To calculate c, we first calculate x and y.

$$\begin{aligned} \cot\alpha &= \frac{x}{h} & \cos\beta &= \frac{y}{a} \\ x &= h\cot\alpha & y &= a\cos\beta \\ &\approx 11.7\cot 31° & &= 14\cos 57° \\ &\approx 11.7(1.664) & &\approx 14(0.5446) \\ x &\approx 19.4 & y &\approx 7.6 \end{aligned}$$

Then $c = x + y \approx 19.4 + 7.6 \approx 27.0$. •

In the next section simpler procedures for finding the unknown parts of an oblique triangle will be developed.

EXERCISES 6.3

Solve each of the following right triangles.

1. $\alpha = 35°$, $c = 10$
2. $\alpha = 31°$, $a = 65$
3. $a = 17$, $b = 24$
4. $\alpha = 56°$, $a = 73$
5. $a = 13.2$, $b = 20.4$
6. $a = 33$, $c = 40$
7. $\beta = 63°$, $b = 7.3$
8. $\beta = 63°10'$, $c = 17.9$
9. $b = 6.13$, $\alpha = 62°20'$
10. $b = 14$, $c = 32$

11. If a rectangle is 7 meters by 11 meters, find the length of the diagonal and the angles it makes with the sides of the rectangle.

12. If the diagonal of a rectangle is 32.4 meters long and makes an angle of 29° with the longer side, find the dimensions of the rectangle.

13. Find the height of a utility pole which casts a shadow 70 feet long when the angle of elevation of the sun (angle of elevation of the top of the pole as observed from the tip of its shadow) is 47°.

14. The height of clouds above a weather station is determined by focusing a searchlight beam directly overhead. Find the height of the clouds if from a point 150 meters away from the searchlight the angle of elevation of the spot of light is 61°.

15. Two points are on the face of a building and one is directly above the other. Their respective angles of elevation are 65°20′ and 53°10′ from a point 50 meters horizontally from the base of the building. How far apart are the points?

16. In the figure, an observer at O is at an elevation of 1000 feet above the level of the point P. The angle of depression of P as seen from O is 20°. What is the straight-line distance of P from O?

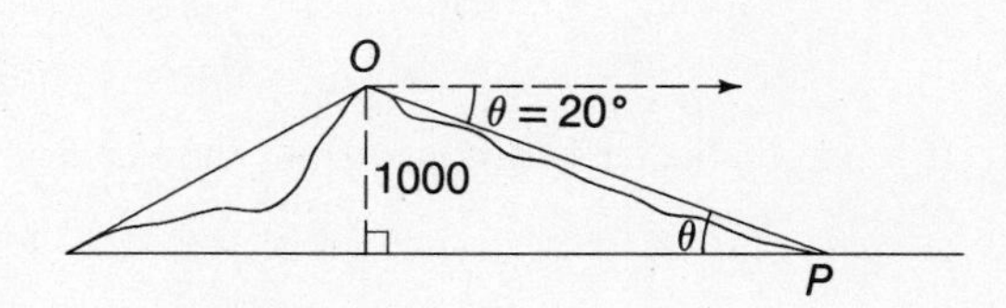

17. In the figure, AB represents a tunnel through a mountain. If the point P at the top of the mountain is 1 mile (5280 feet) along the slope from A and the angles of slope of the mountain are as shown, then (a) how far is the tunnel below the point P? (b) How far is the other entrance B down the mountain from P? (c) How long is the tunnel? (d) How far is point C from the entrance at A?

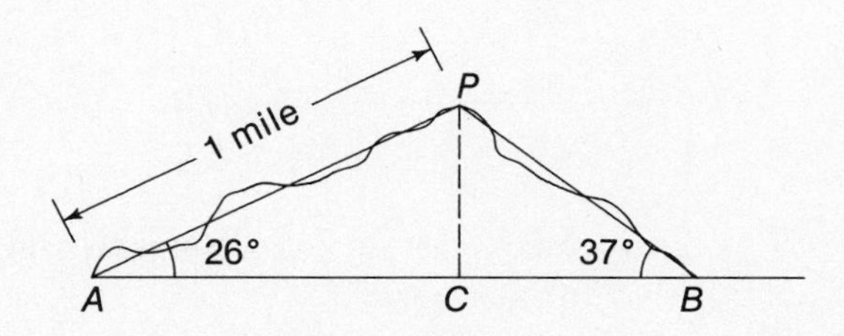

18. Let a, b, and γ be two sides and the included angle of any triangle.

(a) Show that $A = (1/2)ab \sin \gamma$ is a formula for the area of the triangle. [*Hint:* Let h be the length of the altitude upon the side b, express h as a trigonometric ratio, and then use the formula $A = (1/2)bh$.]

(b) Use this formula to calculate the area of a parallelogram whose sides are 12 and 17 and meet at the angle of 41°. [*Hint:* Draw a diagonal.]

6.4 Oblique Triangles: The Law of Sines

In Example 6 of Section 6.3 we solved an oblique triangle by first decomposing it into two right triangles. The procedure there was tedious, and fortunately it is possible to derive formulas for working directly with an oblique triangle.

Law of Sines Figure 6.18 shows oblique triangles labeled in the conventional way. Drop a perpendicular from the vertex of γ to the opposite side (or that side extended, as in (b)). In either case,

$$\sin \alpha = \frac{h}{b} \quad \text{and} \quad \sin \beta = \frac{h}{a}.$$

For Figure 6.18b we use the fact that $\sin(180° - \beta) = \sin \beta$. Then we have $h = b \sin \alpha$ and $h = a \sin \beta$, so

$$b \sin \alpha = a \sin \beta,$$

$$\frac{a}{\sin \alpha} = \frac{b}{\sin \beta}. \tag{1}$$

If a perpendicular is dropped from the vertex of each of the other angles in turn, we obtain

$$\frac{b}{\sin \beta} = \frac{c}{\sin \gamma} \quad \text{and} \quad \frac{a}{\sin \alpha} = \frac{c}{\sin \gamma}. \tag{2}$$

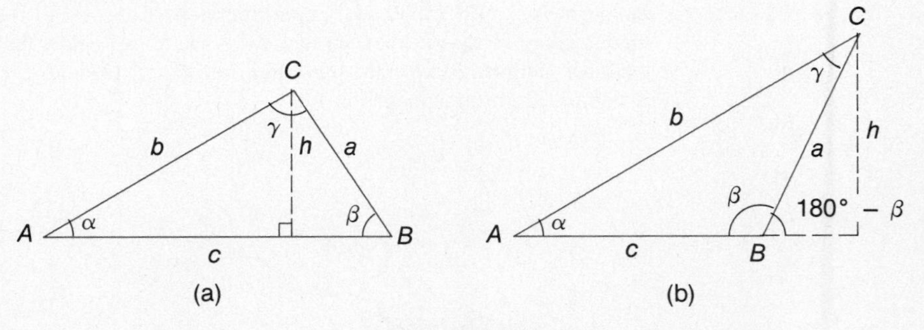

Figure 6.18

Combining these results, we get the following **law of sines,** written in compact form.

THE LAW OF SINES

$$\frac{a}{\sin \alpha} = \frac{b}{\sin \beta} = \frac{c}{\sin \gamma}$$

That is, the ratio of the length of any side of a triangle to the sine of the angle opposite is the same as the corresponding ratio for any other side and angle opposite.

The law of sines is equivalent to the *three* statements in (1) and (2) above. Each of them involves *two sides and the two angles opposite them*. Thus, if a side and the angle opposite it are among the given parts of a triangle, another side (or angle) can be calculated from the law of sines.

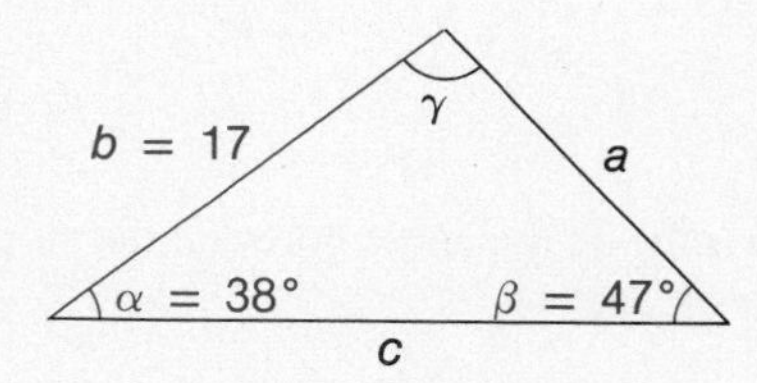

Figure 6.19

EXAMPLE 1 Given $b = 17$, $\alpha = 38°$, and $\beta = 47°$, solve the triangle.

SOLUTION See Figure 6.19. Since α and β are known, γ can be found by subtraction.

$$\begin{aligned} \gamma &= 180° - (\alpha + \beta) \\ &= 180° - (38° + 47°) = 95°. \end{aligned}$$

Then, the law of sines is used twice to find the other sides.

$$\frac{a}{\sin \alpha} = \frac{b}{\sin \beta}$$

$$\begin{aligned} a &= \frac{b \sin \alpha}{\sin \beta} = \frac{17 \sin 38°}{\sin 47°} \\ &\approx \frac{17(0.616)}{0.731} \approx 14.3 \end{aligned}$$

$$\frac{b}{\sin \beta} = \frac{c}{\sin \gamma}$$

$$c = \frac{b \sin \gamma}{\sin \beta} = \frac{17 \sin 95°}{\sin 47°}$$

Since $\sin 95° = \sin(180° - 95°) = \sin 85°$,

$$\begin{aligned} c &= \frac{17 \sin 85°}{\sin 47°} \\ &\approx \frac{17(0.996)}{0.731} \approx 23.2. \end{aligned}$$

●

EXAMPLE 2 Suppose a tree grows vertically on a hillside having an inclination of 15° with the horizontal. If the tree casts a shadow 150 feet long directly down the slope of the hillside when the angle of elevation of the sun is 36°; how tall is the tree?

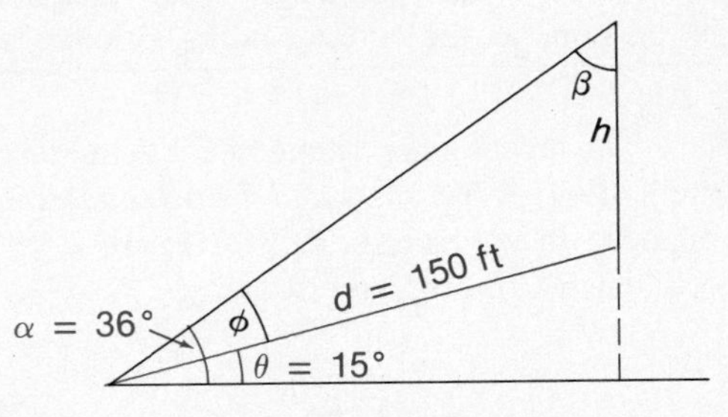

Figure 6.20

SOLUTION Let ϕ (phi) be the angle between the slope of the hill and the line of sight of the sun. In Figure 6.20,

$$\phi = \alpha - \theta = 36° - 15° = 21°,$$
$$\beta = 90° - \alpha = 90° - 36° = 54°.$$

Then, from the law of sines,

$$\frac{h}{\sin \phi} = \frac{d}{\sin \beta}$$
$$h = \frac{d \sin \phi}{\sin \beta}$$
$$= \frac{150 \sin 21°}{\sin 54°}$$
$$\approx \frac{150(0.3584)}{0.8090}$$
$$h \approx 66.5 \text{ (feet)}.$$ ●

The Ambiguous Case The case of two sides and an angle opposite one of them needs separate treatment. This is because there may be either no solution, one solution, or two solutions, depending upon the values of the parts given. To understand the situation, refer to Figure 6.21 and suppose that a, b, and α are given. Then, from the figure, $\sin \alpha = h/b$, so that $h = b \sin \alpha$. Now it may be that $a < h$, $a = h$, or $a > h$. The possible cases are illustrated in the figure and described as follows.

α acute

(a) If $a < b \sin \alpha$, then there is no solution, as seen from part (a) of the figure.

(b) If $a = b \sin \alpha$, then there is exactly one solution. See part (b) of the figure.

(c) If $b \sin \alpha < a < b$, then there are two solutions, as may be seen from part (c) of the figure.

(d) If $b \sin \alpha < b \le a$, then there is one solution. See part (d) of the figure.

α obtuse

(e) If $a \le b$, then there is no solution. See part (e) of the figure.

(f) If $a > b$, there is one solution, as seen in part (f) of the figure.

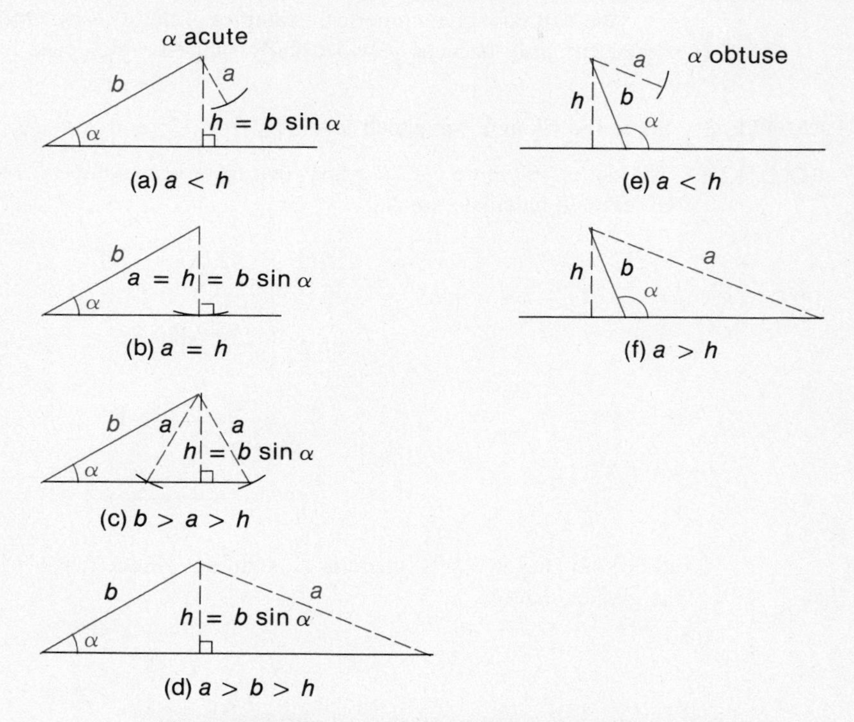

Figure 6.21. The ambiguous case. Given two sides and the angle opposite one of them, there may be no triangle, one triangle, or two triangles.

Before giving examples, let us see what situations we can expect from the numerical calculations. When the law of sines is written as

$$\sin \beta = \frac{b \sin \alpha}{a},$$

it can be seen that $\sin \beta$ may be less than 1, equal to 1, or greater than 1.

(1) Since the sine value is never more than 1, the case $\sin \beta > 1$ can have no solution. This amounts to

$$\frac{h}{a} = \frac{b \sin \alpha}{a} > 1, \quad \text{or} \quad a < h,$$

which is case (a) above.

(2) If $\sin \beta = 1$, then β is a right angle, which is case (b) above.

(3) If $\sin\beta < 1$, there are two possible values of β. This happens because $\sin(180° - \beta) = \sin\beta$, and hence there are two **supplementary** angles (angles whose sum is 180°) having this same sine value. If $\alpha + \beta \geq 180°$, there is no solution since the sum of the angles of a triangle is 180°. See part (e) of the figure. If $\alpha + \beta < 180°$, there is one solution if $a > b$ (see Figure 6.21d) and two solutions if $a < b$ (see Figure 6.21c).

We will now give numerical examples of these cases. One should always sketch a figure carefully because it will usually suggest which case is represented.

EXAMPLE 3 Solve the triangle for which $a = 8.0$, $b = 17$, and $\alpha = 38°$.

SOLUTION The sketch in Figure 6.22 suggests that there is no solution. To verify this, use the law of sines to calculate $\sin\beta$.

$$\frac{\sin\beta}{b} = \frac{\sin\alpha}{a}$$

$$\sin\beta = \frac{b\sin\alpha}{a}$$

$$= \frac{17\sin 38°}{8}$$

$$\approx \frac{17(0.616)}{8} \approx 1.3$$

Since $\sin\beta \approx 1.3 > 1$, there is no solution. Notice that $h = b\sin\alpha = 10.5$, which is greater than a. ●

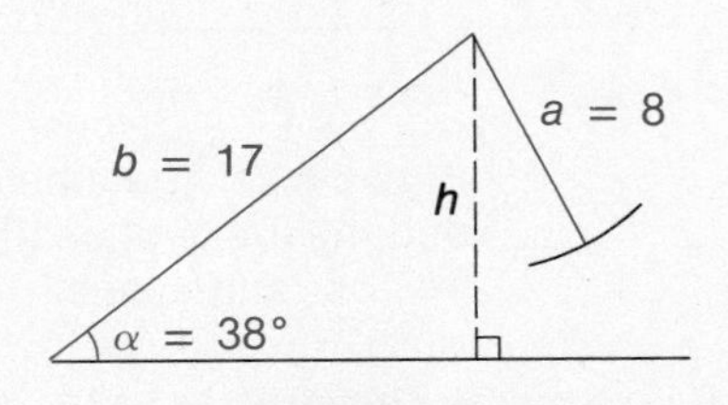

Figure 6.22

EXAMPLE 4 Solve the triangle for which $a = 12$, $b = 17$, and $\alpha = 38°$.

SOLUTION Figure 6.23a indicates that there are two possibilities. To check this, calculate $\sin\beta$.

$$\sin\beta = \frac{b\sin\alpha}{a}$$

$$= \frac{17\sin 38°}{12}$$

$$\approx \frac{17(0.616)}{12} \approx 0.873.$$

$$\beta \approx 61° \text{ or } 119°.$$

Call these two values β_1 and β_2, and the corresponding values for the third angle γ_1 and γ_2. If $\beta_1 = 61°$, then

$$\alpha + \beta_1 = 38° + 61° = 99°,$$
$$\gamma_1 = 180° - (\alpha + \beta_1) = 180° - 99° = 81°.$$

If $\beta_2 = 119°$, then

$$\alpha + \beta_2 = 38° + 119° = 157°,$$
$$\gamma_2 = 180° - (\alpha + \beta_2) = 180° - 157° = 23°.$$

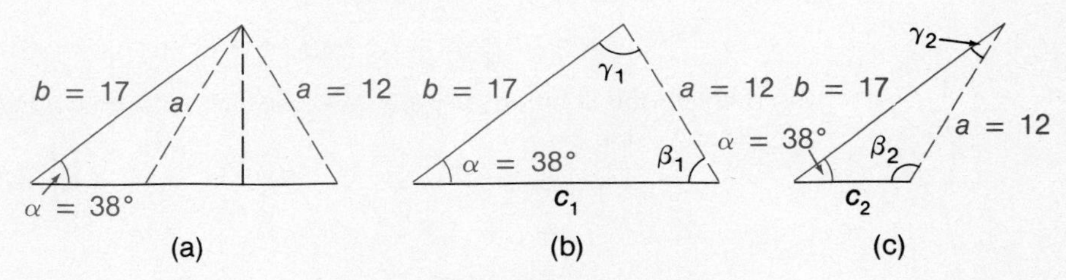

Figure 6.23

We can now calculate c_1 and c_2, the corresponding values of the side c. (Remember that these values of β and γ are approximations.)

$$\frac{c_1}{\sin \gamma_1} = \frac{a}{\sin \alpha} \qquad\qquad \frac{c_2}{\sin \gamma_2} = \frac{a}{\sin \alpha}$$

$$c_1 = \frac{a \sin \gamma_1}{\sin \alpha} \qquad\qquad c_2 = \frac{a \sin \gamma_2}{\sin \alpha}$$

$$\approx \frac{12 \sin 81°}{\sin 38°} \qquad\qquad \approx \frac{12 \sin 23°}{\sin 38°}$$

$$\approx \frac{12(0.988)}{0.616} \qquad\qquad \approx \frac{12(0.391)}{0.616}$$

$$\approx 19 \qquad\qquad \approx 7.6$$

The two solutions are shown in Figures 6.23b and 6.23c. The approximate values are given below.

$$\beta_1 = 61°, \qquad \gamma_1 = 81°, \qquad c_1 = 19$$
$$\beta_2 = 119°, \qquad \gamma_2 = 23°, \qquad c_2 = 7.6 \qquad \bullet$$

EXAMPLE 5 Solve the triangle with $b = 17$, $a = 20$, and $\alpha = 38°$.

SOLUTION The triangle is sketched in Figure 6.24. In this case, there appears to be only one solution. We first calculate $\sin \beta$.

$$\begin{aligned} \sin \beta &= \frac{b \sin \alpha}{a} \\ &= \frac{17 \sin 38°}{20} \\ &\approx \frac{17(0.616)}{20} \approx 0.524 \\ \beta &\approx 32° \end{aligned}$$

Since $\alpha = 38°$, $\alpha + \beta \approx 70°$, so that

$$\gamma = 180° - (\alpha + \beta) \approx 180° - 70° = 110°.$$

To complete the solution, use the law of sines with a, α, and the value for γ just calculated to find c.

$$\begin{aligned} \frac{c}{a} &= \frac{\sin \gamma}{\sin \alpha} \\ c &= \frac{a \sin \gamma}{\sin \alpha} \\ &\approx \frac{20 \sin 110°}{\sin 38°} \\ &\approx \frac{20 \sin 70°}{\sin 38°} \qquad \text{since } \sin 70° = \sin(180° - 110°) \\ &\approx \frac{20(0.940)}{0.616} \\ &\approx 31 \end{aligned}$$

The solution is $\beta = 32°$, $\gamma = 110°$, and $c = 31$ (approximately). •

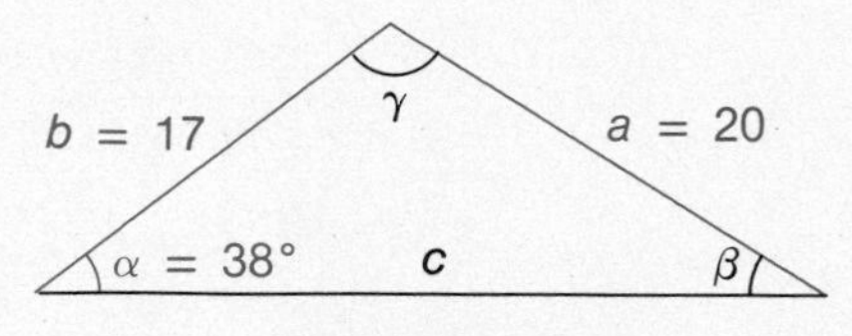

Figure 6.24

The ambiguity in the ambiguous case comes from the fact that the law of sines in this case provides us with the sine of an angle. The value of the angle(s) then depends upon two facts: the sine value is never more than 1, and the sines of supplementary angles are equal. Under these conditions there may be no angle, one angle, or two angles.

The law of sines is frequently used as an identity in proving relationships not associated with numerical solution of triangles, as in the next example.

EXAMPLE 6 Given the angles β and γ and the included side a of a triangle, show that the area is given by

$$\mathbf{A = \frac{a^2 \sin \beta \sin \gamma}{2 \sin(\beta + \gamma)}}.$$

SOLUTION From Exercise 18, Section 6.3, we have the area formula $A = (1/2)ab \sin \gamma$. We start with this and use the law of sines to write a substitution for b.

$$\begin{aligned}
A &= \frac{1}{2}ab \sin \gamma \\
&= \frac{1}{2}a\left(\frac{a \sin \beta}{\sin \alpha}\right) \sin \gamma \\
&= \frac{a^2 \sin \beta \sin \gamma}{2 \sin \alpha} \\
&= \frac{a^2 \sin \beta \sin \gamma}{2 \sin[180^\circ - (\beta + \gamma)]} && \text{since } \alpha + \beta + \gamma = 180^\circ \\
&= \frac{a^2 \sin \beta \sin \gamma}{2 \sin(\beta + \gamma)} && \text{since } \sin(180^\circ - \theta) = \sin \theta
\end{aligned}$$

●

EXAMPLE 7 Find the area of a triangle if two angles and the included side are 40°, 60°, and 120.

SOLUTION Substitute the given values in the above formula for area.

$$\begin{aligned}
A &= \frac{120^2 \sin 40^\circ \sin 60^\circ}{2 \sin 100^\circ} \\
&\approx \frac{14400(0.6428)(0.8660)}{2(0.9848)} && \sin 100^\circ = \sin 80^\circ \\
&\approx 4070
\end{aligned}$$

●

EXERCISES 6.4

Solve the following triangles.

1. $\alpha = 61^\circ,\ \beta = 24^\circ,\ b = 305$
2. $\alpha = 38^\circ,\ \beta = 47^\circ,\ a = 17$
3. $\beta = 42^\circ 10',\ \gamma = 110^\circ 40',\ c = 23.4$
4. $\alpha = 30^\circ 20',\ \gamma = 50^\circ 30',\ c = 11.3$

5. $a = 6.4$, $\alpha = 20°50'$, $\gamma = 50°10'$

6. $b = 131$, $\beta = 62°30'$, $\gamma = 75°40'$

7. $a = 28$, $b = 46$, $\beta = 70°$

8. $a = 42.3$, $b = 81.4$, $\alpha = 37°10'$

9. $\beta = 47°10'$, $a = 21.2$, $b = 28.3$

10. $b = 15.8$, $c = 4.53$, $\gamma = 35°38'$

11. $\gamma = 20°30'$, $a = 24.1$, $\alpha = 45°40'$

12. $b = 25.2$, $a = 24.3$, $\alpha = 30°50'$

Each of the following represents the ambiguous case. Calculate $h = b \sin \alpha$ *and from this determine which of the situations discussed above is involved.*

13. $a = 20$, $b = 24$, $\alpha = 120°$

14. $a = 25$, $b = 24$, $\alpha = 120°$

15. $a = 16$, $b = 32$, $\alpha = 60°$

16. $a = 15$, $b = 24$, $\alpha = 45°$

17. $a = 25$, $b = 24$, $\alpha = 30°$

18. $a = 15$, $b = 24$, $\alpha = 45°$

19. The diagonals of a parallelogram make angles of 33° and 46° with one side of a parallelogram. If this side is 24 inches long, find the length of the diagonals and of the other side of the parallelogram. [Recall that the diagonals of a parallelogram bisect each other.]

20. Two lighthouses A and B are 76 miles apart. Their direction finders determine that a ship is at point C such that angle $ABC = 53°$ and angle $BAC = 81°$. Find the distance of the ship from each lighthouse.

21. A tree stands on the slope of a hill. From a point 72 feet downhill from the base of the tree the line of sight to its top makes an angle of 35° with the slope of the hillside. If the slope of the hillside is 17°, how high is the tree?

22. Find the length of the shorter diagonal and the angle it makes with the shorter side of a parallelogram with sides 8.2 feet and 13.1 feet and an angle of 54°10′ between them.

23. Use the formula of Example 6 to calculate the area of a triangle having two angles 110° and 42°, if the side between the angles is 24 meters.

24. The following is another way to prove the law of sines. From Exercise 18, Section 6.3, the area of a triangle is $A = (1/2)ab \sin \gamma$, where a and b are two sides of the triangle and γ is the included angle. Write the corresponding area formulas for the other two choices of the two sides and included angle. Then equate these three area formulas to obtain the law of sines.

6.5 Oblique Triangles: The Law of Cosines

In Figure 6.25 an oblique triangle is shown with angle α in standard position in the coordinate plane. The following discussion applies equally well to the three possible cases: (a) α acute, (b) α right, (c) α obtuse.

If the coordinates of vertex C are (x, y), then $x = b \cos \alpha$ and $y = b \sin \alpha$. This can be seen in each of the three cases as follows.

(a) $\cos \alpha = \dfrac{x}{b}$ and $\sin \alpha = \dfrac{y}{b}$.

(b) $\alpha = 90°$, so that $x = 0$, $y = b$.

(c) $\dfrac{x}{b} = -\cos(180° - \alpha) = \cos \alpha$, and $\dfrac{y}{b} = \sin(180° - \alpha) = \sin \alpha$.

These expressions for x and y are used to derive a formula for a^2. By the distance formula,

$$a^2 = \overline{BC}^2 = (x - c)^2 + (y - 0)^2.$$

$$\begin{aligned} a^2 &= (b \cos \alpha - c)^2 + (b \sin \alpha - 0)^2 \\ &= (b^2 \cos^2\alpha - 2bc \cos \alpha + c^2) + b^2 \sin^2\alpha \\ &= b^2(\cos^2\alpha + \sin^2\alpha) + c^2 - 2bc \cos \alpha \end{aligned}$$

Since $\cos^2\alpha + \sin^2\alpha = 1$, the result is the following formula.

$$\boldsymbol{a^2 = b^2 + c^2 - 2bc \cos \alpha} \qquad \textbf{(1)}$$

If β instead of α were placed in standard position, we would get, similarly,

$$\boldsymbol{b^2 = a^2 + c^2 - 2ac \cos \beta.} \qquad \textbf{(2)}$$

In the same way we could obtain

$$\boldsymbol{c^2 = a^2 + b^2 - 2ab \cos \gamma.} \qquad \textbf{(3)}$$

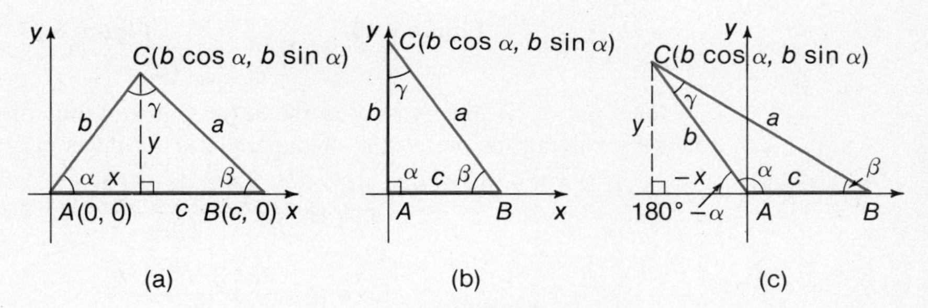

Figure 6.25

Each of these three formulas is a statement of the **law of cosines:**

THE LAW OF COSINES The square of the length of any side of a triangle is equal to the sum of the squares of the other two sides less twice the product of their lengths and the cosine of the included angle.

In the case of a right triangle, one of the angles, say γ, is a right angle. Then, since $\cos 90° = 0$, the law of cosines reduces simply to the Pythagorean theorem, $a^2 + b^2 = c^2$.

EXAMPLE 1 Solve the triangle for which $\alpha = 69°$, $b = 31$, and $c = 23$. See Figure 6.26.

SOLUTION Since α is known, we find side a first.

$$a^2 = b^2 + c^2 - 2bc \cos \alpha$$

$$\begin{aligned} a^2 &= 31^2 + 23^2 - 2(31)(23)\cos 69^\circ \\ &\approx 961 + 529 - 1426(0.358) \\ &\approx 979 \\ a &\approx 31.3 \end{aligned}$$

Now with a and α known we can use the law of sines to calculate the angles β and γ opposite sides b and c. The solutions can be checked in the angle-sum formula $\alpha + \beta + \gamma = 180^\circ$. Details are left to the exercises. ●

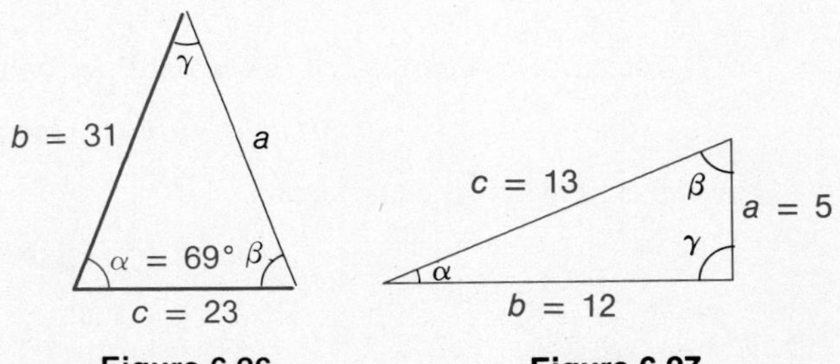

Figure 6.26

Figure 6.27

The law of cosines is also useful in the following alternative forms, which are found by solving for the cosine of the angle in formulas (1), (2), and (3).

$$\cos \alpha = \frac{b^2 + c^2 - a^2}{2bc}$$

$$\cos \beta = \frac{a^2 + c^2 - b^2}{2ac}$$

$$\cos \gamma = \frac{a^2 + b^2 - c^2}{2ab}.$$

From these different versions of the law of cosines we see that, given *two sides and the included angle,* we can calculate the remaining side, and given *three sides,* we can calculate the three angles.

EXAMPLE 2 Calculate the angles of a triangle whose sides are 5, 12, and 13. See Figure 6.27.

SOLUTION Let $a = 5$, $b = 12$, and $c = 13$, and use an alternative form of the law of cosines.

$$\begin{aligned} \cos \alpha &= \frac{b^2 + c^2 - a^2}{2bc} \\ &= \frac{12^2 + 13^2 - 5^2}{2(12)(13)} \\ &= \frac{144 + 169 - 25}{312} \end{aligned}$$

$$= \frac{12}{13} \approx 0.923$$

$$\alpha \approx \arccos 0.923 \approx 22°38'$$

Similarly, we find $\beta \approx 67°22'$ and $\gamma \approx 90°0'$. Actually, $\gamma = 90°$ exactly, since $c^2 = a^2 + b^2$, so that the triangle is a right triangle. In the general three-sided triangle situation one can check by using the angle-sum formula, $\alpha + \beta + \gamma = 180°$. ●

The following summarizes the methods of solution of oblique triangles discussed in the last two sections. The cases are listed according to the parts of the triangle which are given.

1. *Two angles and an included side.* First, the remaining angle is found from the angle-sum formula, $\alpha + \beta + \gamma = 180°$. Then the two remaining sides are calculated by the law of sines.
2. *Two sides and an included angle.* The law of cosines is used to calculate the remaining side and then the law of sines is used for the other two angles.
3. *Three sides.* The alternative form of the law of cosines is used to calculate the three angles.
4. *Two angles and a side opposite one of them.* Use the same procedure as for two angles and an included side.
5. *Two sides and an angle opposite one of them* (the ambiguous case). The law of sines is used to find the angle (none, one, or two) opposite the other side. Then the angle-sum formula is used to find the remaining angle and the law of sines to find the side(s) opposite.

The law of cosines (as well as the law of sines) is a trigonometric identity for any triangle. The following demonstrates its use as an identity. Additional examples are left as exercises at the end of the chapter and many instances occur in advanced mathematics.

EXAMPLE 3 Show that the length m of the median from the vertex A of the triangle ABC to the midpoint of the opposite side is given by

$$m = \sqrt{\frac{1}{2}(b^2 + c^2) - \frac{1}{4}a^2}.$$

SOLUTION The situation is shown in Figure 6.28. From the law of cosines,

$$b^2 = \left(\frac{1}{2}a\right)^2 + m^2 - 2\left(\frac{1}{2}a\right)m \cos \theta$$

$$= \frac{1}{4}a^2 + m^2 - am \cos \theta,$$

and
$$c^2 = \left(\frac{1}{2}a\right)^2 + m^2 - 2\left(\frac{1}{2}a\right)m\cos(180° - \theta)$$
$$= \frac{1}{4}a^2 + m^2 + am\cos\theta,$$

since $\cos(180° - \theta) = -\cos\theta$. Add the expressions for b^2 and c^2 and solve the resulting equation for m.

$$b^2 + c^2 = \frac{1}{2}a^2 + 2m^2,$$
$$m^2 = \frac{1}{2}(b^2 + c^2) - \frac{1}{4}a^2,$$
$$m = \sqrt{\frac{1}{2}(b^2 + c^2) - \frac{1}{4}a^2} \quad \bullet$$

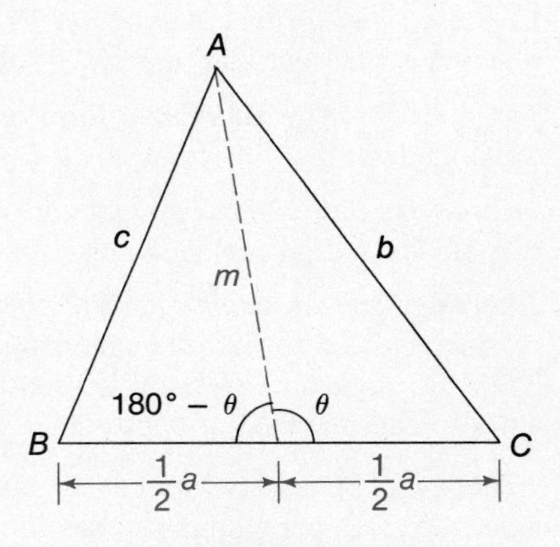

Figure 6.28

EXERCISES 6.5

Solve each of the following triangles.

1. $\gamma = 36°$, $a = 10$, $b = 12$
2. $\beta = 65°$, $a = 100$, $c = 45$
3. $a = 37$, $b = 41$, $\gamma = 35°$
4. $b = 12$, $c = 7$, $\alpha = 21°$
5. $a = 3.6$, $c = 12.5$, $\beta = 46.9°$
6. $a = 17$, $b = 28$, $\gamma = 47°$
7. $a = 13.7$, $b = 17.4$, $c = 15.6$
8. $a = 14.3$, $b = 18.6$, $c = 17.1$

9. $a = 23.5,\ b = 44.2,\ c = 30.1$ **10.** $a = 41.2,\ b = 95.3,\ c = 106$

11. To determine the distance across a lake, AB in the figure, a surveyor made the indicated measurements. Use these data and the law of cosines to calculate AB.

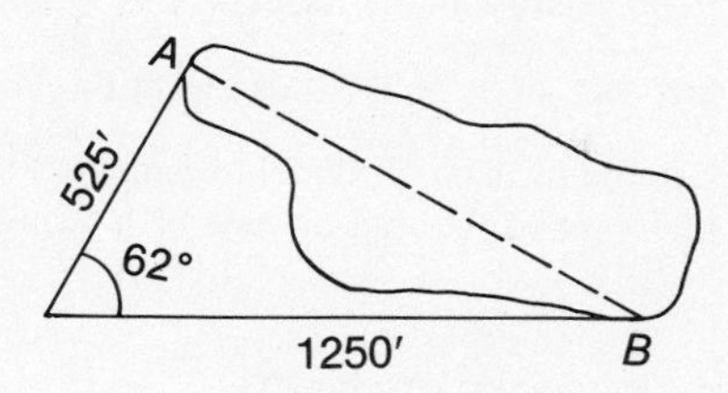

12. Find the length of the shorter diagonal and the angle it makes with the shorter side of a parallelogram, given that the sides measure 7 feet and 12 feet and the angle between these sides is 57°.

13. The diagonals of a parallelogram are 52 feet and 43 feet long and intersect at an angle of 129°. Find the dimensions of the parallelogram.

14. Two boats leave a point at the same time in directions making an angle of 65° with each other. If one boat moves 19 mph and the other 23 mph, how far apart are they after 2 hours?

15. The original vertical height of the leaning tower of Pisa was 179 feet. Recent measurements have shown it leaning 5°27′ from the vertical. Suppose that an observer standing directly in the plane of the tilt of the tower toward him notes that the angle of elevation of the sun is 53°. How long is the shadow of the tower? See the figure.

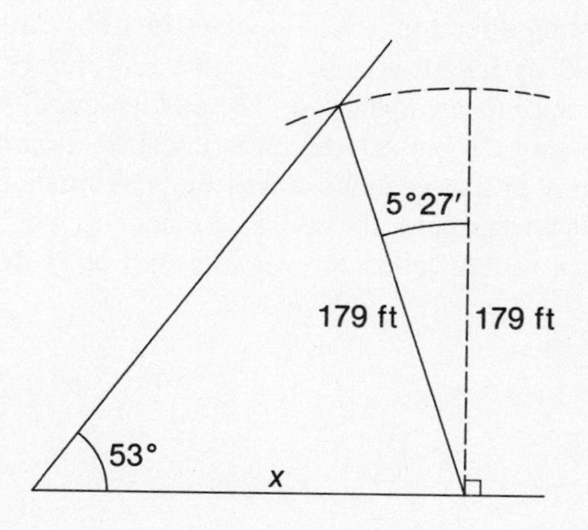

16. Apply the law of cosines to show that

$$1 - \cos\alpha = \frac{(b + c + a)(b + c - a)}{2bc}.$$

The function $f(\alpha) = 1 - \cos\alpha$ is called the *versine* α and is important in the trigonometry of navigation.

17. Show that if the three sides of a triangle are a, b, and c, then the area is given by

$$A = \frac{1}{4}\sqrt{4a^2b^2 - (a^2 + b^2 - c^2)^2}.$$

[*Hint:* Start with the formula $A = \frac{1}{2}ab \sin \gamma$, from Exercise 18, Section 6.3, and use the fact that $\sin^2\gamma = 1 - \cos^2\gamma$ and the law of cosines to substitute for $\cos \gamma$.]

18. Use the formula derived in Example 3 to calculate the length of the median drawn to the midpoint of the longest side of a triangle whose sides are 20, 30, and 40.

6.6 Vector Applications

Vector Quantities In the preceding sections the trigonometric ratios and methods of triangle solving were used primarily to solve surveying problems in which the size of one or more angles or the length of one or more sides of a triangle were determined. In these problems it was unnecessary to specify the direction in which measurement was made. For example, the side of a triangle has no direction, only magnitude (length).

Many quantities, however, can be described completely only by specifying both magnitude *and* direction. For example, when speaking of wind a weather forecaster usually mentions not only magnitude, or speed, but direction as well—15 miles per hour from the northwest. Another example is force which is exerted with a given magnitude in a given direction—10 pounds at an angle of 30° with the ground. Quantities having both magnitude and direction are called **vector quantities.** Other examples are velocity, acceleration, and displacement.

A vector quantity is represented by a directed line segment with an arrow indicating direction. (See Figure 6.29.) The directed segment is called a **geometric vector** and its length is a nonnegative real number. A vector is denoted by $\mathbf{v}$ or, when its endpoints are specified, $\overrightarrow{AB}$, and its length by $|\mathbf{v}|$ or $|\overrightarrow{AB}|$. In physics a quantity having magnitude but not direction is called a *scalar* and, in the context of vectors, the real number that is a measure of the magnitude of a vector is often called a **scalar** as well. As an example, the above reference to wind motion, $\mathbf{v}$ = 15 mph from the northwest, is a vector called the *velocity,* but $|\mathbf{v}|$ = 15 mph is a scalar called the *speed.*

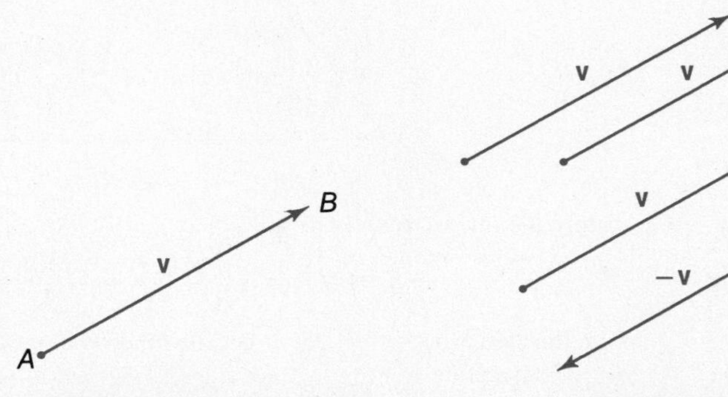

Figure 6.29 A geometric vector.

Figure 6.30

Since a vector is determined by a direction and a length, we may suppose that all directed segments of the same length and direction are geometric vector representations of the *same* vector quantity. A vector having the same length as $\mathbf{v}$ but direction opposite that of $\mathbf{v}$ is called the **negative of v** and written $-\mathbf{v}$. See Figure 6.30.

Composition of Vectors Vectors are combined in ways that agree with the physical situations they represent. If two forces, for instance, are applied at the same point, the combined effect is called the **resultant** (or **sum**) of these forces. The two separate forces are called **components.** In Figure 6.31a, $\mathbf{v}_1$ and $\mathbf{v}_2$ represent forces in the same direction, while $\mathbf{v}_3$ represents a force in the opposite direction. The resultant of $\mathbf{v}_1$ and $\mathbf{v}_2$, denoted $\mathbf{v}_1 + \mathbf{v}_2$, is in the same direction as the component forces and has magnitude $|\mathbf{v}_1 + \mathbf{v}_2|$ equal to the sum of the magnitudes of the components, as shown in (b). The resultant of $\mathbf{v}_1$ and $\mathbf{v}_3$ in (c) is in the direction of $\mathbf{v}_1$, which has the greater magnitude, and its magnitude is the difference of the magnitudes of the components.

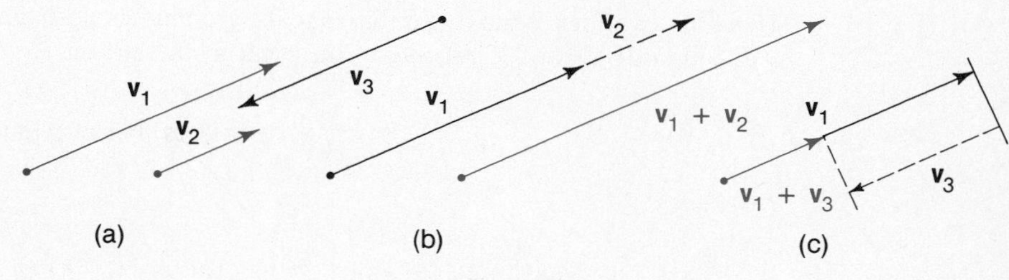

Figure 6.31

EXAMPLE 1 Suppose a plane has an airspeed (speed in still air) of 500 miles per hour and flies in a wind of 30 miles per hour. Find (a) its speed with the wind; (b) its speed against the wind.

SOLUTION (a) If the plane goes in the same direction as the wind, then its resultant speed is the sum of the two speeds, that is,

$$|\mathbf{v}_1 + \mathbf{v}_2| = |\mathbf{v}_1| + |\mathbf{v}_2|$$
$$= 500 + 30 = 530 \text{ (miles per hour).}$$

(b) If the plane flies in the direction opposite that of the wind, the resultant speed is the difference of the speeds of the plane and the wind, that is,

$$|\mathbf{v}_1 + \mathbf{v}_2| = |\mathbf{v}_1| - |\mathbf{v}_2|$$
$$= 500 - 30 = 470 \text{ (miles per hour).}$$ ●

If two vectors are perpendicular, their resultant is the diagonal of the rectangle having the two component vectors as adjacent sides.

EXAMPLE 2 A swimmer swimming at right angles to the current in a stream will be carried downstream as he goes across the stream. Suppose the current moves 7 miles per hour and the swimmer's speed in still water is 3 miles per hour. Find his speed and direction with respect to the bank of the stream.

SOLUTION In Figure 6.32, $\mathbf{v_1}$ represents the velocity of the current and $\mathbf{v_2}$ the swimmer's motion in still water. The combined effect of the two motions is the vector represented by the diagonal of the rectangle shown. The magnitude of the resultant vector $\mathbf{v_1} + \mathbf{v_2}$ is the swimmer's speed. By the Pythagorean theorem,

$$|\mathbf{v_1} + \mathbf{v_2}|^2 = 3^2 + 7^2 = 58,$$
$$|\mathbf{v_1} + \mathbf{v_2}| = \sqrt{58} \approx 7.6.$$

Also,

$$\tan \theta = \frac{3}{7} \approx 0.4286,$$
$$\theta \approx \arctan 0.4286 \approx 23°.$$

Thus, the swimmer is moving (relative to the ground) at approximately 7.6 mph and in a direction of about 23° relative to the bank of the stream. ●

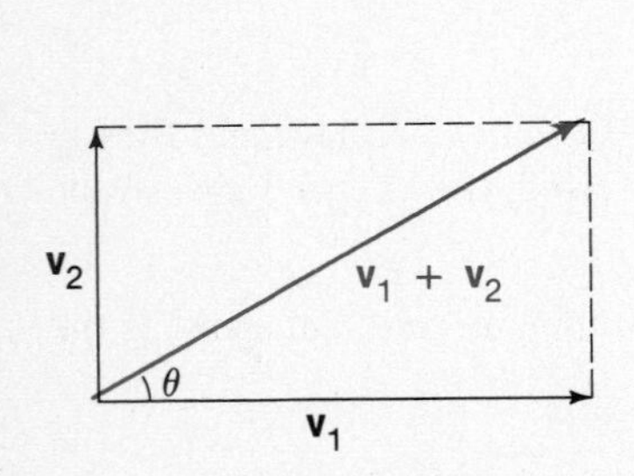

Figure 6.32

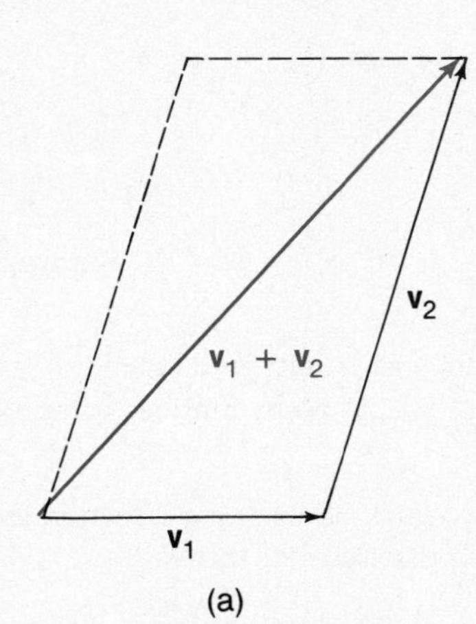

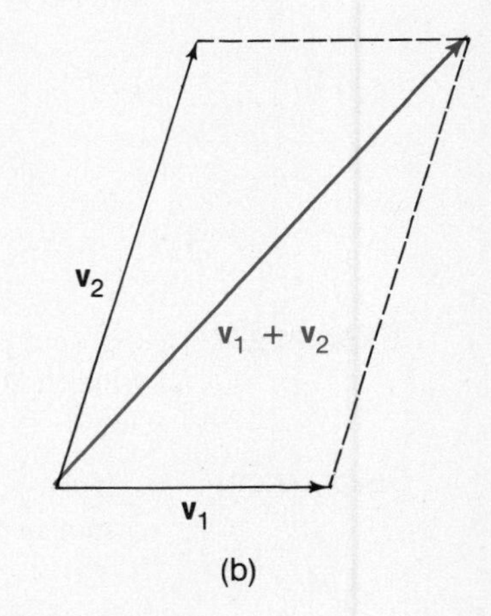

Figure 6.33

The following definition extends the results of the preceding examples to any two vectors with a common initial point.

DEFINITION 6.3 The **resultant** of two geometric vectors having a common initial point is a geometric vector having the same initial point and represented by the diagonal of the parallelogram having the two component vectors as adjacent sides.

The above definition is often called the **parallelogram law** for vectors.

For example, let an object be displaced in the direction and with the force indicated by $\mathbf{v_1}$ in Figure 6.33a and then let it be further displaced as indicated by $\mathbf{v_2}$. The net effect is the same as that of a single displacement represented by $\mathbf{v_1} + \mathbf{v_2}$. Alternatively, the parallelogram law may be illustrated by placing the vectors as shown in Figure 6.33b.

The **difference** $\mathbf{v}_2 - \mathbf{v}_1$ of two vectors is defined as $\mathbf{v}_2 + (-\mathbf{v}_1)$, as shown in Figure 6.34a. It is also useful to define a **zero vector 0** such that $\mathbf{0} + \mathbf{v} = \mathbf{v}$. The zero vector is represented by a point (length zero and direction undefined).

We may refer to a vector, (say, a force) as being twice another. For example, if a force of 10 pounds downward is exerted at an angle of 60° with the ground, twice this force is 20 pounds in the same direction. In general, if $\mathbf{v}$ is any given vector and a any real number, then the vector $a\mathbf{v}$ is called a **scalar multiple of v.** Its magnitude is $|a|\,|\mathbf{v}|$ and its direction the same as that of $\mathbf{v}$. See Figure 6.34b.

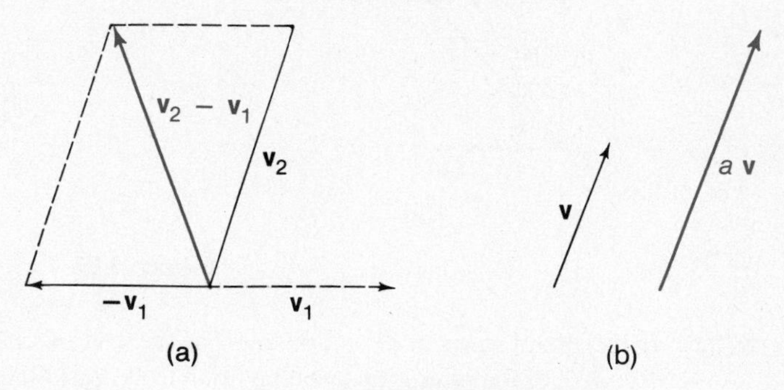

Figure 6.34

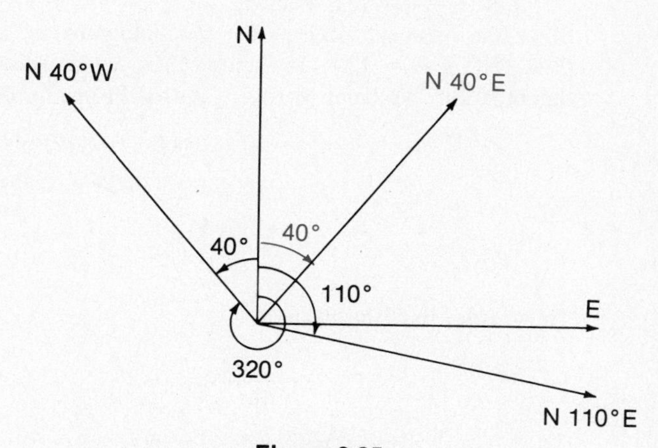

Figure 6.35

Before working the next example, we introduce a conventional notation used in dealing with motions of ships and planes. On a map we take "up" as north and "to the right" as east. If we look north and then sight toward the right along a line at an angle of 40° from the north, as shown in Figure 6.35, we denote this direction N 40°E (read "40° east of north"). A similar meaning is attached to N 40°W, N 110°E, and so on. One can also indicate direction by measuring in degrees from the north clockwise up to 360°. The figure shows a direction having a **bearing** of 320°. This is the same as the direction N 40°W.

EXAMPLE 3 Suppose a plane moves 600 miles per hour (in still air) in the direction N 30°E, as indicated in Figure 6.36a. Also suppose there is a wind of 40 miles per hour directly east. Find the plane's groundspeed (speed relative to the ground) and direction.

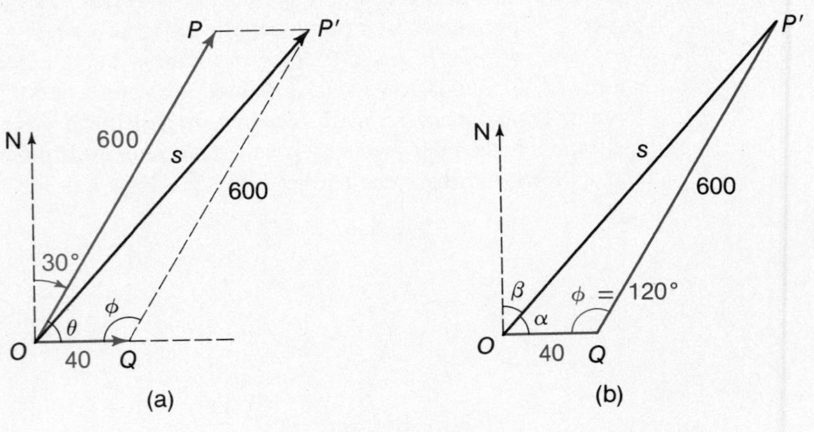

Figure 6.36

SOLUTION If the plane starts at O, its position at the end of one hour in still air would be at P. Because of the wind, its actual position is P', just 40 miles directly east of P. Then $\overrightarrow{OP'}$ represents the plane's actual motion relative to the ground.

Let $s = |\overrightarrow{OP'}|$, $\theta =$ angle $QOP = 90° - 30° = 60°$, and $\phi =$ angle OQP'. Since the adjacent angles of a parallelogram are supplementary (have sum 180°), $\phi = 180° - \theta = 120°$. In Figure 6.36b, which shows the triangle OQP' representing the situation, we want to find s and α. From the law of cosines,

$$\begin{aligned} s^2 &= 40^2 + 600^2 - 2(40)(600)\cos 120° \\ &= 1600 + 360000 + 24000 \\ &= 385600, \\ s &\approx 621. \end{aligned}$$

Then, from the law of sines,

$$\begin{aligned} \frac{600}{\sin\alpha} &= \frac{s}{\sin\phi}, \\ \sin\alpha &= \frac{600\sin\phi}{s} \\ &\approx \frac{600\sin 120°}{621} \\ &\approx \frac{300\sqrt{3}}{621} \\ &\approx 0.8367, \\ \alpha &\approx \arcsin 0.8367 \approx 57°. \end{aligned}$$

The direction β is given by $\beta = 90° - \alpha \approx 33°$. Thus, the plane has a ground speed of about 621 miles per hour in the direction N 33°E. •

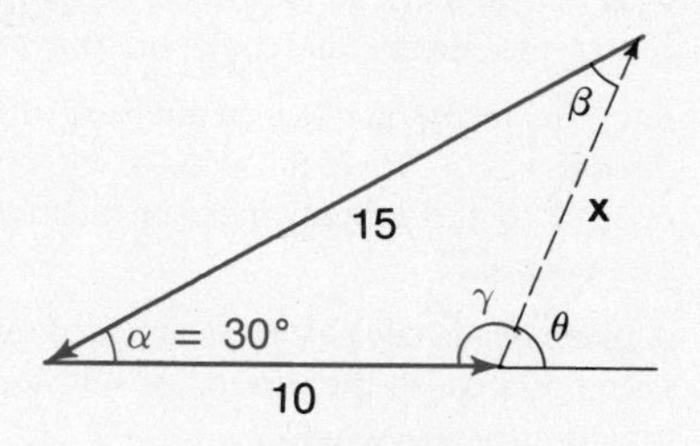

Figure 6.37

Instead of calculating the resultant vector of two component vectors, it may be that we have the resultant and one component vector and wish the other component vector (subtraction of vectors). The following illustrates this.

EXAMPLE 4 Suppose forces of 15 pounds and 10 pounds make an angle of 30° with each other, as shown in Figure 6.37. Find the force such that the resultant of it and the 10-pound force is the 15-pound force.

SOLUTION From the figure and the law of cosines,

$$\begin{aligned} x^2 &= 10^2 + 15^2 - 2(10)(15)\cos 30° \\ &= 100 + 225 - 300\left(\frac{1}{2}\sqrt{3}\right) \\ &\approx 65.2, \\ x &\approx 8.07. \end{aligned}$$

Now we use the law of sines to calculate β.

$$\begin{aligned} \frac{x}{\sin\alpha} &= \frac{10}{\sin\beta} \\ \sin\beta &= \frac{10\sin 30°}{x} \\ &\approx \frac{10\left(\frac{1}{2}\right)}{8.07} \approx 0.6196 \\ \beta &\approx 38° \end{aligned}$$

The direction of **x** is given by θ, which is the supplement of the third angle γ. Since $\gamma = 180° - (\alpha + \beta)$, $\theta = \alpha + \beta \approx 68°$. If the 10-pound force is in a horizontal direction the second component force makes an angle of 68° with the horizontal. •

EXERCISES 6.6

1. A swimmer swims at a uniform rate of 3.5 miles an hour directly across a stream 2100 feet wide. If the current flows 4.8 miles per hour, find the actual direction of the swimmer across the stream and how far downstream from his starting point he reaches shore.

2. Suppose an airplane has an airspeed of 450 miles per hour, and its pilot wishes to fly in the direction N 57°E. If the wind is blowing 43 miles per hour directly east, what direction relative to the ground must be maintained to fly in the desired direction? What is the groundspeed?

3. A plane heads due east at an airspeed of 420 mph, but because of the wind it has a resultant speed relative to the ground of 430 mph in the direction N 80°E. What is the speed and direction of the wind?

4. An aircraft carrier is moving in the direction N 35°W at a rate of 17 knots. A man walks to the starboard (right) across the flight deck at right angles to the direction of the ship's motion at 2.5 knots. In what direction is the man moving with respect to the surface of the water and how fast is he moving in that direction?

5. A kite is rising at 15 feet per second while a wind is moving it horizontally 25 feet per second. Find the angle of the actual path of the kite with the ground and its speed along that path.

6. A force of 124 pounds acts at an angle of 37° with the horizontal. Find the horizontal and vertical components.

7. Suppose a force of 10 pounds is exerted along the handle of a lawnmower, which makes an angle of 31° with the horizontal. Find the component of force exerted vertically (providing friction with the ground) and the one exerted horizontally (providing the forward motion).

8. Two forces of 52 and 81 pounds act simultaneously upon an object in directions at an angle of 63° with one another. Find the magnitude of the resultant force and the direction it makes with each of the component forces.

9. An airplane descends in a glide at a constant angle of 32° with the horizontal. How much distance directly forward does it go in descending 5000 feet?

10. Given forces of 40 pounds and 25 pounds making an angle of 36° with each other, find the force such that the resultant of it and the 25 pound force is the 40-pound force.

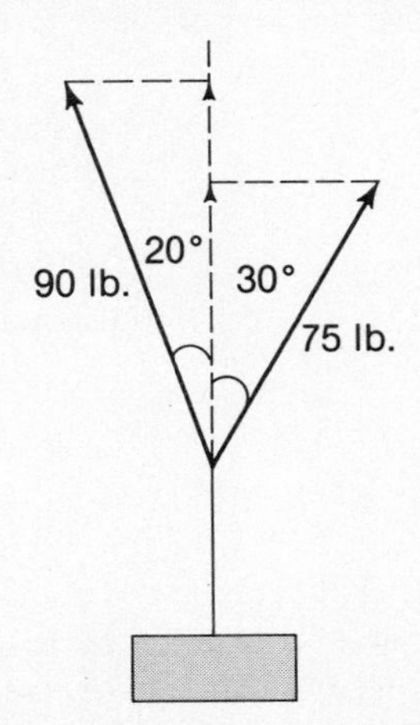

11. The figure on the preceding page shows two ropes, at respective angles of 20° and 30° with the vertical, supporting a crate of ship's freight. The forces along the ropes are 90 pounds and 75 pounds, respectively. What is the weight of the freight (the sum of the vertical components of the forces along the ropes)?

12. A man rolls a barrel weighing 510 pounds up a ramp inclined at an angle of 36°. (See the figure.) What magnitude of force is necessary to keep the barrel from rolling backward down the ramp (force F in the figure equal to the component of the weight force parallel to the ramp)?

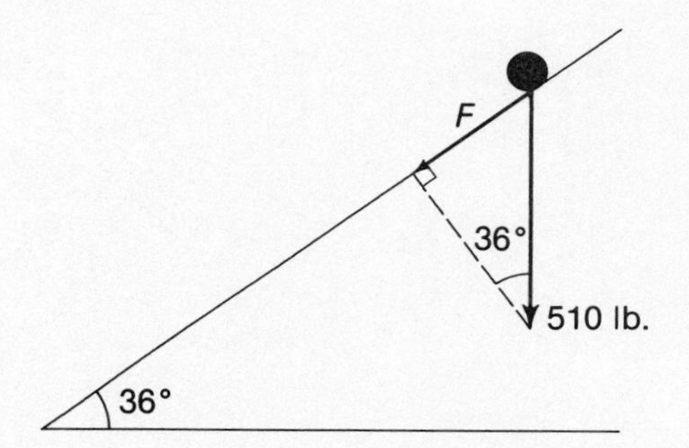

6.7 Polar Coordinates

Coordinate Systems An important use of the trigonometric ratios is in establishing a special coordinate system called *polar coordinates,* which will be introduced here but used more extensively in the next chapter. The Cartesian coordinate system establishes a one-to-one correspondence between the points of the Cartesian plane and the set of ordered pairs of real numbers. These coordinates represent distances from two mutually perpendicular lines. However, in many situations it is useful to identify a point not only by *a distance* but by *a direction* as well. This is the basis of polar coordinates.

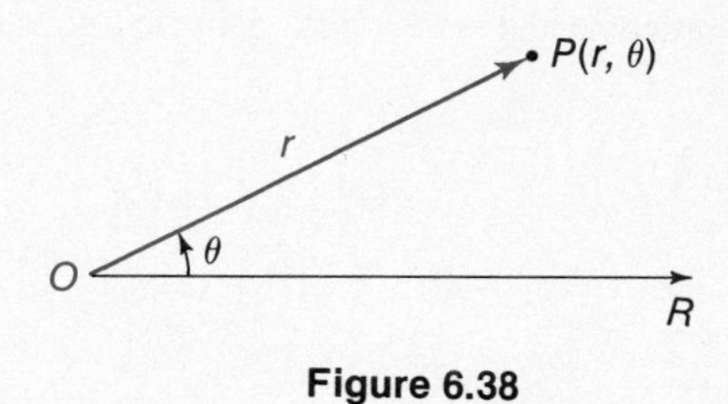

Figure 6.38

Polar Coordinates In Figure 6.38, an arbitrary point O and an arbitrary ray (directed line segment) $\overrightarrow{OR}$ are shown. Any point in the plane, such as P, can be identified by its distance from O and its direction relative to OR. We call O the **pole** and $\overrightarrow{OR}$ the **polar axis.** If r is the directed distance $\overline{OP}$ and θ the directed angle ROP, we call r and θ the **polar coordinates** of P. Observe that $\mathbf{r}$ (or $\overrightarrow{OP}$) is a *vector representation* of the point $P(r, \theta)$.

In Figure 6.38, note that r is measured in a positive direction *from O to P* along the terminal side of θ. The opposite direction is negative, so that in Figure 6.39 the points $(1, 45°)$ and $(-1, 45°)$ are symmetric with respect to the pole O. Also, since the angles in a polar coordinate system are directed angles, in Figure 6.39 the points $(2, \pi/3)$ and $(2, -\pi/3)$ are symmetric with respect to the polar axis.

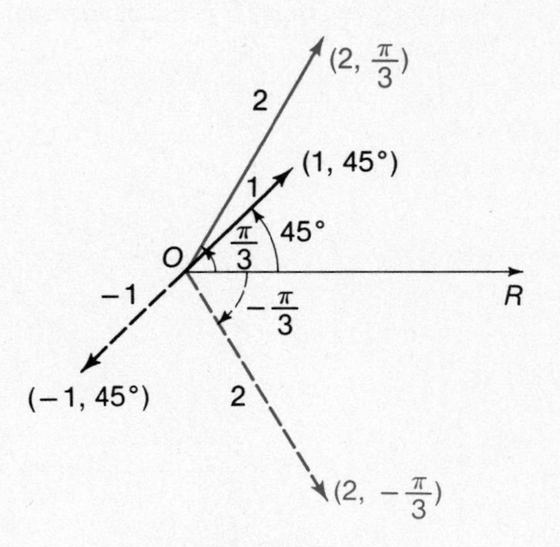

Figure 6.39

Polar coordinates have a feature which requires careful use. In the Cartesian plane there is a one-to-one correspondence between the points of the plane and their pairs of Cartesian coordinates. However, since any one of a set of coterminal angles (Section 6.1) has the same pair of sides, each point in the polar plane has an infinite number of paired polar coordinates. The point associated with $(2, \pi/3)$, for instance, could equally well be represented by $(2, 7\pi/3)$, $(2, -5\pi/3)$, $(-2, 4\pi/3)$, or in general, by $(2, \pi/3 + 2k\pi)$ or by $(-2, \pi/3 + (2k + 1)\pi)$. See Figure 6.40. No polar angle is assigned to the pole. It is identified simply by $r = 0$.

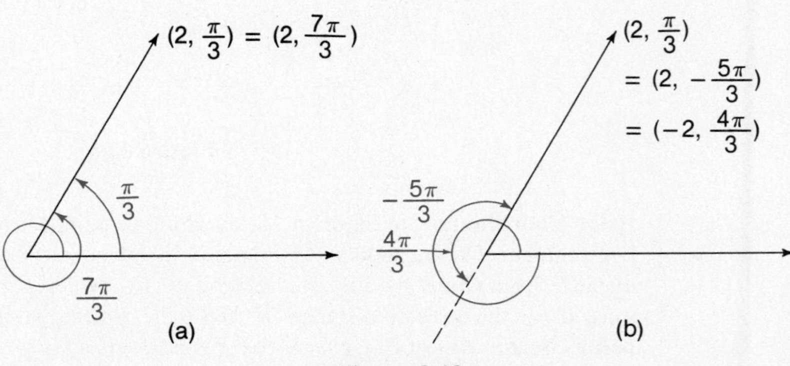

Figure 6.40

Relationships between Polar and Cartesian Coordinates Depending upon the situation, sometimes Cartesian coordinates and sometimes polar coordinates are preferred. There are simple formulas for changing from one system to the other. Using trigonometric ratios and the Pythagorean theorem with Figure 6.41, we get the following conversion formulas.

$$\frac{x}{r} = \cos\theta \quad \text{or} \quad x = r\cos\theta \qquad\qquad r^2 = x^2 + y^2$$

$$\text{and}$$

$$\frac{y}{r} = \sin\theta \quad \text{or} \quad y = r\sin\theta \qquad\qquad \tan\theta = \frac{y}{x}$$

The first pair of formulas determines the Cartesian coordinates (x, y) in terms of the polar coordinates (r, θ). The second pair is used to calculate the polar coordinates (r, θ) from given Cartesian coordinates (x, y). Because of periodicity of the tangent function, the number y/x does not determine a unique angle θ. So one must be careful to choose a value of θ in the correct quadrant.

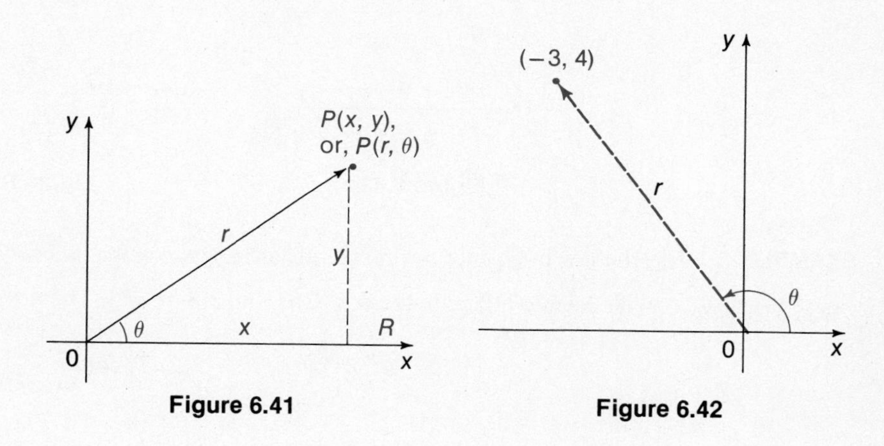

Figure 6.41 **Figure 6.42**

EXAMPLE 1 Find the polar coordinates of the point $(-3, 4)$. See Figure 6.42.

SOLUTION Since the Cartesian coordinates are given, we use the second pair of formulas.

$$r^2 = x^2 + y^2 = (-3)^2 + 4^2 = 25$$

$$r = 5 \quad \text{or} \quad -5$$

and

$$\tan\theta = \frac{y}{x} = -\frac{4}{3}$$

$$\theta \approx 127°,\ 307°,\ \text{and so on.}$$

The point $(-3, 4)$ is in the second quadrant, so we may write its polar coordinates as $(5, 127°)$ or $(-5, 307°)$. We usually choose one with positive r. ●

EXAMPLE 2 Calculate the Cartesian coordinates of the point (2, 60°). See Figure 6.43.

SOLUTION In this case, the first pair of formulas is used to find the Cartesian coordinates.

$$x = r \cos\ \theta = 2 \cos 60° = 2\left(\frac{1}{2}\right) = 1$$

and

$$y = r \sin \theta = 2 \sin 60° = 2\left(\frac{1}{2}\sqrt{3}\right) = \sqrt{3}$$

Thus, the Cartesian coordinates are $(1, \sqrt{3})$. ●

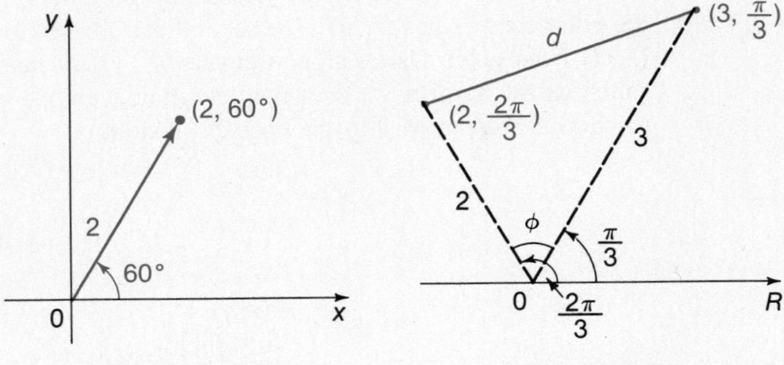

Figure 6.43

Figure 6.44

EXAMPLE 3 Use the law of cosines to find the distance between the points $(2, 2\pi/3)$ and $(3, \pi/3)$.

SOLUTION As seen in Figure 6.44, $\phi = 2\pi/3 - \pi/3 = \pi/3$. Then, from the law of cosines,

$$\begin{aligned} d^2 &= 2^2 + 3^2 - 2(2)(3) \cos \frac{\pi}{3} \\ &= 4 + 9 - 12\left(\frac{1}{2}\right) \\ &= 7, \\ d &= \sqrt{7}. \end{aligned}$$

●

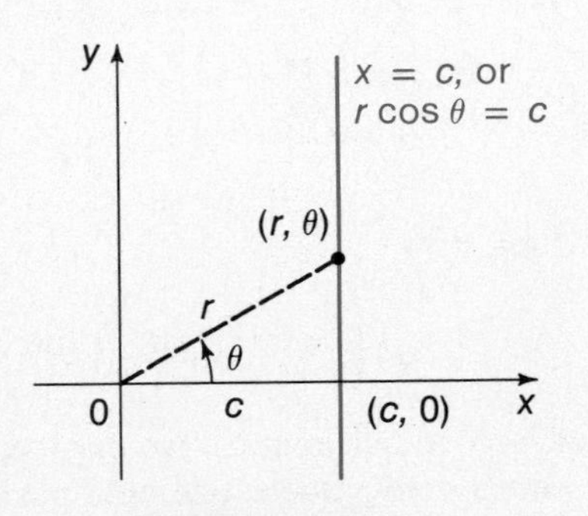

Figure 6.45

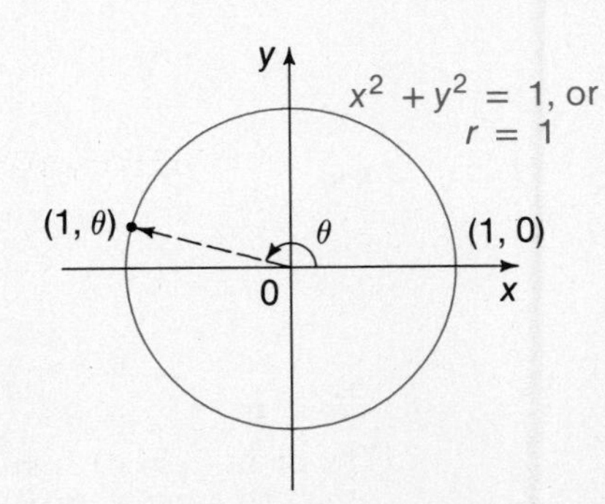

Figure 6.46

We have already seen that the Cartesian equation of a vertical line is $x = c$. Since the polar equivalent of x is $r \cos \theta$, the polar equation of a vertical line is $r \cos \theta = c$, or $r = c \sec \theta$. See Figure 6.45.

EXAMPLE 4 Find the polar equation of the standard unit circle.

SOLUTION Since the Cartesian equation of the unit circle is $x^2 + y^2 = 1$ and since $x^2 + y^2 = r^2$, the equation is $r^2 = 1$, or simply $r = 1$ or $r = -1$. See Figure 6.46. ●

In the next chapter polar coordinates will be used to provide relatively simple equations for curves whose Cartesian equations are not so simple.

EXERCISES 6.7

1. Plot the point whose Cartesian coordinates are (1, 2). Then plot the point whose polar coordinates are (1, 2). Are they the same point? Explain.

Plot the points having the following polar coordinates.

2. $(3, 30°)$ **3.** $\left(1, \frac{1}{4}\pi\right)$ **4.** $\left(-1, \frac{1}{2}\pi\right)$ **5.** $(2, 135°)$

6. $(-1, 300°)$ **7.** $\left(-2, \frac{2\pi}{3}\right)$ **8.** $\left(3, \frac{1}{2}\pi\right)$ **9.** $(3, 2)$

10. $(2, 0)$ **11.** $(0, 3)$ **12.** $\left(1, -\frac{\pi}{4}\right)$ **13.** $\left(-1, -\frac{\pi}{4}\right)$

14. $\left(2, -\frac{2\pi}{3}\right)$ **15.** $\left(-2, -\frac{\pi}{6}\right)$ **16.** $(-1, -1)$

17. For Exercises 2–6 above, give four sets of polar coordinates, two with positive r and two with negative r.

Convert each of the following polar forms to the equivalent Cartesian form.

18. $(2, 60°)$ **19.** $(1, 45°)$ **20.** $\left(\frac{5}{2}, \frac{\pi}{3}\right)$ **21.** $\left(-1, \frac{\pi}{6}\right)$

22. $(-2, 135°)$ **23.** $\left(3, \frac{5\pi}{6}\right)$ **24.** $(1, 180°)$ **25.** $(-2, 0)$

26. $\left(\frac{3}{2}, 210°\right)$ **27.** $\left(1, \frac{4\pi}{3}\right)$ **28.** $r = 3 \sec \theta$ **29.** $\theta = \frac{1}{4}\pi$

30. $r = 2$ **31.** $r = \cos \theta$

Convert each of the following Cartesian coordinates and Cartesian equations to an equivalent polar form with $r > 0$ and $0 \le \theta < 2\pi$.

32. $(2, -2\sqrt{3})$ **33.** $(1, 1)$ **34.** $\left(\frac{1}{2}, \frac{1}{2}\sqrt{3}\right)$ **35.** $\left(-\frac{1}{2}\sqrt{3}, \frac{1}{2}\right)$

36. $(0, -1)$ **37.** $(\sqrt{3}, -1)$ **38.** $(-1, 1)$ **39.** $(\sqrt{2}, -\sqrt{2})$

40. (2, 0) **41.** $\left(\frac{1}{2}\sqrt{2}, \frac{1}{2}\sqrt{2}\right)$ **42.** $x = 2$ **43.** $x = -1$

44. $y = 3$ **45.** $y = 0$ **46.** $x^2 + y^2 = 4$

Use the law of cosines to find the distance between each of the following pairs of points.

47. $\left(1, \frac{1}{2}\pi\right)$ and $\left(2, \frac{\pi}{6}\right)$ **48.** (3, 45°) and (2, 105°)

49. $\left(2, \frac{\pi}{3}\right)$ and $(3, \pi)$ **50.** $\left(1, \frac{2\pi}{3}\right)$ and $\left(3, \frac{4\pi}{3}\right)$

51. (2, 0) and $\left(\frac{3}{2}, 240°\right)$

Historical Notes

Much of the subject matter of this chapter is related to developments in mathematics that span the period from the ancient Greeks and Hindus to relatively modern times. The word "trigonometry" itself means "triangle measurement." The subject was developed as a tool to aid in the study of astronomy, which meant that spherical triangles (triangles on the surface of a sphere) were studied before much was done in the area of plane trigonometry.

The Greek scholars at the ancient university center in Alexandria systematically studied triangles in connection with indirect measurement. They even had theorems which amount to the law of cosines. The first trigonometric tables were constructed in the second century A.D. but, since the Greek concept of number was geometric, their mathematics was somewhat awkwardly expressed.

Analytic geometry and trigonometry had their beginnings with the development of arithmetic and algebra by the Hindus and Arabs. All six of the trigonometric functions were used by Arab astronomers by the tenth century, but were not generally known in Europe for several more centuries. The first text on trigonometry as a subject separate from astronomy was published in the sixteenth century, but used only the sine and cosine functions. It was at about this time that the trigonometric functions were first treated as ratios of sides of a triangle. One mathematician of this period spent 12 years with the help of hired assistants to construct two tables, one a 10-place table of all six functions and another a 15-place table of sines. In Section 5.12 we saw how approximating polynomials and computers simplify this task today.

Modern analytic trigonometry (as distinguished from strictly computational work) began with Descartes' introduction of the coordinate plane. The discipline became firmly established in the middle 18th century as a result of the work of the great Swiss mathematician Euler. He used the abbreviations sin, cos, tang, cot, sec, and csec, and generally systematized the notation of trigonometry. He also developed the notion of periodicity, although it had been vaguely recognized somewhat earlier.

Chapter 6 Review Exercises

Angles are commonly measured in degrees, radians, or revolutions. Convert each of the following angle measurements to its equivalent in each of the other two systems of units.

1. $510°$ **2.** $\frac{7\pi}{3}$ **3.** $140°$ **4.** $\frac{5\pi}{9}$ **5.** 1.4 rev. **6.** 0.72 rev.

7. Calculate the angle, in radians and then in degrees, through which the minute hand of a clock turns in 1 hour and 20 minutes.

8. If the hour hand of a clock is 6 inches long, how far does its tip move in 1 hour and 40 minutes?

Evaluate each of the following.

9. $\sin 315°$ **10.** $\cos 90°$ **11.** $\tan 150°$ **12.** $\cot 240°$ **13.** $\sec 210°$ **14.** $\csc 300°$

Solve for all positive angles θ less than 180°.

15. $\sin^2\theta + \sin\theta = 2$

16. $\cos 2\theta = \cos\theta$

17. $\sec^2\theta + 3\tan^2\theta = 5$

18. From the top of a fire observation tower, a forest ranger measures the angle of depression to the point of an observed fire as 0.5°. The tower is 250 feet high. How far away is the fire (from the foot of the tower)?

19. Calculate the length of the base of an isosceles triangle if the congruent sides are 12 inches long and the congruent angles are 70°. See the figure.

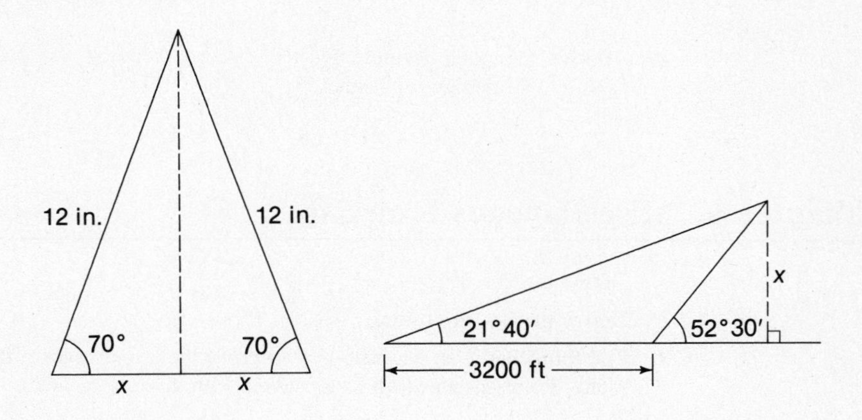

20. From a certain point on a road, a tree appears ahead and to the left in a direction 21°40′ with the road. See the figure. From a point 3200 feet farther down the road, the tree appears to the left at an angle of 52°30′ with the road. How far is the tree from the road?

21. Calculate the least angle of the triangle whose vertices are $(-1, 1)$, $(0, 4)$, and $(3, 2)$. First calculate the lengths of the sides and then use the law of cosines.

22. Calculate the remaining parts of a triangle with two sides respectively 850 cm. and 1200 cm. and a 38°-angle opposite the side of 850 cm.

23. Two streets intersect at an angle of 60°. A corner lot is 100 feet along one street and 150 feet along the other. Calculate the length of the third side of the lot.

24. Two forces acting at right angles to one another have a resultant of 34 pounds. If one of the forces is 16 pounds, calculate the magnitude of the other force and the direction it makes with the first force.

25. The diagonals of a parallelogram are 17 meters and 14 meters long and intersect at an angle of 127°. Find the dimensions of the parallelogram.

Convert each of the following point coordinates to the alternate form (polar to Cartesian and Cartesian to polar).

26. $(1, \sqrt{3})$

27. $(4, \sqrt{3})$

28. $(-1, 45°)$

29. $(2, 210°)$

30. $(3, 120°)$

31. $(-3, 0)$

32. $(3\sqrt{2}, 3\sqrt{2})$

33. $(-2, 300°)$

Use the relationships between polar and Cartesian coordinates to convert each of the following equations to the alternate form.

34. $y = 2$

35. $y = x$

36. $x^2 + y^2 = 5$

37. $x^2 = y$

38. $r = 3$

39. $r = \csc \theta$

40. $\theta = \dfrac{1}{4}\pi$

41. $r \tan \theta \sec \theta = 1$

42. Derive the polar formula for the distance between the points (r_1, θ_1) and (r_2, θ_2). [*Hint:* Use the law of cosines.]

Chapter 6 Miscellaneous Exercises

1. Derive the law of sines.

2. Derive the law of cosines.

3. The following is an alternate way to prove the law of sines. The figure shows a circle of radius r circumscribed about a triangle with sides a, b, and c. First show that

$$r = \frac{\frac{1}{2}a}{\sin' \alpha}, \qquad r = \frac{\frac{1}{2}b}{\sin \beta}, \qquad \text{and} \qquad r = \frac{\frac{1}{2}c}{\sin \gamma},$$

so that

$$2r = \frac{a}{\sin \alpha} = \frac{b}{\sin \beta} = \frac{c}{\sin \gamma}.$$

[*Hint:* Use the fact from geometry that in the figure $\alpha = \alpha'$, and so on.]

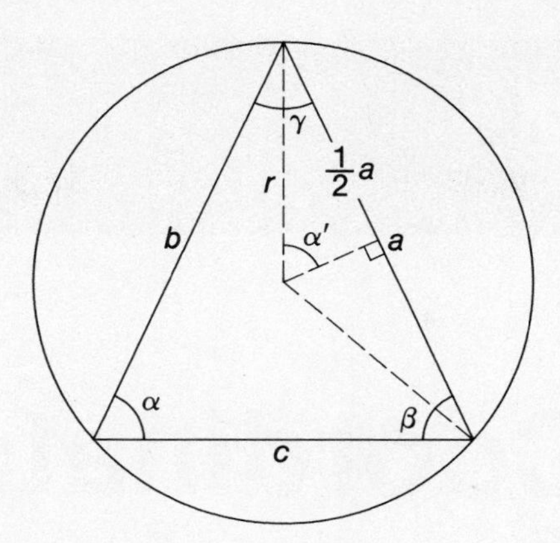

4. Let r be the radius of a circle and $s = (\frac{1}{2})(a + b + c)$ be the **semiperimeter** (one-half the perimeter) of the circumscribed triangle. Show that the area of the triangle is rs. [*Hint:* Use the figure in the preceding exercise.]

5. Use the law of cosines to show that

(a) $\dfrac{\cos \alpha}{a} + \dfrac{\cos \beta}{b} + \dfrac{\cos \gamma}{c} = \dfrac{a^2 + b^2 + c^2}{2abc}$.

(b) $a^2 + b^2 + c^2 = 2(bc \cos \alpha + ac \cos \beta + ab \cos \gamma)$.

6. With the usual convention for labeling a triangle, show that for any triangle $c = b \cos \alpha + a \cos \beta$. [*Hint:* Drop a line from the vertex perpendicular to the side c.]

7. If α, β, and γ are the angles the diagonal of a rectangular box makes with its edges, show that

$$\sin^2\alpha + \sin^2\beta + \sin^2\gamma = 2.$$

[*Hint:* Sketch the box and call the lengths of the edges x, y, and z, and the length of the diagonal d. Then use the fact that $d^2 = x^2 + y^2 + z^2$.]

7

Coordinate Geometry in the Plane

The ancient Greeks as long as 2000 years ago had developed geometry extensively. However, the Greek scholars did not have a corresponding interest in computation, so their arithmetic was limited and awkward. Consequently, they never developed algebra, which is a natural extension of arithmetic. On the other hand, Hindu and Arab scholars cultivated the ideas of number, arithmetic, and algebra. In the early seventeenth century Descartes linked geometry and algebra in such a way as to strengthen both. The union of these two branches of mathematics is called **analytic** or **coordinate geometry.** Geometric problems impossible for the ancient Greek scholars could now be solved. Likewise, ideas of algebra were made clearer by geometric interpretations and illustrations. Concerning coordinate geometry, one writer has said: "Though the idea behind it all is childishly simple, yet the method . . . is so powerful that very ordinary boys of seventeen can use it to prove results which would have baffled the greatest of the Greek geometers."*

Chapters 1 and 2 contained brief discussions of how to use the coordinate plane to apply algebra to geometry. In some of the later chapters, graphs were used to study functions geometrically. In this chapter these methods will be continued in the study of lines, circles, parabolas, ellipses, hyperbolas, and their applications.

7.1 The Straight Line

In Section 3.2 the theorem that every oblique line is identified uniquely by a linear equation (or function) was stated. At that point the linear equation (algebra) was used

*Eric T. Bell, *Men of Mathematics* (New York: Simon and Schuster, 1937), p. 7.

to obtain the line graphically (geometry). In this section the approach is reversed by starting with the line and determining its equation.

Slope The notion of the slope of a straight line introduced briefly in Section 3.2 will now be considered more fully. A line in the coordinate plane may be horizontal, vertical, or oblique. The lines in Figure 7.1 differ in their "steepness." The measure of this steepness is called the *slope* of the line.

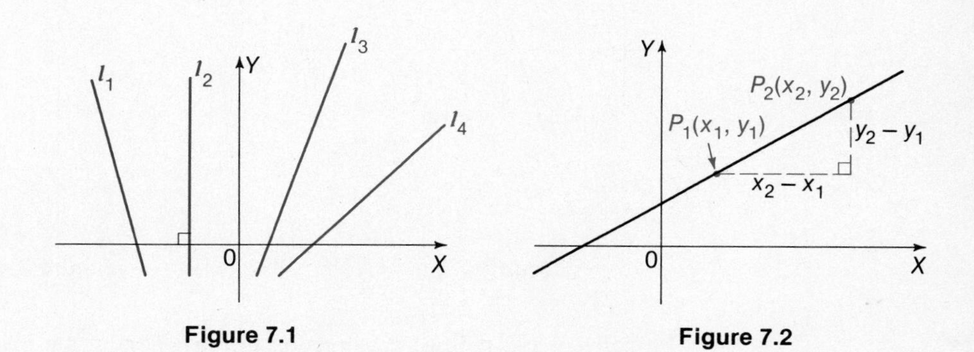

Figure 7.1 **Figure 7.2**

In Figure 7.2, let $P_1(x_1, y_1)$ and $P_2(x_2, y_2)$ be *any* two points on a line. For the given line, the ratio

$$\frac{y_2 - y_1}{x_2 - x_1}, \quad \text{or} \quad \frac{y_1 - y_2}{x_1 - x_2},$$

is unchanged when different choices of (x_1, y_1) and (x_2, y_2) are made.

DEFINITION 7.1 If (x_1, y_1) and (x_2, y_2) are any two points on a nonvertical line, the **slope** of the line is the ratio

$$m = \frac{y_2 - y_1}{x_2 - x_1}.$$

The slope of a vertical line is undefined.

If the line rises to the right, then when $x_2 > x_1$ we have $y_2 > y_1$, so that both $x_2 - x_1 > 0$ and $y_2 - y_1 > 0$ and the slope is positive. Similarly, if the line falls to the right, the slope is negative. The slope of a horizontal line is zero.

EXAMPLE 1 Draw a line having slope 3 through the point $(1, -2)$.

SOLUTION The slope is

$$m = \frac{3}{1} = \frac{y_2 - y_1}{x_2 - x_1}.$$

Then, as in Figure 7.3, we start with the point $(1, -2)$ and move **1** unit to the right and **3** units upward, to the point $(2, 1)$. The slope of this line is 3. ●

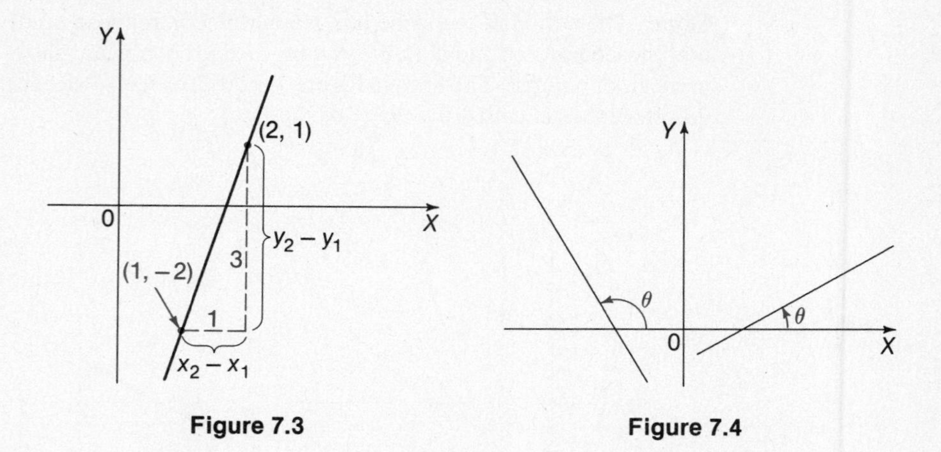

Figure 7.3

Figure 7.4

In highway and railroad construction the steepness of the roadbed is called the **grade.** It is the number of feet of "rise" for each foot of "run." For example, a road with a 3% grade rises 0.03 foot for each foot of horizontal distance, or 3 feet for each 100 feet. The steepness of a stairway and the pitch of a roof are further examples of the notion of *slope*.

The directed angle θ from the positive x-axis counterclockwise to a line is called the **inclination** of the line. See Figure 7.4. Observe that

$$m = \frac{y_2 - y_1}{x_2 - x_1} = \tan \theta,$$

that is, *the slope of a line is the tangent of its inclination*. If the line rises to the right, then $0 < \theta < 90°$ and $m = \tan \theta > 0$. If the line falls to the right, then $90° < \theta < 180°$ and $m = \tan \theta < 0$. For a horizontal line, $\theta = 0°$ or $180°$ and so $m = \tan \theta = 0$. The slope of a vertical line is undefined since, in this case, $\theta = 90°$.

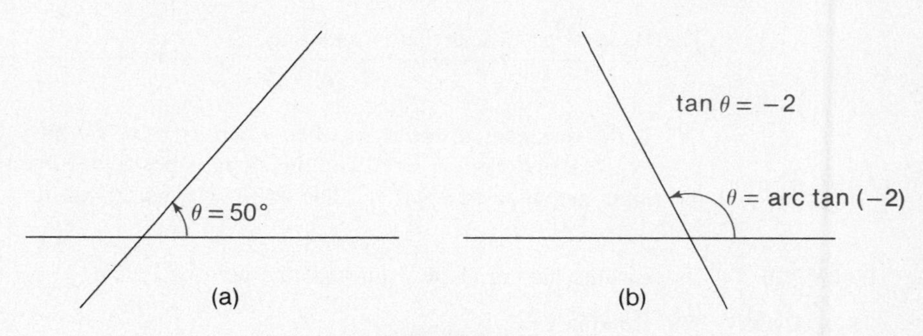

Figure 7.5

EXAMPLE 2 (a) Find the slope m of a line having an inclination of 50°.
(b) Find the inclination θ of a line whose slope is -2.

SOLUTION (a) The line, shown in Figure 7.5a, has slope $m = \tan 50° \approx 1.19$.

(b) See Figure 7.5b. Since $m = -2$ and $\tan\theta = m$, $\tan\theta = -2$, so that $\theta \approx 117°$. ●

Equations of a Line The equation of any oblique line can be written in a variety of forms, each particularly useful in certain situations.

Consider the line through the two points $P_1(x_1, y_1)$ and $P_2(x_2, y_2)$ in Figure 7.6, and let $P(x, y)$ be *any* point on this line. From similar triangles in the figure, we see that

$$\frac{y - y_1}{x - x_1} = \frac{y_2 - y_1}{x_2 - x_1} \quad \text{or} \quad y - y_1 = \frac{y_2 - y_1}{x_2 - x_1}(x - x_1).$$

This is called the **two-point form** of the equation of a line.

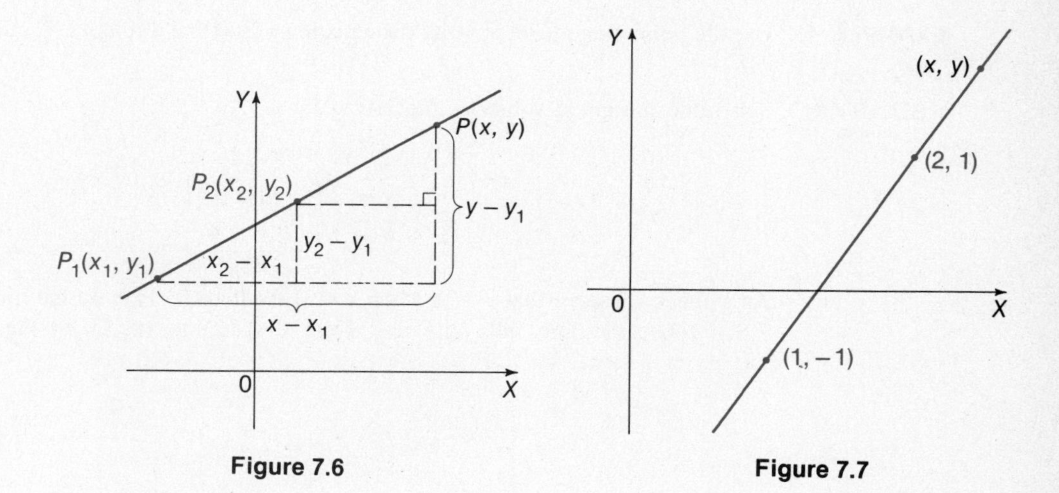

Figure 7.6

Figure 7.7

EXAMPLE 3 Write the equation of the line through the points $(1, -1)$ and $(2, 1)$. Use the two-point form of the equation. (See Figure 7.7.)

SOLUTION Let $(x_1, y_1) = (2, 1)$ and $(x_2, y_2) = (1, -1)$. Then, substitute these values in the formula.

$$\frac{y - y_1}{x - x_1} = \frac{y_2 - y_1}{x_2 - x_1}$$

$$\frac{y - 1}{x - 2} = \frac{-1 - 1}{1 - 2} = 2$$

$$y - 1 = 2(x - 2)$$
$$y - 1 = 2x - 4$$
$$2x - y - 3 = 0$$

If we took $(1, -1)$ as (x_1, y_1) and $(2, 1)$ as (x_2, y_2) we would have

$$\frac{y + 1}{x - 1} = \frac{1 + 1}{2 - 1} = 2,$$

which also reduces to $2x - y - 3 = 0$. ●

It can be seen from Figure 7.6 that the slope is $m = \dfrac{y_2 - y_1}{x_2 - x_1}$, so that the two-point form can be written as

$$\boldsymbol{y - y_1 = m(x - x_1).}$$

In this latter form it is called the **point-slope form** of the equation of a line.

EXAMPLE 4 Use the point-slope form to write the equation of the line through $(-1, 2)$ having slope -1.

SOLUTION Substitute the given values in the formula.

$$y - y_1 = m(x - x_1)$$
$$y - 2 = -1(x + 1)$$
$$x + y - 1 = 0 \quad ●$$

As a check, observe that $y = 1$ when $x = 0$, so that $(0, 1)$ is on the line. (See Figure 7.8.) Thus, the line falls one unit from $(-1, 2)$ to $(0, 1)$ so that the slope is $1/(-1) = -1$.

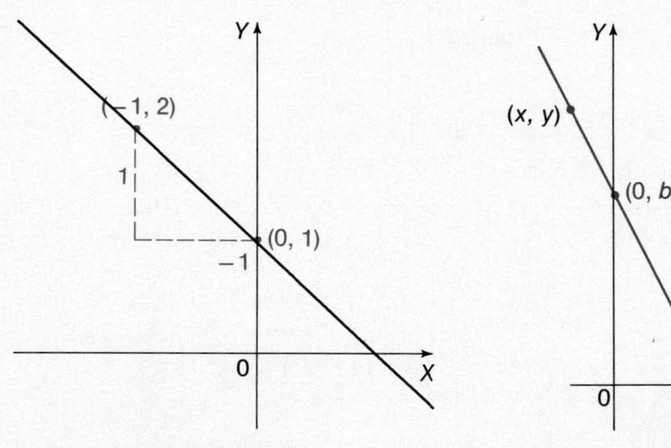

Figure 7.8

Figure 7.9

Next, suppose that the y-intercept of the line in Figure 7.9 with slope m is b, so that the line intersects the y-axis at $(0, b)$.

$$m = \frac{y - b}{x - 0}$$

$$y - b = mx$$

$$y = mx + b$$

This last form is called the **slope-intercept form** of the equation of a line. It is the form used in dealing with graphs of linear functions $f(x) = ax + b$ in Section 3.2. The simplest case is $y = mx$, the equation of a line through the origin with slope m.

EXAMPLE 5 Write the equation of the line having slope 2 and y-intercept 3 and graph.

SOLUTION Substitute 2 for m and 3 for b in the slope-intercept form to get

$$y = 2x + 3.$$

The line goes through $(0, 3)$. To help locate it properly, find another point by letting $y = 0$ and solving for x. The x-intercept is $-3/2$, so $(-3/2, 0)$ is on the line. See Figure 7.10. ●

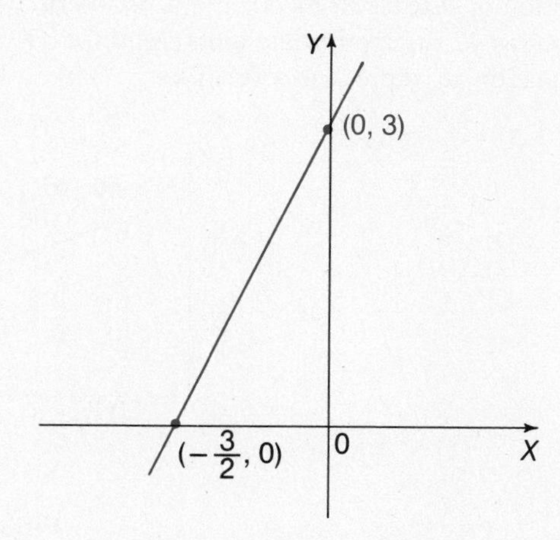

Figure 7.10

Which form to use in a particular situation depends upon the given information. To use the two-point form, two points must be known. The point-slope form is used when one point and the slope are known, and the slope-intercept form if the slope and y-intercept are known.

EXAMPLE 6 Write the equation of the line in Example 3 in the point-slope form.

SOLUTION First find the slope:

$$m = \frac{y_2 - y_1}{x_2 - x_1} = \frac{-1 - 1}{1 - 2} = 2.$$

Then either of the given points may be used in the point-slope form. If $(1, -1)$ is chosen, we have

$$y - y_1 = m(x - x_1),$$
$$y + 1 = 2(x - 1),$$

or $2x - y - 3 = 0$.

Using $(2, 1)$ as the given point, we may write

$$y - y_1 = m(x - x_1)$$
$$y - 1 = 2(x - 2),$$

or $2x - y - 3 = 0$, as before. ●

A simplified illustration of the use of the two-point formula is the following example of developing prediction formulas from observational or experimental data.

EXAMPLE 7 It was found in a botanical study that a certain photochemical reaction in the broad-leaved cattail is more efficient at higher altitudes. Specifically, the reaction (y) is a decreasing linear function of the number of frost-free days (x) at its location (if this period is at least 50 days). For a period of 100 frost-free days the reaction was measured as 42 (in appropriate units), and for 300 such days it was 21 (units). Find a linear equation to express this relation.

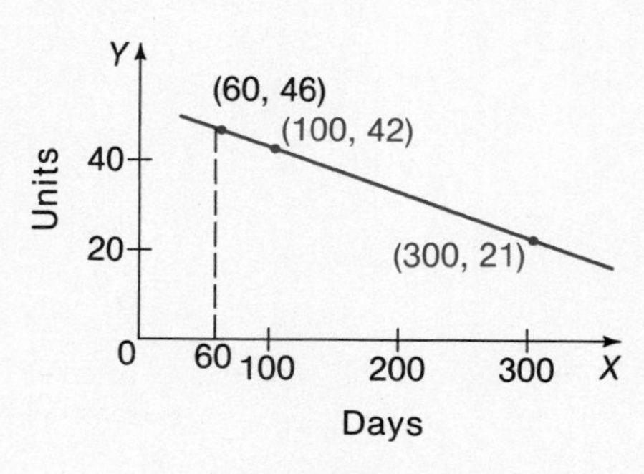

Figure 7.11

SOLUTION The pairs $(100, 42)$ and $(300, 21)$ can be thought of as points on the linear graph of this relation. See Figure 7.11. If the two-point form is used, the equation of this line is

$$\frac{y - 21}{x - 300} = \frac{42 - 21}{100 - 300} = \frac{21}{-200} \approx -0.105,$$

$$y - 21 = -0.105(x - 300),$$
$$= -0.105x + 31.5,$$
$$y = -0.105x + 52.5, \quad 50 \leq x \leq 365.$$

This formula describes the relation between the number of units of reaction (y) and the number of frost-free days (x), and we can use it to predict values of the reaction y for particular values of the time period x. For example, to find the magnitude of the reaction if there are 60 frost-free days, substitute $x = 60$:

$$y = -0.105(60) + 52.5 = -6.3 + 52.5 = 46.2. \quad \bullet$$

The following familiar formulas are equations of lines in slope-intercept form.

$\mathbf{d = rt}$, r constant	distance = rate × time
$\mathbf{f = kx}$, k constant	Hooke's law for a stretched spring
$F = \frac{9}{5}C + 32$ $C = \frac{5}{9}F - \frac{160}{9}$	the formulas relating temperatures in the Fahrenheit and Celsius (centigrade) scales

Just as the grade (slope) of a railroad is a measure of the uniform rate of rise (or fall) of the roadbed, so in the above equations the slope measures a uniform rate of change of one variable relative to another. In $d = rt$, the speed r is the uniform rate at which the distance covered increases with time, and in $f = ks$, the constant k tells how much the force must be increased (uniformly) for each unit of stretch of the spring. Similarly, in $F = (9/5)C + 32$, the slope 9/5 corresponds to the fact that each increase or decrease of one degree Celsius means an increase or decrease of 9/5 degrees Fahrenheit.

Each of the forms of the equation of an oblique line that we have derived may be written in the form $ax + by + c = 0$. For example, $y = mx + b$ may be rewritten

$$mx + (-1)y + b = 0,$$

and $y - y_1 = m(x - x_1)$ may be rewritten

$$mx + (-1)y + (y_1 - mx_1) = 0.$$

Furthermore, if a line is horizontal, its equation is $y = c$, or

$$0 \cdot x + 1 \cdot y - c = 0,$$

and if a line is vertical, its equation is $x = c$, or

$$1 \cdot x + 0 \cdot y - c = 0.$$

Thus, the equation of any line may be written in the **general form**

$$\mathbf{ax + by + c = 0.}$$

The converse is also true, that is, *any equation of the form $ax + by + c = 0$ is the equation of a straight line*. For, solving for y, we get

$$y = \left(-\frac{a}{b}\right)x + \left(-\frac{c}{b}\right),$$

which is the equation of a line with slope $-a/b$ and y-intercept $-c/b$, *provided $b \neq 0$*.

If $b = 0$, then the general equation reduces to $x = -c/a$, the equation of a vertical line. The important result we have just discussed is summed up in the following theorem.

THEOREM 7.1 The equation of every straight line can be written in the general form $ax + by + c = 0$ and, conversely, the graph of every equation $ax + by + c = 0$ is a straight line.

EXAMPLE 8 Write the general form of the equation of the line determined by the points $(-1, 2)$ and $(3, -2)$. Then calculate its slope and inclination. See Figure 7.12.

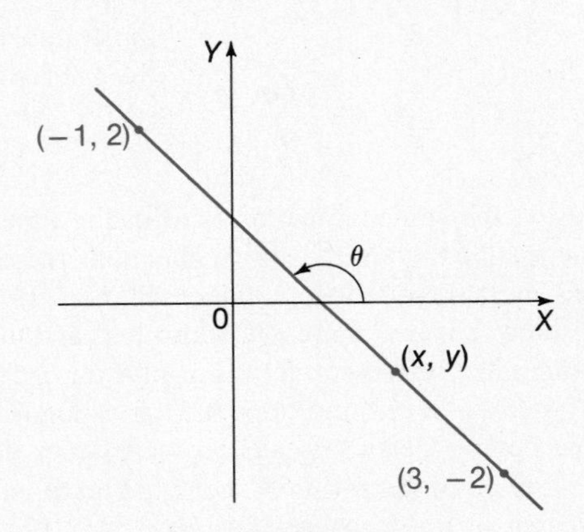

Figure 7.12

SOLUTION Use the two-point form.

$$\frac{y - y_1}{x - x_1} = \frac{y_2 - y_1}{x_2 - x_1}$$

$$\frac{y - 2}{x + 1} = \frac{-2 - 2}{3 + 1} = -1$$

$$y - 2 = -(x + 1)$$

$$x + y - 1 = 0$$

This is the general form. The slope is best found from the slope-intercept form, $y = mx + b$. From $x + y - 1 = 0$, we have $y = -x + 1$. So the slope is -1.

The inclination, θ, is 135°, since $\tan 135° = -1$. (Remember that the angle of inclination is restricted to the interval $0 \leq \theta < 180°$. Why?) ●

EXAMPLE 9 Write the general equation $2x - 6y + 9 = 0$ in the slope-intercept form.

SOLUTION In the slope-intercept form, $y = mx + b$, y is expressed as a function of x. So we solve the given equation for y in terms of x.

$$2x - 6y + 9 = 0$$
$$-6y = -2x - 9$$
$$y = \frac{1}{3}x + \frac{3}{2}$$

In this form, we see that the slope is 1/3 and the y-intercept is 3/2. The equation is graphed in Figure 7.13. ●

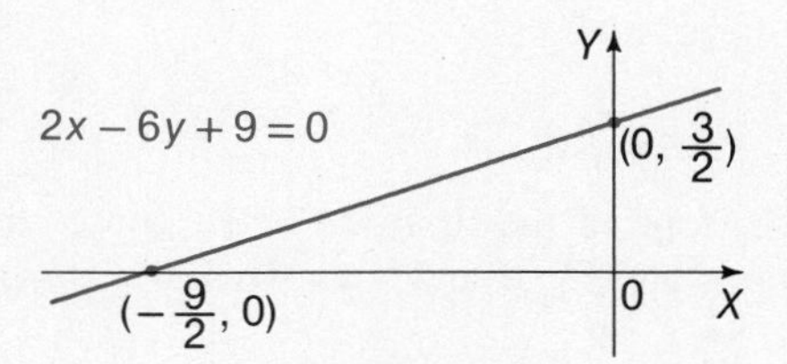

Figure 7.13

In summary, we have the following forms of the equation of a straight line:

two-point form: $\dfrac{y - y_1}{x - x_1} = \dfrac{y_2 - y_1}{x_2 - x_1}, \quad x_1 \neq x_2$

point-slope form: $y - y_1 = m(x - x_1)$

slope-intercept form: $y = mx + b$

general form: $ax + by + c = 0, \quad a, b$ **not both 0**

The slope-intercept form is particularly useful. Other forms of the equation of a line will be covered in the exercises.

Parallel and Perpendicular Lines We know from elementary geometry that the angles θ_1 and θ_2 in Figure 7.14 are equal if and only if lines l_1 and l_2 are parallel. In

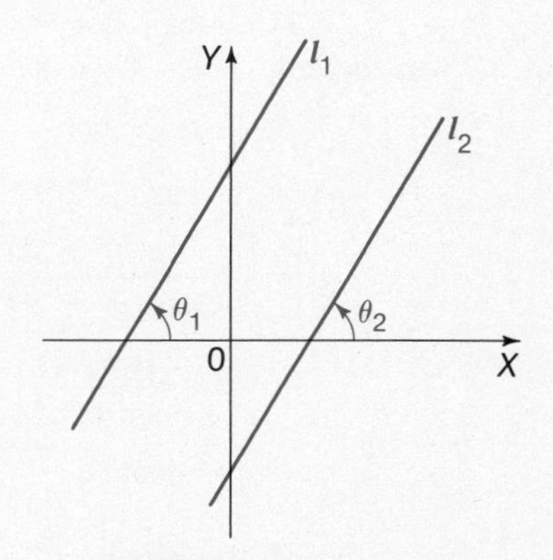

Figure 7.14

this case slope m_1 = slope m_2, and so lines which are parallel all have the same slope. Conversely, lines with equal slopes are parallel.

EXAMPLE 10 Show that the lines $2x - 3y + 1 = 0$ and $4x - 6y - 5 = 0$ are parallel.

SOLUTION First we write each equation in the slope-intercept form.

$$2x - 3y + 1 = 0 \qquad\qquad 4x - 6y - 5 = 0$$

$$-3y = -2x - 1 \qquad\qquad -6y = -4x + 5$$

$$y = \frac{2}{3}x + \frac{1}{3} \qquad\qquad y = \frac{-4}{-6}x - \frac{5}{6}$$

The slope of the first line is 2/3 and of the second line is 4/6. Since 4/6 = 2/3, the slopes are equal and thus the lines are parallel. Their graphs are shown in Figure 7.15. ●

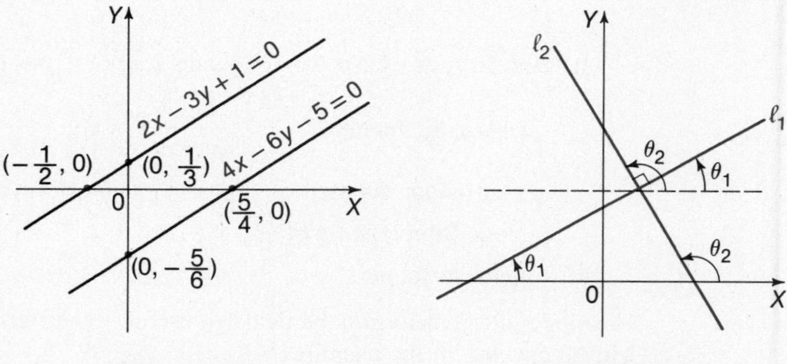

Figure 7.15 **Figure 7.16**

Suppose that l_1 and l_2 are *any* two perpendicular lines neither of which is horizontal. (See Figure 7.16.) Call their slopes m_1 and m_2, respectively, and their inclinations θ_1 and θ_2, respectively. Then, from the figure we see that

$$\theta_2 = \theta_1 + 90°.$$

$$\begin{aligned} \tan\theta_2 &= \tan(\theta_1 + 90°) \\ &= \frac{\sin(\theta_1 + 90°)}{\cos(\theta_1 + 90°)} \\ &= \frac{\cos\theta_1}{-\sin\theta_1} \qquad \text{from Section 5.5} \\ &= -\cot\theta_1 \\ &= -\frac{1}{\tan\theta_1} \qquad = -\frac{1}{m_1} \end{aligned}$$

Thus, $$m_2 = -\frac{1}{m_1}, \quad \text{or} \quad m_1m_2 = -1.$$

This result is stated in the following theorem.

THEOREM 7.2 Two lines are perpendicular if and only if their slopes are negative reciprocals of each other, that is, the product of their slopes is -1.

EXAMPLE 11 Show that the line $3x + 2y + 5 = 0$ is perpendicular to the line $2x - 3y + 1 = 0$.

SOLUTION First write the equations in the slope-intercept form.

$$3x + 2y + 5 = 0, \quad \text{or} \quad y = -\frac{3}{2}x - \frac{5}{2}$$

$$2x - 3y + 1 = 0, \quad \text{or} \quad y = \frac{2}{3}x + \frac{1}{3}$$

Then $m_1 = -3/2$ and $m_2 = 2/3$, so that $m_1m_2 = -1$, and the lines are perpendicular. ●

EXAMPLE 12 Use slopes to show that the triangle whose vertices are $A(-1, -1)$, $B(1, 0)$, and $C(-2, 6)$ is a right triangle.

SOLUTION See Figure 7.17. We want to show that BC is perpendicular to AB.

$$\text{Slope of } AB: \frac{0 + 1}{1 + 1} = \frac{1}{2}$$

$$\text{Slope of } BC: \frac{6 - 0}{-2 - 1} = -2$$

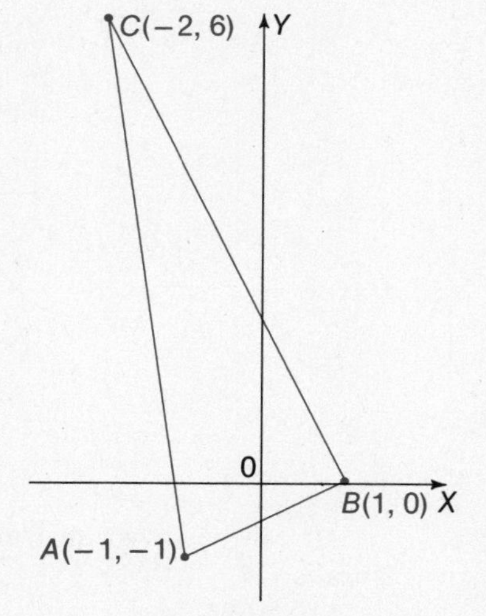

Figure 7.17

Since $(1/2)(-2) = -1$, AB is perpendicular to BC and triangle ABC is a right triangle. ●

EXAMPLE 13 Show that the points (1, 6), (4, 5), (1, −4), and (−2, −3) are vertices of a rectangle by showing that opposite sides are parallel and adjacent sides perpendicular.

SOLUTION Refer to Figure 7.18, where the sides are labeled a, b, c, and d.

$$\text{Slope of } a\text{:} \quad \frac{6+3}{1+2} = 3$$

$$\text{Slope of } c\text{:} \quad \frac{5+4}{4-1} = 3$$

Since slope a = slope c, a and c are parallel. Similarly,

$$\text{slope of } b\text{:} \quad \frac{-3+4}{-2-1} = -\frac{1}{3}$$

and

$$\text{slope of } d\text{:} \quad \frac{6-5}{1-4} = -\frac{1}{3},$$

so b is parallel to d.

The slopes of the adjacent sides are 3 and $-1/3$, which are negative reciprocals, and thus adjacent sides are perpendicular. Since opposite sides are parallel and adjacent sides are perpendicular, the figure is a rectangle. ●

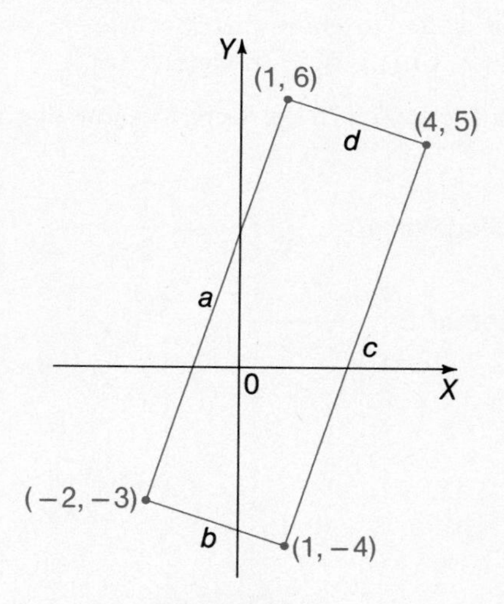

Figure 7.18

It can be shown that the general formula for the angle of intersection of two lines is

$$\tan \phi = \frac{m_2 - m_1}{1 + m_1 m_2},$$

where m_1 and m_2 are the slopes of lines l_1 and l_2, respectively, and ϕ (the Greek letter *phi*) is the angle from l_1 to l_2, measured counterclockwise. The details of the derivation are outlined in the exercises. Note here that if $m_1 = m_2$, then $\tan \phi = 0$, $\phi = 0°$, and the lines are parallel. If $m_1 m_2 = -1$, then $\tan \phi$ is undefined, $\phi = 90°$, and so the lines are perpendicular.

EXERCISES 7.1

Determine the equation of the line in each of the following cases and then reduce it to the general form. Sketch the line.

1. Through (1, 5) and (3, 7)

2. Through (3, −2) and (5, 0)

3. Through (2, −4) with slope 5

4. Through (−2, 2) with slope $-\frac{1}{2}$

5. Through (1, 2) with slope −1

6. Through (−2, −2) with slope 2

7. Having slope 1 and y-intercept 2

8. Having slope −2 and y-intercept −1

9. Having slope 2 and y-intercept 0

10. Having slope $-\frac{1}{3}$ and y-intercept 1

11. Through (−1, 2) with inclination 30°

12. Through (1, 1) with inclination 135°

13. Through (2, 3) with inclination 0°

14. Through the origin with inclination 60°

Graph each of the following by first determining the x- and y-intercepts.

15. $2x + y - 1 = 0$ **16.** $x - 2y - 6 = 0$ **17.** $3x + 4y = 12$ **18.** $5x - 3y = 15$

Put each of the following equations into the slope-intercept form. Then identify the slope and calculate the inclination.

19. $4x - y = 7$

20. $4x + 2y + 3 = 0$

21. $2x + 5y - 3 = 0$

22. $5x - 2y + 10 = 0$

First determine the slopes of each of the following pairs of lines. Then identify pairs which are either parallel or perpendicular. Sketch the lines.

23. $2x + 5y = 3$ and $9 - 4x = 10y$

24. $2x + 3y = 5$ and $8y - 12x = 1$

25. $3x - 5y - 15 = 0$ and $2x + y - 4 = 0$

26. $4x + y = 2$ and $2x - 8y = 3$

27. $3x + 7y - 2 = 0$ and $6x + 14y = 0$

28. $3x + 4y - 5 = 0$ and $6x + 8y = 3$

29. $4x + 3y + 2 = 0$ and $3x - 4y = 2$

30. $3x - 2y - 1 = 0$ and $x - 5y + 1 = 0$

Draw the polygon whose vertices are given. From the figure, guess whether it is a right triangle, rectangle, or parallelogram. Then check your guess by calculating slopes of the sides of each polygon.

31. $A(-4, 1)$, $B(-2, 4)$, $C(6, 2)$

32. $A(6, -5)$, $B(1, 5)$, $C(-2, -1)$

33. $A(4, 1)$, $B(-3, 2)$, $C(-2, -3)$, $D(5, 4)$

34. $A(1, 0)$, $B(-4, 5)$, $C(-6, 3)$, $D(-1, -2)$

35. $A(-2, 3)$, $B(2, 1)$, $C(3, 3)$, $D(-1, 5)$

36. $A(-3, -2)$, $B(2, 0)$, $C(3, 3)$, $D(-2, -1)$

Use the point-slope form to write the equations of the lines through the given point and respectively parallel and perpendicular to the given line.

37. $(2, -4)$ and $3x - y = 6$

38. $(0, 5)$ and $3x - 4y + 7 = 0$

39. $(6, 3)$ and $2x - 6y - 9 = 0$

40. (h, k) and $ax + by + c = 0$

Determine the equation of the perpendicular bisector of the line segment whose endpoints are given. [Hint: As one method, (a) find the midpoint of the segment and then use the point-slope form of the equation of a line. Then as a second method, (b) use the fact that any point on the perpendicular bisector of a line segment is equidistant from the endpoints of the segment.]

41. $(1, -1)$ and $(2, 2)$

42. $(-1, -1)$ and $(3, 0)$

43. $(-2, 2)$ and $(0, 0)$

44. $(-1, 2)$ and $(5, 1)$

Graph each of the following. [Hint: Use the fact that $|a| = a$ if $a \geq 0$, and $|a| = -a$ if $a < 0$.]

45. $y = |x - 1|$

46. $|x| + y = 1$

Write the equations of the sides of the triangles whose vertices are given.

47. $(1, 1)$, $(-2, 3)$, $(-4, 0)$

48. $(2, -1)$, $(-3, -1)$, $(2, 2)$

49–50. The segment joining a vertex of a triangle to the midpoint of the opposite side is called a median. Write the equations of the medians of each of the triangles in exercises 47–48.

51. A printer quotes a price of $1200 for printing a minimum of 100 copies of a pamphlet and $2000 for 500 copies. Assuming the relation between the number of copies (x) and the cost (y) is linear, (a) what is the cost per copy above the minimum, and (b) what is the cost for 750 copies?

52. A moving company has an hourly rate together with a flat charge service fee. If the charge is $120.50 for 2 hours and $187.50 for 5 hours, (a) write the equation of this linear relationship, (b) calculate the hourly rate and the service charge, (c) what is the total charge for 9 hours?

53. The relationship between the length of a stretched spring and the amount of force producing the stretch is a linear one. If a force of 2 pounds stretches a 15-inch spring to 18 inches and a force of 4 pounds stretches it to 21 inches, what is the length of the spring if a force of 5 pounds is applied?

54. A research study showed that if the Democrats win 45% of the two-party vote, they win 42.5% of the seats in the House of Representatives. If the Democrats win 55% of the two-party vote, they win 67.5% of the seats. Assuming the relation between number of seats (y) won and the percent of the vote won is linear, write this linear relation and use it to predict the number of seats the Democrats will win if they get 50% of the vote.

55. Use the figure to derive the formula

$$\tan \phi = \frac{m_2 - m_1}{1 + m_1 m_2}.$$

[*Hint:* Note that $\phi = \theta_2 - \theta_1$ and use the tangent difference formula.]

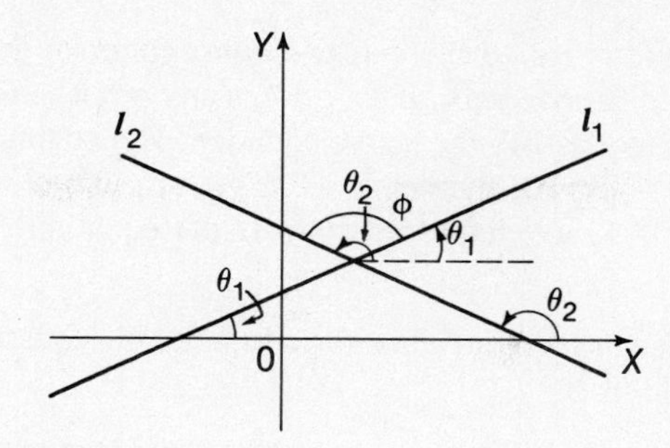

Use the formula of the preceding exercise to calculate the angle from the first line to the second line in each of the following.

56. $2x + 3y = 12$ and $3x - 4y - 12 = 0$

57. $2x - y = 5$ and $4x + y = 2$

58. $2y + 3 = 0$ and $2x + 4y = 1$

59. $x = 3$ and $2x + y = 3$

7.2 The Circle

The equation of a circle was introduced in Section 1.4 and the unit circle $x^2 + y^2 = 1$ was used in the study of the circular functions in Chapter 5. In this section, circles will be discussed in considerable detail.

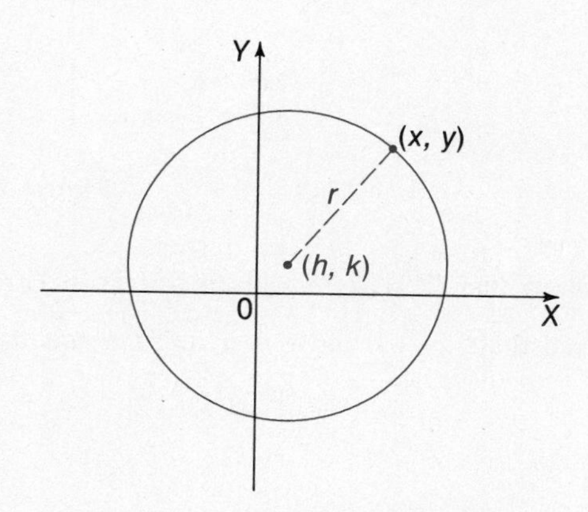

Figure 7.19

Consider any circle in the coordinate plane and call its center (h, k) and its radius r. (See Figure 7.19.) If (x, y) is *any* point on this circle, then from the distance formula and the definition of a circle,

$$(x - h)^2 + (y - k)^2 = r^2.$$

This is called the **center-radius form** of the equation of a circle.

Conversely, if (x', y') is any point in the plane *not* on this circle, its coordinates do not satisfy the above equation. Hence this is the equation of the circle. Of course, if $h = k = 0$, the center is at the origin and the equation reduces to $x^2 + y^2 = r^2$. If $r = 1$, we get $x^2 + y^2 = 1$, the equation of the standard unit circle.

EXAMPLE 1 Use the center-radius form to write the equation of the circle with center at $(-1, 2)$ and radius 3.

SOLUTION Substitute the given values in the equation.

$$(x - h)^2 + (y - k)^2 = r^2$$
$$(x - (-1))^2 + (y - 2)^2 = 3^2$$
$$(x + 1)^2 + (y - 2)^2 = 9$$

The graph is shown in Figure 7.20. ●

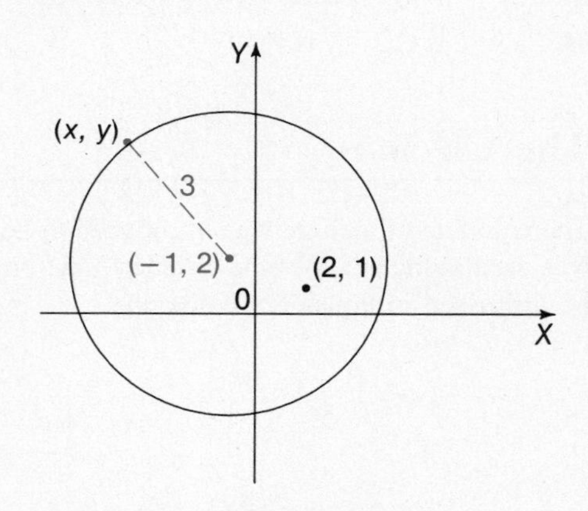

Figure 7.20

EXAMPLE 2 Show that (2, 1) is *not* on the circle of Example 1.

SOLUTION Substitute $x = 2$ and $y = 1$ into the equation of the circle.

$$\begin{aligned}(x + 1)^2 + (y - 2)^2 &= (2 + 1)^2 + (1 - 2)^2 \\ &= 3^2 + (-1)^2 \\ &= 10 \neq 9.\end{aligned}$$

Since (2, 1) does not satisfy the equation of the circle, it is not on the circle. See Figure 7.20. ●

The center-radius form of the equation of the circle can be rewritten, as follows:

$$(x - h)^2 + (y - k)^2 = r^2,$$
$$(x^2 - 2hx + h^2) + (y^2 - 2ky + k^2) = r^2,$$

$$x^2 + y^2 - 2hx - 2ky + (h^2 + k^2 - r^2) = 0,$$

or

$$\mathbf{x^2 + y^2 + Dx + Ey + F = 0,} \tag{1}$$

where $D = -2h$, $E = -2k$, and $F = h^2 + k^2 - r^2$. Equation (1) is called **the general form of the equation of a circle.**

We have shown above that the equation of any circle can be reduced to the general form. Conversely, every equation of this form is the equation of a circle. Consider a particular example.

EXAMPLE 3 Rewrite the equation $x^2 + y^2 - 2x + 2y - 2 = 0$ in the center-radius form.

SOLUTION Begin by completing the square on the x-terms and the y-terms separately (see Section 3.3).

$$\begin{aligned} x^2 + y^2 - 2x + 2y - 2 &= 0 \\ (x^2 - 2x \qquad) + (y^2 + 2y \qquad) &= 2 \\ (x^2 - 2x + 1) + (y^2 + 2y + 1) &= 4 \\ (x - 1)^2 + (y + 1)^2 &= 2^2 \end{aligned}$$

This is the equation of the circle with center at $(1, -1)$ and radius 2, graphed in Figure 7.21. ●

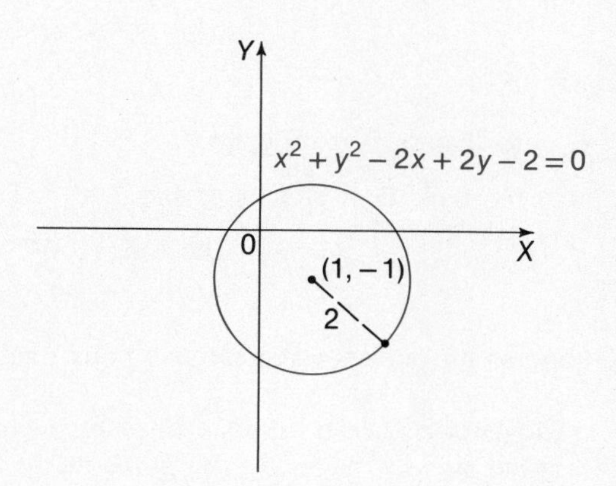

Figure 7.21

The general case is treated in the same way.

$$\begin{aligned} x^2 + y^2 + Dx + Ey + F &= 0, \\ (x^2 + Dx) + (y^2 + Ey) &= -F, \\ \left(x^2 + Dx + \frac{1}{4}D^2\right) + \left(y^2 + Ey + \frac{1}{4}E^2\right) &= -F + \frac{1}{4}D^2 + \frac{1}{4}E^2, \\ \left(x + \frac{1}{2}D\right)^2 + \left(y + \frac{1}{2}E\right)^2 &= \frac{1}{4}(D^2 + E^2 - 4F). \end{aligned}$$

There are three cases to consider:

(1) If $D^2 + E^2 - 4F > 0$, then the above is the equation of a circle with center $(-\frac{1}{2}D, -\frac{1}{2}E)$ and radius equal to $\frac{1}{2}\sqrt{D^2 + E^2 - 4F}$.

(2) If $D^2 + E^2 - 4F = 0$, the equation reduces to $(x + \frac{1}{2}D)^2 + (y+\frac{1}{2}E)^2 = 0$, which is satisfied only by the single point $(-\frac{1}{2}D, -\frac{1}{2}E)$. We then call the graph a *point circle*.

(3) If $D^2 + E^2 - 4F < 0$, then no point (x, y) satisfies the equation, since the sum of the squares on the left side cannot be negative. In this case there is no graph.

EXAMPLE 4 For each of the following, calculate the value of $D^2 + E^2 - 4F$ and tell whether the equation represents a circle, a point circle, or no graph. (a) $x^2 + y^2 + 2x + 4y + 3 = 0$, (b) $x^2 + y^2 + 2x + 4y + 13 = 0$, (c) $x^2 + y^2 + 2x + 4y + 5 = 0$.

SOLUTION (a) Here $D = 2$, $E = 4$, and $F = 3$. Then

$$\begin{aligned} D^2 + E^2 - 4F &= 2^2 + 4^2 - 4(3) \\ &= 8 > 0, \end{aligned}$$

and so a real circle is represented.

(b) In this case, $D = 2$, $E = 4$, and $F = 13$. Then

$$\begin{aligned} D^2 + E^2 - 4F &= 2^2 + 4^2 - 4(13) \\ &= -32 < 0, \end{aligned}$$

so that there is no graph.

(c) We have $D = 2$, $E = 4$, and $F = 5$. Then

$$\begin{aligned} D^2 + E^2 - 4F &= 2^2 + 4^2 - 4(5) \\ &= 0, \end{aligned}$$

and so this equation represents a point circle. ●

Tangents to Circles From elementary geometry, the line tangent to a circle at a point on the circle is perpendicular to the radius drawn to the point of contact.

EXAMPLE 5 Find the equation of the line tangent to the circle $x^2 + y^2 + 2x - 4y = 8$ at the point $(1, -1)$.

SOLUTION First, verify that the point is on the circle, as in Example 2. Then write the equation in the center-radius form by completing the square.

$$x^2 + y^2 + 2x - 4y = 8$$

$$(x + 1)^2 + (y - 2)^2 = 13$$

Thus, the center is $(-1, 2)$. The slope of the radius drawn from the center to the point of contact $(1, -1)$ is

$$m = \frac{2 + 1}{-1 - 1} = -\frac{3}{2}.$$

So the slope of the tangent line is 2/3, since it is perpendicular to the radius. Now use $(1, -1)$ and $m = 2/3$ in the point-slope form of the equation of a line to find the equation of the tangent line.

$$y + 1 = \frac{2}{3}(x - 1)$$

$$2x - 3y - 5 = 0$$

The graphic representation of this is shown in Figure 7.22. ●

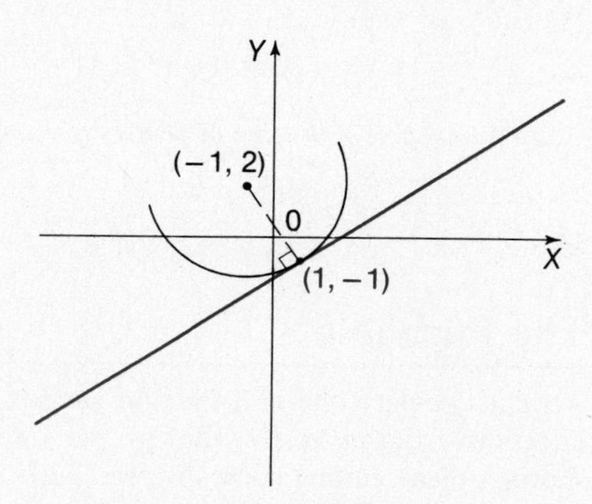

Figure 7.22

EXERCISES 7.2

Draw the graphs and write the equations of the following circles.

1. With center $(2, -3)$ and radius 5
2. With center $(0, 2)$ and radius 2
3. With center $\left(-\frac{1}{2}, 2\right)$ and radius $\frac{1}{2}$
4. With center $(-1, 2)$ and radius $\sqrt{5}$
5. With center $(2, 3)$ and radius 3
6. With center $(4, 0)$ and passing through the origin
7. With $(-1, 2)$ and $(4, -3)$ as endpoints of a diameter
8. With $(-1, 1)$ and $(1, 2)$ as endpoints of a diameter
9. With center $(2, 1)$ and tangent to the x-axis
10. With center $(-1, 2)$ and tangent to the y-axis

By completing the square, reduce each equation to the center-radius form. Identify the type of circle.

11. $x^2 + y^2 - 4x + 6y - 3 = 0$
12. $x^2 + y^2 + 4x + 2y + 1 = 0$
13. $x^2 + y^2 - 6x - 4y + 13 = 0$
14. $x^2 + y^2 - 6x + 4y + 15 = 0$

15. $x^2 + y^2 - 6x = 11$

16. $x^2 + y^2 - x + 3y + 3 = 0$

17. $2x^2 + 2y^2 - x + 3y + 3 = 0.$ [*Hint:* First divide by 2.]

18. $3x^2 + 3y^2 - 2x - 4 = 0$

Find the equation of the tangent line to the given circle at the given point.

19. $x^2 + y^2 = 25$ at $(-3, 4)$

20. $x^2 + y^2 - 2x + 4y + 4 = 0$ at $(1, -1)$

21. $x^2 + y^2 - 6x + 2y = 26$ at $(-3, -1)$

22. $x^2 + y^2 + 6x - 20 = 0$ at $(-5, 5)$

Find the equation of the **line of centers** *(the line through the centers) of each pair of circles.*

23. $x^2 + y^2 + 2x - 4y = 4$ and $x^2 + y^2 - 6x - 4y = 3$

24. $x^2 + y^2 + 6x - 4y + 13 = 0$ and $x^2 + y^2 + 4x + 2y + 1 = 0$

7.3 The Parabola

A right circular cone is a familiar surface. If a conical surface is extended in both directions from its vertex, the two parts of the cone are called **nappes.** Figure 7.23 shows a plane cutting such a conical surface in several ways. The curves in which a plane may cut a conical surface are called **conic sections,** or **conics.** The ancient Greeks studied these curves extensively and discovered a great many of their properties. In more modern times the conic sections have been found to be important in a great variety of ways, some of which will be discussed later.

The conic sections are of three distinct types, represented in Figure 7.23: (a) If a plane cuts only *one nappe* and is *not parallel* to the surface of the cone, the section is an oval-shaped curve called an **ellipse.** In particular, if the section cuts *one nappe*

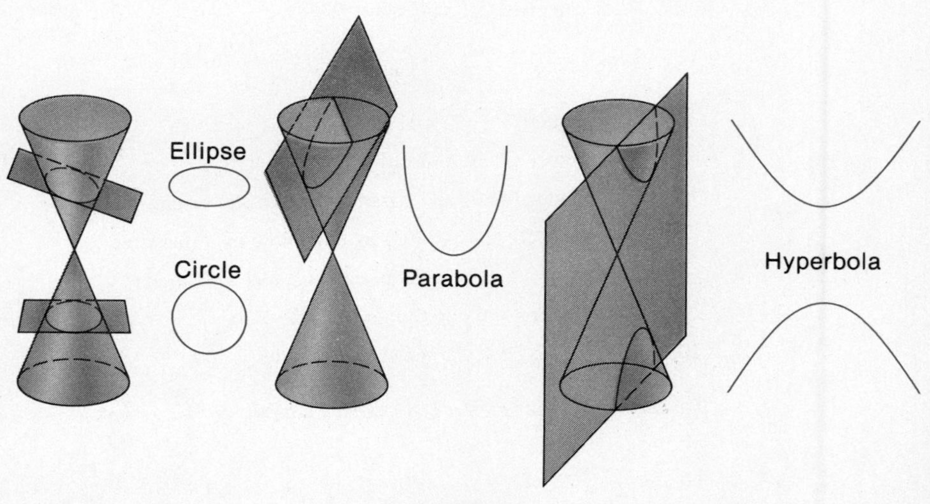

Figure 7.23

perpendicular to the axis of the cone the section is a **circle,** a special case of an ellipse; (b) if the plane cuts only *one nappe* of the cone and is *parallel* to the surface of the cone, the section is called a **parabola;** and (c) if the plane cuts *both nappes,* the section is a curve with two branches, called a **hyperbola.**

If the plane passes through the vertex of the cone but touches it nowhere else, the section is a single point. A plane tangent to the surface of the cone touches it in the points of a single line. And, a plane passing through the vertex and cutting both nappes of the cone will yield a section of two intersecting lines. All of these cases are **degenerate conics.**

The ancient Greek scholars studied these curves using only the methods of elementary geometry. However, with the introduction of coordinate geometry about 1600 and the subsequent derivation of algebraic equations for these curves, they began to be studied more efficiently. In modern times a great number and variety of applications of the properties of the conics have been found. These include paths of projectiles, bridge architecture, systems of navigation, reflecting telescope mirrors, and paths of physical particles, astronomical bodies, and spacecraft, among others. Several of these will be described as we proceed. We begin with the parabola.

The Parabola The parabola is the section of one nappe of a cone cut by a plane parallel to a line on the surface of the cone. See Figure 7.23b. The Greek scholars found that every point on a parabola is equidistant from a given line and a given point. The point is on the axis of symmetry (Section 3.4) and the line is perpendicular to the axis of symmetry. The following definition is based on this property.

DEFINITION 7.2 A **parabola** is the set of points equidistant from a fixed point, called the **focus,** and a fixed line, called the **directrix.**

Thus, in Figure 7.24, F is the focus, l is the directrix, and $d_1 = d_2$, $d_1{}' = d_2{}'$, and so on.

To simplify the derivation of an equation for the parabola, the focus is taken at $F(0, p)$ and the directrix as the line $y = -p$ in the coordinate plane. See Figure 7.25.

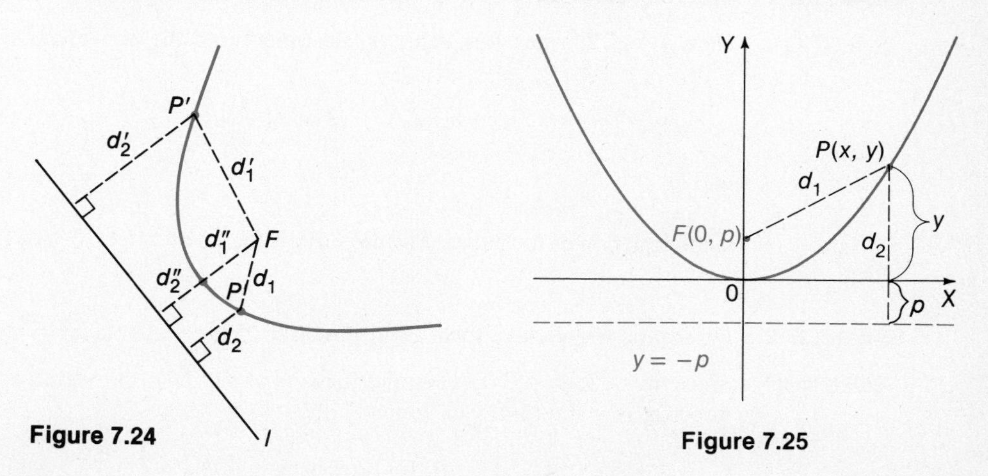

Figure 7.24

Figure 7.25

Then for *any* point $P(x, y)$ on the parabola,

$$d_1 = d_2,$$
$$(x - 0)^2 + (y - p)^2 = (y + p)^2,$$
$$x^2 + (y^2 - 2py + p^2) = y^2 + 2py + p^2,$$
$$x^2 = 4py, \quad \text{or} \quad y = \left(\frac{1}{4p}\right)x^2.$$

This is just another form of the general equation of curves $y = ax^2$ studied in Section 3.4. Such a curve is symmetric to the y-axis (its **axis of symmetry**) and has its lowest point at the origin O. This point is called the **vertex** of the parabola. *The constant p is the distance from the focus to the vertex,* or half the distance from the focus to the directrix.

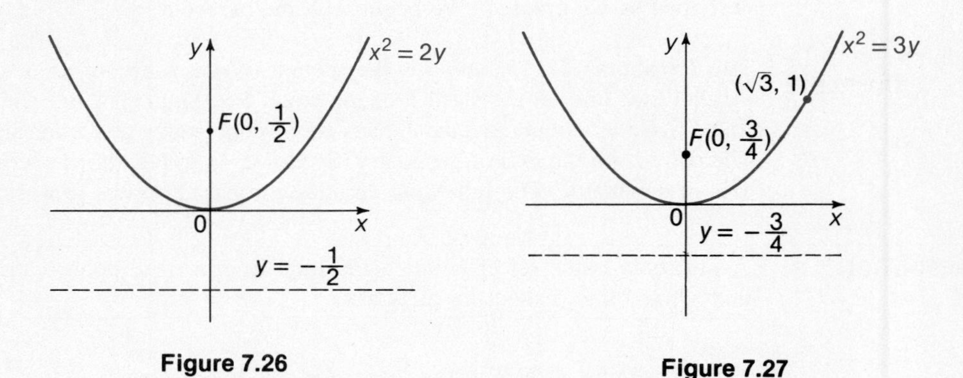

Figure 7.26

Figure 7.27

EXAMPLE 1 Find the equation of the parabola having focus (0, 1/2) and vertex at the origin.

SOLUTION Here $p = 1/2$; when this value is substituted into the above equation,

$$x^2 = 4\left(\frac{1}{2}\right)y,$$

and

$$x^2 = 2y$$

is the equation of the curve. The directrix is the line $y = -1/2$. See Figure 7.26. ●

EXAMPLE 2 Determine the vertex, focus, and directrix of the parabola $x^2 = 3y$.

SOLUTION Since $4p = 3$, $p = 3/4$. Then the focus is at (0, 3/4), the vertex is at (0, 0), and the directrix is $y = -3/4$. See Figure 7.27. ●

In the same way, we can find the equations of parabolas with vertex at the origin and (a) focus at $(p, 0)$, (b) focus at $(0, -p)$, and (c) focus at $(-p, 0)$. See Figure 7.28. In each of the four cases we have considered, the vertex is at the origin and the parabola is symmetric to a coordinate axis. The corresponding equations are called **first standard forms** of the equation of a parabola.

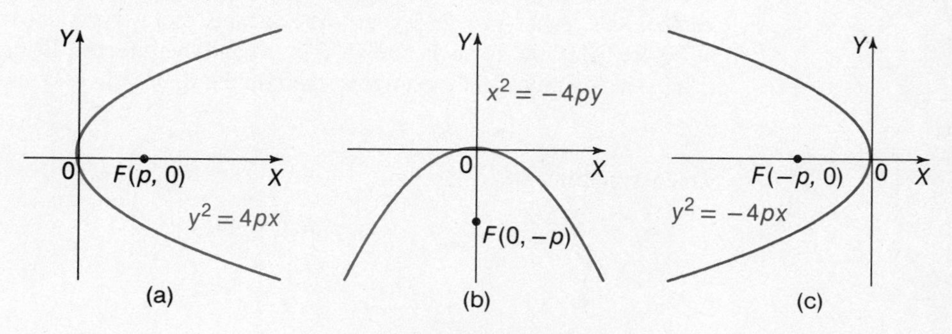

Figure 7.28

EXAMPLE 3 Write the equation of the parabola symmetric to the x-axis, with vertex at the origin, and passing through $(-2, 1)$.

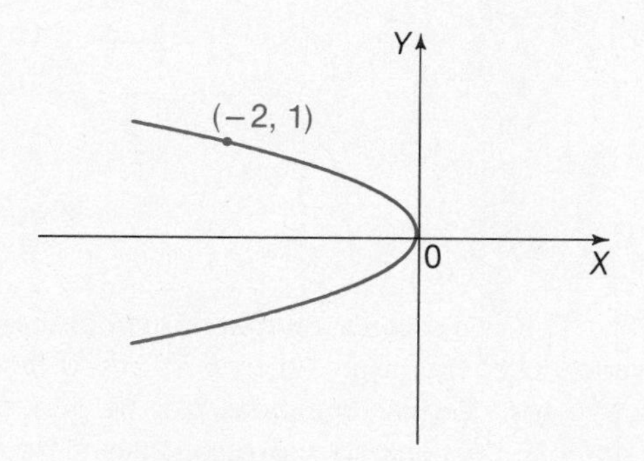

Figure 7.29

SOLUTION As seen in Figure 7.29, this parabola opens to the left. Thus, the standard form of the equation (see Figure 7.28c) is

$$y^2 = -4px.$$

Since the curve passes through $(-2, 1)$, the values $x = -2$ and $y = 1$ satisfy this equation.

$$1^2 = -4p(-2) = 8p,$$

$$p = \frac{1}{8}$$

Substitute this value for p into the standard form.

$$y^2 = -4\left(\frac{1}{8}\right)x = -\frac{1}{2}x$$

The equation is $y^2 = -(1/2)x$. ●

Second Standard Form Now suppose that a parabola has vertex at $V(h, k)$ and a vertical axis. As before, let p denote the distance from vertex to focus. Then, as shown in Figure 7.30, the focus is at $F(h, k + p)$ and the directrix is the line $y = k - p$. If $P(x, y)$ is any point on the parabola, then (in the figure) $d_1 = d_2$, so that $d_1^2 = d_2^2$, or

$$(x - h)^2 + [y - (k + p)]^2 = [y - (k - p)]^2,$$

which simplifies to

$$\mathbf{(x - h)^2 = 4p(y - k).}$$

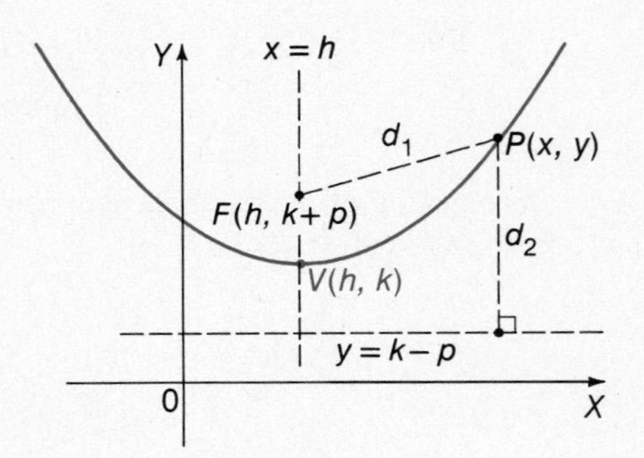

Figure 7.30

This is a **second standard form** of the equation of a parabola. Observe that if the vertex is at the origin, then $h = k = 0$, and this reduces to the first standard form, $x^2 = 4py$. Similar equations can be derived for the other three orientations of the parabola. In general, the second standard form of the equation of a parabola with vertex at (h, k) and distance $|p|$ from focus to vertex is either

$$\mathbf{(x - h)^2 = 4p(y - k)} \quad \text{or} \quad \mathbf{(y - k)^2 = 4p(x - h).}$$

The parabola has the first equation if its axis is parallel to the y-axis, and it opens upward if p is positive and downward if p is negative. The parabola has the second equation if its axis is parallel to the x-axis, and it opens to the right if p is positive and to the left if p is negative. Details are left to the exercises.

EXAMPLE 4 Write the equation of the parabola with focus (2, 3) and vertex (2, 2), and determine the directrix.

SOLUTION The equation has the form $(x - h)^2 = 4p(y - k)$. Since the distance from the focus to the vertex is $3 - 2 = 1$, then $p = 1$. So the equation of the parabola is

$$(x - 2)^2 = 4(y - 2).$$

The directrix is the line $y = 1$, since the distance of this line from the vertex is the same as the distance of the focus from the vertex. This parabola is shown in Figure 7.31. ●

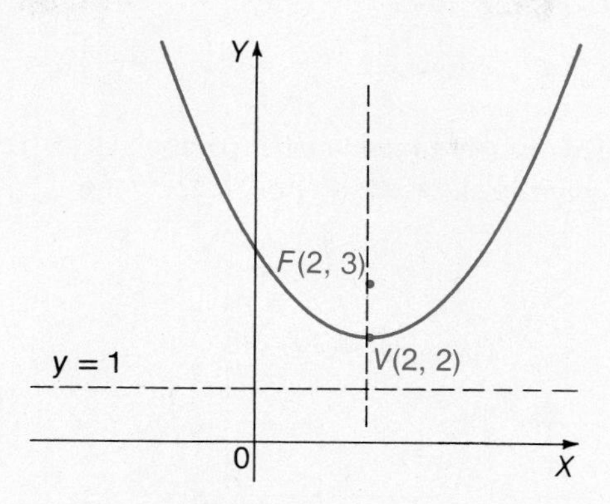

Figure 7.31

General Form At this point, it would be helpful to review Section 3.4. Consider the family of parabolas with vertical axis. Every member of this family is represented by the equation $(x - h)^2 = 4p(y - k)$, which may be rewritten as follows.

$$(x - h)^2 = 4p(y - k)$$
$$x^2 - 2hx + h^2 = 4py - 4pk$$

Solving for y gives

$$y = \left(\frac{1}{4p}\right)x^2 + \left(\frac{h}{-2p}\right)x + \frac{h^2 + 4pk}{4p},$$

or

$$\boldsymbol{y = Ax^2 + Dx + F}, \quad A \neq 0.$$

This is the **general form** of the equation of a parabola with vertical axis. Here $y = f(x)$, where f is the quadratic function discussed in Sections 3.3 and 3.4. The statement made there that the graph of every quadratic function is a parabola can now be proved. First, consider a particular example.

EXAMPLE 5 Show that the graph of the quadratic function $f(x) = 2x^2 + 3x - 1$ is a parabola.

SOLUTION Let $y = 2x^2 + 3x - 1$ and complete the square on the right. Then

$$\begin{aligned} y &= 2\left(x^2 + \frac{3}{2}x \qquad\right) - 1 \\ &= 2\left(x^2 + \frac{3}{2}x + \frac{9}{16}\right) - 1 - \frac{9}{8} \\ &= 2\left(x + \frac{3}{4}\right)^2 - \frac{17}{8}. \end{aligned}$$

Now write this equation in standard form.

$$y + \frac{17}{8} = 2\left(x + \frac{3}{4}\right)^2$$

$$\left(x + \frac{3}{4}\right)^2 = \frac{1}{2}\left(y + \frac{17}{8}\right)$$

This is the equation of a parabola with vertical axis and vertex $(-3/4, -17/8)$. Its graph is shown in Figure 7.32. ●

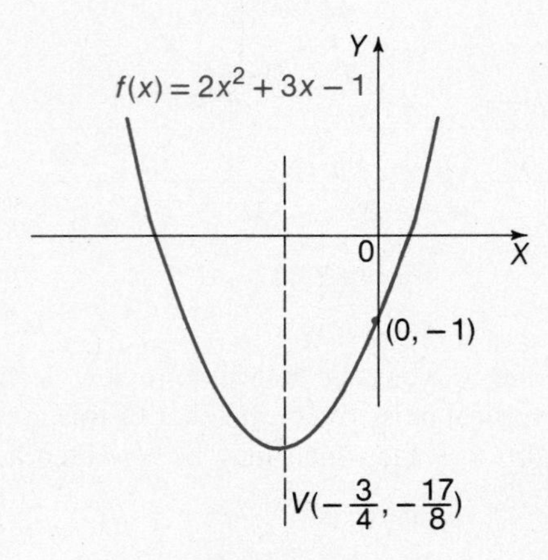

Figure 7.32

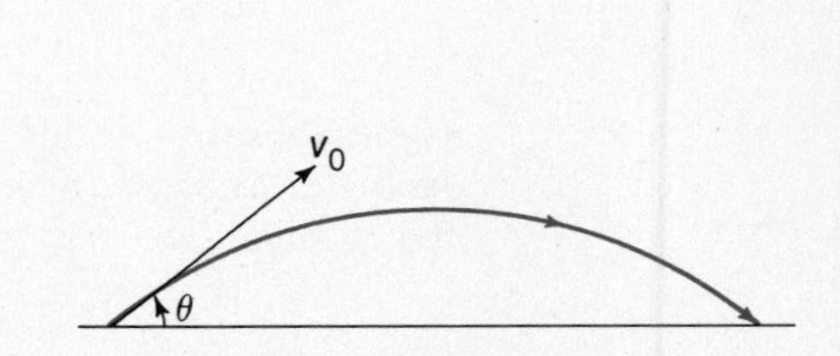

Figure 7.33. Path of a projectile

Now use the same procedure with the general quadratic (Example 8, Section 3.3). First, complete the square.

$$\begin{aligned} y &= ax^2 + bx + c \\ &= a\left(x + \frac{b}{2a}\right)^2 + \frac{4ac - b^2}{4a} \end{aligned}$$

Now, write the equation in the form

$$\left(x + \frac{b}{2a}\right)^2 = \frac{1}{a}\left[y - \frac{4ac - b^2}{4a}\right].$$

This is the standard form of the equation of a parabola with vertical axis, $(x - h)^2 = 4p(y - k)$, where $h = -b/2a$, $k = (4ac - b^2)/4a$, and $p = 1/4a$. That is, *the graph of any quadratic function is a parabola* with vertical axis.

Some Applications The parabola appears in many situations in nature and in mathematics. For example, if a projectile is fired at an angle with the horizontal (see Figure 7.33) and if, with the exception of gravity, all forces (such as air resistance) are ignored, the path of its motion is a parabola. In particular, the paths of a thrown ball and the stream of water from a hose are approximately parabolic.

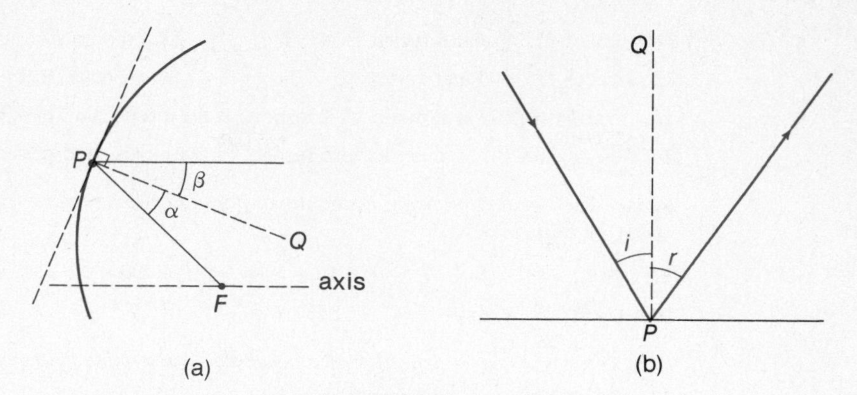

Figure 7.34. Reflection property

A special property of the parabola, called the **reflection property** (or **optical property**), is used in the design of searchlights and reflecting telescopes. In Figure 7.34, P is any point on a parabola for which F is the focus, and PQ is perpendicular to the line tangent to the parabola at P. It can be shown that the **focal radius** FP makes the same angle with PQ as does the line through P parallel to the axis of the parabola. That is, $\alpha = \beta$ in the figure. From optics we know that the *angle of incidence* i is equal to the *angle of reflection* r (see part (b) of the figure). Now think of the parabola in the figure as a cross-section of the surface generated by revolving a parabola about its axis of symmetry (a **paraboloid**). Then if a source of light is placed at F, the paraboloidal surface will reflect all the light in the direction parallel to this axis. This is the principle of the searchlight. (In practice, the surface is only approximately paraboloidal because of the expense of construction.)

The above principle is also used in reflecting telescopes. A star is at such a great distance that light from all its points comes in essentially parallel directions. So if a paraboloidal mirror is directed so that its axis is in the direction of the star, all the light from it will be reflected to the focal point of the mirror, where it can be viewed with an eyepiece. In the same way the paraboloidal dish of the receiving antenna of a radio telescope collects radio wave radiations.

Parabolas are used in innumerable ways in design situations. For instance, they appear in the design of TV field microphones, such as those used in broadcasting sporting events. One method of collecting and directing the energy of the sun is to use a trough-like surface with a parabolic cross-section.

EXERCISES 7.3

Sketch the graph and identify the focus and directrix of each of the following parabolas.

1. $y^2 = 8x$ **2.** $y^2 = -6x$ **3.** $y^2 - x = 0$ **4.** $y^2 + 4x = 0$

5. $x^2 = 4y$ **6.** $x^2 = -6y$ **7.** $x^2 + y = 0$ **8.** $x^2 - 2y = 0$

Write the equation and sketch the graph of the following parabolas.

9. Focus (0, 1) and directrix $y = -1$ **10.** Focus (4, 0) and directrix $x = -4$

11. Focus (2, 0) and directrix $x = -2$

12. Focus (0, −3) and directrix $y = 3$

13. Focus (2, −4) and vertex (2, −6)

14. Focus (0, 0) and vertex (2, 0)

15. Vertex (0, 0), symmetric with respect to the x-axis, and passing through (1, 2)

16. Vertex (0, 0), symmetric with respect to the y-axis, and passing through (−3, −2)

Reduce each of the following equations of parabolas to standard form and sketch, indicating vertex and focus.

17. $4x^2 - 8x - 2y + 7 = 0$

18. $y = -2x^2 + 4x + 13$

19. $2y^2 - 8y + 4x + 1 = 0$

20. $3y^2 + 9y + 3x = 5$

21. $x = y^2 + 6y + 5$

22. $x^2 - 2x - 4y + 5 = 0$

Use the focus-directrix definition of a parabola to determine the equation of the parabola with

23. Focus (3, 4) and directrix $y = -1$

24. Focus (−2, 1) and directrix the x-axis

25. A parabolic stone arch is represented in the figure. The inside span is 40 feet and the height 10 feet. How high is the arch above the level of its base at a point 10 feet from the vertical axis of the parabola? [*Hint:* The equation of the parabola is $x^2 = -4py$ and $h = 10 + y$.]

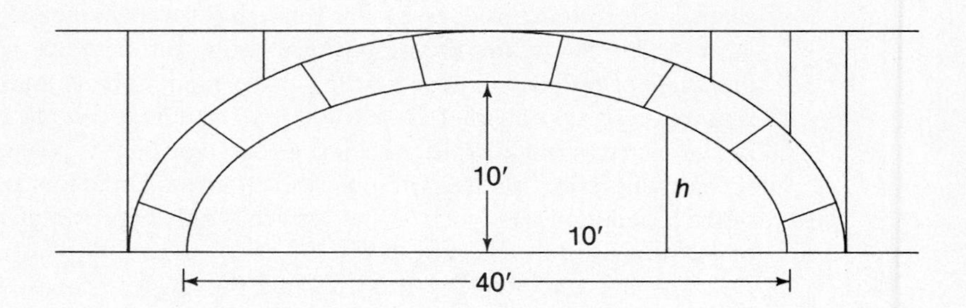

The path of a projectile, ignoring all forces except that of gravity, is the parabola $y = (\tan\theta)x - (16\sec^2\theta/v_0^2)x^2$, where θ is the angle of projection with the horizontal and v_0 is the initial velocity of projection. The vertex is the highest point in the path.

26. Let $\theta = 45°$, $v_0^2 = 10$ and calculate the highest point of the path of the projectile.

27. Repeat for $\theta = 30°$.

28. Repeat for $\theta = 60°$.

29. Compare the results in the three cases above.

30. Derive the equations of the parabolas shown in Figure 7.28.

31. As suggested in the text just before Example 4, derive the other versions of the second standard form of the equation of a parabola.

7.4 The Ellipse

An ellipse is the section of a conical surface obtained when a plane cuts one nappe of the cone (see Figure 7.23a). The Greeks proved geometrically that there are two points inside the ellipse such that the sum of the distances to these points from *any* point on the ellipse is constant. This property makes possible the construction of an elliptical flower bed, as follows: Drive two stakes into the ground and loop a string around them so as to have some slack. See Figure 7.35. Then with a pointed stick hold the string taut while moving along the ground to mark the curve. Ellipses of different sizes and shapes are obtained by varying either the distance separating the stakes or the slack in the string. The definition of the ellipse derives from this property.

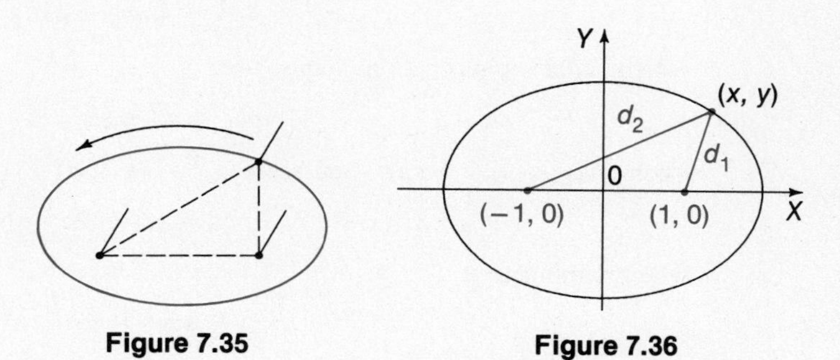

Figure 7.35

Figure 7.36

DEFINITION 7.3 An **ellipse** is the set of points the sum of whose distances from two fixed points is constant. Each of the two points is called a **focus** (plural, **foci**) of the ellipse.

EXAMPLE 1 Find the equation of the set of points P (an ellipse) such that the sum of the distances of P from $(1, 0)$ and $(-1, 0)$ is constantly 4. See Figure 7.36.

SOLUTION In the figure, $d_1 + d_2 = 4$, so

$$\sqrt{(x-1)^2 + y^2} + \sqrt{(x+1)^2 + y^2} = 4,$$

$$\sqrt{(x-1)^2 + y^2} = 4 - \sqrt{(x+1)^2 + y^2}.$$

Now square both sides.

$$\begin{aligned}
(x^2 - 2x + 1) + y^2 &= 16 - 8\sqrt{(x+1)^2 + y^2} + (x^2 + 2x + 1 + y^2) \\
8\sqrt{(x+1)^2 + y^2} &= 4x + 16 \\
64(x^2 + 2x + 1 + y^2) &= 16x^2 + 128x + 256 \\
64x^2 + 128x + 64 + 64y^2 &= 16x^2 + 128x + 256 \\
48x^2 + 64y^2 &= 192 \\
\frac{x^2}{4} + \frac{y^2}{3} &= 1
\end{aligned}$$

●

To obtain a general equation for the ellipse, we introduce a coordinate system into the plane of the ellipse. To simplify the procedure, take the line through the foci as the x-axis and the origin midway between them. See Figure 7.37. Now let $(c, 0)$ and $(-c, 0)$ denote the foci and let $2a$ be the sum of the focal distances from a point on the curve. Then, if $P(x, y)$ is any point on the ellipse, we have

$$d_1 + d_2 = 2a,$$
$$d_1 = 2a - d_2,$$
$$d_1^2 = 4a^2 - 4ad_2 + d_2^2.$$

From Figure 7.37, $d_1^2 = (x - c)^2 + y^2$ and $d_2^2 = (x + c)^2 + y^2$. Thus,

$$(x - c)^2 + y^2 = 4a^2 - 4ad_2 + [(x + c)^2 + y^2],$$

which, after simplification, reduces to

$$ad_2 = a^2 + cx.$$

Then $a^2d_2^2 = (a^2 + cx)^2$, and since $d_2^2 = (x + c)^2 + y^2$, this becomes

$$a^2[(x + c)^2 + y^2] = a^4 + 2a^2cx + c^2x^2,$$

which simplifies to

$$a^2x^2 + a^2c^2 + a^2y^2 = a^4 + c^2x^2,$$

or

$$(a^2 - c^2)x^2 + a^2y^2 = a^2(a^2 - c^2).$$

From elementary geometry we know that the sum of the lengths of any two sides of a triangle is greater than the length of the third side. Thus, in triangle PF_1F_2 (in Figure 7.37), $\overline{PF_1} + \overline{PF_2} > \overline{F_1F_2}$, that is, $2a > 2c$, $a > c$. Hence $a^2 > c^2$, so $a^2 - c^2$ is a positive number, and we may let $a^2 - c^2 = b^2$. As a result, the above equation becomes simply $b^2x^2 + a^2y^2 = a^2b^2$ or, after dividing through by a^2b^2,

$$\frac{x^2}{a^2} + \frac{y^2}{b^2} = 1.$$

This is called the **first standard form** of the equation of an ellipse.

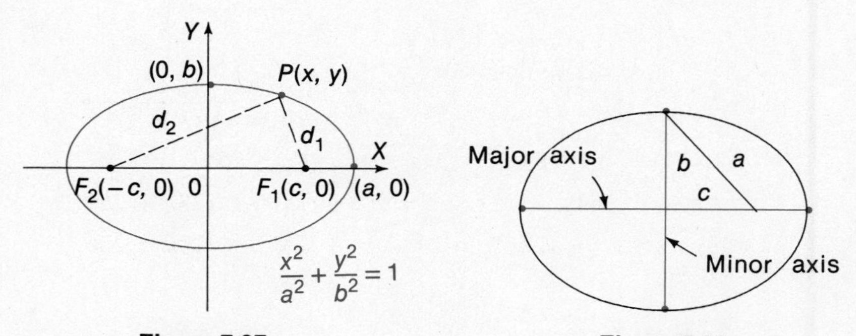

Figure 7.37 **Figure 7.38**

Since the x and y terms have even powers, the graph of this equation is symmetric with respect to both coordinate axes. Starting with $b^2x^2 + a^2y^2 = a^2b^2$ and solving for x^2,

$$b^2x^2 + a^2y^2 = a^2b^2,$$

$$x^2 + \frac{a^2}{b^2}y^2 = a^2,$$

$$x^2 = a^2 - \frac{a^2}{b^2}y^2,$$

so that $x^2 \leq a^2$, or $-a \leq x \leq a$. Similarly, $-b \leq y \leq b$.

The points where the ellipse intersects the x-axis, $(a, 0)$ and $(-a, 0)$, are called **vertices** and the segment connecting them is called the **major axis** of the ellipse. Its midpoint is the **center** of the ellipse. Similarly, the segment connecting $(0, b)$ and $(0, -b)$ is called the **minor axis.** See Figure 7.38. Because $b^2 = a^2 - c^2$, $b^2 \leq a^2$, and thus $b \leq a$.

If $a = b$, the equation of the ellipse reduces to simply $x^2 + y^2 = a^2$, the equation of a circle. In this case, $c = 0$, and the foci coincide. Thus, both algebraically and geometrically, the circle is a particular case of the ellipse. It is the section of a cone cut by a plane perpendicular to the axis of the cone. (See Figure 7.23a.)

EXAMPLE 2 Determine the vertices, foci, and lengths of major and minor axes of the ellipse $4x^2 + 9y^2 = 36$.

SOLUTION Dividing both sides of the equation by 36, we get the standard form

$$\frac{x^2}{9} + \frac{y^2}{4} = 1.$$

Then $a^2 = 9$, $b^2 = 4$, and $c^2 = a^2 - b^2 = 5$, so that $a = 3$, $b = 2$, and $c = \sqrt{5}$. The vertices are $(\pm 3, 0)$, the foci are $(\pm\sqrt{5}, 0)$, the length of the major axis is $2a = 6$, and the length of the minor axis is $2b = 4$. See Figure 7.39. ●

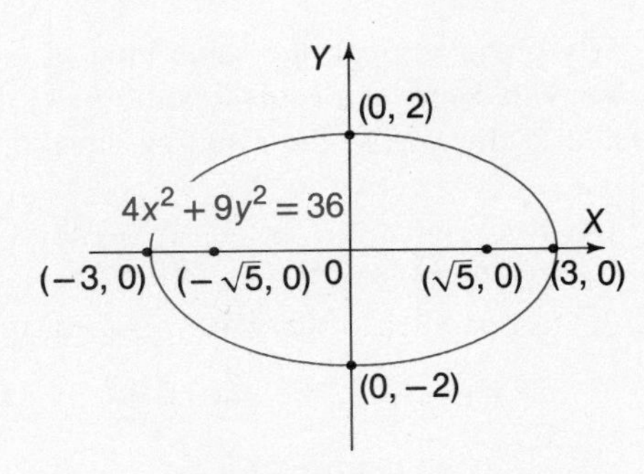

Figure 7.39

EXAMPLE 3 Find the equation of the ellipse whose foci are $(\pm 4, 0)$ and whose major axis has length 12, and sketch the graph.

SOLUTION Here $c = 4$ and $a = 6$, since the length of the major axis, $2a$, is 12. Then

$b^2 = a^2 - c^2 = 36 - 16 = 20$, so that the equation of the ellipse is

$$\frac{x^2}{36} + \frac{y^2}{20} = 1, \text{ or}$$

$$5x^2 + 9y^2 = 180.$$

See Figure 7.40 for the graph. Since $b^2 = 20$, $b = 2\sqrt{5}$, and the y-intercepts are $2\sqrt{5}$ and $-2\sqrt{5}$. ●

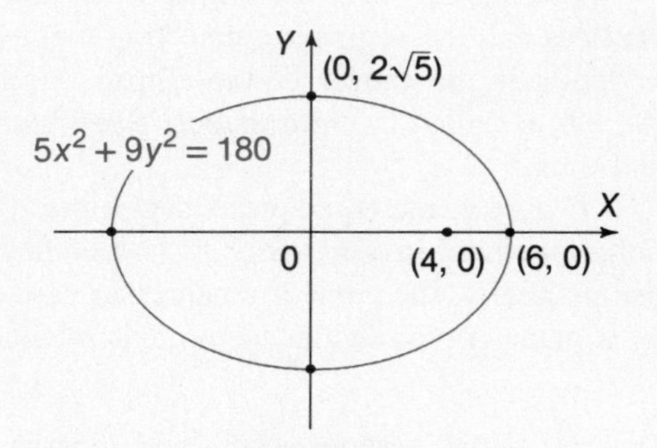

Figure 7.40

If we take the foci on the y-axis and call them $(0, c)$ and $(0, -c)$, the major axis is vertical instead of horizontal. Let the length of the major axis be $2a$. Then we can show that the equation of the ellipse is

$$\frac{x^2}{b^2} + \frac{y^2}{a^2} = 1,$$

where $b^2 = a^2 - c^2$ as before, and the distance $2b$ between the x-intercepts is the minor axis.

By going through the same kind of calculation, we can get the equation of an ellipse with center at a general point (h, k) in the coordinate plane. If the major axis is parallel to the x-axis, the resulting equation is

$$\frac{(x - h)^2}{a^2} + \frac{(y - k)^2}{b^2} = 1,$$

and if it is parallel to the y-axis, the equation is

$$\frac{(x - h)^2}{b^2} + \frac{(y - k)^2}{a^2} = 1.$$

Each of these equations is called a **second standard form** of the equation of an ellipse. Details are left as an exercise.

Applications The ellipse appears in a great variety of situations. About 1600 the German astronomer Johannes Kepler discovered from tedious analysis of thousands of observations that each of the planets moves in an elliptic orbit about the sun, with the sun at one focus of the orbit. The orbit of the earth is very nearly circular. (See Figure 7.41.) Periodic comets also have elliptic orbits about the sun, and the moon moves about the earth in an elliptic path. Artificial satellites orbiting the earth follow elliptic paths. These facts are used in predicting positions of astronomical bodies, and of man-made satellites and spacecraft.

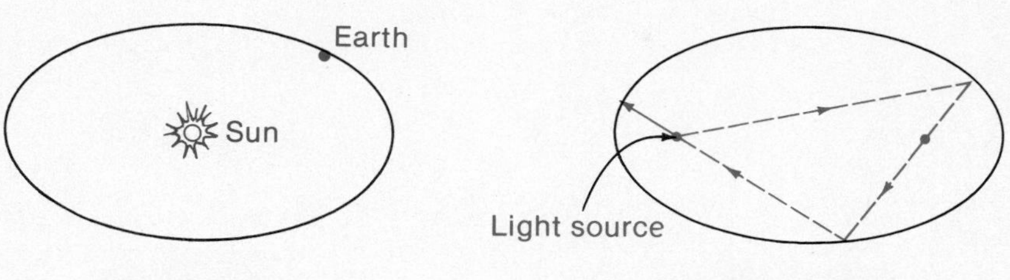

Figure 7.41 **Figure 7.42**

Isaac Newton proved mathematically that the elliptical orbits of the planets are a consequence of the combined effect of the law of gravitation and the laws of motion. He derived a formula showing that the force of attraction between two bodies is inversely proportional to the square of the distance between them. In physics other examples of inverse-square relations arise in relation to electrostatic and electromagnetic force fields.

The earth is not exactly spherical but actually "flattened" at the poles. That is, it is the shape generated by revolving a nearly circular ellipse about its major axis. Thus, the meridians of longitude on the earth are not quite circular but slightly elliptical.

The ellipse has an *optical property* just as does the parabola. It can be shown that if a point source of light is placed at one focus of an ellipsoidal reflecting surface (Figure 7.42), a beam of light reflects off the surface and then passes through the other focus of the ellipse. This principle is used in connection with some laser beam experiments. It also applies to sound and is used in the construction of some auditoriums and so-called whispering galleries.

Concrete and stone bridge and building arches are sometimes elliptical. Elliptical gears are used in machines to obtain a slow, powerful thrust with a rapid return, such as in a hay press, heavy duty punch, planer, and so on.

Eccentricity Since $2a$ measures the major axis and $2b$ the minor axis of an ellipse, the ratio b/a measures the relative "flatness" or "roundness" of an ellipse. The smaller the value of b relative to that of a the more elongated the ellipse. The more nearly b is equal to a the more nearly circular the ellipse. Specifically, the **eccentricity** e of an ellipse is defined by

$$e = \frac{c}{a} = \frac{1}{a}\sqrt{a^2 - b^2}.$$

Thus, the *larger* b is relative to a, the *smaller* the eccentricity e, and vice versa. Observe that, since $c < a$, then $e < 1$ for any ellipse.

EXAMPLE 4 Calculate the eccentricity of the ellipse whose equation is $4x^2 + 9y^2 = 36$.

SOLUTION Dividing both members by 36, we have

$$\frac{x^2}{9} + \frac{y^2}{4} = 1.$$

Then $a^2 = 9$, $b^2 = 4$, so that

$$e = \frac{c}{a} = \frac{1}{a}\sqrt{a^2 - b^2}$$
$$= \frac{1}{3}\sqrt{9 - 4}$$
$$= \frac{1}{3}\sqrt{5} \approx 0.7. \quad \bullet$$

Now imagine that the major axis is kept fixed, so that a remains constant. As the foci move toward the center, c approaches 0. When the foci coincide at the center, $c = 0$, and so eccentricity $e = 0$. The circle is thus an ellipse of eccentricity zero. The fact that the elliptic orbit of the earth about the sun has a small eccentricity (0.0168) means that its orbit is nearly circular.

With the major axis fixed, imagine the foci moving out from the center, so that c gets larger. As c approaches a, b approaches 0 and e approaches 1. Thus, the eccentricity e of an ellipse varies within the interval $0 \leq e < 1$. Some comets have highly elongated orbits, that is, the eccentricity of the orbit is nearly 1.

If we expand the second standard form of the equation of an ellipse and simplify, we get the **general equation**

$$Ax^2 + Cy^2 + Dx + Ey + F = 0,$$

where $A = b^2$, $C = a^2$, $D = -2b^2h$, $E = -2a^2k$, and $F = b^2h^2 + a^2k^2 - a^2b^2$.

EXERCISES 7.4

Rewrite each of the following equations in standard form, sketch the graph, and identify vertices, foci, and center.

1. $9x^2 + 25y^2 = 225$

2. $x^2 + 4y^2 = 4$

3. $3x^2 + 2y^2 = 6$

4. $9x^2 + 4y^2 = 36$

5. $25x^2 + 16y^2 = 400$

6. $x^2 + 2y^2 = 2$

7. $x^2 + 2y^2 - 4x + 2y + 4 = 0$

8. $3x^2 + y^2 - 6x + 6y = 0$

9. $4x^2 - 24x + 9y^2 - 90y + 225 = 0$

10. $9x^2 + 36x + 4y^2 - 24y + 36 = 0$

Calculate the eccentricity of each of the following ellipses.

11. $9x^2 + 16y^2 = 144$

12. $x^2 + 25y^2 = 25$

13. $5x^2 + 9y^2 = 45$

14. $2x^2 + 3y^2 = 6$

15. The figure represents an elliptical stone arch, with the dimensions indicated. Find the height y of the arch at a distance of 6 feet from the center of the base.

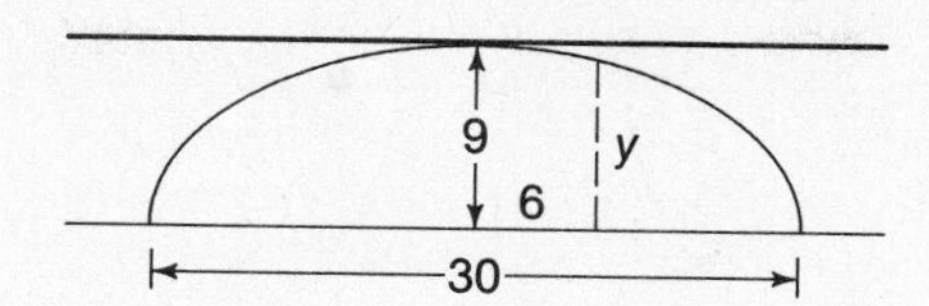

16. In the arch of the bridge of exercise 15 above, there are two points which are 6 feet from the ground level. How far are they from the vertical axis of symmetry of the arch, that is, if $y = 6$, what is x?

17. Suppose an elliptical arch, such as in exercise 15 above, is 12 feet above the ground at its highest point and 20 feet across at the base. How high above ground is a point on the arch 8 feet from the vertical axis of symmetry?

18. Find the equation of the set of points P such that the sum of the distances of P from $(2, 0)$ and $(-2, 0)$ is 6.

19. Find the equation of the set of points P such that the sum of the distances of P from $(0, 8)$ and $(0, -8)$ is 20.

20. The earth's orbit about the sun is an ellipse of eccentricity 0.0168, with the sun at one focus of the ellipse. If the average distance of the earth from the sun, that is, half the length of the major axis, is approximately 92 million miles, what is the least and greatest distance of the earth from the sun?

21. Calculate the equation of the ellipse having foci $(\pm 3, 0)$ and major axis 10, following the pattern of the text development for Example 1.

22. Derive the second standard form of the equation of an ellipse, as suggested in the text.

23. Verify the details for the general equation of an ellipse.

7.5 The Hyperbola

The hyperbola is the conic section obtained when the sectioning plane cuts both nappes of the cone (Figure 7.23c). The ancient Greek mathematicians proved geometrically the general property of the hyperbola which is now given as the definition.

DEFINITION 7.4 A **hyperbola** is the set of points the difference of whose distances from two fixed points is a positive constant. The fixed points are called the **foci.**

By this definition, as shown in Figure 7.43, on the following page, $|d_2 - d_1| = 2a$, $|d_2' - d_1'| = 2a$, and so on.

EXAMPLE 1 Find the equation of the set of points P (a hyperbola) such that the difference of the distances of P from $(5, 0)$ and $(-5, 0)$ is constantly 8.

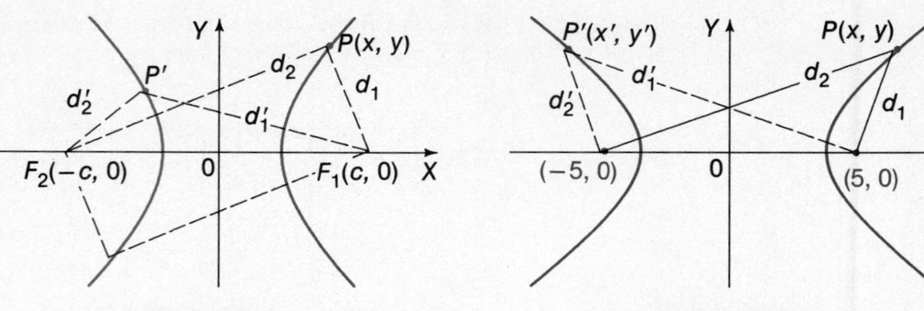

Figure 7.43 **Figure 7.44**

SOLUTION In Figure 7.44,

$$
\begin{aligned}
|d_1 - d_2| &= 8, \\
d_1 - d_2 &= \pm 8, \\
d_1 &= d_2 \pm 8, \\
\sqrt{(x-5)^2 + y^2} &= \sqrt{(x+5)^2 + y^2} \pm 8.
\end{aligned}
$$

Squaring both sides, we get

$$
\begin{aligned}
x^2 - 10x + 25 + y^2 &= (x^2 + 10x + 25 + y^2) \pm 16\sqrt{(x+5)^2 + y^2} + 64, \\
\pm 16\sqrt{(x+5)^2 + y^2} &= 20x + 64.
\end{aligned}
$$

Squaring again,

$$
\begin{aligned}
256[(x+5)^2 + y^2] &= (20x + 64)^2, \\
256(x^2 + 10x + 25 + y^2) &= 400x^2 + 2560x + 4096, \\
256x^2 + 2560x + 6400 + 256y^2 &= 400x^2 + 2560x + 4096, \\
144x^2 - 256y^2 &= 2304, \\
9x^2 - 16y^2 &= 144, \\
\frac{x^2}{16} - \frac{y^2}{9} &= 1. \quad \bullet
\end{aligned}
$$

This same procedure is now used to derive the standard equation for a hyperbola. Take the foci at $(c, 0)$ and $(-c, 0)$ and call the constant distance difference $2a$. Then (in Figure 7.43),

$$
\begin{aligned}
|d_1 - d_2| &= 2a, \\
d_1 - d_2 &= \pm 2a, \\
d_1 &= \pm 2a + d_2, \\
d_1^2 &= 4a^2 \pm 4ad_2 + d_2^2.
\end{aligned}
$$

Then substituting $(x - c)^2 + y^2$ and $(x + c)^2 + y^2$ (from the distance formula) for d_1^2 and d_2^2 and simplifying, we obtain

$$\pm ad_2 = -a^2 - cx.$$

If we square both sides of this equation and substitute $(x + c)^2 + y^2$ for $d_2{}^2$ again, we obtain

$$x^2(c^2 - a^2) - a^2y^2 = a^2(c^2 - a^2).$$

Since the sum of the lengths of any two sides of a triangle is greater than the length of the third side, $\overline{F_1F_2} + d_2 > d_1$ in Figure 7.43. Then,

$$\begin{aligned} \overline{F_1F_2} &> d_1 - d_2, \\ 2c &> 2a, \\ c &> a, \\ c^2 &> a^2, \\ c^2 - a^2 &> 0. \end{aligned}$$

If $c^2 - a^2$ is replaced by b^2, the above equation becomes simply $b^2x^2 - a^2y^2 = a^2b^2$, or

$$\frac{x^2}{a^2} - \frac{y^2}{b^2} = 1.$$

This is called the **first standard form** of the equation of a hyperbola.

To help in sketching hyperbolas, we need some further properties of the hyperbola. Since the powers of x and of y are both even, the graph is symmetric with respect to both coordinate axes. The x-intercepts are a and $-a$, but there are no y-intercepts. Solving the standard equation for y, we obtain

$$y = \pm\frac{b}{a}\sqrt{x^2 - a^2}.$$

It is clear from this equation that

1. $|x| \geq a$, so that $x \geq a$ or $x \leq -a$.
2. $y = 0$ whenever $x = \pm a$.
3. y increases in absolute value as x increases in absolute value.

The hyperbola is an asymptotic curve and the asymptotes are helpful in sketching the graph. The following sequence of equations shows that for the hyperbola $b^2x^2 - a^2y^2 = a^2b^2$, the asymptotes are the lines $y = \dfrac{b}{a}x$ and $y = -\dfrac{b}{a}x$. From the above equation,

$$\begin{aligned} y &= \pm\frac{b}{a}\sqrt{x^2 - a^2}, \\ &= \pm\frac{b}{a}\sqrt{x^2\left(\frac{x^2 - a^2}{x^2}\right)} && \text{Multiplying and dividing the radicand by } x^2 \\ &= \pm\frac{b}{a}\sqrt{x^2\left(1 - \frac{a^2}{x^2}\right)} \\ &= \pm\frac{b}{a}x\sqrt{1 - \frac{a^2}{x^2}}. \end{aligned}$$

From this we can see that as x increases without bound, a^2/x^2 approaches zero and thus y gets arbitrarily close to $\pm(b/a)x$. That is, the lines $y = \pm(b/a)x$ are asymptotes to the hyperbola. Note that these equations for the asymptotes can be calculated simply by replacing 1 in the equation of the hyperbola by 0, since $b^2x^2 - a^2y^2 = 0$ is equivalent to $y = \pm(b/a)x$. The slopes of the asymptotes are $\pm(b/a)$. Now we can complete the sketch of the graph, as in Figure 7.45.

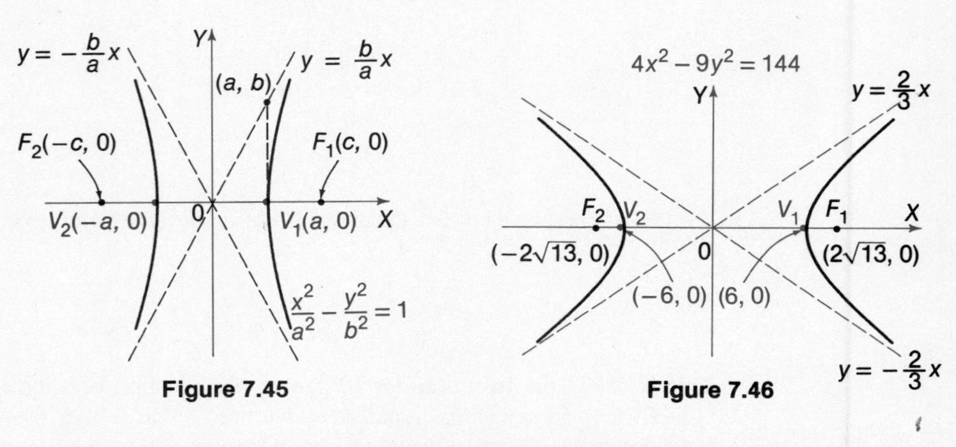

Figure 7.45

Figure 7.46

The two parts of the graph are called **branches** of the hyperbola. The points where the branches intersect the x-axis, $(\pm a, 0)$, are its **vertices,** the segment joining them is the **transverse axis,** and its midpoint is the **center** of the hyperbola.

EXAMPLE 2 Sketch the hyperbola whose equation is $4x^2 - 9y^2 = 144$. Identify the vertices and the foci.

SOLUTION To put the equation in standard form, divide by the constant term, 144.

$$4x^2 - 9y^2 = 144$$

$$\frac{x^2}{36} - \frac{y^2}{16} = 1$$

From the equation, $a^2 = 36$ and $b^2 = 16$, so that $a = 6$ and $b = 4$. The vertices are $(\pm a, 0) = (\pm 6, 0)$. Since $c^2 = a^2 + b^2 = 16 + 36 = 52$, $c = \sqrt{52} = \sqrt{4(13)} = 2\sqrt{13}$. The foci are $(\pm c, 0) = (\pm 2\sqrt{13}, 0)$.

To aid in sketching the hyperbola, we first find the asymptotes. Their equations are

$$y = \pm\frac{b}{a}x,$$

$$y = \pm\frac{4}{6}x = \pm\frac{2}{3}x.$$

We can now draw the curve through the vertices $(\pm 6, 0)$ and asymptotic to $y = \pm(2/3)x$. The result is shown in Figure 7.46. ●

EXAMPLE 3 Find the equation of the hyperbola having vertices $(\pm 2, 0)$ and foci $(\pm 4, 0)$.

SOLUTION Here $a = 2$ and $c = 4$, so that $b^2 = c^2 - a^2 = 16 - 4 = 12$. Substitute these values for a and b in the standard form.

$$\frac{x^2}{a^2} - \frac{y^2}{b^2} = 1$$

$$\frac{x^2}{4} - \frac{y^2}{12} = 1$$

We can obtain the equations of the asymptotes directly by solving for y as follows.

$$\frac{x^4}{4} - \frac{y^2}{12} = 0$$

$$\frac{x^2}{4} = \frac{y^2}{12}$$

$$y^2 = 3x^2$$

$$y = \pm x\sqrt{3}$$

Then, as in the preceding example, we can use the vertices and the asymptotes to sketch the curve properly. See Figure 7.47. ●

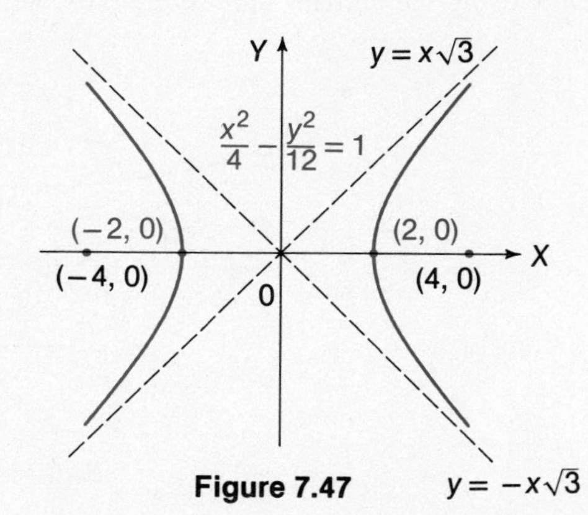

Figure 7.47

An important special case of the hyperbola is that in which $a = b$. Then the equation is $x^2 - y^2 = a^2$ and the asymptotes are the lines $y = \pm x$. These asymptotes are perpendicular and the hyperbola is called a **rectangular,** or **equilateral** hyperbola.

If the foci are taken on the y-axis, the equation is

$$b^2y^2 - a^2x^2 = a^2b^2, \text{ or}$$

$$\frac{y^2}{a^2} - \frac{x^2}{b^2} = 1.$$

In this case, the asymptotes have slope $\pm a/b$, and equations $y = (\pm a/b)x$.

EXAMPLE 4 Write the equation of the hyperbola whose vertices are $(0, \pm 2)$ and whose foci are $(0, \pm 2\sqrt{5})$.

SOLUTION Here $a = 2$ and $c = 2\sqrt{5}$, so that $b^2 = c^2 - a^2 = 20 - 4 = 16$ and $b = 4$. Then the equation is found as follows.

$$\frac{y^2}{a^2} - \frac{x^2}{b^2} = 1$$

$$\frac{y^2}{4} - \frac{x^2}{16} = 1$$

$$16y^2 - 4x^2 = 4(16)$$

$$4y^2 - x^2 = 16$$

To find the asymptotes, we write

$$4y^2 - x^2 = 0,$$

$$4y^2 = x^2,$$

$$2y = \pm x,$$

$$y = \pm\frac{1}{2}x.$$

Now using the vertices and asymptotes, we sketch the curve. See Figure 7.48. ●

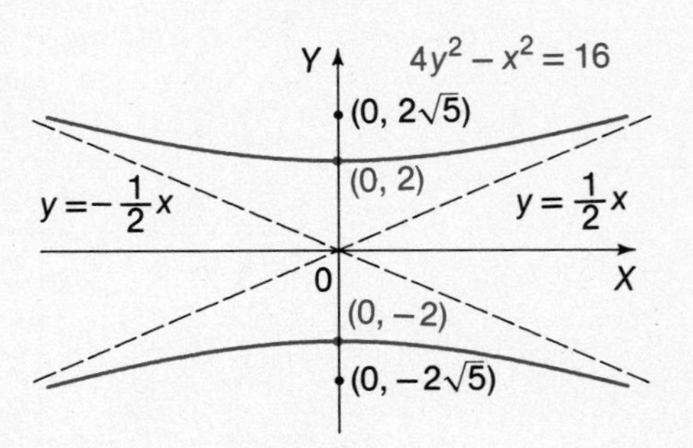

Figure 7.48

Just as in the case of the ellipse, we can derive the equation of a hyperbola with center (h, k) to obtain the **second standard forms,**

$$\frac{(x-h)^2}{a^2} - \frac{(y-k)^2}{b^2} = 1 \quad \text{and} \quad \frac{(y-k)^2}{a^2} - \frac{(x-h)^2}{b^2} = 1.$$

EXAMPLE 5 Graph the hyperbola $9(x - 2)^2 - 16(y - 1)^2 = 144$, and identify the center, vertices, foci, and asymptotes.

SOLUTION First we write the equation in the standard form.

$$\frac{(x-2)^2}{16} - \frac{(y-1)^2}{9} = 1$$

From this we have $a^2 = 16$ and $b^2 = 9$, so that $a = 4$, $b = 3$, $c^2 = a^2 + b^2 = 25$ and $c = 5$. The center is (2, 1). Since $a = 4$, the vertices are $(2 \pm 4, 1)$, or $(-2, 1)$ and (6, 1). Also, since $c = 5$, the foci are $(2 \pm 5, 1)$, or $(-3, 1)$ and (7, 1).

The slopes of the asymptotes are $\pm b/a = \pm 3/4$, and they pass through the center (2, 1). So their equations are $y - 1 = \pm(3/4)(x - 2)$. With this information we can sketch the graph, as shown in Figure 7.49. ●

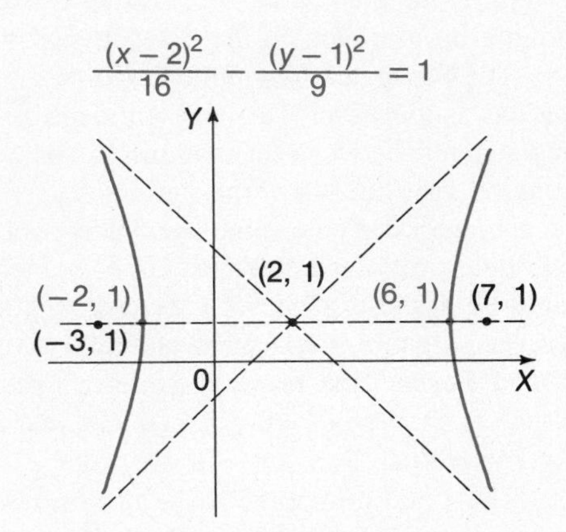

Figure 7.49

As in the case of an ellipse, the ratio $e = c/a$ for a hyperbola is called its **eccentricity.** Note that

$$e^2 = \frac{c^2}{a^2} = \frac{a^2 + b^2}{a^2} = 1 + \frac{b^2}{a^2} > 1,$$

so that $e > 1$ for any hyperbola. (Recall that for any ellipse, $0 < e < 1$.) Since b/a is the slope of the asymptotes, the smaller the angle between the asymptotes, the smaller b/a and so the smaller the eccentricity. Thus, eccentricity of a hyperbola measures the "broadness" of the spread of a branch of the curve.

If we expand the second standard form of the equation of a hyperbola and simplify, we get the **general equation of a hyperbola:**

$$Ax^2 + Cy^2 + Dx + Ey + F = 0,$$

where

$$A = b^2, \quad C = -a^2, \quad D = -2b^2h, \quad E = 2a^2k, \quad \text{and} \quad F = b^2h - a^2k - a^2b^2.$$

Applications A comet generally moves in an elliptic orbit about the sun. However, if its speed as it nears the sun increases sufficiently, it will move off along a branch of a hyperbola, never to return to the vicinity of the sun. If two like-charged particles are propelled toward one another, they will repel each other and each move away along a branch of a hyperbola having the location of the other particle as a focus.

LORAN (LOng RAnge Navigation) is a system for "blind" flying. Two fixed radar sending stations (at S_1 and S_2 in Figure 7.50) send signals simultaneously and these are received by an airplane P at different times. The receiving instrument on the plane is so calibrated that it registers the difference in times as a difference in distances (PS_1 and PS_2 in the figure). Since the sending stations are fixed, the particular difference in distances places the plane on a particular hyperbola having the station locations as foci. If the plane now flies so that it maintains this difference constantly, it will be moving on a branch of the given hyperbola (at P or Q in Figure 7.50).

The navigator is supplied with a map showing a set of hyperbolas having the stations as their foci. He can then identify which hyperbolic path he is on, as follows: The distance between the stations S_1 and S_2 is $2c$ and the difference in distances (read from the instruments) of the stations from the plane is $2a$. Since $c^2 = a^2 + b^2$, using c the navigator can determine b and hence the particular hyperbolic path on the map. The path being followed could be either branch of the given hyperbola, that is, the plane could be at either position P or Q in Figure 7.50. To determine which is the correct position, signals from a third station S_3 are used. From the same procedure as above, S_3 and S_1 can now be used to determine a second hyperbola on the map (only one branch is shown in Figure 7.51). The plane is then at position P of intersection of the two hyperbolas.

The same principle can be used for artillery rangefinding. Three listening posts are in contact by telephone. When the artillery piece is fired each of them notes the time it is heard. The difference in times the sound is received at any two of the listening posts places the gun on a fixed hyperbola. Any other pair of the stations similarly determine its location on a second hyperbola. The gun must then be at the intersection of these two hyperbolas.

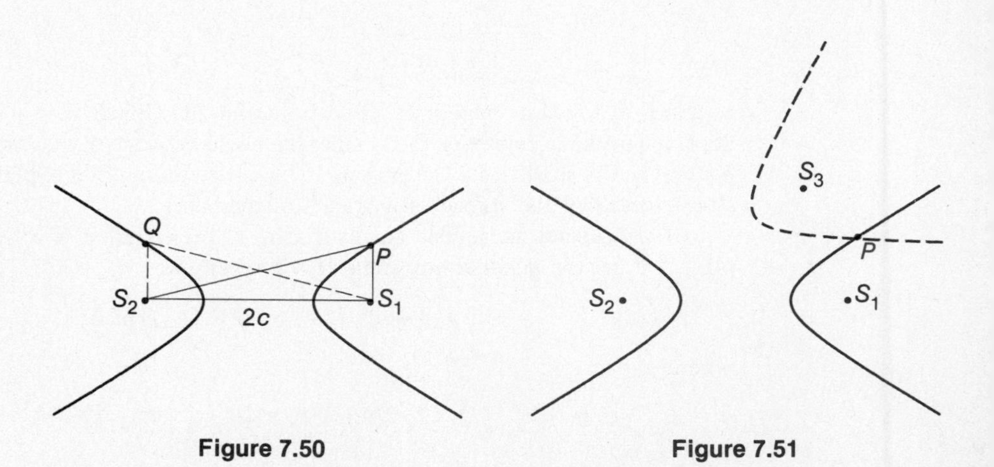

Figure 7.50

Figure 7.51

EXERCISES 7.5

Write the equation and sketch the graph of the following hyperbolas.

1. Foci $(\pm 5, 0)$ and vertices $(\pm 3, 0)$

2. Foci $(\pm 2, 0)$ and vertices $(\pm 1, 0)$

3. Foci $(\pm 4, 0)$ and vertices $(\pm 3, 0)$

4. Foci $(0, \pm 8)$ and vertices $(0, \pm 5)$

5. Foci $(0, \pm 5)$ and vertices $(0, \pm 4)$

6. Foci $(0, \pm\sqrt{5})$ and vertices $(0, \pm 2)$

7. Vertices $(\pm 1, 0)$ and asymptotes $y = \pm 2x$

8. Vertices $(\pm 4, 0)$ and asymptotes $3y = \pm 2x$

9. Vertices $(0, 1)$, $(2, 1)$ and focus $(3, 1)$

10. Vertices $(0, -1)$, $(0, 3)$, and focus $(0, 4)$

Graph each of the following hyperbolas. Identify the foci, vertices, center, and asymptotes. In Exercises 21-24, first complete the squares as for circles in Section 7.2.

11. $4x^2 - 9y^2 = 36$

12. $4x^2 - y^2 = 4$

13. $9x^2 - 16y^2 = 144$

14. $x^2 - 2y^2 = 6$

15. $x^2 - y^2 = 4$

16. $y^2 - x^2 = 1$

17. $8y^2 - 4x^2 = 8$

18. $9y^2 - x^2 = 9$

19. $\dfrac{(x+1)^2}{9} - \dfrac{(y-2)^2}{4} = 1$

20. $\dfrac{(x-2)^2}{16} - \dfrac{(y-1)^2}{9} = 1$

21. $x^2 - 4y^2 - 4x + 8y - 4 = 0$

22. $9x^2 + 54x - 16y^2 + 64y = 127$

23. $y^2 - x^2 + 8y + 8x - 4 = 0$

24. $4y^2 - 9x^2 - 12y - 6x = 28$

25. Following the same pattern as used in the text to derive the first standard form of the equation of a hyperbola, derive the equation of the hyperbola with foci $(\pm 2, 0)$ and transverse axis of length $\sqrt{2}$.

26. Complete the algebraic details of the derivation of the first standard form of the equation of a hyperbola at the beginning of this section.

27. Show that the asymptotes of a rectangular hyperbola are perpendicular.

28. Calculate the eccentricity of the hyperbolas of Exercises 1–3 above.

29. Show that the eccentricity of an equilateral hyperbola is $\sqrt{2}$.

30. Verify the details of the general equation of a hyperbola.

7.6 Transformation of Coordinates (Optional)

In the preceding sections we have seen that each of the conic curves has an equation that can be written in the form

$$Ax^2 + Cy^2 + Dx + Ey + F = 0, \tag{1}$$

a second degree equation in two variables. The equation is that of:

a line, if $A = C = 0$;

a circle, if $A = C$;

a parabola, if either $A = 0$ or $C = 0$;

an ellipse, if A and C have the same sign; and

a hyperbola, if A and C have different signs.

The most general second degree equation in two variables has an additional xy term:

$$Ax^2 + Bxy + Cy^2 + Dx + Ey + F = 0. \tag{2}$$

We now show that it is the equation of one of the conics. In case $B = 0$, we have equation (1) above, which, as we have already seen, represents one of the conic curves in a standard location. The equation of any curve depends upon the location of the axes. We will now see how to change this location to make the above equation take the simplest form possible, thus eliminating the xy-term. We do this by means of *translation of axes* and *rotation of axes*.

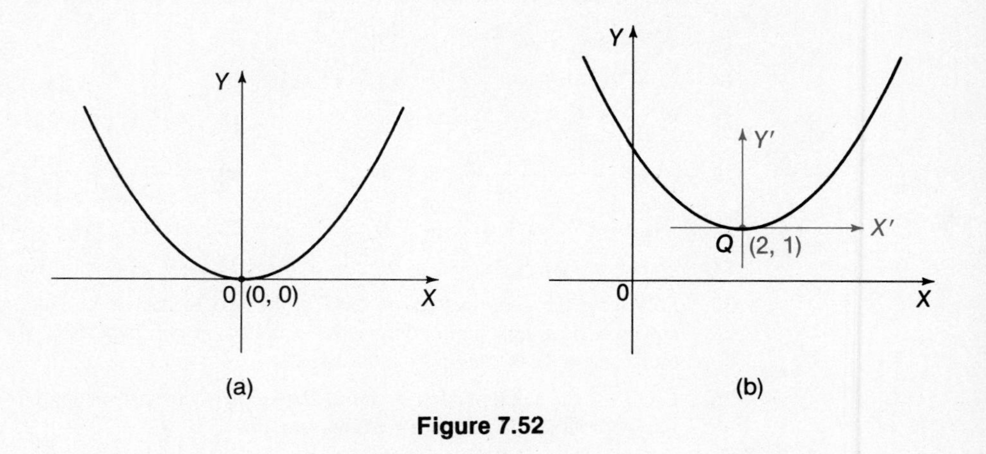

Figure 7.52

Translation of Axes Figure 7.52 shows the same parabola in two locations relative to the coordinate axes. We may think of this situation in either of two different ways:

(1) As in the preceding sections, we may think of the parabola in Figure 7.52b as having been translated two units to the right and one unit upward from the old location in (a).

(2) As shown in Figure 7.52b, we may imagine that the axes x, y, rather than the parabola, have been translated to new locations x', y'.

First, let us see what happens to the coordinates of a point under a translation of axes (case (2) above). In Figure 7.53, suppose the origin for the $x'y'$-axes to be (h, k) relative to the xy-axes. The coordinates of P are then related by the following **translation equations.**

$$\begin{cases} x' = x - h \\ y' = y - k \end{cases} \quad \text{or} \quad \begin{cases} x = x' + h \\ y = y' + k \end{cases}$$

EXAMPLE 1 What are the coordinates of the point (3, 2) relative to the new axes when the origin is translated to $O'(2, 4)$?

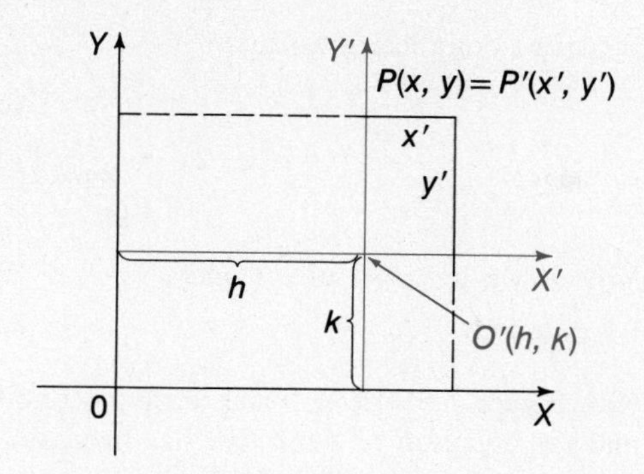

Figure 7.53. Translation of axes

SOLUTION The new coordinates (x', y') are found by using the first pair of equations.

$$x' = x - h = 3 - 2 = 1$$
$$y' = y - k = 2 - 4 = -2$$

See Figure 7.54. ●

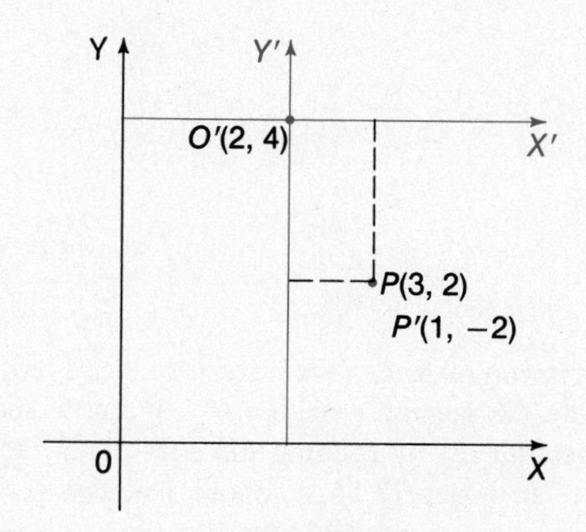

Figure 7.54

To see what happens to the equation of a conic under a translation, refer to Figure 7.52b. The equation of this parabola relative to the xy-axes is

$$y - 1 = (x - 2)^2.$$

If we let $x' = x - 2$ and $y' = y - 1$, the equation becomes

$$y' = x'^2.$$

Relative to the $x'y'$-axes this is the first standard form.

EXAMPLE 2 Transform the equation $x^2 + y^2 - 2x + 4y - 1 = 0$ into a form having no first degree terms.

SOLUTION First, we complete the square.

$$x^2 + y^2 - 2x + 4y - 1 = 0$$
$$(x^2 - 2x + 1) + (y^2 + 4y + 4) = 6$$
$$(x - 1)^2 + (y + 2)^2 = 6$$

Now if we let $x' = x - 1$ and $y' = y + 2$, the equation becomes

$$x'^2 + y'^2 = 6.$$

(See Figure 7.55.) The linear terms $-2x$ and $+4y$ have been removed by a translation and the equation of the curve has been simplified. ●

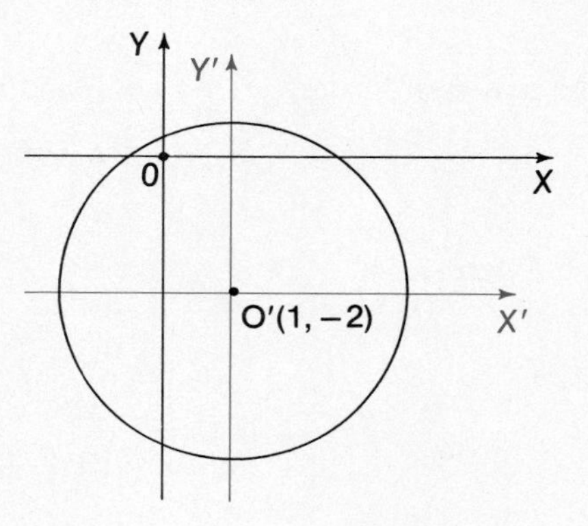

Figure 7.55

Rotation of Axes We have seen that a proper choice of axes removes the linear terms from the general equation (1). We now show that the xy-term can be removed from equation (2) by rotating the axes about the origin through an appropriate angle.

In Figure 7.56, suppose that the xy-axis system is rotated about the origin O through an angle θ into the $x'y'$-axis system. The relationship between the two coordinate systems is expressed by the following pairs of equations.

$$\begin{cases} x' = \overline{OB} = d \cos \phi \\ y' = \overline{BP} = d \sin \phi \end{cases} \quad \text{and} \quad \begin{cases} x = \overline{OA} = d \cos(\theta + \phi) \\ y = \overline{AP} = d \sin(\theta + \phi) \end{cases}$$

Expand the second pair of equations by the addition formulas for sine and cosine.

$$\begin{aligned} x = d \cos(\theta + \phi) &= d(\cos \theta \cos \phi - \sin \theta \sin \phi) \\ &= (d \cos \phi) \cos \theta - (d \sin \phi) \sin \theta \\ &= x' \cos \theta - y' \sin \theta. \\ y = d \sin(\theta + \phi) &= d(\sin \theta \cos \phi + \cos \theta \sin \phi) \\ &= (d \cos \phi) \sin \theta + (d \sin \phi) \cos \theta \\ &= x' \sin \theta + y' \cos \theta. \end{aligned}$$

This gives the **rotation equations:**

$$\boldsymbol{x = x' \cos \theta - y' \sin \theta}$$
$$\boldsymbol{y = x' \sin \theta + y' \cos \theta.}$$

EXAMPLE 3 Transform the equation $xy = 1$ by rotating the axes through an angle of 45° counterclockwise.

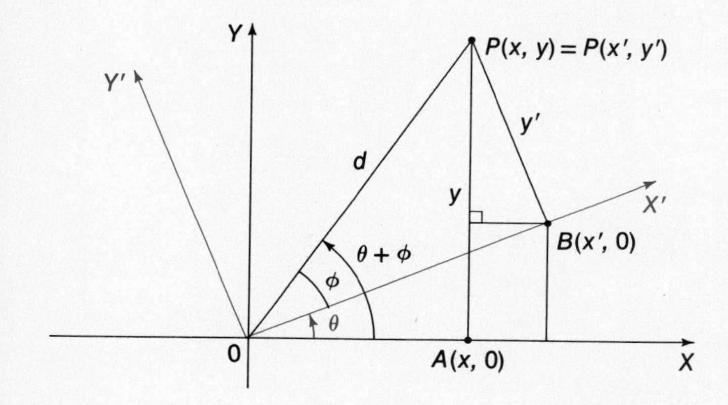

Figure 7.56. Rotation of axes

Figure 7.57

SOLUTION Use the above rotation equations.

$$x = x' \cos 45° - y' \sin 45° = \frac{1}{2}\sqrt{2}(x' - y')$$

$$y = x' \sin 45° + y' \cos 45° = \frac{1}{2}\sqrt{2}(x' + y')$$

Then $xy = 1$ becomes

$$\frac{1}{2}\sqrt{2}(x' - y')\frac{1}{2}\sqrt{2}(x' + y') = 1,$$

$$\frac{1}{2}(x'^2 - y'^2) = 1,$$

$$x'^2 - y'^2 = 2.$$

This is the equation of an equilateral hyperbola (Section 7.5) whose asymptotes are the x- and y-axes. See Figure 7.57. ●

EXAMPLE 4 Transform the equation $x^2 - 4xy + 4y^2 = 4$ by rotating the axes through the angle $\theta = \arctan 1/2$.

SOLUTION Since $\theta = \arctan 1/2$, $\tan \theta = 1/2$ and θ is in the first quadrant. We can find $\sin \theta$ and $\cos \theta$ from the right triangle in Figure 7.58.

$$c^2 = a^2 + b^2 = 1 + 4 = 5$$
$$c = \sqrt{5}$$
$$\sin \theta = \frac{1}{\sqrt{5}}$$
$$\cos \theta = \frac{2}{\sqrt{5}}$$

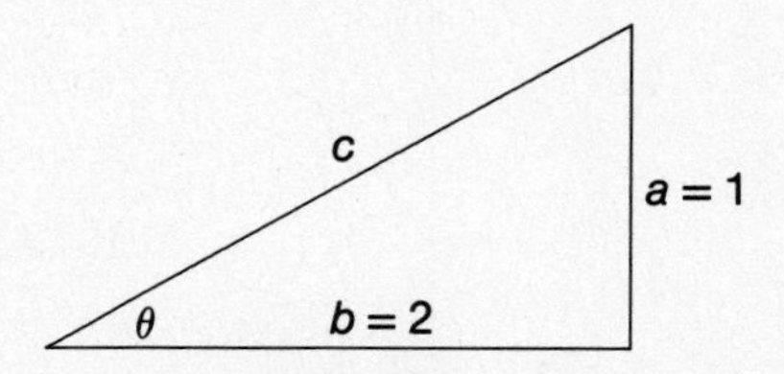

Figure 7.58

The rotation equations are

$$\begin{aligned} x &= x' \cos \theta - y' \sin \theta \\ &= x'\left(\frac{2}{\sqrt{5}}\right) - y'\left(\frac{1}{\sqrt{5}}\right) = \frac{1}{\sqrt{5}}(2x' - y') \\ y &= x' \sin \theta + y' \cos \theta \\ &= x'\left(\frac{1}{\sqrt{5}}\right) + y'\left(\frac{2}{\sqrt{5}}\right) = \frac{1}{\sqrt{5}}(x' + 2y'). \end{aligned}$$

Now, substituting these expressions for x and y in $x^2 - 4xy + 4y^2 = 4$, we get

$$\frac{1}{5}(2x' - y')^2 - 4\left(\frac{1}{\sqrt{5}}\right)(2x' - y')\frac{1}{\sqrt{5}}(x' + 2y') + 4\left(\frac{1}{5}\right)(x' + 2y')^2 = 4.$$

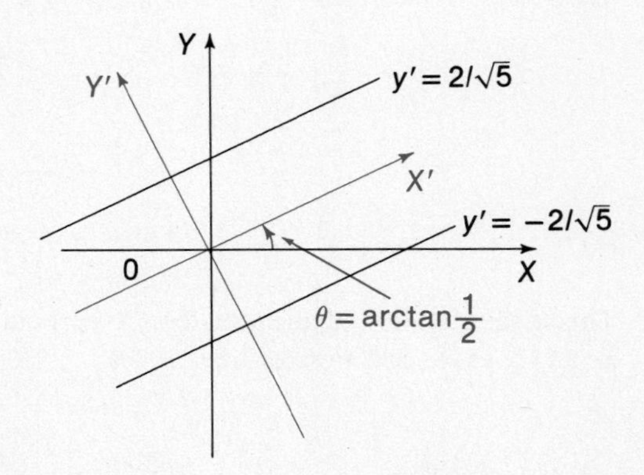

Figure 7.59

Multiplying both sides by 5 simplifies this to

$$(2x' - y')^2 - 4(2x' - y')(x' + 2y') + 4(x' + 2y')^2 = 20.$$

After expanding and simplifying, this equation becomes

$$5y'^2 = 4,$$

or

$$y' = \pm\frac{2}{\sqrt{5}},$$

the equations of the two parallel lines shown in Figure 7.59. ●

In Examples 3 and 4 the rotation transformed the equation into one without an xy-term. A formula for calculating the appropriate angle for this transformation is derived as follows. Substitute the rotation equations

$$\begin{cases} x = x' \cos\theta - y' \sin\theta \\ y = x' \sin\theta + y' \cos\theta \end{cases}$$

into the general equation of the second degree in x and y,

$$Ax^2 + Bxy + Cy^2 + Dx + Ey + F = 0.$$

It then becomes

$$A'x'^2 + B'x'y' + C'y'^2 + D'x' + E'y' + F' = 0,$$

where

$$\begin{aligned}
A' &= A\cos^2\theta + B\sin\theta\cos\theta + C\sin^2\theta,\\
B' &= B(\cos^2\theta - \sin^2\theta) + 2(C - A)\sin\theta\cos\theta,\\
C' &= A\sin^2\theta - B\sin\theta\cos\theta + C\cos^2\theta,\\
D' &= D\cos\theta + E\sin\theta,\\
E' &= -D\sin\theta + E\cos\theta,\\
F' &= F.
\end{aligned}$$

Now remove the $x'y'$-term above by letting $B' = 0$.

$$\begin{aligned}
B' = B(\cos^2\theta - \sin^2\theta) + 2(C - A)\sin\theta\cos\theta &= 0\\
B\cos 2\theta + (C - A)\sin 2\theta &= 0\\
B\cos 2\theta &= (A - C)\sin 2\theta\\
\frac{\cos 2\theta}{\sin 2\theta} &= \frac{A - C}{B}
\end{aligned}$$

$$\cot 2\theta = \frac{A - C}{B} \quad \text{or} \quad \tan 2\theta = \frac{B}{A - C}, \quad A \neq C$$

(From the second form, if $A = C$, then $\tan 2\theta$ is undefined, and $2\theta = 90°$, $\theta = 45°$. In this case, we know in advance that a rotation of 45° is indicated. See Example 3, for instance.)

EXAMPLE 5 Calculate the angle(s) through which the coordinate axes should be rotated to remove the xy-term from the equation $x^2 - 4xy + 4y^2 = 4$. (See Example 4.)

SOLUTION Here $A = 1$, $B = -4$, $C = 4$, $D = 0$, $E = 0$, and $F = -4$. Then

$$\cot 2\theta = \frac{A - C}{B} = \frac{1 - 4}{-4} = \frac{3}{4}.$$

We take $\tan 2\theta = 4/3$ and use the tangent double-angle formula.

$$\tan 2\theta = \frac{4}{3}$$

$$\frac{2 \tan \theta}{1 - \tan^2\theta} = \frac{4}{3}$$

$$6 \tan \theta = 4 - 4 \tan^2\theta$$

$$2 \tan^2\theta + 3 \tan \theta - 2 = 0$$

$$(2 \tan \theta - 1)(\tan \theta + 2) = 0$$

If $2 \tan \theta - 1 = 0$, $\tan \theta = 1/2$, and if $\tan \theta + 2 = 0$, $\tan \theta = -2$. The value $\tan \theta = 1/2$ was used in Example 4. Since $\tan(\theta + 90°) = -\cot \theta = -1/(\tan \theta) = -2$, the second angle above simply represents a rotation an additional 90° beyond arctan 1/2. In Figure 7.59, x' would move to y' and y' to the negative direction of x'. Thus either angle may be used for the rotation. Because of this, θ is usually chosen so that $0° < \theta \leq 90°$. ●

EXAMPLE 6 Transform the equation $9x^2 - 24xy + 16y^2 - 80x - 60y + 200 = 0$ by an appropriate rotation to remove the xy-term, and then sketch the transformed equation.

SOLUTION We have $A = 9$, $B = -24$, and $C = 16$. Then

$$\tan 2\theta = \frac{B}{A - C} = \frac{-24}{9 - 16} = \frac{24}{7},$$

$$\frac{2 \tan \theta}{1 - \tan^2\theta} = \frac{24}{7},$$

$$24 \tan^2\theta + 14 \tan \theta - 24 = 0,$$

$$12 \tan^2 \theta + 7 \tan \theta - 12 = 0,$$

$$(4 \tan \theta - 3)(3 \tan \theta + 4) = 0.$$

If we take $4 \tan \theta - 3 = 0$, then $\tan \theta = 3/4$, and from the triangle in Figure 7.60, we read $\sin \theta = 3/5$ and $\cos \theta = 4/5$. The rotation equations are

$$x = x' \cos \theta - y' \sin \theta$$

$$= \frac{4}{5}x' - \frac{3}{5}y' = \frac{1}{5}(4x' - 3y'),$$

and

$$\begin{aligned} y &= x' \sin \theta + y' \cos \theta \\ &= \frac{3}{5}x' + \frac{4}{5}y' \\ &= \frac{1}{5}(3x' + 4y'). \end{aligned}$$

Now, substituting the rotation equations in the given equation,

$$9x^2 - 24xy + 16y^2 - 80x - 60y + 200 = 0,$$

we get, after expansion and simplification, the parabola

$$y'^2 = 4x' - 8.$$

This is graphed as shown in Figure 7.61. •

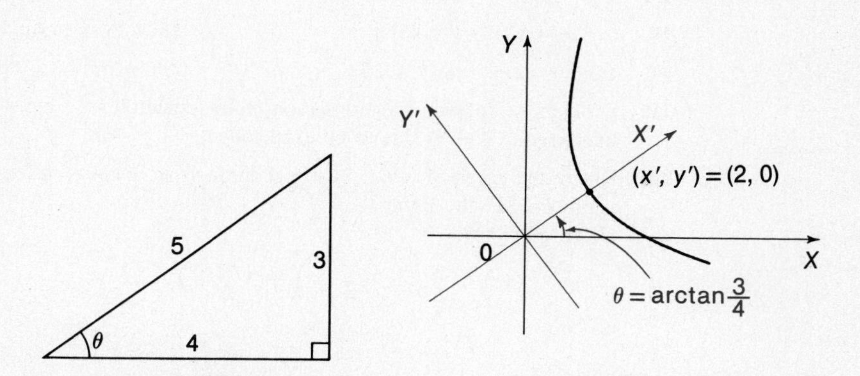

Figure 7.60

Figure 7.61

EXERCISES 7.6

By first completing the square, determine the translation which will remove the linear terms and write the transformed equation.

1. $x^2 + y^2 - 2x - 4y = 0$
2. $2x^2 + 2y^2 - 4x + 6y = 13$
3. $2x^2 - 8x + y^2 + 6y + 11 = 0$
4. $3x^2 + 6x + 2y^2 - 4y = 0$
5. $9x^2 + 4y^2 + 54x - 32y + 1 = 0$
6. $4x^2 + y^2 - 2y = 19$
7. $4x^2 - y^2 - 8x + 4y + 4 = 0$
8. $5x^2 - y^2 + 40x + 35 = 0$

Perform the indicated rotation transformation for each of the following, and then sketch the graph.

9. $xy = -4;\ \theta = 135°$

10. $8x^2 + 4xy + 5y^2 = 9;\ \theta = \arctan \dfrac{1}{2}$

11. $x^2 + xy + y^2 = \tfrac{1}{2};\ \theta = 45°$

12. $x^2 - 2xy + y^2 - 4\sqrt{2}x - 4\sqrt{2}y = 0;\ \theta = 45°$

13. $x^2 - 4xy + y^2 = 5;\ \theta = 135°$

14. $9x^2 - 24xy + 16y^2 - 80x - 60y = 0;\ \theta = \arctan \dfrac{3}{4}$

Determine the angle of rotation to remove the xy-term in each of the following; then transform the equation and sketch the graph.

15. $x^2 - 2xy + y^2 - 8x = 0$

16. $5x^2 + 8xy + 5y^2 = 36$

17. $4x^2 - 4xy + y^2 = 45$

18. $11x^2 - 6xy + 19y^2 = 20$

19. $17x^2 + 12xy + 8y^2 = 20$

20. $16x^2 - 24xy + 9y^2 - 90x - 120y = 0$

21. Perform the rotation transformation on the equation $x^2 + y^2 = a^2$ and note that it is left unchanged. Why is this to be expected?

22. Rotate the graph of $\sqrt{x} + \sqrt{y} = 1$ through an angle of 45° to show that it represents a parabola. See the figure.
[*Hint:* First write

$$\begin{aligned} \sqrt{x} + \sqrt{y} &= 1, \\ \sqrt{y} &= 1 - \sqrt{x}, \\ y &= 1 - 2\sqrt{x} + x, \\ 2\sqrt{x} &= 1 + x - y. \end{aligned}$$

Now square both sides to remove the radical. Finally, perform the rotation transformation.]

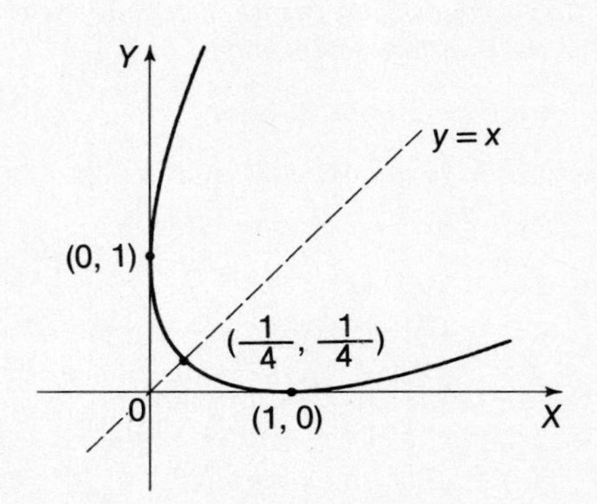

7.7 Equations in Polar Coordinates

In Section 6.7, polar coordinates in the plane were introduced. These are very convenient in the study of many important curves. In particular, they are useful in certain situations with the conic sections.

The relationships between polar and Cartesian coordinates developed earlier are restated here (refer to Figure 7.62).

$$\begin{cases} x = r \cos \theta \\ y = r \sin \theta \end{cases} \quad \text{and} \quad \begin{cases} r^2 = x^2 + y^2 \\ \tan \theta = \dfrac{y}{x} \end{cases}$$

Useful variations on these are:

$$\cos \theta = \frac{x}{r} = \frac{x}{\sqrt{x^2 + y^2}}, \qquad r = \sqrt{x^2 + y^2},$$

and

$$\sin \theta = \frac{y}{r} = \frac{y}{\sqrt{x^2 + y^2}}, \qquad \theta = \arctan \frac{y}{x}, \quad \text{or} \quad \theta = \pi + \arctan \frac{y}{x}.$$

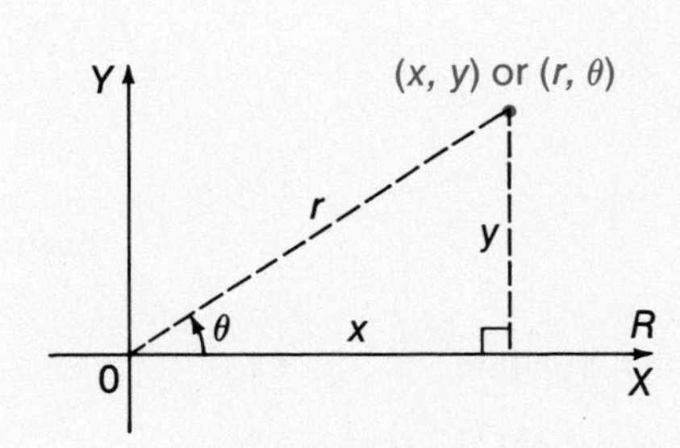

Figure 7.62

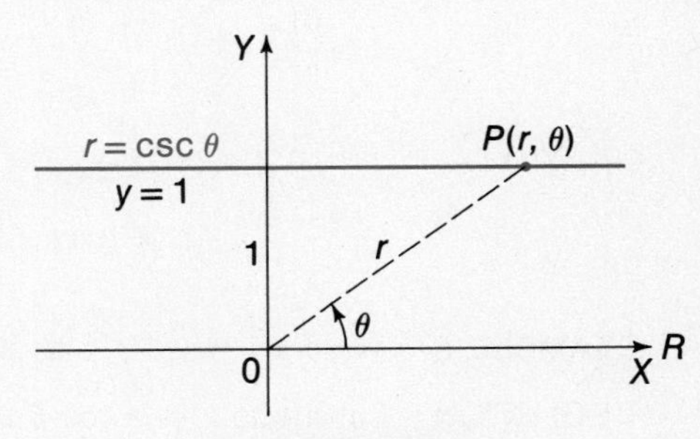

Figure 7.63

These relationships are called **transformation equations.** The choice of θ is determined by the quadrant in which (x, y) is located.

It is often useful to rewrite a Cartesian equation $y = f(x)$ as a polar equation $r = g(\theta)$. The transformation equations are used to do this.

EXAMPLE 1 Write $r = \csc \theta$ as a Cartesian equation.

SOLUTION Multiply both sides by $\sin \theta$ and then use the transformation equation $y = r \sin \theta$.

$$\begin{aligned} r &= \csc \theta \\ r \sin \theta &= \csc \theta \sin \theta \\ y &= 1 \qquad \text{since } \csc \theta \sin \theta = 1 \end{aligned}$$

So $r = \csc \theta$ is the polar equation of the horizontal line $y = 1$. See Figure 7.63. ●

EXAMPLE 2 Write $r = 2 \cos \theta$ as a Cartesian equation.

SOLUTION To use the transformations, we first multiply both sides by r to obtain

$$r^2 = 2r \cos \theta.$$

Substitution for r^2 and $r \cos \theta$ gives

$$x^2 + y^2 = 2x.$$

By completing the square we can write this as

$$(x - 1)^2 + y^2 = 1,$$

which is the equation of a unit circle with center at (1, 0). See Figure 7.64. ●

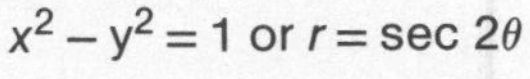

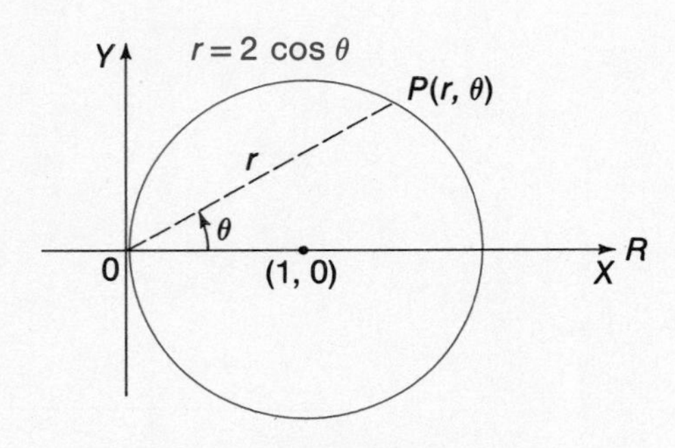

Figure 7.64

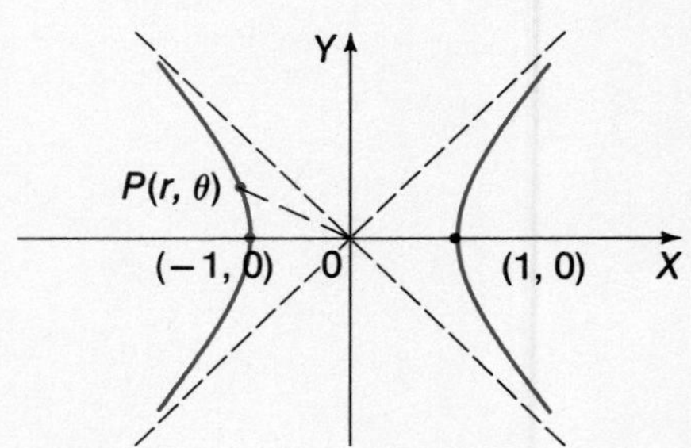

Figure 7.65

EXAMPLE 3 Write $x^2 - y^2 = 1$ as a polar equation.

SOLUTION Substitute $x = r \cos \theta$ and $y = r \sin \theta$.

$$\begin{aligned}(r \cos \theta)^2 - (r \sin \theta)^2 &= 1,\\ r^2(\cos^2\theta - \sin^2\theta) &= 1,\\ r^2 \cos 2\theta &= 1,\\ r^2 &= \sec 2\theta.\end{aligned}$$

This is the equation of the rectangular hyperbola shown in Figure 7.65. ●

Polar Equations of Lines The polar equation of a line through the pole is given by $\theta = c$, c a constant. Such a line is shown in Figure 7.66a. Figure 7.66b shows a line passing through $(a, 0)$ and perpendicular to the polar axis. In this case,

$$\begin{aligned}\cos \theta &= \frac{a}{r},\\ r &= \frac{a}{\cos \theta},\\ r &= a \sec \theta.\end{aligned}$$

Thus, the polar equation of the vertical line with Cartesian equation $x = a$ is $r = a \sec \theta$.

In the same way, we can show that the polar equation of a line parallel to the polar axis and a units from it is $r = a \csc \theta$. See Figure 7.66c.

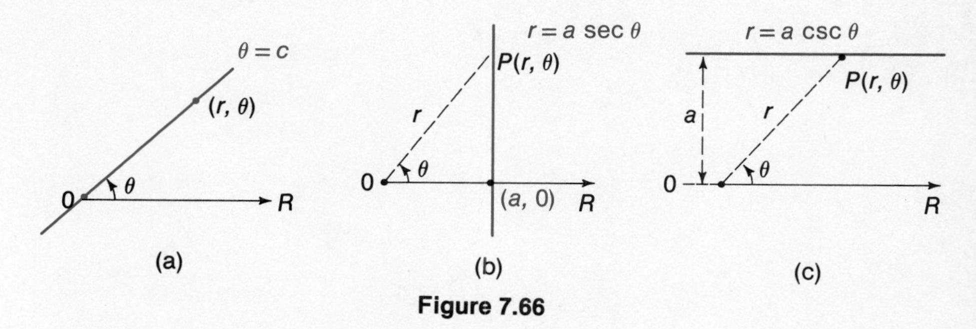

Figure 7.66

Finally, consider any oblique line. In Figure 7.67, let p be the perpendicular distance $\overline{OA}$ of the line from the pole, and α the angle that OA makes with the polar axis. Then

$$\cos(\theta - \alpha) = \frac{p}{r}$$

$$r \cos(\theta - \alpha) = p$$

$$\boldsymbol{r = p \sec(\theta - \alpha).}$$

Thus, $r = p \sec(\theta - \alpha)$ is the polar equation of a straight line. It is left as an exercise to show that the above equations of lines perpendicular or parallel to the polar axis are just particular cases of this equation.

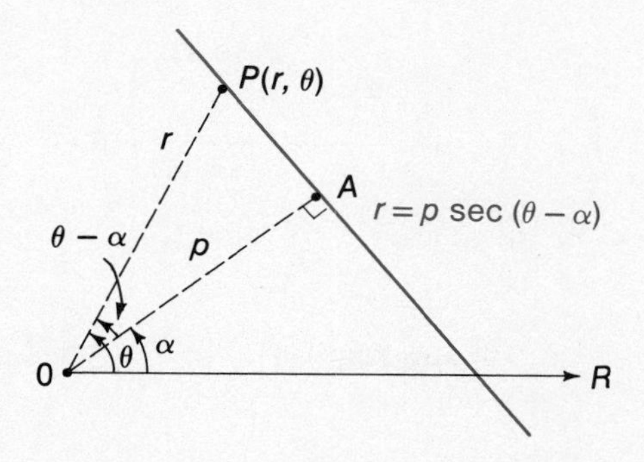

Figure 7.67

EXAMPLE 4 Find the polar equation of the line intersecting the polar axis at (2, 0) at an angle of 60°.

SOLUTION From Figure 7.68,

$$\sin 60° = \frac{p}{2}$$

$$p = 2 \sin 60°$$

$$= 2\left(\frac{1}{2}\sqrt{3}\right) = \sqrt{3}.$$

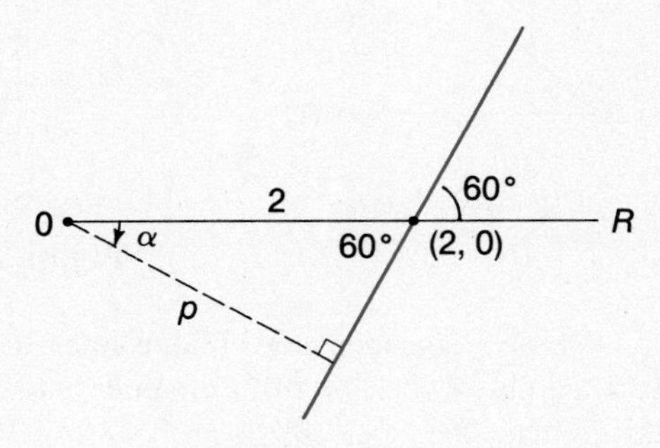

Figure 7.68

If α is the angle that the perpendicular to the line makes with the polar axis, measured clockwise, then

$$\alpha = -(90° - 60°) = -30°.$$

Thus, the equation of the line is $r = p \sec(\theta - \alpha) = \sqrt{3} \sec(\theta + 30°)$. ●

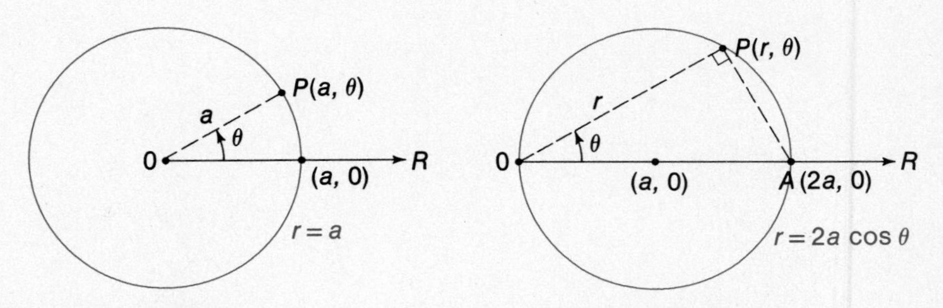

Figure 7.69

Figure 7.70

Polar Equations of Circles The polar equation of a circle of radius a and center at the pole is simply $r = a$. (See Figure 7.69.) Now consider a circle passing through the pole O with center on the polar axis at $(a, 0)$. From elementary geometry, an angle inscribed in a semicircle is a right angle. Thus, in Figure 7.70, angle OPA is a right angle. So

$$\frac{r}{2a} = \cos\theta$$

$$r = 2a\cos\theta.$$

This is the equation of the circle with center at $(a, 0°)$. In the same way, one can show that the circle passing through the pole O with center on the 90°-line at $(a, 90°)$ is $r = 2a\sin\theta$.

Finally, consider *any* circle in the polar coordinate plane. Call its center (c, α) and its radius a. Then from Figure 7.71 and the law of cosines, it follows that $a^2 = r^2 + c^2 - 2cr\cos(\theta - \alpha)$, or

$$\mathbf{r^2 = a^2 - c^2 + 2cr\cos(\theta - \alpha).}$$

EXAMPLE 5 Determine the polar equation of the circle with center $(5, 30°)$ and radius 2.

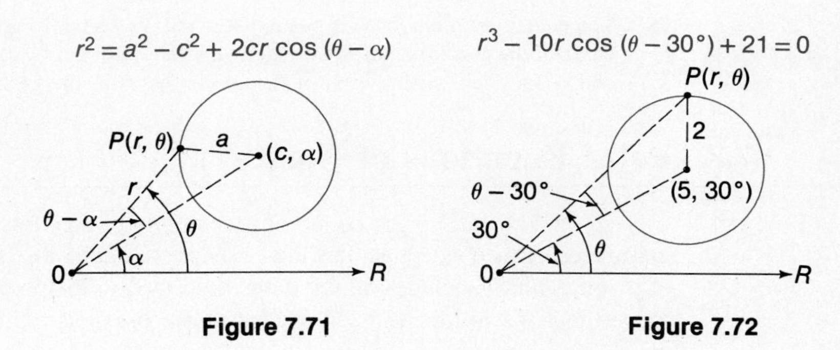

Figure 7.71 **Figure 7.72**

SOLUTION From Figure 7.72,

$$a^2 = r^2 + c^2 - 2cr\cos(\theta - \alpha),$$
$$2^2 = r^2 + 5^2 - 2(5)r\cos(\theta - 30°),$$
$$4 = r^2 + 25 - 10r\cos(\theta - 30°),$$
$$r^2 - 10r\cos(\theta - 30°) + 21 = 0.$$

Note that this is a quadratic equation in r. •

EXERCISES 7.7

Transform the following Cartesian equations to polar form, and graph the curves.

1. $x = a$	**2.** $y = b$	**3.** $y = x$
4. $x + y = 0$	**5.** $y^2 = x$	**6.** $y = x^2$
7. $x^2 + y^2 = 4$	**8.** $x^2 - y^2 = 1$	**9.** $x^2 + y^2 = x$
10. $x^2 + y^2 = y$	**11.** $x^2 + y^2 + 2y = 0$	**12.** $x^2 + y^2 - 2x = 0$

Transform the following polar equations to Cartesian form, and graph the curves.

13. $r = 3$ **14.** $\theta = \dfrac{\pi}{4}$ **15.** $\sin\theta = \dfrac{1}{2}$

16. $\cos\theta = -\dfrac{1}{2}\sqrt{3}$ **17.** $r = 3\sec\theta$ **18.** $r = -2\csc\theta$

19. $r = -2\cos\theta$ **20.** $r = 4\sin\theta$ **21.** $r(1 - \cos\theta) = 1$

22. $r(1 + \sin\theta) = 1$ **23.** $r^2\cos 2\theta = 1$ **24.** $r^2\sin 2\theta = 1$

25. Derive the polar equation of the line parallel to the polar axis and a units above it.

26. Show that the equations of lines parallel or perpendicular to the polar axis are just particular cases of the general polar equation of a line.

27. Derive the polar equation of the circle passing through the pole and with center (a) $(a, \pi/2)$, (b) (a, π), (c) $(a, 3\pi/2)$. [*Hint:* Use the same procedure as used in the text with Figure 7.70.]

28. The equations of circles with center at the pole or else passing through the pole are special cases of the general polar equation of a circle. Show this by identifying the constants a, c, and α and then writing each of the special cases in the general form.

7.8 Polar Equations of Conics

In Sections 7.4 and 7.5 the notion of eccentricity of an ellipse and of a hyperbola was introduced. This concept applies to all conics. Using it, we can give a single definition of a conic which includes all the cases described in the preceding sections. It can be proved that the following is equivalent to the previous definitions.

DEFINITION 7.5 A **conic** is the set of points whose distance from a fixed point is equal to the product of a positive constant e and the distance to a fixed line, not through the fixed point.

The fixed point is called a **focus,** the fixed constant e the **eccentricity,** and the fixed line the **directrix.** The conic is an ellipse if $0 < e < 1$, a parabola if $e = 1$, and a hyperbola if $e > 1$.

The above property of a conic can be used to derive a single general polar equation representing *all* conics, as follows. In Figure 7.73, let the focus be at the pole O and suppose the directrix l is perpendicular to the polar axis. Let $2p$ be the distance from the focus to the directrix and let e be the eccentricity. Then, by the focus-directrix property,

$$\begin{aligned}\overline{OP} &= e(\overline{AP}),\\ r &= e(\overline{AB} + \overline{BP}),\\ &= e(2p + r\cos\theta),\\ &= 2ep + er\cos\theta.\\ r - er\cos\theta &= 2ep\end{aligned}$$

$$r(1 - e \cos \theta) = 2ep$$

$$r = \frac{2ep}{1 - e \cos \theta}.$$

This is the polar equation of an ellipse if $0 < e < 1$, of a parabola if $e = 1$, and of a hyperbola if $e > 1$.

The same procedure can be used to develop the polar equation of a conic if the directrix is vertical and to the right of the pole, or if the directrix is parallel to the polar axis. These cases are left as exercises.

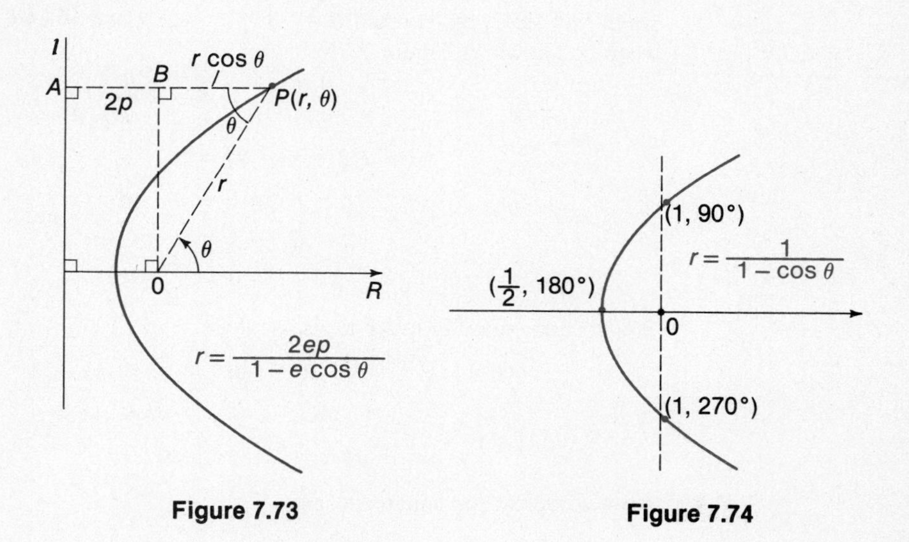

Figure 7.73

Figure 7.74

EXAMPLE 1 Find the polar equation of the parabola with focus at the pole and vertex at (1/2, 180°), and graph.

SOLUTION The eccentricity, e, is 1 and the distance from focus to vertex, p, is 1/2.

$$r = \frac{2(1)\left(\frac{1}{2}\right)}{1 - (1) \cos \theta}$$

$$r = \frac{1}{1 - \cos \theta}$$

is the polar equation.

The vertex is at (1/2, 180°). To complete the graph we find the points of intersection with the 90°-line. When $\theta = 90°$, $r = 1$, and when $\theta = 270°$, $r = 1$. The graph is in Figure 7.74. ●

EXAMPLE 2 Graph $r = 2/(2 - \cos \theta)$, (a) directly from the polar equation and then (b) by first transforming the equation to Cartesian form.

SOLUTION (a) First write the equation in standard polar form for a conic by dividing numerator and denominator by 2.

$$r = \frac{2}{2 - \cos\theta}$$

$$= \frac{1}{1 - \frac{1}{2}\cos\theta}.$$

Since the eccentricity $e = 1/2 < 1$, the curve is an ellipse. To complete the graph, we find intersections with the polar axis and the 90°-line. When $\theta = 0°$, $r = 2$; when $\theta = 90°$, $r = 1$; when $\theta = 180°$, $r = 2/3$; and when $\theta = 270°$, $r = 1$. The graph is shown in Figure 7.75.

(b)

$$r = \frac{2}{2 - \cos\theta}$$

$$r(2 - \cos\theta) = 2$$

$$2r - r\cos\theta = 2$$

$$2r - x = 2$$

$$2r = x + 2$$

Square both sides in order to use $r^2 = x^2 + y^2$.

$$4r^2 = x^2 + 4x + 4$$

$$4(x^2 + y^2) = x^2 + 4x + 4$$

$$3x^2 + 4y^2 - 4x - 4 = 0$$

Now complete the square, to get

$$\frac{\left(x - \frac{2}{3}\right)^2}{\frac{16}{9}} + \frac{y^2}{\frac{4}{3}} = 1.$$

This is the ellipse with center at (2/3, 0) having major axis of length 8/3 and minor axis of length $4/\sqrt{3}$. Compare with Figure 7.75. ●

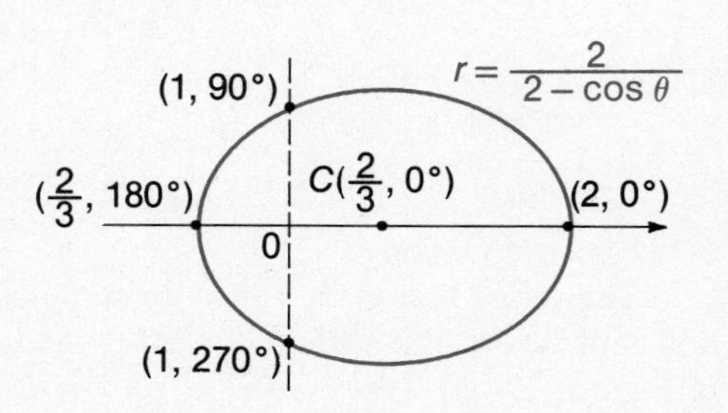

Figure 7.75

Earlier (Section 7.4), we noted that planets, satellites, and comets move in orbits which are conic curves. Since in observing positions in the sky, we cannot measure distance directly but do measure angles (directions), polar coordinates are useful in studying the motion of such bodies.

EXERCISES 7.8

Identify and sketch each of the following.

1. $r(1 + \cos\theta) = 1$

2. $r(1 - \cos\theta) = 1$

3. $r = \dfrac{3}{1 - \frac{1}{2}\cos\theta}$

4. $r = \dfrac{2}{1 + \frac{2}{3}\cos\theta}$

5. $r = \dfrac{6}{2 - \sin\theta}$

6. $r(2 + \cos\theta) = 4$

7. $r(1 + \sin\theta) = 3$

8. Show that $r(\sqrt{2} - 2\cos\theta) = 4\sqrt{2}$ is the equation of a rectangular hyperbola by converting to Cartesian form.

9. Show that the equation of the parabola $r(1 - \cos\theta) = 2p$ can be written in the form $r = p\sec^2\frac{1}{2}\theta$.

Use the procedure of the text in deriving the polar equation of a conic to find the equation of the following curves.

10. Parabola with vertex $(3, 0°)$ and focus at the pole.

11. Parabola with vertex $(2, 270°)$ and focus at the pole.

12. Ellipse having eccentricity $e = 2/3$, directrix $r\cos\theta + 6 = 0$, and focus at the pole.

13. Hyperbola with eccentricity $e = 4/3$, directrix $r\sin\theta = 9$, and focus at the pole.

14. A certain comet moves in a parabolic orbit having the sun at the focus. When the comet is 4×10^7 miles from the sun, the line from the sun to it makes an angle of 60° with the axis of symmetry of the orbit, drawn in the direction in which the curve opens. How near does the comet come to the sun?

15. A comet is observed at two points of its parabolic orbit about the sun. Neither point is the vertex of the orbit. The lines from the sun to these two points make an angle of 90° with each other and have lengths 1×10^7 miles and 2×10^7 miles, respectively. Find the equation of the orbit and determine how near the comet comes to the sun.

16. A conic with focus at the pole and symmetric to the polar axis passes through the two points $(4, 60°)$ and $(2, 90°)$. What is its equation and what kind of conic is it? [*Hint:* Start with the general polar equation $r(1 - e\cos\theta) = 2ep$.]

17. A hyperbola has a focus at the pole and one branch opens to the left about the pole, symmetric to the polar axis (extended to the left). If it passes through the two points $((1/2)\sqrt{2}, 45°)$, and $(\sqrt{2}, 90°)$, find its equation and its eccentricity. [*Hint:* Start with the general equation $r(1 + e\cos\theta) = 2ep$.]

Chapter 7 Review Exercises

Sketch the line determined by the two points $(-3, 4)$ *and* $(2, 1)$. *Then*

1. Derive its equation, using the two-point form.
2. Change the two-point form into the point-slope form (two versions).
3. Write the equation in the slope-intercept form by solving for y in terms of x.
4. Reduce the equation to the general form.
5. Identify the slope, inclination, and intercepts.

Sketch the triangle whose vertices are $(-2, 1)$, $(4, 1)$, *and* $(3, 0)$. *Then*

6. Write the equation of the line from the first vertex to the midpoint of the opposite side.
7. Write the equation of the line through the first vertex perpendicular to the opposite side.
8. Write the equation of the perpendicular bisector of the side opposite the first vertex.
9. Given the line $3x - 2y + 6 = 0$, write the equation of the line through $(1, 1)$ and (a) parallel to the given line, (b) perpendicular to the given line.

Write the equation of the circle

10. Having center $(1, 3)$ and radius 2.
11. With endpoints of diameter $(-1, -1)$ and $(1, 2)$.
12. Write the equations of the tangents to the circle of the preceding exercise at the given points.

Rewrite each of the following equations in the center-radius form and then identify the center and radius, if it is a real circle. If not, identify the kind of "circle."

13. $x^2 + y^2 + 4x - 6y = 12$
14. $x^2 + y^2 - 4x + 2y + 5 = 0$
15. $4x^2 + 4y^2 + 12x + 8y + 3 = 0$
16. $x^2 + y^2 - 2x + 4y + 11 = 0$

Sketch the graph of each of the following conic curves and identify centers, vertices, and foci.

17. $x^2 - 2x - 4y + 5 = 0$
18. $x^2 + 6x - 12y - 15 = 0$
19. $2y^2 + 4y - 3x + 5 = 0$
20. $y^2 + 6y - 8x = 7$
21. $3x^2 + 4y^2 + 6x - 16y + 7 = 0$
22. $2x^2 + y^2 - 4x - 2y = 3$
23. $4x^2 + y^2 - 40x + 96 = 0$
24. $16x^2 + 25y^2 + 32x + 50y = 40$
25. $4x^2 - 5y^2 + 8x + 10y + 19 = 0$
26. $8x^2 - y^2 - 32x - 2y + 23 = 0$
27. The path of a projectile is the parabolic path $y = 40x - 16x^2$. (a) Reduce the equation to standard form and sketch the graph. (b) If the ground is horizontal, how high above the ground does the projectile rise? (c) How far from the initial point does the projectile strike the ground again?
28. Find the equation of the circle through the origin whose center is the focus of the parabola $y^2 = 4x$. Graph.
29. Find the highest point on the curve whose equation is $x^2 + 4y^2 - 8x - 16y + 28 = 0$.
30. Show that the circle which has as a diameter the line segment joining the foci of the hyperbola $5x^2 - 4y^2 = 9$ passes through the foci of the hyperbola $4y^2 - 5x^2 = 9$. (These two hyperbolas are called *conjugates* of one another.) Graph.

Translate the axes so that each of the following equations has no linear terms.

31. $4x^2 + 5y^2 + 8x - 10y = 11$

32. $4x^2 - y^2 + 24x + 4y = 40$

Rotate the axes through the indicated angle to transform the equation into a form without an xy-term. Then sketch the graph.

33. $5x^2 + 8xy + 5y^2 = 36,\ \theta = 45°$

34. $x^2 + 4xy + 4y^2 + 8\sqrt{5}x - 4\sqrt{5}y = 0,\ \theta = \arctan 2$

Determine the angle through which the axes must be rotated to remove the xy-term. Then calculate the transformed equation and sketch the graph.

35. $18x^2 + 8xy + 33y^2 = 34$

36. $8x^2 + 4xy + 5y^2 = 9$

Transform each of the following equations to alternate form (polar to Cartesian and Cartesian to polar). Sketch the curve.

37. $y = 2x$

38. $x^2 + y^2 = 9$

39. $y^2 = 4x + 4$

40. $x^2 + y^2 + 4x = 0$

41. $r = -2 \sec \theta$

42. $r = -2 \cos \theta$

43. $r(1 + \sin \theta) = 1$

44. $r(1 + \cos \theta) = 1$

45. Use the procedure at the beginning of the last section to derive the equation of the conic with focus at the pole and horizontal directrix below the pole.

Chapter 7 Miscellaneous Exercises

A line segment joining a vertex of a triangle with the midpoint of the opposite side is called a median. Determine the equations of the medians of each of the triangles whose vertices are given.

1. $(-2, -1), (4, 3), (2, 7)$

2. $(-1, 4), (-3, -4), (5, -2)$

3. $(2, 2), (3, -3), (6, 4)$

4. $(3, 4), (-5, 4), (3, -1)$

5. Let the equations of two lines be $a_1x + b_1y + c_1 = 0$ and $a_2x + b_2y + c_2 = 0$. Find the relationship between the coefficients of these equations necessary for the two lines to be (a) parallel, (b) perpendicular.

6. Show that if the intercepts of a line are $(a, 0)$ and $(0, b)$, then the equation of the line can be written as $x/a + y/b = 1$. This is called the *intercept form* of the equation of a line.

Find the intercepts of each of the following lines and use them to write the equation in the intercept form.

7. $3x + 4y = 12$

8. $5x - 3y = 15$

9. $x - 2y - 5 = 0$

10. $x + 3y + 4 = 0$

The general equation of a line can be converted into the intercept form as follows:

$$2x - 3y + 6 = 0,$$
$$2x - 3y = -6.$$

Now dividing both sides by -6, *we get*

$$\frac{x}{-3} + \frac{y}{2} = 1.$$

Thus the intercepts are $x = -3$ *and* $y = 2$. *Now use this method to rewrite each of the following equations in the intercept form. Then sketch the graph from this form directly.*

11. $3x - 4y - 12 = 0$

12. $2x + 5y = 10$

13. $x - 2y = 4$

14. $3x + y - 6 = 0$

15. A hyperbola has vertices $(\pm 2, 0)$ and the smaller angle between its asymptotes is 60°. Calculate the equation of the hyperbola and sketch.

16. Describe each of the conic sections.

17. Define each of the conic curves algebraically and use the definition to derive the equation in the first standard form.

Using the procedure of Section 7.8, derive the polar equation of the conic with focus at the pole and

18. Vertical directrix to the right of the pole.

19. Horizontal directrix above the pole.

20. The following is a method (called the *trammel method*) which a draftsman can use to construct accurately any number of points of an ellipse with given axes. Let a and b be the semiaxes of an ellipse. Draw circles centered at the origin, having radii a and b, respectively. See the figure. Now draw any line through the origin intersecting the circles in points B and A. Through B draw a horizontal line and through A draw a vertical line. Call their intersection $P(x, y)$. If the line AB makes an angle ϕ with the x-axis, show that $x = a \cos \phi$, $y = b \sin \phi$, and $b^2x^2 + a^2y^2 = a^2b^2$. Thus, P is a point on the required ellipse. This construction may be repeated any number of times to get any desired number of points.

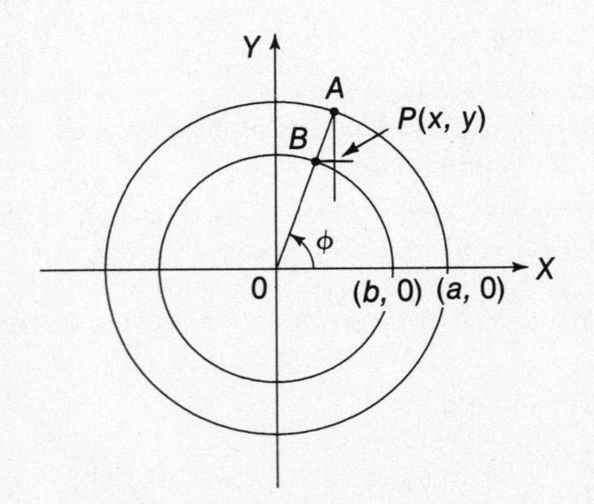

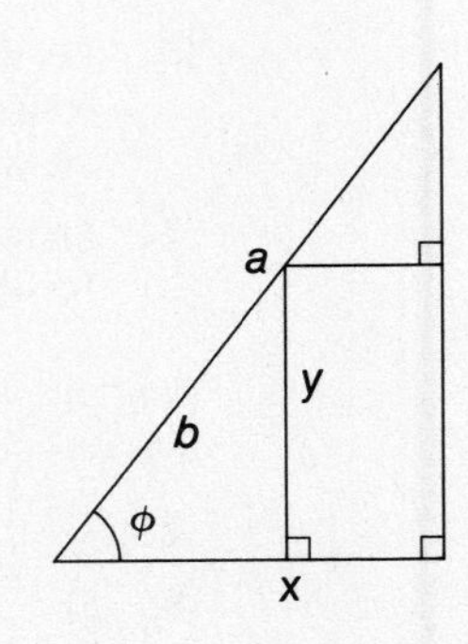

8

Systems of Equations and Inequalities

Often problems are encountered which not only involve a number of variables but require that certain conditions be met as well. For example, if it was desirable to obtain 10 liters of a 12% solution of alcohol by mixing appropriate amounts of 10% and 15% alcohol solutions, two variables would be involved, namely, the amounts of the respective alcohol solutions. Additionally, there would be two conditions to meet, that there be a total of 10 liters in the final mixture and that it be 12% alcohol. To solve this type of problem we frequently make use of two or more equations or inequalities called a *system of equations or inequalities*. This chapter provides an introduction to the methods of dealing with such systems involving several variables and of solving problems like the one above which we will refer to again later on in the chapter.

8.1 Systems of Linear Equations

Solutions of Equations The equation $2x - 3 = 0$, in one variable, is satisfied by $x = 3/2$ and by no other number. The real number 3/2 is called the *solution* of the equation. For equations such as $x + 2y = 4$ and $y^2 = x$ there are an infinite number of *pairs* of numbers (x, y) which satisfy each equation. The tables on the next page list several such pairs of numbers. Each of them is a **solution** of the respective equation.

x	y
0	2
4	0
2	1
-2	3

$x + 2y = 4$

x	y
0	0
1	1
4	2
$\frac{1}{4}$	$-\frac{1}{2}$

$y^2 = x$

Figure 8.1 shows the graphical interpretations of solutions of the equations $x + 2y = 4$ and $y^2 = x$. Each solution gives the coordinates of a point on the corresponding curve. Clearly, there are an infinite number of points on each curve and thus an infinite number of solutions of each equation.

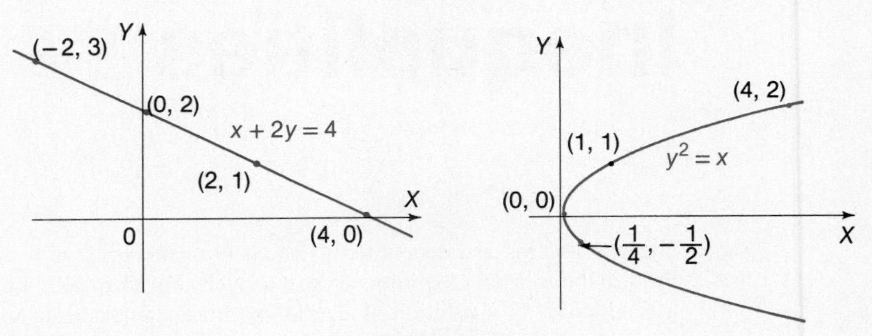

Figure 8.1

Substitution shows that (2, 1) satisfies

$$x + 2y = 4, \qquad 2x + 4y = 8, \qquad \text{and} \qquad -x - 2y = -4.$$

Each of these equations is obtained by multiplying or dividing both sides of another one of the equations by the same number. In fact, every solution of one of these equations is also a solution of the other two. This leads to the following important definition.

DEFINITION 8.1 Two equations having the same set of solutions are called **equivalent equations.**

The equations $x + 2y = 4$, $2x + 4y = 8$, and $-x - 2y = -4$, given above, are equivalent equations.

The equations $\sin^2 x = 1 - \sin x$ and $\sin x = \cos^2 x$ are nonalgebraic examples of equivalent equations, as are $y = e^{x-1}$ and $x = 1 + \ln y$. Each of the transformations made in proving trigonometric identities and solving trigonometric equations (Chapter 5) involved replacing an equation by an equivalent equation.

To solve a system of equations, we use the fact that, given an equation, an equivalent equation is obtained if the following operations are performed:

(1) The same number is added to or subtracted from each side of the equation.

(2) Both sides of the equation are multiplied by or divided by the same number (not zero).

Systems of Equations An equation with more than one variable generally has an infinite set of solutions. Many problems require that a set of equations share a common solution or solutions. For instance, we might wish to know the points of intersection (if any) of the graphs of the line $x - 2y = 0$ and the circle $x^2 + y^2 = 5$. (See Figure 8.2.) Procedures for solving this problem algebraically will be developed later in this chapter. For now, note that both $x = 2$, $y = 1$ and $x = -2$, $y = -1$ satisfy *both* equations. This means that (2, 1) and $(-2, -1)$ are points of intersection of the graphs.

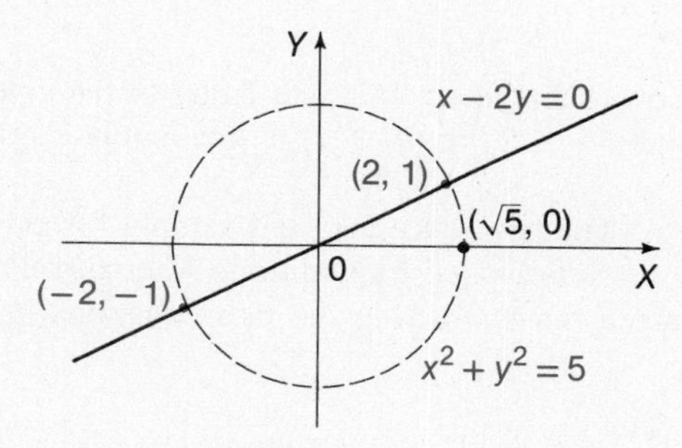

Figure 8.2

The pair of equations

$$\begin{cases} x - 2y = 0 \\ x^2 + y^2 = 5 \end{cases}$$

is called a **system of equations** and the two ordered pairs (2, 1) and $(-2, -1)$ are **simultaneous solutions** (or simply **solutions**) of the system. This is a system of two equations in two variables. In practice, a system may have any number of equations in any number of variables.

DEFINITION 8.2 Two systems of equations having the same set of solutions are called **equivalent systems.**

Solving a system of equations usually involves replacing the given system with an equivalent system. The following transformations result in an equivalent system:

(1) Interchanging any two equations.
(2) Replacing any equation by an equivalent equation.
(3) Adding a multiple of an equation to any other equation of the system.

Solving Systems of Linear Equations We begin with a system of two linear equations in two variables.

EXAMPLE 1 Solve the system

$$\begin{cases} x + 2y = 5 \\ x - 4y = 2. \end{cases}$$

SOLUTION Multiply both sides of the first equation by 2 to obtain the equivalent equation $2x + 4y = 10$. By (2) above, the following system is equivalent to the original system.

$$\begin{cases} 2x + 4y = 10 \\ x - 4y = 2 \end{cases}$$

Adding corresponding sides of the two equations eliminates the variable y and gives

$$3x = 12,$$
$$x = 4.$$

Now, substitute $x = 4$ into either of the given equations to obtain $y = 1/2$. So $x = 4$ and $y = 1/2$, or $(4, 1/2)$ is the simultaneous solution of the given system. ●

The procedure used in Example 1 is called **solution by elimination.** The variable y was eliminated by addition. Alternatively, the same solution could have been obtained by subtracting the two equations in the original system to eliminate x first.

EXAMPLE 2 Solve the system

$$\begin{cases} \dfrac{x}{2} + \dfrac{y}{3} = 1 \\ x + \dfrac{y}{4} = 2. \end{cases}$$

SOLUTION First, multiply both sides of each equation by a common denominator to replace the fractional coefficients by integral coefficients.

$$\begin{cases} 3x + 2y = 6 & \text{multiplying both sides by 6} \\ 4x + y = 8 & \text{multiplying both sides by 4} \end{cases}$$

To eliminate y, multiply both sides of the second equation by 2.

$$\begin{cases} 3x + 2y = 6 \\ 8x + 2y = 16 \end{cases}$$

Subtracting the first equation from the second gives

$$5x = 10,$$
$$x = 2.$$

Substitute 2 for x in either of the given equations to get $y = 0$. The solution is $(2, 0)$. ●

We can now solve the problem posed at the beginning of this chapter.

EXAMPLE 3 How much of each of a 10% solution and a 15% solution of alcohol must be combined to obtain 10 liters of a 12% solution?

SOLUTION Let x be the number of liters of the 10% solution and y the number of liters of the 15% solution. Then

$$\begin{cases} x + y = 10 \\ 0.10x + 0.15y = 0.12(10). \end{cases}$$

As in the preceding example, we first multiply both sides of the second equation by 100 to obtain integral coefficients.

$$\begin{cases} x + \quad y = \quad 10 \\ 10x + 15y = 120 \end{cases}$$

Multiply the first equation by 3 and divide the second equation by 5.

$$\begin{cases} 3x + 3y = 30 \\ 2x + 3y = 24 \end{cases}$$

Subtracting the second equation from the first,

$$x = 6.$$

If $x = 6$, then $y = 4$, and so 6 liters of the 10% solution must be mixed with 4 liters of the 15% solution. ●

A system of equations may represent any one of a variety of situations. One of the simplest is the intersection of two curves represented by a pair of equations. A geometric representation of the system of Example 1 is the pair of lines whose equations are $x + 2y = 5$ and $x - 4y = 2$. The graph is shown in Figure 8.3.

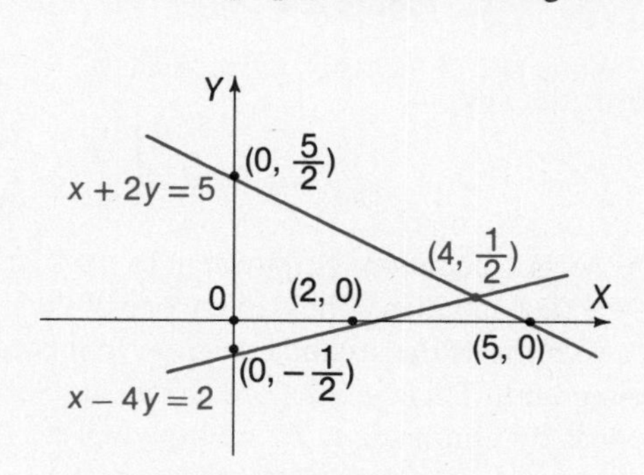

Figure 8.3

A different situation which leads to the same system of equations occurs if $y = -(1/2)x + 5/2$, $x > 0$, is a supply curve, and $y = (1/4)x - 1/2$, $x > 0$, is the corresponding demand curve. (See Section 3.2.) This pair of equations is equivalent to the pair $x + 2y = 5$ and $x - 4y = 2$ of the above example. The intersection of their graphs is the equilibrium point of supply and demand. Here supply equals demand when the price is $y = p(4) = 1/2$.

Corresponding to the fact that two lines may be parallel and so have no point of intersection, a system of two linear equations in two variables may have no solution.

EXAMPLE 4 Solve the system

$$\begin{cases} x - 2y = 6 \\ 2x - 4y = -4. \end{cases}$$

SOLUTION Multiply both sides of the first equation by -2 to obtain the following equivalent system.

$$\begin{cases} -2x + 4y = -12 \\ 2x - 4y = -4 \end{cases}$$

Adding these two equations gives $0 = -16$. Since this is impossible, the given system has no solution. Such a system is **inconsistent.** The graphical situation is shown in Figure 8.4. The graphs are parallel lines. ●

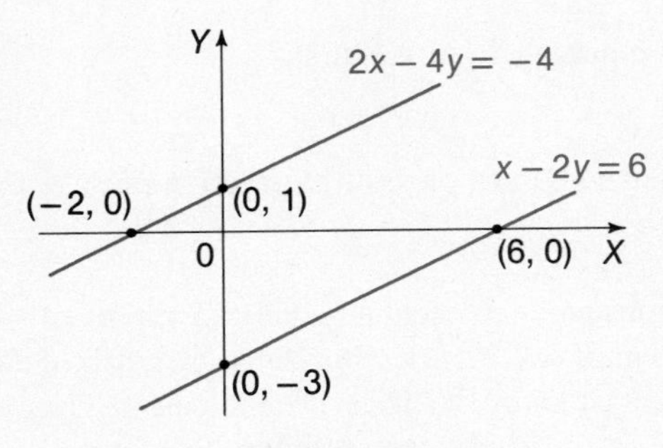

Figure 8.4

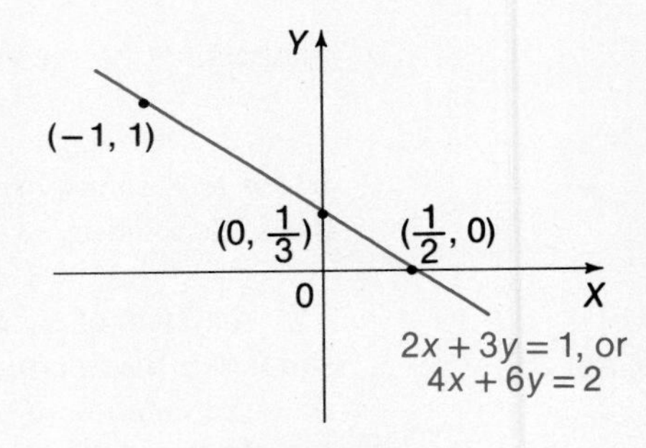

Figure 8.5

EXAMPLE 5 Solve the system

$$\begin{cases} 2x + 3y = 1 \\ 4x + 6y = 2. \end{cases}$$

SOLUTION The second equation is equivalent to the first one, since its coefficients are twice those of the first equation. Thus, every one of the infinite set of solutions of the first equation is a solution of the second equation. In this case we say the equations of the system are **dependent.** Their graphs are identical. See Figure 8.5. If we attempt to solve this system by elimination, by multiplying the first equation by 2, we get

$$\begin{cases} 4x + 6y = 2 \\ 4x + 6y = 2, \end{cases}$$

and subtraction gives $0 = 0$.

Since $2x + 3y = 1$, $y = (1 - 2x)/3$, and every solution is of the form $(x, (1 - 2x)/3)$. Taking $x = -1$, 0, and 1/2, respectively, gives us the solutions $(-1, 1)$, $(0, 1/3)$, and $(1/2, 0)$, as seen in the figure. ●

There is a simple algebraic test to determine whether a system is consistent or not. Consider the general system of two equations in two variables.

$$\begin{cases} a_1x + b_1y = c_1, & b_1 \neq 0 \\ a_2x + b_2y = c_2, & b_2 \neq 0 \end{cases}$$

To solve by elimination, first obtain an equivalent system.

$$\begin{cases} a_1b_2x + b_1b_2y = b_2c_1 & \text{multiplying by } b_2 \\ a_2b_1x + b_1b_2y = b_1c_2 & \text{multiplying by } b_1 \end{cases}$$

Now eliminate y by subtraction,

$$a_1b_2x - a_2b_1x = b_2c_1 - b_1c_2$$
$$x(a_1b_2 - a_2b_1) = b_2c_1 - b_1c_2$$
$$x = \frac{b_2c_1 - b_1c_2}{a_1b_2 - a_2b_1}.$$

There are three possibilities:

1. If the denominator $a_1b_2 - a_2b_1 \neq 0$, there is exactly one solution and the given system is consistent.
2. If the denominator $a_1b_2 - a_2b_1 = 0$ but the numerator $b_2c_1 - b_1c_2 \neq 0$, there is no solution and the system is inconsistent.
3. If both numerator and denominator are zero, then from $a_1b_2 - a_2b_1 = 0$ and $b_2c_1 - b_1c_2 = 0$,

$$\frac{a_1}{a_2} = \frac{b_1}{b_2} \quad \text{and} \quad \frac{b_1}{b_2} = \frac{c_1}{c_2}.$$

In this case, the coefficients of corresponding terms in the given equations are proportional. The two equations are equivalent and any solution of one is a solution of the other. Such a system is dependent and has an infinite number of solutions.

Each of the above algebraic possibilities has a graphical counterpart. To see this, write each of the given equations in the slope-intercept form (solve for y) to obtain an equivalent system.

$$\begin{cases} y = -\dfrac{a_1}{b_1}x + \dfrac{c_1}{b_1} \\[2ex] y = -\dfrac{a_2}{b_2}x + \dfrac{c_2}{b_2} \end{cases}$$

The slopes of the two lines are $m_1 = -a_1/b_1$ and $m_2 = -a_2/b_2$. In the first case above, the lines do not have the same slopes and so they intersect. Example 1 (Figure 8.3) illustrates this case. In the second case, the slopes are equal and the lines are parallel. (See Example 4 and Figure 8.4.) In the third case, the equations represent the same line, as in Example 5 and Figure 8.5.

In summary, there are three possibilities for a system of two equations in two variables:

(1) The system is consistent (has a unique solution) and is represented by two intersecting lines.

(2) The system is inconsistent (has no solution) and is represented by two parallel lines.

(3) The system is dependent (has an infinite number of solutions) and is represented by two coinciding lines (a single line).

Nonlinear Systems A **nonlinear system** is one in which at least one of the equations is not linear. Methods for solving such systems are illustrated in the following examples.

EXAMPLE 6 Find two numbers whose sum is 2, such that one number is the square of the other.

SOLUTION If we let x and y be the numbers, then

$$\begin{cases} x + y = 2 \\ \quad\;\; y = x^2 \end{cases}$$

is the resulting system. Substituting $y = x^2$ from the second equation into the first equation reduces the system to the single equation

$$x + x^2 = 2.$$

This may be solved by factoring, as follows.

$$x^2 + x - 2 = 0$$

$$(x + 2)(x - 1) = 0$$

$$x + 2 = 0 \quad \text{or} \quad x - 1 = 0$$

$$x = -2 \quad \text{or} \quad x = 1$$

Substituting each of these values for x in either of the original equations gives the corresponding values of y: if $x = -2$, $y = 4$, and if $x = 1$, $y = 1$. The solution consists of the two pairs $(-2, 4)$ and $(1, 1)$. Note that these results do actually satisfy the conditions of the given problem. For, in each case the sum of the two numbers is 2 and the second one is the square of the first one. The check here is essential because our procedure only shows that these results follow *if* there is a solution; it does not show that the results satisfy the problem statement. As shown in Figure 8.6, these solutions are the points of intersection of the line $x + y = 2$ and the parabola $y = x^2$. The procedure used in the solution of this system is called **solution by substitution.** ●

EXAMPLE 7 Find the points of intersection of the ellipse $2x^2 + y^2 = 8$ and the hyperbola $x^2 - y^2 = 1$.

SOLUTION We must find simultaneous solutions of the following system.

$$\begin{cases} 2x^2 + y^2 = 8 \\ x^2 - y^2 = 1 \end{cases}$$

Add the two equations.

$$3x^2 = 9$$
$$x^2 = 3$$
$$x = \pm\sqrt{3}$$

From the second equation, if x is either $+\sqrt{3}$ or $-\sqrt{3}$ then

$$y^2 = x^2 - 1$$
$$= 3 - 1 = 2,$$
$$y = \pm\sqrt{2}.$$

So there are four points of intersection, namely, $(\sqrt{3}, \sqrt{2})$, $(-\sqrt{3}, \sqrt{2})$, $(-\sqrt{3}, -\sqrt{2})$, and $(\sqrt{3}, -\sqrt{2})$, as seen in Figure 8.7. ●

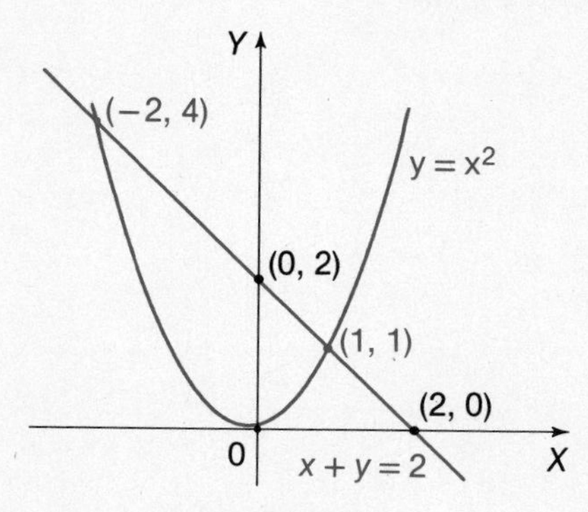

Figure 8.6

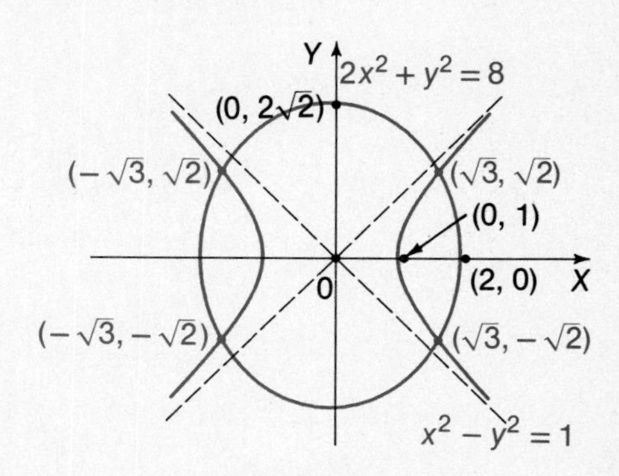

Figure 8.7

EXERCISES 8.1

Solve each of the following systems of linear equations and graph.

1. $\begin{cases} y = x \\ x + y = 4 \end{cases}$

2. $\begin{cases} x + y = 9 \\ x - y = 5 \end{cases}$

3. $\begin{cases} x - y = 7 \\ 3x + y = 9 \end{cases}$

4. $\begin{cases} x + 4y = 1 \\ x + 3y = 0 \end{cases}$

5. $\begin{cases} x + y + 3 = 0 \\ 2x - y + 3 = 0 \end{cases}$

6. $\begin{cases} 3x + 2y + 6 = 0 \\ 5x - 2y + 10 = 0 \end{cases}$

7. $\begin{cases} x - 3y = 4 \\ -2x + 6y = 3 \end{cases}$

8. $\begin{cases} -3x + 2y = 6 \\ 6x - 4y = 5 \end{cases}$

9. $\begin{cases} 2x - 3y = 5 \\ -6x + 9y = 8 \end{cases}$

10. $\begin{cases} 3x + y = -2 \\ 6x + 2y = 3 \end{cases}$

11. $\begin{cases} x - 3y = 1 \\ 2x - 6y = 2 \end{cases}$

12. $\begin{cases} -3x + 6y = 3 \\ x - 2y = -1 \end{cases}$

Solve the following systems of equations.

13. $\begin{cases} 3x + 4y = 6 \\ 6x - y = 1 \end{cases}$

14. $\begin{cases} 2x - 3y = 8 \\ 3x + 5y = 12 \end{cases}$

15. $\begin{cases} x + \dfrac{5}{2}y = 2 \\ \dfrac{1}{2}x + y = \dfrac{1}{2} \end{cases}$

16. $\begin{cases} \dfrac{x}{3} + \dfrac{y}{6} = \dfrac{1}{2} \\ \dfrac{x}{2} + \dfrac{3y}{10} = 1 \end{cases}$

17. $\begin{cases} 0.4x + 0.3y = 0.5 \\ 0.2x + 0.1y = 0.2 \end{cases}$
[*Hint:* First multiply each equation by 10 to remove the decimals.]

18. $\begin{cases} 0.03x - 0.02y + 0.13 = 0 \\ 0.02x + 0.05y = 0.04 \end{cases}$

The following pairs of equations represent either a line and a conic curve or two conic curves. Find the points of intersection and graph.

19. $\begin{cases} x + y = 7 \\ x^2 = 4y \end{cases}$

20. $\begin{cases} 4x - 3y = 0 \\ x^2 + y^2 = 25 \end{cases}$

21. $\begin{cases} x - y + 2 = 0 \\ y = x^2 - 6x + 8 \end{cases}$

22. $\begin{cases} 2x - y = 6 \\ x^2 + y^2 - 6x = 5 \end{cases}$

23. $\begin{cases} 5x - 3y + 2 = 0 \\ xy = 1 \end{cases}$

24. $\begin{cases} y = x^2 \\ x^2 = 2 - y \end{cases}$

25. $\begin{cases} x^2 + y = 19 \\ x^2 + y^2 = 25 \end{cases}$

26. $\begin{cases} x^2 = 2y + 10 \\ x^2 + y^2 = 10 \end{cases}$

27. $\begin{cases} 3x^2 + 4y^2 = 192 \\ 3x^2 - y^2 = 12 \end{cases}$

28. $\begin{cases} x^2 + y^2 = 4 \\ (x - 2)^2 + y^2 = 4 \end{cases}$

29. A child opened a piggy bank and found 34 coins, consisting of only dimes and quarters. If the total value was \$5.35, how many of each kind of coin was in the bank?

30. Two trains start at the same time and travel in opposite directions from the same station. Each goes at a constant speed but one travels 15 miles per hour faster than the other. At the end of 2 hours they are 210 miles apart. What is the speed of each train?

31. The total amount of a loan is \$1000. Part of this amount is loaned at 12% per year and the rest at 15% per year. If the total interest for one year is \$138, how much is loaned at each rate?

32. One hundred liters of a 20% acid solution is obtained by mixing some 10% solution with some 25% solution. How much of each is used?

33. Find two positive numbers which differ by 3 and whose product is 108.

34. The perimeter of a rectangle is 70 meters and its area is 304 square meters. What are its dimensions?

35. The area of a right triangle is 210 square meters and its hypotenuse is 37 meters long. How long are its legs?

36. Find two numbers whose sum is 21 and the sum of whose squares is 245.

The figure represents forces F_1 and F_2 in balance on a lever. The laws of physics require that $x_1F_1 = x_2F_2$, or $x_1F_1 - x_2F_2 = 0$. Find the position of the fulcrum for balance in the following cases.

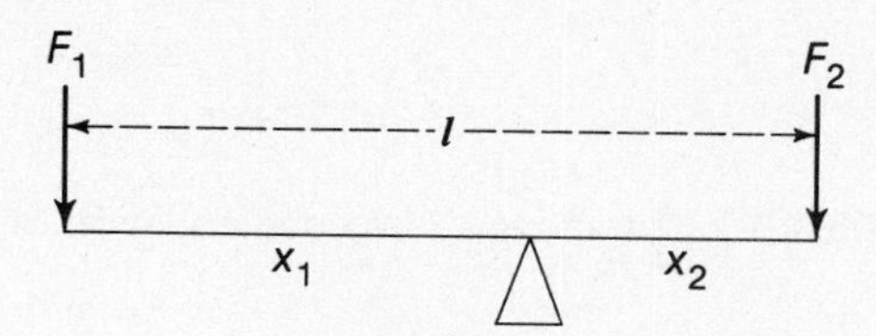

37. $F_1 = 8$, $F_2 = 6$, $l = 7$

38. $F_1 = 10$, $F_2 = 4$, $l = 8.75$

39. Prove algebraically that the circle $x^2 + y^2 = 1$ and the line $x + y = 3$ do not intersect. Illustrate graphically.

40. Prove algebraically that the ellipse $2x^2 + y^2 = 2$ and the line $x - y = 10$ do not intersect. Illustrate graphically.

8.2 Systems of Equations in Three Variables

The equation $3x - 2y + z = 2$ in three variables is satisfied by $x = 1, y = 2, z = 3$, by $x = -2, y = -4, z = 0$, and by infinitely many other **ordered triples** (x, y, z). To see this, simply substitute *any* two numbers for x and y and then solve the resulting single equation in one variable for z. For example, if $x = 2$ and $y = 3$,

$$3(2) - 2(3) + z = 2,$$
$$z = 2.$$

Thus, the triple (2, 3, 2) is also a solution.

In the last section simultaneous solutions to systems of two equations in two variables were found by eliminating one of the variables. This procedure is extended below to systems of equations in three variables. The idea is to select two pairs of equations and eliminate the same variable from each pair. The method is best explained by an example.

EXAMPLE 1 Solve the system

$$\begin{cases} x + y + z = 6 \\ x - 3y - 2z = -9 \\ 2x - 2y + 3z = 1. \end{cases}$$

SOLUTION We begin by pairing the first two equations and eliminating x from the second equation. Subtracting the first equation from the second one gives the equivalent system

$$\begin{cases} x + y + z = 6 \\ \quad - 4y - 3z = -15 \\ 2x - 2y + 3z = 1. \end{cases}$$

Now multiplying the first equation by -2 and adding it to the third equation gives

$$\begin{cases} x + y + z = 6 \\ \quad - 4y - 3z = -15 \\ \quad - 4y + z = -11. \end{cases}$$

Finally, subtracting the second equation from the third one gives

$$\begin{cases} x + y + z = 6 & \textbf{(1)} \\ \quad - 4y - 3z = -15 & \textbf{(2)} \\ \qquad 4z = 4. & \textbf{(3)} \end{cases}$$

From the last equation,

$$4z = 4,$$
$$z = 1.$$

From equation (2), substituting $z = 1$,

$$-4y - 3z = -15,$$
$$-4y - 3 = -15,$$
$$-4y = -12,$$
$$y = 3.$$

Finally, from equation (1), substituting $y = 3$, $z = 1$,

$$x + y + z = 6,$$
$$x + 3 + 1 = 6,$$
$$x = 2.$$

The solution is $x = 2$, $y = 3$, $z = 1$, or $(2, 3, 1)$. This solution should be checked by substitution into the given system of equations. ●

The sequence of steps here is important. The first equation was left intact and used to eliminate the variable x from the first and second equations. Then both the first and second equations were left unchanged while the second equation was used to eliminate the variable y from the third equation. The resulting equivalent system consists of a first equation in three variables, a second equation in two variables, and a third equation in one variable. Then, starting with the last equation, a succession of equations in one variable was solved for the respective variables. This systematic procedure is called **row reduction by elimination.** It is also called reduction to **triangular form.**

It is shown in 3-dimensional analytic geometry that a linear equation in three variables is represented graphically by a plane. The unique solution (2, 3, 1) of the above example then represents the single common point of intersection of the three planes corresponding to the three given equations, as illustrated in Figure 8.8a.

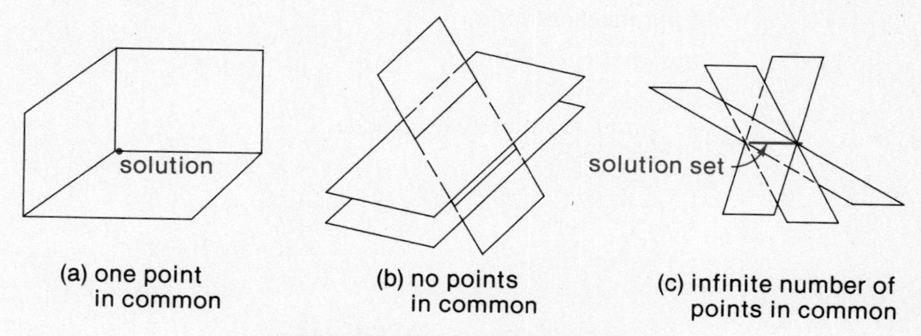

Figure 8.8. Intersection of planes

It may happen that there is no point common to all three planes. One such situation is shown in Figure 8.8b. It is also possible for three given planes to intersect along a common line, as in Figure 8.8c. The three geometric possibilities in the figure correspond to the following cases for solutions of a system of three linear equations in three variables:

(1) There is a unique solution (consistent system); the three planes intersect in one point.

(2) There is no solution (inconsistent system); the three planes do not all intersect.

(3) There are infinitely many solutions (a dependent system); the three planes intersect in one line.

EXAMPLE 2 Solve the system

$$\begin{cases} x + y + z = 1 \\ x - 2y + 2z = 4 \\ 2x - y + 3z = 5. \end{cases}$$

SOLUTION To eliminate x from the second equation, subtract the first equation from the second equation. The following system results.

$$\begin{cases} x + y + z = 1 & \text{keeping the first equation} \\ -3y + z = 3 & \text{replacing the second equation} \\ 2x - y + 3z = 5 & \text{by an equivalent equation} \end{cases}$$

Now we eliminate x from the last equation by multiplying the first equation by -2 and adding the result to the third equation.

$$\begin{cases} x + y + z = 1 \\ -3y + z = 3 \\ -3y + z = 3 \end{cases} \quad \text{new third equation}$$

The last two equations above are the same, so there is not a single solution. Solve one of the equations for z,

$$z = 3y + 3,$$

and substitute in the first equation.

$$\begin{aligned} x + y + z &= 1 \\ x + y + (3y + 3) &= 1 \\ x + 4y &= -2 \\ x &= -4y - 2 \end{aligned}$$

Both x and z are now expressed in terms of y.

$$\begin{aligned} x &= -4y - 2 \\ z &= 3y + 3 \end{aligned}$$

We can substitute *any* value of y in these equations to obtain corresponding values of x and z. Thus, there are an infinite number of solutions and the system is dependent. All the solutions are of the form $(-4y - 2, y, 3y + 3)$. If we let $y = 1$, for example, then $x = -6$ and $z = 6$, so that $(-6, 1, 6)$ is a solution. ●

In the above example, we eliminated x, but we could equally well eliminate any one of the variables first and express the solutions in terms of either of the two remaining variables.

EXAMPLE 3 Solve the system

$$\begin{cases} x + y + 2z = 1 \\ -x + 2y + z = 3 \\ y + z = 1. \end{cases}$$

SOLUTION Eliminate x from the second equation by adding the first equation to it. The resulting system is

$$\begin{cases} x + y + 2z = 1 \\ 3y + 3z = 4 \\ y + z = 1. \end{cases}$$

The last two equations are inconsistent, since the last one says $y + z = 1$ while the second one says that $y + z = 4/3$. If we tried to eliminate y from the third equation by subtracting $1/3$ times the second from it, we would get $0 = -1/3$, which is impossible. So the system has no solution (is inconsistent). ●

EXAMPLE 4 Determine the equation of the parabola having a vertical axis which passes through the three points (0, 3), (1, 1), and (2, 2).

SOLUTION The equation can be written in the form $y = ax^2 + bx + c$ (see Section 3.4). Substitution of the coordinates of each of the three points in turn into this equation gives the following equations.

$$\begin{cases} 3 = a\cdot 0 + b\cdot 0 + c & \text{letting } x = 0,\ y = 3 \\ 1 = a\cdot 1 + b\cdot 1 + c & \text{letting } x = 1,\ y = 1 \\ 2 = a\cdot 4 + b\cdot 2 + c & \text{letting } x = 2,\ y = 2 \end{cases}$$

This is a system of three equations in three variables. It may be written in the following more convenient form.

$$\begin{cases} c = 3 \\ a + b + c = 1 \\ 4a + 2b + c = 2 \end{cases}$$

Substituting from the first equation into the second and third equations gives

$$\begin{cases} a + b = -2 \\ 4a + 2b = -1. \end{cases}$$

The solution of this system (as you can show) is $a = 3/2$, $b = -7/2$, $c = 3$. The equation of the parabola is then

$$\begin{aligned} y &= ax^2 + bx + c \\ &= \frac{3}{2}x^2 - \frac{7}{2}x + 3, \qquad \text{or} \qquad 2y = 3x^2 - 7x + 6. \end{aligned}$$

We can complete the square to obtain

$$\left(x - \frac{7}{6}\right)^2 = \frac{2}{3}\left(y - \frac{23}{24}\right).$$

The graph is shown in Figure 8.9. Check that the three given pairs satisfy the equation we found, that is, all three points lie on the parabola. ●

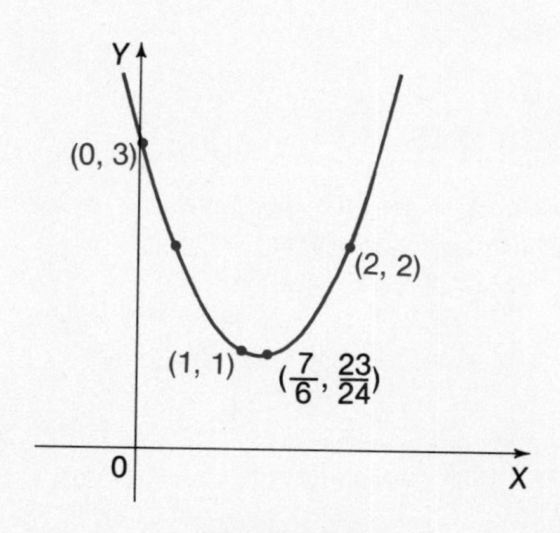

Figure 8.9

EXAMPLE 5 Write the rational expression

$$\frac{x^2 - 4x - 2}{x^3 - x^2 - 2x}$$

as a sum of partial fractions. (See Section 3.8.)

SOLUTION First factor the denominator.

$$\frac{x^2 - 4x - 2}{x^3 - x^2 - 2x} = \frac{x^2 - 4x - 2}{x(x^2 - x - 2)} = \frac{x^2 - 4x - 2}{x(x + 1)(x - 2)}$$

Each factor is the denominator of one of the partial fractions.

$$\frac{x^2 - 4x - 2}{x^3 - x^2 - 2x} = \frac{A}{x} + \frac{B}{x + 1} + \frac{C}{x - 2}$$

Multiply both sides by $x(x + 1)(x - 2)$.

$$\begin{aligned} x^2 - 4x - 2 &= A(x + 1)(x - 2) + Bx(x - 2) + Cx(x + 1) \\ &= A(x^2 - x - 2) + B(x^2 - 2x) + C(x^2 + x) \\ &= x^2(A + B + C) + x(-A - 2B + C) - 2A \end{aligned}$$

For this equation to hold, coefficients of corresponding powers of x must be equal, that is, we must have the following system of equations.

$$\begin{cases} A + B + C = 1 \\ -A - 2B + C = -4 \\ -2A = -2 \end{cases}$$

The solution is $A = 1$, $B = 1$, and $C = -1$ (check this), and thus

$$\begin{aligned} \frac{x^2 - 4x - 2}{x^3 - x^2 - 2x} &= \frac{A}{x} + \frac{B}{x + 1} + \frac{C}{x - 2} \\ &= \frac{1}{x} + \frac{1}{x + 1} - \frac{1}{x - 2}. \end{aligned}$$ ●

EXERCISES 8.2

Solve each of the following systems of equations by elimination. Identify each as consistent, dependent, or inconsistent.

1. $\begin{cases} -x + 2y + 2z = 2 \\ x + 4y - z = -3 \\ -3x + 2y + 2z = 0 \end{cases}$

2. $\begin{cases} x + 2y - 7z = 1 \\ 2x + y + z = 0 \\ 6x - 2y + 3z = 1 \end{cases}$

3. $\begin{cases} x + 7y - 3z = -14 \\ 2x - 3y + 2z = 3 \\ 4x + 8y + z = 2 \end{cases}$

4. $\begin{cases} x - 3y - 4z = 5 \\ 3x + y + 4z = 0 \\ 5x + y + 3z = 1 \end{cases}$

5. $\begin{cases} x + y + z = 1 \\ 2y - z = 0 \\ x - 2y = 0 \end{cases}$

6. $\begin{cases} 6x - 12y = 0 \\ 8y + z = 0 \\ 9x - z = 12 \end{cases}$

7. $\begin{cases} x + 2y + 3z = 8 \\ 3x - y + 2z = -5 \\ -2x - 4y - 6z = 5 \end{cases}$

8. $\begin{cases} x + y + z = 1 \\ x - 2y + 2z = 4 \\ 2x - y + 3z = 5 \end{cases}$

9. $\begin{cases} x + y - 5z = 0 \\ x - y + 3z = 4 \\ 2x - y + z = 5 \end{cases}$

10. $\begin{cases} x + y + 2z = 1 \\ -x + 2y + z = 3 \\ y + z = 1 \end{cases}$

11. $\begin{cases} x - y + 4z = 0 \\ 2x + y - z = 0 \\ x + y - 2z = 0 \end{cases}$

12. $\begin{cases} x + 3y + z = 0 \\ x + y - z = 0 \\ x - 2y - 4z = 0 \end{cases}$

13. Determine the three numbers whose sum is 3, such that three times the first plus the sum of the other two is 7, and twice the sum of the first two numbers plus the third number is -6.

14. A collection of nickels, dimes, and quarters has the value \$3.40. The total number of coins is 32, and there are twice as many dimes as quarters. How many of each kind are there? [*Hint:* Let x = the number of nickels, and so on.]

Find the equation of the parabola with vertical axis and passing through the given points. [*Hint:* See Example 4.]

15. (0, 0), (1, 0), and (3, 6)

16. $(-2, 0)$, $(0, -3)$, and (1, 0)

Find the equation of the parabola with horizontal axis and passing through the given points. [*Hint:* Begin with $x = Cy^2 + Ey + F$, and use the procedure of Example 4.]

17. (0, 3), (1, 1), (2, 2)

18. (2, 0), $(1, -1)$, $(-1, 2)$

Find the equation of the circle determined by each of the following sets of three points. [*Hint:* Begin with the general equation $x^2 + y^2 + Dx + Ey + F = 0$ and use the same procedure as in Example 4.]

19. (0, 1), $(-1, -1)$, (2, 0)

20. (1, 3), (1, 1), $(-2, 3)$

21. (1, 6), (2, 5), $(-6, -1)$

22. (4, 6), $(-2, -2)$, $(-4, 2)$

Decompose each of the following rational fractions into partial fractions. [*Hint:* See Example 5.]

23. $\dfrac{1}{(x - 1)(x - 2)(x + 1)}$

24. $\dfrac{x^2 - 3x + 7}{(x - 1)(x - 2)(x - 3)}$

25. $\dfrac{2x + 1}{x^3 - x}$

26. $\dfrac{1}{16x - x^3}$

27. $\dfrac{x^2 + 1}{x^3 - 4x}$

28. $\dfrac{-7x - 2}{x^3 + x - 2x}$

29. The figure shows an electrical circuit with two power sources of voltages E_1 and E_2 (volts) and three resistances R_1, R_2, and R_3 (ohms). The laws for such electrical circuits require that the current values i_1, i_2, and i_3 satisfy the following set of equations. Solve this system for the current values (amperes).

$$\begin{cases} i_1 - i_2 - i_3 = 0 \\ 5i_1 + 5i_3 = 10 \\ 10i_2 - 5i_3 = 20 \end{cases}$$

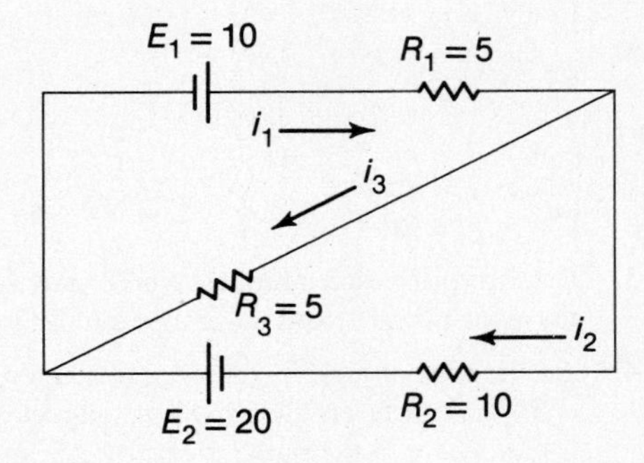

30. The figure shows another electrical circuit with the values of the voltages and resistances indicated. The current values must satisfy the following system of equations. Calculate them.

$$\begin{cases} i_1 + i_2 + i_3 = 0 \\ 2i_1 - 5i_2 = 3 \\ 9i_2 - 5i_3 = 10 \end{cases}$$

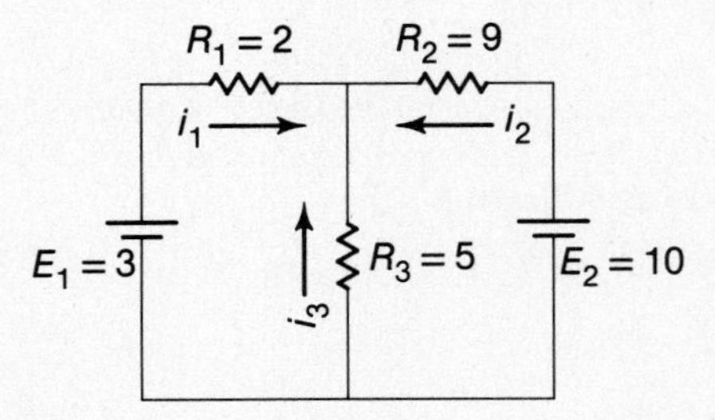

31. Solve the system in Example 2 by first eliminating (a) y; (b) z. Note that this does not change the set of solutions.

8.3 Matrices and Solutions of Systems

Some of the linear systems in the preceding section were solved by the method of row reduction by elimination. The method is tedious, particularly if there are more than three variables. It is possible to simplify this method by noting that we worked only with the coefficients and the constant terms in the equations. The variables essentially served only to keep the coefficients in their proper places. In this section we will develop a way to work with the coefficients alone. From this we will get a shortcut procedure that can be applied to systems of any number of equations in any number of variables.

To illustrate, let us go back to the system of Example 1 in the preceding section.

$$\begin{cases} x + y + z = 6 \\ x - 3y - 2z = -9 \\ 2x - 2y + 3z = 1 \end{cases}$$

Two rectangular arrays may be written from the numbers of the system. The first array consists of the coefficients of the variables. The second array also contains the constant terms, separated from the coefficients by a vertical line.

$$\begin{bmatrix} 1 & 1 & 1 \\ 1 & -3 & -2 \\ 2 & -2 & 3 \end{bmatrix} \qquad \left[\begin{array}{ccc|c} 1 & 1 & 1 & 6 \\ 1 & -3 & -2 & -9 \\ 2 & -2 & 3 & 1 \end{array}\right]$$

A rectangular array of numbers such as these is called a **matrix** (plural, **matrices**). The one on the left above is called the **matrix of the coefficients** of the system, and the one on the right is the **augmented matrix** of the system.

The row reduction method is shown step by step below, with the corresponding augmented matrix alongside.

$$\begin{cases} x + y + z = 6 \\ x - 3y - 2z = -9 \\ 2x - 2y + 3z = 1 \end{cases} \qquad \left[\begin{array}{ccc|c} 1 & 1 & 1 & 6 \\ 1 & -3 & -2 & -9 \\ 2 & -2 & 3 & 1 \end{array}\right]$$

First, subtract each element of the first row from the corresponding element of the second row.

$$\begin{cases} x + y + z = 6 \\ \quad - 4y - 3z = -15 \\ 2x - 2y + 3z = 1 \end{cases} \qquad \left[\begin{array}{ccc|c} 1 & 1 & 1 & 6 \\ 0 & -4 & -3 & -15 \\ 2 & -2 & 3 & 1 \end{array}\right]$$

Next, subtract twice the elements of the first row from the corresponding elements of the third row.

$$\begin{cases} x + y + z = 6 \\ \quad - 4y - 3z = -15 \\ \quad - 4y + z = -11 \end{cases} \qquad \left[\begin{array}{ccc|c} 1 & 1 & 1 & 6 \\ 0 & -4 & -3 & -15 \\ 0 & -4 & 1 & -11 \end{array}\right]$$

Finally, subtract the elements of the second row from the corresponding elements of the third row.

$$\begin{cases} x + y + z = 6 \\ \quad - 4y - 3z = -15 \\ \qquad 4z = 4 \end{cases} \qquad \left[\begin{array}{ccc|c} 1 & 1 & 1 & 6 \\ 0 & -4 & -3 & -15 \\ 0 & 0 & 4 & 4 \end{array}\right]$$

The above steps show that we could have worked with the *matrix alone*. After the final reduction the last system of equations could be written from the final matrix. The rest of the solution process goes exactly as in the last section.

The final matrix above has 0's below the diagonal line of coefficients beginning in the upper left-hand corner of the matrix. This is called the **echelon form** of the matrix. We will now use it to repeat the solutions of the systems of Examples 2 and 3 in Section 8.2.

EXAMPLE 1 Solve the system of Example 2, Section 8.2, by matrices.

SOLUTION There we started with the following system.

$$\begin{cases} x + y + z = 1 \\ x - 2y + 2z = 4 \\ 2x - y + 3z = 5 \end{cases}$$

The corresponding augmented matrix is shown below.

$$\left[\begin{array}{ccc|c} 1 & 1 & 1 & 1 \\ 1 & -2 & 2 & 4 \\ 2 & -1 & 3 & 5 \end{array}\right]$$

Subtracting the first row from the second row and twice the first row from the third row results in the following matrix.

$$\left[\begin{array}{ccc|c} 1 & 1 & 1 & 1 \\ 0 & -3 & 1 & 3 \\ 0 & -3 & 1 & 3 \end{array}\right]$$

Then subtract the second row of this matrix from the third row.

$$\left[\begin{array}{ccc|c} 1 & 1 & 1 & 1 \\ 0 & -3 & 1 & 3 \\ 0 & 0 & 0 & 0 \end{array}\right]$$

From this we have the following system of equations, equivalent to the original given system.

$$\begin{cases} x + y + z = 1 \\ \quad - 3y + z = 3 \end{cases}$$

The solution is then completed in the same way as before, by solving for x and z in terms of y. ●

EXAMPLE 2 Solve the system of Example 3, Section 8.2, by matrices.

SOLUTION There we started with the following system.

$$\begin{cases} x + y + 2z = 1 \\ -x + 2y + z = 3 \\ \quad y + z = 1 \end{cases}$$

Its augmented matrix is shown below.

$$\begin{bmatrix} 1 & 1 & 2 & 1 \\ -1 & 2 & 1 & 3 \\ 0 & 1 & 1 & 1 \end{bmatrix}$$

Then, replace the second row by the result of adding the first row to the second row.

$$\left[\begin{array}{ccc|c} 1 & 1 & 2 & 1 \\ 0 & 3 & 3 & 4 \\ 0 & 1 & 1 & 1 \end{array}\right]$$

Next, subtract one-third the second row from the third row.

$$\left[\begin{array}{ccc|c} 1 & 1 & 2 & 1 \\ 0 & 3 & 3 & 4 \\ 0 & 0 & 0 & -\frac{1}{3} \end{array}\right]$$

The equation corresponding to the last row is $0 \cdot x + 0 \cdot y + 0 \cdot z = -\frac{1}{3}$, which has no solution. Thus, the system is inconsistent, as previously stated. ●

The row reduction of the matrix of coefficients to echelon form can be applied to any number of equations in any number of variables. The following example illustrates this for a system of three equations in four variables.

EXAMPLE 3 Solve the following system.

$$\begin{cases} x + y + z = 2 \\ 2x - z + w = 3 \\ 3x + 2y + w = 4 \end{cases}$$

SOLUTION The following is the augmented matrix of the system.

$$\left[\begin{array}{cccc|c} 1 & 1 & 1 & 0 & 2 \\ 2 & 0 & -1 & 1 & 3 \\ 3 & 2 & 0 & 1 & 4 \end{array}\right]$$

Subtracting twice the first row from the second row and three times the first row from the third row reduces the matrix as shown below.

$$\left[\begin{array}{cccc|c} 1 & 1 & 1 & 0 & 2 \\ 0 & -2 & -3 & 1 & -1 \\ 0 & -1 & -3 & 1 & -2 \end{array}\right]$$

Since interchanging any two of the equations in a system does not affect its solution, for convenience we interchange the second and third rows in this matrix.

$$\left[\begin{array}{rrrr|r} 1 & 1 & 1 & 0 & 2 \\ 0 & -1 & -3 & 1 & -2 \\ 0 & -2 & -3 & 1 & -1 \end{array}\right]$$

Now subtract twice the second row from the third row.

$$\left[\begin{array}{rrrr|r} 1 & 1 & 1 & 0 & 2 \\ 0 & -1 & -3 & 1 & -2 \\ 0 & 0 & 3 & -1 & 3 \end{array}\right]$$

The following is the corresponding system of equations equivalent to the original set.

$$\begin{cases} x + y + z = 2 \\ -y - 3z + w = -2 \\ 3z - w = 3 \end{cases}$$

The third equation may be solved for z.

$$3z - w = 3$$
$$z = \frac{1}{3}(w + 3)$$

Solve the second equation for y and then substitute the above expression for z.

$$-y - 3z + w = -2$$
$$y = -3z + w + 2$$
$$= -(w + 3) + w + 2$$
$$y = -1$$

Now use the the first equation to find x.

$$x + y + z = 2$$
$$x = -y - z + 2$$
$$= 1 - \frac{1}{3}(w + 3) + 2$$
$$x = -\frac{1}{3}w + 2$$

The solution is $(-w/3 + 2, -1, w/3 + 1, w)$. Here the solution is expressible in terms of w. Since values may be assigned arbitrarily to w, there are infinitely many solutions. The given system of equations is dependent. Check the above solution in the original system. ●

The dependency of the above system is a consequence of the fact that there are fewer equations than variables. If we were to solve the following system in which a fourth equation has been added to the above system,

$$\begin{cases} x + y + z = 2 \\ 2x - z + w = 3 \\ 3x + 2y + w = 4 \\ 3y + 2z + 2w = 7, \end{cases}$$

we would get the solution $(1, -1, 2, 3)$. Thus, this system of four equations in four variables is consistent.

Homogeneous Systems A system of linear equations in which all the constant terms are 0 is called a **homogeneous system.** It is clear that $(0, 0, 0, \ldots)$ is a solution of a homogeneous system. It is called the **trivial solution.** A homogeneous system may also have nontrivial solutions, as in the following example.

EXAMPLE 4 Solve the homogeneous system

$$\begin{cases} x - y + 2z = 0 \\ 3x + y - z = 0 \\ 2x + 2y - 3z = 0. \end{cases}$$

SOLUTION The matrix of the system is shown below.

$$\left[\begin{array}{rrr|r} 1 & -1 & 2 & 0 \\ 3 & 1 & -1 & 0 \\ 2 & 2 & -3 & 0 \end{array}\right]$$

Multiply the first row by -3 and add it to the second.

$$\left[\begin{array}{rrr|r} 1 & -1 & 2 & 0 \\ 0 & 4 & -7 & 0 \\ 2 & 2 & -3 & 0 \end{array}\right]$$

Now, multiply the first row by -2 and add it to the third row.

$$\left[\begin{array}{rrr|r} 1 & -1 & 2 & 0 \\ 0 & 4 & -7 & 0 \\ 0 & 4 & -7 & 0 \end{array}\right]$$

Finally, subtract the second row from the third row.

$$\left[\begin{array}{rrr|r} 1 & -1 & 2 & 0 \\ 0 & 4 & -7 & 0 \\ 0 & 0 & 0 & 0 \end{array}\right]$$

The corresponding system of equations is shown below.

$$\begin{cases} x - y + 2z = 0 \\ 4y - 7z = 0 \end{cases}$$

The following is obtained from the second equation.

$$4y = 7z$$

$$y = \frac{7}{4}z$$

Now substitute into the first equation.

$$\begin{aligned} x - y + 2z &= 0 \\ x &= y - 2z \\ &= \frac{7}{4}z - 2z \\ x &= -\frac{1}{4}z \end{aligned}$$

The solution is the set of triples $(-z/4, 7z/4, z)$ or more simply $(-z, 7z, 4z)$, which also satisfies the system. Any value may be assigned to z and so there is an infinite number of solutions, in fact, as many solutions as there are real numbers. ●

By a careful study of the echelon form of the matrix of a system of linear equations it is possible to determine whether the system is consistent, dependent, or inconsistent, without actually carrying through the solution.

This is illustrated as follows with the examples worked out at the beginning of this section. Three systems and the *echelon* form of the corresponding matrices are shown.

$$(1)\ \begin{cases} x + y + z = 6 \\ x - 3y - 2z = -9 \\ 2x - 2y + 3z = 1 \end{cases} \qquad \left[\begin{array}{ccc|c} 1 & 1 & 1 & 6 \\ 0 & -4 & -3 & -15 \\ 0 & 0 & 4 & 4 \end{array}\right]$$

$$(2)\ \begin{cases} x + y + z = 1 \\ x - 2y + 2z = 4 \\ 2x - y + 3z = 5 \end{cases} \qquad \left[\begin{array}{ccc|c} 1 & 1 & 1 & 1 \\ 0 & -3 & 1 & 3 \\ 0 & 0 & 0 & 0 \end{array}\right]$$

$$(3)\ \begin{cases} x + 2y + 2z = 1 \\ -x + 2y + z = 3 \\ y + z = 1 \end{cases} \qquad \left[\begin{array}{ccc|c} 1 & 1 & 2 & 1 \\ 0 & 3 & 3 & 4 \\ 0 & 0 & 0 & -\frac{1}{3} \end{array}\right]$$

The first system is consistent, the second dependent, and the third inconsistent. The dependency in (2) results from the fact that the echelon form has all 0's in the last row, so that there are fewer equations (here, two) than in the original system (here, three). In (3) the augmented matrix has a nonzero element only in the last position in the last row, corresponding to a reduced equation of the form $0 \cdot z = a$. Since this is impossible, the system is inconsistent.

Thus we see that a system is consistent if the last row of the echelon form of its augmented matrix has nonzero elements in the last two columns, dependent if the last row has all zeros, or inconsistent if the last row has a nonzero element only in the last column.

EXERCISES 8.3

Reduce each of the following matrices to echelon form.

1. $\begin{bmatrix} 1 & -2 & 0 \\ 2 & 3 & -1 \\ -4 & 1 & 2 \end{bmatrix}$

2. $\begin{bmatrix} 1 & -1 & 2 \\ 3 & 1 & -1 \\ 2 & 2 & -3 \end{bmatrix}$

3. $\begin{bmatrix} -1 & 3 & -1 \\ 2 & 2 & 0 \\ -6 & 2 & 2 \end{bmatrix}$

4. $\begin{bmatrix} 2 & -1 & 1 \\ -2 & 1 & -2 \\ 4 & 1 & -2 \end{bmatrix}$

5. $\begin{bmatrix} 1 & 1 & 1 & 1 \\ 1 & -1 & 1 & 1 \\ 1 & -1 & -1 & 1 \\ 1 & -1 & -1 & -1 \end{bmatrix}$

6. $\begin{bmatrix} 1 & 1 & -1 & 1 & 0 \\ 0 & 0 & 0 & 1 & 1 \\ 2 & 2 & -1 & 0 & 1 \\ 1 & 1 & -2 & 0 & -1 \\ -1 & -1 & 2 & -3 & 1 \end{bmatrix}$

7. $\begin{bmatrix} 2 & -1 & 1 \\ -4 & 2 & 2 \end{bmatrix}$

8. $\begin{bmatrix} 4 & -2 & 2 & 1 \\ 2 & -1 & -3 & 1 \end{bmatrix}$

9. $\begin{bmatrix} 1 & -1 & 1 \\ 1 & 1 & 1 \\ -1 & 0 & 1 \\ 1 & 2 & 1 \end{bmatrix}$

10. $\begin{bmatrix} 2 & -1 \\ 1 & -1 \\ 1 & 4 \end{bmatrix}$

11. $\begin{bmatrix} 2 & -1 \\ 1 & 4 \\ 0 & 0 \\ 3 & -3 \end{bmatrix}$

12. $\begin{bmatrix} 1 & 1 & 2 & 0 \\ 1 & 2 & 3 & 1 \\ 2 & 1 & 3 & 4 \end{bmatrix}$

Use matrices to solve the following systems of equations.

13. $\begin{cases} x - y + 7z = 1 \\ 2x - y + 8z = 2 \\ 3x + y - z = 0 \end{cases}$

14. $\begin{cases} x + y + 2z = 5 \\ 2x + y + z = 5 \\ x + 3y + 2z = 6 \end{cases}$

15. $\begin{cases} x + y + 2z = 1 \\ 2x + 3y - 4z = 2 \\ 5x + 6y + 2z = 5 \end{cases}$

16. $\begin{cases} x + y + 3z = 3 \\ x - y + 3z = 2 \\ 2x + 3y + 4z = 8 \end{cases}$

17. $\begin{cases} x + y + 2z = 3 \\ 2x - y + 3z = 3 \\ x + 3y - 4z = 5 \end{cases}$

18. $\begin{cases} x + 2y + 3z = 0 \\ x - y + z = 1 \\ 4x + 5y + 10z = 2 \end{cases}$

19. $\begin{cases} x + y + z + w = 4 \\ x - y + z - w = 0 \\ 2x + y - z + 2w = 3 \\ x - y + 2z - w = 1 \end{cases}$

20. $\begin{cases} x + y + z = 2 \\ 2x - z + w = 3 \\ 3x + 2y + w = 4 \\ 3y + 2z + 2w = 7 \end{cases}$

21. $\begin{cases} x + y + z = 5 \\ 2x - y + 3z = 4 \end{cases}$

22. $\begin{cases} x + y = 5 \\ 2x - y = 7 \\ x + 2y = 3 \end{cases}$

23. $\begin{cases} 2x + y + w = 2 \\ 3x + 3y + 3z + 5w = 4 \\ 3x - 3z - 2w = 3 \end{cases}$

24. $\begin{cases} x + 2y - z = -3 \\ 2x + y + 3z = 7 \\ -3x - 2y + z = 1 \\ 4x + 2y - 3z = -4 \end{cases}$

25. A grocer wishes to mix three kinds of miniature candy bars to obtain 20 pounds of a mixture to sell at \$2.69 per pound. The prices of the candy bars separately are \$2.60, \$2.70, and \$2.90 per pound. If there is to be as much of the least expensive kind as of the other two kinds combined, how many pounds of each kind should be used?

26. It is desired to make a 20% alcohol solution using a mixture of 10% solution, 15% solution, and 25% solution. If there is to be three times as much of the 10% solution as of the 15% solution, how much of each must be used to have 10 liters of the 20% solution?

27. A collection of nickels, dimes, and quarters has the value \$2.60. There are twice as many nickels as quarters and 26 coins in all. How many of each kind are there?

Given the three points $(-3, 2)$, $(1, 1)$, *and* $(-1, -1)$, *find the equation of the curve indicated below through these points.*

28. A circle

29. A parabola with vertical axis

30. A parabola with horizontal axis

31. Is it possible to find three numbers a, b, and c such that $x^3 = ax^2 + bx + c$ for all values of x? Prove that the answer is "No" by showing that it is generally not true for even a set of four arbitrarily chosen values of x. [*Suggestion:* Choose $x = 0$, 1, 2, and 3 and substitute in the above equation to obtain four equations in the three unknowns a, b, and c. Then show that this system has no solution by showing that a solution of the first three equations fails to satisfy the fourth.]

8.4 Determinants and Cramer's Rule

Matrices arise quite naturally in connection with solutions of systems of linear equations. There are many other applications of matrices which will not be considered here. However, **square matrices** (those with the same number of rows and columns) are especially important. Associated with each square matrix there is a number called the **determinant of the matrix.**

DEFINITION 8.3 The **determinant D of the matrix** $\begin{bmatrix} a & b \\ c & d \end{bmatrix}$ is defined as

$$\mathbf{D} = \begin{vmatrix} \mathbf{a} & \mathbf{b} \\ \mathbf{c} & \mathbf{d} \end{vmatrix} = \mathbf{ad - bc}.$$

(The vertical bars used to denote a determinant do not refer to absolute value.)

EXAMPLE 1 Evaluate the determinants of the following matrices. (a) $\begin{bmatrix} 1 & -2 \\ 5 & -3 \end{bmatrix}$; (b) $\begin{bmatrix} 2 & -3 \\ -4 & 6 \end{bmatrix}$

SOLUTION (a) $\begin{vmatrix} 1 & -2 \\ 5 & -3 \end{vmatrix} = 1(-3) - 5(-2) = 7$

(b) $\begin{vmatrix} 2 & -3 \\ -4 & 6 \end{vmatrix} = 2(6) - (-4)(-3) = 0$ •

Note that in Example (1b) the rows are proportional (that is, $2/-4 = -3/6$). In this case, the determinant is always equal to 0. This is shown as follows.

$$\begin{vmatrix} a & b \\ ka & kb \end{vmatrix} = a(kb) - (ka)b = 0$$

The number of rows (or columns) of a square matrix is called its **order.** The above are determinants of second order matrices. To deal systematically with matrices of any order and their determinants a standard notation is needed. For example, general matrices of order two, three, and four are often written as follows.

$$\begin{bmatrix} a_1 & b_1 \\ a_2 & b_2 \end{bmatrix} \qquad \begin{bmatrix} a_1 & b_1 & c_1 \\ a_2 & b_2 & c_2 \\ a_3 & b_3 & c_3 \end{bmatrix} \qquad \begin{bmatrix} a_1 & b_1 & c_1 & d_1 \\ a_2 & b_2 & c_2 & d_2 \\ a_3 & b_3 & c_3 & d_3 \\ a_4 & b_4 & c_4 & d_4 \end{bmatrix}$$

The determinant of a matrix of any given order is defined in terms of determinants of matrices of the next lower order. For example, the determinant of a third-order matrix is defined as follows.

DEFINITION 8.4

$$\begin{vmatrix} a_1 & b_1 & c_1 \\ a_2 & b_2 & c_2 \\ a_3 & b_3 & c_3 \end{vmatrix} = a_1 \begin{vmatrix} b_2 & c_2 \\ b_3 & c_3 \end{vmatrix} - a_2 \begin{vmatrix} b_1 & c_1 \\ b_3 & c_3 \end{vmatrix} + a_3 \begin{vmatrix} b_1 & c_1 \\ b_2 & c_2 \end{vmatrix}$$

$$= a_1(b_2c_3 - b_3c_2) - a_2(b_1c_3 - b_3c_1) + a_3(b_1c_2 - b_2c_1)$$

$$= a_1b_2c_3 - a_1b_3c_2 - a_2b_1c_3 + a_2b_3c_1 + a_3b_1c_2 - a_3b_2c_1.$$

Instead of trying to remember this complicated formula, we can use a very simple pattern for this definition. Below, an element in the first column has been circled and the rest of the row and column in which it appears has been crossed out.

$$\begin{bmatrix} \circled{a_1} & \not{b_1} & \not{c_1} \\ \not{a_2} & b_2 & c_2 \\ \not{a_3} & b_3 & c_3 \end{bmatrix} \qquad \begin{bmatrix} \not{a_1} & b_1 & c_1 \\ \circled{a_2} & \not{b_2} & \not{c_2} \\ \not{a_3} & b_3 & c_3 \end{bmatrix} \qquad \begin{bmatrix} \not{a_1} & b_1 & c_1 \\ \not{a_2} & b_2 & c_2 \\ \circled{a_3} & \not{b_3} & \not{c_3} \end{bmatrix}$$

The determinant of the 2×2-array remaining in each case is called the **minor** of the corresponding circled element. For example, the minor of a_2 is the determinant

$$\begin{vmatrix} b_1 & c_1 \\ b_3 & c_3 \end{vmatrix}.$$

Now notice that the determinant of a third-order matrix is defined as the sum of the products of the elements of the first column (multiplied alternately by 1 and -1)

with their minors. We call this the **expansion of the determinant by minors** of the elements of the first column. (For simplicity, we will sometimes refer to the "rows" or "columns" of a determinant, rather than to the rows of the corresponding matrix.)

EXAMPLE 2 Evaluate the following determinant by expanding by minors of the first column.

SOLUTION

$$\begin{vmatrix} -1 & -2 & -3 \\ 2 & 4 & 1 \\ -1 & 2 & 0 \end{vmatrix} = -1\begin{vmatrix} 4 & 1 \\ 2 & 0 \end{vmatrix} - 2\begin{vmatrix} -2 & -3 \\ 2 & 0 \end{vmatrix} + (-1)\begin{vmatrix} -2 & -3 \\ 4 & 1 \end{vmatrix}$$

$$= -1(0 - 2) - 2(0 + 6) - 1(-2 + 12)$$

$$= 2 - 12 - 10 = -20. \quad \bullet$$

The same result would be obtained by expanding by minors of any row or column, provided appropriate signs are used for the "circled" elements. The choice of signs is given by the following array.

$$\begin{matrix} + & - & + \\ - & + & - \\ + & - & + \end{matrix}$$

The rule is: Start with + in the upper left-hand column and alternate − and + along any row or column in the array.

EXAMPLE 3 Expand the determinant of Example 2 by minors of (a) the second row; (b) the third column.

SOLUTION (a)

$$\begin{vmatrix} -1 & -2 & -3 \\ 2 & 4 & 1 \\ -1 & 2 & 0 \end{vmatrix} = -2\begin{vmatrix} -2 & -3 \\ 2 & 0 \end{vmatrix} + 4\begin{vmatrix} -1 & -3 \\ -1 & 0 \end{vmatrix} - 1\begin{vmatrix} -1 & -2 \\ -1 & 2 \end{vmatrix}$$

$$= -2(0 + 6) + 4(0 - 3) - (-2 - 2)$$

$$= -12 - 12 + 4 = -20$$

This is the result obtained in Example 2, as expected.

(b)

$$\begin{vmatrix} -1 & -2 & -3 \\ 2 & 4 & 1 \\ -1 & 2 & 0 \end{vmatrix} = -3\begin{vmatrix} 2 & 4 \\ -1 & 2 \end{vmatrix} - 1\begin{vmatrix} -1 & -2 \\ -1 & 2 \end{vmatrix} + 0\begin{vmatrix} -1 & -2 \\ 2 & 4 \end{vmatrix}$$

$$= -3(4 + 4) - 1(-2 - 2) + 0(-4 + 4)$$

$$= -24 + 4 + 0 = -20 \quad \bullet$$

A similar method can be used to evaluate the determinant of any n by n matrix by minors of any row or column, and an extension of the same pattern of signs applies.

Matrices may also be written with a double-subscript notation. For example, a third-order matrix may be written as

$$\begin{bmatrix} a_{11} & a_{12} & a_{13} \\ a_{21} & a_{22} & a_{23} \\ a_{31} & a_{32} & a_{33} \end{bmatrix}.$$

The elements are read as "a-one-one, a-one-two, a-two-three," and so on. The first subscript denotes the row and the second subscript the column of the element. The number in row i and column j is a_{ij}, and its minor is denoted by A_{ij}. When this notation is used, the expansion of the above determinant is written compactly as

$$+a_{11}A_{11} - a_{12}A_{12} + a_{13}A_{13}.$$

Note that an element is multiplied by $+1$ if the sum of its subscripts is even and by -1 if the sum of its subscripts is odd.

This notation may be used to write the determinant of the general nth order matrix and its expansion by elements of the first row as follows.

$$\begin{vmatrix} a_{11} & a_{12} & a_{13} & a_{14} & \cdots & a_{1n} \\ a_{21} & a_{22} & a_{23} & a_{24} & \cdots & a_{2n} \\ a_{31} & a_{32} & a_{33} & a_{34} & \cdots & a_{3n} \\ \vdots & \vdots & \vdots & \vdots & & \vdots \\ a_{n1} & a_{n2} & a_{n3} & a_{n4} & \cdots & a_{nn} \end{vmatrix}$$

$$= a_{11}A_{11} - a_{12}A_{12} + a_{13}A_{13} - a_{14}A_{14} + \cdots \pm a_{1n}A_{1n}$$

The use of this double-subscript notation is essential in the theory of determinants and matrices.

Area It can be shown that the area of the triangle having vertices (x_1, y_1), (x_2, y_2), and (x_3, y_3) is the absolute value of

$$\frac{1}{2}\begin{vmatrix} 1 & x_1 & y_1 \\ 1 & x_2 & y_2 \\ 1 & x_3 & y_3 \end{vmatrix}.$$

EXAMPLE 4 Calculate the area of the triangle whose vertices are (1, 0), (2, 3), and (−2, 2). See Figure 8.10.

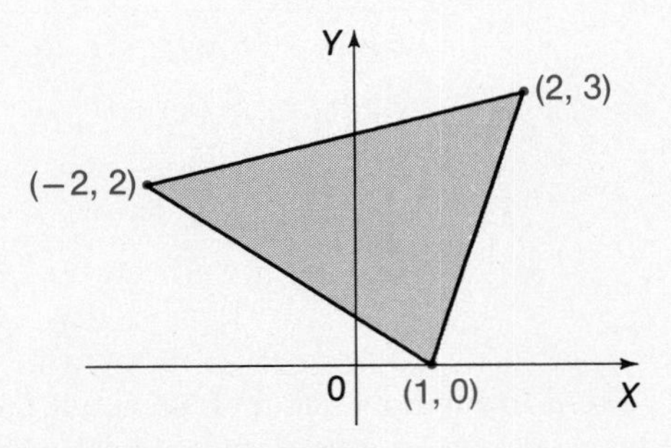

Figure 8.10

SOLUTION The area is given by the absolute value of A, where

$$2A = \begin{vmatrix} 1 & 1 & 0 \\ 1 & 2 & 3 \\ 1 & -2 & 2 \end{vmatrix} = 1\begin{vmatrix} 2 & 3 \\ -2 & 2 \end{vmatrix} - 1\begin{vmatrix} 1 & 0 \\ -2 & 2 \end{vmatrix} + 1\begin{vmatrix} 1 & 0 \\ 2 & 3 \end{vmatrix}$$

$$= (4 + 6) - (2 + 0) + (3 - 0) = 11.$$

So area $A = \frac{1}{2}(11) = 5\frac{1}{2}$. ●

Solution of Systems of Linear Equations by Determinants To see how determinants can be used to solve systems, we first solve the general system of two equations in two variables by elimination as in Section 8.1.

$$\begin{cases} a_1x + b_1y = c_1 \\ a_2x + b_2y = c_2 \end{cases}$$

To eliminate y, multiply the first equation by b_2 and the second by $-b_1$ and then add. The resulting equation is then solved for x.

$$\begin{aligned} a_1b_2x + b_1b_2y &= c_1b_2 \\ -a_2b_1x - b_2b_1y &= -c_2b_1 \\ \hline a_1b_2x - a_2b_1x &= c_1b_2 - c_2b_1 \\ x(a_1b_2 - a_2b_1) &= c_1b_2 - c_2b_1 \end{aligned}$$

$$x = \frac{c_1b_2 - c_2b_1}{a_1b_2 - a_2b_1} = \frac{\begin{vmatrix} c_1 & b_1 \\ c_2 & b_2 \end{vmatrix}}{\begin{vmatrix} a_1 & b_1 \\ a_2 & b_2 \end{vmatrix}} \qquad a_1b_2 - a_2b_1 \neq 0$$

Similarly, we can eliminate x to obtain a formula for y. Multiply the first equation by $-a_2$ and the second equation by a_1 and then add.

$$\begin{aligned} -a_1a_2x - b_1a_2y &= -c_1a_2 \\ a_2a_1x + b_2a_1y &= c_2a_1 \\ \hline b_2a_1y - b_1a_2y &= c_2a_1 - c_1a_2 \\ y(b_2a_1 - b_1a_2) &= c_2a_1 - c_1a_2 \end{aligned}$$

$$y = \frac{c_2a_1 - c_1a_2}{b_2a_1 - b_1a_2} = \frac{\begin{vmatrix} a_1 & c_1 \\ a_2 & c_2 \end{vmatrix}}{\begin{vmatrix} a_1 & b_1 \\ a_2 & b_2 \end{vmatrix}} \qquad a_1b_2 - a_2b_1 \neq 0$$

From Examples 4 and 5 of Section 8.1 it is clear that if $a_1b_2 - a_2b_1 = 0$, then the two equations represent straight lines which are either parallel or identical. In this case they do not have a unique point of intersection.

Note the pattern of the determinants above.

$$D = \begin{vmatrix} a_1 & b_1 \\ a_2 & b_2 \end{vmatrix}$$

is the **determinant of the coefficients** of the system of equations.

$$D_x = \begin{vmatrix} c_1 & b_1 \\ c_2 & b_2 \end{vmatrix}$$

is obtained from D by replacing the coefficients of x by the corresponding c's.

$$D_y = \begin{vmatrix} a_1 & c_1 \\ a_2 & c_2 \end{vmatrix}$$

is obtained from D by replacing the coefficients of y by the corresponding c's.

The solutions found above may be written in terms of the above determinants, as follows:

$$x = \frac{D_x}{D} \quad \text{and} \quad y = \frac{D_y}{D}.$$

EXAMPLE 5 Use determinants to solve $\begin{cases} 3x - y = -6 \\ 2x + 3y = 7 \end{cases}$.

SOLUTION From above:

$$D = \begin{vmatrix} 3 & -1 \\ 2 & 3 \end{vmatrix} = 3(3) - 2(-1) = 11,$$

$$D_x = \begin{vmatrix} -6 & -1 \\ 7 & 3 \end{vmatrix} = -6(3) - 7(-1) = -11,$$

$$D_y = \begin{vmatrix} 3 & -6 \\ 2 & 7 \end{vmatrix} = 3(7) - 2(-6) = 33.$$

Now use the formulas:

$$x = \frac{D_x}{D} = \frac{-11}{11} = -1 \quad \text{and} \quad y = \frac{D_y}{D} = \frac{33}{11} = 3. \quad \bullet$$

The procedure used above is an illustration of **Cramer's Rule.** It can be applied to any system of n equations in n variables. The theorem is stated here for the case $n = 3$.

CRAMER'S RULE Let $D \neq 0$ be the determinant of the coefficients of a system of three equations in three variables. Also, let D_x, D_y, and D_z be the determinants obtained by substituting the constant terms for the coefficients of x, y, and z, respectively, in D. Then

$$x = \frac{D_x}{D}, \quad y = \frac{D_y}{D}, \quad z = \frac{D_z}{D}$$

is the unique solution of the system of equations.

EXAMPLE 6 Use Cramer's Rule to solve the following system.

$$\begin{cases} x + y - z = 3 \\ 2x - 2y + z = -2 \\ x + 2y + 3z = 5 \end{cases}$$

SOLUTION The required determinants are shown below.

$$D = \begin{vmatrix} 1 & 1 & -1 \\ 2 & -2 & 1 \\ 1 & 2 & 3 \end{vmatrix} \qquad D_x = \begin{vmatrix} 3 & 1 & -1 \\ -2 & -2 & 1 \\ 5 & 2 & 3 \end{vmatrix}$$

$$D_y = \begin{vmatrix} 1 & 3 & -1 \\ 2 & -2 & 1 \\ 1 & 5 & 3 \end{vmatrix} \qquad D_z = \begin{vmatrix} 1 & 1 & 3 \\ 2 & -2 & -2 \\ 1 & 2 & 5 \end{vmatrix}$$

Verify that

$$D = -19, \qquad D_x = -19, \qquad D_y = -38, \qquad D_z = 0.$$

Then

$$x = \frac{D_x}{D} = \frac{-19}{-19} = 1, \qquad y = \frac{D_y}{D} = \frac{-38}{-19} = 2, \qquad z = \frac{D_z}{D} = \frac{0}{-19} = 0.$$

The solution, (1, 2, 0), should be checked by substitution. •

Cramer's Rule can be used whenever the number of equations is the *same* as the number of variables.

EXERCISES 8.4

Evaluate each of the following determinants.

1. $\begin{vmatrix} 2 & -3 \\ 4 & -2 \end{vmatrix}$ **2.** $\begin{vmatrix} 2 & 4 \\ -1 & 3 \end{vmatrix}$ **3.** $\begin{vmatrix} 1 & -1 \\ 2 & 3 \end{vmatrix}$ **4.** $\begin{vmatrix} 2 & 3 \\ -2 & 1 \end{vmatrix}$

Solve the following equations.

5. $\begin{vmatrix} 2 & 3 \\ 4 & x \end{vmatrix} = 0$ **6.** $\begin{vmatrix} x & 13 \\ 2 & 3 \end{vmatrix} = 0$

7. $\begin{vmatrix} 4 - x & 2 \\ -1 & 1 - x \end{vmatrix} = 0$ **8.** $\begin{vmatrix} x - 10 & 2 \\ x + 2 & 2 + x \end{vmatrix} = 0$

Evaluate each of the following determinants by expanding by minors of any row (or column).

9. $\begin{vmatrix} 1 & 2 & 3 \\ 0 & 0 & -1 \\ 4 & -2 & 0 \end{vmatrix}$ **10.** $\begin{vmatrix} -1 & -2 & -3 \\ 2 & 4 & 1 \\ -1 & 2 & 0 \end{vmatrix}$

11. $\begin{vmatrix} 1 & 1 & 1 \\ 1 & 0 & -1 \\ 0 & 1 & -1 \end{vmatrix}$

12. $\begin{vmatrix} 2 & -1 & 3 \\ 0 & 1 & 2 \\ 3 & -2 & 1 \end{vmatrix}$

13. $\begin{vmatrix} -1 & 2 & -3 \\ 2 & 1 & 1 \\ 1 & 3 & 2 \end{vmatrix}$

14. $\begin{vmatrix} 2 & -1 & 3 \\ 1 & 2 & 3 \\ 3 & -2 & 1 \end{vmatrix}$

15. $\begin{vmatrix} 1 & x & -2 \\ 2 & 3 & 2 \\ 2 & 6 & -4 \end{vmatrix}$

16. $\begin{vmatrix} 3x & 3 & 2 \\ 1 & x & 2 \\ -1 & -1 & 1 \end{vmatrix}$

17. $\begin{vmatrix} -2 & -1 & -2 \\ 4 & x & 1 \\ 6 & 3 & x \end{vmatrix}$

18. $\begin{vmatrix} x & y & y \\ y & x & y \\ y & y & x \end{vmatrix}$

19. $\begin{vmatrix} 2 & -1 & 0 & 3 \\ 1 & 2 & -1 & -3 \\ 3 & 0 & 2 & 0 \\ 4 & 3 & 1 & 2 \end{vmatrix}$

20. $\begin{vmatrix} 2 & 1 & 4 & 0 \\ 3 & 2 & 6 & 1 \\ 4 & 3 & 8 & -1 \\ -1 & 4 & -2 & 2 \end{vmatrix}$

21. $\begin{vmatrix} 1 & -1 & 3 & 2 \\ 2 & -2 & 2 & -1 \\ 3 & 0 & 0 & -2 \\ 4 & 3 & 1 & 1 \end{vmatrix}$

22. $\begin{vmatrix} 1 & 1 & 1 & 1 \\ 1 & 0 & -1 & 0 \\ 0 & 1 & 1 & -1 \\ 2 & 0 & -1 & -3 \end{vmatrix}$

Use Cramer's Rule to solve each of the following systems.

23. $\begin{cases} x - y = 3 \\ x + y = 5 \end{cases}$

24. $\begin{cases} x + y = 2 \\ 2x - y = 1 \end{cases}$

25. $\begin{cases} 2x + 3y = -6 \\ x - 3y = 6 \end{cases}$

26. $\begin{cases} 5x - 2y = 4 \\ 2x + 3y = 13 \end{cases}$

27. $\begin{cases} 2x - y = 2 \\ 6x - 3y = 6 \end{cases}$

28. $\begin{cases} x + 2y = -1 \\ 3x + 6y = -3 \end{cases}$

29. $\begin{cases} x + y + z = 4 \\ 2x - y - 2z = -1 \\ x - 2y - z = 1 \end{cases}$

30. $\begin{cases} x + 7y - 3z = -14 \\ x + 3y + z = 2 \\ 4x + 8y + z = 3 \end{cases}$

31. $\begin{cases} x - 2y + 3z = 4 \\ 2x + y - 4z = 3 \\ -3x + 4y - z = -2 \end{cases}$

32. $\begin{cases} 2x - y + z = 4 \\ 2x + 2y + 3z = 3 \\ 6x - 9y - 2z = 17 \end{cases}$

33. $\begin{cases} 3x - y - 3z = 1 \\ -x - 4y - 2z = 1 \\ 3x - y - z = 1 \end{cases}$

34. $\begin{cases} 4x + 2y - z = 0 \\ x + 3y + 2z = 0 \\ x + y + 3z = 4 \end{cases}$

35. $\begin{cases} 2x + y + 3z = 1 \\ -x + 4y + 2z = 0 \\ 3x + y + z = -1 \end{cases}$

36. $\begin{cases} x + y + z = 0 \\ 2y - 3z = -1 \\ 3y + 5z = 22 \end{cases}$

37. $\begin{cases} x + y - 2z = 3 \\ -y + 3z = -1 \\ 2x + 5z = 0 \end{cases}$

38. $\begin{cases} 2x + 3y - 4z = 2 \\ x + y + 2z = 1 \\ 5x + 6y + 2z = 5 \end{cases}$

39. $\begin{cases} x \qquad\quad - 2z + 2w = 1 \\ \quad y + z - w = 0 \\ -2x + 3y + 4z \qquad = -1 \\ 3x + y - 2z - w = 3 \end{cases}$

40. $\begin{cases} x + 2y - z - 4w = -2 \\ -x \qquad + 2z + 6w = -1 \\ \quad - 4y - 2z - 8w = 10 \\ 3x - 2y \qquad + 5w = 12 \end{cases}$

Calculate the area of the triangle whose vertices are given.

41. $(-1, -2), (3, -2), (1, 5)$

42. $(-2, 3), (-6, -3), (1, 1)$

43. $(1, 0), (0, -5), (-4, 2)$

44. $(-4, -3), (6, -5), (-2, 4)$

45. It can be proved that the rows and columns of any matrix may be interchanged without affecting the value of its determinant. Show this directly for matrices of orders 2 and 3. That is, show that

$$\begin{vmatrix} a & b \\ c & d \end{vmatrix} = \begin{vmatrix} a & c \\ b & d \end{vmatrix} \quad \text{and} \quad \begin{vmatrix} a_1 & b_1 & c_1 \\ a_2 & b_2 & c_2 \\ a_3 & b_3 & c_3 \end{vmatrix} = \begin{vmatrix} a_1 & a_2 & a_3 \\ b_1 & b_2 & b_3 \\ c_1 & c_2 & c_3 \end{vmatrix}.$$

46. Solve the following rotation equations (Section 7.6)

$$\begin{cases} x' \cos\theta - y' \sin\theta = x \\ x' \sin\theta + y' \cos\theta = y \end{cases}$$

for x' and y'. This gives the inverse rotation transformation. [*Hint:* Use determinants and note that $D = 1$.]

8.5 Systems of Inequalities

Linear inequalities in one variable were studied in Chapter 1. In this section that discussion is extended to linear and quadratic inequalities in two variables, together with their graphs. Simple examples of these inequalities are $x - 2y < 1$, $x^2 - 2y > 4$, and $x^2 + y^2 \leq 1$. Graphs will be used to describe solutions of such inequalities.

EXAMPLE 1 Graph the inequality $y < x$.

SOLUTION First graph the line $y = x$. Any point on this line has coordinates (p, p). If (p, q) is any point below this line, then $q < p$, which satisfies the inequality $y < x$. Thus, the *shaded region* below the line in Figure 8.11 is the *graph of the solution set* of $y < x$. Since $y = x$ is not part of this region, it is drawn as a dashed line. Note that above the line all points satisfy the inequality $y > x$. Thus, the plane has been partitioned into three sets of points: the **half-plane** $y < x$, the line $y = x$, and the **half-plane** $y > x$. ●

EXAMPLE 2 Graph the inequality $2x + y \geq 3$.

SOLUTION First, write this inequality as $y \geq -2x + 3$, then graph the line $y = -2x + 3$. The reasoning used in Example 1 shows that all points *on or above* the line $y = -2x + 3$

satisfy the inequality $y \geq -2x + 3$. The graph of the line in Figure 8.12 is solid to indicate that its points are part of the graph of the solution.

It is a good idea to test some points by substitution to see if the inequality is satisfied. For example, since $2 > -2(1) + 3$, the point $(1, 2)$ satisfies $2x + y \geq 3$, as does any point on the same side of the line as $(1, 2)$. The point $(-1, 4)$ does not satisfy the inequality, nor does any point on the same side of the line as $(-1, 4)$. ●

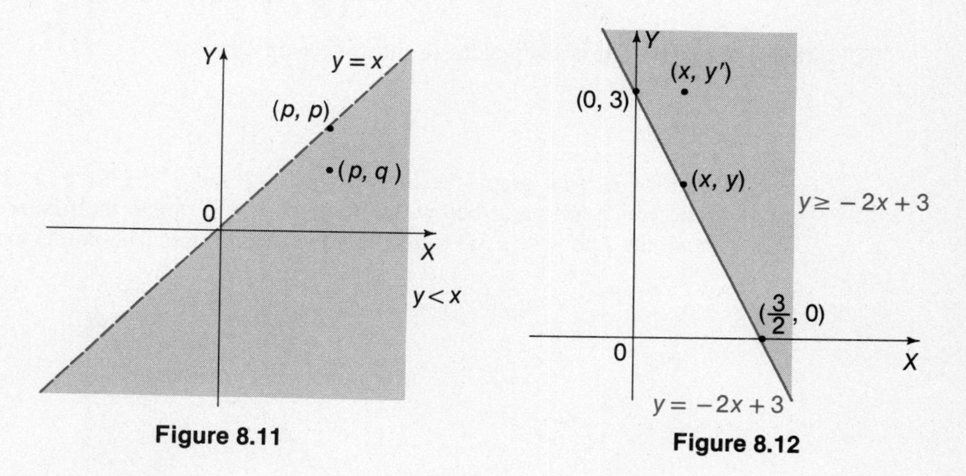

Figure 8.11 **Figure 8.12**

EXAMPLE 3 Graph $y > x^2$.

SOLUTION First graph the parabola $y = x^2$. All points above this curve satisfy $y > x^2$. This region is shaded in Figure 8.13. The curve itself is drawn with a dashed line since its points do not satisfy the inequality. ●

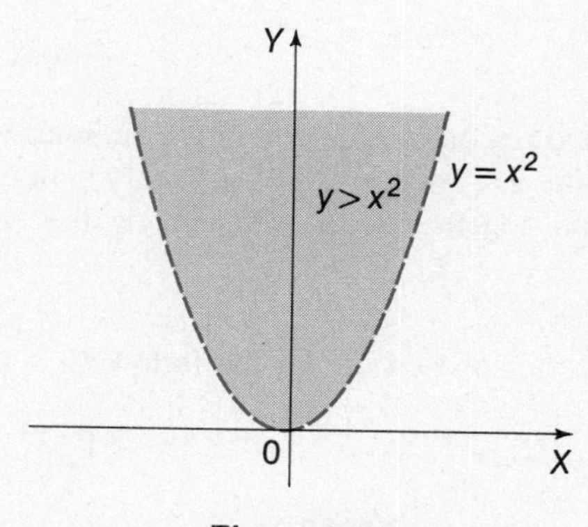

Figure 8.13

Observe that in each of the examples above, the curve partitions the plane into three sets of points: those for which $y < f(x)$, $y = f(x)$, and $y > f(x)$.

Systems of Inequalities Systems of inequalities are treated similarly to systems of equations. The method of solution is illustrated in the following examples.

EXAMPLE 4 Find the points which simultaneously satisfy the following pair of inequalities.

$$\begin{cases} x - y \geq -2 \\ 3x + y > 3 \end{cases}$$

SOLUTION This system is equivalent to the following system.

$$\begin{cases} y \leq x + 2 \\ y > -3x + 3 \end{cases}$$

To solve it, first graph the lines $y = x + 2$ and $y = -3x + 3$. Those points in the plane which are on or below the line $y = x + 2$ satisfy the first inequality, and those above the line $y = -3x + 3$ satisfy the second one. The points located in the doubly

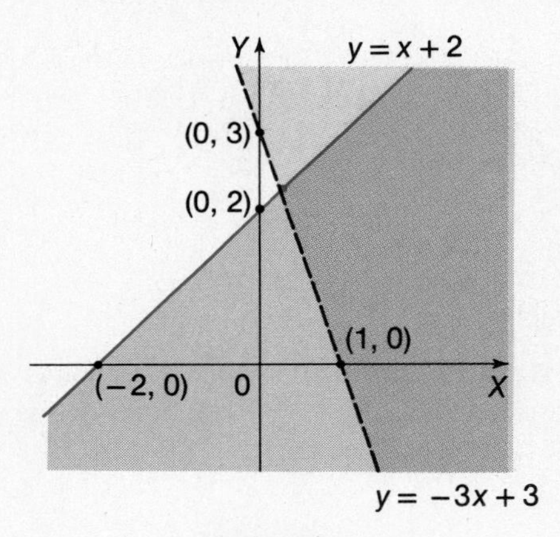

Figure 8.14

shaded region in Figure 8.14 are simultaneously below the line $y = x + 2$ and above the line $y = -3x + 3$ and so satisfy both inequalities. The solution set consists of this region together with the part of the line $y = x + 2$ which is above the line $y = -3x + 3$. ●

EXAMPLE 5 Find the solution set for the following system.

$$\begin{cases} x \geq -1 \\ x - y \leq 0 \\ x + y \leq 2 \end{cases}$$

SOLUTION This system is equivalent to the system below.

$$\begin{cases} x \geq -1 \\ y \geq x \\ y \leq 2 - x \end{cases}$$

The equations $x = -1$, $y = x$, and $y = 2 - x$ are graphed in Figure 8.15. The solution set consists of portions of these lines and the points simultaneously to the right of the line $x = -1$, above the line $y = x$, and below the line $y = 2 - x$. This region is shaded in the figure. ●

EXAMPLE 6 Solve the system

$$\begin{cases} x^2 < y + 1 \\ y^2 \leq 3 - x^2. \end{cases}$$

SOLUTION This system of inequalities is equivalent to the system

$$\begin{cases} y > x^2 - 1 \\ x^2 + y^2 \leq 3. \end{cases}$$

Figure 8.16 shows the graphs of the parabola $y = x^2 - 1$ and the circle $x^2 + y^2 = 3$. Since the points (x, y) for which $y = x^2 - 1$ lie on the parabola, the points satisfying the first inequality lie above this parabola. Recall that $x^2 + y^2$ is the square of the distance of (x, y) from the origin O. For points on the circle, $x^2 + y^2 = 3$, and for points inside the circle, $x^2 + y^2 < 3$. So the second inequality is satisfied by points *on and inside* the circle. The solution for the system of inequalities is shaded in the figure. Note that the circle is solid and the parabola is dashed. ●

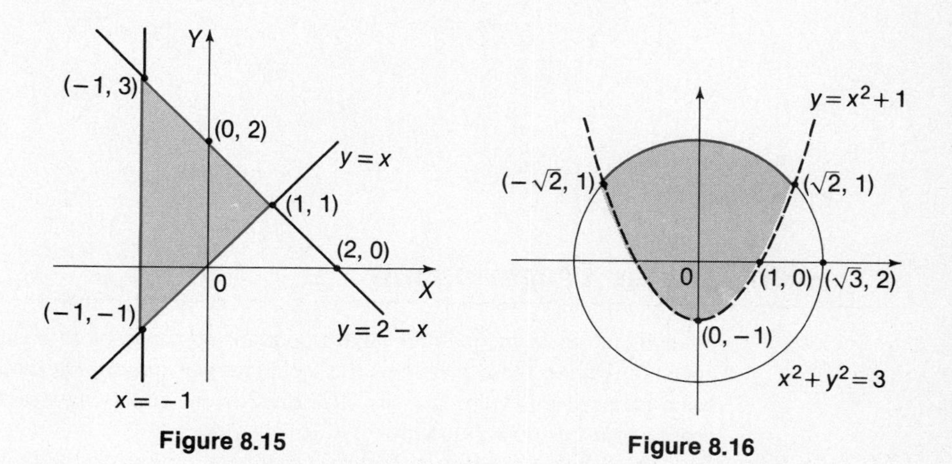

Figure 8.15 **Figure 8.16**

There is a very wide variety of applications of systems of inequalities. In the area of business and sociology systems of linear inequalities play an important role. Some of these applications will be discussed in the next section.

EXERCISES 8.5

Graph the region representing the solution set for each of the following inequalities.

1. $y < 2x + 3$ **2.** $2y - 4 \leq x$ **3.** $3x - 2y < 6$

4. $4x + 3y > 12$
5. $x^2 + y^2 \geq 1$
6. $x^2 + y^2 < 4$
7. $(x - 1)^2 + (y + 1)^2 \leq 5$
8. $(x - 2)^2 + (y - 3)^2 \leq 6$
9. $y^2 \leq x$
10. $y \geq (x - 2)^2$
11. $(x + 1)^2 \leq -2y$
12. $x^2 - 2x - y > 1$
13. $y \leq e^x$
14. $y > \ln x$
15. $y > e^x - 1$
16. $y \leq 1 + \ln x$
17. $y \geq \sin x$
18. $y \geq \sec x$
19. $y \geq \tan x$
20. $y \leq \cot x$

Solve the following systems of inequalities by graphing the solution set in each case.

21. $\begin{cases} y < 1 - x \\ 4 - y < 2 - x \end{cases}$
22. $\begin{cases} x + y \geq 4 \\ y \leq 2x + 3 \end{cases}$
23. $\begin{cases} y \geq 2x - 4 \\ x + y \leq 4 \end{cases}$
24. $\begin{cases} 2y + x \leq 4 \\ 3y - 2x < 6 \end{cases}$
25. $\begin{cases} y \leq 2 \\ y \geq x^2 \end{cases}$
26. $\begin{cases} y - 1 < x^2 \\ y + 1 > x^2 \end{cases}$
27. $\begin{cases} x^2 + y^2 \leq 4 \\ x^2 + y^2 \geq 1 \end{cases}$
28. $\begin{cases} x^2 + y^2 \geq 4 \\ x^2 + y^2 \leq 1 \end{cases}$
29. $\begin{cases} y \geq x^2 \\ x^2 + y^2 \leq 1 \end{cases}$
30. $\begin{cases} x^2 + y^2 \leq 1 \\ x + y \geq 1 \end{cases}$
31. $\begin{cases} x + y < 2 \\ x > 0 \\ y > 0 \end{cases}$
32. $\begin{cases} 2x + y < 2 \\ -x + y < 1 \\ x > 0 \end{cases}$
33. $\begin{cases} 0 \leq x \leq 2 \\ 0 \leq y \leq 3 \end{cases}$
34. $\begin{cases} x + y < 2 \\ 0 \leq x \leq 1 \\ y \geq 0 \end{cases}$
35. $\begin{cases} y \leq e^x \\ y \geq \ln x, \ x > 0 \end{cases}$
36. $\begin{cases} y \leq \cos x \\ y \geq \sin x, \ -\frac{1}{2}\pi \leq x \leq \frac{1}{2}\pi \end{cases}$
37. $\begin{cases} y \geq \tan x \\ y \geq \cot x, \ 0 < x < \frac{1}{2}\pi \end{cases}$
38. $\begin{cases} y \leq 1 + e^x \\ y \geq \ln(x + 1), \ x > 0 \end{cases}$

8.6 Linear Programming

Systems of linear inequalities have important applications to business and social science. In solving these problems the techniques of *linear programming* are used. The basic ideas of this technique are illustrated in the following example on maximizing profits in a business situation.

The Smith Company makes two products, tape decks and amplifiers. Each tape deck gives a profit of \$3, while each amplifier earns \$7. The company must manufacture at least one tape deck per day to satisfy one of its customers, but no more than five because of production problems. Also, the number of amplifiers produced cannot exceed six per day. As a further requirement, the number of tape decks cannot exceed the number of amplifiers. How many of each should the company manufacture in order to obtain the maximum profit?

To begin, translate the statements of the problem into symbols by assuming

$$x = \text{number of tape decks to be produced daily}$$

$$y = \text{number of amplifiers to be produced daily.}$$

According to the statement of the problem given above, the company must produce at least one tape deck (one or more), so that

$$x \geq 1.$$

No more than 5 tape decks may be produced: $x \leq 5$.

Not more than 6 amplifiers may be made in one day: $y \leq 6$.

The number of tape decks may not exceed the number of amplifiers: $x \leq y$.

The number of tape decks and of amplifiers cannot be negative:

$$x \geq 0 \quad \text{and} \quad y \geq 0.$$

All the restrictions, or **constraints,** that are placed on production are stated below.

$$x \geq 1, \quad x \leq 5, \quad y \leq 6, \quad x \leq y, \quad x \geq 0, \quad y \geq 0.$$

We need to find the maximum possible profit that the company can make, subject to these constraints. To begin, we sketch the graph of each constraint, as in Figure 8.17. (Since $x \geq 1$ implies that $x \geq 0$, the graph of $x \geq 0$ is not shown.)

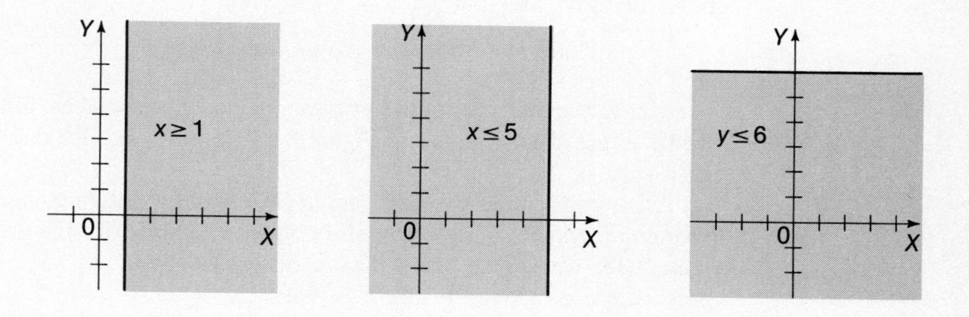

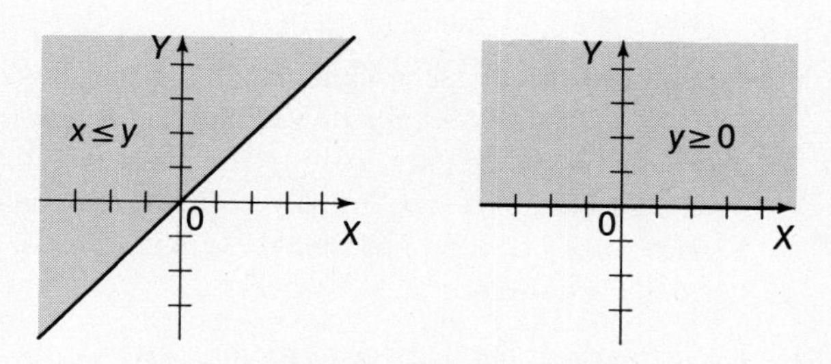

Figure 8.17

The **feasible** values of x and y are those that satisfy all constraints. They are the values which lie in the intersection of the graphs of the constraints. The intersection is shown in Figure 8.18 on the next page. Any point lying inside or on the boundary of the region in Figure 8.18 satisfies the restrictions on the number of tape decks and amplifiers that may be produced.

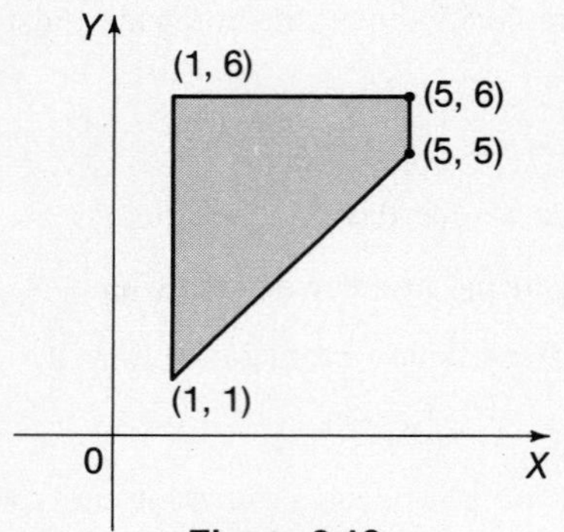

Figure 8.18

Point	Profit = $3x + 7y$
(1, 1)	$3(1) + 7(1) = 10$ ← minimum
(1, 6)	$3(1) + 7(6) = 45$
(5, 5)	$3(5) + 7(5) = 50$
(5, 6)	$3(5) + 7(6) = 57$ ← maximum

Since each tape deck gives a profit of \$3, the daily profit from the production of x decks is $3x$ dollars. Also, the profit from the production of y amplifiers will be $7y$ dollars per day. The total daily profit is thus

$$\text{profit} = 3x + 7y.$$

The problem of the Smith Company may now be stated as follows: find values of x and y in the shaded region of Figure 8.18 that will produce the maximum possible value of $3x + 7y$.

A basic principle of linear programming says that in a case such as this the maximum (or minimum) profit will be found at a corner of the graph. By looking at Figure 8.18, we see that the corner points are (1, 1), (1, 6), (5, 5), and (5, 6). Now check each corner point to find the profit.

Maximum profit of \$57 per day will be obtained if five tape decks and six amplifiers are made each day.

EXAMPLE 1 Robin, who is ill, takes vitamin pills. Each day, she must have at least 16 units of Vitamin A, at least 5 units of Vitamin B_1, and at least 20 units of Vitamin C. She can choose between red pills, costing 10¢ each, which contain 8 units of A, 1 of B_1, and 2 of C; and blue pills, costing 20¢ each, which contain 2 units of A, 1 of B_1, and 7 of C. How many of each pill should she buy in order to minimize her cost and yet fulfill her daily requirements?

SOLUTION Let x represent the number of red pills to buy, and let y represent the number of blue pills to buy. Then the cost in pennies per day is given by

$$\text{cost} = 10x + 20y,$$

since Robin buys x of the 10¢ pills and y of the 20¢ pills. Vitamin A comes as follows: 8 units from each red pill and 2 units from each blue pill. Altogether, she gets $8x + 2y$ units of A per day. Since she must get at least 16 units,

$$8x + 2y \geq 16.$$

Figure 8.19

Point	Cost = $10x + 20y$	
(10, 0)	$10(10) + 20(0) = 100$	
(3, 2)	$10(3) + 20(2) = 70$	← minimum
(1, 4)	$10(1) + 20(4) = 90$	
(0, 8)	$10(0) + 20(8) = 160$	

Each red pill and each blue pill supplies 1 unit of Vitamin B_1. Robin needs at least 5 units per day:

$$x + y \geq 5.$$

For Vitamin C we get the inequality $2x + 7y \geq 20$.

Also, $x \geq 0$ and $y \geq 0$, since Robin cannot buy negative numbers of the pills.

Again, we minimize total cost of the pills by finding the solution of the system of inequalities formed by the constraints. (See Figure 8.19.) Check the corners to find the lowest cost.

Robin's best bet is to buy 3 red pills and 2 blue ones, for a total cost of 70¢ per day. She receives just the minimum amounts of Vitamins B_1 and C, but an excess of Vitamin A. Even though she has an excess of A, this is still the best buy. ●

EXERCISES 8.6

Exercises 1–4 show regions of feasible solutions. Find the maximum and minimum values of the given expressions.

1. $3x + 5y$

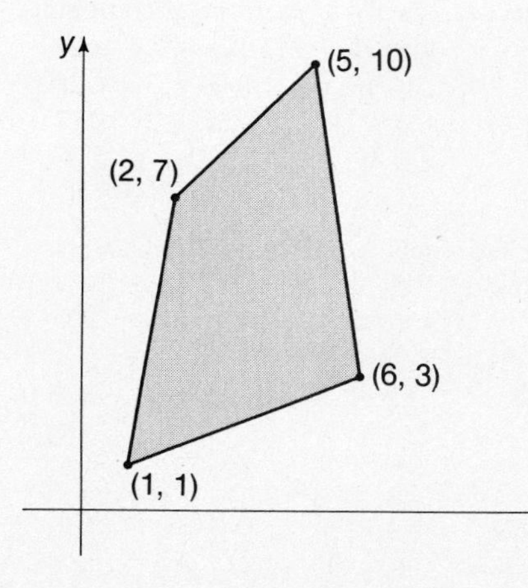

2. $6x + y$

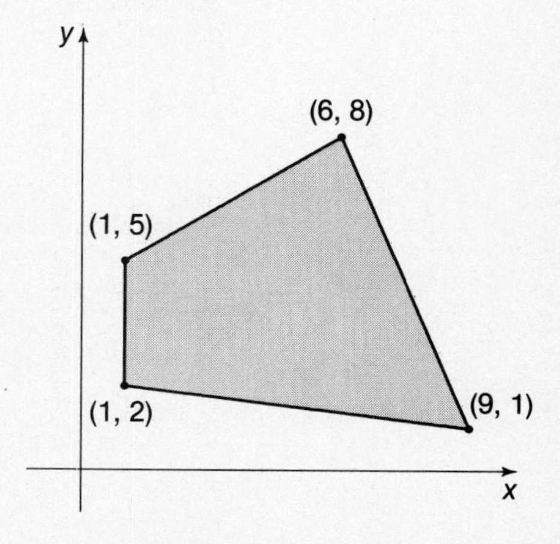

3. $40x + 75y$

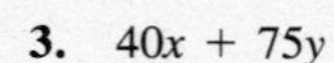

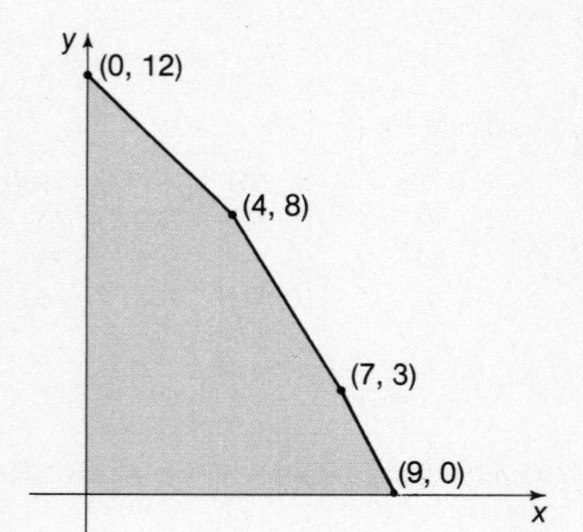

4. $35x + 125y$

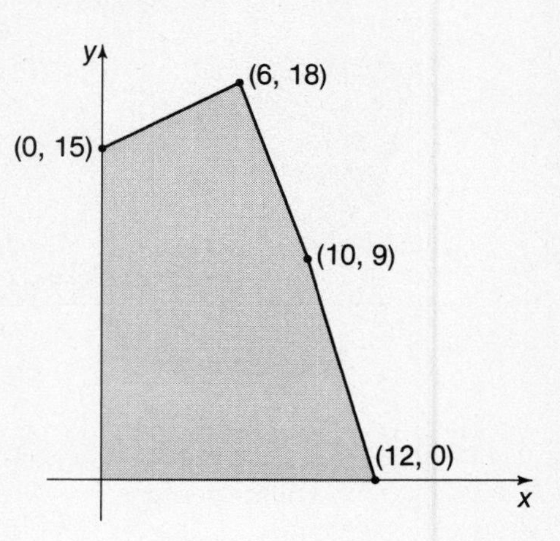

In Exercises 5–10, use graphical methods to find values of x and y satisfying the given conditions. Find the value of the maximum or minimum.

5. Find $x \geq 0$ and $y \geq 0$ such that
$$2x + 3y \leq 6$$
$$4x + y \leq 6$$
and $5x + 2y$ is maximized.

6. Find $x \geq 0$ and $y \geq 0$ such that
$$x + y \leq 10$$
$$5x + 2y \geq 20$$
$$2y \geq x$$
and $x + 3y$ is minimized.

7. Find $x \geq 2$ and $y \geq 5$ such that
$$3x - y \geq 12$$
$$x + y \leq 15$$
and $2x + y$ is minimized.

8. Find $x \geq 10$ and $y \geq 20$ such that
$$2x + 3y \leq 100$$
$$5x + 4y \leq 200$$
and $x + 3y$ is maximized.

9. Find $x \geq 0$ and $y \geq 0$ such that
$$x - y \leq 10$$
$$5x + 3y \leq 75$$
and $4x + 2y$ is maximized.

10. Find $x \geq 0$ and $y \geq 0$ such that
$$10x - 5y \leq 100$$
$$20x + 10y \geq 150$$
and $4x + 5y$ is minimized.

11. Maximize $10x + 12y$ subject separately to each set of constraints.

(a) $x + y \leq 20$, $x + 3y \leq 24$, $x \geq 0$, $y \geq 0$

(b) $3x + y \leq 15$, $x + 2y \leq 18$, $x \geq 0$, $y \geq 0$

(c) $2x + 5y \geq 22$, $4x + 3y \leq 28$, $2x + 2y \leq 17$, $x \geq 0$, $y \geq 0$

12. Minimize $3x + 2y$ subject separately to each set of constraints.

(a) $10x + 7y \leq 42$, $4x + 10y \geq 35$, $x \geq 0$, $y \geq 0$

(b) $6x + 5y \geq 25$, $2x + 6y \geq 15$, $x \geq 0$, $y \geq 0$

(c) $x + 2y \geq 10$, $2x + y \geq 12$, $x - y \leq 8$, $x \geq 0$, $y \geq 0$

Solve each of the following linear programming problems.

13. Farmer Jones raises only pigs and geese. He wants to raise no more than 16 animals with no more than 12 geese. He spends \$5 to raise a pig and \$2 to raise a goose. He has \$50 available for this purpose. Find the maximum profit he can make if he makes a profit of \$8 per goose and \$4 per pig.

14. A wholesaler of party goods wishes to display her products at a convention of social secretaries in such a way that she gets the maximum number of inquiries about her whistles and hats. Her booth at the convention has 12 square meters of floor space to be used for display purposes. A display unit for hats requires 2 square meters; and for whistles, 4 square meters. Experience tells the wholesaler that she should never have more than a total of 5 units of whistles and hats on display at one time. If she receives three inquiries for each unit of hats and two inquiries for each unit of whistles on display, how many of each should she display in order to get the maximum number of inquiries?

15. An office manager wants to buy some filing cabinets. She knows that cabinet #1 costs \$10 each, requires 6 square feet of floor space, and holds 8 cubic feet of files. On the other hand, cabinet #2 costs \$20 each, requires 8 square feet of floor space, and holds 12 cubic feet. She can spend no more than \$140 due to budgetary limitations, while her office has room for no more than 72 square feet of cabinets. She desires the maximum storage capacity within the limitations imposed by funds and space. How many of each type of cabinet should she buy?

16. The manufacturer of certain souvenir ashtrays and cufflinks requires two machines, a drill and a saw. Each ashtray needs one minute on the drill and two minutes on the saw; each cufflink set needs two minutes on the drill and one minute on the saw. The drill is available at most 12 minutes per day, and the saw is available no more than 15 minutes per day. Each product provides a profit of \$1. How many of each should be manufactured in order to maximize profit?

Chapter 8 Review Exercises

Solve the following systems of equations by elimination.

1. $\begin{cases} 3x + 2y = 5 \\ x - y = 5 \end{cases}$

2. $\begin{cases} 2x - 3y = 13 \\ 5x + 2y = 4 \end{cases}$

3. $\begin{cases} \frac{1}{2}x + \frac{1}{3}y = 10 \\ 5x - 3(y + 1) = 21 \end{cases}$

4. $\begin{cases} 0.3x + 0.8y = 3 \\ 0.9x + 3.4y = 12 \end{cases}$

5. $\begin{cases} x - 3y = 4 \\ 3x - 9y = 12 \end{cases}$

6. $\begin{cases} 2x - 4y = 3 \\ -3x + 6y = 2 \end{cases}$

7. $\begin{cases} 0.5x + 0.7y = 0.6 \\ x - 0.3y = 4.6 \end{cases}$

8. $\begin{cases} \frac{x}{2} + \frac{y}{3} = 8 \\ \frac{2x}{3} + \frac{3y}{2} = 17 \end{cases}$

9. $\begin{cases} x + 3y - 6z = -7 \\ 2x - y + z = 8 \\ x + 2y + 2z = -10 \end{cases}$

10. $\begin{cases} 2x + 5y + 2z = 9 \\ 3x - 8y - 2z = -7 \\ 4x - 7y - 3z = -6 \end{cases}$

Use row reduction to solve the following systems of equations.

11. $\begin{cases} 4x + 8y + z = 2 \\ x + 7y - 3z = -14 \\ 2x - 3y + 2z = 3 \end{cases}$

12. $\begin{cases} 2x - 5y - z = 3 \\ 5x - 14y - z = 18 \\ 7x + 9y - 2z = 2 \end{cases}$

13. $\begin{cases} x + 2y - z = 0 \\ 2x - y + z = 0 \\ 3x + 2y - 2z = 0 \end{cases}$

14. $\begin{cases} x + y + z = 0 \\ 2x - y - 2z = 0 \\ x - 2y - z = 0 \end{cases}$

15. $\begin{cases} x + 2y - 3z = -14 \\ y - z - 4w = -4 \\ -2x + y + 2w = 3 \\ 3x - 2y - 6w = 13 \end{cases}$

16. $\begin{cases} -x - 3y + 2z - w = -1 \\ 2x - 5y - 3z + 2w = 3 \\ -3x + 4y + 8z - 2w = 4 \\ 6x - y - 6z + 4w = 2 \end{cases}$

Solve each of the following systems of nonlinear equations by substitution and represent the solutions graphically.

17. $\begin{cases} x^2 = 4y \\ x + 2y = 4 \end{cases}$

18. $\begin{cases} x^2 + y^2 = 25 \\ 3x - 4y = 0 \end{cases}$

19. $\begin{cases} xy = 12 \\ x + y = 7 \end{cases}$

20. $\begin{cases} 4x^2 + 25y^2 = 100 \\ 6x - 5y = 10 \end{cases}$

Use row reduction to reduce each of the following matrices to echelon form.

21. $\begin{bmatrix} 1 & -2 & 3 \\ 2 & 1 & -4 \\ -3 & 4 & -1 \end{bmatrix}$

22. $\begin{bmatrix} 1 & 3 & 1 \\ 1 & 1 & -1 \\ 1 & -2 & -4 \end{bmatrix}$

23. $\begin{bmatrix} 1 & -1 & 4 \\ 2 & 1 & -1 \\ -1 & -1 & 2 \end{bmatrix}$

24. $\begin{bmatrix} 1 & 0 & -7 \\ 2 & -1 & 0 \\ 0 & 5 & 3 \end{bmatrix}$

25. $\begin{bmatrix} 1 & 2 & -1 & -3 \\ 1 & 1 & 3 & -2 \\ 3 & 1 & -2 & -1 \\ 4 & -3 & -1 & -2 \end{bmatrix}$

26. $\begin{bmatrix} 1 & 0 & -2 & 2 \\ 0 & 1 & 1 & -1 \\ -2 & 3 & 4 & 0 \\ 3 & 1 & -2 & -1 \end{bmatrix}$

27–32. Evaluate the determinant of each of the matrices in exercises 21–26 above by expansion by minors.

Use Cramer's Rule to solve each of the following systems of linear equations.

33. $\begin{cases} x - 2y = -4 \\ 3x + 4y = 3 \end{cases}$

34. $\begin{cases} 2x + 8y = 20 \\ 3x - 5y = -21 \end{cases}$

35. $\begin{cases} x + y + z = 9 \\ x + 2y + 3z = 9 \\ x + 3y + 6z = 3 \end{cases}$

36. $\begin{cases} 2x - 3y + z = 4 \\ x + y - z = 2 \\ 4x - y + 3z = 1 \end{cases}$

37. $\begin{cases} 2x + 3y - z = 1 \\ 4x + 6y - 2z = 2 \\ x - y + z = 2 \end{cases}$

38. $\begin{cases} 3x - 8y + 6z = 16 \\ 2x + 4y - 3z = 17 \\ 8x - 2y - 9z = 41 \end{cases}$

39. $\begin{cases} x + 2z + 2w = -7 \\ 3x + y + 4z + 2w = -9 \\ 2x - 3y - z + 5w = -6 \\ x + 2y - w = 6 \end{cases}$

40. $\begin{cases} x + y - z + w = 2 \\ x + 3z + 2w = 5 \\ x - y + z + w = 4 \\ -2x + y + 2z - w = -5 \end{cases}$

Represent graphically the solution set of each of the following inequalities.

41. $2x + y \le 1$

42. $4x - 2y \le 3$

43. $y \le x^2 + 1$

44. $x^2 + y^2 \ge 3$

Solve the following systems of inequalities by graphing the solution set in each case.

45. $\begin{cases} x - 2y + 4 \le 0 \\ x \ge -4 \end{cases}$

46. $\begin{cases} 3x - y \le 0 \\ x - 2y \le 8 \end{cases}$

47. $\begin{cases} y \le x^2 + 1 \\ y \le 3 \end{cases}$

48. $\begin{cases} x^2 + y^2 \le 2 \\ x + y \le 0 \end{cases}$

49. Find the maximum value of $x + 3y$ if $x \ge 0$, $y \ge 0$, $x + 2y \le 11$, $4x + 3y \le 24$.

50. Find the maximum value of $3x + y$ if $x \ge 0$, $y \ge 0$, $2x + y \ge 8$, $x + y \le 0$.

51. Find the minimum value of $2x + y$ if $x \ge 0$, $y \ge 0$, $4x + y \ge 16$, $3x + 2y \ge 24$.

52. Find the minimum value of $2.5x + y$ if $x \ge 0$, $y \ge 0$, $2x + y \ge 12$, $x + 2y \ge 12$.

53. A walnut rancher has 2400 boxes of walnuts to be shipped to markets in Chicago and Dallas. The Chicago market needs at least 1000 boxes, while the Dallas market must have at least 800 boxes. How many boxes should be shipped to each city to minimize shipping costs if it costs \$18 to ship a box to Chicago and \$16 to ship a box to Dallas?

54. The Puwati Indians make woven blankets and skirts. Each blanket requires 24 hours for spinning the yarn, 4 hours for dying the yarn, and 15 hours for weaving. Skirts require 12, 3, and 9 hours respectively. There are 216, 44, and 147 hours available for spinning, dying, and weaving respectively. How many of each item should be made to maximize profit, if the profit is \$40 on a blanket and \$25 on a skirt?

Chapter 8 Miscellaneous Exercises

The following are useful properties of determinants. Prove each of them for determinants of second and third order matrices.

1. A factor common to the elements of a row or a column may be factored outside the determinant. That is,

$$\begin{vmatrix} ka & b \\ kc & d \end{vmatrix} = k \begin{vmatrix} a & b \\ c & d \end{vmatrix} \quad \text{and} \quad \begin{vmatrix} a_1 & kb_1 & c_1 \\ a_2 & kb_2 & c_2 \\ a_3 & kb_3 & c_3 \end{vmatrix} = k \begin{vmatrix} a_1 & b_1 & c_1 \\ a_2 & b_2 & c_2 \\ a_3 & b_3 & c_3 \end{vmatrix}.$$

2. Interchanging adjacent rows or columns changes the sign of the determinant. That is,

$$\begin{vmatrix} c & d \\ a & b \end{vmatrix} = -\begin{vmatrix} a & b \\ c & d \end{vmatrix} \quad \text{and} \quad \begin{vmatrix} b_1 & a_1 & c_1 \\ b_2 & a_2 & c_2 \\ b_3 & a_3 & c_3 \end{vmatrix} = -\begin{vmatrix} a_1 & b_1 & c_1 \\ a_2 & b_2 & c_2 \\ a_3 & b_3 & c_3 \end{vmatrix}.$$

3. If two rows or columns are proportional, the value of the determinant is zero. That is,

$$\begin{vmatrix} a & b \\ ka & kb \end{vmatrix} = 0 \quad \text{and} \quad \begin{vmatrix} a_1 & b_1 & kb_1 \\ a_2 & b_2 & kb_2 \\ a_3 & b_3 & kb_3 \end{vmatrix} = 0.$$

4. Prove the following result for the determinant of a second order matrix.

$$\begin{vmatrix} a + a' & b \\ c + c' & d \end{vmatrix} = \begin{vmatrix} a & b \\ c & d \end{vmatrix} + \begin{vmatrix} a' & b \\ c' & d \end{vmatrix}$$

If the three vertices of a triangle are thought of as changing position continuously until they lie on a straight line, the area of the triangle shrinks to the value 0. This is the intuitive basis for the fact that three points (x_1, y_1), (x_2, y_2), and (x_3, y_3) lie on a straight line if and only if the determinant representing the area of a triangle is zero (see Example 4, Section 8.4), that is, if

$$\begin{vmatrix} 1 & x_1 & y_1 \\ 1 & x_2 & y_2 \\ 1 & x_3 & y_3 \end{vmatrix} = 0.$$

Use this fact to determine which of the following sets of points are collinear.

5. $(-3, -3)$, $(3, 5)$, $(6, 9)$

6. $(0, 1)$, $(-1, 3)$, $(2, -3)$

7. $(0, 5/2)$, $(1, 5)$, $(-2, -1)$

8. $(0, 1)$, $(-1, 3)$, $(2, -3)$

9. $(1, 2)$, $(2, 6)$, $(4, 14)$

10. $(6, -5)$, $(1, 5)$, $(-2, -1)$

Any point (x, y) on a line is collinear with any other two points of the line. Thus, from above, the equation of the line determined by (x_1, y_1) and (x_2, y_2) is given by

$$\begin{vmatrix} 1 & x & y \\ 1 & x_1 & y_1 \\ 1 & x_2 & y_2 \end{vmatrix} = 0.$$

Use this fact to write the equation of the line determined by each of the following pairs of points.

11. $(2, 3)$ and $(5, 6)$

12. $(2, -3)$ and $(-3, -4)$

13. $(2, -3)$ and $(2, 5)$

14. $(3, -2)$ and $(-2, 4)$

9

Complex Numbers

In Chapter 1 the set of real numbers and some of its subsets—integers, rational numbers, and irrational numbers—were introduced. In this chapter, the *complex numbers,* mentioned in Chapter 3, are formally defined and operations and properties of real numbers are extended to these new numbers.

9.1 Complex Numbers and Their Operations

Each of the subsets of the real numbers above is related to the solution of a particular type of algebraic equation. For example, the equation $x - 5 = 0$ has the *integral* solution $x = 5$, the equation $3x - 4 = 0$ has the *rational* solution $x = 4/3$, and the equation $x^2 - 2 = 0$ has the *irrational* solutions $x = \sqrt{2}$ and $x = -\sqrt{2}$. However, the equation $x^2 + 1 = 0$ has no real number solution, since it is equivalent to $x^2 = -1$, and we know that the square of any real number is nonnegative.

To provide for a solution to this equation as well as many others, the real number system is extended to include a number whose square is -1. This number is called i; it has the property that

$$i^2 = -1 \qquad \text{or} \qquad i = \sqrt{-1}.$$

If the ordinary laws of exponents are applied, the first eight powers of i are:

$$\begin{aligned} i &= \sqrt{-1} & i^5 &= i^4 \cdot i = 1 \cdot i = i \\ i^2 &= -1 & i^6 &= i^4 \cdot i^2 = 1 \cdot (-1) = -1 \\ i^3 &= i^2 \cdot i = -1 \cdot i = -i & i^7 &= i^4 \cdot i^3 = 1 \cdot (-i) = -i \\ i^4 &= i^2 \cdot i^2 = (-1)(-1) = 1 & i^8 &= i^4 \cdot i^4 = 1 \cdot 1 = 1. \end{aligned}$$

Clearly, further powers will repeat the same values and thus the powers of i are, in succession, the four numbers i, -1, $-i$, and 1. This fact simplifies many calculations with complex numbers.

The number i is the simplest of all the complex numbers. The **system of complex numbers** is constructed by using the familiar algebraic properties and rules of operations with i. First, we use i to write the square roots of negative numbers. For example,

$$\sqrt{-3} = \sqrt{-1}\sqrt{3} = i\sqrt{3},$$
$$\sqrt{-4} = \sqrt{-1}\sqrt{4} = 2i,$$
$$\sqrt{-12} = \sqrt{-4}\sqrt{3} = 2i\sqrt{3}.$$

In general, if a is any positive real number,

$$\sqrt{-a} = \sqrt{-1}\sqrt{a} = i\sqrt{a}.$$

The numbers above are of the form bi, where b is a real number. Real numbers are combined with numbers of the form bi, according to the following definition, to get a new set of numbers.

DEFINITION 9.1 If a and b are any two real numbers, then $\mathbf{a + bi}$, where $i^2 = -1$, is a **complex number.**

We agree to take $0i = 0$. Then, if $b = 0$, $a + bi$ is the real number a. If $a = 0$ but $b \neq 0$, then $a + bi$ becomes bi, which is called a **(pure) imaginary number.** Finally, if $a = 0$ and $b = 0$, then $a + bi$ becomes $0 + 0i = 0$. Thus, both real numbers and pure imaginary numbers are particular cases of complex numbers.

When a complex number is written in the form $a + bi$, it is said to be in **standard form.** Then a is called its **real part** and b its **imaginary part.** The number i is called the **imaginary unit.** Note that in the examples above, the symbol i was written in front of the radical to prevent confusion about what is under the radical.

The term *imaginary* was introduced at a time when these numbers were not well understood. However, there are many "real" applications for these numbers in physics, electrical engineering, and advanced mathematics.

Two important concepts are frequently used in connection with complex numbers. The first is an extension of the notion of *absolute value* for real numbers:

> The **absolute value** of the complex number $a + bi$, written $|a + bi|$, is the *real number* $\sqrt{a^2 + b^2}$.

A geometric interpretation of the absolute value of a complex number similar to the one for real numbers will be given in Section 9.2. For now, note that this definition is equivalent to the one for real numbers in case $b = 0$.

Another important term used with complex numbers is the following:

> The **conjugate** of $a + bi$ is the complex number $a - bi$.

From the definition, a real number is its own conjugate, and the conjugate of an imaginary number bi is its **negative,** $-bi$. Note also that a complex number and its conjugate have the same absolute value.

Sometimes, a single letter is used to stand for a complex number. For example, we may use z for $a + bi$ or w for $c + di$. With this notation, $\bar{z}$ denotes the conjugate and $|z|$ the absolute value. That is, if $z = a + bi$, then $\bar{z} = a - bi$ and $|z| = \sqrt{a^2 + b^2}$.

The diagram in Figure 9.1 shows the relationships among the various sets of numbers we now know.

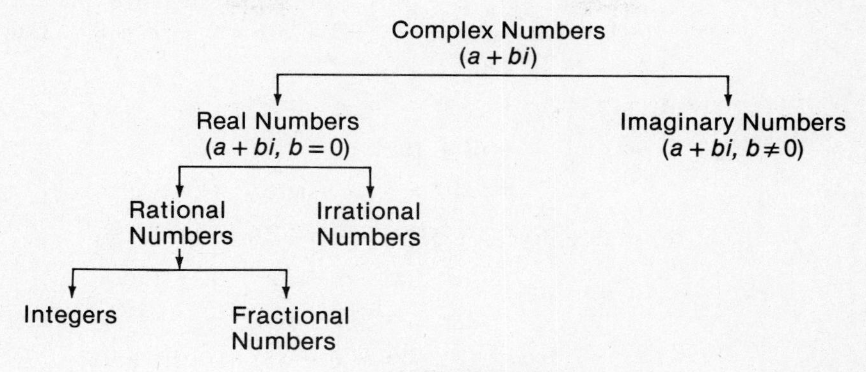

Figure 9.1

Before discussing operations on complex numbers it is necessary to have a definition of **equality** of complex numbers.

DEFINITION 9.2 Given the two complex numbers $a + bi$ and $c + di$,

$$a + bi = c + di \quad \textbf{if and only if} \quad a = c \textbf{ and } b = d.$$

Thus, two complex numbers are equal if and only if their *real parts are equal* and their *imaginary parts are equal*.

EXAMPLE 1 (a) If $x + yi = 2 - 3i$, then $x = 2$ and $y = -3$.

(b) If $(x + 3) + (y - 2)i = 1 + 2i$, then

$$x + 3 = 1 \quad \text{and} \quad y - 2 = 2,$$
$$x = -2 \quad \text{and} \quad y = 4. \quad \bullet$$

Operations on Complex Numbers The rules for operating on complex numbers are the same as the familiar algebraic rules for real numbers, with the additional fact that $i^2 = -1$. The following examples show how the rules are used.

EXAMPLE 2 (a) $(-2 + 3i) + (3 + 5i) = -2 + 3i + 3 + 5i$

$$= (-2 + 3) + (3i + 5i)$$
$$= 1 + 8i$$

(b) $(-2 + 3i) - (3 + 5i) = -2 + 3i - 3 - 5i$

$$= (-2 - 3) + (3i - 5i)$$
$$= -5 - 2i$$

(c) $-1(-2 + 3i) = (-1)(-2) + (-1)(3i)$
$= 2 - 3i$

In part (c), $2 - 3i$ is called the **negative** of $-2 + 3i$, since $(2 - 3i) + (-2 + 3i) = (2 - 2) + (-3 + 3)i = 0 + 0i = 0$. Thus, if $z = -2 + 3i$, then $-z = 2 - 3i$.

(d) $2i(3 + 5i) = 2i(3) + 2i(5i)$
$= 6i + 10i^2$
$= -10 + 6i$, since $i^2 = -1$.

(e) $(-2 + 3i) \cdot (3 + 5i) = -2(3 + 5i) + 3i(3 + 5i)$
$= -6 - 10i + 9i + 15i^2$
$= -6 - 10i + 9i - 15$, since $i^2 = -1$,
$= (-6 - 15) + (-10 + 9)i$
$= -21 - i$

(f) $(a + bi)(a - bi) = a^2 - (bi)^2$
$= a^2 - b^2i^2$
$= a^2 + b^2$, since $i^2 = -1$. •

Example 2(f) shows that *the product of a complex number and its conjugate is the square of their absolute value*. Observe that the result of each of the above operations on two complex numbers is a complex number.

To divide two complex numbers, for example, $(-2 + 3i) \div (3 + 5i)$, first write the quotient as a fraction:

$$(-2 + 3i) \div (3 + 5i) = \frac{-2 + 3i}{3 + 5i}.$$

To reduce this fractional expression to the standard form $a + bi$, both the numerator and denominator are multiplied by the same number. Since the product of conjugate complex numbers $a + bi$ and $a - bi$ is the real number $a^2 + b^2$, as shown in Example 2(f), this suggests multiplying numerator and denominator by $3 - 5i$, the conjugate of the denominator. The steps are given in part (a) of the next example.

EXAMPLE 3 (a)

$$(-2 + 3i) \div (3 + 5i) = \frac{-2 + 3i}{3 + 5i}$$
$$= \frac{(-2 + 3i)(3 - 5i)}{(3 + 5i)(3 - 5i)}$$
$$= \frac{-6 - 15i^2 + 9i + 10i}{9 - 25i^2}$$
$$= \frac{-6 + 15 + (9 + 10)i}{9 + 25}$$

$$= \frac{9 + 19i}{34}$$

$$= \frac{9}{34} + \frac{19}{34}i.$$

(b) $1 \div (-3 + 2i) = \dfrac{1}{-3 + 2i}$

$$= \frac{1}{-3 + 2i} \cdot \frac{-3 - 2i}{-3 - 2i}$$

$$= \frac{-3 - 2i}{9 - 4i^2}$$

$$= \frac{-3 - 2i}{9 + 4}, \qquad \text{since } i^2 = -1$$

$$= -\frac{3}{13} - \frac{2}{13}i \qquad \bullet$$

Example 3 illustrates that the quotient of two complex numbers is also a complex number.

The procedures used in the examples above can be carried out for any two complex numbers. This is shown as follows:

(1) $$(a + bi) + (c + di) = a + bi + c + di$$
$$= (a + c) + (b + d)i$$

(2) $$(a + bi) - (c + di) = a + bi - c - di$$
$$= (a - c) + (b - d)i$$

$$(a + bi) \cdot (c + di) = a(c + di) + bi(c + di) \qquad \text{distributive property}$$
$$= ac + adi + bci + bdi^2$$
$$= ac + (ad + bc)i - bd, \qquad \text{since } i^2 = -1,$$
$$= (ac - bd) + (ad + bc)i$$

(4) $$(a + bi) \div (c + di) = \frac{a + bi}{c + di} = \frac{(a + bi)(c - di)}{(c + di)(c - di)}$$
$$= \frac{ac - adi + bci - bdi^2}{c^2 - d^2i^2}$$
$$= \frac{(ac + bd) + (bc - ad)i}{c^2 + d^2}$$
$$= \frac{ac + bd}{c^2 + d^2} + \frac{bc - ad}{c^2 + d^2}i$$

As shown above, we add (or subtract) complex numbers by adding (or subtracting) the real parts and the imaginary parts separately. It is not advisable to use the general rules for multiplication and division in actual calculations. Rather, we proceed directly as in Example 2(c)–(f) and Example 3. The formulas are needed only for more general discussions of the operations. For instance, using them we can show that the usual commutative, associative, and distributive properties hold for these operations just as they do for the operations on real numbers.

In Section 3.3, the quadratic formula for solving quadratic equations was developed:

$$\text{If } ax^2 + bx + c = 0, \text{ then } x = \frac{-b \pm \sqrt{b^2 - 4ac}}{2a}.$$

There we were restricted to the cases $b^2 - 4ac \geq 0$, so that x would be real. Now this restriction can be removed.

EXAMPLE 4 Solve $x^2 + 2x + 5 = 0$.

SOLUTION If $x^2 + 2x + 5 = 0$, then by the quadratic formula

$$\begin{aligned} x &= \frac{-2 \pm \sqrt{2^2 - 4(5)}}{2(1)} \\ &= \frac{-2 \pm \sqrt{-16}}{2} \\ &= \frac{-2 \pm 4i}{2}. \end{aligned}$$

$$x = -1 + 2i \quad \text{or} \quad x = -1 - 2i.$$

Observe that the roots are *conjugate* complex numbers. This is always the case when there are complex roots, as the quadratic formula shows. •

The extension of our number system to complex numbers provides the result that every quadratic equation has exactly two roots, real or complex. Indeed, a similar fact holds for every polynomial equation, as stated in the following two theorems.

THEOREM 9.1 **Fundamental Theorem of Algebra**

Every polynomial $f(x)$ of degree $n > 0$ with real or complex coefficients has at least one zero, real or complex.

This theorem was first proved in 1799 by the great German mathematician Gauss. Its proof is too advanced to be given here.

It can also be shown that the division algorithm, the remainder theorem, and the factor theorem (Section 3.5) can be extended to polynomials with complex coefficients. As a consequence of these results, we have the following theorem.

THEOREM 9.2 If the polynomial $f(x)$ of degree $n > 0$ has complex coefficients and each zero of multiplicity k is counted k times, then $f(x)$ has *exactly* n zeros, real or complex.

From this theorem, we know, without actually solving, that the equation $x^3 - 4x^2 + 7x - 6 = 0$ has exactly three roots, real or complex.

EXAMPLE 5 Solve the equation $x^3 - 4x^2 + 7x - 6 = 0$.

SOLUTION By using synthetic division we find that $x = 2$ is a root and that $x^3 - 4x^2 + 7x - 6 = (x - 2)(x^2 - 2x + 3)$. The other two roots are found by solving the equation

$$x^2 - 2x + 3 = 0.$$

$$x = \frac{2 \pm \sqrt{(-2)^2 - 4(1)(3)}}{2(1)}$$

$$= \frac{2 \pm \sqrt{-8}}{2} = \frac{2 \pm 2i\sqrt{2}}{2}$$

$$= 1 \pm i\sqrt{2} \qquad \bullet$$

As seen in Examples 4 and 5, the complex roots of a polynomial equation with real coefficients *always* occur in conjugate pairs.

EXERCISES 9.1

Express each of the following powers of i in the simplest form as i, −i, 1, or −1.

1. i^3 **2.** i^5 **3.** i^6 **4.** i^7 **5.** i^8 **6.** i^9

7. i^{10} **8.** i^{11} **9.** i^{15} **10.** i^{17} **11.** i^{20} **12.** i^{42}

For each of the following, (a) write in standard form; (b) give the conjugate.

13. $1 + \sqrt{-4}$ **14.** $2 - \sqrt{-9}$ **15.** $5 + \sqrt{-3}$ **16.** $2 + \sqrt{-5}$

17. $-1 + \sqrt{-8}$ **18.** $3 + \sqrt{-16}$ **19.** $\sqrt{-9} + 2$ **20.** $\sqrt{-1} + 3$

21. $3i^3 - 2i^2$ **22.** $4i^2 + 5i^3$ **23.** $\sqrt{-24}$ **24.** 3

Use the definition of equality of complex numbers to solve for x and y.

25. $x + yi = -2 + 5i$ **26.** $x + yi = -1 + 4i$ **27.** $x - 2yi = 4$

28. $-x + yi = -3i$ **29.** $(x - 2) + (y + 1)i = 1 - i$

30. $(x + 1) + (y - 2)i = 2 + i$ **31.** $(x - y) + (x + y)i = 1 + 5i$

32. $(x + 2y) + (3x - y)i = 1 - 4i$ **33.** $(2x + 3y) + (5x - 2y)i = 13 + 4i$

34. $(3x - 3y) + (3x + y)i = 1 + 7i$

Perform the indicated operations and reduce the result to the standard form.

35. $(3 + 3i) + (6 - i)$ **36.** $(4 - i) + (7 + 4i)$ **37.** $(1 - i) + (4 + 3i)$

38. $(-2 + 3i) + (5 - 5i)$ **39.** $(3 - \sqrt{-1}) + (6 + \sqrt{-4})$

40. $(6 + \sqrt{-9}) + (2 - \sqrt{-16})$ **41.** $(-2 + 3i) - (5 - 5i)$

42. $(2 - 3i) - (4 + i)$ **43.** $2i - (3 + 6i)$ **44.** $(2 - i) - 3i$

45. $(1 + i)(3 - 4i)$ **46.** $(7 + 2i)(1 - i)$ **47.** $(1 - 6i)(3 + i)$

48. $(5 + 4i)(3 + 2i)$ **49.** $(2 - i)(2 + i)$ **50.** $(-3 + 2i)(-3 - 2i)$

51. $(3 + 2i) \cdot i$ **52.** $-3(1 - 2i)$ **53.** $(2 + i) \div (3 + i)$

54. $(3 - 2i) \div (3 + 2i)$ **55.** $(2 + 7i) \div (1 - 3i)$ **56.** $(3 + i) \div (4 + i)$

57. $\dfrac{1}{2 + i}$ **58.** $\dfrac{3 + i}{2}$ **59.** $\dfrac{1 - i}{1 + i}$ **60.** $\dfrac{3 - i}{i}$

Solve the following quadratic equations and reduce the complex roots to standard form.

61. $x^2 + 4 = 0$ **62.** $2x^2 + 1 = 0$ **63.** $x^2 - x + 1 = 0$

64. $x^2 + x + 1 = 0$ **65.** $x^2 + 2x + 4 = 0$ **66.** $x^2 + 4x + 3 = 0$

67. $2x^2 - 2x + 5 = 0$ **68.** $2x^2 + 2x + 1 = 0$ **69.** $2x^2 + 2x + 2 = 0$

70. $3x^2 + 2x + 1 = 0$

Use the procedure of Example 5 to solve each of the following equations.

71. $x^3 - 2x^2 + x - 2 = 0$ **72.** $x^3 + 4x^2 + 8x + 8 = 0$

73. $4x^4 + x^2 - 3x + 1 = 0$ **74.** $6x^4 - 11x^3 + 15x^2 - 22x + 6 = 0$

75. Show that a complex number z, its negative $-z$, and its conjugate $\bar{z}$ all have the same absolute value.

76. Show that the product of a pair of conjugate complex numbers is the square of their absolute value.

77. Show that if a complex number is real, it is equal to its conjugate and, conversely, if a complex number is equal to its conjugate, it is real.

78. Show that the conjugate of the sum of two complex numbers is the sum of their conjugates.

79. Show that a pure imaginary complex number is equal to the negative of its conjugate and, if a complex number is equal to the negative of its conjugate, it is a pure imaginary number.

80. If a and b are nonnegative real numbers, then a law of radicals states that $\sqrt{a} \cdot \sqrt{b} = \sqrt{ab}$. Show that, however, $\sqrt{-1} \cdot \sqrt{-1} \neq \sqrt{(-1)(-1)}$.

9.2 Geometric Representation of Complex Numbers

Any complex number may be represented by the form $a + bi$, where a and b are real numbers. That is, a complex number has associated with it a pair of real numbers, a and b. Thus, it may be pictured graphically in the Cartesian plane by the point with coordinates (a, b). Conversely, each point (a, b) can be associated with a unique complex number $a + bi$. Thus, *there is a one-to-one correspondence between the set of complex numbers and the set of points in the Cartesian plane*. In this context the plane is called the **complex plane.** See Figure 9.2. In the complex plane the real numbers are graphed on the x-axis, which is called the **real axis.** All imaginary numbers are graphed on the y-axis, and so it is called the **imaginary axis.**

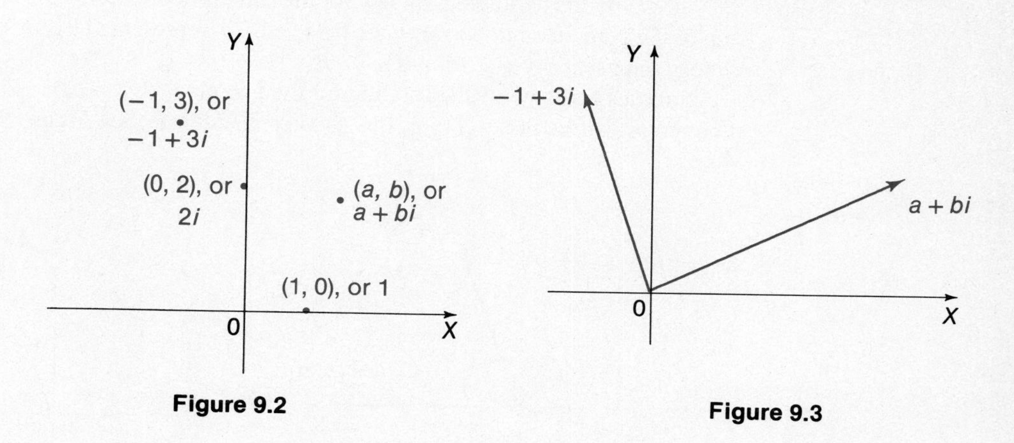

Figure 9.2

Figure 9.3

It is sometimes useful to represent complex numbers geometrically as vectors, as in Figure 9.3. This allows a simple geometric representation of sums and differences of complex numbers.

When vectors are used to represent complex numbers, the rule for adding complex numbers is equivalent to the parallelogram law for adding vectors (Section 6.6). To see this, let vectors $\overrightarrow{OP}$, $\overrightarrow{OQ}$, and $\overrightarrow{OR}$ represent the complex numbers $a + bi$, $c + di$, and $(a + c) + (b + d)i$, respectively, as in Figure 9.4. We will show that $OPRQ$ is a parallelogram by showing that its opposite sides are parallel.

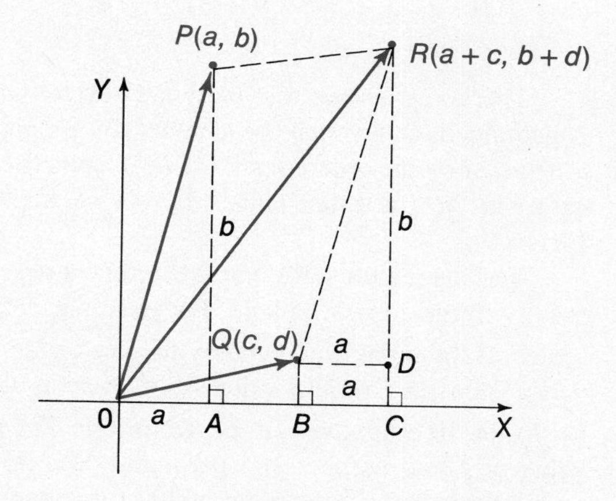

Figure 9.4

From the figure, the slope of OP is b/a. Since $\overline{OC} = a + c$ and $\overline{OB} = c$, then $\overline{BC} = \overline{OC} - \overline{OB} = (a + c) - c = a$, so that $\overline{QD}$ is also equal to a. Similarly, $\overline{DR} = \overline{CR} - \overline{CD} = (b + d) - d = b$, so that the slope of QR is b/a. Therefore, OP is parallel to QR, since their slopes are equal. In the same way, we can show that OQ is parallel to PR. Hence $OPRQ$ is a parallelogram, since its opposite sides are parallel.

Since (by the parallelogram law for the sum of two vectors) $\overrightarrow{OR} = \overrightarrow{OP} + \overrightarrow{OQ}$, the sum of the complex numbers $a + bi$ and $c + di$ is represented by the vector sum of the vectors representing $a + bi$ and $c + di$.

Subtraction is equivalent to adding the negative, that is, $z - u = z + (-u)$, so subtraction can be done graphically as vector addition. See Figure 9.5.

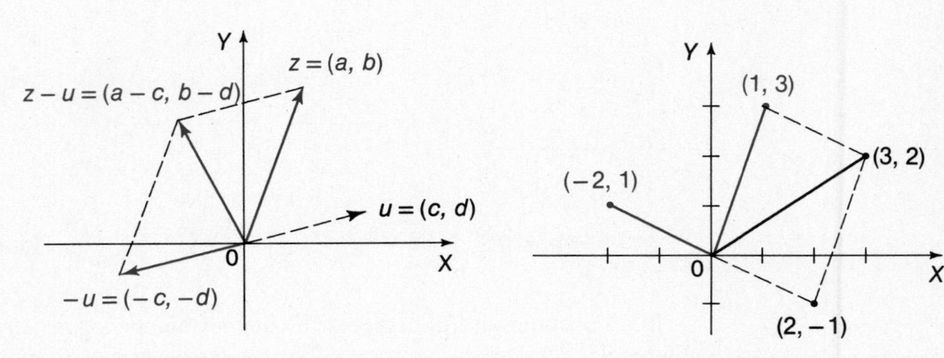

Figure 9.5 **Figure 9.6**

EXAMPLE 1 Represent $(1 + 3i) - (-2 + i)$ graphically.

SOLUTION In Figure 9.6 the numbers $1 + 3i$, $-2 + i$, and $-(-2 + i)$, or $2 - i$, are shown in the complex plane. When the parallelogram having adjacent sides $1 + 3i$ and $2 - i$ is completed, the endpoint of its diagonal is $(3, 2)$. Verify algebraically that $(1 + 3i) - (-2 + i) = 3 + 2i$. ●

The above shows that there are several ways of representing complex numbers. Algebraically, they may be denoted by a single letter, say z, by the standard form $a + bi$, or by the ordered pair (a, b). Geometrically they may be represented by a point (a, b) in the Cartesian plane or by a vector whose components are a and b, as in Figure 9.3.

Various features of complex numbers described earlier may be exhibited geometrically in the complex plane. For example, let the complex number $a + bi$ be represented as the point (a, b) and by the vector $\overrightarrow{OP}$ with terminal point (a, b). (See Figure 9.7.) Then its absolute value $\sqrt{a^2 + b^2}$ is simply the distance from the origin to (a, b), or the magnitude of the vector $\overrightarrow{OP}$. The geometric interpretation of the absolute value of a real number is a particular case of this, since it is the distance along the x-axis from the origin to the point corresponding to the real number.

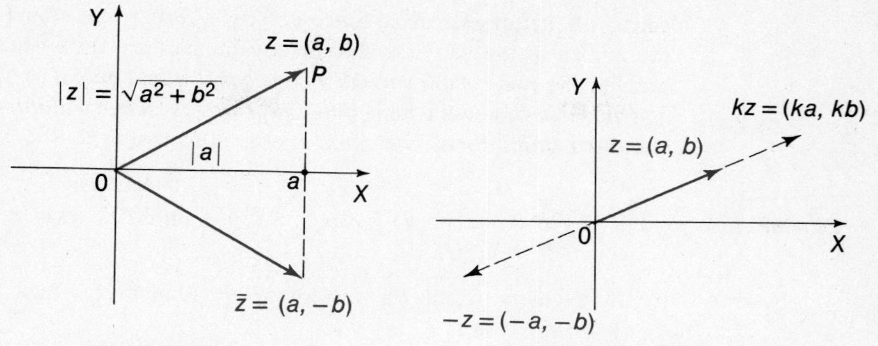

Figure 9.7 **Figure 9.8**

As Figure 9.7 shows, conjugate complex numbers $a + bi$ and $a - bi$ are represented by points symmetric to one another with respect to the real axis (x-axis). For k real and z complex, $kz = k(a + bi) = ka + (kb)i$, which is associated with the point (ka, kb). So the vector representing kz is in the same direction as z but has magnitude k times that of z. (See Figure 9.8.) In particular, if $k = -1$, then $kz = -z = -a - bi$, the negative of z. It is represented graphically by the point $(-a, -b)$, which is symmetric to z with respect to the origin.

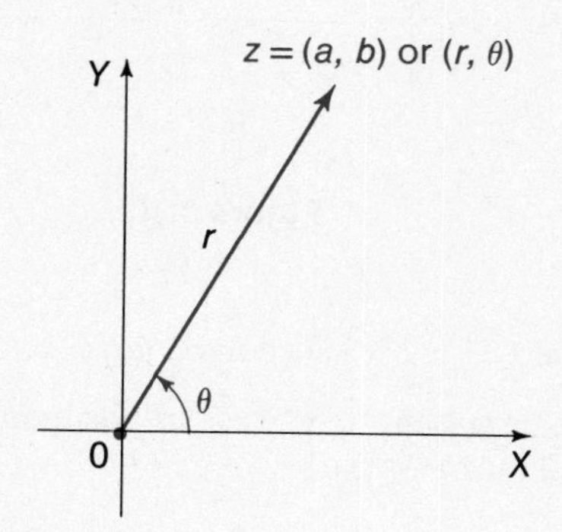

Figure 9.9

Polar Form of Complex Numbers If the point (a, b) is represented by its polar coordinates, as in Figure 9.9, then $a = r \cos \theta$ and $b = r \sin \theta$. The complex number $z = a + bi$ associated with (a, b) is given by

$$\begin{aligned} a + bi &= r \cos \theta + (r \sin \theta)i \\ &= r(\cos \theta + i \sin \theta). \end{aligned}$$

(This latter form is sometimes abbreviated $\boldsymbol{r}$ **cis** $\boldsymbol{\theta}$, suggested by **c**osine + $\boldsymbol{i}$ **s**ine.)

This form is called the **polar** (or **trigonometric**) **form** of the complex number $a + bi$. The form $a + bi$ will be referred to as the **Cartesian** or **algebraic form.** Here, the positive real number r, which is equal to $|z|$, is called the **modulus** of z and θ is called the **argument** of z. Since $\cos(\theta + 2k\pi) = \cos\theta$ and $\sin(\theta + 2k\pi) = \sin\theta$, the angles in the set $\{\theta + 2k\pi \mid k \text{ any integer}\}$ are arguments of the same complex number. So *two complex numbers are equal if and only if they have the same modulus and have arguments which differ at most by some multiple of 2π, or $360°$*. As a **standard polar form,** we often choose θ so that $0 \le \theta < 2\pi$.

EXAMPLE 2 (a) If $r(\cos\theta + i\sin\theta) = 2(\cos 30° + i\sin 30°)$, then $r = 2$ and $\theta = 30°$, or $\theta = 30° + k(360°)$.

(b) If $r(\cos\theta + i\sin\theta) = 3(\cos\pi/2 + i\sin\pi/2)$, then $r = 3$ and $\theta = \pi/2 + 2k\pi$. ●

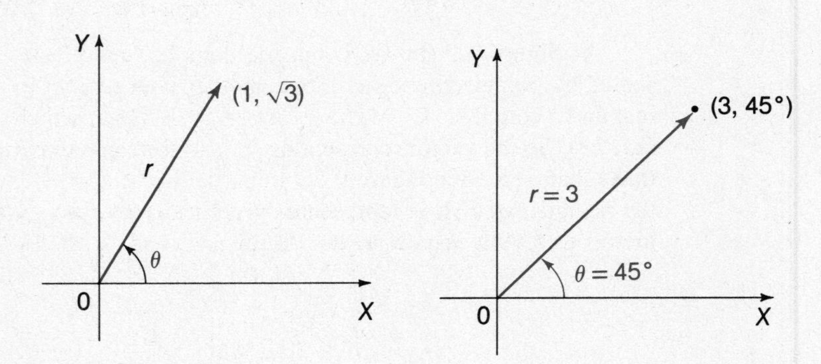

Figure 9.10

Figure 9.11

EXAMPLE 3 Convert $1 + i\sqrt{3}$ to polar form.

SOLUTION First, represent $1 + i\sqrt{3}$ as the point $(1, \sqrt{3})$, as shown in Figure 9.10. Then $r = |1 + i\sqrt{3}|$, so

$$r^2 = 1^2 + (\sqrt{3})^2 = 1 + 3 = 4,$$
$$r = 2.$$

Also,

$$\tan\theta = \frac{\sqrt{3}}{1} = \sqrt{3},$$

$$\theta = 60°, \text{ or } \frac{\pi}{3}.$$

Thus,

$$1 + i\sqrt{3} = r(\cos\theta + i\sin\theta)$$
$$= 2\left(\cos\frac{\pi}{3} + i\sin\frac{\pi}{3}\right).$$

As a check, observe that

$$2\left(\cos\frac{\pi}{3} + i\sin\frac{\pi}{3}\right) = 2\left(\frac{1}{2} + \frac{1}{2}i\sqrt{3}\right) = 1 + i\sqrt{3},$$

since $\cos \pi/3 = 1/2$ and $\sin \pi/3 = \frac{1}{2}\sqrt{3}$. ●

EXAMPLE 4 Convert $3(\cos 45° + i \sin 45°)$ to Cartesian form. (Refer to Figure 9.11.)

SOLUTION We know that $\cos 45° = \frac{1}{2}\sqrt{2}$ and $\sin 45° = \frac{1}{2}\sqrt{2}$, and so

$$3(\cos 45° + i\sin 45°) = 3\left(\frac{1}{2}\sqrt{2} + \frac{1}{2}i\sqrt{2}\right)$$
$$= \frac{3}{2}\sqrt{2} + \frac{3}{2}i\sqrt{2}. \quad ●$$

In the general case,

$$a + bi = r(\cos\theta + i\sin\theta)$$
$$= \sqrt{a^2+b^2}\left(\frac{a}{\sqrt{a^2+b^2}} + \frac{b}{\sqrt{a^2+b^2}}i\right),$$

where $r = \sqrt{a^2+b^2}$, $\cos\theta = a/\sqrt{a^2+b^2}$, $\sin\theta = b/\sqrt{a^2+b^2}$, and $\tan\theta = b/a$, all of which can be seen from Figure 9.12. We know that $\tan(\theta + \pi) = \tan\theta$, from Section 5.2. So, if $\tan\theta = b/a$, then $\theta = \arctan b/a$ or $\pi + \arctan b/a$. The proper choice will be clear from the quadrant in which θ is located.

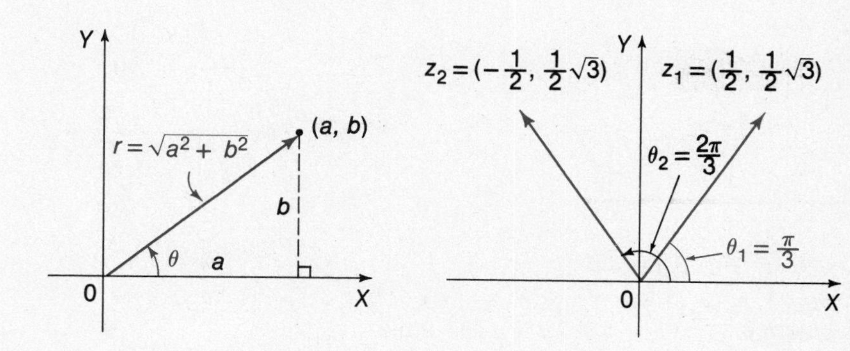

Figure 9.12

Figure 9.13

For example, as pictured in Figure 9.13, let

$$z_1 = \frac{1}{2} + \frac{1}{2}i\sqrt{3} \quad \text{and} \quad z_2 = -\frac{1}{2} + \frac{1}{2}i\sqrt{3},$$

and let θ_1 be the angle associated with z_1 and θ_2 the angle associated with z_2. Then

$$\tan\,\theta_1 = \frac{\frac{1}{2}\sqrt{3}}{\frac{1}{2}} = \sqrt{3},$$

$$\theta_1 = \frac{\pi}{3} \quad \text{or} \quad \frac{4\pi}{3}.$$

Since z_1 is in the first quadrant, we choose $\theta_1 = \pi/3$, or arctan $\sqrt{3}$. Similarly,

$$\tan\,\theta_2 = \frac{\frac{1}{2}\sqrt{3}}{-\frac{1}{2}} = -\sqrt{3},$$

$$\theta_2 = \frac{2\pi}{3} \quad \text{or} \quad \frac{5\pi}{3}.$$

Since z_2 is in the second quadrant, we choose $\theta_2 = 2\pi/3$. Note that $2\pi/3 \neq \arctan(-\sqrt{3})$, but rather $2\pi/3 = \pi + \arctan(-\sqrt{3})$. Thus, the right choice of θ can always be made from the graph. Just note in which quadrant the vector is located.

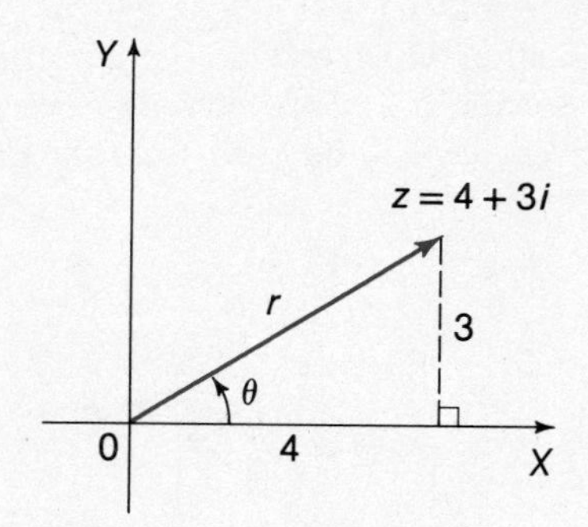

Figure 9.14

EXAMPLE 5 Write $4 + 3i$ in polar form.

SOLUTION As shown in Figure 9.14,

$$r^2 = 4^2 + 3^2 = 25 \qquad \qquad \tan\,\theta = \frac{3}{4} = 0.75$$
$$\text{and}$$
$$r = 5, \qquad \qquad 6 = 0.64 \text{ (or } 36.9^\circ\text{), approximately.}$$

Since $a + bi = r(\cos\,\theta + i\,\sin\,\theta)$,

$$4 + 3i = 5(\cos 0.64 + i \sin 0.64), \text{ approximately.}$$ ●

Multiplication and Division of Complex Numbers with the Polar Form The polar form of complex numbers is especially useful in multiplication and division and in raising to powers and extracting roots. These applications are based upon the following theorem.

THEOREM 9.3 If $u = r_1(\cos\theta + i\sin\theta)$ and $v = r_2(\cos\phi + i\sin\phi)$, then

(1) $$uv = r_1r_2[\cos(\theta + \phi) + i\sin(\theta + \phi)]$$

and

(2) $$\frac{u}{v} = \frac{r_1}{r_2}[\cos(\theta - \phi) + i\sin(\theta - \phi)].$$

Proof: (1) $r_1(\cos\theta + i\sin\theta) \cdot r_2(\cos\phi + i\sin\phi)$

$$= r_1r_2[\cos\theta\cos\phi + \cos\theta(i\sin\phi) + (i\sin\theta)\cos\phi + (i\sin\theta)(i\sin\phi)]$$

$$= r_1r_2[\cos\theta\cos\phi + i\cos\theta\sin\phi + i\sin\theta\cos\phi + i^2\sin\theta\sin\phi]$$

$$= r_1r_2[(\cos\theta\cos\phi - \sin\theta\sin\phi) + i(\cos\theta\sin\phi + \sin\theta\cos\phi)]$$

$$= r_1r_2[\cos(\theta + \phi) + i\sin(\theta + \phi)].$$

(2) $r_1(\cos\theta + i\sin\theta) \div r_2(\cos\phi + i\sin\phi)$

$$= \frac{r_1(\cos\theta + i\sin\theta)}{r_2(\cos\phi + i\sin\phi)}$$

$$= \frac{r_1}{r_2} \cdot \frac{(\cos\theta + i\sin\theta)(\cos\phi - i\sin\phi)}{(\cos\phi + i\sin\phi)(\cos\phi - i\sin\phi)}$$

$$= \frac{r_1}{r_2} \cdot \frac{(\cos\theta\cos\phi + \sin\theta\sin\phi) + i(\sin\theta\cos\phi - \cos\theta\sin\phi)}{\cos^2\phi - i^2\sin^2\phi}$$

$$= \frac{r_1}{r_2} \cdot \frac{\cos(\theta - \phi) + i\sin(\theta - \phi)}{\cos^2\phi + \sin^2\phi}$$

$$= \frac{r_1}{r_2}[\cos(\theta - \phi) + i\sin(\theta - \phi)], \text{ since } \sin^2\phi + \cos^2\phi = 1.$$

Theorem 9.3 says that (1) the modulus of the product of two complex numbers is the product of their moduli and the argument of the product is the sum of the arguments, and (2) the modulus of the quotient of two complex numbers is the quotient of their moduli, and the argument of the quotient is the difference of the arguments. Briefly, to multiply two complex numbers in polar form, multiply their moduli and add their arguments; to divide them, divide their moduli and subtract their arguments.

EXAMPLE 6 (a) $2(\cos 30° + i \sin 30°) \cdot 3(\cos 120° + i \sin 120°)$

$$= (2)(3)[\cos(30° + 120°) + i \sin(30° + 120°)]$$
$$= 6(\cos 150° + i \sin 150°)$$

(b) $4\left(\cos \frac{\pi}{6} + i \sin \frac{\pi}{6}\right) \cdot 2\left(\cos \frac{\pi}{4} + i \sin \frac{\pi}{4}\right)$

$$= 4(2)\left[\cos\left(\frac{\pi}{6} + \frac{\pi}{4}\right) + i \sin\left(\frac{\pi}{6} + \frac{\pi}{4}\right)\right]$$
$$= 8\left(\cos \frac{5\pi}{12} + i \sin \frac{5\pi}{12}\right) \quad \bullet$$

EXAMPLE 7 (a) $4(\cos 115° + i \sin 115°) \div 2(\cos 70° + i \sin 70°)$

$$= \frac{4}{2}[\cos(115° - 70°) + i \sin(115° - 70°)]$$
$$= 2(\cos 45° + i \sin 45°)$$

(b) $2\left(\cos \frac{\pi}{6} + i \sin \frac{\pi}{6}\right) \div 3\left(\cos \frac{\pi}{4} + i \sin \frac{\pi}{4}\right)$

$$= \frac{2}{3}\left[\cos\left(\frac{\pi}{6} - \frac{\pi}{4}\right) + i \sin\left(\frac{\pi}{6} - \frac{\pi}{4}\right)\right]$$
$$= \frac{2}{3}\left[\cos\left(-\frac{\pi}{12}\right) + i \sin\left(-\frac{\pi}{12}\right)\right]$$
$$= -\frac{2}{3}\left[\cos\left(\frac{11\pi}{12}\right) + i \sin\left(\frac{11\pi}{12}\right)\right],$$

or

$$\frac{2}{3}\left[\cos\left(\frac{23\pi}{12}\right) + i \sin\left(\frac{23\pi}{12}\right)\right] \quad \bullet$$

EXERCISES 9.2

For each of the following complex numbers z, graph z, its negative $-z$, and its conjugate $\bar{z}$ on the same figure. Also calculate its absolute value $|z|$.

1. 3 **2.** $1 - i$ **3.** $-3 - 2i$ **4.** $2i$ **5.** $2 + 3i$

6. $5 + 2i$ **7.** $-3i$ **8.** $-4 + 2i$ **9.** $-3 + 4i$

Represent each of the following sums and differences geometrically.

10. $(2 + i) + (3 - 2i)$

11. $(4 - 3i) + (1 - 2i)$

12. $(4 + 3i) - (1 - i)$

13. $(5 - 2i) - (2 + 3i)$

14. $(2 - 3i) + (1 + i)$

15. $(5 + 2i) + (3 - 4i)$

16. $(3 - 4i) - (1 + i)$

17. $(7 + 5i) - (2 + 4i)$

For each of the following complex numbers, (a) graph and (b) convert into Cartesian form.

18. $2(\cos 30° + i \sin 30°)$ **19.** $3(\cos 45° + i \sin 45°)$ **20.** $4\left(\cos \frac{5\pi}{3} + i \sin \frac{5\pi}{3}\right)$

21. $\cos \pi + i \sin \pi$ **22.** $-2(\cos 270° + i \sin 270°)$ **23.** $3(\cos 60° + i \sin 60°)$

24. $2(\cos 135° + i \sin 135°)$ **25.** $5\left(\cos \frac{5\pi}{6} + i \sin \frac{5\pi}{6}\right)$ **26.** $-(\cos 0° + i \sin 0°)$

27. $\cos\left(-\frac{1}{2}\pi\right) + i \sin\left(-\frac{1}{2}\pi\right)$

Convert each of the following complex numbers to polar form, with $0 \le \theta < 360°$, using a graph as in the text examples.

28. 2 **29.** $2 + 2i$ **30.** $2 - 2i$ **31.** -3 **32.** $-2 + 2i$ **33.** $\frac{1}{2} - \frac{1}{2}i\sqrt{3}$

34. $3i$ **35.** $-1 + i$ **36.** $1 - 3i$ **37.** $-2i$ **38.** $1 - i$ **39.** $-1 + 3i$

Write in Cartesian form. Use decimal approximations.

40. $2(\cos 40° + i \sin 40°)$ **41.** $\cos 65° + i \sin 65°$ **42.** $\cos \frac{5\pi}{12} + i \sin \frac{5\pi}{12}$

43. $3(\cos 20° + i \sin 20°)$ **44.** $\cos 37° + i \sin 37°$ **45.** $\cos \frac{3\pi}{8} + i \sin \frac{3\pi}{8}$

Use the conditions for equality of complex numbers in polar form to solve for r and θ, $r > 0$ and $0 \le \theta < 2\pi$.

46. $r(\cos \theta + i \sin \theta) = 3(\cos 45° + i \sin 45°)$

47. $r(\cos \theta + i \sin \theta) = -2(\cos 30° + i \sin 30°)$

48. $r(\cos 2\theta + i \sin 2\theta) = 4\left(\cos \frac{2\pi}{3} + i \sin \frac{2\pi}{3}\right)$

49. $r(\cos 3\theta + i \sin 3\theta) = 2\left(\cos \frac{5\pi}{3} + i \sin \frac{5\pi}{3}\right)$

50. Show that if $z = r(\cos \theta + i \sin \theta)$, then its conjugate $\bar{z} = r[\cos(-\theta) + i \sin(-\theta)]$.

51. Every real number has the same argument. What is it? Repeat for every imaginary number.

Perform the indicated operations and leave the result in polar form.

52. $3\left(\cos \frac{\pi}{6} + i \sin \frac{\pi}{6}\right) \cdot 2\left(\cos \frac{\pi}{3} + i \sin \frac{\pi}{3}\right)$

53. $2(\cos 70° + i \sin 70°) \cdot 3(\cos 20° + i \sin 20°)$

54. $3(\cos 10° + i \sin 10°) \cdot 4(\cos 125° + i \sin 125°)$

55. $4(\cos 2\pi + i \sin 2\pi) \cdot 2\left(\cos \frac{\pi}{2} + i \sin \frac{\pi}{2}\right)$

56. $(\cos 20° + i \sin 20°)(\cos 30° + i \sin 30°)(\cos 40° + i \sin 40°)$

57. $\left(\cos \frac{\pi}{12} + i \sin \frac{\pi}{12}\right)\left(\cos \frac{\pi}{6} + i \sin \frac{\pi}{6}\right)\left(\cos \frac{\pi}{4} + i \sin \frac{\pi}{4}\right)$

58. $4(\cos 70° + i \sin 70°) \div 2(\cos 40° + i \sin 40°)$

59. $(\cos 130° + i \sin 130°) \div 4(\cos 85° + i \sin 85°)$

60. $-8\left(\cos \frac{\pi}{3} + i \sin \frac{\pi}{3}\right) \div 4\left(\cos \frac{\pi}{6} + i \sin \frac{\pi}{6}\right)$

61. $6\left(\cos \frac{3\pi}{4} + i \sin \frac{3\pi}{4}\right) \div 2\left(\cos \frac{\pi}{4} + i \sin \frac{\pi}{4}\right)$

62. $2\left(\cos \frac{2\pi}{3} + i \sin \frac{2\pi}{3}\right) \div 3\left(\cos \frac{\pi}{2} + i \sin \frac{\pi}{2}\right)$

63. $5\left(\cos \frac{\pi}{3} + i \sin \frac{\pi}{3}\right) \div \left(\cos \frac{\pi}{4} + i \sin \frac{\pi}{4}\right)$

9.3 De Moivre's Theorem: Powers and Roots

The last section showed the simplicity of multiplication and division of complex numbers in the polar form. A further use for this form is the calculation of powers and roots of complex numbers.

EXAMPLE 1 Find the square of $3\left(\cos \frac{\pi}{6} + i \sin \frac{\pi}{6}\right)$.

SOLUTION

$$\begin{aligned}\left[3\left(\cos \frac{\pi}{6} + i \sin \frac{\pi}{6}\right)\right]^2 &= 3\left(\cos \frac{\pi}{6} + i \sin \frac{\pi}{6}\right) \cdot 3\left(\cos \frac{\pi}{6} + i \sin \frac{\pi}{6}\right)\\ &= 3 \cdot 3\left[\cos\left(\frac{\pi}{6} + \frac{\pi}{6}\right) + i \sin\left(\frac{\pi}{6} + \frac{\pi}{6}\right)\right]\\ &= 3^2\left[\cos 2\left(\frac{\pi}{6}\right) + i \sin 2\left(\frac{\pi}{6}\right)\right]\end{aligned}$$ •

In general,

$$\begin{aligned}[r(\cos \theta + i \sin \theta)]^2 &= r^2[\cos(\theta + \theta) + i \sin(\theta + \theta)]\\ &= r^2(\cos 2\theta + i \sin 2\theta).\end{aligned}$$

Similarly, we could show that

$$[r(\cos \theta + i \sin \theta)]^3 = r^3(\cos 3\theta + i \sin 3\theta).$$

These are instances of the following theorem.

THEOREM 9.4 **De Moivre's Theorem**

If k is any integer, then

$$[r(\cos \theta + i \sin \theta)]^k = r^k(\cos k\theta + i \sin k\theta).$$

The pattern for proving this theorem for positive integral values of k is shown in the examples above. The case for $k = 0$, which follows, is very simple.

$$[r(\cos \theta + i \sin \theta)]^0 = 1,$$

by definition of $a^0 = 1$. Also,

$$r^0[\cos 0(\theta) + i \sin 0(\theta)] = 1(\cos 0 + i \sin 0) = 1 \cdot 1 = 1.$$

So

$$[r(\cos \theta + i \sin \theta)]^0 = r^0[\cos 0(\theta) + i \sin 0(\theta)],$$

which is De Moivre's Thereom for $k = 0$.

As an example with k negative, the particular case $k = -1$ is proved below.

$$\begin{aligned}
[r(\cos \theta + i \sin \theta)]^{-1} &= \frac{1}{r(\cos \theta + i \sin \theta)} \\
&= \frac{1(\cos 0 + i \sin 0)}{r(\cos \theta + i \sin \theta)} \qquad \text{since } \cos 0 + i \sin 0 = 1 \\
&= \frac{1}{r}[\cos(0 - \theta) + i \sin(0 - \theta)] \\
&= \frac{1}{r}[\cos(-\theta) + i \sin(-\theta)] \\
&= r^{-1}[\cos(-1)\theta + i \sin(-1)\theta]
\end{aligned}$$

The above result is in agreement with De Moivre's Theorem. Other cases for k, a positive integer, are left to the exercises.

EXAMPLE 2 Use De Moivre's theorem to find (a) $(1 + i)^3$, (b) $(-1 + i\sqrt{3})^{-2}$.

SOLUTION (a)

$$\begin{aligned}
(1 + i)^3 &= \left[\sqrt{2}\left(\cos \frac{\pi}{4} + i \sin \frac{\pi}{4}\right)\right]^3 \\
&= (\sqrt{2})^3\left(\cos \frac{3\pi}{4} + i \sin \frac{3\pi}{4}\right) \\
&= 2\sqrt{2}\left(-\frac{1}{2}\sqrt{2} + \frac{1}{2}i\sqrt{2}\right) \\
&= -2 + 2i
\end{aligned}$$

(b)

$$\begin{aligned}
(-1 + i\sqrt{3})^{-2} &= \left[2\left(\cos \frac{2\pi}{3} + i \sin \frac{2\pi}{3}\right)\right]^{-2} \\
&= 2^{-2}\left[\cos(-2)\frac{2\pi}{3} + i \sin(-2)\frac{2\pi}{3}\right] \\
&= \frac{1}{4}\left[\cos\left(-\frac{4\pi}{3}\right) + i \sin\left(-\frac{4\pi}{3}\right)\right]
\end{aligned}$$

$$= \frac{1}{4}\left(-\frac{1}{2} + \frac{1}{2}i\sqrt{3}\right)$$

$$= -\frac{1}{8} + \frac{1}{8}i\sqrt{3} \qquad \bullet$$

EXAMPLE 3 Use De Moivre's Theorem to derive the double-value formulas for the sine and cosine functions. (These formulas are discussed in Section 5.6.)

SOLUTION

$$\begin{aligned}(\cos\theta + i\sin\theta)^2 &= \cos^2\theta + 2i\sin\theta\cos\theta + i^2\sin^2\theta \\ &= (\cos^2\theta - \sin^2\theta) + i(2\sin\theta\cos\theta)\end{aligned}$$

By De Moivre's Theorem,

$$(\cos\theta + i\sin\theta)^2 = \cos 2\theta + i\sin 2\theta.$$

From the last two equations,

$$\cos 2\theta + i\sin 2\theta = (\cos^2\theta - \sin^2\theta) + i(2\sin\theta\cos\theta).$$

The desired formulas follow from the conditions for equality of complex numbers.

$$\cos 2\theta = \cos^2\theta - \sin^2\theta \text{ and } \sin 2\theta = 2\sin\theta\cos\theta \qquad \bullet$$

Formulas for $\sin 3\theta$, $\cos 3\theta$, $\sin 4\theta$, $\cos 4\theta$, and so on, can be similarly derived from De Moivre's Theorem.

Roots of Complex Numbers To introduce this idea, we again start with some examples.

EXAMPLE 4 Calculate the square roots of $-1 + i\sqrt{3}$.

SOLUTION Begin by finding the polar form of $-1 + i\sqrt{3}$.

$$r^2 = a^2 + b^2 = (-1)^2 + (\sqrt{3})^2 = 1 + 3 = 4$$

$$r = 2$$

Also,

$$\tan\theta = \frac{b}{a} = \frac{\sqrt{3}}{-1} = -\sqrt{3}$$

$$\theta = \frac{2\pi}{3},$$

since the complex number $-1 + i\sqrt{3}$ is in the second quadrant. Now, let z be the square root of $-1 + i\sqrt{3}$, so that

$$z^2 = -1 + i\sqrt{3} = 2\left(\cos\frac{2\pi}{3} + i\sin\frac{2\pi}{3}\right)$$

in polar form. Now, if z is represented by $s(\cos\phi + i\sin\phi)$, then

$$z^2 = [s(\cos \phi + i \sin \phi)]^2$$
$$= s^2(\cos 2\phi + i \sin 2\phi). \quad \text{by De Moivre's theorem}$$

$$s^2 = 2 \qquad 2\phi = \frac{2\pi}{3} + 2k\pi$$

and

$$s = \sqrt{2} \qquad \phi = \frac{\pi}{3} \text{ and } \frac{\pi}{3} + \pi = \frac{4\pi}{3},\ k = 0, 1$$

Thus, if we let z_0 and z_1 be the values of the roots of $-1 + i\sqrt{3}$ corresponding to the two values of ϕ,

$$z_0 = \sqrt{2}\left(\cos \frac{\pi}{3} + i \sin \frac{\pi}{3}\right) \quad \text{and} \quad z_1 = \sqrt{2}\left(\cos \frac{4\pi}{3} + i \sin \frac{4\pi}{3}\right)$$
$$= \sqrt{2}\left(\frac{1}{2} + \frac{1}{2}i\sqrt{3}\right) \qquad = \sqrt{2}\left(-\frac{1}{2} - \frac{1}{2}i\sqrt{3}\right)$$
$$= \frac{1}{2}\sqrt{2} + \frac{1}{2}i\sqrt{6} \qquad = -\frac{1}{2}\sqrt{2} - \frac{1}{2}i\sqrt{6}.$$

Check these results by squaring. ●

In the above development we used the identities $\cos 2\pi/3 = \cos(2\pi/3 + 2k\pi)$ and $\sin 2\pi/3 = \sin(2\pi/3 + 2k\pi)$ with the values $k = 0$ and $k = 1$. Since k may be any integer, we may write

$$2\phi = \frac{2\pi}{3} + 2k\pi, \quad \text{or} \quad \phi = \frac{\pi}{3} + k\pi.$$

Then

$$\text{for } k = 0, \quad \phi_0 = \frac{\pi}{3};$$
$$\text{for } k = 1, \quad \phi_1 = \frac{\pi}{3} + \pi = \frac{4\pi}{3};$$
$$\text{for } k = 2, \quad \phi_2 = \frac{\pi}{3} + 2\pi = \frac{7\pi}{3};$$
$$\text{for } k = 3, \quad \phi_3 = \frac{\pi}{3} + 3\pi = \frac{10\pi}{3},$$

and so on.

Now,

$$\sin \frac{7\pi}{3} = \sin \frac{\pi}{3} \text{ and } \cos \frac{7\pi}{3} = \cos \frac{\pi}{3},$$
$$\sin \frac{10\pi}{3} = \sin \frac{4\pi}{3} \text{ and } \cos \frac{10\pi}{3} = \cos \frac{4\pi}{3},$$

and so on. The only *distinct* roots are obtained from $\phi_0 = \pi/3$ and $\phi_1 = 4\pi/3$.

Observe that, just as in the case of real numbers, this complex number has *two* square roots, which are *negatives* of one another. Graphically, they appear at the same distance from the origin and are located symmetrically with respect to the origin. See Figure 9.15.

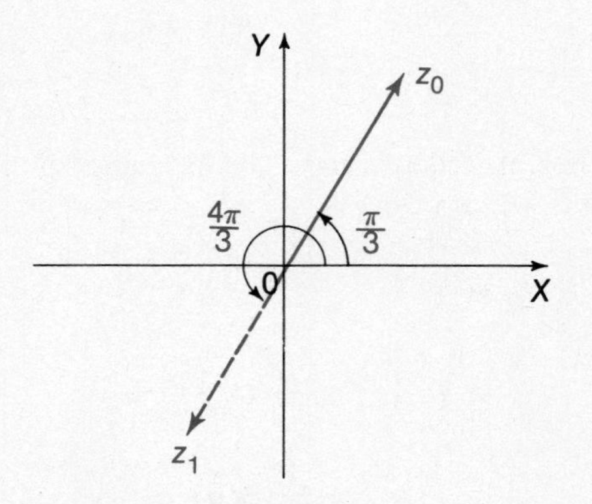

Figure 9.15

EXAMPLE 5 Calculate the cube roots of 1.

SOLUTION Let $z^3 = 1$, where $z = r(\cos\theta + i\sin\theta)$. Then

$$[r(\cos\theta + i\sin\theta)]^3 = 1$$

$$r^3(\cos 3\theta + i\sin 3\theta) = \cos 0 + i\sin 0,$$

since $\cos 0 + i\sin 0 = 1 + i(0) = 1$,

$$= \cos 2k\pi + i\sin 2k\pi.$$

Thus, $r^3 = 1$, $r = 1$, and $3\theta = 2k\pi$, $\theta = 2k\pi/3$.

Then, for $k = 0$, $\theta_0 = 0$, for $k = 2$, $\theta_2 = \dfrac{4\pi}{3}$,

for $k = 1$, $\theta_1 = \dfrac{2\pi}{3}$, for $k = 3$, $\theta_3 = \dfrac{6\pi}{3} = 2\pi$,

and so on. Since $\cos 2\pi = \cos 0$ and $\sin 2\pi = \sin 0$, the only *distinct* values of the cube roots $z = r(\cos\theta + i\sin\theta)$ are those for $\theta = 0$, $\theta = 2\pi/3$, and $\theta = 4\pi/3$. Corresponding to them we get

$$z_0 = \cos 0 + i\sin 0 = 1,$$

$$z_1 = \cos\frac{2\pi}{3} + i\sin\frac{2\pi}{3} = -\frac{1}{2} + \frac{1}{2}\sqrt{3},$$

$$z_2 = \cos\frac{4\pi}{3} + i\sin\frac{4\pi}{3} = -\frac{1}{2} - \frac{1}{2}\sqrt{3}.$$

Graphically, the roots are represented by three equally-spaced points on the standard unit circle. See Figure 9.16. ●

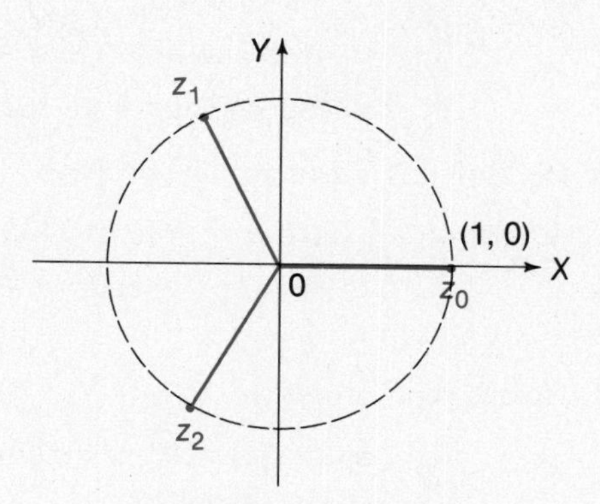

Figure 9.16

The results of the above example may be obtained algebraically as follows. Let

$$x^3 = 1,$$
$$x^3 - 1 = 0,$$
$$(x - 1)(x^2 + x + 1) = 0,$$

by factoring the difference of cubes (Section 3.1).

$$x - 1 = 0 \qquad \text{or} \qquad x^2 + x + 1 = 0$$

$$x = 1 \qquad\qquad x = \frac{-1 \pm \sqrt{1^2 - 4(1)(1)}}{2}$$

$$x = \frac{-1 \pm \sqrt{-3}}{2} = \frac{-1 \pm i\sqrt{3}}{2},$$

so that the cube roots of 1 are 1, $(-1 + i\sqrt{3})/2$, and $(-1 - i\sqrt{3})/2$.

EXAMPLE 6 Calculate the fourth roots of $-8 - 8i\sqrt{3}$.

SOLUTION We first need to convert to the polar form.

$$r^2 = a^2 + b^2 = (-8)^2 + (-8\sqrt{3})^2$$
$$= 64 + 192 = 256$$
$$r = 16$$

Also,

$$\tan\theta = \frac{b}{a} = \frac{-8\sqrt{3}}{-8} = \sqrt{3},$$
$$\theta = 240°,$$

since $-8 - 8i\sqrt{3}$ is in the third quadrant. We now let z denote a fourth root of $-8 - 8i\sqrt{3}$.

$$\begin{aligned} z^4 &= -8 - 8i\sqrt{3} \\ &= 16(\cos 240^\circ + i \sin 240^\circ) \\ &= 16[(\cos 240^\circ + k \cdot 360^\circ) + i \sin(240^\circ + k \cdot 360^\circ)] \end{aligned}$$

By De Moivre's Theorem, this becomes

$$r^4(\cos 4\theta + i \sin 4\theta) = 16[(\cos 240^\circ + k \cdot 360^\circ) + i \sin(240^\circ + k \cdot 360^\circ)]$$

$$r^4 = 16 \quad \text{and} \quad 4\theta = 240^\circ + k \cdot 360^\circ$$

$$r = 2 \qquad \theta = 60^\circ + k \cdot 90^\circ.$$

The fourth roots are then

$$z_k = 2[\cos(60^\circ + k \cdot 90^\circ) + i \sin(60^\circ + k \cdot 90^\circ)],$$

for $k = 0, 1, 2,$ and 3.

$$\text{For } k = 0, \quad z_0 = 2(\cos 60^\circ + i \sin 60^\circ) = 2\left(\frac{1}{2} + \frac{1}{2} i\sqrt{3}\right) = 1 + i\sqrt{3}.$$

$$\text{For } k = 1, \quad z_1 = 2[\cos(60^\circ + 90^\circ) + i \sin(60^\circ + 90^\circ)] = 2(\cos 150^\circ + i \sin 150^\circ) = 2\left(-\frac{1}{2}\sqrt{3} + \frac{1}{2} i\right) = -\sqrt{3} + i.$$

$$\text{For } k = 2, \quad z_2 = 2[\cos(60^\circ + 180^\circ) + i \sin(60^\circ + 180^\circ)] = 2(\cos 240^\circ + i \sin 240^\circ) = 2\left(-\frac{1}{2} - \frac{1}{2} i\sqrt{3}\right) = -1 - i\sqrt{3}.$$

$$\text{For } k = 3, \quad z_3 = 2[\cos(60^\circ + 270^\circ) + i \sin(60^\circ + 270^\circ)] = 2(\cos 330^\circ + i \sin 330^\circ) = 2\left(\frac{1}{2}\sqrt{3} - \frac{1}{2} i\right) = \sqrt{3} - i.$$

Check that for $k = 4$, $\theta = 420^\circ$, which will give the first root z_0 again. Thus, the four complex fourth roots of $-8 - 8i\sqrt{3}$ are $1 + i\sqrt{3}$, $-1 - i\sqrt{3}$, $-\sqrt{3} + i$, and $\sqrt{3} - i$. Observe that they occur in pairs of negatives of one another. Check these results by calculating the fourth power of each of them. [*Hint:* First square and then square again.] ●

Roots of Unity The complex roots of 1 are called **roots of unity.** They are very useful in working with the complex roots of any real number. The following roots of 1 have already been found.

The square roots of 1: 1 and -1 **(1)**

The cube roots of 1: $1, \quad -\frac{1}{2} + \frac{1}{2}i\sqrt{3}, \quad -\frac{1}{2} - \frac{1}{2}i\sqrt{3}$ **(2)**

The fourth roots of 1: $1, \quad -1, \quad i, \quad -i$ **(3)**

We will calculate the fifth roots of unity to help us see the situation in general for the nth roots of unity. Now if $z^5 = 1$ and $z = r(\cos\theta + i\sin\theta)$,

$$\begin{aligned} z^5 &= r^5(\cos 5\theta + i \sin 5\theta) = 1 \\ &= 1 \cdot (\cos 2k\pi + i \sin 2k\pi). \end{aligned}$$

Then, $r = 1$ and $5\theta = 2k\pi$, for $k = 0, 1, 2, 3, 4$, so that $r = 1$ and $\theta = 2k\pi/5$.

Now,

$$\begin{aligned} &\text{for } k = 0, \quad z_0 = \cos 0 + i \sin 0 = 1, \\ &\text{for } k = 1, \quad z_1 = \cos\frac{2\pi}{5} + i\sin\frac{2\pi}{5}, \\ &\text{for } k = 2, \quad z_2 = \cos\frac{4\pi}{5} + i\sin\frac{4\pi}{5}, \\ &\text{for } k = 3, \quad z_3 = \cos\frac{6\pi}{5} + i\sin\frac{6\pi}{5}, \\ &\text{for } k = 4, \quad z_4 = \cos\frac{8\pi}{5} + i\sin\frac{8\pi}{5}. \end{aligned}$$

If larger values of k are used, the values of z_k already obtained are simply repeated.

This same procedure may be used to calculate the nth roots of unity for any positive integral value of n. They are

$$\mathbf{z_k = \cos\frac{2k\pi}{n} + i\sin\frac{2k\pi}{n}, \quad k = 0, 1, 2, \ldots, n - 1.}$$

This formula enables us to write down directly the nth roots of unity for any positive integral value of n. The following examples show how to use the roots of unity to write the roots of any real number.

EXAMPLE 7 Calculate the three complex cube roots of 27.

SOLUTION

$$\begin{aligned} 27 = 27 \cdot 1 &= 27(\cos 0 + i \sin 0) \\ &= 27(\cos 2k\pi + i \sin 2k\pi). \end{aligned}$$

Then if $z^3 = 27$, the cube roots z of 27 are

$$z = 27^{1/3}\left(\cos\frac{2k\pi}{3} + i\sin\frac{2k\pi}{3}\right)$$

$$= 3\left(\cos\frac{2k\pi}{3} + i\sin\frac{2k\pi}{3}\right),\ k = 0, 1, 2. \qquad \bullet$$

Observe that these three complex numbers are just the three numbers obtained by multiplying the three complex roots of unity, namely, $\cos 2k\pi/3 + i \sin 2k\pi/3$, by the *real* cube root of 27, that is, 3. At the beginning of this discussion from (2), we saw that the cube roots of unity are 1, $-1/2 + \frac{1}{2}i\sqrt{3}$, and $-1/2 - \frac{1}{2}i\sqrt{3}$. Hence the cube roots of 27 are 3, $-3/2 + \frac{3}{2}i\sqrt{3}$, and $-3/2 - \frac{3}{2}i\sqrt{3}$.

EXAMPLE 8 Write directly the four complex fourth roots of 16.

SOLUTION The four fourth roots of unity have already been found: they are $1, -1, i$, and $-i$. Since the *real* fourth root of 16 is 2, the complex fourth roots are $2, -2, 2i$, and $-2i$. $\bullet$

We will now discuss the general case of the nth roots of a complex number. This most important result is expressed by the following theorem.

THEOREM 9.5 Every complex number has exactly n nth roots.

We can see this as follows.

Let z be an nth root of the complex number $r(\cos\theta + i\sin\theta)$, so that

$$z^n = r(\cos\theta + i\sin\theta), \text{ or}$$

$$z^n = r[\cos(\theta + 2k\pi) + i\sin(\theta + 2k\pi)],$$

$$z = r^{1/n}\left[\cos\left(\frac{\theta + 2k\pi}{n}\right) + i\sin\left(\frac{\theta + 2k\pi}{n}\right)\right].$$

Then for $k = 0$, $z_0 = r^{1/n}\left(\cos\frac{\theta}{n} + i\sin\frac{\theta}{n}\right)$;

for $k = 1$, $z_1 = r^{1/n}\left[\cos\left(\frac{\theta}{n} + \frac{2\pi}{n}\right) + i\sin\left(\frac{\theta}{n} + \frac{2\pi}{n}\right)\right]$;

for $k = 2$, $z_2 = r^{1/n}\left[\cos\left(\frac{\theta}{n} + \frac{4\pi}{n}\right) + i\sin\left(\frac{\theta}{n} + \frac{4\pi}{n}\right)\right]$;

for $k = n - 1$, $z_{n-1} = r^{1/n}\left[\cos\left(\frac{\theta}{n} + \frac{(n-1)2\pi}{n}\right) + i\sin\left(\frac{\theta}{n} + \frac{(n-1)2\pi}{n}\right)\right]$;

for $k = n$, we get $z_n = r^{1/n}\left[\cos\left(\frac{\theta}{n} + 2\pi\right) + i\sin\left(\frac{\theta}{n} + 2\pi\right)\right]$;

and the sequence of roots begins to repeat as in the preceding examples. One may check by applying De Moivre's Theorem that these are actually nth roots.

Since no two of the arguments of the above complex roots differ by a multiple of 2π, they are all *distinct*. Furthermore, there are no additional nth roots, since the moduli and arguments must satisfy De Moivre's Theorem. Thus, the theorem stated above has been proven.

EXERCISES 9.3

Calculate the indicated powers, using De Moivre's Theorem.

1. $\left[2\left(\cos\frac{2\pi}{3} + i\sin\frac{2\pi}{3}\right)\right]^2$

2. $\left[-\left(\cos\frac{\pi}{4} + i\sin\frac{\pi}{4}\right)\right]^3$

3. $\left[\frac{1}{2}(\cos 30° + i\sin 30°)\right]^4$

4. $\left(\cos\frac{\pi}{3} + i\sin\frac{\pi}{3}\right)^{-1}$

5. $\left(\cos\frac{\pi}{4} + i\sin\frac{\pi}{4}\right)^{-2}$

6. $\left[\cos\left(-\frac{\pi}{6}\right) + i\sin\left(-\frac{\pi}{6}\right)\right]^{-3}$

7. $(\cos 17° + i\sin 17°)^0$

8. $(\cos 18° + i\sin 18°)^5$

Convert each of the following from Cartesian to polar form. Use De Moivre's Theorem to raise to the indicated power, and give results in Cartesian form.

9. $(1 - i)^3$

10. $(-1 - i\sqrt{3})^5$

11. $(1 + i\sqrt{3})^6$

12. $(1 + i)^{-2}$

13. $\left(\frac{1}{2} + \frac{1}{2}i\sqrt{3}\right)^{-1}$

14. $(2\sqrt{3} - 2i)^3$

15. $(1 + i)^4$

16. $(-\sqrt{3} - i)^6$

17. $(\sqrt{2} + i\sqrt{2})^4$

18. $(1 - i)^{-3}$

19. $\left(\frac{1}{2} - \frac{1}{2}i\sqrt{3}\right)^{-4}$

20. $(-3\sqrt{3} + 3i)^3$

Use the procedure of Example 3 to derive formulas for the following.

21. $\sin 3\theta$ in terms of $\sin\theta$

22. $\cos 3\theta$ in terms of $\cos\theta$

23. $\sin 4\theta$ in terms of $\sin\theta$ and $\cos\theta$

24. $\cos 4\theta$ in terms of $\cos\theta$

Following the procedure of the text examples, prove De Moivre's Theorem for the following cases.

25. $k = 3$

26. $k = 4$

27. $k = 5$

28. $k = -2$

Check the results of the following text examples by actually raising them to the appropriate powers.

29. Example 4, square roots of $-1 + i\sqrt{3}$

30. Example 5, cube roots of 1

31. Example 6, fourth roots of $-8 - 8i\sqrt{3}$

Calculate each of the following.

32. Square roots of $-2 + 2i\sqrt{3}$

33. Square roots of $2 - 2i\sqrt{3}$

34. Square roots of $4i$

35. Square roots of $-4i$

36. Cube roots of $-i$

37. Cube roots of $8i$

38. Cube roots of $-8i$

39. Cube roots of $-64i$

40. Cube roots of $27i$

41. Cube roots of $-27i$

42. Cube roots of $-4 - 4i\sqrt{3}$

43. Cube roots of $-4 + 4i\sqrt{3}$

44. Fourth roots of $-8 + 8i\sqrt{3}$

45. Fourth roots of $8 + 8i\sqrt{3}$

46. Fourth roots of $4(\cos 120° + i\sin 120°)$

47. Cube roots of $8(\cos 135° + i \sin 135°)$

48. Fifth roots of $32(\cos 150° + i \sin 150°)$

49. Sixth roots of $\cos 240° + i \sin 240°$

Solve the following equations for all complex roots.

50. $z^3 + 1 = 0$ **51.** $z^4 - 16 = 0$ **52.** $z^3 = 27$ **53.** $z^2 + 8 + 8i\sqrt{3} = 0$

54. $z^3 - 1 = 0$ **55.** $z^4 + 16 = 0$ **56.** $z^4 = 81$ **57.** $z^3 + 4 + 4i\sqrt{3} = 0$

Use the roots of unity given in the text to calculate each of the following.

58. The square roots of -4

59. The cube roots of 8

60. The cube roots of -8

61. The fourth roots of -1

62. The fourth roots of 16

63. The fourth roots of 81

64. The fourth roots of -81

65. The fifth roots of -1

66. The fifth roots of 32

67. The fifth roots of -32

Chapter 9 Review Exercises

Use the definition of equality of complex numbers to find x and y in each of the following.

1. $x + yi = -1 + 4i$

2. $(x + 2) + (y - 1)i = 3i.$

3. $(x + y) - (x - y)i = 7 + i$

4. $(2x - 3y) + (5x + 2y)i = 13 + 4i$

For each of the following pairs of complex numbers, find the (a) sum, (b) difference, (c) product, and (d) quotient.

5. $1 + i$ and $4 - 2i$

6. $2 - 3i$ and $3 + 2i$

7. $2 + 5i$ and $2 - 5i$

8. $3 - 4i$ and $4 + 3i$

Calculate each of the following products and quotients and leave in polar form.

9. $6(\cos 80° + i \sin 80°) \cdot 2(\cos 40° + i \sin 40°)$

10. $6(\cos 60° + i \sin 60°) \div 2(\cos 20° + i \sin 20°)$

11. $5\left(\cos \frac{5\pi}{6} + i \sin \frac{5\pi}{6}\right) \div \left(\cos \frac{\pi}{3} + i \sin \frac{\pi}{3}\right)$

12. $-3\left(\cos \frac{\pi}{8} + i \sin \frac{\pi}{8}\right) \cdot 4\left(\cos \frac{3\pi}{4} + i \sin \frac{3\pi}{4}\right)$

Graph each of the following complex numbers. Also name and graph on the same figure with each number its negative and its conjugate.

13. $\cos \frac{2\pi}{3} + i \sin \frac{2\pi}{3}$

14. $\cos 330° + i \sin 330°$

15. $2(\cos 210° + i \sin 210°)$

16. $3\left(\cos \frac{1}{4}\pi + i \sin \frac{1}{4}\pi\right)$

Convert each of the following complex numbers into the alternate form (Cartesian to polar and polar to Cartesian). For each state the argument and modulus.

17. $2(\cos 30° + i \sin 30°)$

18. $3\left(\cos \frac{1}{2}\pi + i \sin \frac{1}{2}\pi\right)$

19. $1 + i\sqrt{3}$

20. $-\sqrt{2} + i\sqrt{2}$

Represent graphically.

21. $(1 + i) + (2 - 3i)$

22. $(3 - 2i) - (1 - i)$

Calculate each of the following powers and roots and write the result in Cartesian form.

23. $(\cos 60° + i \sin 60°)^3$

24. $[2(\cos 135° + i \sin 135°)]^4$

25. $(1 - i\sqrt{3})^5$

26. $(\sqrt{2} + i\sqrt{2})^4$

27. Sixth roots of -64

28. Cube roots of $64i$

29. Square roots of $4 + 4i\sqrt{3}$

30. Fourth roots of $-32 - 32i\sqrt{3}$

Chapter 9 Miscellaneous Exercises

1. Show that the conjugate of the difference of two complex numbers is the difference of their conjugates.

2. Show that the conjugate of the product of two complex numbers is the product of their conjugates.

3. Show that the reciprocal of the conjugate of a nonzero complex number is the conjugate of its reciprocal.

4. Show that the conjugate of the quotient of two complex numbers is the quotient of their conjugates.

5. If the product of two complex numbers is zero, show that at least one of them is zero.

6. If z is a complex number, show that its reciprocal is its conjugate if and only if its absolute value is 1, that is $1/z = \bar{z}$ if and only if $|z| = 1$.

7. Prove that any two complex numbers $a + bi$ and $c + di$ lie on the same straight line through the origin if and only if their quotient is a real number.

8. Prove De Moivre's Theorem for k *any negative integer*. [*Hint:* Let $k = -c$, where $c > 0$. Then use De Moivre's Theorem for $c > 0$, together with the text result for the case $k = -1$ and the law of exponents $(z^a)^b = z^{ab}$. For instance, $[r(\cos \theta + i \sin \theta)]^{-3} = [r(\cos \theta + i \sin \theta)^3]^{-1}$.

9. Show that the cube roots of 8 are located at the vertices of an equilateral triangle inscribed in a circle, center at the origin and radius 2.

10. Show that the fourth roots of 16 are located at the vertices of a square inscribed in a circle, center at the origin and radius 2.

11. Two of the cube roots of unity are imaginary. Show that each of them is the square of the other.

12. Calculate the sixth roots of unity and write them in the standard form $a + bi$. Identify among them both the square roots and the cube roots of unity. Why is this to be expected? Graph.

13. Calculate the eighth roots of unity and write in the standard form $a + bi$. Identify among them both the square roots and the fourth roots of unity. Why is this to be expected? Graph.

14. Calculate the twelfth roots of unity and write them in the standard form $a + bi$. Identify among them the square roots, the cube roots, the fourth roots, and the sixth roots of unity. Why is this to be expected? Graph.

15. Prove that the product of any two nth roots of unity is also a root of unity.

The reciprocal of an nth root of unity is also an nth root of unity. Show that this is true for the following cases.

16. The square roots of 1.

17. The cube roots of 1.

18. The fourth roots of 1.

Use the basic definitions of equality, addition, and multiplication of complex numbers (Section 9.1) to prove that

19. Addition is commutative

20. Addition is associative

21. Multiplication is commutative

22. Multiplication is associative

23. Multiplication is distributive over addition

10

Sequences and Series

This chapter studies a special kind of function, a *sequence,* that is important both for applications and for the development of other mathematical ideas. Some of these ideas and applications related to sequences and series are also discussed.

10.1 Sequences and Series

Sequences The following example introduces some of the basic notions of this section.

EXAMPLE 1 Give the domain and range of each of the functions f defined as follows:

(a) $f(1) = 2, f(2) = 4, f(3) = 6$;
(b) $f(1) = 1, f(2) = 3, f(3) = 5, \ldots, f(k) = 2k - 1, \ldots$.

SOLUTION (a) The domain of f is the set of the first three positive integers and its range is the set of the first three positive even integers.

(b) In this case, the domain is the set of all positive integers and the range is the set of all positive odd integers. ●

Note that the domains of both functions in Example 1 are sets of positive integers rather than continuous intervals of real numbers. A function whose domain is a set of positive integers 1, 2, 3, . . . is a **sequence function.**

Since the domain of a sequence function is a set of positive integers, the members of the range of the function fall into a natural order: $f(1), f(2), f(3)$, and so on. This leads to the following definition.

DEFINITION 10.1 The range of a sequence function, when arranged in the order $f(1)$, $f(2)$, $f(3)$, and so on, is called a **sequence.** That is, a sequence is any set of numbers arranged in a definite order.

The sequence of Example 1(a),

$$2, 4, 6,$$

is a **finite sequence,** and that of Example 1(b),

$$1, 3, 5, \ldots, 2k - 1, \ldots,$$

is an **infinite sequence** (usually simply called a **sequence**).

If we denote the infinite sequence function by $a(n)$ and write a_n for $a(n)$ we have the following conventional notation for the general infinite sequence:

$$a_1, a_2, a_3, \ldots, a_k, \ldots.$$

We call a_1 the **first term,** a_2 the **second term,** and a_k the **general** or **kth term** of the sequence. For Example 1(b) above,

$$a_1 = f(1) = 1, \quad a_2 = f(2) = 3, \quad a_3 = f(3) = 5, \quad \ldots, \quad a_k = f(k) = 2k - 1,$$

and so on. Using the formula $a_k = 2k - 1$ for the general term, with the understanding that $k = 1, 2, 3, \ldots$, we can write any desired term of the sequence. For example,

$$a_7 = 2(7) - 1 = 13 \qquad \text{and} \qquad a_{13} = 2(13) - 1 = 25.$$

EXAMPLE 2 Write all the terms of the finite sequence whose general term is k^2 and whose domain is the set of the first five positive integers.

SOLUTION In this case, $a_k = k^2$, so

$$a_1 = 1^2 = 1, \qquad a_2 = 2^2 = 4, \qquad a_3 = 3^2 = 9,$$
$$a_4 = 4^2 = 16, \qquad a_5 = 5^2 = 25.$$

This may be written simply as

$$1, 4, 9, 16, 25. \quad \bullet$$

EXAMPLE 3 Write the first four terms of the infinite sequence whose general term is (a) $1/k$; (b) $(-1)^k/k^2$.

SOLUTION (a) Let $k = 1, 2, 3, 4$ successively to get $1, \dfrac{1}{2}, \dfrac{1}{3}, \dfrac{1}{4}, \ldots.$

(b) Let $k = 1, 2, 3, 4$. This sequence is $-1, \dfrac{1}{4}, -\dfrac{1}{9}, \dfrac{1}{16}, \ldots.$ •

We may be given some terms of a sequence and asked to determine a general term.

EXAMPLE 4 Find a general term for the following sequence.

$$0, 1, 2, 3, \ldots$$

SOLUTION By inspection we see that each of the above terms is 1 less than the corresponding term in the sequence of integers 1, 2, 3, 4. Thus, they can be represented by $k - 1$ for $k = 1, 2, 3, 4$. If we assume that the same pattern continues, the general term would be $a_k = k - 1$. ●

It may be that there is no formula for the general term of a sequence, since *any* ordered set of numbers may be considered a sequence. For example, the prime numbers, in order, are 2, 3, 5, 7, 11, 13, ... However, there is no known expression representing all primes.

Furthermore, it is not necessary that terms of a sequence be distinct. For example,

$$1, -1, 1, -1, \ldots$$

and

$$1, 2, 3, 1, 2, 3, \ldots$$

are sequences. In the first case, $a_1 = a_3 = a_5 = \ldots = 1$, while $a_2 = a_4 = a_6 = \ldots = -1$. A similar situation holds in the second case. As another illustration, let $a_k = k^2 - 6k + 8$. Then for $k = 1, 2, 3, 4, 5$, we get

$$3, 0, -1, 0, 3,$$

whose terms are not distinct.

The terms of a sequence are not necessarily determined by the pattern of the first few terms. For example, given

$$1, 4, 7, 8, 5, -2, \ldots,$$

the formula $a_k = 3k - 2$ generates the first three terms but not any of the following terms. Thus, before concluding anything about a sequence we need either a general term or other information that will completely describe the sequence.

If the first terms of a sequence are given, there may be more than one formula for a general term that describes these first terms.

EXAMPLE 5 Write the first five terms corresponding to the general term (a) $a_k = 1/k$; (b) $a_k = 1/k + (k - 1)(k - 2)(k - 3)$.

SOLUTION (a) Substituting, in turn, 1, 2, 3, 4, and 5 for k, we get the first five terms:

$$1, \frac{1}{2}, \frac{1}{3}, \frac{1}{4}, \frac{1}{5}.$$

(b) When 1, 2, 3, 4, and 5 are substituted for k, the terms are

$$1, \frac{1}{2}, \frac{1}{3}, \frac{25}{4}, \frac{121}{5}.$$

The first three terms of this sequence are the same as in part (a). This is because for $k = 1$, 2, or 3, the product $(k - 1)(k - 2)(k - 3)$ is zero. Many similar examples could be given. So we use care in determining the general term of a sequence.

Series The indicated sum of the terms of a finite sequence is called a **series.** For example, the series associated with the finite sequence 1, 3, 5, 7 is

$$1 + 3 + 5 + 7.$$

The series formed by taking the first term, or the first two, three, four, and so on, terms of a sequence is called a **partial sum.** For the above sequence, there are four partial sums:

$$S_1 = 1,$$
$$S_2 = 1 + 3 = 4,$$
$$S_3 = 1 + 3 + 5 = 9,$$
$$S_4 = 1 + 3 + 5 + 7 = 16.$$

These make up the **sequence of partial sums** S_1, S_2, S_3, S_4. Similarly, for a general sequence

$$a_1, a_2, a_3, \ldots, a_n,$$

the general or kth partial sum S_k is given by

$$S_k = a_1 + a_2 + a_3 + \cdots + a_k.$$

In dealing with series it is often useful to use **summation notation.** To illustrate, the partial sums associated with the general sequence above are

$$S_1 = a_1$$
$$S_2 = a_1 + a_2 = \sum_{k=1}^{2} a_k,$$
$$S_3 = a_1 + a_2 + a_3 = \sum_{k=1}^{3} a_k,$$

and so on. The symbol Σ is the Greek letter sigma corresponding to S, and so we sometimes call this the **sigma notation.**

In the summation notation above, the letter k is called the **summation variable.** Any other letter could be used for this variable. Thus, $\Sigma_{i=1}^{n} a_i$ describes the same series as $\Sigma_{k=1}^{n} a_k$. Also, it is not necessary that the summation start with 1. It can begin with any positive integral value.

EXAMPLE 6 Calculate $\Sigma_{n=2}^{5} n(n - 1)$.

SOLUTION In this case, the summation begins with $n = 2$.

$$\begin{aligned}\sum_{n=2}^{5} n(n-1) &= 2(2-1) + 3(3-1) + 4(4-1) + 5(5-1)\\ &= 2(1) + 3(2) + 4(3) + 5(4)\\ &= 40 \quad \bullet\end{aligned}$$

Summation notation can also be used to write the general polynomial

$$P_n(x) = a_0 + a_1x + a_2x^2 + \cdots + a_nx^n$$

as

$$P_n(x) = \sum_{k=0}^{n} a_kx^k.$$

The above are examples of finite series. The notion of infinite series will be discussed later.

EXERCISES 10.1

Write the first five terms of the sequence whose general term is given.

1. $3n + 2$ **2.** k^3 **3.** $\frac{1}{n+1}$ **4.** $\frac{n}{n+2}$ **5.** $(-1)^k$

6. $\frac{2^n}{n}$ **7.** $\frac{1}{k!}$ **8.** $\frac{n^2}{n!}$ **9.** $\frac{(-1)^n}{n!}$ **10.** $\frac{2^k}{k+1}$

Write a general term suggested by each of the following first few terms of a sequence.

11. $3, 8, 13, 18, \ldots$

12. $-2, -5, -8, -11, \ldots$

13. $5, 7, 9, 11, \ldots$

14. $\frac{1}{1(2)}, \frac{1}{2(3)}, \frac{1}{3(4)}, \ldots$

15. $\frac{1}{2}, \frac{2}{3}, \frac{3}{4}, \frac{4}{5}, \ldots$

16. $2, 4, 8, 16, \ldots$

17. $-1, 4, -27, 256, \ldots$

18. $1, \frac{1}{2}, \frac{1}{4}, \frac{1}{8}, \ldots$

19. $x^2, x^4, x^6, x^8, \ldots$

20. $-x, x^3, -x^5, x^7, \ldots$

21. $\frac{1}{x}, \frac{1}{2x}, \frac{1}{3x}, \frac{1}{4x}, \ldots$

22. $x - 1, 2x - 1, 3x - 1, \ldots$

Write each of the following summations as an indicated sum. Simplify terms where possible.

23. $\sum_{k=1}^{5} k$ **24.** $\sum_{k=0}^{4} 2k$ **25.** $\sum_{k=1}^{4} k(k+2)$ **26.** $\sum_{k=2}^{6} (k-1)(k-2)$

27. $\sum_{k=1}^{3} \frac{k+1}{k!}$ **28.** $\sum_{k=1}^{5} \frac{1}{\sqrt{k}}$ **29.** $\sum_{k=1}^{4} \sqrt[3]{k}$ **30.** $\sum_{n=2}^{5} (-1)^n n^2$

31. $\sum_{n=1}^{3} \frac{1}{n^3}$ **32.** $\sum_{n=2}^{7} n\sqrt{n}$ **33.** $\sum_{n=1}^{5} nx^n$ **34.** $\sum_{n=1}^{4} (n+1)x^{n-1}$

35. $\sum_{n=1}^{6} n!x^n$ **36.** $\sum_{n=0}^{4} (-x)^n$

Determine a formula for the general term and then use summation notation to write the following finite series.

37. The sum of the first seven multiples of 2

38. The sum of the first ten multiples of 3

39. The sum of the first five powers of 2

40. The sum of the first six powers of -1

41. $4 + 7 + 10 + \cdots$ to nine terms

42. $11 + 5 - 1 - 7 - \cdots$ to eleven terms

43. $1 - 1 - 3 - 5 - \cdots$ to five terms

44. $\frac{1}{3} + \frac{1}{9} + \frac{1}{27} + \frac{1}{81} + \cdots$ to six terms

45. $-1 + \frac{1}{2} - \frac{1}{3} + \frac{1}{4} - \cdots$ to eight terms

46. $1! + 2! + 3! + 4! + \cdots$ to seven terms

47. $x + x^2 + x^3 + x^4 + \cdots$ to twelve terms

48. $x^3 + x^6 + x^9 + x^{12} + \cdots$ to five terms

49. The distance d traveled by a car moving uniformly at 40 miles per hour is given by $d = 40t$, where t is the time in hours. Write the sequence representing the total distance covered at the end of each of the first five hours.

50. A freely falling body drops s feet in t seconds, where $s = 16t^2$. Write the sequence representing (a) the total distance covered at the end of each of the first five seconds, (b) the distance the body falls *during* each of the first five seconds.

51. The Fahrenheit temperature F in terms of the Celsius temperature C is given by $F = \frac{9}{5}C + 32$. Write the sequence of Fahrenheit temperatures corresponding to Celsius temperatures of 0°, 5°, 10°, 15°, 20°.

10.2 Arithmetic Sequences and Series

Arithmetic Sequences In the sequence of odd integers

$$1, 3, 5, 7, \ldots,$$

any two successive terms differ by the same number, 2. The same is true of the sequence of even integers. Similarly, any two successive terms of the sequence

$$2, 5, 8, 11, \ldots,$$

differ by the same number, 3. Such sequences are called *arithmetic sequences* or *arithmetic progressions*.

DEFINITION 10.2 An **arithmetic sequence** is a sequence of numbers such that any two successive terms differ by the same number, called the **common difference.**

In the illustrations above, each term is greater than the preceding one and the difference is positive. The terms of an arithmetic sequence may also have a negative common difference. In this case the terms decrease rather than increase. For example,

$$5, 1, -3, -7, \ldots,$$

is an arithmetic sequence with common difference -4.

If a_1 denotes the first term of an arithmetic sequence and d the common difference, then the terms of the sequence are

$$a_1, a_1 + d, a_1 + 2d, \ldots, a_1 + (n - 1)d.$$

The nth term is $a_n = a_1 + (n - 1)d$. Clearly, given any three of the quantities a_1, d, n, and a_n, we can solve for the remaining one.

EXAMPLE 1 Write the first four terms and the nth term of the arithmetic sequence with (a) first term 3 and common difference 4; (b) first term 2 and common difference -1.

SOLUTION (a) Use the formula for the nth term, and let $a_1 = 3$ and $d = 4$.

$$a_n = a_1 + (n - 1)d$$
$$a_n = 3 + (n - 1)4$$

Substitute 1, 2, 3, and 4, in turn, for n in the above formula to get the first four terms. The sequence is

$$3, 7, 11, 15, \ldots, 3 + (n - 1)4.$$

(b) Use the formula for the nth term with $a_1 = 2$ and $d = -1$.

$$2, 1, 0, -1, \ldots, 2 + (n - 1)(-1).$$ ●

EXAMPLE 2 If an arithmetic sequence has $a_1 = 3$, $a_n = -13$, and $n = 5$, find d.

SOLUTION Substitute the given values in the formula and solve for d.

$$a_n = a_1 + (n - 1)d$$
$$-13 = 3 + 4d$$
$$4d = -16$$
$$d = -4$$ ●

EXAMPLE 3 Find the tenth term of the arithmetic sequence whose first four terms are 1, 4, 7, 10.

SOLUTION Here $a_1 = 1$, $d = 3$, and $n = 10$ (we want the tenth term).

$$a_n = a_1 + (n - 1)d$$
$$a_{10} = 1 + 9(3) = 28$$ ●

Arithmetic Series For each finite arithmetic sequence there is a corresponding arithmetic series. That is, given the finite sequence

$$a_1, a_1 + d, a_1 + 2d, \ldots, a_1 + (n - 1)d,$$

the series

$$S_n = a_1 + (a_1 + d) + (a_1 + 2d) + \cdots + [a_1 + (n - 1)d]$$

is associated with it. In summation notation, this series is written

$$\sum_{k=1}^{n} (a_1 + (k - 1)d).$$

There is a formula for calculating the sum of an arithmetic series without actually adding up all the terms. To see how the formula is derived, consider the series

$$\sum_{k=1}^{100} k = 1 + 2 + 3 + \cdots + 100,$$

which represents the sum of the first 100 integers. Write out the series, and below it, write out the terms of the series in reverse order.

$$\begin{aligned} S_{100} &= 1 + 2 + 3 + \cdots + 98 + 99 + 100 \\ S_{100} &= 100 + 99 + 98 + \cdots + 3 + 2 + 1 \end{aligned}$$

Adding these two series term by term, we get

$$\begin{aligned} 2S_{100} &= 101 + 101 + 101 + \cdots + 101 + 101 + 101 \\ &= 100(101) \\ S_{100} &= 50(101) \qquad \text{there are 100 terms} \\ &= 5050. \end{aligned}$$

(This method of summing an arithmetic series was supposedly discovered by the mathematician Carl Friedrich Gauss when he was ten years old.) The general formula is derived in the same way. First, write S_n in the two orders shown above.

$$S_n = a_1 + (a_1 + d) + (a_1 + 2d) + \cdots + (a_n - d) + a_n,$$

and

$$S_n = a_n + (a_n - d) + (a_n - 2d) + \cdots + (a_1 + d) + a_1.$$

Adding term by term, we get

$$\begin{aligned} 2S_n &= (a_1 + a_n) + (a_1 + a_n) + \cdots + (a_1 + a_n) \\ &= n(a_1 + a_n) \qquad \text{there are } n \text{ terms} \\ S_n &= \frac{n}{2}(a_1 + a_n) \end{aligned}$$

$$\boldsymbol{S_n = n\left(\frac{a_1 + a_n}{2}\right).}$$

In words, the sum of the terms of an arithmetic series is the product of the number of terms, n, and the average of the first and last terms, $(a_1 + a_n)/2$.

Another form of the formula for S_n may be found by substituting $a_1 + (n - 1)d$ for a_n in the formula just derived.

$$S_n = \frac{1}{2}n(a_1 + a_n)$$

$$= \frac{1}{2}n[a_1 + a_1 + (n - 1)d]$$

$$\boldsymbol{S_n = \frac{n}{2}[2a_1 + (n - 1)d]}$$

EXAMPLE 4 Find the sum of (a) the arithmetic series with five terms whose first term is 2 and last term is 14; (b) the first twenty odd integers.

SOLUTION (a) Here $n = 5$, $a_1 = 2$, and $a_n = 14$.

$$S_n = \frac{n}{2}(a_1 + a_n) = \frac{1}{2}(5)(2 + 14) = 40$$

(b) In this case, $n = 20$, $a_1 = 1$, and $d = 2$.

$$S_n = \frac{n}{2}[2a_1 + (n - 1)d] = \frac{1}{2}(20)[2 + (19)2] = 400$$

●

EXAMPLE 5 A ladder has 5 rungs, each shorter by the same amount than the one below it. If the bottom rung is 18 inches long and the top one 8 inches long, what is the total length of the rungs (see Figure 10.1).

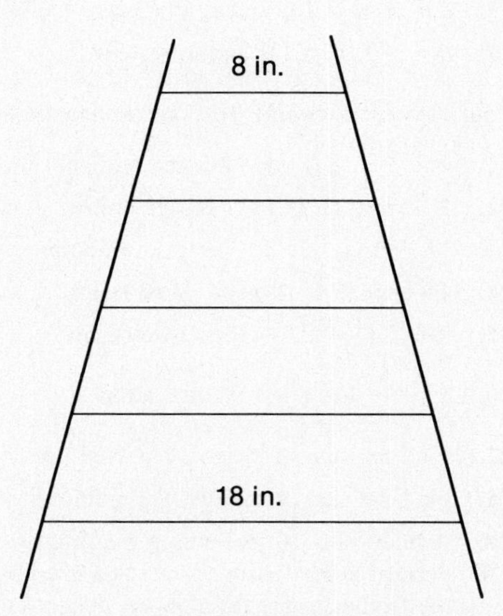

Figure 10.1

SOLUTION The lengths of the rungs are in arithmetic progression, with $a_1 = 18$, $a_5 = 8$, and $n = 5$. Then

$$S_5 = \frac{1}{2}(5)(18 + 8)$$
$$= 65 \text{ (inches)}. \quad \bullet$$

EXERCISES 10.2

Find the common difference and the general term suggested by the first terms of the given arithmetic sequence.

1. $3, 5, 7, 9, \ldots$

2. $4, 7, 10, 13, \ldots$

3. $-11, -8, -5, -2, \ldots$

4. $10, 7, 4, 1, \ldots$

5. $2, \frac{7}{2}, 5, \frac{13}{2}, \ldots$

6. $7.2, 10.4, 13.6, 16.8, \ldots$

7. $x, x + y, x + 2y, x + 3y, \ldots$

8. $x, x - y, x - 2y, x - 3y, \ldots$

9. $2x - 1, 2x - 3, 2x - 5, 2x - 7, \ldots$

10. $3 + x, 3 + \frac{x}{2}, 3, 3 - \frac{x}{2}, \ldots$

In each of the following, three of the numbers a_1, d, a_n, n, and S_n are given. Find the remaining ones.

11. $a_1 = 5$, $a_n = 25$, and $d = 2$

12. $a_1 = 7$, $a_n = 75$, and $n = 18$

13. $a_1 = 8$, $a_n = 38$, and $n = 16$

14. $a_1 = 5$, $a_n = 68$, and $d = 3$

15. $d = 4$, $n = 12$, and $S_n = 300$

16. $d = 3$, $n = 13$, and $a_n = 236$

17. $d = 5$, $n = 13$, and $a_n = 63$

18. $a_1 = 6$, $S_n = 720$, and $a_n = 74$

19. $d = -3$, $n = 13$, and $a_n = -16$

20. $a_1 = 2$, $S_n = 572$, and $a_n = 86$

Find the common difference, the general term, and the sum of the given arithmetic series.

21. $2 + 5 + 8 + \cdots$; thirteen terms

22. $3 + 7 + 11 + \cdots$; sixteen terms

23. $11 + 5 - 1 - 7 - \cdots$; seven terms

24. $14 + 16.5 + 19 + \cdots$; ten terms

25. $3 + 2.6 + 2.2 + \cdots$; seven terms

26. $1 + \frac{5}{3} + \frac{7}{3} + \cdots$; five terms

27. Find the sum of the first ten even positive integers.

28. Find the sum of the first ten multiples of 3.

29. A body falls 16 feet during the first second, 48 feet during the second second, 80 feet during the third second, and so on in arithmetic progression. (a) How far does it fall during the tenth second? (b) How far does it fall in 10 seconds?

10.3 Geometric Sequences and Series

Geometric Sequences In the sequence

$$2,\ 6,\ 18,\ 54,\ \ldots,$$

each term (after the first one) is obtained from the preceding term by multiplying it by 3. This is an example of a *geometric sequence* with *common ratio* 3. (The ratio of each term to the preceding one is 3.) Similarly, the sequence

$$12,\ -6,\ 3,\ -\frac{3}{2},\ \ldots$$

is a geometric sequence with common ratio $-1/2$.

DEFINITION 10.3 A **geometric sequence** is one in which each term after the first is obtained by multiplying the preceding term by a fixed number r, called the **common ratio.**

If the first term is denoted by a_1, the common ratio by r, the nth term by a_n, and the number of terms by n, the general (finite) geometric sequence is

$$a_1,\quad a_1r,\quad (a_1r)r = a_1r^2,\quad (a_1r^2)r = a_1r^3,\ \ldots,\ a_1r^{n-1},$$

where

$$a_k = a_1r^{k-1} \qquad \text{for} \qquad 1 \le k \le n.$$

EXAMPLE 1 Write (a) the finite geometric sequence of five terms whose first term is 3 and whose common ratio is 2; (b) the first four terms and the nth term of the geometric sequence whose first term is 6 and whose common ratio is 2/3.

SOLUTION (a) According to the above, $a_2 = 3 \cdot 2 = 6$, $a_3 = 3 \cdot 2^2 = 12$, $a_4 = 3 \cdot 2^3 = 24$, and $a_5 = 3 \cdot 2^4 = 48$. Thus, the sequence is

$$3,\ 6,\ 12,\ 24,\ 48.$$

(b) The terms are 6, $6(2/3)$, $6(2/3)^2$, $6(2/3)^3$, and the nth term is $6(2/3)^{n-1}$. Thus, the sequence is

$$6,\ 4,\ \frac{8}{3},\ \frac{16}{9},\ \ldots,\ 6\left(\frac{2}{3}\right)^{n-1},\ \ldots. \qquad \bullet$$

EXAMPLE 2 Given the sequence 16, 8, 4, . . . , find (a) the common ratio; (b) the tenth term.

SOLUTION (a) $r = \dfrac{8}{16} = \dfrac{4}{8} = \dfrac{1}{2}$.

(b) If $n = 10$, then (substituting in the above formula)

$$a_{10} = a_1r^{10-1}$$

$$= 16\left(\frac{1}{2}\right)^9 = 16\left(\frac{1}{512}\right)$$

$$= \frac{1}{32}. \quad \bullet$$

Geometric Series The indicated sum of the terms of a finite geometric sequence is called a **geometric series.** Given the sequence

$$a_1, a_1r, a_1r^2, \ldots, a_1r^{n-1},$$

the corresponding geometric series is

$$S_n = a_1 + a_1r + a_1r^2 + \cdots + a_1r^{n-1}.$$

S_n is also the nth partial sum of the corresponding infinite sequence.

First we derive a formula for S_n in terms of a_1, r, and n. If

$$S_n = a_1 + a_1r + a_1r^2 + \cdots + a_1r^{n-1},$$

then

$$rS_n = a_1r + a_1r^2 + \cdots + a_1r^{n-1} + a_1r^n.$$

Subtract the last line from the preceding one.

$$S_n - rS_n = a_1 - a_1r^n$$

$$S_n(1 - r) = a_1(1 - r^n)$$

$$S_n = \frac{a_1(1 - r^n)}{1 - r} \qquad \text{provided } r \neq 1$$

or

$$S_n = \frac{a_1 - a_1r^n}{1 - r} \qquad r \neq 1$$

EXAMPLE 3 Find the sum of the first five powers of 3.

SOLUTION The first five powers of 3 are

$$3^1 = 3,\ 3^2 = 9,\ 3^3 = 27,\ 3^4 = 81,\ 3^5 = 243.$$

Here $n = 5$, $a_1 = 3$, and $r = 3$. Then

$$S_n = \frac{a_1(1 - r^n)}{1 - r};$$

$$S_5 = \frac{3(1 - 3^5)}{1 - 3}$$

$$= \frac{3(1 - 243)}{-2}$$

$$= 363.$$

Or, by the second formula above,

$$S_n = \frac{a - ar^n}{1 - r}$$

$$= \frac{3 - 3(243)}{1 - 3}$$

$$= 363. \quad \bullet$$

EXAMPLE 4 Each person has two parents, four grandparents, and so on. How many persons are there in a family tree of ten generations consisting of you, your parents, and the eight preceding generations of ancestors.

SOLUTION The number of people at the end of ten generations is

$$1 + 2 + 4 + 8 + 16 + 32 + 64 + 128 + 256 + 512.$$

Here $a_1 = 1$, $r = 2$, and $n = 10$.

$$S_n = \frac{a_1 - a_1 r^n}{1 - r}$$

$$S_{10} = \frac{1 - 1(2^{10})}{1 - 2}$$

$$= 1023 \quad \bullet$$

Infinite Geometric Series The above are illustrations of sums for finite geometric series. Under appropriate conditions it is possible to define the sum of an infinite number of terms of a geometric series.

First, recall some terminology used in earlier chapters. The expression "$x \to a$", read "x approaches a," means that x varies in such a way that its difference from a becomes as small as we wish (arbitrarily small). The expression "$x \to \infty$", read "x increases without bound," or "x increases infinitely," means that x varies in such a way that it eventually becomes larger than any number however large.

Given the infinite geometric series

$$a_1 + a_1 r + a_1 r^2 + \cdots + a_1 r^{n-1} + a_1 r^n + \cdots,$$

we will investigate how the partial sum of n terms,

$$a_1 + a_1 r + a_1 r^2 + \cdots + a_1 r^{n-1},$$

varies as n increases without bound. We have

$$S_n = \frac{a_1(1 - r^n)}{1 - r}$$

$$= \frac{a_1 - a_1 r^n}{1 - r}$$

$$= \frac{a_1}{1 - r} - \left(\frac{a_1}{1 - r}\right) r^n.$$

It is shown in the calculus that, if $-1 < r < 1$, then as $n \to \infty$, $r^n \to 0$. That is, r^n becomes arbitrarily small as n increases without bound. Then, as $n \to \infty$,

$$\left(\frac{a_1}{1-r}\right)r^n \to 0 \qquad \text{and} \qquad S_n \to \frac{a_1}{1-r}.$$

We let

$$\mathbf{S = \frac{a_1}{1-r} \qquad -1 < r < 1}$$

and call S the **sum** of the infinite geometric series. Note that, since we cannot add an infinite number of terms, this is an extension of the usual idea of addition.

EXAMPLE 5 Find the sums of the following series.
(a) $2 + 1 + 1/2 + 1/4 + \cdots$; (b) $1 - 2/3 + 4/9 - 8/27 + \cdots$.

SOLUTION (a) Here $r = 1/2$ and $a_1 = 2$.

$$\begin{aligned} S &= \frac{a_1}{1-r} \\ &= \frac{2}{1-\frac{1}{2}} \\ &= 4 \end{aligned}$$

(b) In this case, $r = -2/3$ and $a_1 = 1$.

$$\begin{aligned} S &= \frac{a_1}{1-r} \\ &= \frac{1}{1+\frac{2}{3}} \\ &= \frac{3}{5} \end{aligned}$$

In summation notation, these series are written as follows.

(a) $$\sum_{k=1}^{\infty} 2\left(\frac{1}{2}\right)^{k-1} = 4, \quad \text{or} \quad \sum_{k=0}^{\infty} 2\left(\frac{1}{2}\right)^{k} = 4$$

(b) $$\sum_{k=1}^{\infty} 1\left(-\frac{2}{3}\right)^{k-1} = \frac{3}{5}, \quad \text{or} \quad \sum_{k=0}^{\infty} 1\left(-\frac{2}{3}\right)^{k} = \frac{3}{5}$$ •

It is instructive to look at the number line representation of the sum of an infinite geometric series. As an example, consider

$$\frac{1}{2} + \frac{1}{4} + \frac{1}{8} + \cdots + \left(\frac{1}{2}\right)^k + \cdots.$$

First, we have

$$S = \frac{a_1}{1 - r} = \frac{\frac{1}{2}}{1 - \frac{1}{2}} = 1.$$

The first four partial sums are

$$S_1 = \frac{1}{2},$$

$$S_2 = \frac{1}{2} + \frac{1}{4} = \frac{3}{4},$$

$$S_3 = \frac{1}{2} + \frac{1}{4} + \frac{1}{8} = \frac{7}{8},$$

$$S_4 = \frac{1}{2} + \frac{1}{4} + \frac{1}{8} + \frac{1}{16} = \frac{15}{16}.$$

These are shown on the number line in Figure 10.2. It is clear that each successive partial sum is nearer to 1 than the preceding one and that S_k can be as close to 1 as we wish if k is taken large enough. This is what is meant by the notation "$S_k \to 1$ as $n \to \infty$" and why $S = 1$ is called the "sum."

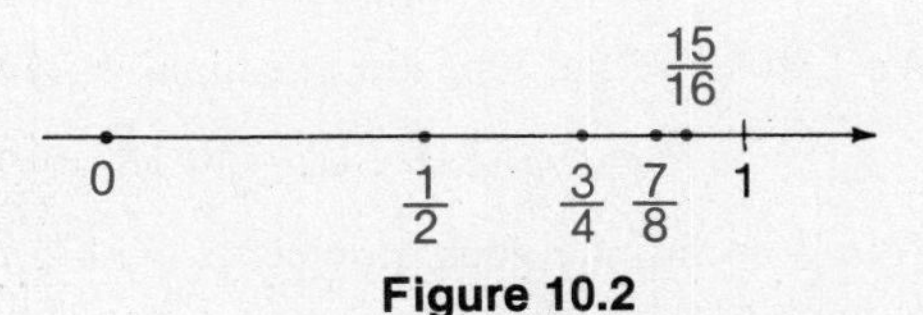

Figure 10.2

The process of "adding" the terms of this infinite series corresponds to repeatedly cutting a piece of wire into halves. The sum of n terms in the series represents the total length cut off after n cuts. The length of the part left gets smaller and smaller, approaching zero, and the total of the parts cut off approaches the total original length.

EXAMPLE 6 Find $\Sigma_{k=0}^{\infty} 3(1/4)^k$.

SOLUTION Here $a_1 = 3$ and $r = 1/4$.

$$S = \frac{a_1}{1 - r} = \frac{3}{1 - \frac{1}{4}} = 4$$

Therefore,

$$\sum_{k=0}^{\infty} 3\left(\frac{1}{4}\right)^k = 4. \quad \bullet$$

If $|r| \geq 1$, then the sequence of partial sums does not approach any real number, and we say that the series does not have a sum. For example, if $r = 1$, the series is

$$a_1 + a_1 + a_1 + \cdots,$$

and $S_n = na_1$, which clearly gets larger and larger as n increases without bound. If $r = -1$, then the series is

$$a_1 - a_1 + a_1 - a_1 + \cdots.$$

In this case, if n is even, then $S_n = 0$, and if n is odd, then $S_n = a_1$. Since the terms continue to alternate in this fashion, S_n does not approach any fixed number.

As an example for $|r| > 1$, suppose $r = 2$. Then the terms of the series after the first are all larger than the corresponding terms when $r = 1$. Since the series with $r = 1$ did not have a sum, clearly a series with even larger terms does not have a sum. The same kind of argument can be used for negative r when $|r| > 1$.

Repeating Decimals Infinite geometric series can be used to show that every repeating decimal represents a rational number (Section 1.1). The following examples illustrate this.

EXAMPLE 7 Find the fractional number corresponding to 0.181818 . . . .

SOLUTION We first write the repeating decimal $0.\overline{18}$ as an infinite series.

$$0.181818\ldots = 0.18 + 0.0018 + 0.000018 + \cdots$$

This is an infinite geometric series with $a_1 = 0.18$ and $r = 0.01$. Since $|r| \leq 1$, this series has a sum S, and

$$\begin{aligned} S &= \frac{a_1}{1 - r} \\ &= \frac{0.18}{1 - 0.01} \\ &= \frac{0.18}{0.99} = \frac{18}{99} = \frac{2}{11}. \end{aligned}$$

Thus, $0.181818\ldots = 2/11$. This can be checked by actually dividing 2 by 11. •

EXAMPLE 8 Find the fractional number equivalent to 1.666 . . . .

SOLUTION First, we write

$$1.666\ldots = 1 + (0.6 + 0.06 + 0.006 + \cdots).$$

The expression in the parentheses is a geometric series with $a_1 = 0.6$ and $r = 0.1$. Then

$$1.666\ldots = 1 + \frac{0.6}{1 - 0.1}$$
$$= 1 + \frac{0.6}{0.9}$$
$$= 1 + \frac{6}{9} = 1 + \frac{2}{3}$$
$$= \frac{5}{3}. \quad \bullet$$

Other Sequences and Series The discussion so far has been limited to arithmetic and geometric sequences and series. Many other kinds of sequences could have been treated. A simple example is the **harmonic sequence** whose terms are the reciprocals of the terms of an arithmetic sequence. Examples of harmonic sequences are

$$1, \frac{1}{2}, \frac{1}{3}, \frac{1}{4}, \ldots \quad \text{and} \quad 2, \frac{2}{3}, \frac{2}{5}, \frac{2}{7}, \ldots.$$

A sequence of both historical and practical importance is the famous **Fibonacci sequence:**

$$1, 1, 2, 3, 5, 8, 13, \ldots.$$

The first and second terms are 1; each term thereafter is the sum of the two preceding terms. This sequence was discovered in the Middle Ages by the mathematician Fibonacci as the answer to a question about the population growth of a family of rabbits. It appears in a great variety of areas: mathematics, physics, chemistry, art, music, and nature. For instance, it is related to patterns of arrangements of plant leaves and seeds.

The subject of infinite sequences and series in general is a major topic of calculus. The approximating polynomials treated at the ends of Chapters 4 and 5 were finite series, extracted from infinite series, in a variable x. For example, to say that

$$e^x = 1 + x + \frac{x^2}{2} + \frac{x^3}{3!} + \cdots, \quad \text{for all } x,$$

means that if any real number, say 1.5, is substituted for x in this series, then the sequence of partial sums S_n approaches $e^{1.5}$ as $n \to \infty$.

EXERCISES 10.3

Find the common ratio and then write the next two terms and the general term of the given geometric sequence.

1. $3, 6, 12, \ldots$

2. $6, -3, \frac{3}{2}, \ldots$

3. $2, 6, 18, \ldots$

4. $2, 1, \frac{1}{2}, \ldots$ **5.** $8, 6, \frac{9}{2}, \ldots$ **6.** $12, -18, 27, \ldots$

7. $1, \sqrt{2}, 2, \ldots$ **8.** $1, \frac{1}{2}, \frac{1}{4}, \ldots$ **9.** $x, xy, xy^2, \ldots$ **10.** $x, -x^2, x^3, \ldots$

For the given geometric sequence, find the common ratio r and the tenth term.

11. $2, 6, 18, \ldots$ **12.** $3, -6, 12, \ldots$ **13.** $3, -\frac{3}{2}, \frac{3}{4}, \ldots$

14. $1, 3^x, 3^{2x}, \ldots$ **15.** $\frac{x}{v}, \frac{xz}{v}, \frac{xz^2}{v}, \ldots$

For the given geometric series, find (a) the common ratio; (b) the sum of the first five terms; (c) the sum of the infinite series.

16. $8 + 4 + 2 + \cdots$ **17.** $1 - \frac{1}{2} + \frac{1}{4} - \cdots$ **18.** $1 - \frac{1}{2}\sqrt{2} + \frac{1}{2} - \cdots$

19. $3 + 2 + \frac{4}{3} + \cdots$ **20.** $16 + 12 + 9 + \cdots$ **21.** $5 - 3 + \frac{9}{5} - \cdots$

22. $1 - \frac{1}{3} + \frac{1}{9} - \cdots$ **23.** $4 + 0.4 + 0.04 + \cdots$ **24.** $4 + 1.2 + 0.36 + \cdots$

25. $200 - 20 + 2 - \cdots$

In each of the following, some of the quantities a_1, r, n, a_n, and S_n are given. Find the remaining quantities.

26. $a_1 = 512$, $r = \frac{1}{2}$, and $a_n = 16$ **27.** $a_1 = 36$, $r = \frac{1}{3}$, and $a_n = \frac{4}{9}$

28. $S_n = 248$, $a_n = 128$, and $r = 2$ **29.** $r = -\frac{1}{2}$, $n = 5$, and $a_n = \frac{1}{4}$

30. $a_1 = 1$, $r = -\sqrt{3}$, and $r = 5$

Calculate the fractional number equivalent to each of the following repeating decimals.

31. 0.777 . . . **32.** 0.353535 . . . **33.** 0.090909 . . .

34. 0.123123123 . . . **35.** 0.737373 . . . **36.** 0.010101 . . .

37. 0.142857142857142857 . . . **38.** 1.444 . . . **39.** 2.050505 . . .

40. 3.373737 . . .

Find a general term for each of the following series.

41. $e^x = 1 + x + \frac{x^2}{2} + \frac{x^3}{3!} + \cdots$ **42.** $e^{-x} = 1 - x + \frac{x^2}{2} - \frac{x^3}{3!} + \cdots$

43. $\ln(1 + x) = x - \frac{x^2}{2} + \frac{x^3}{3} - \frac{x^4}{4} + \cdots$

44. $\ln(1 - x) = -x - \frac{x^2}{2} - \frac{x^3}{3} - \frac{x^4}{4} - \cdots$

45. $\sin x = x - \dfrac{x^3}{3!} + \dfrac{x^5}{5!} - \cdots$

46. $\cos x = 1 - \dfrac{x^2}{2!} + \dfrac{x^4}{4!} - \cdots$

47. $\arctan x = x - \dfrac{x^3}{3} + \dfrac{x^5}{5} - \cdots$

48. $\dfrac{1}{1 - x} = 1 + x + x^2 + x^3 + \cdots$

In Section 10.4 you will see how to prove that each of the following statements is true for every positive integral value of n. Test each of them for the cases n = 1, 2, 3.

49. $1^2 + 2^2 + 3^2 + \cdots + n^2 = \dfrac{n}{6}(n + 1)(2n + 1)$

50. $1 \cdot 3 + 2 \cdot 4 + 3 \cdot 5 + \cdots + n(n + 2) = \dfrac{n}{6}(n + 1)(2n + 7)$

51. Suppose a dropped ball always rebounds one half the height it falls. If it is dropped from a height of 128 feet, (a) how far has it traveled when it reaches the top of the fifth bounce? (b) If it could continue to bounce forever, what total distance would it travel?

52. A person receives a salary of \$8000 the first year and a 10% raise every year. What is his salary for the fifth year?

53. A tank full of alcohol is emptied of 1/3 of its contents and then filled with water and mixed. If this is done six times, what fraction of the original alcohol remains?

54. Suppose a house costing \$50,000 appreciates in value 10% each year. What is it worth at the end of 5 years?

55. Show that the reciprocals of the terms of a geometric sequence also form a geometric sequence.

10.4 Mathematical Induction

In Section 4.4, we proved that $\log xy = \log x + \log y$, but we *assumed* that $\log xyz = \log x + \log y + \log z$. To prove such general theorems as, for instance, $\log x_1x_2x_3 \cdots x_n = \log x_1 + \log x_2 + \log x_3 + \cdots + \log x_n$, we use the **principle of mathematical induction.** This principle is nicely illustrated as follows. Suppose, as shown in Figure 10.3, that we have an endless row of dominoes standing on end. In (a), to tip over every domino would require a separate action for each of the

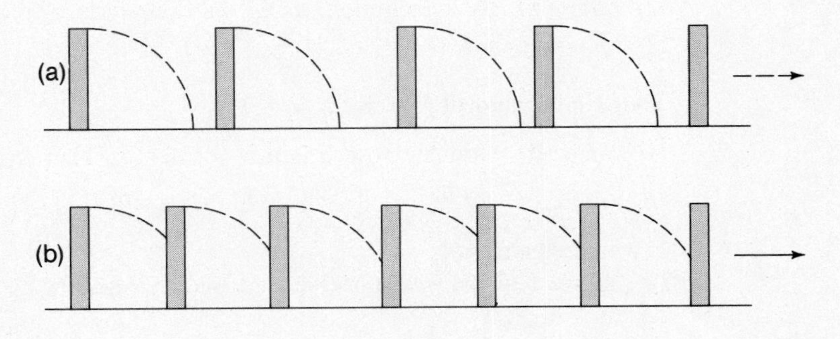

Figure 10.3

dominoes (an infinite number of actions). However, if they are lined up as in (b), when the first one is tipped over, it falls against the second one, which in turn falls against the third one, and so on infinitely. Only one action is needed for the entire operation. This is the sort of thing that is commonly referred to as "the domino effect."

How this idea is used in a mathematical situation is shown by the following. Observe that

$$1 = 1^2,$$
$$1 + 3 = 4 = 2^2,$$
$$1 + 3 + 5 = 9 = 3^2,$$
$$1 + 3 + 5 + 7 = 16 = 4^2,$$

and so on. From this it appears that the sum of the first n positive odd integers is equal to n^2. Indeed, it appears even more likely if testing continues for $n = 5, 6$, and so on. But, proving a general statement for any finite number of cases, however large, does not prove it for *every* value of n, since it is possible that the statement is not true for some large value of n not yet tested. To prove it in every case, the following important principle is used.

The Principle of Mathematical Induction A statement involving a general positive integer n is true for every $n \geq 1$ provided the following are true:
(A) The statement is true for $n = 1$.
(B) Whenever the statement is true for $n = k$, it is also true for $n = k + 1$.

First, let us observe that proving a statement by mathematical induction requires two separate steps: (A) By direct substitution, show that the statement is true for $n = 1$, and (B) show that *if* the statement is true for $n = k$, *then* it is necessarily true for $n = k + 1$.

EXAMPLE 1 Prove $1 + 3 + 5 + \cdots + (2n - 1) = n^2$.

SOLUTION Note that $2n - 1$ represents the nth positive odd integer. Also notice that in this formula, when $n = 1$, there is one term on the left side of the equals sign, when $n = 2$, there are two terms on the left side, and so on.

Step (A). By substitution, using only one term of the left side,

$$1 = 1^2.$$

Thus, the statement is true for $n = 1$.

Step (B). Now suppose it is true for $n = k$. That is, assume that

$$1 + 3 + 5 + \cdots + (2k - 1) = k^2$$

is a true statement.

If we add the *next* positive odd integer, namely $2k + 1$, to both sides of the equality, we get

$$\begin{aligned} 1 + 3 + 5 + \cdots + (2k - 1) + (2k + 1) &= k^2 + (2k + 1) \\ &= k^2 + 2k + 1, \end{aligned}$$

or, $$1 + 3 + 5 + \cdots + (2k - 1) + (2k + 1) = (k + 1)^2.$$

Here the left member is the sum of the first $k + 1$ positive odd integers and the right member is the square of $k + 1$. So the statement is true for $n = k + 1$ and part (B) is proved. Steps A and B have both been carried out, so by the principle of mathematical induction, we have proved that

$$1 + 3 + 5 + \cdots + (2n - 1) = n^2.$$

That is, *the sum of the first n positive odd integers is equal to the square of n.*

Observe that in (B) we do not prove that the theorem *is* true for $n = k + 1$. Rather, we show that *if* it is true for $n = k$, *then* it is true for $n = k + 1$. In Step (A), it was shown to be true for $n = 1$. By Step (B), it must then be true for $n = 2$ (since $2 = 1 + 1$). But now that we know it to be true for $n = 2$, (B) shows that it must be true for $n = 3$ (since $3 = 2 + 1$), and so on. ●

EXAMPLE 2 Prove that $1 + 2 + 4 + \cdots + 2^{n-1} = 2^n - 1$.

SOLUTION *Step (A).* Substituting $n = 1$, we get $1 = 2^1 - 1$, so that the statement is true for the case $n = 1$.

Step (B). Now suppose that for some k,

$$1 + 2 + 4 + \cdots + 2^{k-1} = 2^k - 1.$$

Adding 2^k to both sides gives

$$\begin{aligned} 1 + 2 + 4 + \cdots + 2^{k-1} + 2^k &= (2^k - 1) + 2^k \\ &= 2(2^k) - 1 \end{aligned}$$

or

$$1 + 2 + 4 + \cdots + 2^{k-1} + 2^k = 2^{k+1} - 1.$$

This is precisely the statement of the theorem for the case $n = k + 1$. Since we have shown both parts (A) and (B) of the principle of mathematical induction, we have proved that

$$1 + 2 + 4 + 8 + \cdots + 2^{n-1} = 2^n - 1$$

for every positive integral value of n. ●

EXAMPLE 3 Prove the following general law of logarithms:

$$\log x_1x_2x_3 \cdots x_n = \log x_1 + \log x_2 + \log x_3 + \cdots + \log x_n.$$

SOLUTION *Step (A).* Since $\log x_1 = \log x_1$, the statement is true for the case $n = 1$.

Step (B). Now suppose that it is true for some $n = k$, that is,

$$\log x_1x_2x_3 \cdots x_k = \log x_1 + \log x_2 + \cdots + \log x_k.$$

We know that

$$\log(x_1x_2 \cdots x_k)x_{k+1} = \log(x_1x_2 \cdots x_k) + \log x_{k+1},$$

because of the law of logarithms, $\log ab = \log a + \log b$, which was proved in Section 4.4 (let $x_1x_2 \cdots x_k = a$ and $x_{k+1} = b$). But $x_1x_2x_3 \cdots x_{k+1} = (x_1x_2 \cdots x_k)x_{k+1}$, so

$$\begin{aligned}\log x_1x_2 \cdots x_{k+1} &= \log(x_1x_2 \cdots x_k)x_{k+1}\\ &= \log(x_1x_2 \cdots x_k) + \log x_{k+1}\\ &= (\log x_1 + \log x_2 + \cdots + \log x_k) + \log x_{k+1}\end{aligned}$$

or, $\log x_1x_2 \cdots x_{k+1} = \log x_1 + \log x_2 + \cdots + \log x_{k+1}$. This is the theorem for the case $n = k + 1$. Since we have established both (A) and (B) of the principle of mathematical induction, we have proved the theorem:

$$\log x_1x_2x_3 \cdots x_n = \log x_1 + \log x_2 + \cdots + \log x_n$$

for every positive integral value of n.

●

EXAMPLE 4 Prove De Moivre's Theorem, $[r(\cos\theta + i\sin\theta)]^n = r^n(\cos n\theta + i\sin n\theta)$, for every positive integral value of n.

SOLUTION To use mathematical induction to prove the theorem for every *positive* integral value of n, we must show that:

Step (A). The theorem is true for $n = 1$;

Step (B). If the theorem is true for $n = k$, then it is true for $n = k + 1$.

The proof of (A) is immediate, for

$$[r(\cos\theta + i\sin\theta)]^1 = r^1(\cos 1\cdot\theta + i\sin 1\cdot\theta).$$

To prove (B) we must suppose it is true that

$$[r(\cos\theta + i\sin\theta)]^k = r^k(\cos k\theta + i\sin k\theta),$$

for some k, and show that

$$[r(\cos\theta + i\sin\theta)]^{k+1} = r^{k+1}[\cos(k + 1)\theta + i\sin(k + 1)\theta].$$

Given

$$[r(\cos\theta + i\sin\theta)]^k = r^k(\cos k\theta + i\sin k\theta),$$

we multiply both sides by $r(\cos\theta + i\sin\theta)$ to obtain

$$[r(\cos\theta + i\sin\theta)]^{k+1} = r^k(\cos k\theta + i\sin k\theta)\cdot r(\cos\theta + i\sin\theta).$$

Expanding the product on the right gives the result

$$\begin{aligned}[r(\cos\theta + i\sin\theta)]^{k+1} &= r^k\cdot r\cdot(\cos k\theta + i\sin k\theta)\cdot(\cos\theta + i\sin\theta)\\ &= r^{k+1}(\cos k\theta\cos\theta + \cos k\theta\cdot i\sin\theta\\ &\qquad + i\sin k\theta\cdot\cos\theta + i^2\sin k\theta\cdot\sin\theta)\\ &= r^{k+1}[(\cos k\theta\cos\theta - \sin k\theta\sin\theta)\\ &\qquad + i(\sin k\theta\cos\theta + \cos k\theta\sin\theta)]\end{aligned}$$

$$= r^{k+1}[\cos(k\theta + \theta) + i \sin(k\theta + \theta)]$$
$$= r^{k+1}[\cos(k + 1)\theta + i \sin(k + 1)\theta].$$

This completes the proof of part (B). By the principle of mathematical induction we have proved De Moivre's Theorem for all positive integral values of n. ●

EXAMPLE 5 Prove that $2^n > n$ for every positive integral value of n.

SOLUTION *Step (A).* Clearly, $2^1 > 1$, so the theorem is true for the case $n = 1$.

Step (B). Suppose that $2^k > k$ for some positive integral value of k. We must show that $2^{k+1} > k + 1$. Under the assumption that $2^k > k$, multiplying both sides by 2 gives

$$2(2^k) > 2(k)$$
$$2^{k+1} > 2k.$$

But $2k > k + 1$ when $k > 1$, and therefore,

$$2^{k+1} > k + 1.$$

This proves (B), so the theorem is true by the principle of mathematical induction. ●

It is essential to remember that proof by mathematical induction requires showing both parts (A) and (B) of the principle. In the example of the dominoes, unless one domino is tipped over, the following ones will not fall, even if properly spaced. And, even if some first domino is tipped over, not all the rest will fall unless each is spaced properly relative to its immediate neighbors. The following is a mathematical illustration of the second situation.

Consider the statement

$$2 + 4 + 6 + \cdots + 2n = n(n + 1) + (n - 1).$$

Step (A). For $n = 1$, this becomes $2 = 1(1 + 1) + 0$, which is true.

Step (B). Assume the truth of

$$2 + 4 + 6 + \cdots + 2k = k(k + 1) + (k - 1),$$

for some k. Now add the next even integer, $2(k + 1)$, to both sides.

$$2 + 4 + 6 + \cdots + 2k + 2(k + 1) = k(k + 1) + (k - 1) + 2(k + 1)$$
$$= (k + 2)(k + 1) + (k - 1)$$

However, if we substitute $n = k + 1$ in the original statement, we get

$$2 + 4 + 6 + \cdots + 2(k + 1) = (k + 1)(k + 1 + 1) + (k + 1 - 1)$$
$$= (k + 2)(k + 1) + k,$$

which is not the same as the preceding result. So part (B) is not true. In fact, the statement is not true for $n \geq 2$.

Mathematical induction should not be confused with the more familiar *empirical* induction used to determine scientific "laws" from a series of observations. These latter are actually conjectures which cannot be proved. They are used until and unless it is found that a different conjecture fits the observed data better.

A good illustration of this kind of reasoning with mathematical observations is given, for example, by the formula $p(n) = n^2 - n + 41$. Substitution of $n = 1, 2, 3, 4$, and 5 yields 41, 43, 47, 53, and 61 as values of $p(n)$. Each of these is a *prime** number. Larger values of n produce prime numbers also. We might easily conjecture that this formula produces primes for all positive integral values of n. However, when we reach $n = 41$, we get $p(41) = 41^2 - 41 + 41 = 41^2$, which clearly is not prime. Hence the conjecture is not valid.

Empirical or scientific induction is sometimes called "incomplete induction" to contrast it with mathematical induction, or "complete induction." Mathematical induction is a fundamental principle of mathematics. It is a powerful tool for proving an infinite number of statements at once, without having to test each of them separately (which would be impossible).

EXERCISES 10.4

By substitution, verify each of the following statements for the cases $n = 1$, $n = 2$, and $n = 3$. Then use mathematical induction to prove the statement for all positive integral values of n.

1. $1 + 2 + 3 + \cdots + n = \frac{n}{2}(n + 1)$

2. $2 + 4 + 6 + \cdots + 2n = n(n + 1)$

3. $4 + 9 + 12 + \cdots + 4n = 2n(n + 1)$

4. $2 + 5 + 8 + \cdots + (3n - 1) = \frac{n}{2}(3n + 1)$

5. $1^2 + 2^2 + 3^2 + \cdots + n^2 = \frac{n}{6}(n + 1)(2n + 1)$

6. $1^3 + 2^3 + 3^3 + \cdots + n^3 = \frac{n^2}{4}(n + 1)^2$

7. $2 + 2^2 + 2^3 + \cdots + 2^n = 2(2^n - 1)$

8. $3 + 3^2 + 3^3 + \cdots + 3^n = \frac{3(3^n - 1)}{2}$

9. $1(2) + 2(3) + 3(4) + \cdots + n(n + 1) = \frac{n}{3}(n + 1)(n + 2)$

10. $\left(1 + \frac{1}{1}\right)\left(1 + \frac{1}{2}\right)\left(1 + \frac{1}{3}\right) \cdots \left(1 + \frac{1}{n}\right) = n + 1$

11. $1 + r + r^2 + \cdots + r^{n-1} = \frac{r^n - 1}{r - 1}$

12. $a + (a + 2) + (a + 4) + \cdots + (a + 2n - 2) = 2(a + n - 1)$

*A **prime** number is an integer whose only factors are itself and 1.

13. $\dfrac{1}{1(2)} + \dfrac{1}{2(3)} + \dfrac{1}{3(4)} + \cdots + \dfrac{1}{n(n+1)} = \dfrac{n}{n+1}$

14. $\dfrac{1}{1(3)} + \dfrac{1}{3(5)} + \dfrac{1}{5(7)} + \cdots + \dfrac{1}{(2n-1)(2n+1)} = \dfrac{n}{2n+1}$

15. $\sum_{k=1}^{n} 3k = \dfrac{3n(n+1)}{2}$

16. $\sum_{k=1}^{n} 4k = 2k(k+1)$

17. $\sum_{k=1}^{n} (3k-2) = \dfrac{n}{2}(3n-1)$

18. $\sum_{k=1}^{n} 2(3^{n-1}) = 3^n - 1$

19. $3^n > n$

20. $3n < 3^n$

21. $3^n < 3^{n+1}$

22. $3^n > 2^n + 10n,\ n \geq 4$

23. $1 \cdot 2 \cdot 3 \cdot 4 \cdots n > n^3,\ n \geq 6$

Use mathematical induction to prove the following formulas.

24. The formula for the sum of an arithmetic series:

$$\sum_{k=1}^{n} (a + kd) = \frac{n}{2}(2a + (n-1)d).$$

25. The formula for the sum of a geometric series:

$$\sum_{k=1}^{n} ar^{k-1} = \frac{a(1-r^n)}{1-r}.$$

Use mathematical induction to prove the following laws of exponents.

26. $(ab)^n = a^n b^n$

27. $\left(\dfrac{a}{b}\right)^n = \dfrac{a^n}{b^n}$

Use mathematical induction to prove the following identities.

28. $\sin(\theta + n\pi) = (-1)^n \sin \theta$

29. $\cos(\theta + n\pi) = (-1)^n \cos \theta$

30. Use mathematical induction to prove the **generalized distributive law** for multiplication:

$$a(b_1 + b_2 + b_3 + \cdots + b_n) = ab_1 + ab_2 + ab_3 + \cdots + ab_n.$$

31. Prove that the sum of the interior angles of a polygon of n sides ($n \geq 3$) is $(n-2)180°$. [*Hint:* Draw all the diagonals from a single vertex and use the fact that the sum of the interior angles of a triangle is 180°.]

10.5 The Binomial Theorem

In Exercises 3.1, several of the following formulas, called **binomial expansions,** were calculated using multiplication.

$$(x + y)^0 = 1$$
$$(x + y)^1 = x + y$$
$$(x + y)^2 = x^2 + 2xy + y^2$$

$$(x + y)^3 = x^3 + 3x^2y + 3xy^2 + y^3$$
$$(x + y)^4 = x^4 + 4x^3y + 6x^2y^2 + 6xy^3 + y^4$$
$$(x + y)^5 = x^5 + 5x^4y + 10x^3y^2 + 10x^2y^3 + 5xy^4 + y^5$$

However, calculating these expansions by multiplication becomes quite tedious and impractical for large powers. Fortunately, there is a simple way to obtain them. To help find it, arrange just the above coefficients in the triangular pattern shown below.

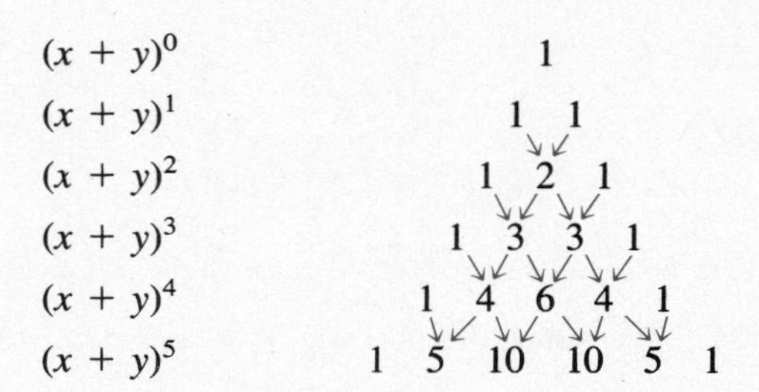

The above pattern is called **Pascal's triangle.** (Pascal was a 17th-century French mathematician, but this scheme was known much earlier). The following are some of the features of the pattern:

1. The first and last number in each row is 1.
2. The number of coefficients in each row is one more than the corresponding power.
3. Each number is the sum of the two adjacent numbers in the row just above (as shown by the arrows in the diagram).
4. The second and second to last coefficients in each row are the same, as are the third and third to last, and so on. That is, the coefficients are symmetric about the middle of the row.

Now going back to the given binomial expansions, note that:

5. The first term is x^n and the last term is y^n.
6. The exponent on x decreases by 1 and the exponent on y increases by 1 in successive terms from left to right.
7. The sum of the exponents on x and y is n.

EXAMPLE 1 Use Pascal's triangle to write the expansions for $(x + y)^6$ and $(x + y)^7$.

SOLUTION Continue the pattern described above to write the coefficients in the next rows of the triangle as follows.

$(x + y)^5$ 1 5 10 10 5 1
$(x + y)^6$ 1 6 15 20 15 6 1
$(x + y)^7$ 1 7 21 35 35 21 7 1

From this, we can write

$$(x + y)^6 = x^6 + 6x^5y + 15x^4y^2 + 20x^3y^3 + 15x^2y^4 + 6xy^5 + y^6$$

and

$$(x + y)^7 = x^7 + 7x^6y + 21x^5y^2 + 35x^4y^3 + 35x^3y^4 + 21x^2y^5 + 7xy^6 + y^7.$$

●

To write the binomial coefficients from Pascal's triangle it is necessary to have at hand (or else develop) the preceding row of the triangle. A table of these coefficients is given below for $n = 0$ to $n = 10$. For convenience the triangle is arranged differently than before. Pascal's triangle (binomial coefficients) is useful in mathematics in many ways other than in the binomial expansion.

n											
0	1										
1	1	1									
2	1	2	1								
3	1	3	3	1							
4	1	4	6	4	1						
5	1	5	10	10	5	1					
6	1	6	15	20	15	6	1				
7	1	7	21	35	35	21	7	1			
8	1	8	28	56	70	56	28	8	1		
9	1	9	36	84	126	126	84	36	9	1	
10	1	10	45	120	210	252	210	120	45	10	1

To expand a binomial without the use of the table, the following theorem is used.

THEOREM 10.1 **Binomial Theorem** For any positive integer n,

$$(x + y)^n = x^n + nx^{n-1}y + \frac{n(n-1)}{2}x^{n-2}y^2 + \frac{n(n-1)(n-2)}{2 \cdot 3}x^{n-3}y^3 + \cdots + nxy^{n-1} + y^n.$$

EXAMPLE 2 Expand $(x + y)^5$ by using the binomial theorem.

SOLUTION

$$(x + y)^5 = x^5 + 5x^4y + \frac{5(5-1)}{2}x^3y^2 + \frac{5(5-1)(5-2)}{2 \cdot 3}x^2y^3 + \frac{5(5-1)(5-2)(5-3)}{2 \cdot 3 \cdot 4}xy^4 + \frac{5(5-1)(5-2)(5-3)(5-4)}{2 \cdot 3 \cdot 4 \cdot 5}y^5$$

$$= x^5 + 5x^4y + \frac{5 \cdot 4}{2}x^3y^2 + \frac{5 \cdot 4 \cdot 3}{2 \cdot 3}x^2y^3$$

$$+ \frac{5 \cdot 4 \cdot 3 \cdot 2}{2 \cdot 3 \cdot 4}xy^4 + \frac{5 \cdot 4 \cdot 3 \cdot 2 \cdot 1}{2 \cdot 3 \cdot 4 \cdot 5}y^5$$

$$= x^5 + 5x^4y + 10x^3y^2 + 10x^2y^3 + 5xy^4 + y^5$$ ●

There is a very useful scheme for writing down the coefficients without substituting in the formula and then simplifying, as was done above. A careful look at the last line above will show that each coefficient can be obtained from the preceding one by the following rule:

Multiply the exponent on x in the preceding term by its coefficient and then divide by the number of the preceding term.

Thus, to obtain the second coefficient above, note that the exponent on x in the first term is 5 and the coefficient of the first term is 1. The second coefficient is

$$\frac{5(1)}{1}, \quad \text{or} \quad 5.$$

Similarly, the third coefficient is 4(5)/2, or 10, the fourth is 3(10)/3, or 10, and so on. Check your understanding of the procedure by checking the coefficients in each of the expansions given above.

The same pattern used for expanding $(x + y)^n$ can be used for the expansion of $(x - y)^n$ or $(ax \pm by)^n$, as shown by the following examples.

EXAMPLE 3 Expand and simplify $(x - y)^6$.

SOLUTION

$$(x - y)^6 = [x + (-y)]^6$$

$$= x^6 + 6x^5(-y) + 15x^4(-y)^2 + 20x^3(-y)^3 + 15x^2(-y)^4 + 6x(-y)^5 + (-y)^6$$

$$= x^6 - 6x^5y + 15x^4y^2 - 20x^3y^3 + 15x^2y^4 - 6xy^5 + y^6$$

Observe that the effect of the negative sign on y is to make the signs alternately positive and negative. ●

EXAMPLE 4 Expand and simplify $(2x + y)^5$.

SOLUTION

$$(2x + y)^5 = (2x)^5 + 5(2x)^4y + 10(2x)^3y^2 + 10(2x)^2y^3 + 5(2x)y^4 + y^5$$

$$= 32x^5 + 80x^4y + 80x^3y^2 + 40x^2y^3 + 10xy^4 + y^5$$ ●

The binomial expansion can be used for expanding numerical expressions as well as algebraic expressions.

EXAMPLE 5 Evaluate $(1.01)^3$ by using the binomial expansion.

SOLUTION

$$\begin{aligned}(1.01)^3 &= (1 + 0.01)^3 \\ &= 1^3 + 3(1^2)(0.01) + 3(1)(0.01)^2 + (0.01)^3 \\ &= 1 + 0.03 + 0.0003 + 0.000001 = 1.030301.\end{aligned}$$

Proof of the Binomial Theorem Mathematical induction may be used as follows to prove the binomial theorem.

Proof: *Step (A).* The theorem is true for $n = 1$, since $(x + y)^1 = x + y = x^1 + y^1$.

Step (B). Assume that the theorem is true for $n = k$.

$$\begin{aligned}(x + y)^k = x^k + kx^{k-1}y &+ \frac{k(k-1)}{2}x^{k-2}y^2 \\ &+ \frac{k(k-1)(k-2)}{2 \cdot 3}x^{k-3}y^3 + \cdots + kxy^{k-1} + y^k\end{aligned}$$

Next, the rules for exponents and the distributive property are used with the above assumption to expand $(x + y)^{k+1}$.

$$\begin{aligned}(x + y)^{k+1} &= (x + y)(x + y)^k \\ &= x(x + y)^k + y(x + y)^k \\ &= x(x^k + kx^{k-1}y + \frac{k(k-1)}{2}x^{k-2}y^2 \\ &\quad + \frac{k(k-1)(k-2)}{2 \cdot 3}x^{k-3}y^3 + \cdots + kxy^{k-1} + y^k) \\ &\quad + y(x^k + kx^{k-1}y + \frac{k(k-1)}{2}x^{k-2}y^2 \\ &\quad + \frac{k(k-1)(k-2)}{2 \cdot 3}x^{k-3}y^3 + \cdots + kxy^{k-1} + y^k).\end{aligned}$$

Now combine terms in order of decreasing powers of x to obtain

$$\begin{aligned}(x + y)^{k+1} &= x^{k+1} + x^k(ky + y) + x^{k-1}\left(\frac{k(k-1)}{2}y^2 + ky^2\right) \\ &\quad + x^{k-2}\left(\frac{k(k-1)(k-2)}{2 \cdot 3}y^3 + \frac{k(k-1)}{2}y^3\right) \\ &\quad + \cdots + x(y^k + ky^k) + y^{k+1} \\ &= x^{k+1} + x^k y(k + 1) + x^{k-1}y^2\left(\frac{k(k-1)}{2} + k\right) \\ &\quad + x^{k-2}y^3\left(\frac{k(k-1)(k-2)}{2 \cdot 3} + \frac{k(k-1)}{2}\right) + \cdots \\ &\quad + xy^k(k + 1) + y^{k+1}\end{aligned}$$

$$= x^{k+1} + (k + 1)x^k y + \frac{(k + 1)k}{2}x^{k-1}y^2$$

$$+ \frac{(k + 1)(k)(k - 1)}{2 \cdot 3}x^{k-2}y^3 + \cdots + (k + 1)xy^k + y^{k+1}.$$

This is the statement of the binomial theorem for $n = k + 1$. Hence, by the principle of mathematical induction, the theorem is true.

Binomial Coefficients The binomial coefficients appear in many different contexts in mathematics. For instance, they play an important role in probability and statistics. For this reason, it is useful to have simple formulas for them. First let us recall the *factorial notation* introduced in Section 5.12.

$$n! = n(n - 1)(n - 2) \cdots 3 \cdot 2 \cdot 1$$

or

$$n! = 1 \cdot 2 \cdot 3 \cdots (n - 2)(n - 1)n$$

By the above, $1! = 1$, and it is useful to define $0! = 1$ also. Then, for example,

$$4! = 4 \cdot 3 \cdot 2 \cdot 1 \quad \text{or} \quad 4! = 4(3 \cdot 2 \cdot 1) = 4(3!),$$

$$5! = 5 \cdot 4 \cdot 3 \cdot 2 \cdot 1 \quad \text{or} \quad 5! = 5 \cdot 4 \cdot (3 \cdot 2 \cdot 1) = 5 \cdot 4(3!).$$

Expressed differently,

$$\frac{4!}{3!} = 4 \quad \text{and} \quad \frac{5!}{3!} = 5 \cdot 4.$$

In general,

$$n! = n \cdot (n - 1)! \quad \text{or} \quad n! = n(n - 1) \cdot (n - 2)!,$$

and

$$\frac{n!}{n} = (n - 1)!, \qquad \frac{n!}{(n - 1)!} = n, \qquad \frac{n!}{(n - 2)!} = n(n - 1),$$

and so on. As an example using this notation,

$$(x + y)^5 = x^5 + 5x^4y + \frac{5 \cdot 4}{2}x^3y^2 + \frac{5 \cdot 4 \cdot 3}{2 \cdot 3}x^2y^3$$

$$+ \frac{5 \cdot 4 \cdot 3 \cdot 2}{2 \cdot 3 \cdot 4}xy^4 + \frac{5 \cdot 4 \cdot 3 \cdot 2 \cdot 1}{2 \cdot 3 \cdot 4 \cdot 5}y^5$$

$$= \frac{x^5}{0!} + \frac{5}{1!}x^4y + \frac{5 \cdot 4}{2!}x^3y^2 + \frac{5 \cdot 4 \cdot 3}{3!}x^2y^3$$

$$+ \frac{5 \cdot 4 \cdot 3 \cdot 2}{4!}xy^4 + \frac{5 \cdot 4 \cdot 3 \cdot 2 \cdot 1}{5!}y^5$$

$$= \frac{5!}{5!0!}x^5 + \frac{5!}{4!1!}x^4y + \frac{5!}{3!2!}x^3y^2 + \frac{5!}{2!3!}x^2y^3$$

$$+ \frac{5!}{1!4!}xy^4 + \frac{5!}{5!}y^5.$$

Note the pattern of the coefficients: the coefficient of x^4y is $5!/(4!1!)$ and that of x^2y^3 is $5!/(2!3!)$. In general, the coefficient of x^py^q in the expansion of $(x + y)^n$ is $n!/(p!q!)$. Expressed differently, the coefficient of the term involving x^k is

$$\frac{n!}{k!(n-k)!}.$$

That of y^k is the same because of the symmetric pattern of the coefficients.

Now, combining summation notation (Section 10.1) and factorial notation, we can write the expansion of $(x + y)^5$ concisely as

$$(x+y)^5 = \sum_{k=0}^{5} \frac{5!}{k!(5-k)!} x^{5-k}y^k.$$

The general formula for the binomial expansion is

$$(x+y)^n = \sum_{k=0}^{n} \frac{n!}{k!(n-k)!} x^{n-k}y^k.$$

An alternative notation for these coefficients is introduced in the Miscellaneous Exercises for Chapter 10.

To calculate powers of complex numbers algebraically, we need the binomial power $(a + bi)^n$.

EXAMPLE 6 Expand $(1 + 2i)^5$ and reduce to the standard form $a + bi$.

SOLUTION Use the binomial expansion.

$$\begin{aligned}(1+2i)^5 &= 1^5 + 5(1^4)(2i) + 10(1^3)(2i)^2 \\ &\quad + 10(1^2)(2i)^3 + 5(1)(2i)^4 + (2i)^5 \\ &= 1 + 10i + 40i^2 + 80i^3 + 80i^4 + 32i^5\end{aligned}$$

Since $i^2 = -1$, $i^3 = -i$, and so on, this reduces to

$$\begin{aligned}(1+2i)^5 &= 1 + 10i - 40 - 80i + 80 + 32i \\ &= 41 - 38i. \quad \bullet\end{aligned}$$

Binomial Series Consider the binomial expansion

$$\begin{aligned}(1+x)^n &= 1 + nx + \frac{n(n-1)}{2}x^2 + \frac{n(n-1)(n-2)}{3!}x^3 + \cdots \\ &\quad + \frac{n(n-1)(n-2)\cdots(n-k+1)}{k!}x^k + \cdots.\end{aligned}$$

In the preceding discussion n was always a positive integer, and the expansions terminated naturally with the term containing x^n. This is because if $k > n$ the coefficient of x^k contains the factor zero in the numerator. For example, if $k = n + 1$, the factor $n - k + 1$ in the numerator is $n - (n + 1) + 1 = 0$.

If n is not a positive integer, say a negative integer or a fraction, the expansion does not terminate but becomes an infinite series. For example,

$$(1 + x)^{\frac{1}{2}} = 1 + \frac{x}{2} + \frac{\frac{1}{2}\left(\frac{1}{2} - 1\right)}{2}x^2 + \frac{\frac{1}{2}\left(\frac{1}{2} - 1\right)\left(\frac{1}{2} - 2\right)}{2 \cdot 3}x^3 + \cdots$$

$$= 1 + \frac{x}{2} - \frac{x^2}{8} + \frac{x^3}{16} - \cdots.$$

This is called a **binomial series.** In calculus it is proved that this infinite series has a "sum" (as described in Section 10.3), provided that $|x| < 1$. Thus,

$$f(x) = \sqrt{1 + x} = (1 + x)^{\frac{1}{2}} = 1 + \frac{x}{2} - \frac{x^2}{8} + \frac{x^3}{16} - \cdots,$$

gives polynomial approximations for $\sqrt{1 + x}$. (See Sections 4.5 and 5.12).

EXAMPLE 7 Use the binominal expansion to approximate $\sqrt{1.01}$.

SOLUTION If the first four terms are used,

$$\sqrt{1.01} = (1 + 0.01)^{\frac{1}{2}}$$

$$= 1 + \frac{1}{2}(0.01) - \frac{1}{8}(0.01)^2 + \frac{1}{16}(0.01)^3 - \cdots$$

$$= 1 + 0.005 - 0.0000125 + \cdots$$

$$\approx 1.005.$$

A check with the calculator shows that this result is correct to three decimal places. ●

EXAMPLE 8 Use the first three terms of the binomial expansion to approximate $\sqrt[3]{7}$.

SOLUTION First, we write 7 in a form involving $1 + x$, where $|x| < 1$, so that we can use the expansion for $(1 + x)^{1/3}$. Thus,

$$\sqrt[3]{7} = \sqrt[3]{8 - 1} = \sqrt[3]{8(1 - 1/8)}$$

$$= 2(1 - 1/8)^{1/3} = 2\left(1 + \left(-\frac{1}{8}\right)\right)^{1/3}$$

$$= 2\left[1 + \frac{1}{3}\left(-\frac{1}{8}\right) + \frac{\frac{1}{3}\left(\frac{1}{3} - 1\right)}{2}\left(-\frac{1}{8}\right)^2 + \cdots\right]$$

$$\approx 2\left[1 - \frac{1}{24} - \frac{1}{9}\left(\frac{1}{64}\right)\right]$$

$$\approx 2(0.957) \approx 1.91.$$

The calculator gives $7^{1/3} \approx 1.91$ (rounded to two decimal places). ●

EXAMPLE 9 Use the binomial expansion to find a series representation for $1/(1 - x)$.

SOLUTION

$$\frac{1}{1-x} = (1 - x)^{-1}$$

$$= 1 - (-1)x + \frac{(-1)(-2)}{2}x^2 - \frac{(-1)(-2)(-3)}{2 \cdot 3}x^3 + \cdots$$

$$= 1 + x + x^2 + x^3 + \cdots. \quad \bullet$$

Binomial expansions for negative and fractional powers were used by Newton (1642–1727), but the need for the restriction $|x| < 1$ was not known until years later.

EXERCISES 10.5

Expand and simplify, using the table of binomial coefficients (Pascal's triangle).

1. $(x + y)^6$ **2.** $(x + y)^7$ **3.** $(x + y)^8$ **4.** $(x - y)^9$ **5.** $(x - y)^{10}$

6. $(x - y)^{11}$ **7.** $(x + 2y)^5$ **8.** $(3x + y)^4$ **9.** $(2x - 3y)^3$

10. $\left(\frac{x}{2} - y\right)^6$

Expand and simplify, using the short-cut rule of the text to determine the binomial coefficients.

11. $(x + y)^3$ **12.** $(x + y)^4$ **13.** $(x + y)^6$ **14.** $(x + y)^7$

15. $(x - y)^8$ **16.** $(x - y)^9$ **17.** $(x - y)^{10}$ **18.** $(x - y)^{11}$

Expand and simplify, using the statement of the binomial theorem.

19. $(x + y)^3$ **20.** $(x + y)^4$ **21.** $(x + y)^6$

22. $(x + y)^7$ **23.** $(2x + y)^7$ **24.** $(3x - 2y)^8$

Write:

25. The first four terms of $(2x + 3y)^{12}$

26. The first five terms of $(x + y)^{25}$

27. The last three terms of $\left(\frac{x}{2} - 2y\right)^{10}$

28. The last two terms of $(x - 4y)^{13}$

29. The sixth term of $(x + y)^{15}$

30. The ninth term of $(x - y)^{14}$

31. The term of $(x + y)^{16}$ involving x^7

32. The term of $(x - y)^8$ involving y^5

33. The coefficient of x^2y^5 in $(x + y)^7$

34. The coefficient of x^3y^6 in $(x + y)^9$

35. The coefficient of x^4y^4 in $(x - y)^8$

36. The coefficient of x^8y^2 in $(x - y)^{10}$

Expand the following powers of complex numbers and simplify to the standard form $a + bi$.

37. $(2 - i)^6$ **38.** $(3 + 4i)^3$ **39.** $(1 + 3i)^5$

40. $(2 - 3i)^4$ **41.** $(5 + \sqrt{-4})^3$ **42.** $(2 + \sqrt{-3})^4$

43. $(1 + i\sqrt{2})^6$ **44.** $\left(\frac{1}{2}\sqrt{2} + \frac{i}{2}\sqrt{2}\right)^2$

Calculate by first expanding by the binomial expansion.

45. 101^3 [*Hint:* $101 = 100 + 1$.]

46. $(1.02)^4$

47. 99^4

48. $(0.97)^3$

Write the first three terms of the binomial expansion of each of the following.

49. $(1 + x)^{1/4}$

50. $(x - 1)^{-2}$

51. $(1 + x)^{-3}$

52. $(1 - 2x)^{1/3}$

53. $(1 + x^2)^{-3}$

54. $(1 - x)^{-4}$

Use the first three terms of the binomial expansion to approximate the following.

55. $\sqrt{10}$

56. $\sqrt[3]{65}$

57. $(1.02)^{1/4}$

58. $(1.01)^{-5}$

59. Expand $(x + y + z)^3$. [*Hint:* $x + y + z = (x + y) + z$.]

60. Calculate $(2 + \sqrt{3})^4 + (2 - \sqrt{3})^4$. (Note that the result is an integer.)

61. Show that $\dfrac{1}{1 + x^2} = \Sigma_{k=0}^{\infty} (-1)^k x^2$. [*Hint:* The summation is a geometric series.]

Use the binomial expansion to show the following.

62. $(1.1)^{10} > 2.5$

63. $(1.002)^{10} > 1.02$

64. Complete the next two rows ($n = 11$ and $n = 12$) of Pascal's triangle in the table in the text.

65. It can be shown that the number of different ways n people can line up is $n!$. How many different possible arrangements are there at a theater ticket window if there are 7 persons? 10 persons? 12 persons?

Chapter 10 Review Exercises

Write the next three terms and the general term of each of the following sequences.

1. $-1, 2, 5, 8, \ldots$

2. $9, 6, 4, 8/3, \ldots$

3. $1, -1/2\sqrt{2}, 1/3\sqrt{3}, -1/8, \ldots$

4. $\dfrac{1}{2^3}, -\dfrac{2}{3^3}, \dfrac{3}{4^3}, -\dfrac{4}{5^3}, \ldots$

5. $1, -\dfrac{x}{5}, \dfrac{x^2}{5^2}, -\dfrac{x^3}{5^3}, \ldots$

6. $1, \dfrac{x}{3}, \dfrac{x^2}{3^2}, \dfrac{x^3}{3^3}, \ldots$

Find the common difference, the last term, and the sum of each of the following arithmetic series.

7. $2 + 5 + 8 + \cdots$; to eight terms

8. $3 + 7 + 11 + \cdots$; to seven terms

9. $-11 - 8 - 5 - \cdots$; to ten terms

10. $\dfrac{2}{3} + \dfrac{7}{3} + 4 + \cdots$; to six terms

11. $x + \left(x + \dfrac{3}{4}\right) + \left(x + \dfrac{3}{2}\right) + \cdots$; to five terms

12. $2x - \dfrac{1}{2}x - 3x - \cdots$; to nine terms

Find the common ratio, the last term, and the sum of each of the following geometric series, and write in summation notation.

13. $5 + 10 + 20 + \cdots$; five terms

14. $6 + 2 + \frac{2}{3} + \cdots$; six terms

15. $5 + 15 + 45 + \cdots$; six terms

16. $192 + 96 + 48 + \cdots$; five terms

17. $\frac{1}{3} + \frac{2}{9} + \frac{4}{27} + \cdots$; four terms

18. $3 + \frac{3}{2} + \frac{3}{4} + \cdots$; six terms

19. $1 - \frac{1}{2} + \frac{1}{4} - \frac{1}{8} + \cdots$; seven terms

20. $1 - \frac{1}{3} + \frac{1}{9} - \frac{1}{27} + \cdots$; six terms

21. $2 + 0.2 + 0.02 + \cdots$; five terms

22. $1 + 0.01 + 0.0001 + \cdots$; six terms

Use the formula $S = a_1/(1 - r)$ to find the sum of the following infinite geometric series.

23. $\sum_{k=0}^{k=\infty} \frac{1}{4^k}$

24. $\sum_{k=1}^{k=\infty} \frac{1}{2^{k-1}}$

25. $\sum_{k=1}^{k=\infty} \left(\frac{3}{4}\right)^{k-1}$

26. $\sum_{k=0}^{k=\infty} \left(-\frac{2}{3}\right)^{k}$

Calculate the fractional equivalents of the following repeating decimals.

27. 0.282828 . . .

28. 1.513513513 . . .

29. A man is employed at \$8000 base pay per year, with the promise of an annual increase of \$750 per year for each of the next four years. (a) What is his salary for the fifth year? (b) What is the total salary received for five years?

30. A grocer displays cans of soup against a counter in a pyramid arrangement, one row above another. Each row has 2 fewer cans than the row below it. If there are 17 cans in the bottom row and 5 cans in the top row, (a) how many rows are there; (b) how many cans are there altogether?

31. If a person deposits \$1 in savings and then doubles the amount deposited each successive month, how much will he have accumulated at the end of 12 months (exclusive of interest)?

32. At the end of each year the value of a certain machine has depreciated by 20% of the value at the beginning of the year. If it is worth \$1000 at first, how much is it worth at the end of 5 years?

33. Find the sum of the first five multiples of 4.

34. Find the sum of the first six powers of 3.

Use mathematical induction to show that each of the following is true for every positive integral value of n.

35. $3 + 6 + 9 + \cdots + 3n = \frac{3n(n + 1)}{2}$

36. $\frac{1}{2} + \frac{1}{4} + \frac{1}{8} + \cdots + \frac{1}{2^n} = \frac{2^n - 1}{2^n}$

37. $1 + 2 + 2^2 + 2^3 + \cdots + 2^n = 2^{n+1} - 1$

38. $1(3) + 2(4) + 3(5) + \cdots + n(n + 2) = \frac{n}{6}(n + 1)(2n + 7)$

Use the binomial theorem to expand each of the following.

39. $(x + y)^9$ **40.** $(x - y)^7$ **41.** $(x - 3y)^6$ **42.** $(3x + y)^5$

Use the binomial expansion to reduce each of the following to the standard form $a + bi$.

43. $(2 + 3i)^5$ **44.** $(1 - 2i)^6$ **45.** $(1 + i)^4$ **46.** $(-\sqrt{3} + i)^3$

Write the first four terms of each of the following.

47. $\dfrac{1}{1 + x}$ [*Hint:* Write $\dfrac{1}{1 + x}$ as $(1 + x)^{-1}$.]

48. $\sqrt[3]{1 - x}$ [*Hint:* Write $\sqrt[3]{1 - x}$ as $(1 - x)^{1/3}$.]

Use the binomial series to approximate each of the following. Take four terms of the series.

49. $\sqrt{47}$ **50.** $\sqrt[3]{25}$

Chapter 10 Miscellaneous Exercises

The following are very useful facts. Use mathematical induction to prove them.

1. $(1 + a)^n > 1 + na$

2. $x - y$ is a factor of $x^n - y^n$ for all positive integral values of n. [*Hint:* First write $x^{k+1} - y^{k+1} = (x^{k+1} - xy^k) + (xy^k - y^{k+1})$.]

By the binomial theorem, the coefficient of $x^k y^{n-k}$ in $(x + y)^n$ is $n!/k!(n - k)!$. This expression is important in probability and other areas of mathematics. It is sometimes denoted by $C(n, k)$ or ${}_nC_k$, but now more commonly by $\binom{n}{k}$. Evaluate each of the following.

3. $\binom{6}{2}$ **4.** $\binom{5}{3}$ **5.** $\binom{10}{8}$ **6.** $\binom{8}{5}$

7. $\binom{n}{n}$ **8.** $\binom{n}{0}$ **9.** $\binom{n}{1}$ **10.** $\binom{n}{n-1}$

The value of $\binom{n}{k}$ is the number of ways one may select k objects from a set of n objects. Use this fact to find the following.

11. The number of subcommittees of 3 which may be selected from a whole committee of 7

12. The number of different sums of money that can be formed from a penny, a nickel, a dime, and a quarter

13. The number of lines determined by the vertices of a quadrilateral

14. The number of triangles determined by the vertices of a hexagon

15. Prove that $\binom{n}{n-k} = \binom{n}{k}$ and interpret this fact relative to (a) the selection of subcommittees, (b) the binomial coefficients.

16. Let $a_1, a_1r, a_1r^2, \ldots$ be a geometric sequence with $a_1 > 0$ and $r > 0$. Show that $\log a_1, \log a_1r, \log a_1r^2, \ldots$ is an arithmetic sequence.

Appendixes

Appendix 1 Approximate Numbers and Calculation

Small hand calculators are now commonly used for arithmetic and scientific calculations. The exercises in this book may be done with or without a calculator. Tables of function values with instructions for their use are given in Appendix 2. Details on the use of any particular calculator are given in the instruction booklet accompanying it.

In numerical applications we often deal with approximate numbers. This comes from the use of physical measurements, from rounded-off entries in the tables, and from limitations of a particular calculator. For instance, a distance may be measured to the nearest meter, to the nearest centimeter, or to a particular decimal part of any unit. The tables which follow list entries to four or five digits, rounded off (generally) from unending decimals. A hand calculator may display eight or ten digits.

The number of digits in a number (not counting zeros on the left) is a useful measure of the accuracy of data. If the number is written with a decimal point, its number of **significant digits** is the number of digits from the left-most nonzero digit to the right-most digit. The zeros to the right of nonzero digits may or may not be significant digits, depending on the situation. For example, 23.405 has five significant digits, while 0.0076 has two. The zeros in the latter are necessary to place the nonzero digits properly. The distance from the Earth to the Sun is 93,000,000 miles to the *nearest million* miles or 92,900,000 to the *nearest hundred thousand* miles. The zeros on the right are not significant digits. We make the distinction clear when we write these in scientific notation as 9.3×10^7 and 9.29×10^7, respectively. The measure 23.4 means accuracy to the nearest 0.1. If we wish it to mean accuracy to the nearest 0.01, we write it 23.40, and the "0" is a significant digit. In scientific notation we write these as 2.34×10^2 and 2.340×10^2, respectively.

The following are useful guidelines for use in rounding off results of calculations:

(1) In the sum or difference of approximate numbers, retain no more digits than occur in the number having the *smallest number of decimal places*.

(2) In the product or quotient of approximate numbers, retain no more digits than occur in the number having the *smallest number of significant digits*.

In single-operation calculations with a calculator the error in the display is at most ± 1. If several successive operations are performed there may be an accumulated error greater than this. Usually we perform the calculations, retaining at each stage at least one more digit than the above rules suggest, and then round off appropriately. We will generally follow these guidelines in calculations in this book.

Appendix 2 Tables

Table A. Exponential Function Values

Listed in this table are the values of e^x and e^{-x} from $x = 0$ to $x = 10$. For $0 < x < 0.2$ they are given at intervals of 0.01, for $0.2 < x < 3$ at intervals of 0.1, and for $3 < x < 10$ at intervals of 0.5. For example, to find $e^{0.13}$, first locate 0.13 in the column headed x. Then in the same line and in the column headed e^x we read 1.1388. Thus, $e^{0.13} = 1.1388$ (approximately). Similarly, $e^{1.8} = 6.0496$. Since values of e^x are generally irrational, we can write them decimally only approximately. (For convenience, however, we will use the equals sign, =.) Here the values are given to five significant figures.

In the same way, $e^{-0.13}$ is found in the line with $x = 0.13$ and in the column headed e^{-x}. So $e^{-0.13} = 0.87810$. To find e^x and e^{-x} for $x > 5$ we use a law of exponents. Thus, for example, $e^{6.13} = e^{5+1.13} = e^5e^1(e^{0.13})$ (rounded off to five digits) = 1.4841(2.7183)(1.1388) = 459.42.

Interpolation It is not possible to list the values of any function for all values of x. For intermediate values we use a procedure called **linear interpolation.** The idea will be clear from the figure. Suppose we want $e^{1.84}$. Since the exponential function is an increasing function, $e^{1.8} < e^{1.84} < e^{1.9}$. Part (a) in the figure shows the graphical situation. In part (b) of the figure, we have enlarged the portion of the figure including the interval $1.8 < x < 1.9$. From the table, $e^{1.8} = 6.0496$ and $e^{1.9} = 6.6859$. In the figure a straight-line chord connects the points $P(1.8, 6.0496)$ and $Q(1.9, 6.6859)$. For this small interval, the line segment PQ approximates the curve quite closely. (This accounts for the term linear approximation). So its ordinates are good approximations to those of the curve for $1.8 < x < 1.9$. Hence the ordinate of R on the line approximates the ordinate $y = e^{1.84}$ of S on the curve.

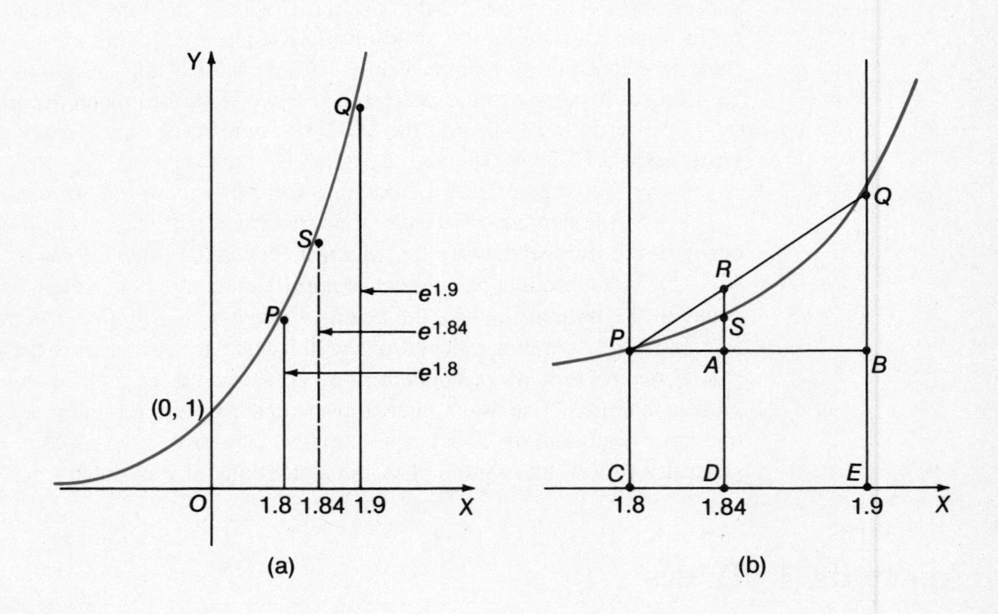

The right triangles PAR and PBQ are similar, so that

$$\frac{\overline{AR}}{\overline{BQ}} = \frac{\overline{PA}}{\overline{PB}}.$$

Here, $\overline{PA} = 0.04$, $\overline{PB} = 0.1$, and

$$\overline{BQ} = \overline{EQ} - \overline{EB}$$

$$
\begin{aligned}
&= \overline{EQ} - \overline{CP} \\
&= e^{1.9} - e^{1.8} \\
&= 6.6859 - 6.0496 = 0.6363.
\end{aligned}
$$

Now let $d = \overline{AR}$, and then

$$
\begin{aligned}
\frac{d}{0.6363} &= \frac{0.04}{0.1} = \frac{4}{10}, \\
10\,d &= 4(0.6363) = 2.5452, \\
d &= 0.2545.
\end{aligned}
$$

So we have

$$
\begin{aligned}
e^{1.84} &= \overline{DS} \approx \overline{DR} \\
&= \overline{DA} + \overline{AR} \\
&= \overline{CP} + d \\
&= e^{1.8} + d \\
&= 6.0496 + 0.2545 = 6.3041.
\end{aligned}
$$

This calculation can be done quite simply if we note the following pattern.

$$
10\left(4\left(\begin{matrix} 1.8 \\ 1.84 \end{matrix} \quad \begin{matrix} 6.0496 \\ e^{1.84} \end{matrix} \right) d \atop \begin{matrix} 1.9 \end{matrix} \quad \begin{matrix} 6.6859 \end{matrix} \right) 0.6363
$$

Here 1.84 is 0.4 the difference from 1.8 to 1.9. So we add the proportional difference $0.4(0.6363) = 0.2545$ to $e^{1.8}$.

EXAMPLE Find $e^{-0.63}$.

SOLUTION From the table,

$$
\left. \begin{aligned} e^{-0.6} &= 0.54881 \\ e^{-0.7} &= 0.49659 \end{aligned} \right) -0.05222
$$

Here e^{-x} is decreasing, so that we add $d = 0.3(-0.05222) = -0.01567$ to $e^{-0.6}$.

$$
\begin{aligned}
e^{-0.63} &= e^{-0.6} + d \\
&= 0.54881 - 0.01567 \\
&= 0.53314 \quad \bullet
\end{aligned}
$$

Table B. Common Logarithms

First, we will show that only the decimal part of common logarithms need be listed in the tables. For example,

$$
\begin{aligned}
6.53 &= 6.53 \times 10^0, \\
65.3 &= 6.53 \times 10^1, \\
653 &= 6.53 \times 10^2,
\end{aligned}
$$

$$6530 = 6.53 \times 10^3, \text{ and so on.}$$
$$0.653 = 6.53 \times 10^{-1},$$
$$0.0653 = 6.53 \times 10^{-2}, \text{ and so on.}$$

Now,

$$\log N \cdot 10^k = \log N + \log 10^k = \log N + k.$$

So if we are given $\log 6.53 = 0.8149$, we have

$$\log 0.0653 = \log 6.53 - 2 = 0.8149 - 2,$$
$$\log 0.653 = \log 6.53 - 1 = 0.8149 - 1,$$
$$\log 6.53 = \log 6.53 + 0 = 0.8149 + 0,$$
$$\log 65.3 = \log 6.53 + 1 = 0.8149 + 1,$$
$$\log 653 = \log 6.53 + 2 = 0.8149 + 2,$$
$$\log 6530 = \log 6.53 + 3 = 0.8149 + 3, \text{ and so on.}$$

If the context is clear, it is customary to write $\log x$ for $\log_{10} x$, as above.

From the examples, we see that we can write the common logarithm of any positive number if we have the common logarithms of all numbers *between 1 and 10*. These logarithms are called *mantissas*. Thus, all numbers having the same sequence of digits have the same mantissa. In the above example these digits are 6, 5, and 3, and the mantissa is the same for all the logarithms, namely, 0.8149.

When the decimal part of a logarithm is positive, the integral part of the logarithm is called the *characteristic* of the logarithm. Thus, the characteristic of log 6.53 is 0, that of log 653 is 2, and that of log 0.653 is -1. We may write $\log 0.0653 = 0.8149 - 2 = -1.1851$, which is often a useful form for calculations. However, the decimal part of this latter form is *not* the mantissa and the integral part is *not* the characteristic. For, $-1.1851 = -1 + (-0.1851)$, so that the decimal part is not positive.

We may equally well write

$$\begin{aligned}\log 0.653 &= 0.8149 - 1\\ &= 1.8149 - 2\\ &= 2.8149 - 3\end{aligned}$$

and so on. In each case the integral part (the characteristic) is the same: $0 - 1 = 1 - 2 = 2 - 3 = -1$. The standard convention here is to write in this pattern such that it ends in -10. Thus, $\log 0.653 = 9.8149 - 10$.

To determine the characteristic we need simply first write the number in scientific form and then read off the exponent on 10. So, in practice, we determine the characteristic by inspection and the mantissa from the table.

EXAMPLE Find (a) log 37, (b) log 231, (c) log 0.842.

SOLUTION In Table B we look for the first two digits of the number in the first column, headed N, and for the third digit in the same line in the appropriate column, headed 0, 1, or 2, and so on. For simplicity, the decimal points of mantissas may be omitted in the tabular entries.

(a) To find log 37, we first find 3.7 in the column headed N. In the same line under 0 we see 5682. So the mantissa of log 37 is 0.5682. Since $37 = 3.7 \times 10^1$, the characteristic of the logarithm is 1. Finally, $\log 37 = 1.5682$.

(b) To find log 231, we first locate 2.3 in the column headed N. In the same line under 1 we read the mantissa 0.3636. Since $231 = 2.31 \times 10^2$, then $\log 231 = 2.3636$.

(c) From the table, as in (a) and (b), we find the mantissa for 842 is 0.9253. Since $0.842 = 8.42 \times 10^{-1}$, then $\log 0.842 = 0.9253 - 1$, or $9.9253 - 10$, or -0.0747. The characteristic is -1. ●

In exactly the same way as we did for the exponential values, we interpolate for values of logarithms when the argument is intermediate to two table entries.

Antilogarithms In the illustration above we found $\log 6.53 = 0.8149$. We then call 6.53 the *antilogarithm* of 0.8149.

EXAMPLE Find the antilogarithm of (a) 1.2148, (b) 0.6972.

SOLUTION Each of these numbers is a logarithm. We look for its mantissa (its decimal part) in the body of Table B.

(a) We find 2148 in the line opposite 1.6 and in the column headed 4. So the significant digits of the antilogarithm of 1.2148 are 1, 6, and 4. Since the characteristic of the logarithm is 1, then antilog $1.2148 = 1.64 \times 10^1 = 16.4$.

(b) We find 0.6972 opposite 4.9 in the column headed N and below the heading 8. So the significant digits are 4, 9, and 8. The characteristic of the logarithm is 0. Then antilog $0.6972 = 4.98 \times 10^0 = 4.98$. ●

Calculation with Logarithms Logarithms may be used to simplify the calculation of products, quotients, powers, and roots, using the laws of logarithms (Section 4.4). Historically, this was the reason for the invention of logarithms. And, they are still helpful for this purpose if one does not have a calculator available but does have tables of logarithms. The laws of logarithms apply to any base. However, since our system of numeration is decimal (base 10), the integral parts of logarithms (characteristics) to base 10 are obtained simply by inspection, as we have just seen. For that reason we use common logarithms in our calculations.

EXAMPLE Calculate (a) 8.34×0.652, (b) $\dfrac{8.34 \times 0.652}{376}$, (c) $(8.34)^3$, (d) $\sqrt{8.34}$.

SOLUTION (a) Let $N = 8.34 \times 0.652$. Then

$$\begin{aligned}\log N &= \log 8.34 + \log 0.652 \\ &= 0.9212 + (0.8142 - 1) \\ &= 1.7354 - 1 = 0.7354.\end{aligned}$$

Then from Table B, $N = 5.44$. This result is, of course, approximate, since the logarithms are irrational numbers. Also, since the data was given to only three significant digits, the result is rounded off to three significant digits. In practice, this calculation may be arranged as follows.

$$\begin{aligned} \log 8.34 &= 0.9212 \\ \log 0.652 &= 0.8142 - 1 \\ \hline \log N &= 1.7354 - 1 \\ &= 0.7534 \\ N &= 5.44 \end{aligned}$$

We use similar arrangements for the remaining calculations.

(b) Let $N = (8.34 \times 0.652)/376$. Then $\log N = \log 8.34 + \log 0.652 - \log 376$. We arrange this as follows.

$$\begin{aligned} \log 8.34 &= 0.9212 \\ \log 0.652 &= 0.8142 - 1 \\ \hline &= 1.7354 - 1 && \text{adding} \\ \log 376 &= 2.5752 \\ \hline \log N &= 0.1602 - 2 && \text{subtracting} \\ N &= 0.0145 \end{aligned}$$

(c) Let $N = (8.34)^3$. Then $\log N = 3 \log 8.34$.

$$\begin{aligned} \log 8.34 &= 0.9212 \\ 3 \log 8.34 &= 2.7636 \\ N &= 580 \end{aligned}$$

(d) Let $N = \sqrt{8.34} = (8.34)^{1/2}$. Then $\log N = (1/2) \log 8.34$.

$$\begin{aligned} \log 8.34 &= 0.9212 \\ \frac{1}{2} \log 8.34 &= 0.4606 \\ N &= 2.89 \end{aligned}$$

●

Table C. Natural Logarithms

In this table are given the values of $\ln x$ for $1 \leq x < 5.5$ at intervals of 0.01. Since the base is e, not 10, both the integral and the decimal part of the logarithms must be tabulated.

EXAMPLE Find (a) ln 1.76, (b) ln 4.32.

SOLUTION (a) We locate the first two digits of 1.76 in the first column, headed x. In the same line and in the column headed by the third digit, 6, we find 0.5653. Then $\ln 1.76 = 0.5653$.

(b) Similarly, we find $\ln 4.32 = 1.4633$. ●

If x is outside the interval $1 \leq x < 5.5$, we can either use more extensive tables or else use results of the change-of-base procedures (Section 4.5). From that we know, for example, that $\ln 10 = 2.3026$.

EXAMPLE Find (a) ln 0.325, (b) ln 534.

SOLUTION

(a)
$$\begin{aligned} \ln 0.325 &= \ln(3.25 \times 10^{-1}) \\ &= \ln 3.25 - \ln 10 \\ &= 1.1787 - 2.3026 \\ &= -1.1239 \end{aligned}$$

(b)
$$\begin{aligned} \ln 534 &= \ln(5.34 \times 10^{2}) \\ &= \ln 5.34 + 2 \ln 10 \\ &= 1.6752 + 2(2.3026) \\ &= 1.6752 + 4.6052 \\ &= 6.2804 \end{aligned}$$
●

Table D. Circular Functions of Real Numbers

Here are tabulated values of each of the six circular functions for the interval $0 \leq x \leq 1.57$, at intervals of 0.01. This corresponds to the first-quadrant values of x, since $\frac{1}{2}\pi = 1.57$. For intermediate values we can use linear interpolation, and for arguments outside this interval we use reduction formulas.

Table E. Trigonometric Functions of Angles in Degrees and Radians

This table lists the trigonometric functions of angles θ for $0° \leq \theta \leq 90°$, at intervals of $10'$, with the corresponding radian measures of angles in a parallel column. Angles θ for $0° \leq \theta \leq 45°$ are shown on the left-hand side of the table and the trigonometric functions along the top. For angles θ such that $45° \leq \theta \leq 90°$, we read the angles on the right-hand side and the functions along the bottom of the table. For intermediate values of θ we use linear interpolations, and for angles greater than 90° we use reduction formulas (Section 5.5).

EXAMPLE Find (a) sin 23°40′, (b) cos 61°20′, (c) tan 130°, (d) sec 320°, (e) sin 0.4712, (f) cos 0.9018.

SOLUTION

(a) $\sin 23°40' = 0.4014$, directly from the table.

(b) $\cos 61°20' = 0.4797$, reading 61°40′ on the right side and $\cos \theta$ along the bottom of the table.

(c) $\tan 130° = -\tan(180° - 130°) = -\tan 50° = -1.192$.

(d) $\sec 320° = \sec(360° - 320°) = \sec 40° = 1.305$.

(e) $\sin 0.4712 = 0.4540$

(f) $\cos 0.9018 = 0.6202$ ●

Table A Exponential Function Values

x	e^x	e^{-x}	x	e^x	e^{-x}
0.00	1.00000	1.00000			
0.01	1.01005	0.99004	1.60	4.95302	0.20189
0.02	1.02020	0.98019	1.70	5.47394	0.18268
0.03	1.03045	0.97044	1.80	6.04964	0.16529
0.04	1.04081	0.96078	1.90	6.68589	0.14956
0.05	1.05127	0.95122	2.00	7.38905	0.13533
0.06	1.06183	0.94176			
0.07	1.07250	0.93239	2.10	8.16616	0.12245
0.08	1.08328	0.92311	2.20	9.02500	0.11080
0.09	1.09417	0.91393	2.30	9.97417	0.10025
0.10	1.10517	0.90483	2.40	11.02316	0.09071
			2.50	12.18248	0.08208
0.11	1.11628	0.89583	2.60	13.46372	0.07427
0.12	1.12750	0.88692	2.70	14.87971	0.06720
0.13	1.13883	0.87810	2.80	16.44463	0.06081
0.14	1.15027	0.86936	2.90	18.17412	0.05502
0.15	1.16183	0.86071	3.00	20.08551	0.04978
0.16	1.17351	0.85214			
0.17	1.18530	0.84366	3.50	33.11545	0.03020
0.18	1.19722	0.83527	4.00	54.59815	0.01832
0.19	1.20925	0.82696	4.50	90.01713	0.01111
0.20	1.22140	0.81873	5.00	148.41316	0.00674
0.30	1.34985	0.74081	5.50	224.69193	0.00409
0.40	1.49182	0.67032			
0.50	1.64872	0.60653	6.00	403.42879	0.00248
0.60	1.82211	0.54881	6.50	665.14163	0.00150
0.70	2.01375	0.49658			
0.80	2.22554	0.44932	7.00	1096.63316	0.00091
0.90	2.45960	0.40656	7.50	1808.04241	0.00055
1.00	2.71828	0.36787			
			8.00	2980.95799	0.00034
			8.50	4914.76884	0.00020
1.10	3.00416	0.33287			
1.20	3.32011	0.30119	9.00	8130.08392	0.00012
1.30	3.66929	0.27253	9.50	13359.72683	0.00007
1.40	4.05519	0.24659			
1.50	4.48168	0.22313	10.00	22026.46579	0.00005

Table B Common Logarithms

N	0	1	2	3	4	5	6	7	8	9
1.0	.0000	.0043	.0086	.0128	.0170	.0212	.0253	.0294	.0334	.0374
1.1	.0414	.0453	.0492	.0531	.0569	.0607	.0645	.0682	.0719	.0755
1.2	.0792	.0828	.0864	.0899	.0934	.0969	.1004	.1038	.1072	.1106
1.3	.1139	.1173	.1206	.1239	.1271	.1303	.1335	.1367	.1399	.1430
1.4	.1461	.1492	.1523	.1553	.1584	.1614	.1644	.1673	.1703	.1732
1.5	.1761	.1790	.1818	.1847	.1875	.1903	.1931	.1959	.1987	.2014
1.6	.2041	.2068	.2095	.2122	.2148	.2175	.2201	.2227	.2253	.2279
1.7	.2304	.2330	.2355	.2380	.2405	.2430	.2455	.2480	.2504	.2529
1.8	.2553	.2577	.2601	.2625	.2648	.2672	.2695	.2718	.2742	.2765
1.9	.2788	.2810	.2833	.2856	.2878	.2900	.2923	.2945	.2967	.2989
2.0	.3010	.3032	.3054	.3075	.3096	.3118	.3139	.3160	.3181	.3201
2.1	.3222	.3243	.3263	.3284	.3304	.3324	.3345	.3365	.3385	.3404
2.2	.3424	.3444	.3464	.3483	.3502	.3522	.3541	.3560	.3579	.3598
2.3	.3617	.3636	.3655	.3674	.3692	.3711	.3729	.3747	.3766	.3784
2.4	.3802	.3820	.3838	.3856	.3874	.3892	.3909	.3927	.3945	.3962
2.5	.3979	.3997	.4014	.4031	.4048	.4065	.4082	.4099	.4116	.4133
2.6	.4150	.4166	.4183	.4200	.4216	.4232	.4249	.4265	.4281	.4298
2.7	.4314	.4330	.4346	.4362	.4378	.4393	.4409	.4425	.4440	.4456
2.8	.4472	.4487	.4502	.4518	.4533	.4548	.4564	.4579	.4594	.4609
2.9	.4624	.4639	.4654	.4669	.4683	.4698	.4713	.4728	.4742	.4757
3.0	.4771	.4786	.4800	.4814	.4829	.4843	.4857	.4871	.4886	.4900
3.1	.4914	.4928	.4942	.4955	.4969	.4983	.4997	.5011	.5024	.5038
3.2	.5051	.5065	.5079	.5092	.5105	.5119	.5132	.5145	.5159	.5172
3.3	.5185	.5198	.5211	.5224	.5237	.5250	.5263	.5276	.5289	.5302
3.4	.5315	.5328	.5340	.5353	.5366	.5378	.5391	.5403	.5416	.5428
3.5	.5441	.5453	.5465	.5478	.5490	.5502	.5514	.5527	.5539	.5551
3.6	.5563	.5575	.5587	.5599	.5611	.5623	.5635	.5647	.5658	.5670
3.7	.5682	.5694	.5705	.5717	.5729	.5740	.5752	.5763	.5775	.5786
3.8	.5798	.5809	.5821	.5832	.5843	.5855	.5866	.5877	.5888	.5899
3.9	.5911	.5922	.5933	.5944	.5955	.5966	.5977	.5988	.5999	.6010
4.0	.6021	.6031	.6042	.6053	.6064	.6075	.6085	.6096	.6107	.6117
4.1	.6128	.6138	.6149	.6160	.6170	.6180	.6191	.6201	.6212	.6222
4.2	.6232	.6243	.6253	.6263	.6274	.6284	.6294	.6304	.6314	.6325
4.3	.6335	.6345	.6355	.6365	.6375	.6385	.6395	.6405	.6415	.6425
4.4	.6435	.6444	.6454	.6464	.6474	.6484	.6493	.6503	.6513	.6522
4.5	.6532	.6542	.6551	.6561	.6571	.6580	.6590	.6599	.6609	.6618
4.6	.6628	.6637	.6646	.6656	.6665	.6675	.6684	.6693	.6702	.6712
4.7	.6721	.6730	.6739	.6749	.6758	.6767	.6776	.6785	.6794	.6803
4.8	.6812	.6821	.6830	.6839	.6848	.6857	.6866	.6875	.6884	.6893
4.9	.6902	.6911	.6920	.6928	.6937	.6946	.6955	.6964	.6972	.6981
5.0	.6990	.6998	.7007	.7016	.7024	.7033	.7042	.7050	.7059	.7067
5.1	.7076	.7084	.7093	.7101	.7110	.7118	.7126	.7135	.7143	.7152
5.2	.7160	.7168	.7177	.7185	.7193	.7202	.7210	.7218	.7226	.7235
5.3	.7243	.7251	.7259	.7267	.7275	.7284	.7292	.7300	.7308	.7316
5.4	.7324	.7332	.7340	.7348	.7356	.7364	.7372	.7380	.7388	.7396
N	0	1	2	3	4	5	6	7	8	9

Table B Common Logarithms (Continued)

N	0	1	2	3	4	5	6	7	8	9
5.5	.7404	.7412	.7419	.7427	.7435	.7443	.7451	.7459	.7466	.7474
5.6	.7482	.7490	.7497	.7505	.7513	.7520	.7528	.7536	.7543	.7551
5.7	.7559	.7566	.7574	.7582	.7589	.7597	.7604	.7612	.7619	.7627
5.8	.7634	.7642	.7649	.7657	.7664	.7672	.7679	.7686	.7694	.7701
5.9	.7709	.7716	.7723	.7731	.7738	.7745	.7752	.7760	.7767	.7774
6.0	.7782	.7789	.7796	.7803	.7810	.7818	.7825	.7832	.7839	.7846
6.1	.7853	.7860	.7868	.7875	.7882	.7889	.7896	.7903	.7910	.7917
6.2	.7924	.7931	.7938	.7945	.7952	.7959	.7966	.7973	.7980	.7987
6.3	.7993	.8000	.8007	.8014	.8021	.8028	.8035	.8041	.8048	.8055
6.4	.8062	.8069	.8075	.8082	.8089	.8096	.8102	.8109	.8116	.8122
6.5	.8129	.8136	.8142	.8149	.8156	.8162	.8169	.8176	.8182	.8189
6.6	.8195	.8202	.8209	.8215	.8222	.8228	.8235	.8241	.8248	.8254
6.7	.8261	.8267	.8274	.8280	.8287	.8293	.8299	.8306	.8312	.8319
6.8	.8325	.8331	.8338	.8344	.8351	.8357	.8363	.8370	.8376	.8382
6.9	.8388	.8395	.8401	.8407	.8414	.8420	.8426	.8432	.8439	.8445
7.0	.8451	.8457	.8463	.8470	.8476	.8482	.8488	.8494	.8500	.8506
7.1	.8513	.8519	.8525	.8531	.8537	.8543	.8549	.8555	.8561	.8567
7.2	.8573	.8579	.8585	.8591	.8597	.8603	.8609	.8615	.8621	.8627
7.3	.8633	.8639	.8645	.8651	.8657	.8663	.8669	.8675	.8681	.8686
7.4	.8692	.8698	.8704	.8710	.8716	.8722	.8727	.8733	.8739	.8745
7.5	.8751	.8756	.8762	.8768	.8774	.8779	.8785	.8791	.8797	.8802
7.6	.8808	.8814	.8820	.8825	.8831	.8837	.8842	.8848	.8854	.8859
7.7	.8865	.8871	.8876	.8882	.8887	.8893	.8899	.8904	.8910	.8915
7.8	.8921	.8927	.8932	.8938	.8943	.8949	.8954	.8960	.8965	.8971
7.9	.8976	.8982	.8987	.8993	.8998	.9004	.9009	.9015	.9020	.9025
8.0	.9031	.9036	.9042	.9047	.9053	.9058	.9063	.9069	.9074	.9079
8.1	.9085	.9090	.9096	.9101	.9106	.9112	.9117	.9122	.9128	.9133
8.2	.9138	.9143	.9149	.9154	.9159	.9165	.9170	.9175	.9180	.9186
8.3	.9191	.9196	.9201	.9206	.9212	.9217	.9222	.9227	.9232	.9238
8.4	.9243	.9248	.9253	.9258	.9263	.9269	.9274	.9279	.9284	.9289
8.5	.9294	.9299	.9304	.9309	.9315	.9320	.9325	.9330	.9335	.9340
8.6	.9345	.9350	.9355	.9360	.9365	.9370	.9375	.9380	.9385	.9390
8.7	.9395	.9400	.9405	.9410	.9415	.9420	.9425	.9430	.9435	.9440
8.8	.9445	.9450	.9455	.9460	.9465	.9469	.9474	.9479	.9484	.9489
8.9	.9494	.9499	.9504	.9509	.9513	.9518	.9523	.9528	.9533	.9538
9.0	.9542	.9547	.9552	.9557	.9562	.9566	.9571	.9576	.9581	.9586
9.1	.9590	.9595	.9600	.9605	.9609	.9614	.9619	.9624	.9628	.9633
9.2	.9638	.9643	.9647	.9652	.9657	.9661	.9666	.9671	.9675	.9680
9.3	.9685	.9689	.9694	.9699	.9703	.9708	.9713	.9717	.9722	.9727
9.4	.9731	.9736	.9741	.9745	.9750	.9754	.9759	.9763	.9768	.9773
9.5	.9777	.9782	.9786	.9791	.9795	.9800	.9805	.9809	.9814	.9818
9.6	.9823	.9827	.9832	.9836	.9841	.9845	.9850	.9854	.9859	.9863
9.7	.9868	.9872	.9877	.9881	.9886	.9890	.9894	.9899	.9903	.9908
9.8	.9912	.9917	.9921	.9926	.9930	.9934	.9939	.9943	.9948	.9952
9.9	.9956	.9961	.9965	.9969	.9974	.9978	.9983	.9987	.9991	.9996
N	0	1	2	3	4	5	6	7	8	9

Table C Natural Logarithms

x	0	1	2	3	4	5	6	7	8	9
1.0	0.0000	0.0100	0.0198	0.0296	0.0392	0.0488	0.0583	0.0677	0.0770	0.0862
1.1	0.0953	0.1044	0.1133	0.1222	0.1310	0.1398	0.1484	0.1570	0.1655	0.1740
1.2	0.1823	0.1906	0.1989	0.2070	0.2151	0.2231	0.2311	0.2390	0.2469	0.2546
1.3	0.2624	0.2700	0.2776	0.2852	0.2927	0.3001	0.3075	0.3148	0.3221	0.3293
1.4	0.3365	0.3436	0.3507	0.3577	0.3646	0.3716	0.3784	0.3853	0.3920	0.3988
1.5	0.4055	0.4121	0.4187	0.4253	0.4318	0.4383	0.4447	0.4511	0.4574	0.4637
1.6	0.4700	0.4762	0.4824	0.4886	0.4947	0.5008	0.5068	0.5128	0.5188	0.5247
1.7	0.5306	0.5365	0.5423	0.5481	0.5539	0.5596	0.5653	0.5710	0.5766	0.5822
1.8	0.5878	0.5933	0.5988	0.6043	0.6098	0.6152	0.6206	0.6259	0.6313	0.6366
1.9	0.6419	0.6471	0.6523	0.6575	0.6627	0.6678	0.6729	0.6780	0.6831	0.6881
2.0	0.6932	0.6981	0.7031	0.7080	0.7129	0.7178	0.7227	0.7275	0.7324	0.7372
2.1	0.7419	0.7467	0.7514	0.7561	0.7608	0.7655	0.7701	0.7747	0.7793	0.7839
2.2	0.7885	0.7930	0.7975	0.8020	0.8065	0.8109	0.8154	0.8198	0.8242	0.8286
2.3	0.8329	0.8373	0.8416	0.8459	0.8502	0.8544	0.8587	0.8629	0.8671	0.8713
2.4	0.8755	0.8796	0.8838	0.8879	0.8920	0.8961	0.9002	0.9042	0.9083	0.9123
2.5	0.9163	0.9203	0.9243	0.9282	0.9322	0.9361	0.9400	0.9439	0.9478	0.9517
2.6	0.9555	0.9594	0.9632	0.9670	0.9708	0.9746	0.9783	0.9821	0.9858	0.9895
2.7	0.9933	0.9969	1.0006	1.0043	1.0080	1.0116	1.0152	1.0188	1.0225	1.0260
2.8	1.0296	1.0332	1.0367	1.0403	1.0438	1.0473	1.0508	1.0543	1.0578	1.0613
2.9	1.0647	1.0682	1.0716	1.0750	1.0784	1.0818	1.0852	1.0886	1.0919	1.0953
3.0	1.0986	1.1019	1.1053	1.1086	1.1119	1.1151	1.1184	1.1217	1.1249	1.1282
3.1	1.1314	1.1346	1.1378	1.1410	1.1442	1.1474	1.1506	1.1537	1.1569	1.1600
3.2	1.1632	1.1663	1.1694	1.1725	1.1756	1.1787	1.1817	1.1848	1.1878	1.1909
3.3	1.1939	1.1969	1.2000	1.2030	1.2060	1.2090	1.2119	1.2149	1.2179	1.2208
3.4	1.2238	1.2267	1.2296	1.2326	1.2355	1.2384	1.2413	1.2442	1.2470	1.2499
3.5	1.2528	1.2556	1.2585	1.2613	1.2641	1.2669	1.2698	1.2726	1.2754	1.2782
3.6	1.2809	1.2837	1.2865	1.2892	1.2920	1.2947	1.2975	1.3002	1.3029	1.3056
3.7	1.3083	1.3110	1.3137	1.3164	1.3191	1.3218	1.3244	1.3271	1.3297	1.3324
3.8	1.3350	1.3376	1.3403	1.3429	1.3455	1.3481	1.3507	1.3533	1.3558	1.3584
3.9	1.3610	1.3635	1.3661	1.3686	1.3712	1.3737	1.3762	1.3788	1.3813	1.3838
4.0	1.3863	1.3888	1.3913	1.3938	1.3962	1.3987	1.4012	1.4036	1.4061	1.4085
4.1	1.4110	1.4134	1.4159	1.4183	1.4207	1.4231	1.4255	1.4279	1.4303	1.4327
4.2	1.4351	1.4375	1.4398	1.4422	1.4446	1.4469	1.4493	1.4516	1.4540	1.4563
4.3	1.4586	1.4609	1.4633	1.4656	1.4679	1.4702	1.4725	1.4748	1.4771	1.4793
4.4	1.4816	1.4839	1.4861	1.4884	1.4907	1.4929	1.4951	1.4974	1.4996	1.5019
4.5	1.5041	1.5063	1.5085	1.5107	1.5129	1.5151	1.5173	1.5195	1.5217	1.5239
4.6	1.5261	1.5282	1.5304	1.5326	1.5347	1.5369	1.5390	1.5412	1.5433	1.5454
4.7	1.5476	1.5497	1.5518	1.5539	1.5560	1.5581	1.5602	1.5623	1.5644	1.5665
4.8	1.5686	1.5707	1.5728	1.5748	1.5769	1.5790	1.5810	1.5831	1.5851	1.5872
4.9	1.5892	1.5913	1.5933	1.5953	1.5974	1.5994	1.6014	1.6034	1.6054	1.6074
5.0	1.6094	1.6114	1.6134	1.6154	1.6174	1.6194	1.6214	1.6233	1.6253	1.6273
5.1	1.6292	1.6312	1.6332	1.6351	1.6371	1.6390	1.6409	1.6429	1.6448	1.6467
5.2	1.6487	1.6506	1.6525	1.6544	1.6563	1.6582	1.6601	1.6620	1.6639	1.6658
5.3	1.6677	1.6696	1.6715	1.6734	1.6752	1.6771	1.6790	1.6808	1.6827	1.6845
5.4	1.6864	1.6882	1.6901	1.6919	1.6938	1.6956	1.6974	1.6993	1.7011	1.7029

Table D Circular Functions of Real Numbers

x	sin x	cos x	tan x	cot x	sec x	csc x
.00	.0000	1.0000	.0000	—	1.000	—
.01	.0100	1.0000	.0100	99.997	1.000	100.00
.02	.0200	.9998	.0200	49.993	1.000	50.00
.03	.0300	.9996	.0300	33.323	1.000	33.34
.04	.0400	.9992	.0400	24.987	1.001	25.01
.05	.0500	.9988	.0500	19.983	1.001	20.01
.06	.0600	.9982	.0601	16.647	1.002	16.68
.07	.0699	.9976	.0701	14.262	1.002	14.30
.08	.0799	.9968	.0802	12.473	1.003	12.51
.09	.0899	.9960	.0902	11.081	1.004	11.13
.10	.0998	.9950	.1003	9.967	1.005	10.02
.11	.1098	.9940	.1104	9.054	1.006	9.109
.12	.1197	.9928	.1206	8.293	1.007	8.353
.13	.1296	.9916	.1307	7.649	1.009	7.714
.14	.1395	.9902	.1409	7.096	1.010	7.166
.15	.1494	.9888	.1511	6.617	1.011	6.692
.16	.1593	.9872	.1614	6.197	1.013	6.277
.17	.1692	.9856	.1717	5.826	1.015	5.911
.18	.1790	.9838	.1820	5.495	1.016	5.586
.19	.1889	.9820	.1923	5.200	1.018	5.295
.20	.1987	.9801	.2027	4.933	1.020	5.033
.21	.2085	.9780	.2131	4.692	1.022	4.797
.22	.2182	.9759	.2236	4.472	1.025	4.582
.23	.2280	.9737	.2341	4.271	1.027	4.386
.24	.2377	.9713	.2447	4.086	1.030	4.207
.25	.2474	.9689	.2553	3.916	1.032	4.042
.26	.2571	.9664	.2660	3.759	1.035	3.890
.27	.2667	.9638	.2768	3.613	1.038	3.749
.28	.2764	.9611	.2876	3.478	1.041	3.619
.29	.2860	.9582	.2984	3.351	1.044	3.497
.30	.2955	.9553	.3093	3.233	1.047	3.384
.31	.3051	.9523	.3203	3.122	1.050	3.278
.32	.3146	.9492	.3314	3.018	1.053	3.179
.33	.3240	.9460	.3425	2.920	1.057	3.086
.34	.3335	.9428	.3537	2.827	1.061	2.999
.35	.3429	.9394	.3650	2.740	1.065	2.916
.36	.3523	.9359	.3764	2.657	1.068	2.839
.37	.3616	.9323	.3879	2.578	4.073	2.765
.38	.3709	.9287	.3994	2.504	1.077	2.696
.39	.3802	.9249	.4111	2.433	1.081	2.630
.40	.3894	.9211	.4228	2.365	1.086	2.568
.41	.3986	.9171	.4346	2.301	1.090	2.509
.42	.4078	.9131	.4466	2.239	1.095	2.452
.43	.4169	.9090	.4586	2.180	1.100	2.399
.44	.4259	.9048	.4708	2.124	1.105	2.348
.45	.4350	.9004	.4831	2.070	1.111	2.299
.46	.4439	.8961	.4954	2.018	1.116	2.253
.47	.4529	.8916	.5080	1.969	1.122	2.208
.48	.4618	.8870	.5206	1.921	1.127	2.166
.49	.4706	.8823	.5334	1.875	1.133	2.125
.50	.4794	.8776	.5463	1.830	1.139	2.086
.51	.4882	.8727	.5594	1.788	1.146	2.048
.52	.4969	.8678	.5726	1.747	1.152	2.013
.53	.5055	.8628	.5859	1.707	1.159	1.978
.54	.5141	.8577	.5994	1.668	1.166	1.945
.55	.5227	.8525	.6131	1.631	1.173	1.913
.56	.5312	.8473	.6269	1.595	1.180	1.883
.57	.5396	.8419	.6410	1.560	1.188	1.853
.58	.5480	.8365	.6552	1.526	1.196	1.825
.59	.5564	.8309	.6696	1.494	1.203	1.797
.60	.5646	.8253	.6841	1.462	1.212	1.771
.61	.5729	.8196	.6989	1.431	1.220	1.746
.62	.5810	.8139	.7139	1.401	1.229	1.721
.63	.5891	.8080	.7291	1.372	1.238	1.697
.64	.5972	.8021	.7445	1.343	1.247	1.674
.65	.6052	.7961	.7602	1.315	1.256	1.652
.66	.6131	.7900	.7761	1.288	1.266	1.631
.67	.6210	.7838	.7923	1.262	1.276	1.610
.68	.6288	.7776	.8087	1.237	1.286	1.590
.69	.6365	.7712	.8253	1.212	1.297	1.571
.70	.6442	.7648	.8423	1.187	1.307	1.552
.71	.6518	.7584	.8595	1.163	1.319	1.534
.72	.6594	.7518	.8771	1.140	1.330	1.517
.73	.6669	.7452	.8949	1.117	1.342	1.500
.74	.6743	.7385	.9131	1.095	1.354	1.483
.75	.6816	.7317	.9316	1.073	1.367	1.467
.76	.6889	.7248	.9505	1.052	1.380	1.452
.77	.6961	.7179	.9697	1.031	1.393	1.437
.78	.7033	.7109	.9893	1.011	1.407	1.422
.79	.7104	.7038	1.009	.9908	1.421	1.408

Table D Circular Functions of Real Numbers (Continued)

x	sin x	cos x	tan x	cot x	sec x	csc x
.80	.7174	.6967	1.030	.9712	1.435	1.394
.81	.7243	.6895	1.050	.9520	1.450	1.381
.82	.7311	.6822	1.072	.9331	1.466	1.368
.83	.7379	.6749	1.093	.9146	1.482	1.355
.84	.7446	.6675	1.116	.8964	1.498	1.343
.85	.7513	.6600	1.138	.8785	1.515	1.331
.86	.7578	.6524	1.162	.8609	1.533	1.320
.87	.7643	.6448	1.185	.8437	1.551	1.308
.88	.7707	.6372	1.210	.8267	1.569	1.297
.89	.7771	.6294	1.235	.8100	1.589	1.287
.90	.7833	.6216	1.260	.7936	1.609	1.277
.91	.7895	.6137	1.286	.7774	1.629	1.267
.92	.7956	.6058	1.313	.7615	1.651	1.257
.93	.8016	.5978	1.341	.7458	1.673	1.247
.94	.8076	.5898	1.369	.7303	1.696	1.238
.95	.8134	.5817	1.398	.7151	1.719	1.229
.96	.8192	.5735	1.428	.7001	1.744	1.221
.97	.8249	.5653	1.459	.6853	1.769	1.212
.98	.8305	.5570	1.491	.6707	1.795	1.204
.99	.8360	.5487	1.524	.6563	1.823	1.196
1.00	.8415	.5403	1.557	.6421	1.851	1.188
1.01	.8468	.5319	1.592	.6281	1.880	1.181
1.02	.8521	.5234	1.628	.6142	1.911	1.174
1.03	.8573	.5148	1.665	.6005	1.942	1.166
1.04	.8624	.5062	1.704	.5870	1.975	1.160
1.05	.8674	.4976	1.743	.5736	2.010	1.153
1.06	.8724	.4889	1.784	.5604	2.046	1.146
1.07	.8772	.4801	1.827	.5473	2.083	1.140
1.08	.8820	.4713	1.871	.5344	2.122	1.134
1.09	.8866	.4625	1.917	.5216	2.162	1.128
1.10	.8912	.4536	1.965	.5090	2.205	1.122
1.11	.8957	.4447	2.014	.4964	2.249	1.116
1.12	.9001	.4357	2.066	.4840	2.295	1.111
1.13	.9044	.4267	2.120	.4718	2.344	1.106
1.14	.9086	.4176	2.176	.4596	2.395	1.101
1.15	.9128	.4085	2.234	.4475	2.448	1.096
1.16	.9168	.3993	2.296	.4356	2.504	1.091
1.17	.9208	.3902	2.360	.4237	2.563	1.086
1.18	.9246	.3809	2.427	.4120	2.625	1.082
1.19	.9284	.3717	2.498	.4003	2.691	1.077

x	sin x	cos x	tan x	cot x	sec x	csc x
1.20	.9320	.3624	2.572	.3888	2.760	1.073
1.21	.9356	.3530	2.650	.3773	2.833	1.069
1.22	.9391	.3436	2.733	.3659	2.910	1.065
1.23	.9425	.3342	2.820	.3546	2.992	1.061
1.24	.9458	.3248	2.912	.3434	3.079	1.057
1.25	.9490	.3153	3.010	.3323	3.171	1.054
1.26	.9521	.3058	3.113	.3212	3.270	1.050
1.27	.9551	.2963	3.224	.3102	3.375	1.047
1.28	.9580	.2867	3.341	.2993	3.488	1.044
1.29	.9608	.2771	3.467	.2884	3.609	1.041
1.30	.9636	.2675	3.602	.2776	3.738	1.038
1.31	.9662	.2579	3.747	.2669	3.878	1.035
1.32	.9687	.2482	3.903	.2562	4.029	1.032
1.33	.9711	.2385	4.072	.2456	4.193	1.030
1.34	.9735	.2288	4.256	.2350	4.372	1.027
1.35	.9757	.2190	4.455	.2245	4.566	1.025
1.36	.9779	.2092	4.673	.2140	4.779	1.023
1.37	.9799	.1994	4.913	.2035	5.014	1.021
1.38	.9819	.1896	5.177	.1931	5.273	1.018
1.39	.9837	.1798	5.471	.1828	5.561	1.017
1.40	.9854	.1700	5.798	.1725	5.883	1.015
1.41	.9871	.1601	6.165	.1622	6.246	1.013
1.42	.9887	.1502	6.581	.1519	6.657	1.011
1.43	.9901	.1403	7.055	.1417	7.126	1.010
1.44	.9915	.1304	7.602	.1315	7.667	1.009
1.45	.9927	.1205	8.238	.1214	8.299	1.007
1.46	.9939	.1106	8.989	.1113	9.044	1.006
1.47	.9949	.1006	9.887	.1011	9.938	1.005
1.48	.9959	.0907	10.983	.0910	11.029	1.004
1.49	.9967	.0807	12.350	.0810	12.390	1.003
1.50	.9975	.0707	14.101	.0709	14.137	1.003
1.51	.9982	.0608	16.428	.0609	16.458	1.002
1.52	.9987	.0508	19.670	.0508	19.695	1.001
1.53	.9992	.0408	24.498	.0408	24.519	1.001
1.54	.9995	.0308	32.461	.0308	32.476	1.000
1.55	.9998	.0208	48.078	.0208	48.089	1.000
1.56	.9999	.0108	92.620	.0108	92.626	1.000
1.57	1.0000	.0008	1255.8	.0008	1255.8	1.000

Table E Trigonometric Functions of Angles in Degrees and Radians

θ (degrees)	θ (radians)	sin θ	cos θ	tan θ	cot θ	sec θ	csc θ		
0°00′	.0000	.0000	1.0000	.0000	—	1.000	—	1.5708	**90°00′**
10	.0029	.0029	1.0000	.0029	343.8	1.000	343.8	1.5679	50
20	.0058	.0058	1.0000	.0058	171.9	1.000	171.9	1.5650	40
30	.0087	.0087	1.0000	.0087	114.6	1.000	114.6	1.5621	30
40	.0116	.0116	.9999	.0116	85.94	1.000	85.95	1.5592	20
50	.0145	.0145	.9999	.0145	68.75	1.000	68.76	1.5563	10
1°00′	.0175	.0175	.9998	.0175	57.29	1.000	57.30	1.5533	**89°00′**
10	.0204	.0204	.9998	.0204	49.10	1.000	49.11	1.5504	50
20	.0233	.0233	.9997	.0233	42.96	1.000	42.98	1.5475	40
30	.0262	.0262	.9997	.0262	38.19	1.000	38.20	1.5446	30
40	.0291	.0291	.9996	.0291	34.37	1.000	34.38	1.5417	20
50	.0320	.0320	.9995	.0320	31.24	1.001	31.26	1.5388	10
2°00′	.0349	.0349	.9994	.0349	28.64	1.001	28.65	1.5359	**88°00′**
10	.0378	.0378	.9993	.0378	26.43	1.001	26.45	1.5330	50
20	.0407	.0407	.9992	.0407	24.54	1.001	24.56	1.5301	40
30	.0436	.0436	.9990	.0437	22.90	1.001	22.93	1.5272	30
40	.0465	.0465	.9989	.0466	21.47	1.001	21.49	1.5243	20
50	.0495	.0494	.9988	.0495	20.21	1.001	20.23	1.5213	10
3°00′	.0524	.0523	.9986	.0524	19.08	1.001	19.11	1.5184	**87°00′**
10	.0553	.0552	.9985	.0553	18.07	1.002	18.10	1.5155	50
20	.0582	.0581	.9983	.0582	17.17	1.002	17.20	1.5126	40
30	.0611	.0610	.9981	.0612	16.35	1.002	16.38	1.5097	30
40	.0640	.0640	.9980	.0641	15.60	1.002	15.64	1.5068	20
50	.0669	.0669	.9978	.0670	14.92	1.002	14.96	1.5039	10
4°00′	.0698	.0698	.9976	.0699	14.30	1.002	14.34	1.5010	**86°00′**
10	.0727	.0727	.9974	.0729	13.73	1.003	13.76	1.4981	50
20	.0756	.0756	.9971	.0758	13.20	1.003	13.23	1.4952	40
30	.0785	.0785	.9969	.0787	12.71	1.003	12.75	1.4923	30
40	.0814	.0814	.9967	.0816	12.25	1.003	12.29	1.4893	20
50	.0844	.0843	.9964	.0846	11.83	1.004	11.87	1.4864	10
5°00′	.0873	.0872	.9962	.0875	11.43	1.004	11.47	1.4835	**85°00′**
10	.0902	.0901	.9959	.0904	11.06	1.004	11.10	1.4806	50
20	.0931	.0929	.9957	.0934	10.71	1.004	10.76	1.4777	40
30	.0960	.0958	.9954	.0963	10.39	1.005	10.43	1.4748	30
40	.0989	.0987	.9951	.0992	10.08	1.005	10.13	1.4719	20
50	.1018	.1016	.9948	.1022	9.788	1.005	9.839	1.4690	10
6°00′	.1047	.1045	.9945	.1051	9.514	1.006	9.567	1.4661	**84°00′**
10	.1076	.1074	.9942	.1080	9.255	1.006	9.309	1.4632	50
20	.1105	.1103	.9939	.1110	9.010	1.006	9.065	1.4603	40
30	.1134	.1132	.9936	.1139	8.777	1.006	8.834	1.4573	30
40	.1164	.1161	.9932	.1169	8.556	1.007	8.614	1.4544	20
50	.1193	.1190	.9929	.1198	8.345	1.007	8.405	1.4515	10
		cos θ	sin θ	cot θ	tan θ	csc θ	sec θ	θ (radians)	θ (degrees)

θ (degrees)	θ (radians)	sin θ	cos θ	tan θ	cot θ	sec θ	csc θ		
7°00′	.1222	.1219	.9925	.1228	8.144	1.008	8.206	1.4486	**83°00′**
10	.1251	.1248	.9922	.1257	7.953	1.008	8.016	1.4457	50
20	.1280	.1276	.9918	.1287	7.770	1.008	7.834	1.4428	40
30	.1309	.1305	.9914	.1317	7.596	1.009	7.661	1.4399	30
40	.1338	.1334	.9911	.1346	7.429	1.009	7.496	1.4370	20
50	.1376	.1363	.9907	.1376	7.269	1.009	7.337	1.4341	10
8°00′	.1396	.1392	.9903	.1405	7.115	1.010	7.185	1.4312	**82°00′**
10	.1425	.1421	.9899	.1435	6.968	1.010	7.040	1.4283	50
20	.1454	.1449	.9894	.1465	6.827	1.011	6.900	1.4254	40
30	.1484	.1478	.9890	.1495	6.691	1.011	6.765	1.4224	30
40	.1513	.1507	.9886	.1524	6.561	1.012	6.636	1.4195	20
50	.1542	.1536	.9881	.1554	6.435	1.012	6.512	1.4166	10
9″00′	.1571	.1564	.9877	.1584	6.314	1.012	6.392	1.4137	**81°00′**
10	.1600	.1593	.9872	.1614	6.197	1.013	6.277	1.4108	50
20	.1629	.1622	.9868	.1644	6.084	1.013	6.166	1.4079	40
30	.1658	.1650	.9863	.1673	5.976	1.014	6.059	1.4050	30
40	.1687	.1679	.9858	.1703	5.871	1.014	5.955	1.4021	20
50	.1716	.1708	.9853	.1733	5.769	1.015	5.855	1.3992	10
10″00′	.1745	.1736	.9848	.1763	5.671	1.015	5.759	1.3963	**80°00′**
10	.1774	.1765	.9843	.1793	5.576	1.016	5.665	1.3934	50
20	.1804	.1794	.9838	.1823	5.485	1.016	5.575	1.3904	40
30	.1833	.1822	.9833	.1853	5.396	1.017	5.487	1.3875	30
40	.1862	.1851	.9827	.1883	5.309	1.018	5.403	1.3846	20
50	.1891	.1880	.9822	.1914	5.226	1.018	5.320	1.3817	10
11″00′	.1920	.1908	.9816	.1944	5.145	1.019	5.241	1.3788	**79°00′**
10	.1949	.1937	.9811	.1974	5.066	1.019	5.164	1.3759	50
20	.1978	.1965	.9805	.2004	4.989	1.020	5.089	1.3730	40
30	.2007	.1994	.9799	.2035	4.915	1.020	5.016	1.3701	30
40	.2036	.2022	.9793	.2065	4.843	1.021	4.945	1.3672	20
50	.2065	.2051	.9787	.2095	4.773	1.022	4.876	1.3643	10
12″00′	.2094	.2079	.9781	.2126	4.705	1.022	4.810	1.3614	**78°00′**
10	.2123	.2108	.9775	.2156	4.638	1.023	4.745	1.3584	50
20	.2153	.2136	.9769	.2186	4.574	1.024	4.682	1.3555	40
30	.2182	.2164	.9763	.2217	4.511	1.024	4.620	1.3526	30
40	.2211	.2193	.9757	.2247	4.449	1.025	4.560	1.3497	20
50	.2240	.2221	.9750	.2278	4.390	1.026	4.502	1.3468	10
13″00′	.2269	.2250	.9744	.2309	4.331	1.026	4.445	1.3439	**77°00′**
10	.2298	.2278	.9737	.2339	4.275	1.027	4.390	1.3410	50
20	.2327	.2306	.9730	.2370	4.219	1.028	4.336	1.3381	40
30	.2356	.2334	.9724	.2401	4.165	1.028	4.284	1.3352	30
40	.2385	.2363	.9717	.2432	4.113	1.029	4.232	1.3323	20
50	.2414	.2391	.9710	.2462	4.061	1.030	4.182	1.3294	10
		cos θ	sin θ	cot θ	tan θ	csc θ	sec θ	θ (radians)	θ (degrees)

Table E Trigonometric Functions of Angles in Degrees and Radians (Continued)

θ (degrees)	θ (radians)	sin θ	cos θ	tan θ	cot θ	sec θ	csc θ		
14°00′	.2443	.2419	.9703	.2493	4.011	1.031	4.134	1.3265	**76°00′**
10	.2473	.2447	.9696	.2524	3.962	1.031	4.086	1.3235	50
20	.2502	.2476	.9689	.2555	3.914	1.032	4.039	1.3206	40
30	.2531	.2504	.9681	.2586	3.867	1.033	3.994	1.3177	30
40	.2560	.2532	.9674	.2617	3.821	1.034	3.950	1.3148	20
50	.2589	.2560	.9667	.2648	3.776	1.034	3.906	1.3119	10
15°00′	.2618	.2588	.9659	.2679	3.732	1.035	3.864	1.3090	**75°00′**
10	.2647	.2616	.9652	.2711	3.689	1.036	3.822	1.3061	50
20	.2676	.2644	.9644	.2742	3.647	1.037	3.782	1.3032	40
30	.2705	.2672	.9636	.2773	3.606	1.038	3.742	1.3003	30
40	.2734	.2700	.9628	.2805	3.566	1.039	3.703	1.2974	20
50	.2763	.2728	.9621	.2836	3.526	1.039	3.665	1.2945	10
16°00′	.2793	.2756	.9613	.2867	3.487	1.040	3.628	1.2915	**74°00′**
10	.2822	.2784	.9605	.2899	3.450	1.041	3.592	1.2886	50
20	.2851	.2812	.9596	.2931	3.412	1.042	3.556	1.2857	40
30	.2880	.2840	.9588	.2962	3.376	1.043	3.521	1.2828	30
40	.2909	.2868	.9580	.2994	3.340	1.044	3.487	1.2799	20
50	.2938	.2896	.9572	.3026	3.305	1.045	3.453	1.2770	10
17°00′	.2967	.2924	.9563	.3057	3.271	1.046	3.420	1.2741	**73°00′**
10	.2996	.2952	.9555	.3089	3.237	1.047	3.388	1.2712	50
20	.3025	.2979	.9546	.3121	3.204	1.048	3.356	1.2683	40
30	.3054	.3007	.9537	.3153	3.172	1.049	3.326	1.2654	30
40	.3083	.3035	.9528	.3185	3.140	1.049	3.295	1.2625	20
50	.3113	.3062	.9520	.3217	3.108	1.050	3.265	1.2595	10
18°00′	.3142	.3090	.9511	.3249	3.078	1.051	3.236	1.2566	**72°00′**
10	.3171	.3118	.9502	.3281	3.047	1.052	3.207	1.2537	50
20	.3200	.3145	.9492	.3314	3.018	1.053	3.179	1.2508	40
30	.3229	.3173	.9483	.3346	2.989	1.054	3.152	1.2479	30
40	.3258	.3201	.9474	.3378	2.960	1.056	3.124	1.2450	20
50	.3287	.3228	.9465	.3411	2.932	1.057	3.098	1.2421	10
19°00′	.3316	.3256	.9455	.3443	2.904	1.058	3.072	1.2392	**71°00′**
10	.3345	.3283	.9446	.3476	2.877	1.059	3.046	1.2363	50
20	.3374	.3311	.9436	.3508	2.850	1.060	3.021	1.2334	40
30	.3403	.3338	.9426	.3541	2.824	1.061	2.996	1.2305	30
40	.3432	.3365	.9417	.3574	2.798	1.062	2.971	1.2275	20
50	.3462	.3393	.9407	.3607	2.773	1.063	2.947	1.2246	10
20°00′	.3491	.3420	.9397	.3640	2.747	1.064	2.924	1.2217	**70°00′**
10	.3520	.3448	.9387	.3673	2.723	1.065	2.901	1.2188	50
20	.3549	.3475	.9377	.3706	2.699	1.066	2.878	1.2159	40
30	.3578	.3502	.9367	.3739	2.675	1.068	2.855	1.2130	30
40	.3607	.3529	.9356	.3772	2.651	1.069	2.833	1.2101	20
50	.3636	.3557	.9346	.3805	2.628	1.070	2.812	1.2072	10
		cos θ	sin θ	cot θ	tan θ	csc θ	sec θ	θ (radians)	θ (degrees)

θ (degrees)	θ (radians)	sin θ	cos θ	tan θ	cot θ	sec θ	csc θ		
21°00′	.3665	.3584	.9336	.3839	2.605	1.071	2.790	1.2043	**69°00′**
10	.3694	.3611	.9325	.3872	2.583	1.072	2.769	1.2014	50
20	.3723	.3638	.9315	.3906	2.560	1.074	2.749	1.1985	40
30	.3752	.3665	.9304	.3939	2.539	1.075	2.729	1.1956	30
40	.3782	.3692	.9293	.3973	2.517	1.076	2.709	1.1926	20
50	.3811	.3719	.9283	.4006	2.496	1.077	2.689	1.1897	10
22°00′	.3840	.3746	.9272	.4040	2.475	1.079	2.669	1.1868	**68°00′**
10	.3869	.3773	.9261	.4074	2.455	1.080	2.650	1.1839	50
20	.3898	.3800	.9250	.4108	2.434	1.081	2.632	1.1810	40
30	.3927	.3827	.9239	.4142	2.414	1.082	2.613	1.1781	30
40	.3956	.3854	.9228	.4176	2.394	1.084	2.595	1.1752	20
50	.3985	.3881	.9216	.4210	2.375	1.085	2.577	1.1723	10
23°00′	.4014	.3907	.9205	.4245	2.356	1.086	2.559	1.1694	**67°00′**
10	.4043	.3934	.9194	.4279	2.337	1.088	2.542	1.1665	50
20	.4072	.3961	.9182	.4314	2.318	1.089	2.525	1.1636	40
30	.4102	.3987	.9171	.4348	2.300	1.090	2.508	1.1606	30
40	.4131	.4014	.9159	.4383	2.282	1.092	2.491	1.1577	20
50	.4160	.4041	.9147	.4417	2.264	1.093	2.475	1.1548	10
24°00′	.4189	.4067	.9135	.4452	2.246	1.095	2.459	1.1519	**66°00′**
10	.4218	.4094	.9124	.4487	2.229	1.096	2.443	1.1490	50
20	.4247	.4120	.9112	.4522	2.211	1.097	2.427	1.1461	40
30	.4276	.4147	.9100	.4557	2.194	1.099	2.411	1.1432	30
40	.4305	.4173	.9088	.4592	2.177	1.100	2.396	1.1403	20
50	.4334	.4200	.9075	.4628	2.161	1.102	2.381	1.1374	10
25°00′	.4363	.4226	.9063	.4663	2.145	1.103	2.366	1.1345	**65°00′**
10	.4392	.4253	.9051	.4699	2.128	1.105	2.352	1.1316	50
20	.4422	.4279	.9038	.4734	2.112	1.106	2.337	1.1286	40
30	.4451	.4305	.9026	.4770	2.097	1.108	2.323	1.1257	30
40	.4480	.4331	.9013	.4806	2.081	1.109	2.309	1.1228	20
50	.4509	.4358	.9001	.4841	2.066	1.111	2.295	1.1199	10
26°00′	.4538	.4384	.8988	.4877	2.050	1.113	2.281	1.1170	**64°00′**
10	.4567	.4410	.8975	.4913	2.035	1.114	2.268	1.1141	50
20	.4596	.4436	.8962	.4950	2.020	1.116	2.254	1.1112	40
30	.4625	.4462	.8949	.4986	2.006	1.117	2.241	1.1083	30
40	.4654	.4488	.8936	.5022	1.991	1.119	2.228	1.1054	20
50	.4683	.4514	.8923	.5059	1.977	1.121	2.215	1.1025	10
27°00′	.4712	.4540	.8910	.5095	1.963	1.122	2.203	1.0996	**63°00′**
10	.4741	.4566	.8897	.5132	1.949	1.124	2.190	1.0966	50
20	.4771	.4592	.8884	.5169	1.935	1.126	2.178	1.0937	40
30	.4800	.4617	.8870	.5206	1.921	1.127	2.166	1.0908	30
40	.4829	.4643	.8857	.5243	1.907	1.129	2.154	1.0879	20
50	.4858	.4669	.8843	.5280	1.894	1.131	2.142	1.0850	10
		cos θ	sin θ	cot θ	tan θ	csc θ	sec θ	θ (radians)	θ (degrees)

Table E Trigonometric Functions of Angles in Degrees and Radians (Continued)

θ (degrees)	θ (radians)	$\sin\theta$	$\cos\theta$	$\tan\theta$	$\cot\theta$	$\sec\theta$	$\csc\theta$		
28°00′	.4887	.4695	.8829	.5317	1.881	1.133	2.130	1.0821	**62°00′**
10	.4916	.4720	.8816	.5354	1.868	1.134	2.118	1.0792	50
20	.4945	.4746	.8802	.5392	1.855	1.136	2.107	1.0763	40
30	.4974	.4772	.8788	.5430	1.842	1.138	2.096	1.0734	30
40	.5003	.4797	.8774	.5467	1.829	1.140	2.085	1.0705	20
50	.5032	.4823	.8760	.5505	1.816	1.142	2.074	1.0676	10
29°00′	.5061	.4848	.8746	.5543	1.804	1.143	2.063	1.0647	**61°00′**
10	.5091	.4874	.8732	.5581	1.792	1.145	2.052	1.0617	50
20	.5120	.4899	.8718	.5619	1.780	1.147	2.041	1.0588	40
30	.5149	.4924	.8704	.5658	1.767	1.149	2.031	1.0559	30
40	.5178	.4950	.8689	.5696	1.756	1.151	2.020	1.0530	20
50	.5207	.4975	.8675	.5735	1.744	1.153	2.010	1.0501	10
30°00′	.5236	.5000	.8660	.5774	1.732	1.155	2.000	1.0472	**60°00′**
10	.5265	.5025	.8646	.5812	1.720	1.157	1.990	1.0443	50
20	.5294	.5050	.8631	.5851	1.709	1.159	1.980	1.0414	40
30	.5323	.5075	.8616	.5890	1.698	1.161	1.970	1.0385	30
40	.5352	.5100	.8601	.5930	1.686	1.163	1.961	1.0356	20
50	.5381	.5125	.8587	.5969	1.675	1.165	1.951	1.0327	10
31°00′	.5411	.5150	.8572	.6009	1.664	1.167	1.942	1.0297	**59°00′**
10	.5440	.5175	.8557	.6048	1.653	1.169	1.932	1.0268	50
20	.5469	.5200	.8542	.6088	1.643	1.171	1.923	1.0239	40
30	.5498	.5225	.8526	.6128	1.632	1.173	1.914	1.0210	30
40	.5527	.5250	.8511	.6168	1.621	1.175	1.905	1.0181	20
50	.5556	.5275	.8496	.6208	1.611	1.177	1.896	1.0152	10
32°00′	.5585	.5299	.8480	.6249	1.600	1.179	1.887	1.0123	**58°00′**
10	.5614	.5324	.8465	.6289	1.590	1.181	1.878	1.0094	50
20	.5643	.5348	.8450	.6330	1.580	1.184	1.870	1.0065	40
30	.5672	.5373	.8434	.6371	1.570	1.186	1.861	1.0036	30
40	.5701	.5398	.8418	.6412	1.560	1.188	1.853	1.0007	20
50	.5730	.5422	.8403	.6453	1.550	1.190	1.844	.9977	10
33°00′	.5760	.5446	.8387	.6494	1.540	1.192	1.836	.9948	**57°00′**
10	.5789	.5471	.8371	.6536	1.530	1.195	1.828	.9919	50
20	.5818	.5495	.8355	.6577	1.520	1.197	1.820	.9890	40
30	.5847	.5519	.8339	.6619	1.511	1.199	1.812	.9861	30
40	.5876	.5544	.8323	.6661	1.501	1.202	1.804	.9832	20
50	.5905	.5568	.8307	.6703	1.492	1.204	1.796	.9803	10
34°00′	.5934	.5592	.8290	.6745	1.483	1.206	1.788	.9774	**56°00′**
10	.5963	.5616	.8274	.6787	1.473	1.209	1.781	.9745	50
20	.5992	.5640	.8258	.6830	1.464	1.211	1.773	.9716	40
30	.6021	.5664	.8241	.6873	1.455	1.213	1.766	.9687	30
40	.6050	.5688	.8225	.6916	1.446	1.216	1.758	.9657	20
50	.6080	.5712	.8208	.6959	1.437	1.218	1.751	.9628	10
		$\cos\theta$	$\sin\theta$	$\cot\theta$	$\tan\theta$	$\csc\theta$	$\sec\theta$	θ (radians)	θ (degrees)

θ (degrees)	θ (radians)	$\sin\theta$	$\cos\theta$	$\tan\theta$	$\cot\theta$	$\sec\theta$	$\csc\theta$		
35°00′	.6109	.5736	.8192	.7002	1.428	1.221	1.743	.9599	**55°00′**
10	.6138	.5760	.8175	.7046	1.419	1.223	1.736	.9570	50
20	.6167	.5783	.8158	.7089	1.411	1.226	1.729	.9541	40
30	.6196	.5807	.8141	.7133	1.402	1.228	1.722	.9512	30
40	.6225	.5831	.8124	.7177	1.393	1.231	1.715	.9483	20
50	.6254	.5854	.8107	.7221	1.385	1.233	1.708	.9454	10
36°00′	.6283	.5878	.8090	.7265	1.376	1.236	1.701	.9425	**54°00′**
10	.6312	.5901	.8073	.7310	1.368	1.239	1.695	.9396	50
20	.6341	.5925	.8056	.7355	1.360	1.241	1.688	.9367	40
30	.6370	.5948	.8039	.7400	1.351	1.244	1.681	.9338	30
40	.6400	.5972	.8021	.7445	1.343	1.247	1.675	.9308	20
50	.6429	.5995	.8004	.7490	1.335	1.249	1.668	.9279	10
37°00′	.6458	.6018	.7986	.7536	1.327	1.252	1.662	.9250	**53°00′**
10	.6487	.6041	.7969	.7581	1.319	1.255	1.655	.9221	50
20	.6516	.6065	.7951	.7627	1.311	1.258	1.649	.9192	40
30	.6545	.6088	.7934	.7673	1.303	1.260	1.643	.9163	30
40	.6574	.6111	.7916	.7720	1.295	1.263	1.636	.9134	20
50	.6603	.6134	.7898	.7766	1.288	1.266	1.630	.9105	10
38°00′	.6632	.6157	.7880	.7813	1.280	1.269	1.624	.9076	**52°00′**
10	.6661	.6180	.7862	.7860	1.272	1.272	1.618	.9047	50
20	.6690	.6202	.7844	.7907	1.265	1.275	1.612	.9018	40
30	.6720	.6225	.7826	.7954	1.257	1.278	1.606	.8988	30
40	.6749	.6248	.7808	.8002	1.250	1.281	1.601	.8959	20
50	.6778	.6271	.7790	.8050	1.242	1.284	1.595	.8930	10
39°00′	.6807	.6293	.7771	.8098	1.235	1.287	1.589	.8901	**51°00′**
10	.6836	.6316	.7753	.8146	1.228	1.290	1.583	.8872	50
20	.6865	.6338	.7735	.8195	1.220	1.293	1.578	.8843	40
30	.6894	.6361	.7716	.8243	1.213	1.296	1.572	.8814	30
40	.6923	.6383	.7698	.8292	1.206	1.299	1.567	.8785	20
50	.6952	.6406	.7679	.8342	1.199	1.302	1.561	.8756	10
40°00′	.6981	.6428	.7660	.8391	1.192	1.305	1.556	.8727	**50°00′**
10	.7010	.6450	.7642	.8441	1.185	1.309	1.550	.8698	50
20	.7039	.6472	.7623	.8491	1.178	1.312	1.545	.8668	40
30	.7069	.6494	.7604	.8541	1.171	1.315	1.540	.8639	30
40	.7098	.6517	.7585	.8591	1.164	1.318	1.535	.8610	20
50	.7127	.6539	.7566	.8642	1.157	1.322	1.529	.8581	10
41°00′	.7156	.6561	.7547	.8693	1.150	1.325	1.524	.8552	**49°00′**
10	.7185	.6583	.7528	.8744	1.144	1.328	1.519	.8523	50
20	.7214	.6604	.7509	.8796	1.137	1.332	1.514	.8494	40
30	.7243	.6626	.7490	.8847	1.130	1.335	1.509	.8465	30
40	.7272	.6648	.7470	.8899	1.124	1.339	1.504	.8436	20
50	.7301	.6670	.7451	.8952	1.117	1.342	1.499	.8407	10
		$\cos\theta$	$\sin\theta$	$\cot\theta$	$\tan\theta$	$\csc\theta$	$\sec\theta$	θ (radians)	θ (degrees)

Table E Trigonometric Functions of Angles in Degrees and Radians (Continued)

θ (degrees)	θ (radians)	sin θ	cos θ	tan θ	cot θ	sec θ	csc θ		
42°00′	.7330	.6691	.7431	.9004	1.111	1.346	1.494	.8378	**48°00′**
10	.7359	.6713	.7412	.9057	1.104	1.349	1.490	.8348	50
20	.7389	.6734	.7392	.9110	1.098	1.353	1.485	.8319	40
30	.7418	.6756	.7373	.9163	1.091	1.356	1.480	.8290	30
40	.7447	.6777	.7353	.9217	1.085	1.360	1.476	.8261	20
50	.7476	.6799	.7333	.9271	1.079	1.364	1.471	.8232	10
43°00′	.7505	.6820	.7314	.9325	1.072	1.367	1.466	.8203	**47°00′**
10	.7534	.6841	.7294	.9380	1.066	1.371	1.462	.8174	50
20	.7563	.6862	.7274	.9435	1.060	1.375	1.457	.8145	40
30	.7592	.6884	.7254	.9490	1.054	1.379	1.453	.8116	30
40	.7621	.6905	.7234	.9545	1.048	1.382	1.448	.8087	20
50	.7560	.6926	.7214	.9601	1.042	1.386	1.444	.8058	10
44°00′	.7679	.6947	.7193	.9657	1.036	1.390	1.440	.8029	**46°00′**
10	.7709	.6967	.7173	.9713	1.030	1.394	1.435	.7999	50
20	.7738	.6988	.7153	.9770	1.024	1.398	1.431	.7970	40
30	.7767	.7009	.7133	.9827	1.018	1.402	1.427	.7941	30
40	.7796	.7030	.7112	.9884	1.012	1.406	1.423	.7912	20
50	.7825	.7050	.7092	.9942	1.006	1.410	1.418	.7883	10
		cos θ	sin θ	cot θ	tan θ	csc θ	sec θ	θ (radians)	θ (degrees)

Answers to Odd-Numbered Exercises

CHAPTER 1

Exercises 1.1

3. 0.24 **5.** 0.181818 . . . **7.** 0.3 **9.** 0.111 . . . **11.** 0.454545 . . . **19.** (a) Digits of successive even integers and thus a non-repeating pattern. (b) Digits of successive multiples of 10 and thus a non-repeating pattern. (c) Digits of successive multiples of 11 and thus a non-repeating pattern. **21.** $-2 > -5$ **23.** $6 > -1$ **25.** $3/4 = 9/12$ **27.** $\sqrt{3} < 1.7$ **29.** $0.6 > 3/8$ **31.** Commutative (twice) **33.** Distributive **35.** Associative **37.** Distributive **39.** Inverse **41.** Identity **43.** Inverse **57.** $x < 1$ **59.** $x < 7$ **61.** No solution **63.** $x > -1$ **65.** $x < 16$

Exercises 1.2

1. {California, Arizona, New Mexico, Texas} **3.** {1} **5.** The set of the last three letters of the alphabet. **7.** The first ten positive integers. **9.** {1, 2, 3} **11.** {2, 4} **13.** {−4}

15. $\{x | x \leq 2\}$

17. $x < 2$

19. $-2 < x < 3$

21. $-1 \leq x < 2$

23. $x \leq 0$

25. $x \geq 0$

27. $[-2, 1]$

29. $(0, 3)$

31. $(1, 3]$

33. $[1, \infty)$

35. $(-\infty, -2)$

37. $(-\infty, 1) \cup (3, \infty)$

39. $x > 6$

41. $-3 \leq x < -3/2$

43. $1 < x < 3$

45. $0 < x < 2$

Exercises 1.3

1. 3 **3.** 3 **5.** 0 **7.** $x - y$ if $x \geq y$ and $y - x$ if $x \leq y$ **9.** $x + 2$ if $x \geq -2$ and $-x - 2$ if $x \leq -2$ **11.** 1, 5 **13.** 2/3 **15.** $1 < x < 5$ **17.** $x > 1$ or $x < -3$ **19.** $|x| \geq 1/5$ **21.** $x < -3/2$ or $x > -1/2$ **23.** $x < 0$, or $x > 2/3$ **25.** $x < 2/3$ or $x > 1$ **27.** All real numbers

31. (a) (b) x is between 0 and 2 (c) $|x - 1| < 1$ (d) $(0, 2)$

33. $2 \leq x \leq 5$ (a) (b) Given (c) $|x - 7/2| < 3/2$ (d) $[-2, 5]$

35. $-1 \leq x < 2$ (a)
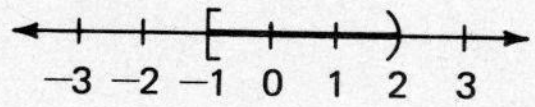

(b) given (c) not applicable because half-open interval (d) $[-1, 2)$

37. $x < -5$ or $x > 1$

(a)
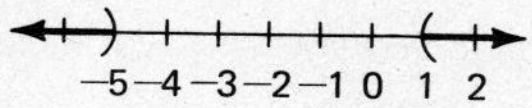

(b) x is less than -5 or greater than 1 (c) $|x + 2| > 3$ (d) $(-\infty, 5)$ or $(1, \infty)$

Exercises 1.4

1. Ordinates all have value -3 **3.** (a) $(2, -3)$ (b) $(-2, -3)$ (c) $(2, 3)$ (d) $(3, -2)$; (a) $(3, 2)$ (b) $(-3, 2)$ (c) $(3, -2)$ (d) $(-2, -3)$ **5.** $(-5, -1)$ **7.** 13 **9.** $\sqrt{37}$ **11.** $\sqrt{x^2 + (y - 8)^2}$ **13.** $\sqrt{a^2 + b^2}$ **15.** No **17.** No **19.** Isosceles **21.** Equilateral **23.** None of these **25.** Right **27.** $(-1, 2)$ **29.** $(2, 5/2)$ **31.** $(2, -3)$ **37.** $(0, 1)$ **39.** $(x + 1)^2 + y^2 = 4$ **41.** $(x - 2)^2 + (y - 3)^2 = 16$

Chapter 1 Review Exercises

1. $x = -1$ or $x = -3$
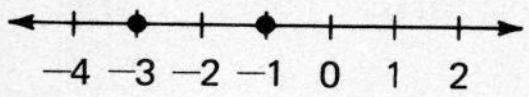

3. $x = 8$ or $x = 4$
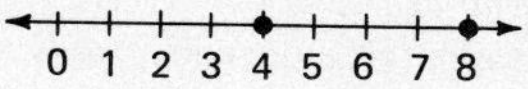

5. $x = 1$ or $x = 0$
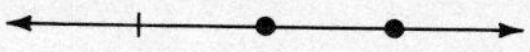

7. $x > -2$
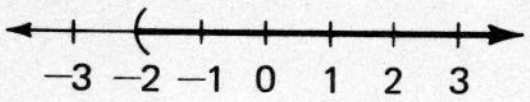

9. $x < -1$
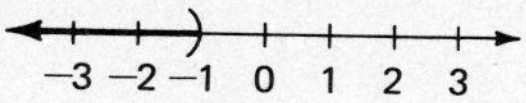

11. $1 < x < 5$
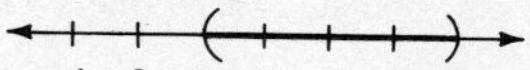

13. $x < -1$ or $x > 5$
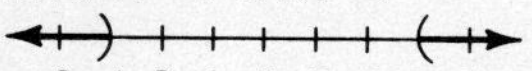

15. $-4 < x < 5$
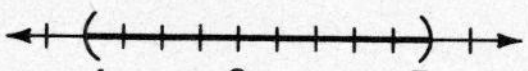

17. $-4 < x < -1$
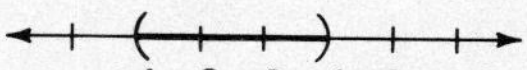

19. $-1 < x < 0$
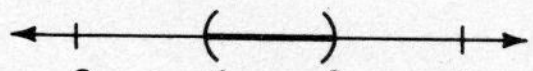

21. $0 < x < 1$, $x > 2$
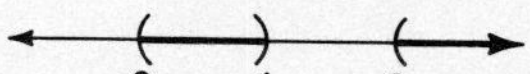

23. III

25. II

27. IV

29. No
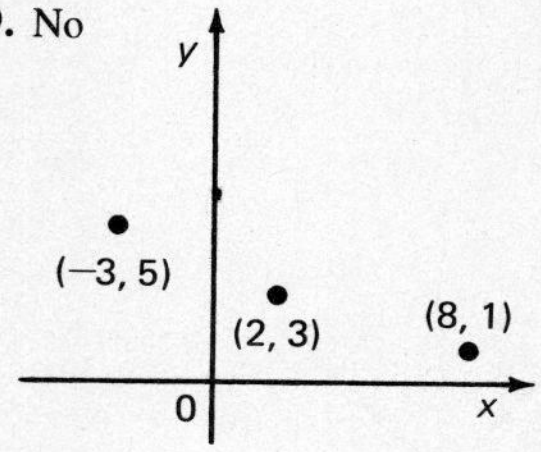

31. No
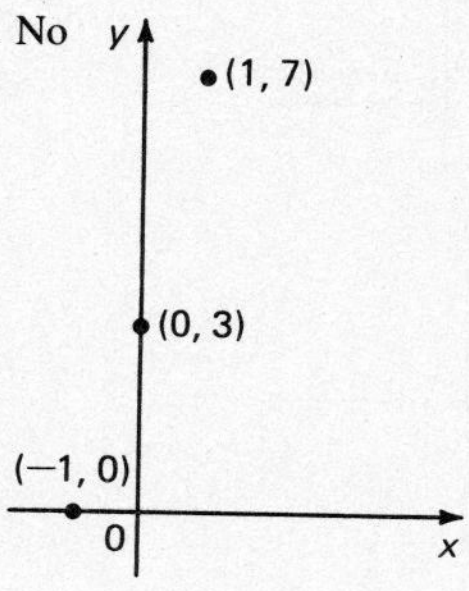

33. Right
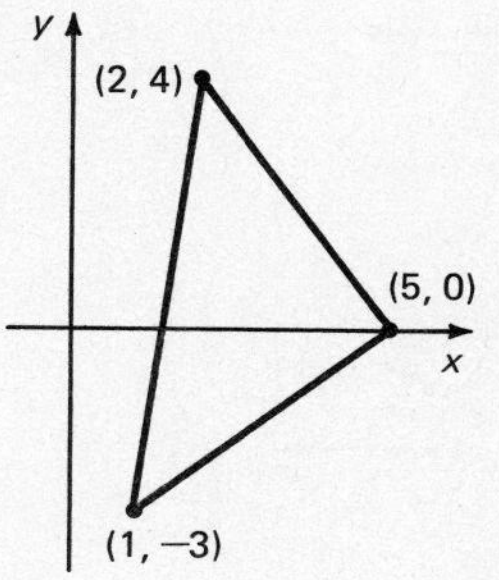

35. None of these
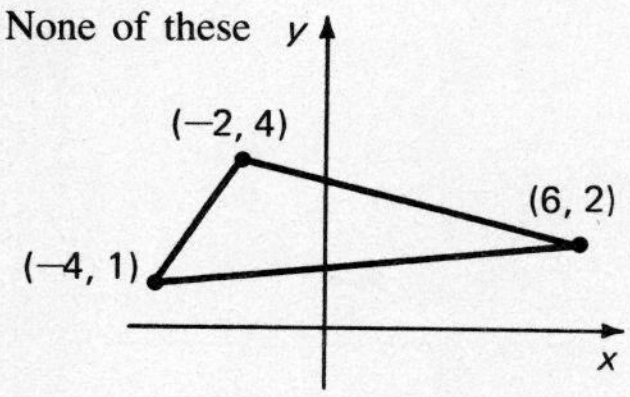

39. Midpoints, $(3, 5)$, $(1, 1)$, and $(0, 3)$; corresponding medians have lengths $\sqrt{61}$, $\sqrt{37}$, and 4, respectively **41.** $c = 2$ or $c = 4$ **43.** $(4, 0)$ **45.** $(0, 5)$ **47.** $y = -2$ **49.** $y = -x$ **51.** $(x - 2)^2 + (y + 1)^2 = 5$ **53.** $(x + 1)^2 + (y - 2)^2 = 9$

CHAPTER 2

Exercises 2.1

1. Yes; domain = set of persons, range = set of corresponding first names **3.** No; to each person corresponds more than one parent **5.** Yes; domain = set of index fingerprints, range = set of persons **7.** Yes; domain = set of Presidents, range = set of states of birth of Presidents **9.** No; to each x corresponds more than one y. **11.** $y = x/2$
13. $y = \pi x^2$ **15.** $y = \sqrt{x}$ **17.** $0, 1, \sqrt{3}, \sqrt{u - 1}, \sqrt{x}$ **19.** 0.40 **21.** 0.41 **23.** 0.42
25. $A(s) = s^2$ **27.** $A(s) = (s^2/4)\sqrt{3}$ **29.** $a = \sqrt{49 - b^2}$ **33.** (a) $x + h$ (b) h (c) 1 **35.** (a) $-2(x + h)^2$ (b) $-4hx - 2h^2$ (c) $-4x - 2h$ **37.** (a) $\sqrt{x + h}$ (b) $\sqrt{x + h} - \sqrt{x}$ (c) $(\sqrt{x + h} - \sqrt{x})/h$ **39.** 212
41. 122 **43.** 77 **45.** $A = P^2/16$ **47.** $4x + 4$

Exercises 2.2

1. Domain = set of all reals, range = set of all reals **3.** Domain = $\{x|x \geq 0\}$, range = $\{y|y \geq 0\}$ **5.** Domain = set of all reals, range = $\{y|y \geq 0\}$ **7.** All reals **9.** $x \geq -1$ **11.** $x \neq 2$ **13.** All reals **15.** $x \neq 1$ and $x \neq -1$ **17.** $x > 1$ **19.** All reals **21.** $y = 1 - x$; yes, because to each x corresponds exactly one y
23. $y = \pm\sqrt{x^2 - 1}$; no, because to each x corresponds more than one y **25.** $y = 1/x^2$; yes, because to each x in the domain corresponds exactly one y **27.** $y = 1/(x - 2)$; yes, because to each x in the domain corresponds exactly one y
29. $y = \pm\sqrt{4 - x}$; no, because to each x corresponds more than one y **31.** $y = 1/(x + 1)$; yes, because to each x corresponds exactly one y **33.** $y \leq x$ or $y \geq -x$; no, because to each x corresponds more than one y
35. $y = (2x + 1)/(1 - x)$; yes, because to each $x \neq 1$ corresponds exactly one y **37.** $y = (x - 1)/(x + 2)$; yes, because to each $x \neq -2$, corresponds exactly one y **39.** The domain of $A(x)$ is $\{x|x > 0\}$, while that of the square function is the set of all reals

Exercises 2.3

1.

3.

5.

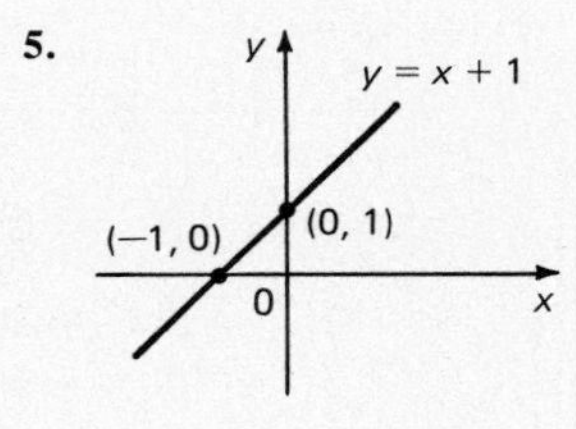

7.

9.

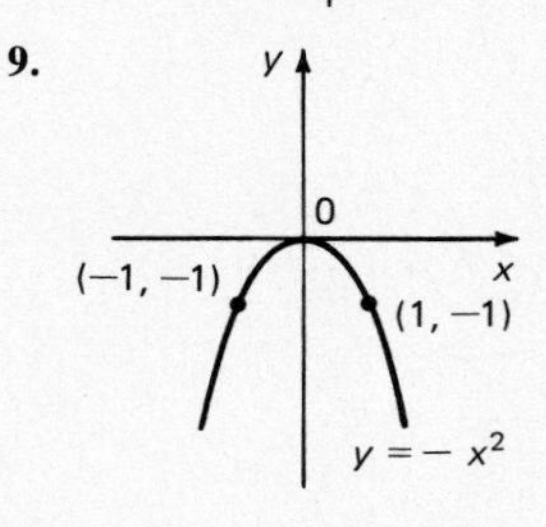

11.

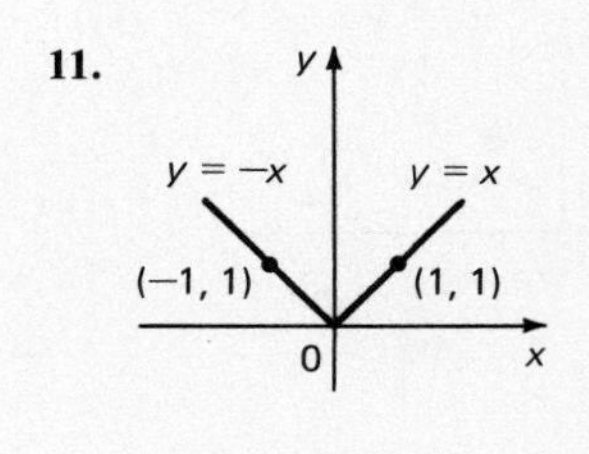

13.

15.

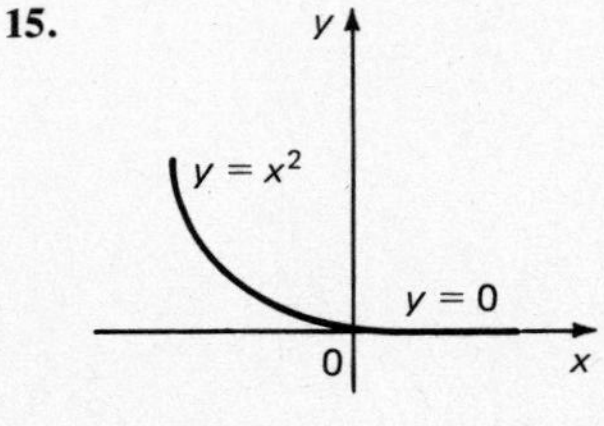

17.

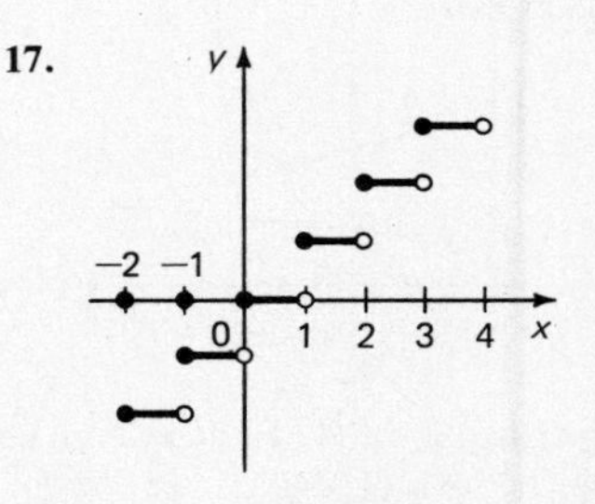

Exercises 2.4

1. The domain of f is the set of all reals and g the set of reals $x \geq 0$. (a) $f(x) + g(x) = x + \sqrt{x}$, all reals $x \geq 0$ (b) $f(x) - g(x) = x - \sqrt{x}$, all reals $x \geq 0$ (c) $f(x)g(x) = x\sqrt{x}$, all reals $x \geq 0$ (d) $f(x)/g(x) = \sqrt{x}$, all reals $x > 0$ **3.** f has domain the set of all reals $x \neq 0$ and g the set of all reals x (a) $f(x) + g(x) = 1/x + x, x \neq 0$ (b) $f(x) - g(x) = 1/x - x, x \neq 0$ (c) $f(x)g(x) = 1, x \neq 0$ (d) $f(x)/g(x) = 1/x^2, x \neq 0$ **5.** f and g both have as domain the set of all reals. (a) $f(x) + g(x) = x + x^2$, all real x (b) $f(x) - g(x) = x - x^2$, all real x (c) $f(x)g(x) = x^3$, all real x (d) $f(x)/g(x) = 1/x, x \neq 0$ **7.** f and g both have as domain the set of all reals (a) $f(x) + g(x) = x^3 + 4$, all real x (b) $f(x) - g(x) = x^3 - 4$, all real x (c) $f(x)g(x) = 4x^3$, all real x (d) $f(x)/g(x) = x^3/4$, all real x **9.** f and g both have as domain the set of all reals (a) $f(x) + g(x) = 4x + 2$, all real x (b) $f(x) - g(x) = 2x - 2$, all real x (c) $f(x)g(x) = 3x(x + 2)$, all real x (d) $f(x)/g(x) = 3x/(x + 2), x \neq -2$ **11.** $f[g(x)] = 1/x^2$ and $g[f(x)] = 1/x^2$, so that $f \circ g = g \circ f, x \neq 0$ **13.** $f[g(x)] = x$ and $g[f(x)] = x$, so that $f \circ g = g \circ f, x \geq 0$ **15.** $f[g(x)] = x^2$ and $g[f(x)] = x^2$, so that $f \circ g = g \circ f, x \geq 0$ **17.** $f[g(x)] = x^2$ and $g[f(x)] = x^2$, so that $f \circ g = g \circ f$, for all x **19.** $f[g(x)] = (1 - x)/(1 + x)$, $x \neq -1, 0$, and $g[f(x)] = (x + 1)/(x - 1), x \neq -1, 0, 1$, so that $f \circ g = g \circ f, x \neq -1, 0, 1$

Exercises 2.5

11. $f^{-1}(x) = -x + 1$ **13.** $f^{-1}(x) = (-1/3)(x - 2)$ **15.** No inverse **17.** $f^{-1}(x) = 3/x$ **19.** No inverse

Exercises 2.6

1. 1/4 **3.** 2/3 **5.** 0 **7.** 0 **9.** 2 **11.** Domain, all real x; range, $0 \leq y \leq 1$ **15.** 1, 0, 0, 1

Exercises 2.7

1. Odd **3.** Neither **5.** Neither **7.** Even **9.** Even **11.** Odd **13.** -1 **15.** $\pm\sqrt{3}$ **17.** 1 **19.** 1/2 **21.** ± 1 **23.** 0

Chapter 2 Review Exercises

1. Yes, because to each x corresponds exactly one y. Domain and range the set of all reals **3.** Yes, because to each species there is exactly one number (of chromosomes). Domain is set of species and range is set of corresponding numbers of chromosomes **5.** No, because to each house corresponds more than one room **7.** No, because to each salesman corresponds more than one state **9.** Yes, because to each person corresponds exactly one shadow **11.** 3 **13.** $2x + h + 2$ **15.** $3x^2 + 3hx + h^2$ **17.** $|x| \geq 1$ **19.** $x \neq 1$ **21.** $x \neq 0, 1$ **23.** $x \neq 1/2$ **25.** Yes, since to each x corresponds exactly one y **27.** No, since to each x corresponds more than one y **29.** $f[g(x)] = 1/\sqrt{x}$, $g[f(x)] = 1/\sqrt{x}$ **31.** $f[g(x)] = g[f(x)] = x, x \neq 0, 1$ **33.** See Figure 2.6 **35.** See Figure 2.8 **37.** See Figure 2.11

39.

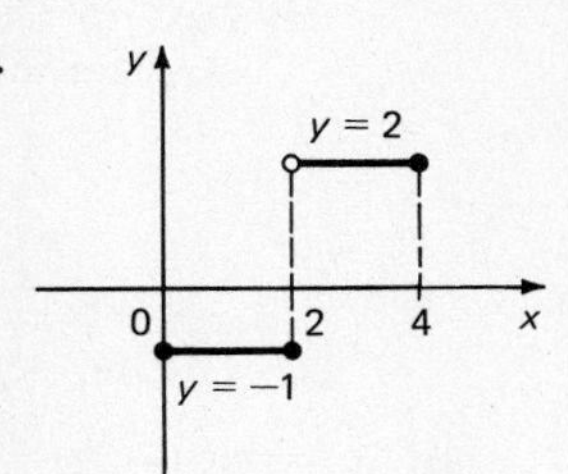

41.

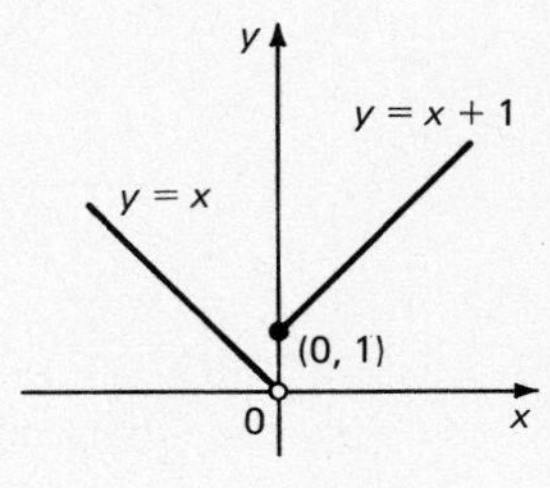

43. $f^{-1}(x) = (x + 1)/3$

45. $f^{-1}(x) = \dfrac{7x - 5}{1 - 2x}$

47.

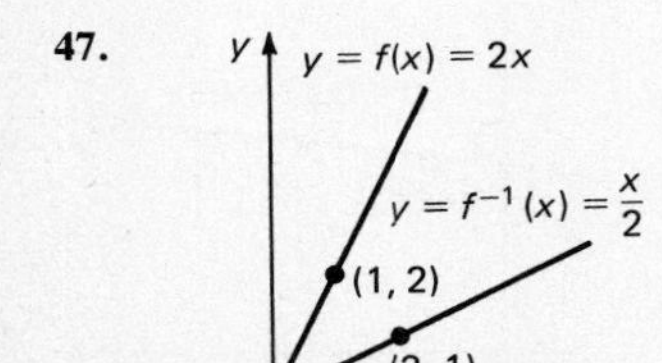

49.

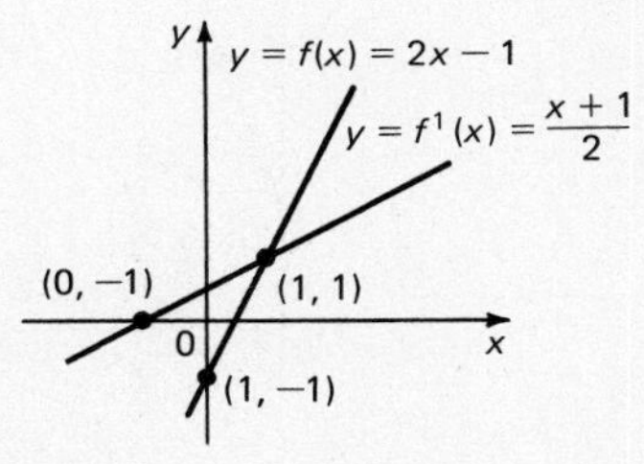

51. Even **53.** Odd **55.** Even

Chapter 2 Miscellaneous Exercises

1. (a) Domain, $-1 \leq x \leq 4$; range, $0 \leq y \leq 1$
(b) $f(-1) = 0, f(0) = 1, f(1/2) = 1, f(3) = 0$
(c) $$f(x) = \begin{cases} x + 1, & -1 \leq x < 0 \\ 1, & 0 \leq x < 1 \\ -x + 2, & 1 \leq x < 2 \\ 0, & 2 \leq x < 4 \end{cases}$$

3. (a) $f(r/2) = c/2r^2$
(b) $f(0) = 0$
(c) $f(3r) = 3c/r^2$

5.

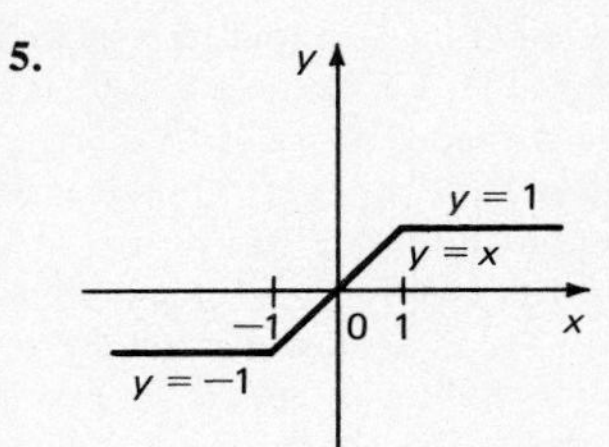

CHAPTER 3

Exercises 3.1

1. Polynomial **3.** Neither **5.** Rational **7.** Neither **9.** Polynomial **11.** Polynomial **13.** Rational **15.** Sum, $5x^2 - x - 7$ Difference, $-x^2 + 11x - 9$ Product, $6x^4 + 3x^3 - 52x^2 + 53x - 8$ Quotient, $2/3 + (27x - 26)/3(3x^2 - 6x + 1)$ **17.** Sum, $2x^4 + x^3 + x^2 - 2x - 4$ Difference, $2x^4 + x^3 - 4x - 6$ Product, $2x^6 + 3x^5 + 3x^4 - 2x^3 - 8x^2 - 8x - 5$ Quotient, $2x^2 - x - 1 - (x + 4)/(x^2 + x + 1)$ **19.** $x^3 + 3x^2y + 3xy^2 + y^3$ **21.** $x^5 + 5x^4y + 10x^3y^2 + 10x^2y^3 + 5xy^4 + y^5$ **23.** $5(4x - 1)$ **25.** $(x - 5)(x + 3)$ **27.** $(x - 5)^2$ **29.** $(4x - 7)(4x + 7)$ **31.** $(2 - 3x)(4 + 6x + 9x^2)$ **33.** $(3x - 4)(x + 3)$ **35.** $(4x + 3)^2$ **37.** $2(x + 2y - 3z)$ **39.** $(3x - 2)(x + 4)$ **41.** $(3x + 1)^2$ **43.** $(5x - y)(5x + y)$ **45.** $3x(4x^2 + 7x - 5)$ **47.** $(3x + 4)(4x - 3)$ **49.** 180 **51.** 132 **53.** -7.52620159 **55.** 514.368

Exercises 3.2

1.

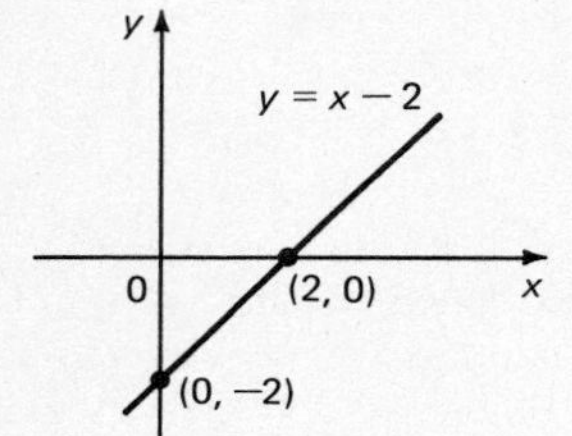

3.

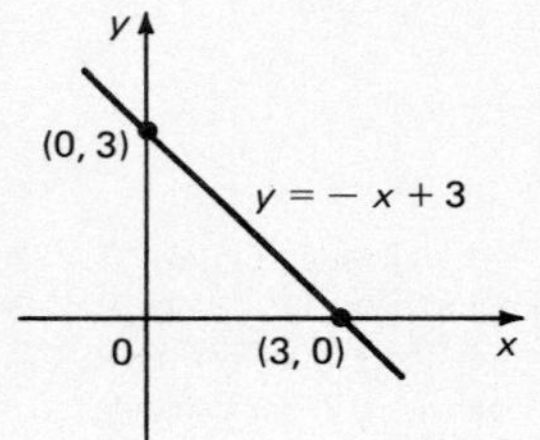

5.

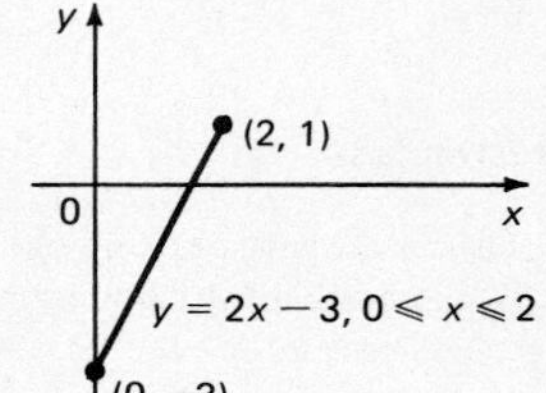

7. 1
9. 5
11. 7/4

13. Slope = 3

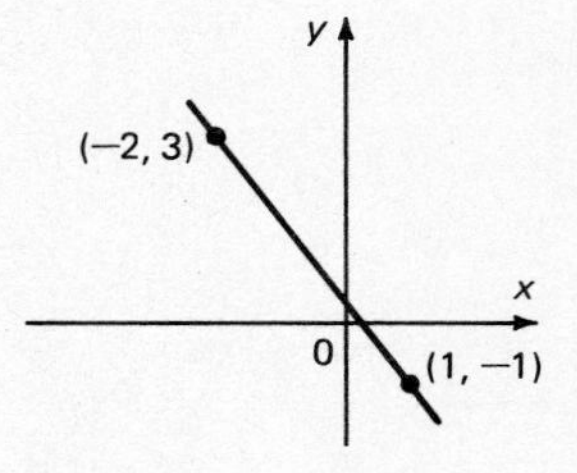

15. Slope = $-4/3$

17. Slope = 3, y-intercept = -4
19. Slope = -1, y-intercept = 0

21. The F-intercept, 32, is the Fahrenheit reading for 0°C, and the C-intercept, $-160/9$, is the Celsius reading for 0°F

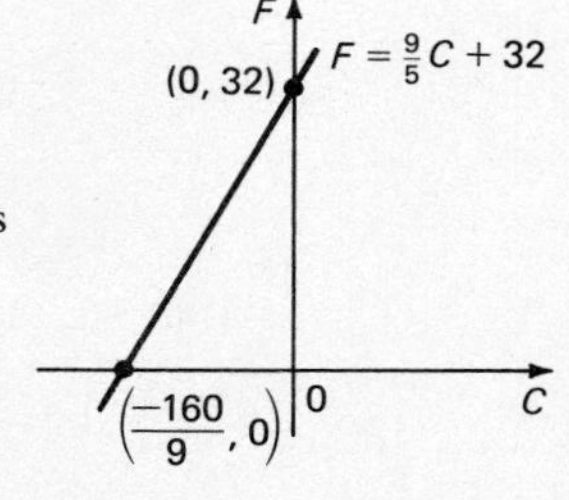

23. $C(x) = 5x + 20$

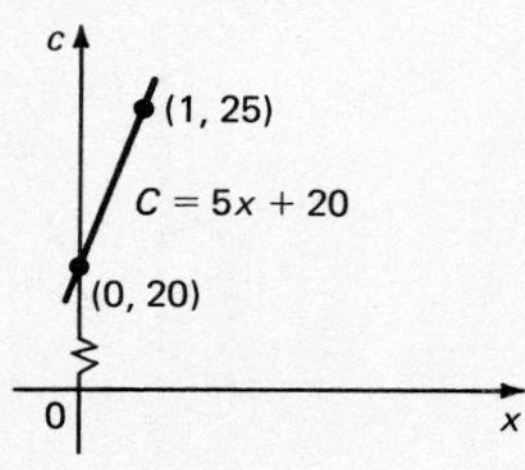

25. $C(x) = 20 + 0.25x$

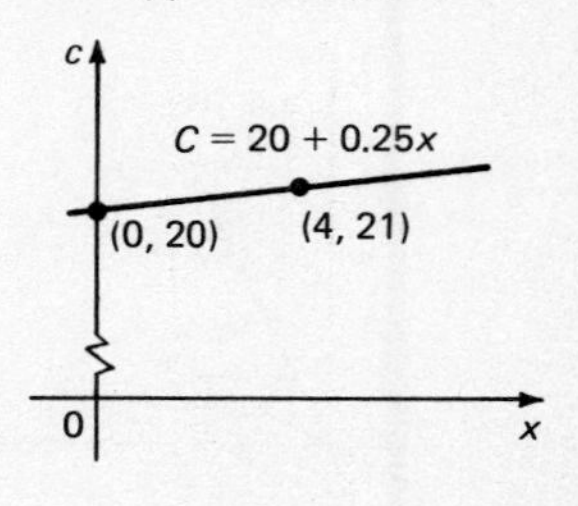

27. (a) $C(x) = 2 + 0.6x$
(b) $C(5) = 5$ (dollars)

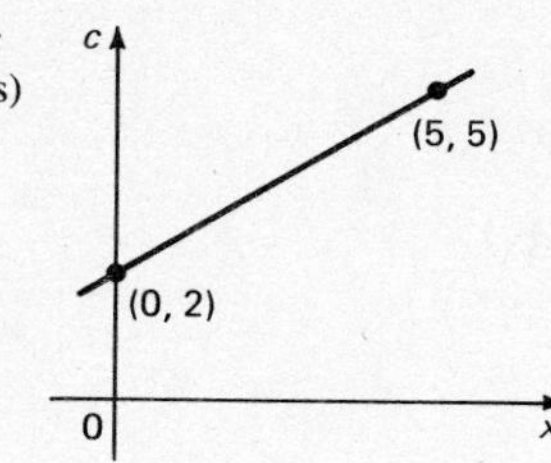

29.

Exercises 3.3

1. $(x - 1)^2 - 1$ **3.** $(x + 5/2)^2 - 25/4$ **5.** $2(x - 5/2)^2 - 3/2$ **7.** $2(x + 1)^2 - 3$ **9.** $5(x - 3/2)^2 - 13/4$ **11.** Least value, -1; range, $f(x) \geq -1$ **13.** Greatest value, 1; range, $f(x) \leq 1$ **15.** Least value, 39/8; range, $f(x) \geq 39/8$ **17.** Greatest value, 1/4; range, $f(x) \leq 1/4$ **19.** Least value, $-3/2$; range, $f(x) \geq -3/2$ **21.** $x = -2, 4$ **23.** $x = -4, 3$ **25.** $x = (15 \pm \sqrt{65})/10$ **27.** $x = (-3 \pm \sqrt{33})/4$ **29.** $x = (-2 \pm \sqrt{6})/2$ **31.** $x < -4$ or $x > 3$ **33.** $-1 < x < 3$ **35.** $x < 3/4$ or $x > 2$ **37.** $f(x) \leq 0$ if $-3 \leq x \leq 3$ and $f(x) \geq 0$ if $|x| \geq 3$ **39.** $f(x) \leq 0$ if $-2 \leq x \leq 4$ and $f(x) \geq 0$ if $x \leq -2$ or $x \geq 4$ **41.** $f(x) \leq 0$ if $-1 \leq x < 3/8$ and $f(x) \geq 0$ if $x \leq -1$ or $x \geq 3/8$ **43.** (a) 16 feet (b) 36 feet (c) 4 feet **45.** $t = -2 + \sqrt{99} \approx 7.95$ (feet) **49.** 4, 5 and -4, -5 **51.** 3 and 3

Exercises 3.4

1. $(0, 0)$; $x = 0$ **3.** $(2, -4)$; $x = 2$ **5.** $(1, 0)$; $x = 1$ **7.** $(-3/4, -9/8)$; $x = -3/4$

9.

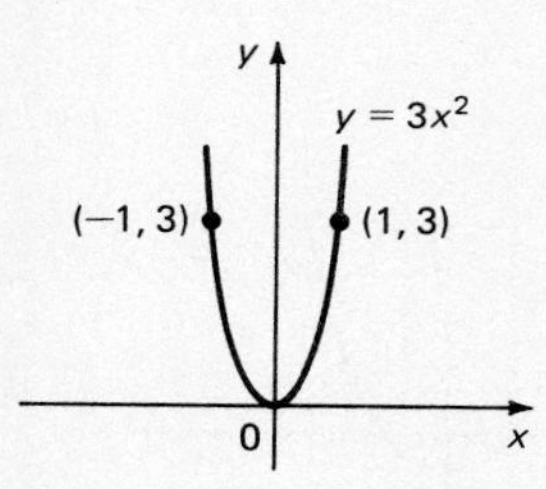

11.

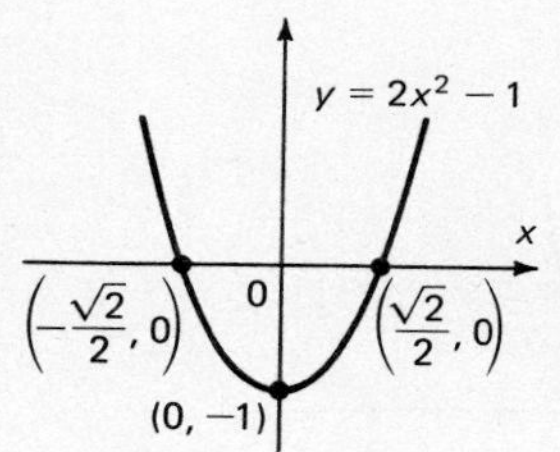

13.

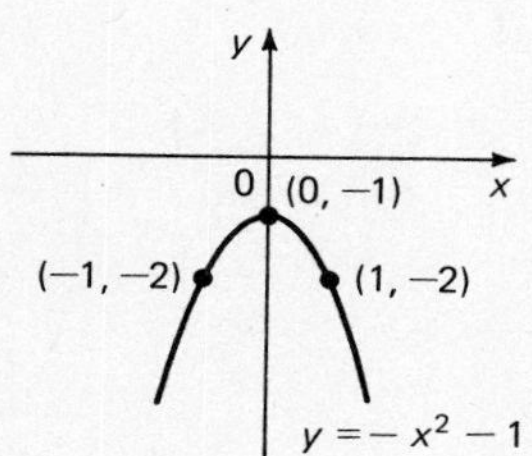

15.

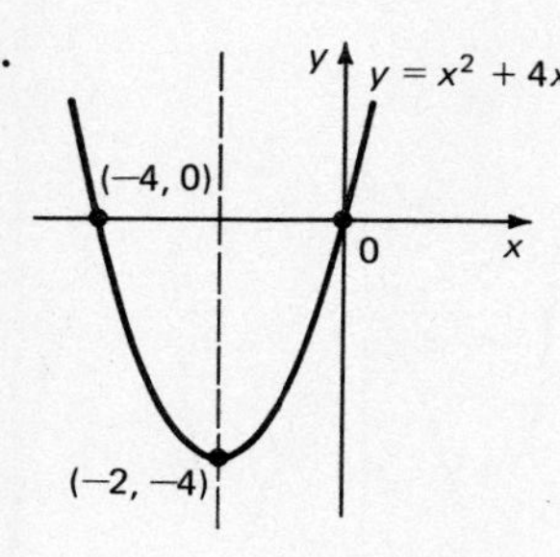

17.

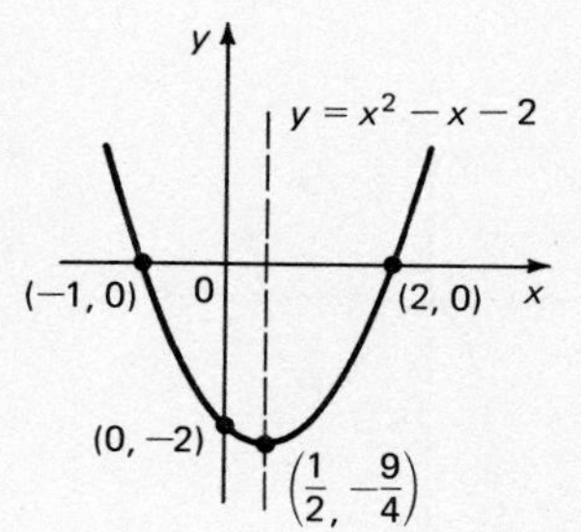

19.

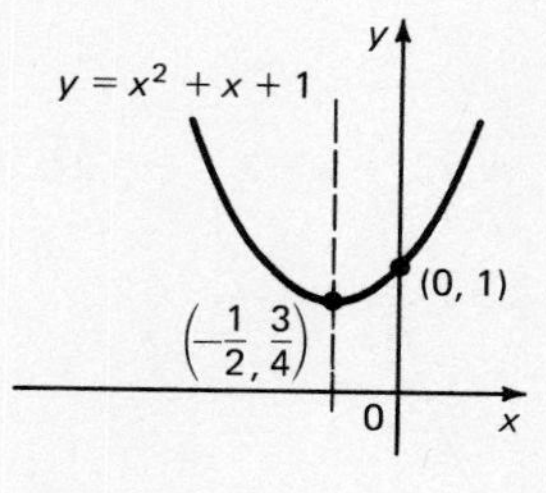

21.

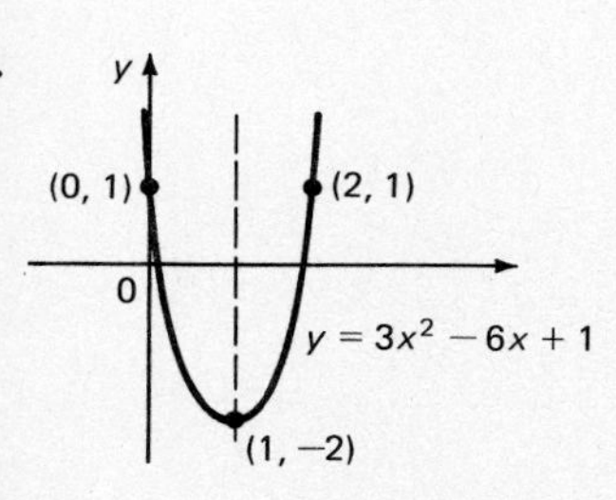

23. (a) $f(x) \geq 0$ for all x (b) $f(x)$ increasing for $x > 0$ and decreasing for $x < 0$
25. (a) $f(x) > 0$ for $0 < x < 3$ and $f(x) < 0$ for $x < 0$ or $x > 3$
(b) $f(x)$ increasing for $x < 3/2$ and decreasing for $x > 3/2$
27. (a) $f(x) > 0$ for $x < 1 - \sqrt{2}/2$ and $f(x) < 0$ for $x > 1 + \sqrt{2}/2$
(b) $f(x)$ increasing for $x > 1$ and decreasing for $x < 1$
29. $y = (x - 2)^2 + (c - 4)$, so that the lowest point is $(2, c - 4)$. Then $c - 4 = 0$, or $c = 4$

Exercises 3.5

1. Quotient, $x - 4$; remainder, 6 **3.** Quotient, $x^2 + 5x + 2$; remainder, 0 **5.** Quotient, $x^3 + 11x^2 - 15x + 4$; remainder, -6 **7.** -6, 6, 84 **9.** 0, -35 **11.** -15, 0, 65 **13.** $(x + 1)(2x^2 - 7x + 6)$ **15.** $(x - 1)(8x^2 - 2x - 3)$ **17.** $(x + 2)(x^3 - 2x^2 + 4x - 8)$ **19.** $(x - 3 - \sqrt{2})(x - 3 + \sqrt{2})$ **21.** $(x - 3/2 + (3/2)\sqrt{2})(x - 3/2 - (3/2)\sqrt{2})$ **23.** $(x + 1/2 - \sqrt{3}/2)(x + 1/2 + \sqrt{3}/2)$ **25.** $c(x^3 - 3x^2 + x)$ **27.** $c(x^2 - 2)$ **29.** $c(x^4 - 6x^3 + 13x^2 - 12x + 4)$

Exercises 3.6

1. No rational zeros **3.** (a) 2, -2, -3, (b) $f(x) = (x - 2)(x + 2)(x + 3)$ **5.** (a) $-1/2$, $-1/2$ (b) $f(x) = (2x + 1)(2x + 1)(x^2 - x - 1)$ **7.** (a) 0, 2, 2 (b) $f(x) = x(x - 2)(x - 2)$ **9.** (a) 2, 2, 2 (b) $f(x) = (x - 2)^3$ **11.** $x = 4$ is the only real root **13.** 0, 2

Exercises 3.7

1.

y
(−1, 1)
0
x
(1, −1)
y = −x³

3.

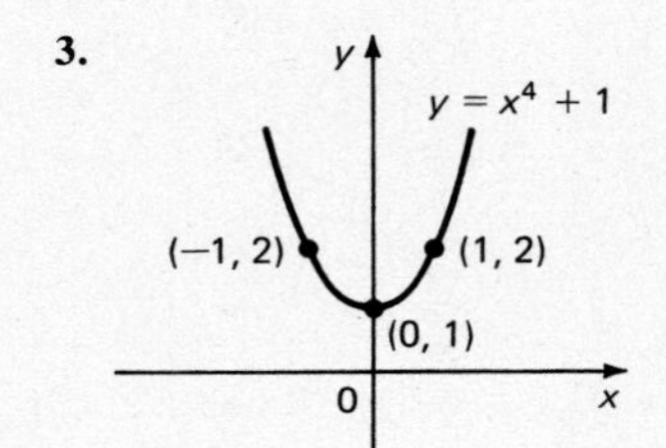

5.

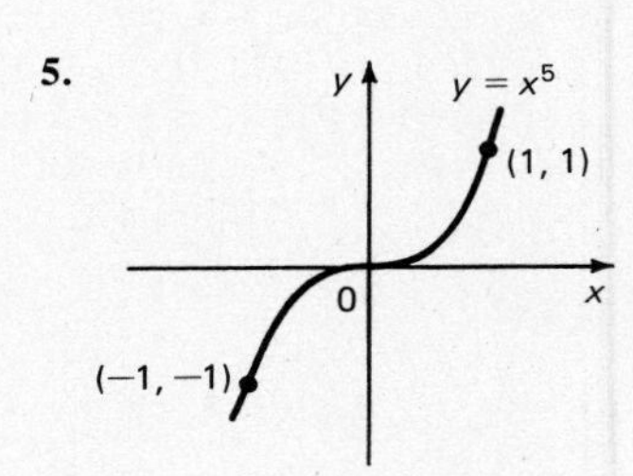

7.

y
y = (x + 1)³
(0, 1)
(−1, 0)
0
x

9.

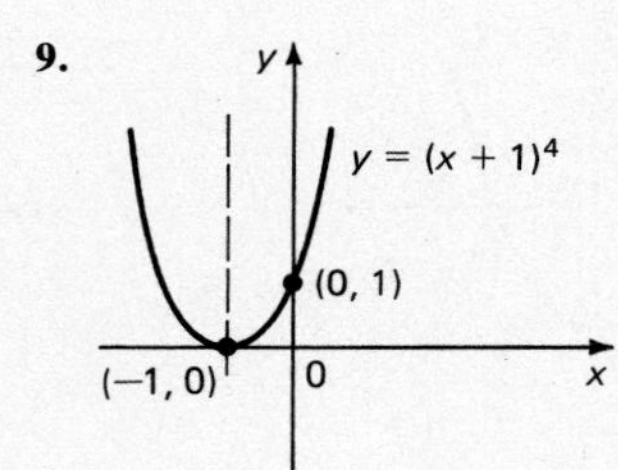

11.

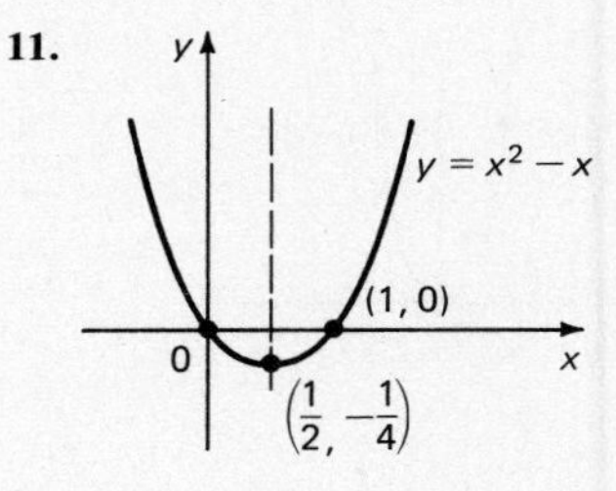

13.

y = x² − x − 2
(−1, 0)
0
(2, 0)
(1/2, −9/4)

15.

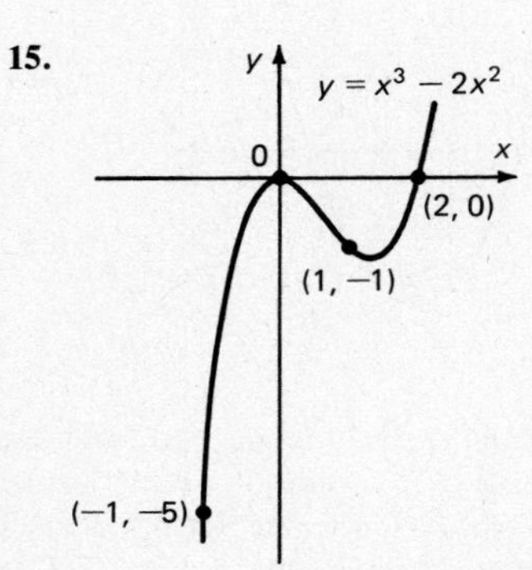

17.

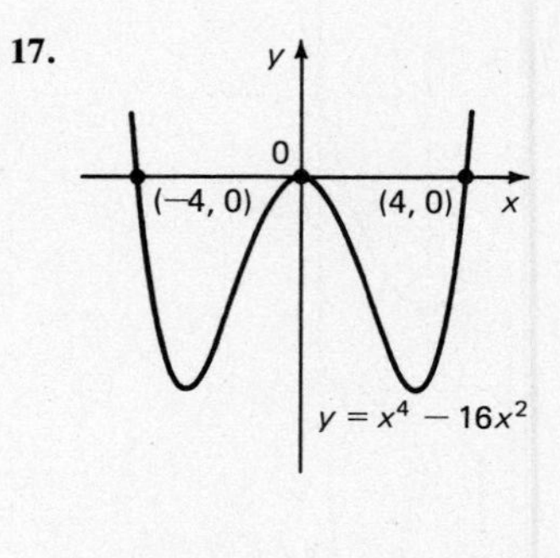

19.

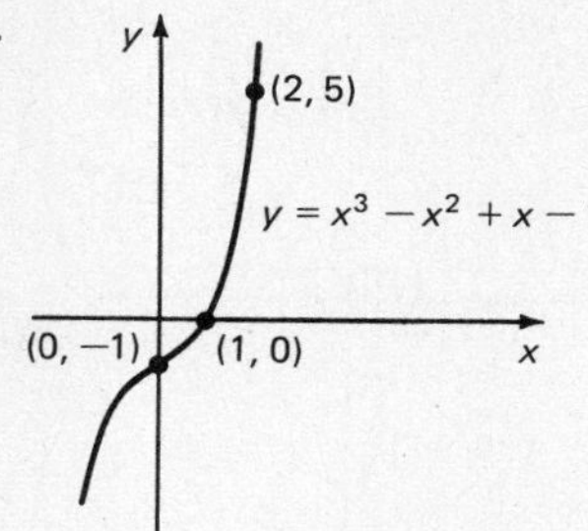

21.

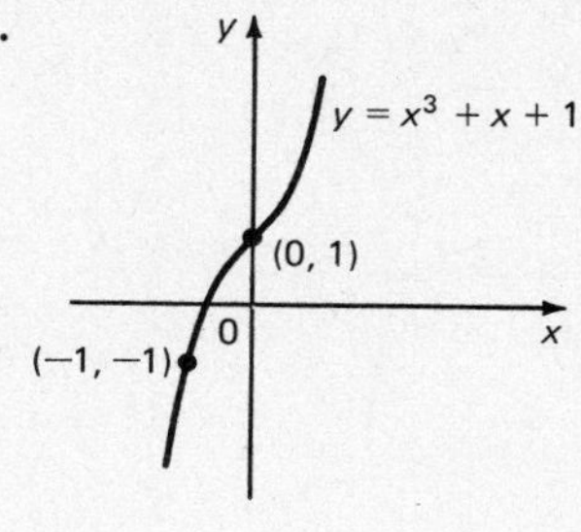

23.

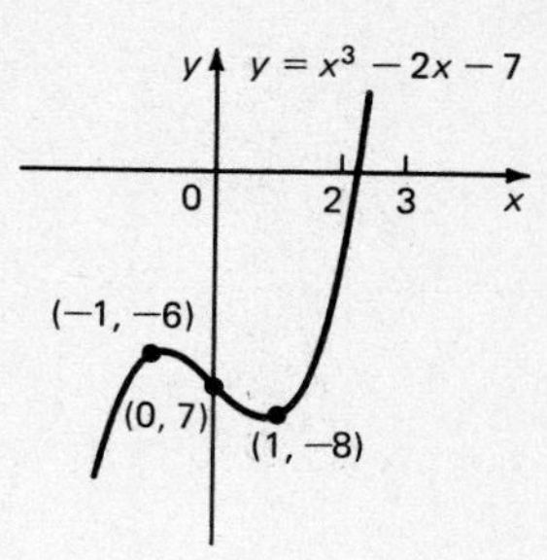

25.

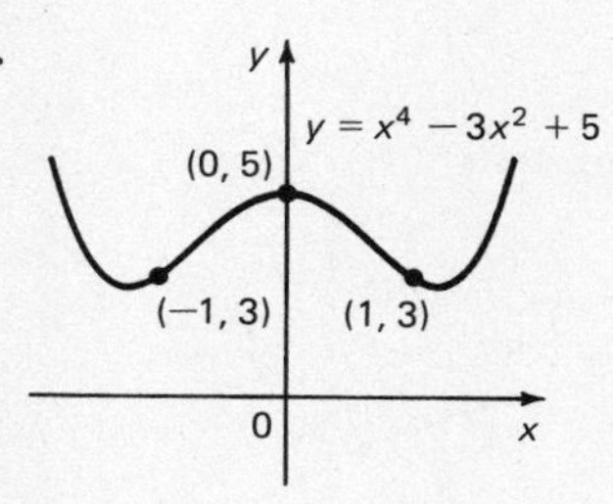

27.

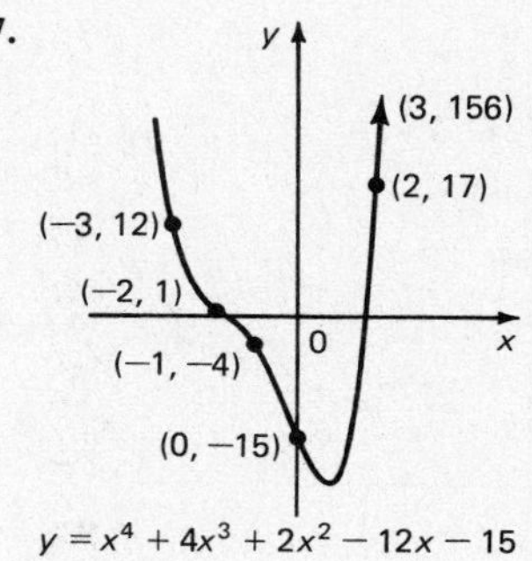

Exercises 3.8

1. $(x + y)/(x + z)$ **3.** $(x - 2y)/(x + y)$ **5.** $(2x + 5)/(x - 2)$ **7.** $2x/(x + 1)(x - 1)$ **9.** $-4x/(x + 1)(x - 1)$
11. $2x(x + y)$ **13.** 1 **15.** $1/9x^2$ **17.** $1/x - 1/(x + 1)$ **19.** $(2/5)/(x + 2) + (3/5)(x - 3)$
21. $(-3/2)/(x + 2) + (5/2)/(x + 4)$ **23.** $1/(x - 3) + 4/(x + 2)$ **25.** $x - 4 + 1/2x + 31/2(x + 4)$
27. $x^2 - 1 + 2/(x + 2)$ **29.** $1 - 1/2(x + 1) + 1/2(x - 1)$

Exercises 3.9

1.

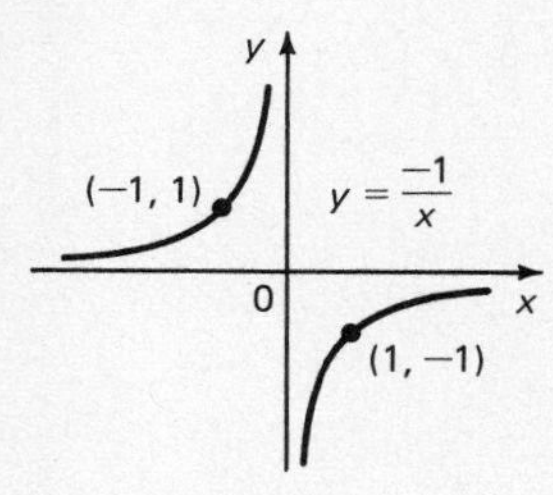

3.

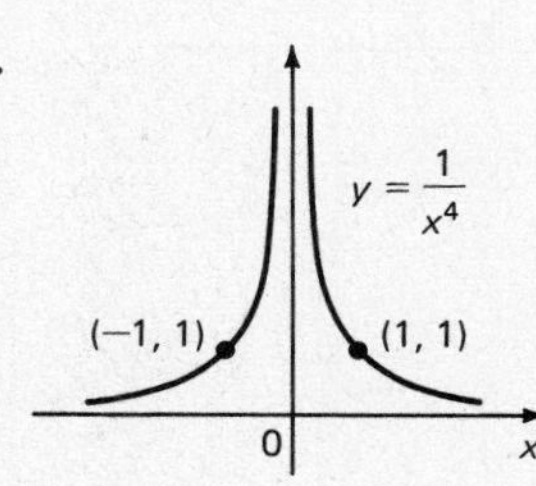

5.

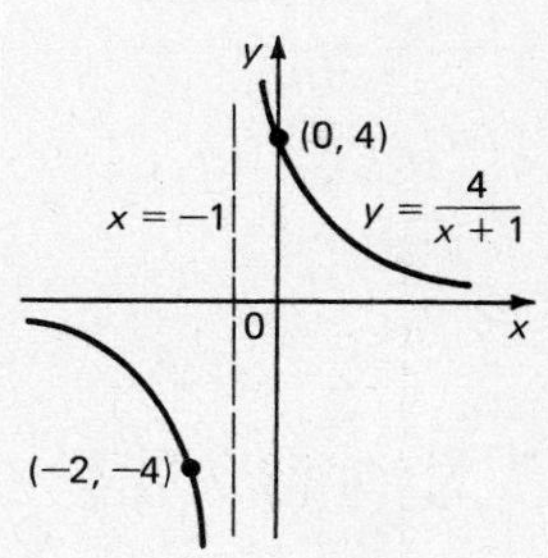

7.

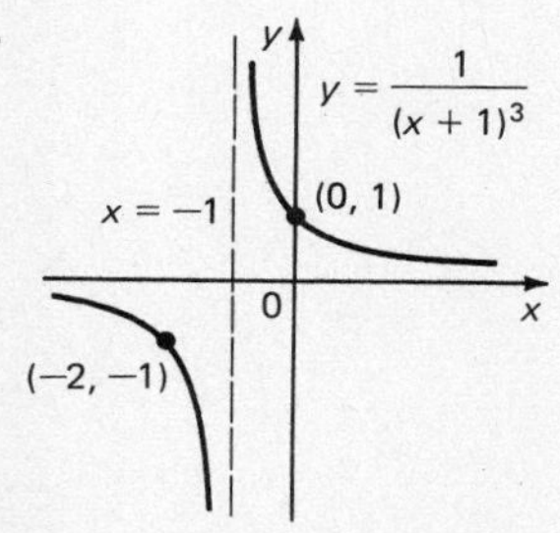

9.

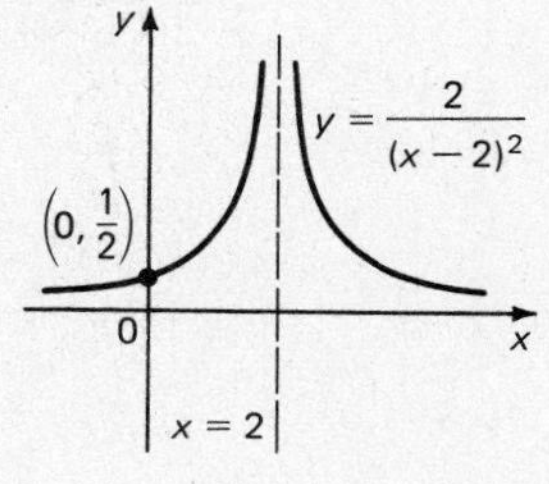

11.

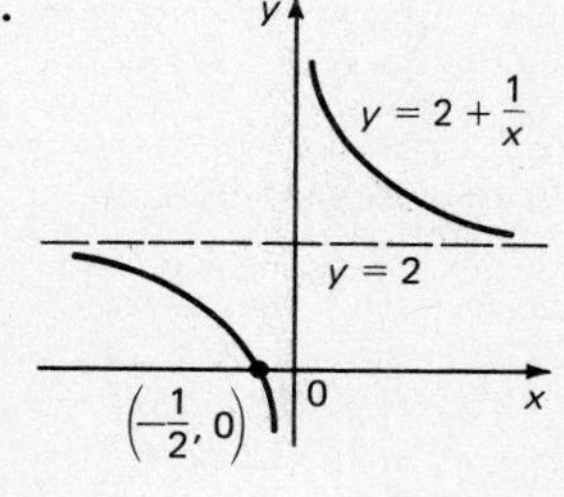

13.

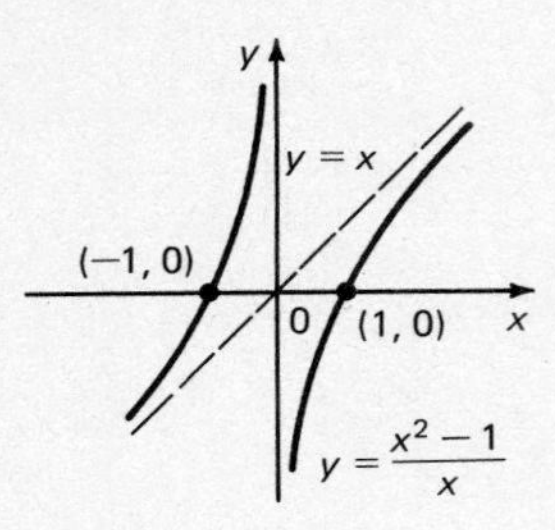

15.

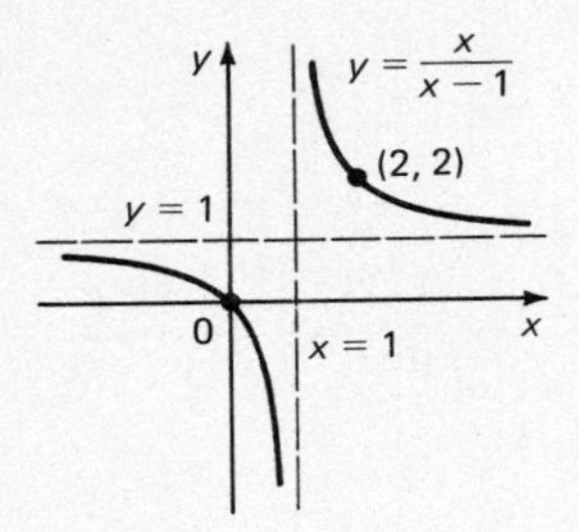

17.

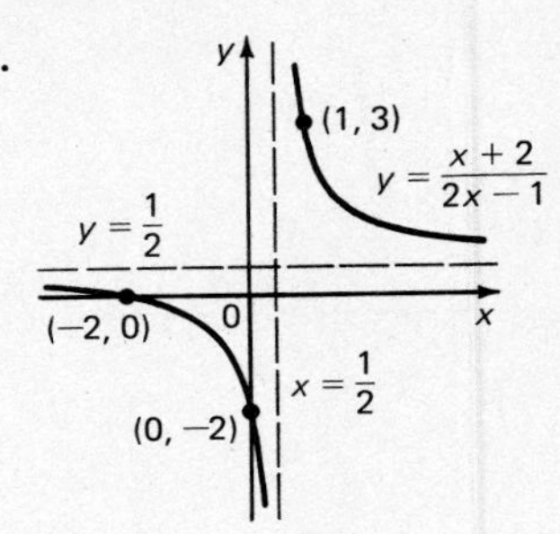

Exercises 3.10

1. Domain, $x \geq 0$

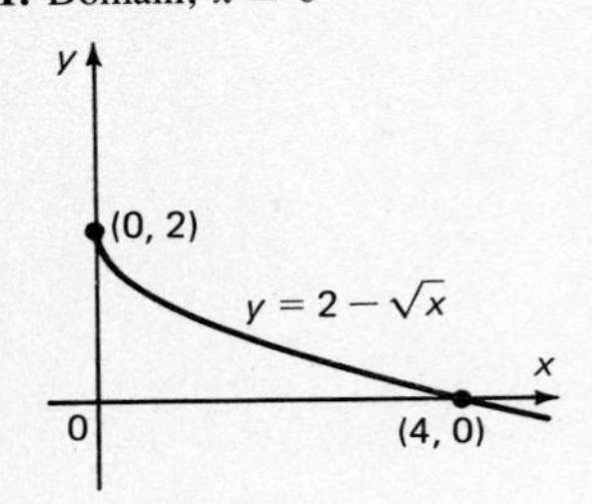

3. Domain, $x \geq 1$

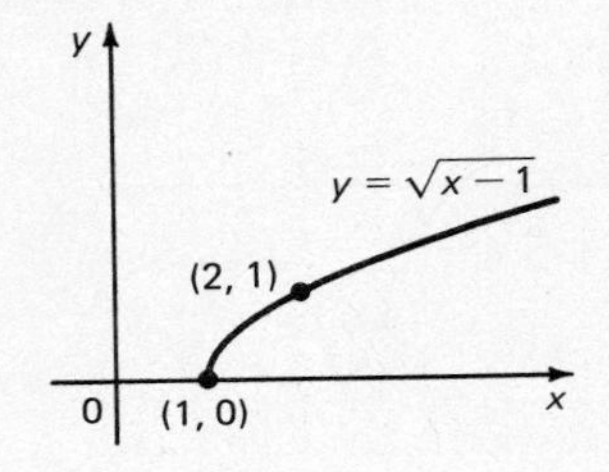

5. Domain, $x \leq 1$

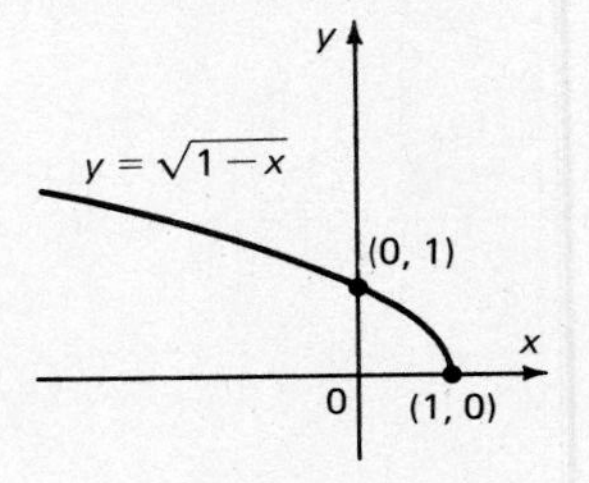

7. Domain, $x > 0$

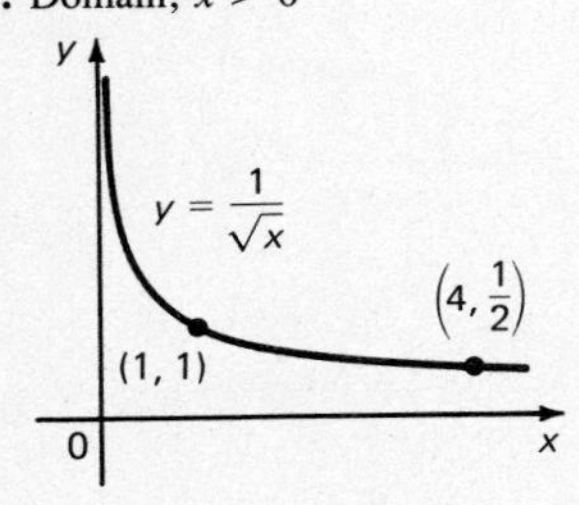

9. Domain, $|x| \geq 1$

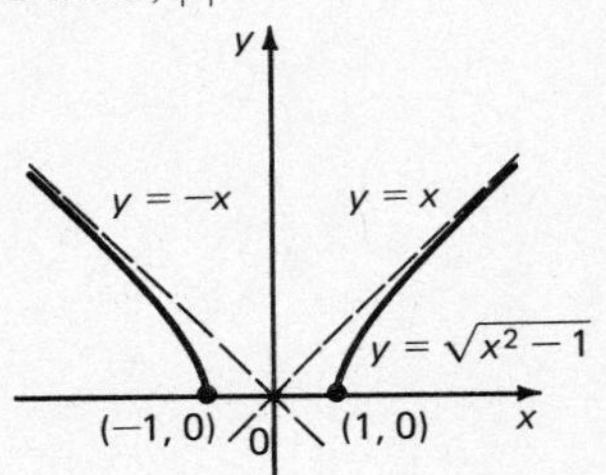

11. Domain, $|x| \geq 2$

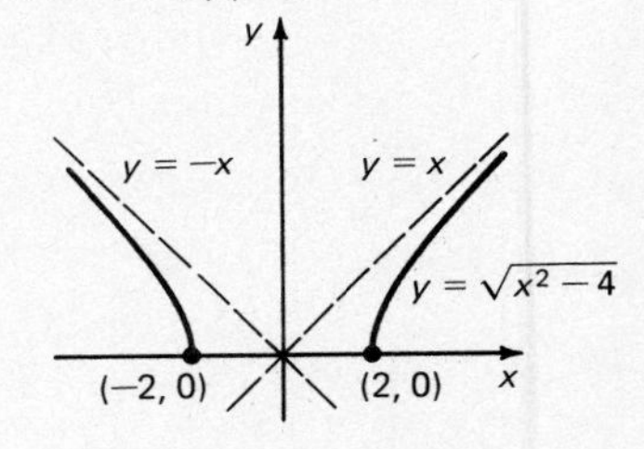

13.

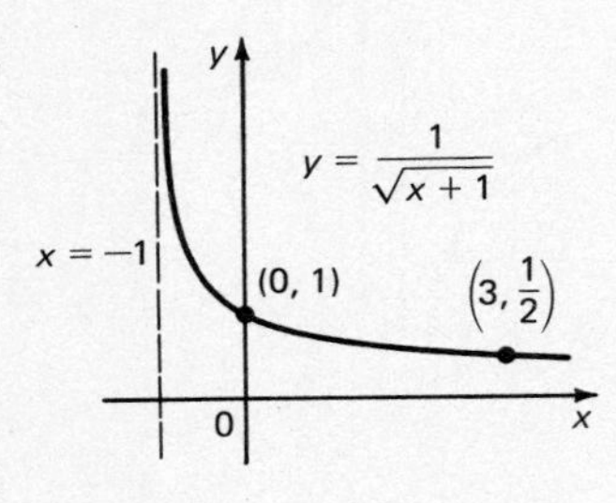

15.

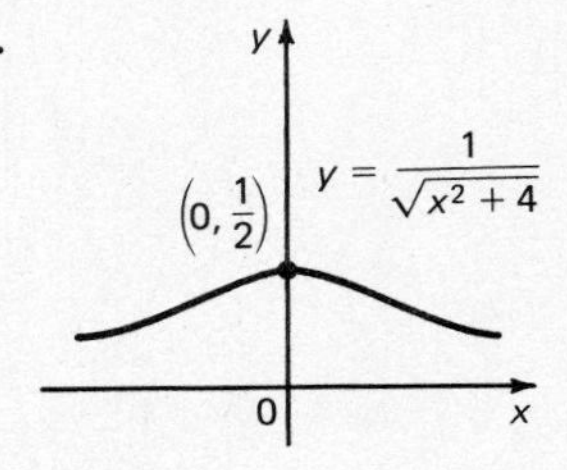

Chapter 3 Review Exercises

1. Polynomial **3.** Polynomial **3.** Rational **7.** $(x - 4)(x + 1)$ **9.** $(-x + 4)(x - 3)$, or $(x - 4)(-x + 3)$
11. $(3x - 2)(x + 4)$ **13.** Least value, -4 **15.** Least value, 3 **17.** Least value, 7/8 **19.** $x = 3, -1$
21. $x = (-2 \pm \sqrt{6})/2$ **23.** No real roots **25.** $1 < x < 2$ **27.** $x \leq -2$ or $x \geq 1/2$
29. Quotient, $x^2 + 1$; remainder, -1

31. Quotient, $x^3 - 3x^2 + 3x/2 - 9/4$; remainder, $59/8$ **33.** 22, −18 **35.** −9, 0
37. 1, 3, −2; $f(x) = (x - 1)(x - 3)(x + 2)$ **39.** −1/2, −1/2; $f(x) = (2x + 1)(2x + 1)(x^2 - x - 1)$

41.
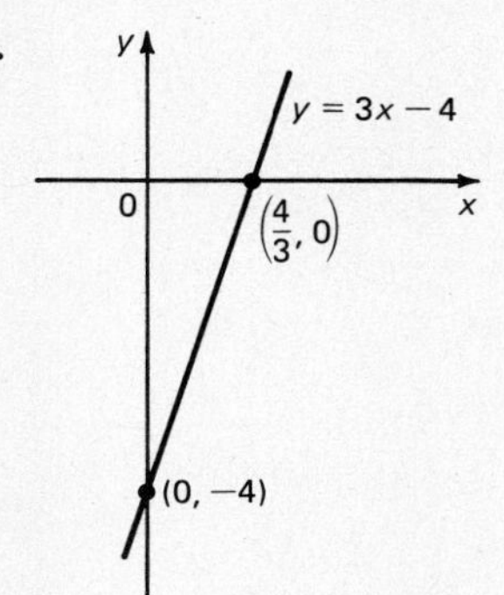

43.
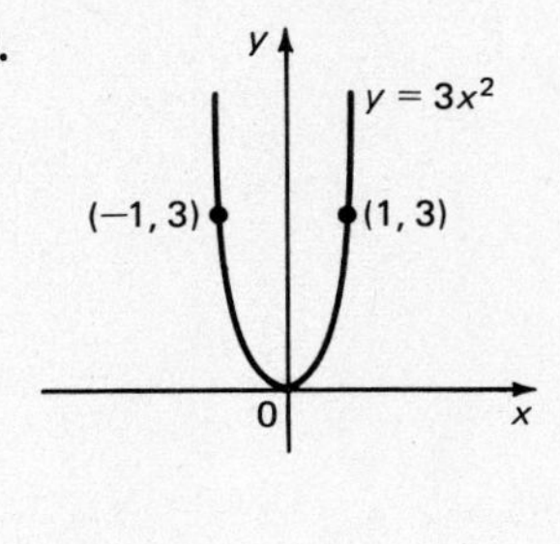

45.
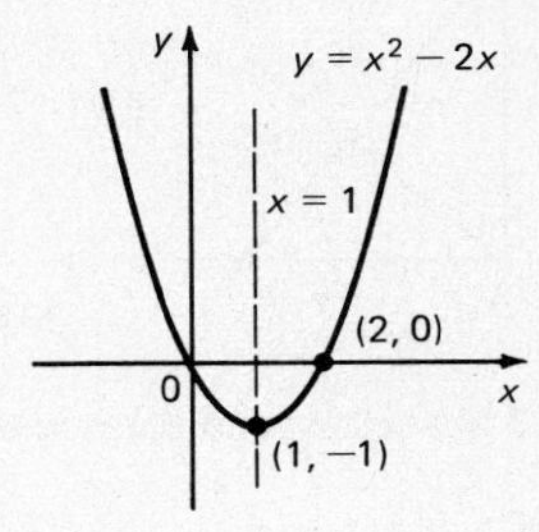

47.
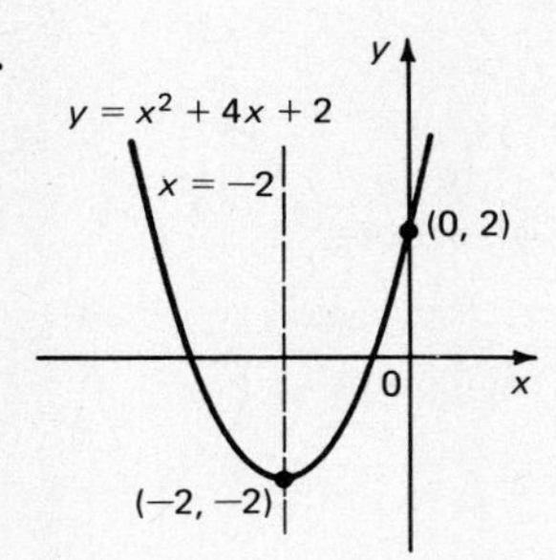

49.
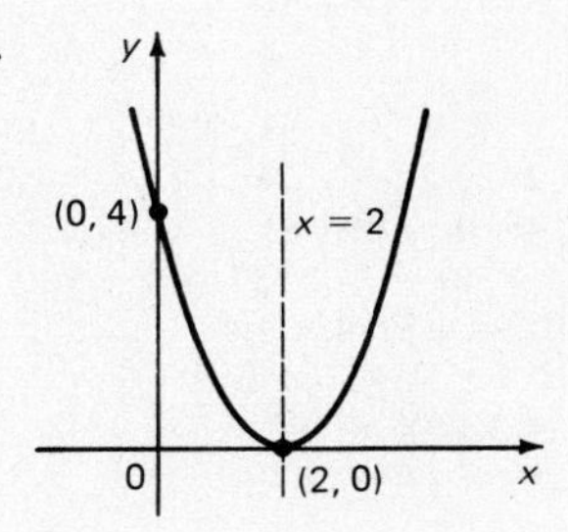

51.
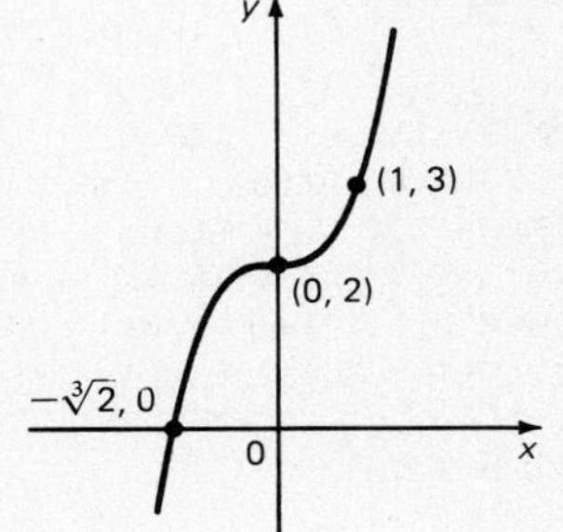

53.
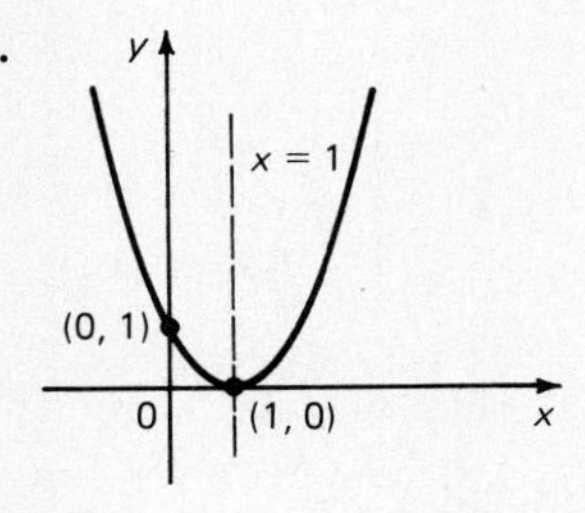

55.
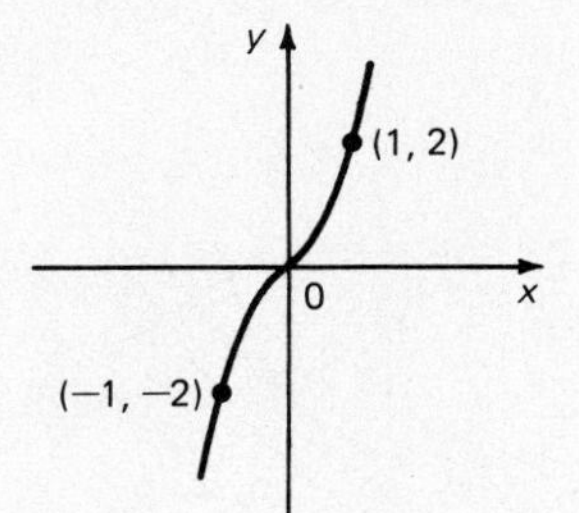

57.
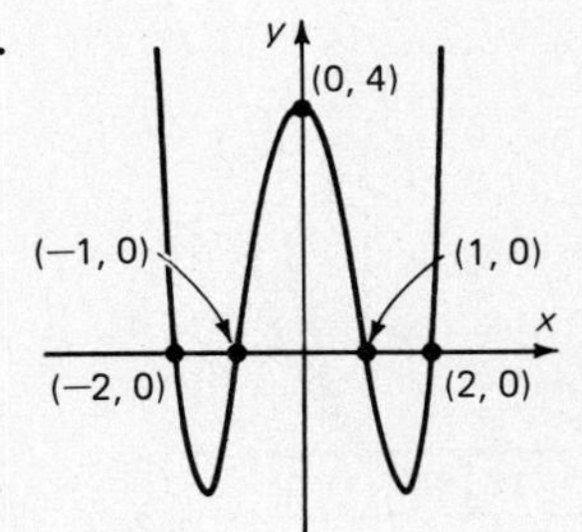

59. $(-1/3)/(x + 1) + (1/3)/(x - 2)$ **61.** $(3/4)/(x + 2) + (5/4)/(x - 2)$ **63.** $x - 1 - 1/(x + 1)$

65.
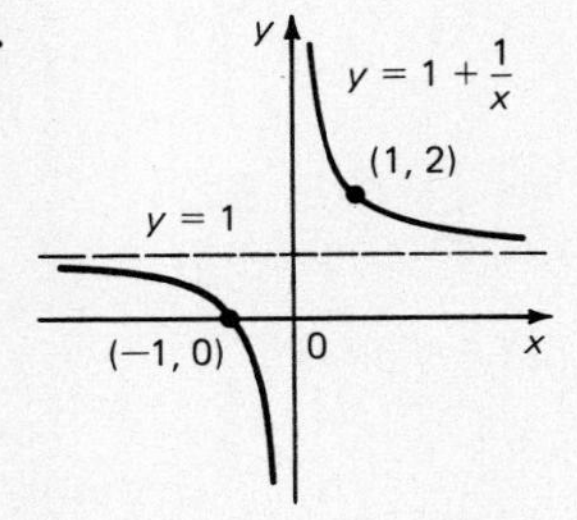

67.
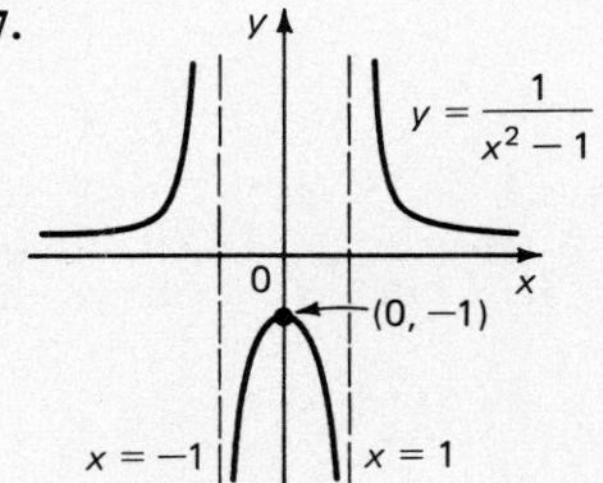

69.
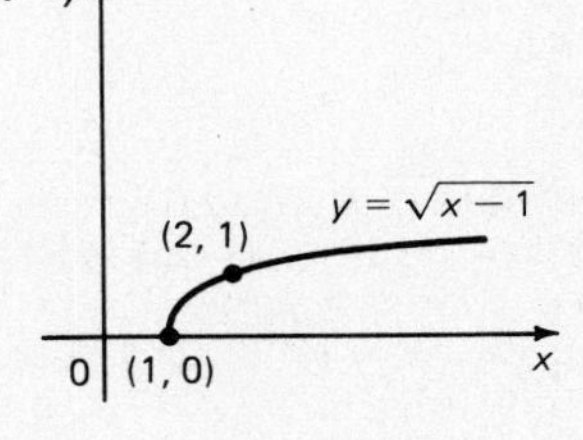

71.

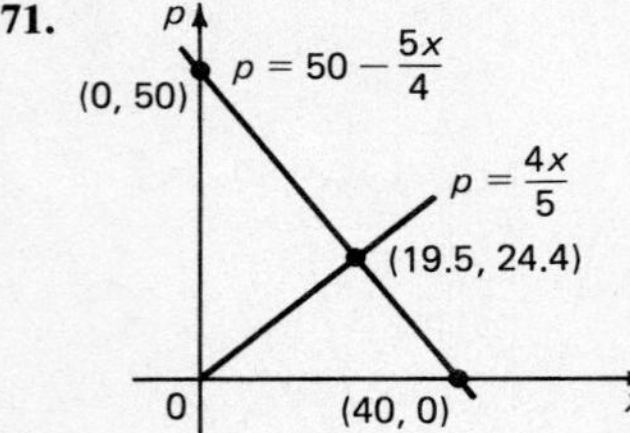

73. 3, −3 **75.** $(3/2)\sqrt{2}$ **77.** 250 yd. by 375 yd. **79.** 1/2 by 1/2

Chapter 3 Miscellaneous Exercises

3. $x_1 + x_2 = 5$, $x_1x_2 = 6$ **5.** $x_1 + x_2 = -11/3$, $x_1x_2 = -4/3$ **7.** $x_1 + x_2 = -19/4$, $x_1x_2 = -5/4$

CHAPTER 4

Exercises 4.1

1. $8x/y$ **3.** x^7x^6 **5.** 1 **7.** x^4y^3 **9.** 9 **11.** $2^{3/4}$ **13.** x/y^3 **15.** $x^{5/6}$ **17.** 3 **19.** 3
21. 1/2 **23.** 8 **25.** 4 **27.** 8 **29.** 1/2 **31.** 1/27 **33.** 10^{-1} **35.** 10^{-2} **37.** 10^1
39. 10^3 **41.** 10,000 **43.** 0.0000001 **45.** 0, 00000367 **47.** 574,000 **49.** 0.000842
51. 1.745×10^{-2} **53.** 4.6×10^0 **55.** 1×10^{-3} **57.** 7.8×10^1
59. $(2 \times 10^2) \times (3.6 \times 10) \times 2.54 = 18.288 \times 10^3 = 1.8288 \times 10^4$
61. $(9.3 \times 10^7) \times (1.609 \times 10^3) = 14.9637 \times 10^{10} = 1.49637 \times 10^{11}$
63. $(8 \times 10^3) \times (3.937 \times 10^{-1}) = 31.496 \times 10^2 = 3.1496 \times 10^3$

Exercises 4.2

1. 7.3891 **3.** 1.6487 **5.** 492.75 **7.** 0.01832 **9.** 4.9530 **11.** 148.41 **13.** 1.1 **15.** 1.0
17. 1.7 **19.** 1.8 **21.** 5.0

23.

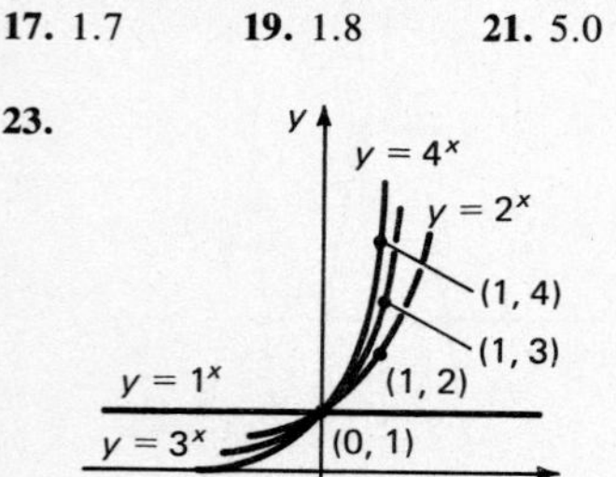

25.

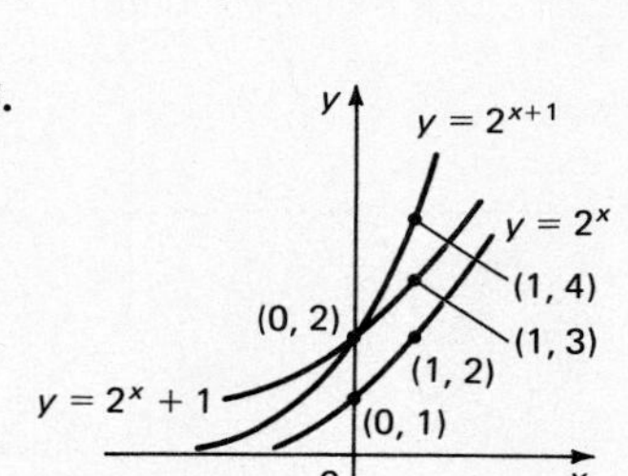

27.

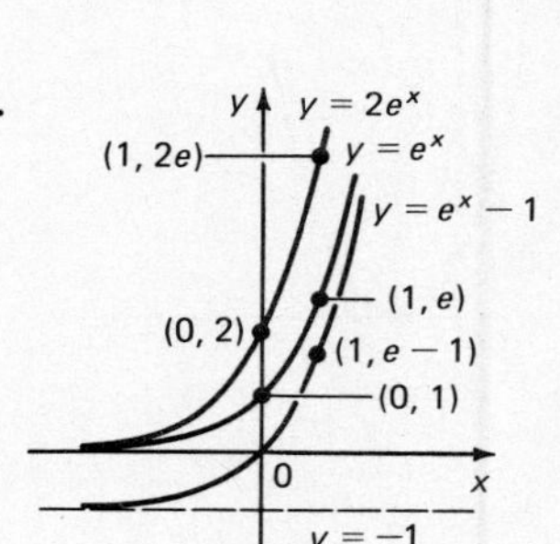

29. 9 min. **31.** 7 min. **33.** $q_o = q(0) = 2^{-10}q = q/1024$, where q is the quantity in a full tank
35. 1.6 (approx.) **37.** No solution, since $e^k > 0$

Exercises 4.3

1. (a) 1000 (b) 3597 (c) 1.08 **3.** 3679 (approx.) **5.** (a) 5.37×10^9 (b) 6.56×10^9 (c) 3.60×10^9
7. (a) 9.9 yr. (b) 495 yr. (c) 231 yr. **9.** 10.0 (approx.) **11.** (a) \$1048.65 (b) \$1268.07 (c) \$1608.01
13. (a) \$1083.00 (b) \$1082.43 (c) \$1081.60 **15.** (a) 10.3 yr. (b) 9.2 yr. (c) 8.7 yr.

Exercises 4.4

1. $2^4 = 16$ **3.** $8^{2/3} = 4$ **5.** $e^0 = 1$ **7.** $10^0 = 1$ **9.** $e^1 = e$ **11.** $\log_2 32 = 5$ **13.** $\log_8 2 = 1/3$
15. $\log_4(1/2) = -1/2$ **17.** $\log_4 64 = 3$ **19.** $\log_3 1 = 0$ **21.** $\log_3 9 = 2$ **23.** $5^2 = 25$
25. $\log_{27} 9 = 2/3$ **27.** $4^3 = 64$ **29.** $\ln 0.223 = -1.5$ **31.** $\ln u = v$

33.

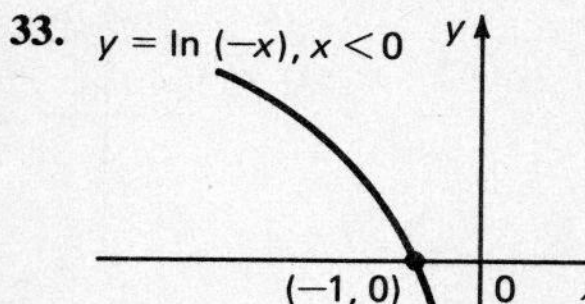

35.

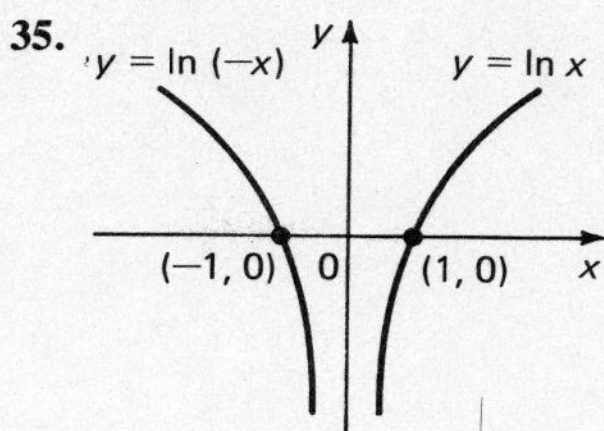

37.

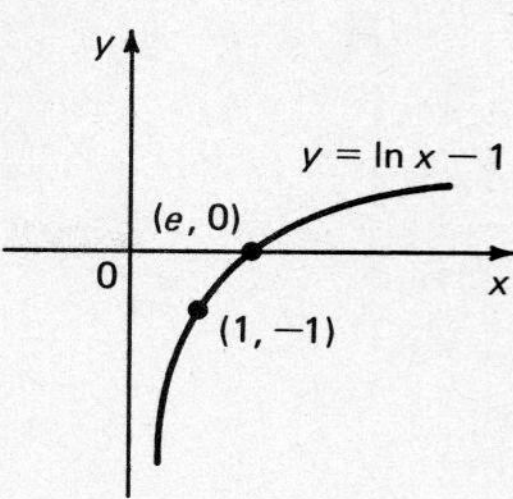

39. $2 \log x + (1/2)\log(1 + x)$ **41.** $\log x + \log y - \log z$ **43.** $(1/2)\log x + (1/2)\log y$
45. $\log x + (1/2)\log y - (3/2)\log z$ **47.** $(1/2)\log x - (1/2)\log y$ **49.** 4 **51.** 3/2 **53.** −1 **55.** 2
57. 0 **59.** 3 **61.** 0.778 **63.** −0.097 **65.** 1.079 **67.** 1.0485 **69.** (a) 0.53 (b) 8.3 **71.** 1.2

Exercises 4.5

1. −2 **3.** 4 **5.** 2 **7.** 32 **9.** 10 **11.** 1.43 **13.** ±2 **15.** 0.697 **17.** 1600
19. 174.6 **21.** 4, −2 **23.** 1.386 **25.** 5.523 **27.** 3.091 **29.** 0.845 **31.** 1.381 **33.** 0.954
35. 1.431 **37.** 1.292 **39.** 1.387 **41.** (a) 6.333 (b) 0.333 **43.** 1.417

Chapter 4 Review Exercises

1. 3 **3.** −1, 2 **5.** −3/2 **7.** 3/4 **9.** ±2 **11.** −1 **13.** 1 **15.** 12
17. $(\ln A - \ln P)/\ln(1 + r)$ **19.** $(\log c - \log p)/\log v$ **21.** $3 \log x + 2 \log y - 5 \log z$
23. $2 \log z + (1/2) \log x - (1/2) \log y$ **25.** $(5/2) \log x - (1/2) \log(x^2 - 1)$ **27.** $\log\sqrt{xy}$ **29.** $\log(xy^3/z^2)$
31. log 3(4)/1.5

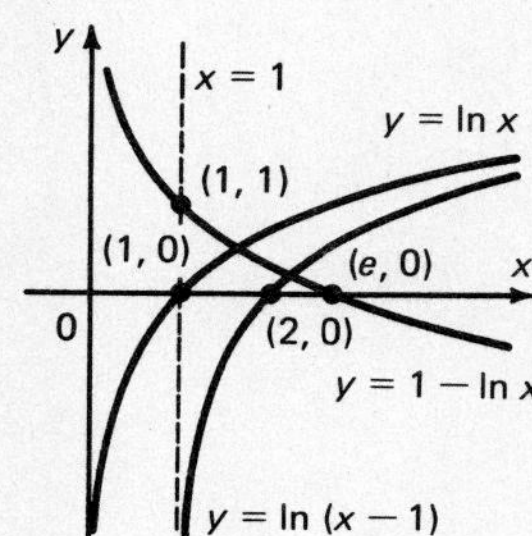

35. 740 grams **37.** (a) 17.3 yr. (b) 11.6 yr. (c) 9.9 yr. (d) 6.9 yr.

39. $1/x$ **41.** $e^{-1/x}$ **43.** $-\ln x$

Chapter 4 Miscellaneous Exercises

3. (a) 0.0349 or 3.5% (b) 5.7×10^7 or 57,000,000 **5.** (a) 9.02×10^9 yr. (b) 1.05×10^{10} yr. **7.** (a) 72 min.
(b) 84 min. **9.** (a) 5 (b) 1.5 (c) $3.1 = 10^{-5}$ **11.** $A = \$1.0832$ or 8.32 (cents per dollar) $\approx$ 8 1/3%

13.

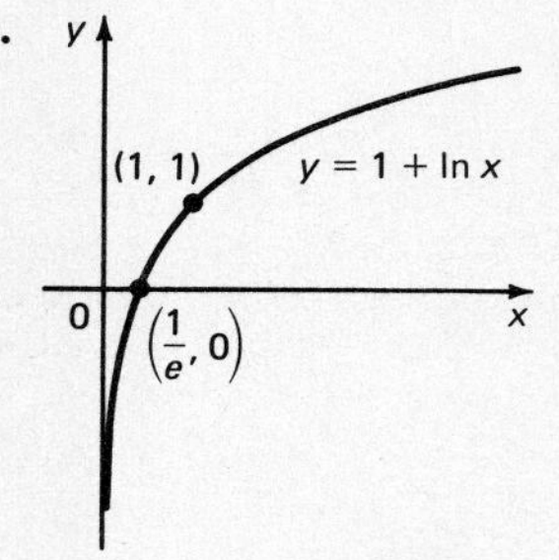

15.

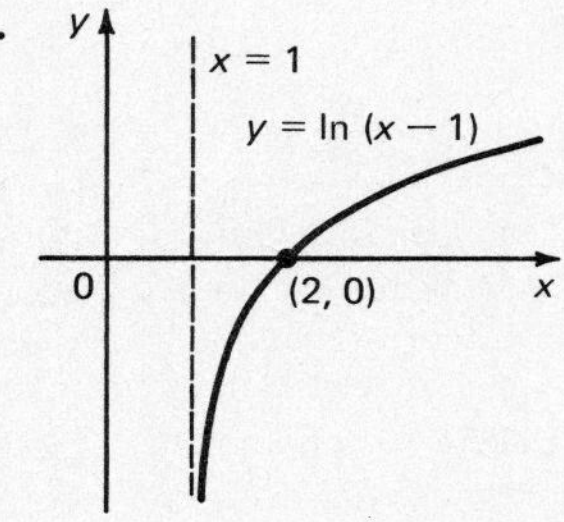

17.

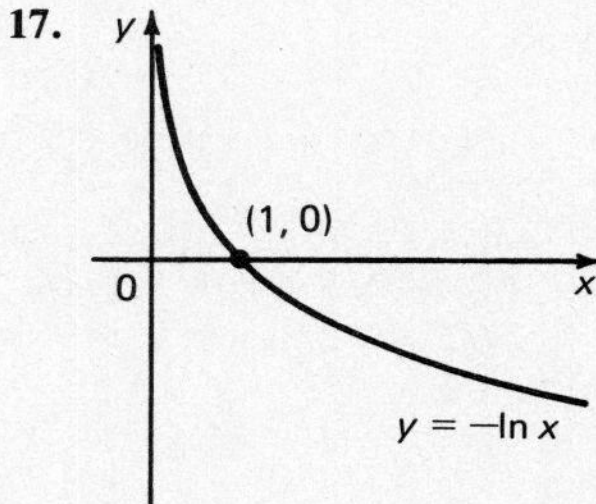

19. $2 \ln x$ **21.** x **23.** $\log K = (\tfrac{1}{2})[\log s + \log(s - a) + \log(s - b) + \log(s - c)]$
25. $\log p = (\log R + \log T) - (\log m + \log v)$ **27.** $p = p_0(1 - \ell z/T_0)^{1/\ell R}$ **29.** $v = (c/p)^{1/n}$ **31.** 13.9 yr.
33. 9.9 yr. **39.** (a) 0 (b) 0.46 **41.** 1.17 **43.** -1.17 **45.** $\ln(y + \sqrt{y^2 + 1})$

CHAPTER 5

Exercises 5.1

1. (a)

u	$\sin u$	$\cos u$
$\pi/2$	1	0
$2\pi/3$	$\sqrt{3}/2$	$-1/2$
$3\pi/4$	$\sqrt{2}/2$	$-\sqrt{2}/2$
$5\pi/6$	$1/2$	$-\sqrt{3}/2$
π	0	-1

(b)

u	$\sin u$	$\cos u$
π	0	-1
$7\pi/6$	$-1/2$	$-\sqrt{3}/2$
$5\pi/4$	$-\sqrt{2}/2$	$-\sqrt{2}/2$
$4\pi/3$	$-\sqrt{3}/2$	$-1/2$
$3\pi/2$	-1	0

(c)

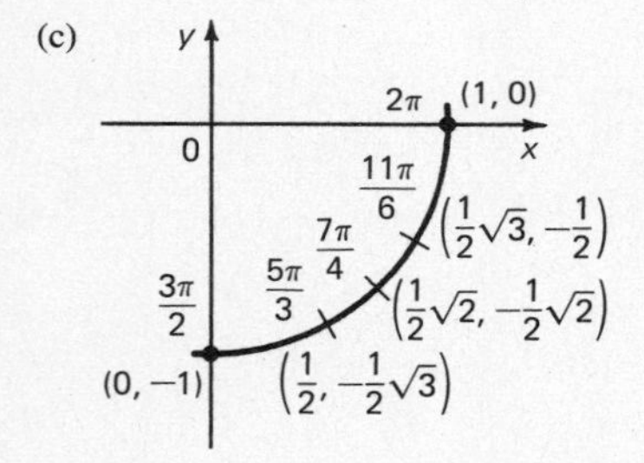

u	$\sin u$	$\cos u$
$3\pi/2$	-1	0
$5\pi/3$	$-\sqrt{3}/2$	$1/2$
$7\pi/4$	$-\sqrt{2}/2$	$\sqrt{2}/2$
$11\pi/6$	$-1/2$	$\sqrt{3}/2$
2π	0	1

3. $\sqrt{2}/2$ **5.** $1/2$ **7.** $-\sqrt{2}/2$ **9.** $-\sqrt{2}/2$ **11.** $\sqrt{2}/2$ **13.** 0 **15.** $1/2$ **17.** I, II
19. III, IV **21.** I **23.** III **25.** $\pi/6$ or $5\pi/6$ **27.** $\pi/6$ or $11\pi/6$ **29.** $\pi/4$ or $7\pi/4$ **31.** 0 or π
33. $\pi/2$ or $3\pi/2$ **35.** $\pm 2\sqrt{2}/3$ **37.** $\pm\sqrt{5}/3$

Exercises 5.2

1.

u	$\sin u$	$\cos u$	$\tan u$	$\cot u$	$\sec u$	$\csc u$
0	0	1	0	Undef.	1	Undef.
$\pi/6$	$1/2$	$\sqrt{3}/2$	$\sqrt{3}/3$	$\sqrt{3}$	$2\sqrt{3}/3$	2
$\pi/4$	$\sqrt{2}/2$	$\sqrt{2}/2$	1	1	$\sqrt{2}$	$\sqrt{2}$
$\pi/3$	$\sqrt{3}/2$	$1/2$	$\sqrt{3}$	$\sqrt{3}/3$	2	$2\sqrt{3}/3$
$\pi/2$	1	0	Undef.	0	Undef.	1

3. $2\sqrt{3}/3$ **5.** 1 **7.** $\sqrt{3}/3$ **9.** -1
11. $\pi/4$ **13.** $\pi/3$ **15.** $0, \pi$
17. $5\pi/6$ **19.** $\pi/2, 3\pi/2, -\pi/2$, for example, since $\tan u = \sin u/\cos u$ and $\cos u = 0$ for each of these numbers **21.** The same as exercise 19, since $\sec u = 1/\cos u$ and $\cos u = 0$ here
23. $1/2$ **25.** $\sqrt{2}/2$ **27.** $1/2$
29. $1/2$ **31.** 1

Exercises 5.4

1. $(\sqrt{2} - \sqrt{6})/4$, or $\sqrt{2}(1 - \sqrt{3})/4$ **3.** $-(\sqrt{2} + \sqrt{6})/4$, or $-\sqrt{2}(1 + \sqrt{3})$ **5.** $(\sqrt{2} + \sqrt{6})/4$, or $\sqrt{2}(1 + \sqrt{3})/4$

Exercises 5.5

1. $(\sqrt{6} + \sqrt{2})/4$, or $\sqrt{2}(\sqrt{3} + 1)/4$ **3.** $(\sqrt{2} - \sqrt{6})/4$, or $\sqrt{2}(1 - \sqrt{3})/4$ **5.** $\sqrt{6} - \sqrt{2}$, or $\sqrt{2}(\sqrt{3} - 1)$
7. $(1 + \sqrt{3})/(1. - \sqrt{3})$, or $-2 - \sqrt{3}$ **9.** $(\sqrt{3} - 1)/(\sqrt{3} + 1)$, or $2 - \sqrt{3}$

Exercises 5.6

1. $\sqrt{2 + \sqrt{3}}/2$ **3.** $\sqrt{2 + \sqrt{3}}/2$ **5.** $\sqrt{2 - \sqrt{2}}/2$ **7.** $\sqrt{2 - \sqrt{2}}/2$ **9.** $\sqrt{2} - 1$ **11.** $\sqrt{2} + 1$

Exercises 5.7

1. $\pi/2, 3\pi/2$ **3.** $3\pi/2$ **5.** $\pi/3, \pi, 5\pi/3$ **7.** $\pi/6, 5\pi/6, 7\pi/6, 11\pi/6$ **9.** $\pi/3, \pi/2, 5\pi/6, 3\pi/2$
11. $\pi/3, \pi, 5\pi/3$ **13.** 0.374, 2.77 (approx.) **15.** $\pi/4, 5\pi/4$ **17.** $0, \pi$ **19.** $0, \pi/6, 5\pi/6, \pi, 7\pi/6, 11\pi/6$
21. $0, \pi/6, 5\pi/6, \pi$ **23.** $\pi/2, 3\pi/2$ **25.** 3.380, 6.045 **27.** 1.89, 5.03 (approx.), $\pi/4, 5\pi/4$

Exercises 5.8

1. Period, 2π
Amplitude, 2
Phase shift, 0

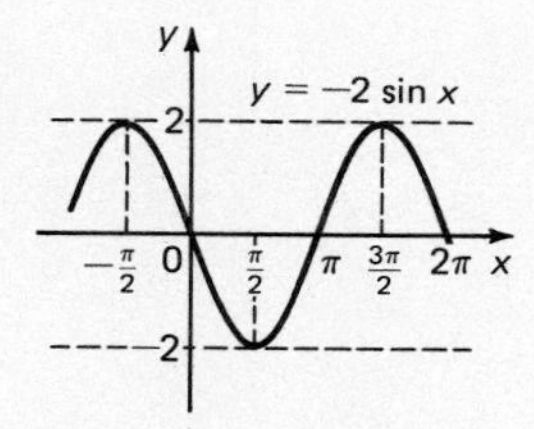

3. Period, 2π
Amplitude, 1/2
Phase shift, 0

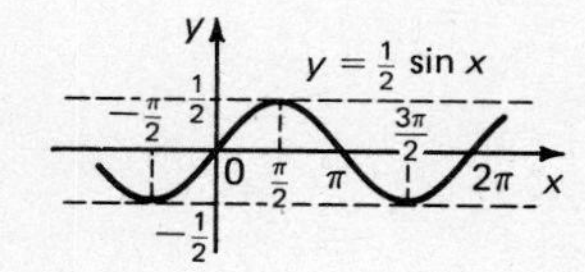

5. Period, $2\pi/3$
Amplitude, 1
Phase shift, 0

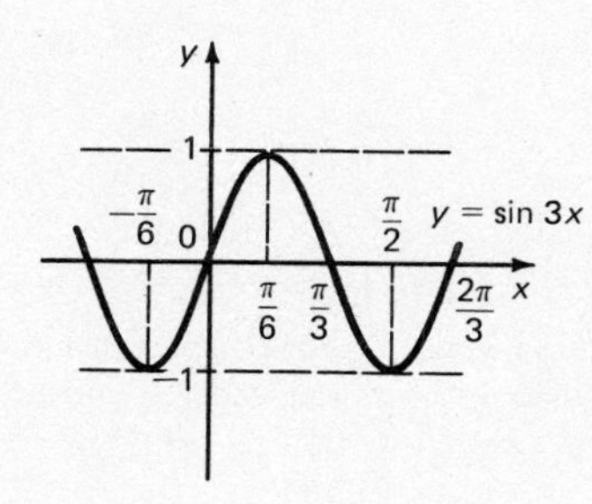

7. Period, 2π
Amplitude, 1
Phase shift, 0

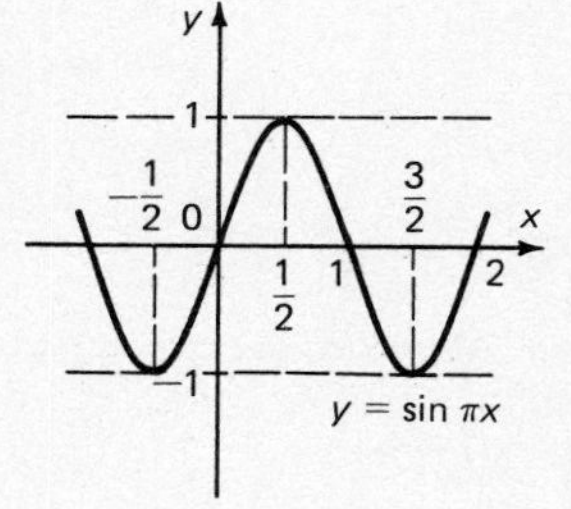

9. Period, 2π
Amplitude, 1
Phase shift,
$\pi/2$ to the left

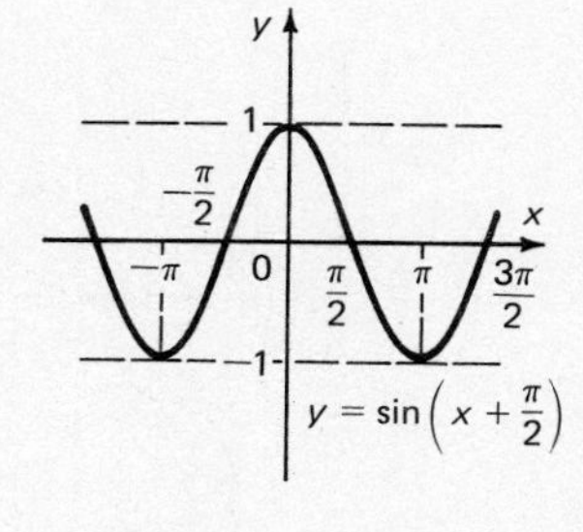

11. Period, 2π
Amplitude, 1
Phase shift,
$\pi/6$ to the left

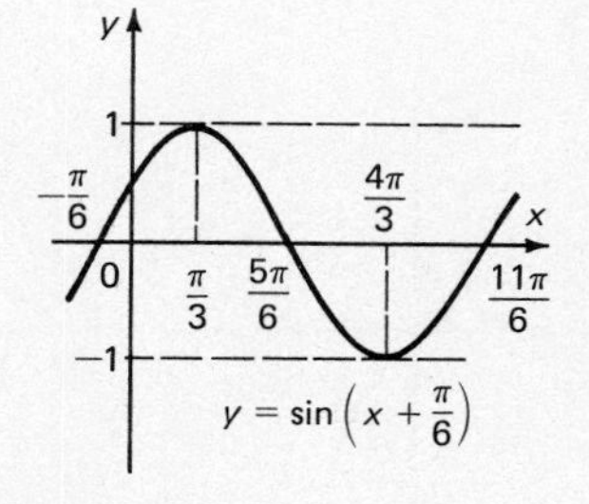

13. Period, 2π
Amplitude, 2
Phase shift,
$\pi/3$ to the left

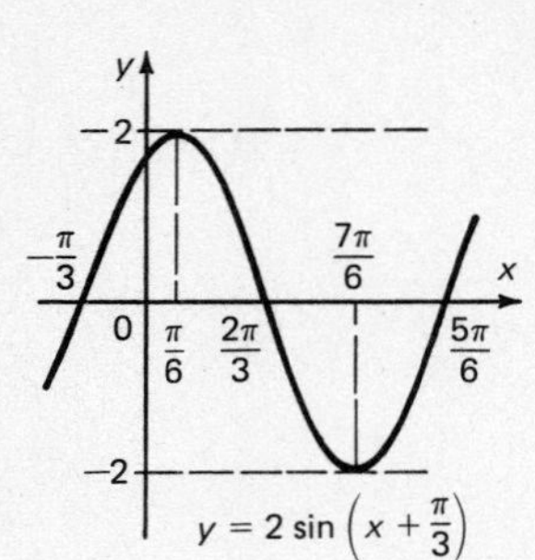

15. Period, 2π
Amplitude, 1
Phase shift, 0

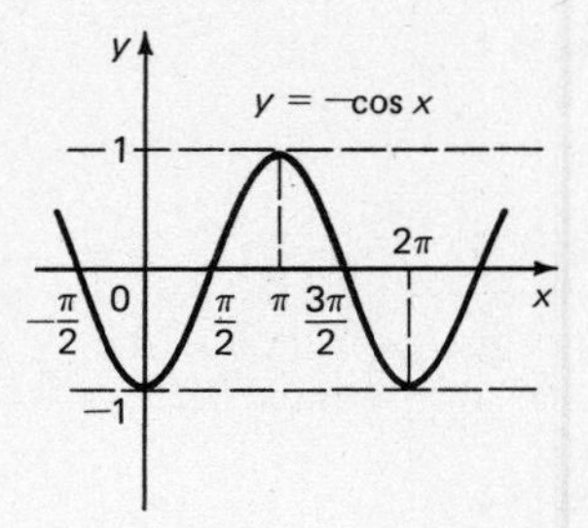

17. Period, $2\pi/3$
Amplitude, 1
Phase shift, 0

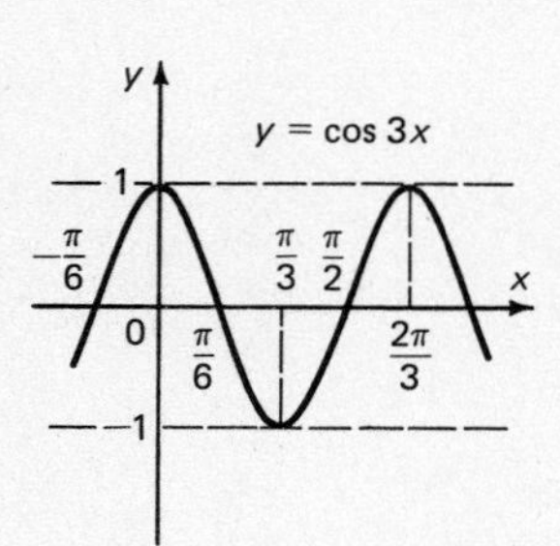

19. Period, 2π
Amplitude, 1
Phase shift,
$\pi/4$ to the right

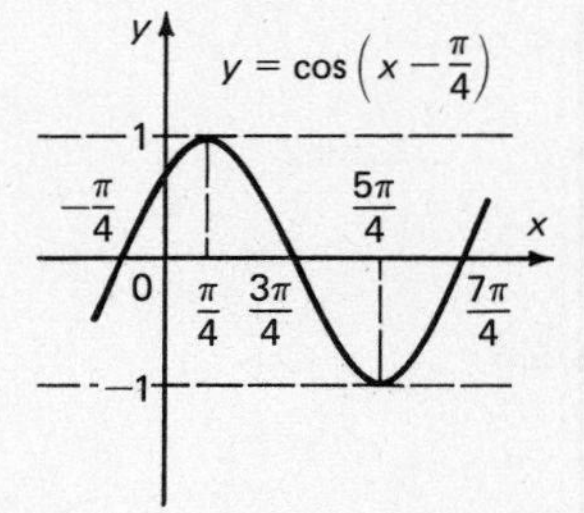

21. Period, 2π
Amplitude, 2
Phase shift,
$\pi/2$ to the left

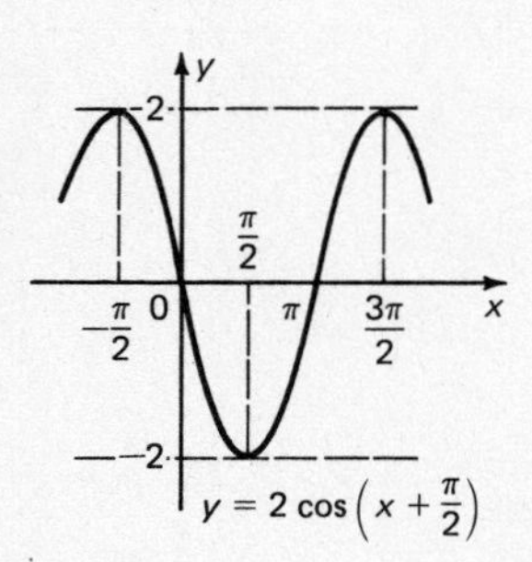

23. $-2, 2$ **25.** $-1, 1$ **27.** $-1/2, 1/2$ **29.** $0, 2$

Exercises 5.9

1. (a) $y = \sin 2t$; amplitude, 1; period, π; frequency, $1/\pi$ (b) $y = \sin 3t$; amplitude, 1; period $2\pi/3$; frequency, $3/2\pi$ (c) $y = \sin 4t$; amplitude, 1; period, $\pi/2$; frequency, $2/\pi$ **3.** $y = 2\sin(t + \pi/3)$; amplitude, 2; period, 2π; frequency, $1/2\pi$ **5.** $8/\pi^2$ (feet) **7.** (a) period, $2\pi\sqrt{m/k}$; frequency, $(1/2\pi)\sqrt{k/m}$, (b) $1/\pi^2$ **9.** $I(t) = 15 \sin 120\pi t$

Exercises 5.10

1.

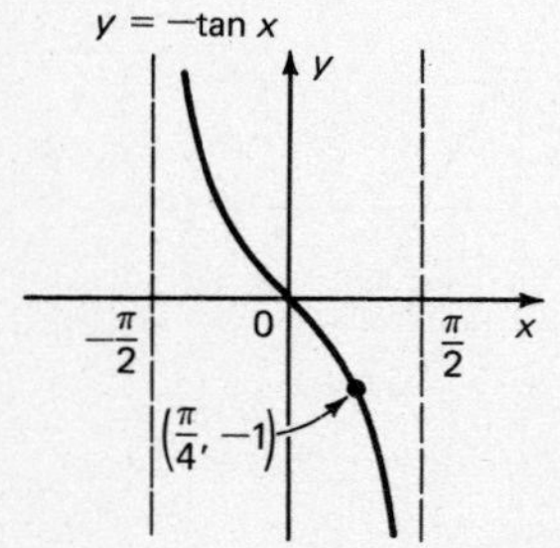

3.

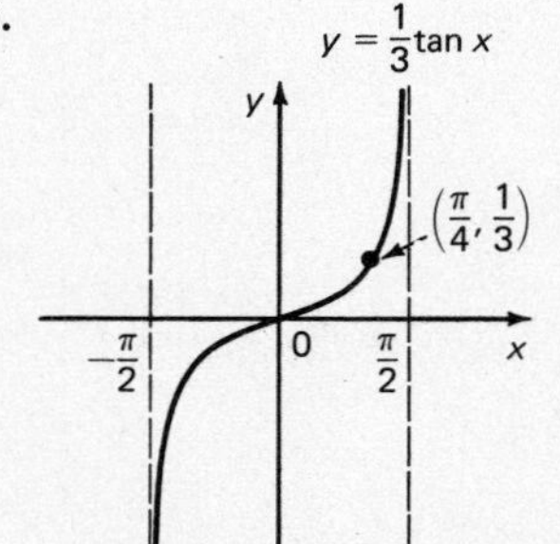

5.

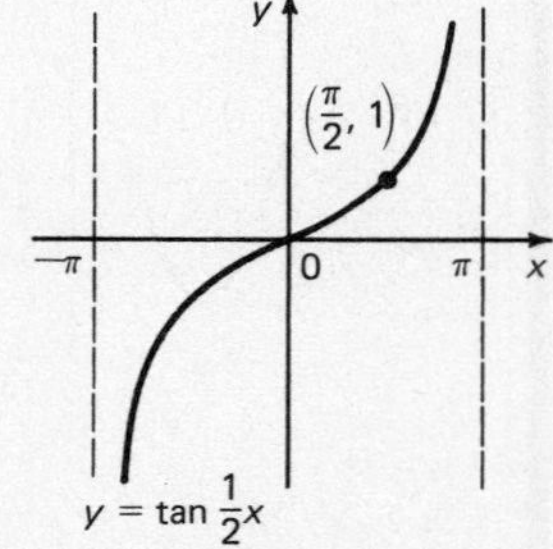

7.

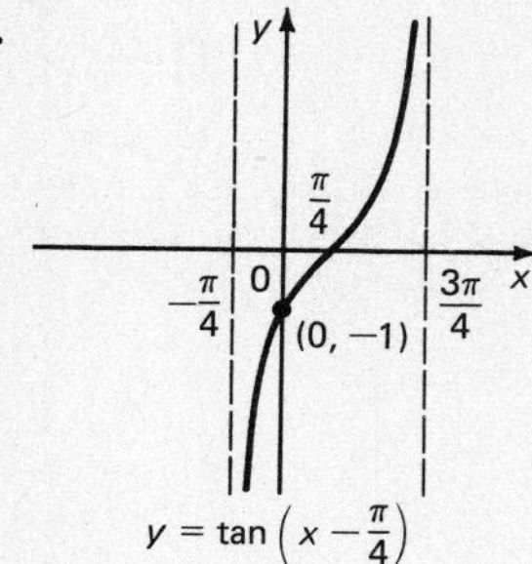

9.

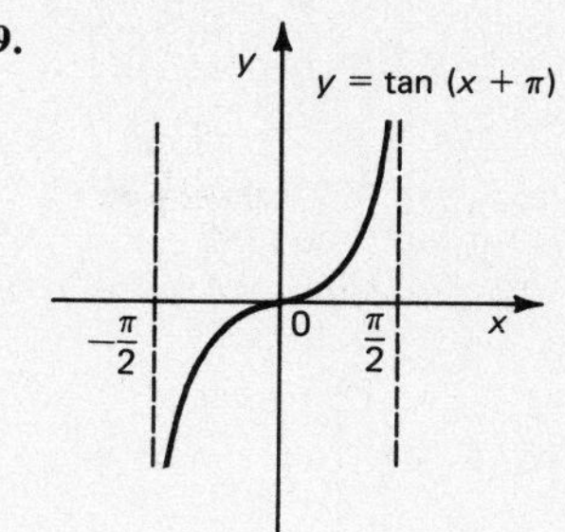

11.

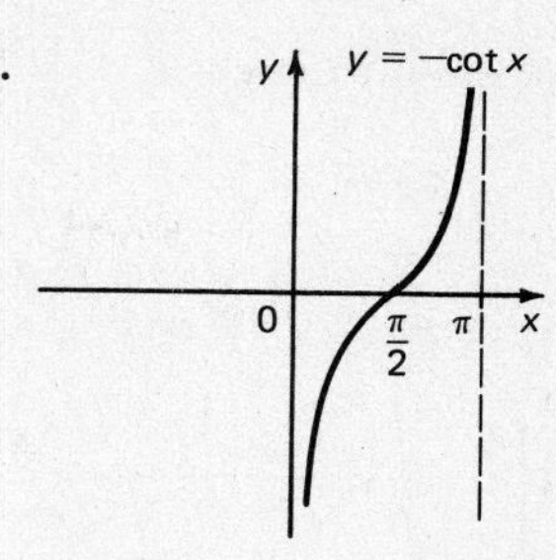

13.

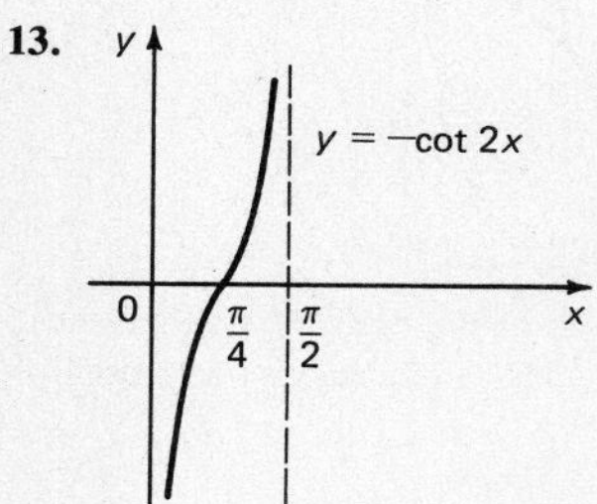

15.

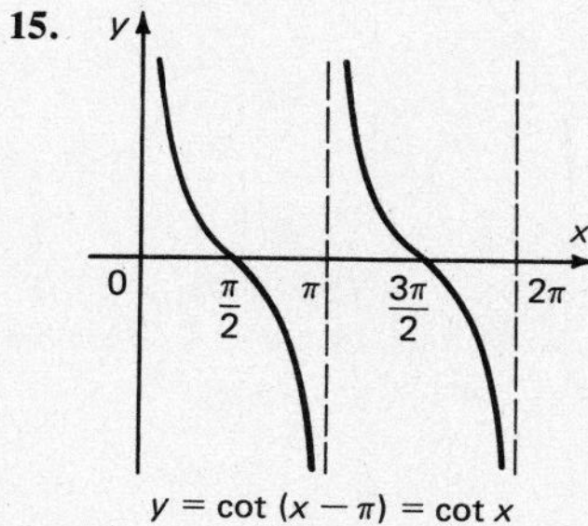

17.

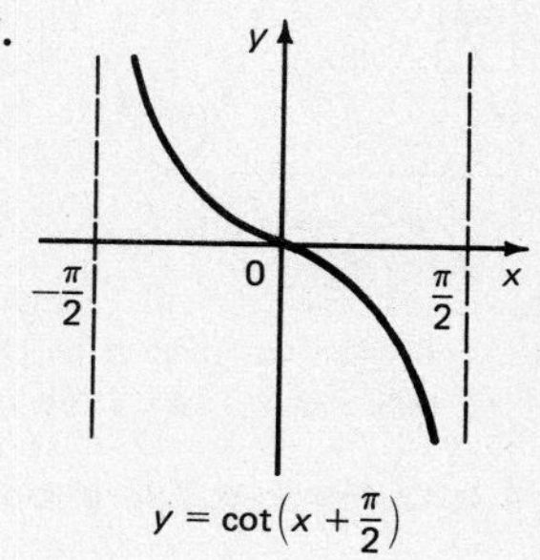

Exercises 5.11

1. $\pi/3$ **3.** $\pi/2$ **5.** $\pi/2$ **7.** $\pi/4$ **9.** 0 **11.** $-\pi/6$ **13.** No solution **15.** $3\pi/4$ **17.** No solution **19.** $\pi/6$ **21.** arcsin 1/4 **23.** $(\cos y)/2$ **25.** $\arccos(-1/3)$ **27.** $(\sin y)/3$ **29.** $\sqrt{3}/2$ **31.** 4/3 **33.** $1/\sqrt{5}$ **35.** x **37.** $1/x$ **39.** x **41.** $1/x$ **43.** $2x\sqrt{1 - x^2}$ **45.** $2x$

47.

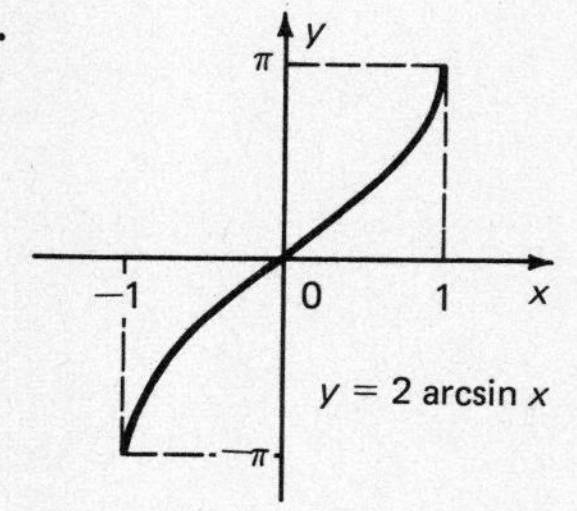

49.

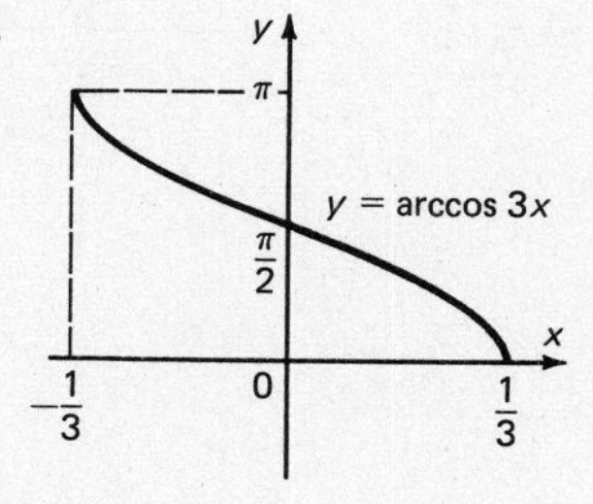

51.

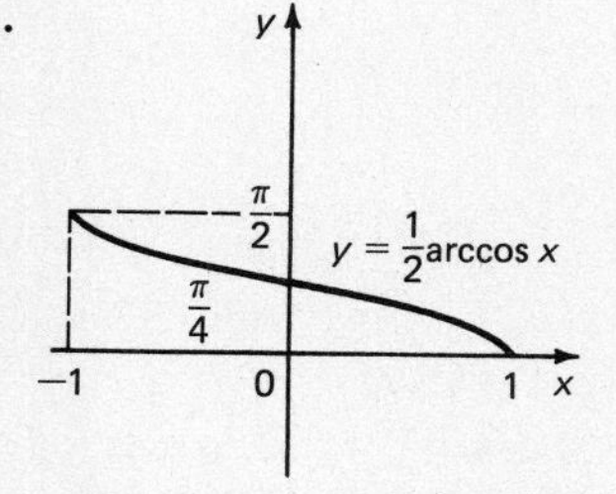

53.

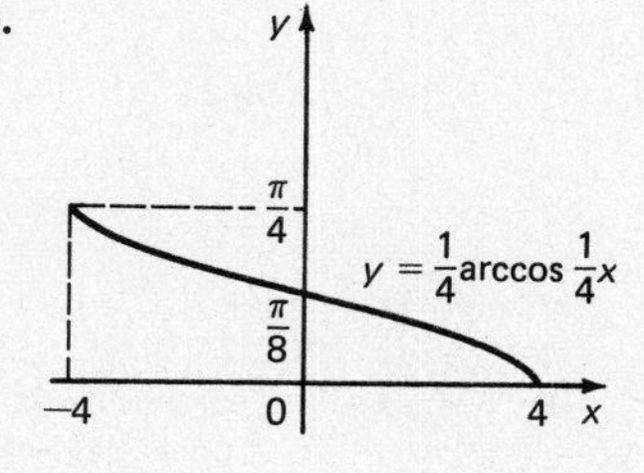

Exercises 5.12

1. 0 **3.** 0.479 **5.** 1 **7.** 0.542 **9.** 0 **11.** 0.965

Chapter 5 Review Exercises

1. $\sqrt{2}/2$ **3.** $\sqrt{3}/2$ **5.** 1 **7.** $-\sqrt{3}$ **9.** -1 **11.** $\pi/3$, $2\pi/3$ **13.** $\pi/6$, $11\pi/6$ **15.** 0, π **17.** $3\pi/4$, $7\pi/4$ **19.** $5\pi/4$, $7\pi/4$ **21.** (a) $\cos u = \sqrt{3}/2$, $\tan u = \sqrt{3}/3$, $\cot u = \sqrt{3}$, $\sec u = 2/\sqrt{3}$, $\csc u = 2$ (b) $\sin u = 2/\sqrt{13}$, $\cos u = -3/\sqrt{13}$, $\cot u = -3/2$, $\sec u = -\sqrt{13}/3$, $\csc u = \sqrt{13}/2$ (c) $\sin u = -\sqrt{3}/2$, $\cos u = -1/2$, $\tan u = \sqrt{3}$, $\cot u = \sqrt{3}/3$, $\csc u = -2/\sqrt{3}$ **31.** $\pi/6$, $5\pi/6$ **33.** $\pi/3$, π, $5\pi/3$ **35.** π, $2\pi/3$

37.

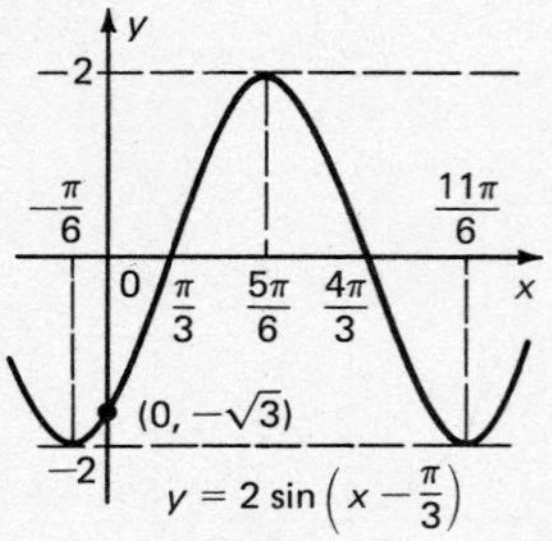

39.

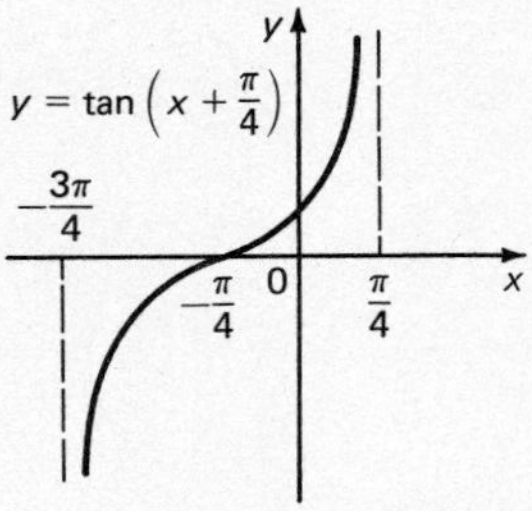

41.

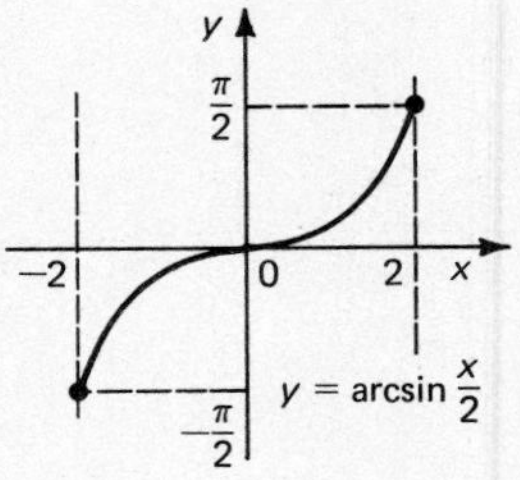

49. $2\pi/3$ **51.** $\pi/4$ **53.** $-\pi/4$ **63.** -0.91 **65.** -0.91 **67.** If the center of the wheel is taken at the origin and $t = 0$ when the given point crosses the positive axis, then $y = (1/2)\sin(8/5)t$. Period, $5\pi/4$; frequency, $4/5\pi$; amplitude, 1/2 **69.** Period, $2\pi\sqrt{2}$; frequency, $1/2\pi\sqrt{2}$; amplitude, 1/4

Chapter 5 Miscellaneous Exercises

1. x **3.** $\ln(\sin x)$ **5.** $e^{\sin x}$ **7.** 0 **9.** 0

17.

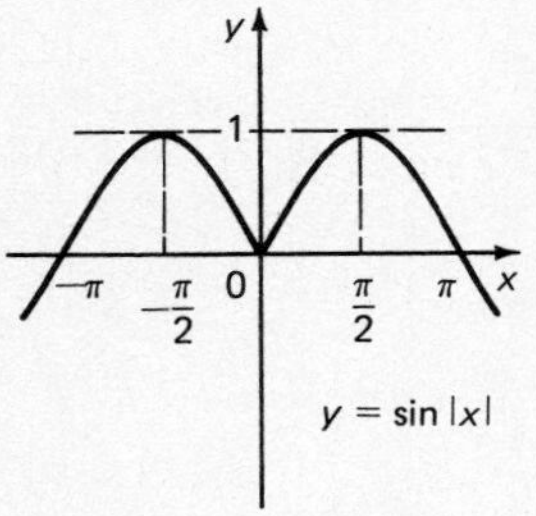

19.

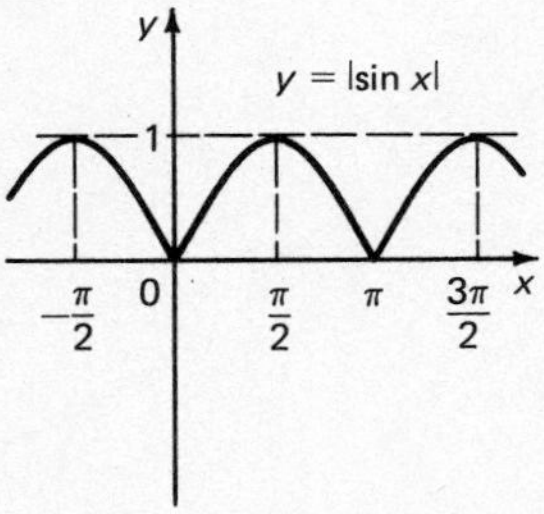

21. $2\sin x$, $\sin 2x$
23. Yes
25. $f[g(x)] = 3\cos x$. Yes

CHAPTER 6

Exercises 6.1

1.

30°	45°	60°	90°	120°	135°	150°	180°
$\pi/6$	$\pi/4$	$\pi/3$	$\pi/2$	$2\pi/3$	$3\pi/4$	$5\pi/6$	π

3. (a) 5 in. (b) 37.5 sq. in. **5.** (a) 2.5 radians (b) 143° (approx.) **7.** 0.5°

Exercises 6.2

1.

0	$\pi/6$	$\pi/4$	$\pi/3$	$\pi/2$	$2\pi/3$	$3\pi/4$	$5\pi/6$	π
0	1/2	$\sqrt{2}/2$	$\sqrt{3}/2$	1	$\sqrt{3}/2$	$\sqrt{2}/2$	1/2	0
1	$\sqrt{3}/2$	$\sqrt{2}/2$	1/2	0	$-1/2$	$-\sqrt{2}/2$	$-\sqrt{3}/2$	-1

3. $\sqrt{3}$ **5.** 1/2 **7.** $-\sqrt{3}/2$ **9.** 2 **11.** $-\sqrt{3}/3$ **13.** 0.9902 **15.** 0.9001 **17.** 0.6841 **19.** 0.7400 **21.** 5.164 **35.** 0° **37.** 30°, 210° **39.** 60°, 300° **41.** 60°, 300° **43.** 30°, 210° **45.** $\pi/6$, $5\pi/6$ **47.** 1.12, 2.45, 3.83, 5.16 (all approx.) **49.** $7\pi/6$, $11\pi/6$ **51.** $\pi/6$, $5\pi/6$, $7\pi/6$, $11\pi/6$ **53.** $\pi/6$, $5\pi/6$, $3\pi/2$

Exercises 6.3

1. $a \approx 5.74$, $b \approx 8.19$, $\beta = 55°$ **3.** $\alpha \approx 35.3°$, $\beta \approx 54.7°$, $c \approx 29.4$ **5.** $c = 24.3$, $\alpha \approx 32.9°$, $\beta \approx 57.1°$ **7.** $\alpha = 27°$, $a \approx 3.7$, $c \approx 8.2$ **9.** $\beta = 27°40'$, $a \approx 11.7$, $c \approx 13.2$ **11.** 13.0 ft., 32.5°, 57.5° **13.** 75.1 ft **15.** 42.1 m. **17.** (a) 0.438 mile, or 2300 ft. (b) 3800 ft. (c) 7800 ft. (d) 4700 ft.

Exercises 6.4

1. $\gamma = 95°$, $a \approx 660$, $c \approx 750$ **3.** $\alpha = 27°10'$, $a \approx 11.4$, $b \approx 16.8$ **5.** $\beta = 109°$, $b \approx 17.0$, $c \approx 13.8$ **7.** $\alpha \approx 34.9°$, $\gamma \approx 75.1°$, $c \approx 47.3$ **9.** $\alpha \approx 33.3°$, $\gamma \approx 99.5°$, $c \approx 38.1$ **11.** $\beta = 113°50'$, $b \approx 30.8$, $c \approx 11.8$ **13.** Case (e) **15.** Case (a) **17.** Case (d) **19.** Diagonals, 35.2 in. and 26.6 in.; side, 19.9 in. (approx.) **21.** 67 ft. **23.** 386 sq. m.

Exercises 6.5

1. $c = 7.06$, $\alpha = 56.4°$, $\beta = 87.5°$ (all approx.) **3.** $c \approx 23.8$, $\alpha \approx 63.3°$, $\beta \approx 81.8°$ **5.** $b \approx 10.4$, $\alpha \approx 14.7°$, $\gamma \approx 118.3°$ **7.** $\alpha \approx 48.7°$, $\beta \approx 72.5°$, $\gamma \approx 58.8°$ **9.** $\alpha \approx 29.9°$, $\beta \approx 110.5°$, $\gamma \approx 39.6°$ **11.** 1105 ft. **13.** 42.9 ft. by 20.8 ft. **15.** 151 ft.

Exercises 6.6

1. 36.1° with the direction of the stream; 2880 ft. **3.** 74.7 mph, N 1.7°W **5.** 31°, 29 ft./sec. **7.** 5.2 lb., 8.6 lb. **9.** 8000 ft. **11.** 68 lb.

Exercises 6.7

1.

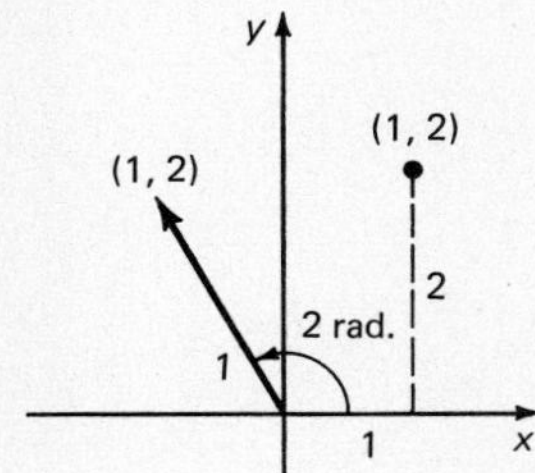

3–15.

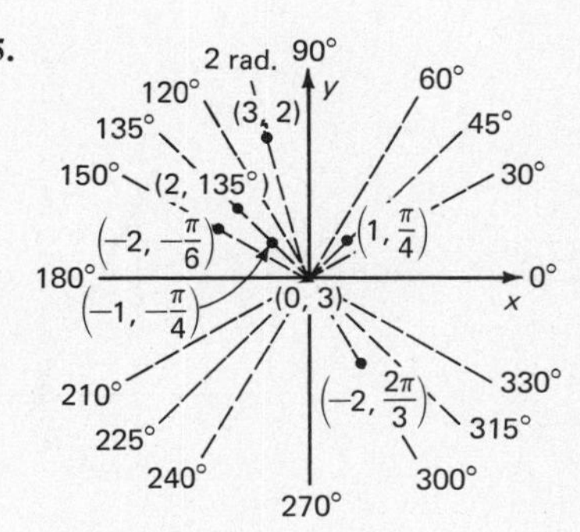

17. (3, 30°); (3, 390°), (3, −330°), (−3, 210°), (−3, −150°). (1, π/4); (1, 9π/4), (1, −7π/4), (−1, 5π/4), (−1, −3π/4). (−1, π/2); (1, 3π/2), (1, −π/2), (−1, 5π/2), (1, 3π/2). (2, 135°); (2, 495°), (2, −225°), (−2, 315°), (−2, −45°). (−1, 300°); (1, 120°), (1, −240°), (−1, −60°), (−1, 660°) **19.** $(\sqrt{2}/2, \sqrt{2}/2)$ **21.** $(-\sqrt{3}/2, -1/2)$ **23.** $(-3\sqrt{3}/2, 3/2)$ **25.** (−2, 0) **27.** $(-1/2, -\sqrt{3}/2)$ **29.** $y = x$ **31.** $x^2 + y^2 = x$ **33.** $(\sqrt{2}, 45°)$ **35.** (1, 150°) **37.** (2, 330°) **39.** (2, 315°) **41.** (1, 45°) **43.** $r \cos \theta + 1 = 0$ **45.** $\theta = 0$ **47.** $\sqrt{3}$ **49.** $\sqrt{19}$ **51.** $\sqrt{37}/2$

Chapter 6 Review Exercises

1. 510° = 17π/6 rad. = 17/12 rev. **3.** 140° = 7π/9 rad. = 7/18 rev. **5.** 1.4 rev. = 504° = 14π/5 rad. **7.** 8π/3 rad., 480° **9.** $-\sqrt{2}/2$ **11.** $-1/\sqrt{3}$ **13.** $-2/\sqrt{3}$ **15.** 90° **17.** 45°, 135° **19.** 8.2 in. **21.** 47.7° **23.** 218 ft. **25.** 13.9 m. by 7.0 m. **27.** $\sqrt{19}$, 23.4° **29.** $(-\sqrt{3}, -1)$ **31.** (3, 180°) **33.** $(-1, \sqrt{3})$ **35.** $\theta = 45°$ **37.** $r \cos^2\theta = \sin \theta$, or $r \cot \theta \cos \theta = 1$ **39.** $y = 1$ **41.** $y(x^2 + y^2) = x^2$

CHAPTER 7

Exercises 7.1

1. $x - y + 4 = 0$

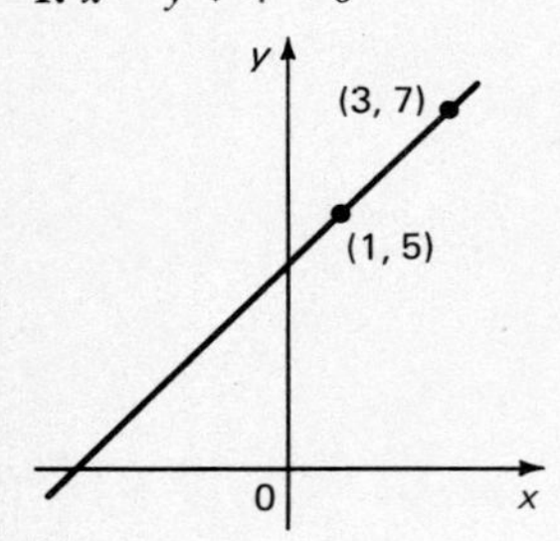

3. $5x - y - 14 = 0$

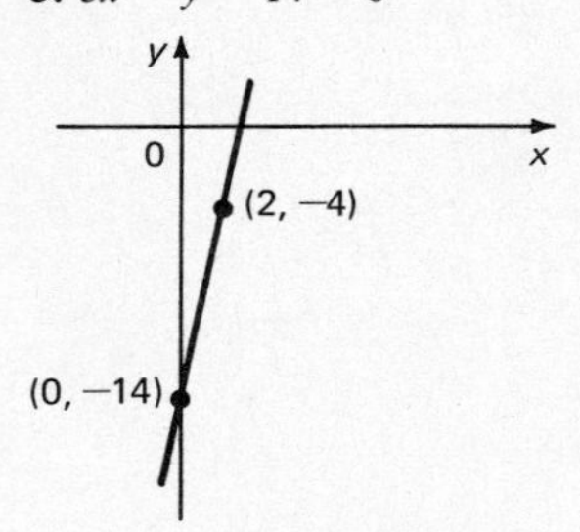

5. $x + y - 3 = 0$

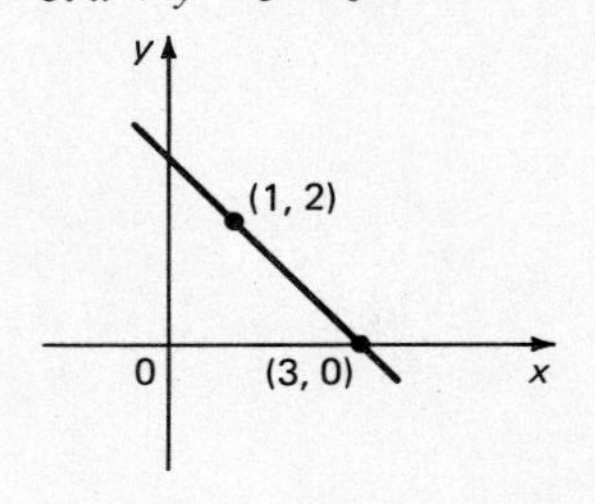

7. $x - y + 2 = 0$

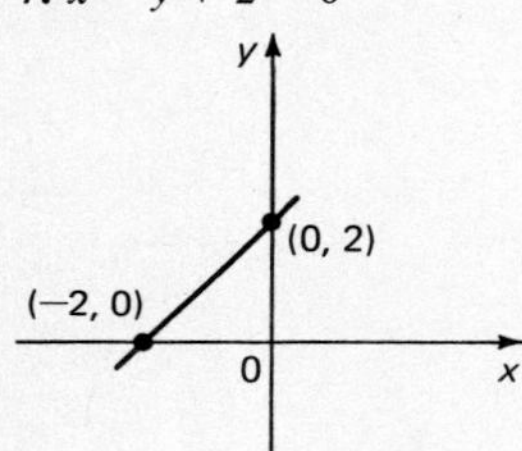

9. $2x - y = 0$

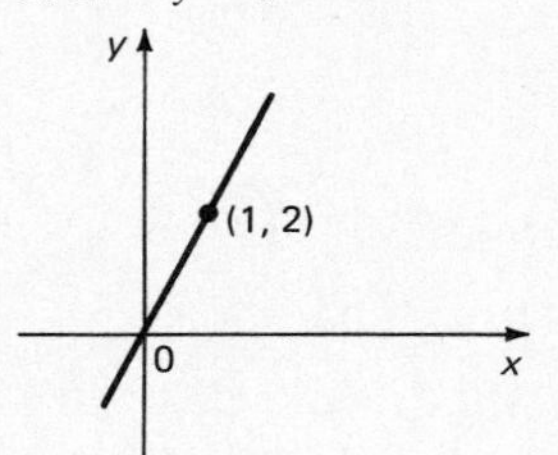

11. $x - y\sqrt{3} + (1 + 2\sqrt{3}) = 0$

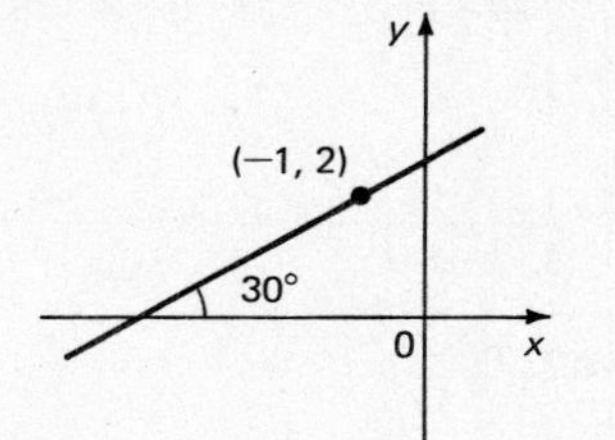

13. $y - 3 = 0$

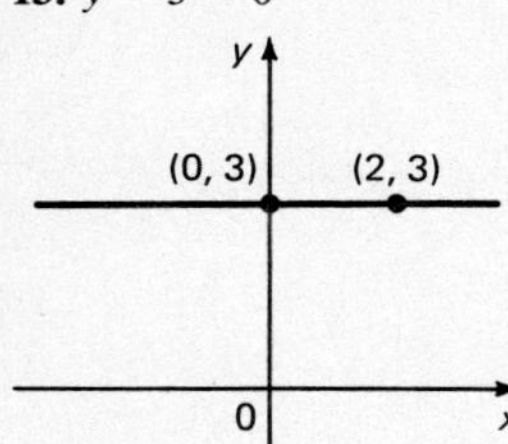

15.

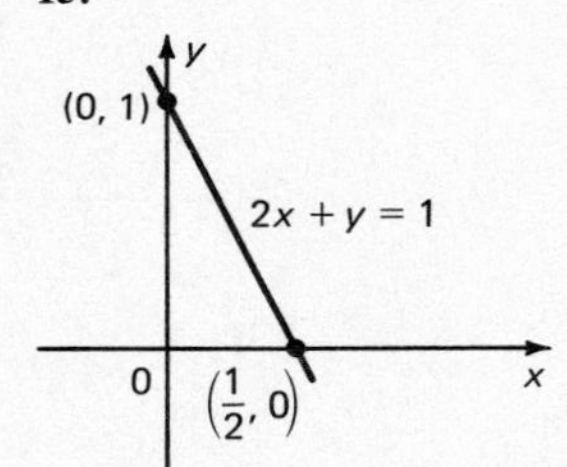

17.

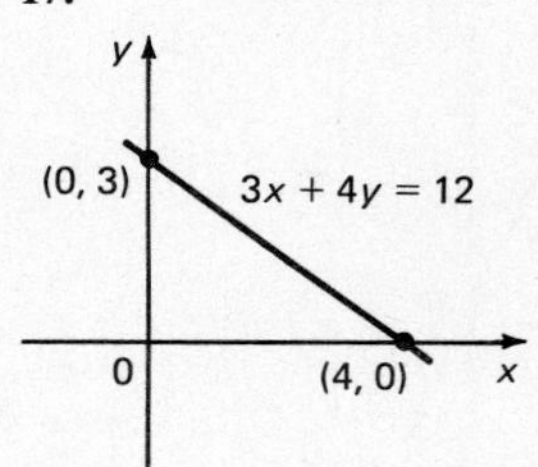

19. $y = 4x - 7$; slope, 4; inclination, arctan $4 \approx 76°$ **21.** $y = (-2/5)x + (3/5)$; slope, $-2/5$; inclination, arctan $(-2/5) \approx 158°$ **23.** $m_1 = m_2 = -2/5$; parallel **25.** $m_1 = 3/5$, $m_2 = -2$; neither parallel nor perpendicular **27.** $m_1 = m_2 = -3/7$; parallel **29.** $m_1 = -4/3$, $m_2 = 3/4$; perpendicular **31.** None of these **33.** None of these **35.** Rectangle **37.** Parallel, $3x - y - 10 = 0$ Perpendicular, $x + 3y + 10 = 0$ **39.** Parallel, $x - 3y + 3 = 0$ Perpendicular, $3x + y - 21 = 0$ **41.** $x + 3y - 3 = 0$ **43.** $x - y + 2 = 0$

45.

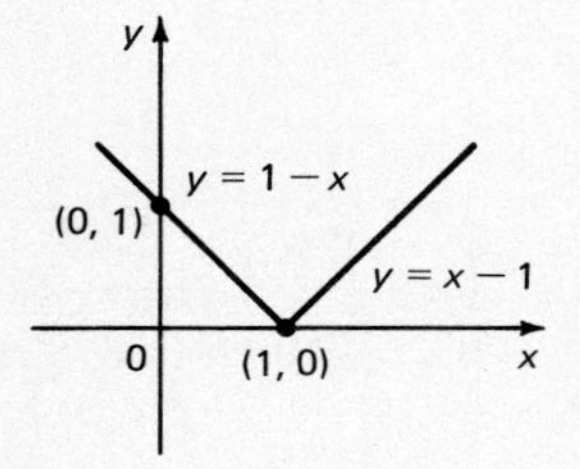

47. $2x + 3y - 5 = 0$, $3x - 2y + 12 = 0$, $x - 5y + 4 = 0$ **49.** $5x + y + 7 = 0$, $4x - 7y + 16 = 0$, $x + 8y - 9 = 0$ **51.** (a) \$2 (b) \$2500 **53.** 7.5 in. **57.** 40.6° **59.** 26.6°

Exercises 7.2

1. $(x - 2)^2 + (y + 3)^2 = 25$

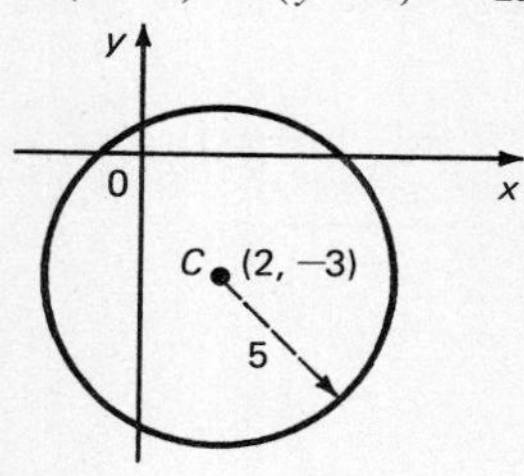

3. $(x + 1/2)^2 + (y - 2)^2 = 1/4$

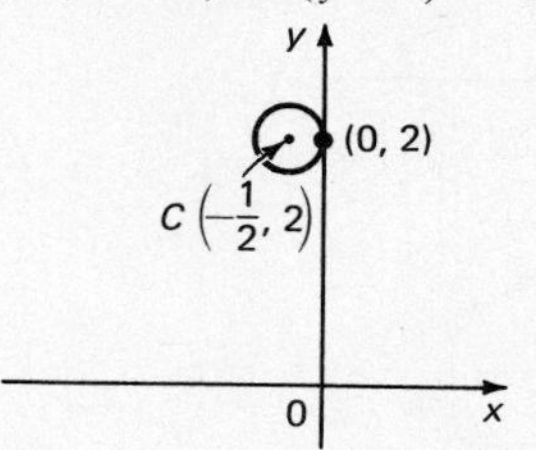

5. $(x - 2)^2 + (y - 3)^2 = 9$

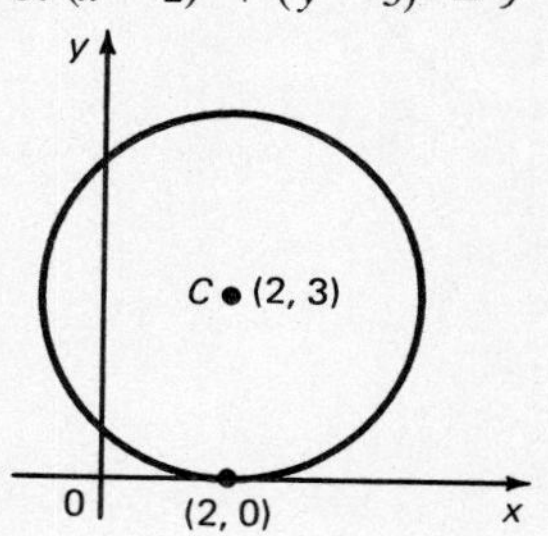

7. $(x - 3/2)^2 + (y + 1/2)^2 = 25/2$

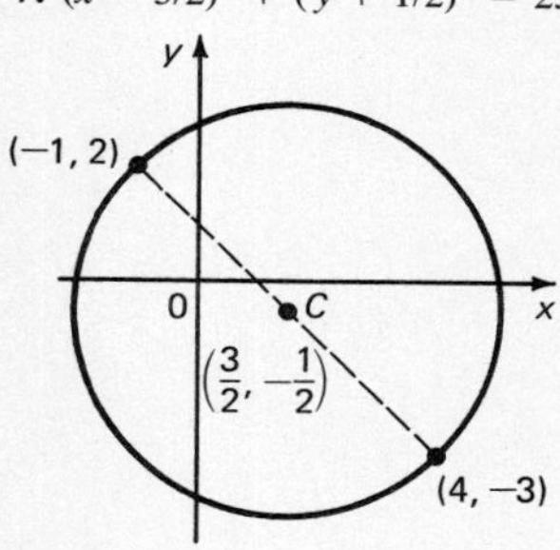

9. $(x - 2)^2 + (y - 1)^2 = 1$

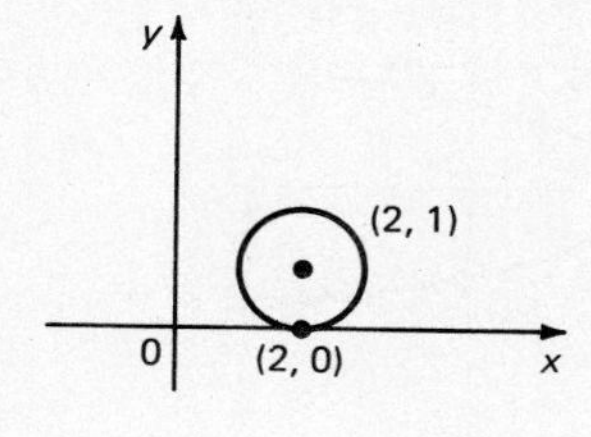

11. $(x - 2)^2 + (y + 3)^2 = 16$
13. $(x - 3)^2 + (y - 2)^2 = 0$, "point circle"
15. $(x - 3)^2 + (y - 0)^2 = 20$
17. $(x - 1/4)^2 + (y + 3/4)^2 = -7/8$, no graph
19. $3x - 4y + 25 = 0$ **21.** $x = -3$
23. $y = 2$

Exercises 7.3

1.

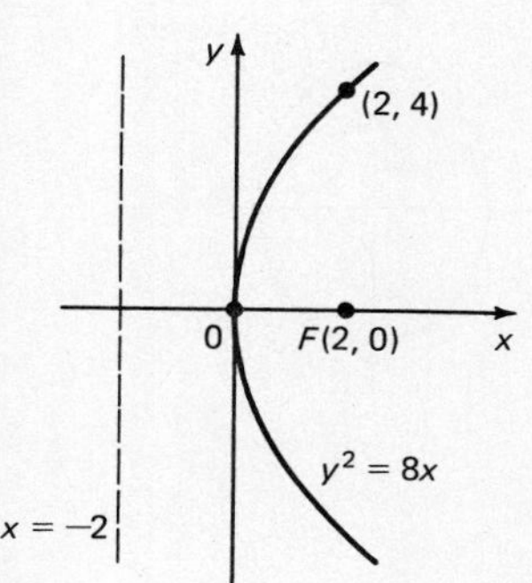

3.

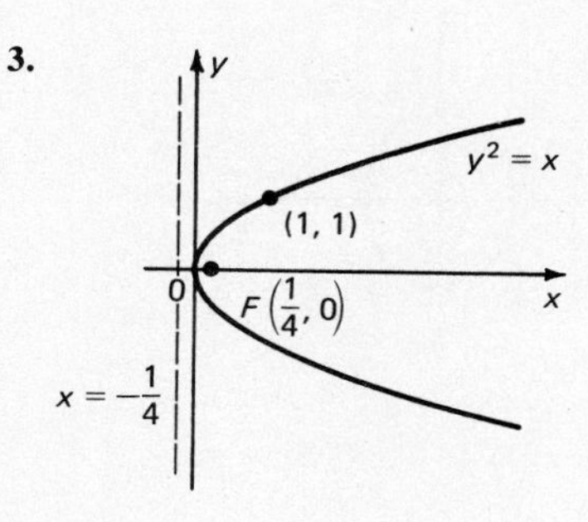

5.

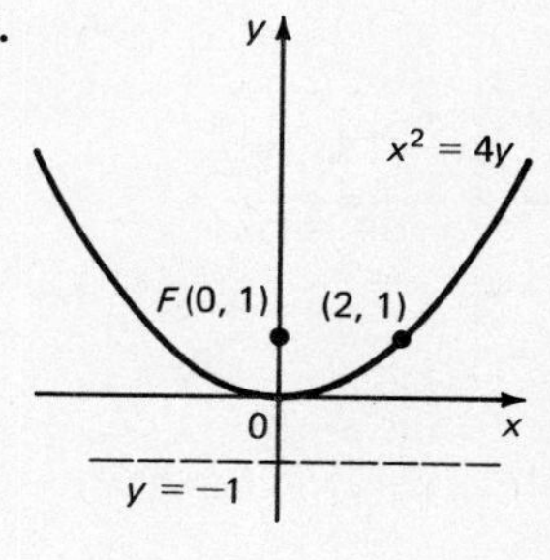

7.

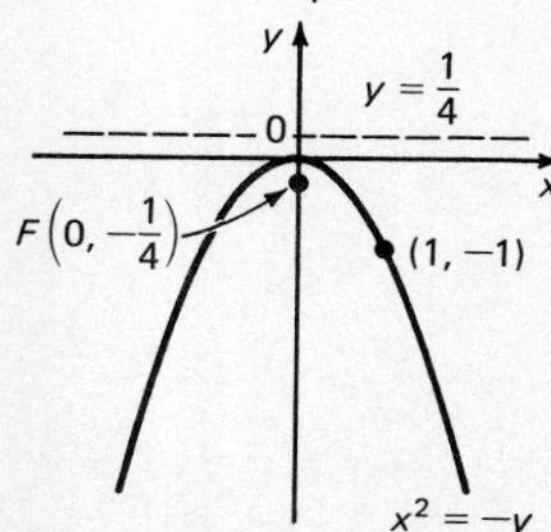

9. $x^2 = 4y$

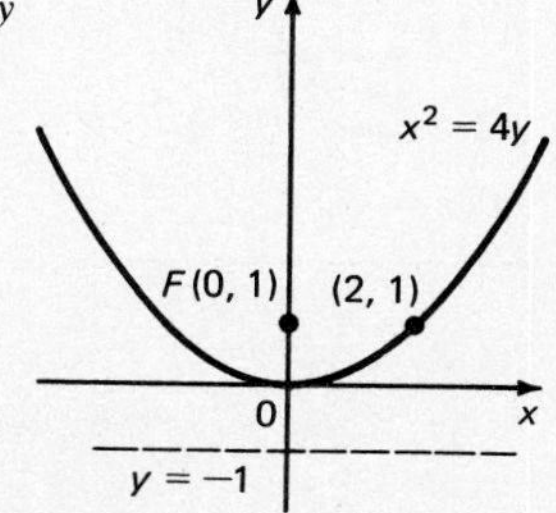

11. $y^2 = 8x$

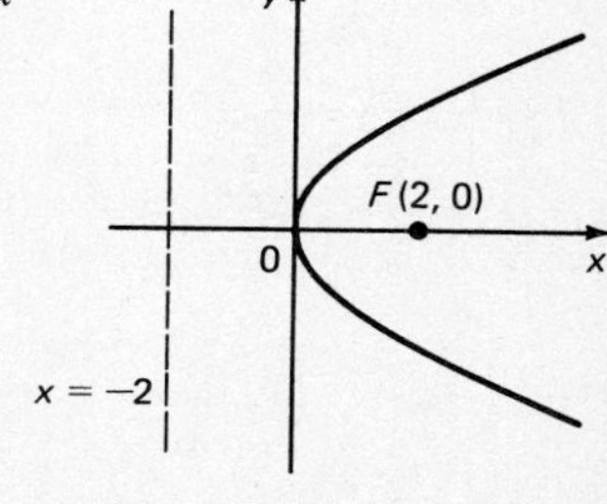

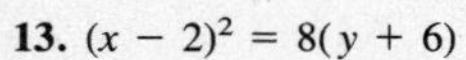

13. $(x - 2)^2 = 8(y + 6)$

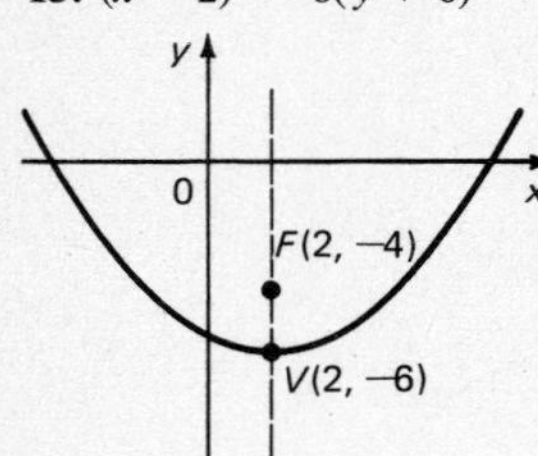

15. $y^2 = 4x$

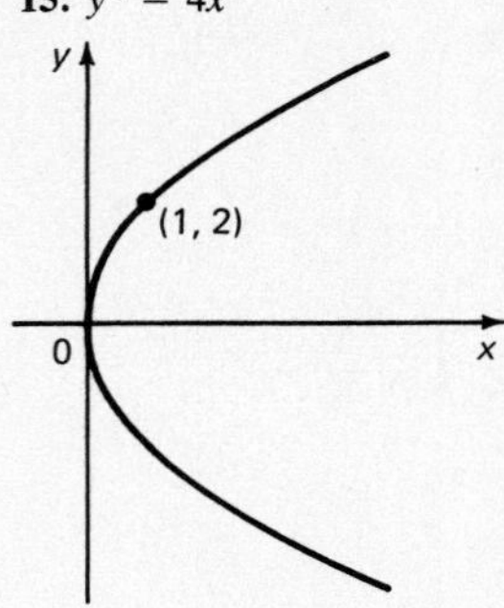

17. $4x^2 - 8x - 2y + 7 = 0$

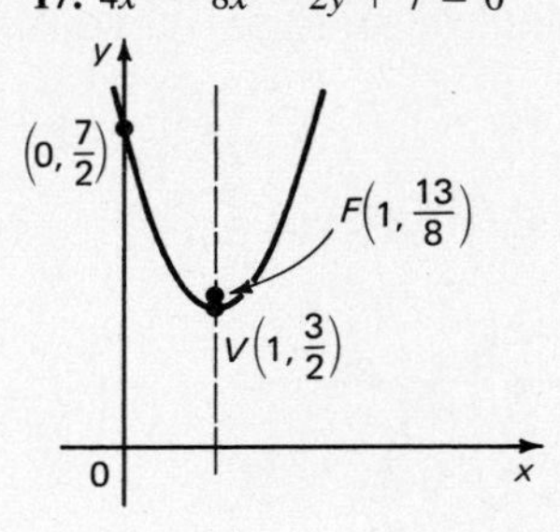

19. $2y^2 - 8y + 4x + 1 = 0$

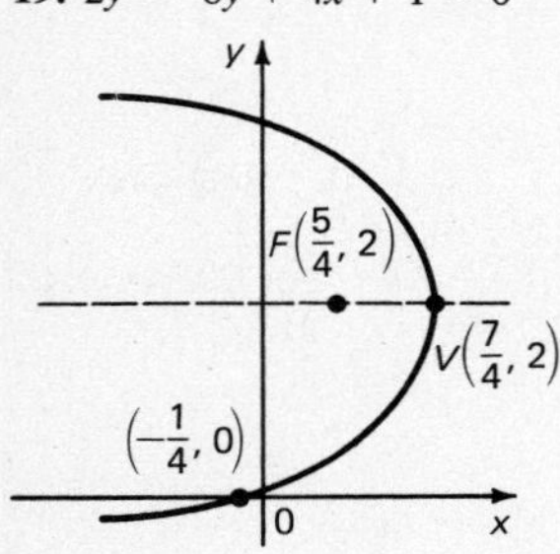

21. $x^2 = y^2 + 6y + 5$

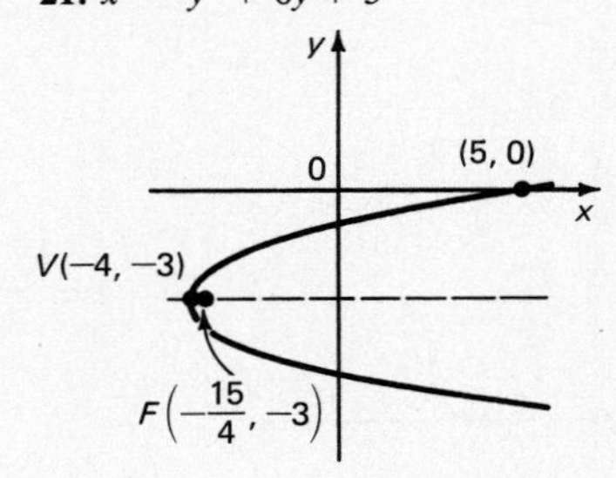

23. $(x - 3)^2 = 10(y - 3/2)$ **25.** 7 1/2 feet **27.** $25/32 \approx 0.78$ ft.

Exercises 7.4

1. $x^2/25 + y^2/9 = 1$

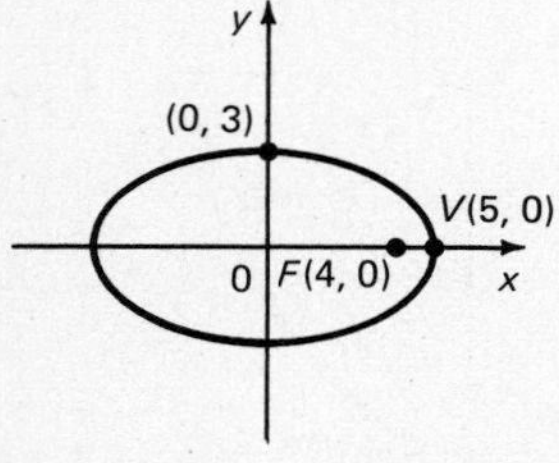

3. $x^2/2 + y^2/3 = 1$

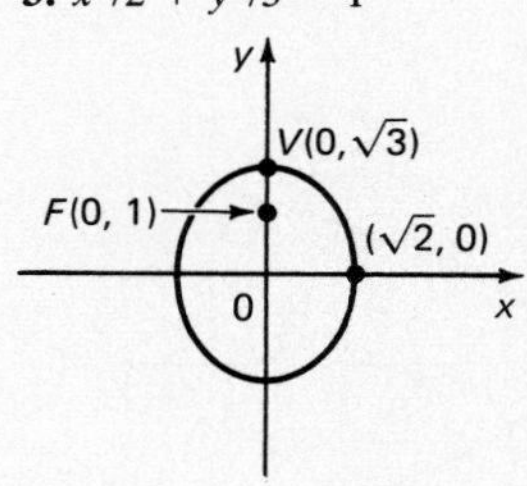

5. $x^2/16 + y^2/25 = 1$

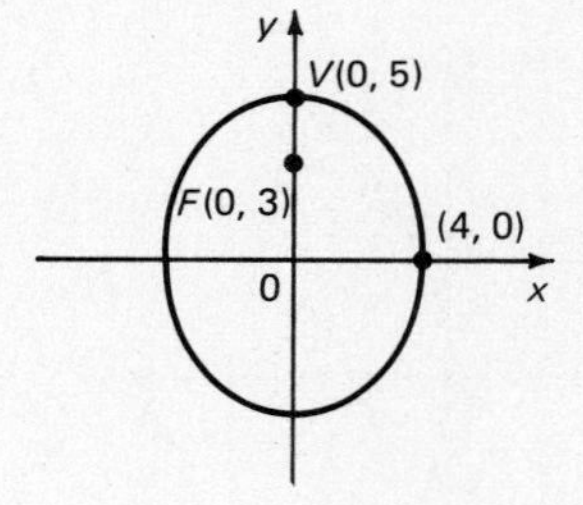

7. $(x - 2)^2/(1/2) + (y + 1/2)^2/(1/4) = 1$

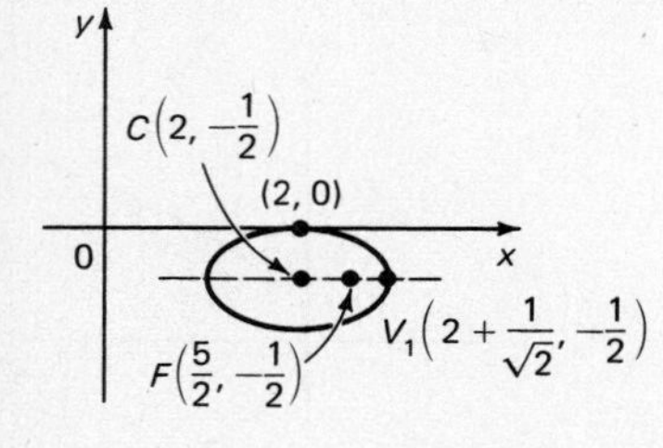

9. $(x - 3)^2/9 + (y - 5)^2/4 = 1$

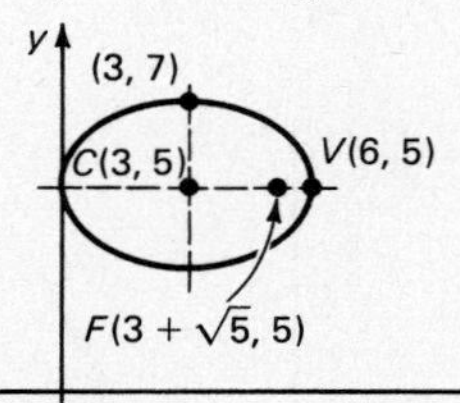

11. $\sqrt{7}/4$ **13.** 2/3 **15.** $9\sqrt{21}/5$ **17.** 7.2 feet **19.** $100x^2 + 36y^2 = 3600$ **21.** $16x^2 + 25y^2 = 400$

Exercises 7.5

1. $16x^2 - 9y^2 = 144$

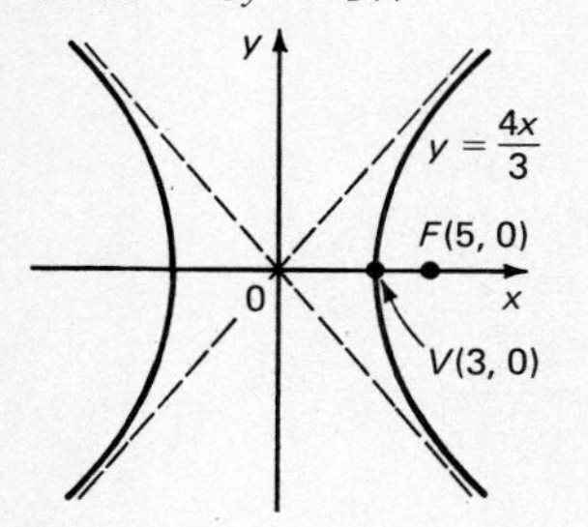

3. $7x^2 - 9y^2 = 63$

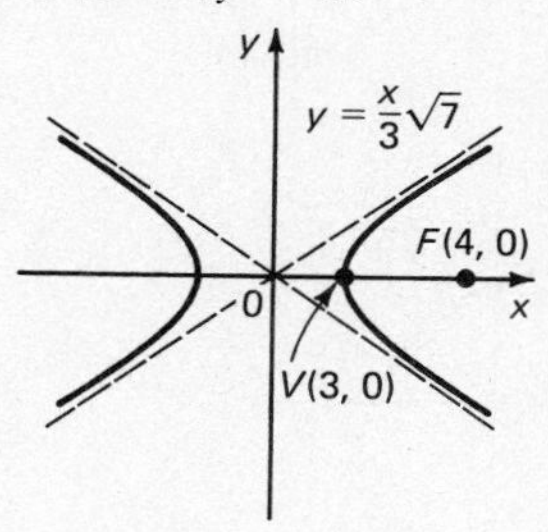

5. $9y^2 - 16x^2 = 144$

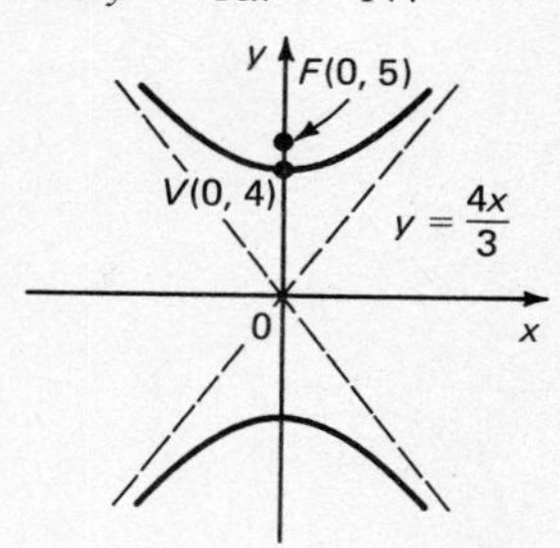

7. $4x^2 - y^2 = 4$

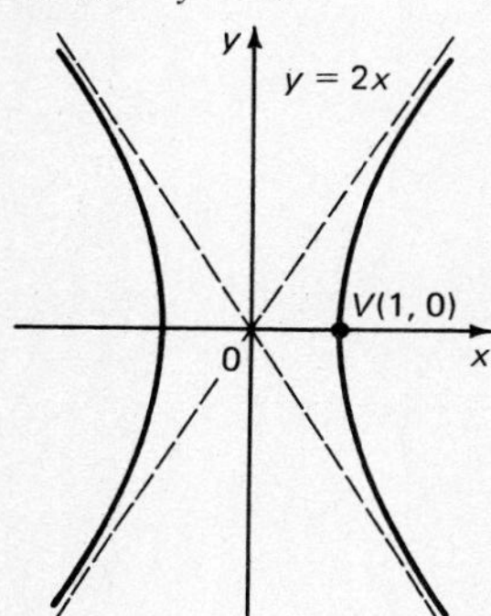

9. $3(x - 1)^2 - (y - 1)^2 = 3$

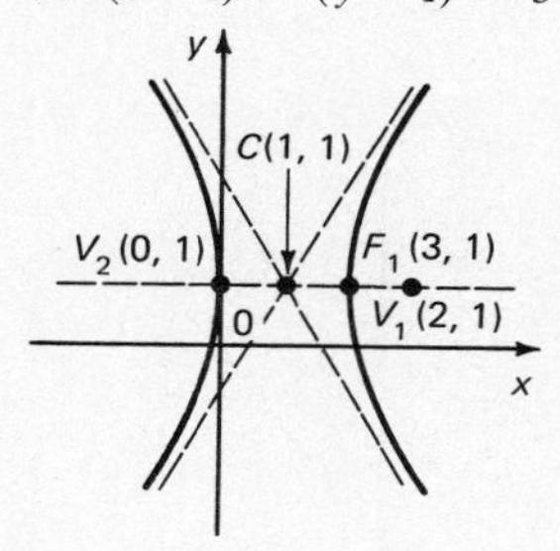

11.

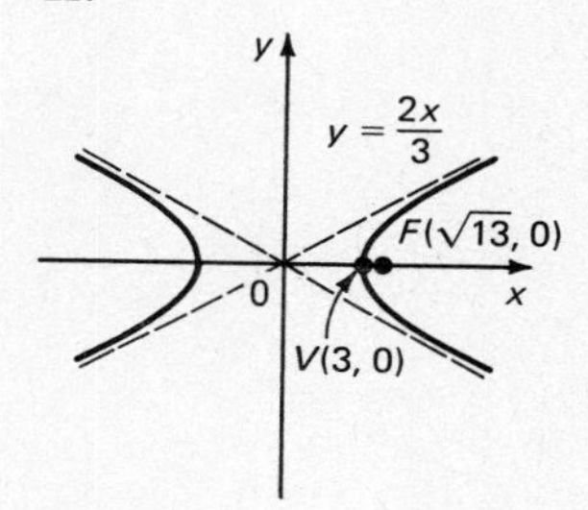

13.

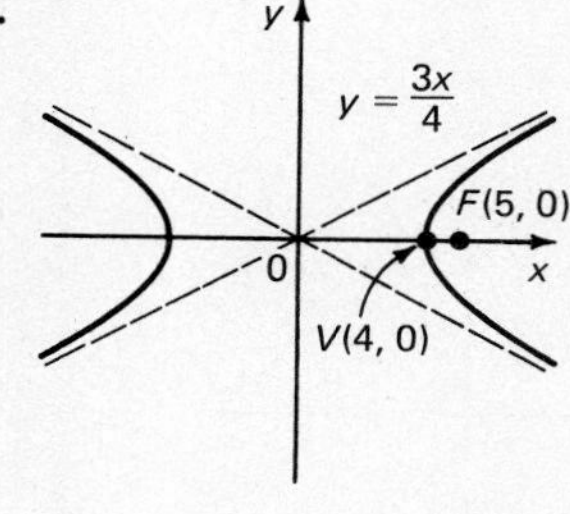

15.

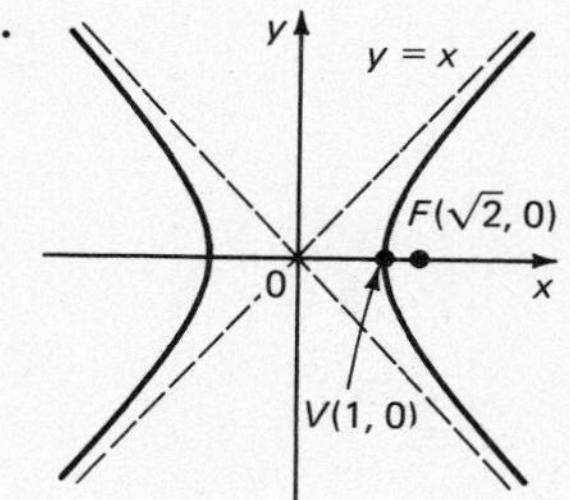

17.

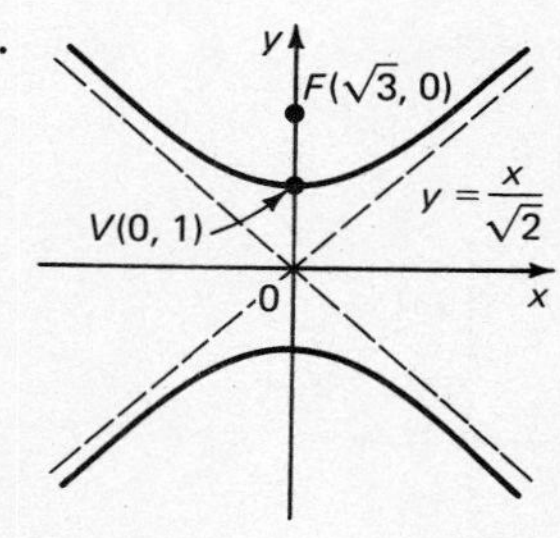

19.

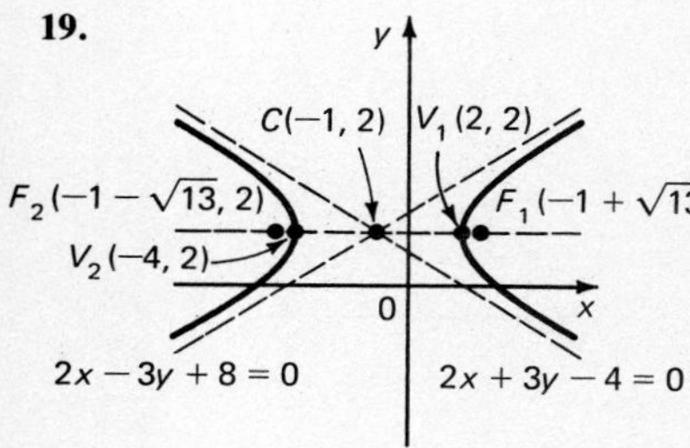

21.

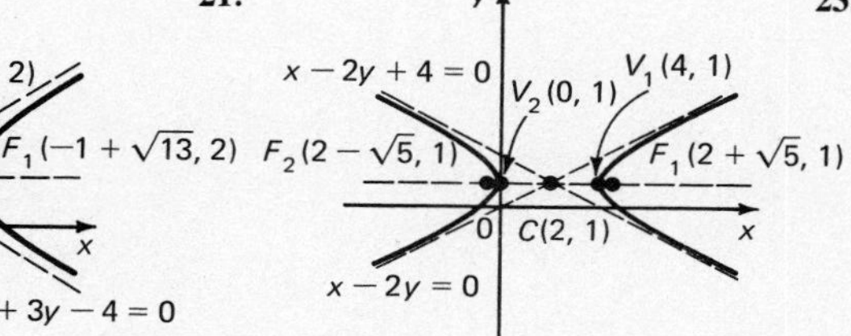

23.

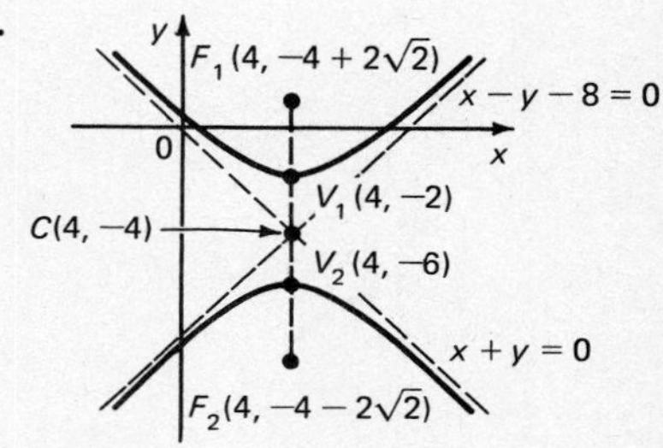

25. $x^2 - y^2 = 2$

Exercises 7.6

1. $x'^2 + y'^2 = 5$ **3.** $2x'^2 + y'^2 = 6$ **5.** $9x'^2 + 4y'^2 = 144$ **7.** $y'^2 - 4x'^2 = 4$

9. $x'^2 - y'^2 = 8$

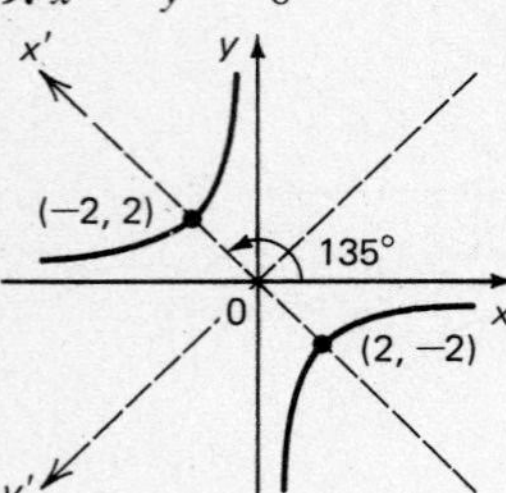

11. $3x'^2 + y'^2 = 1$

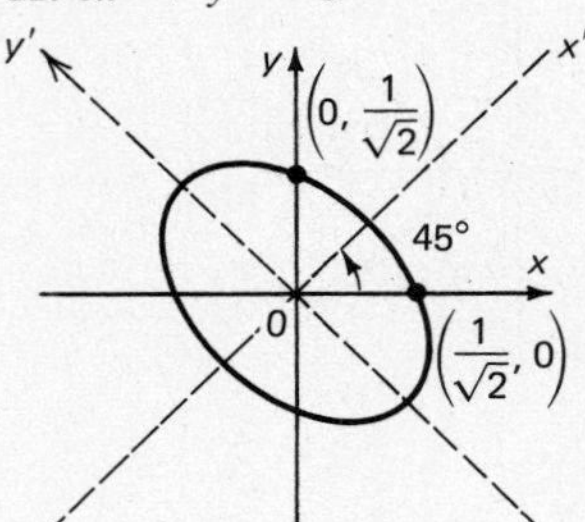

13. $3x'^2 - y'^2 = 5$

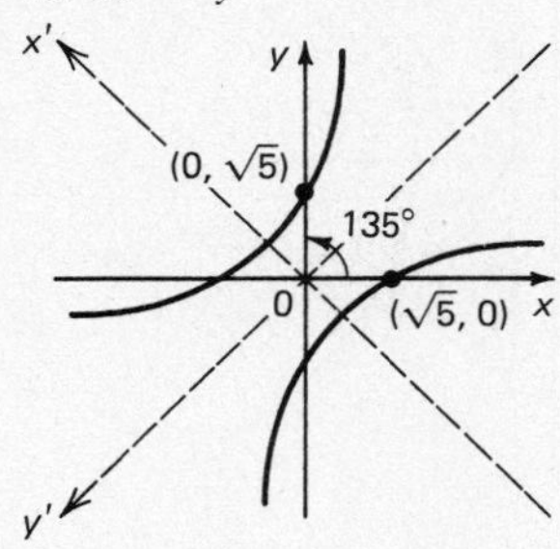

15. $\theta = 45°$, $(y' + \sqrt{2})^2 = 2\sqrt{2}(x' + \sqrt{2}/2)$

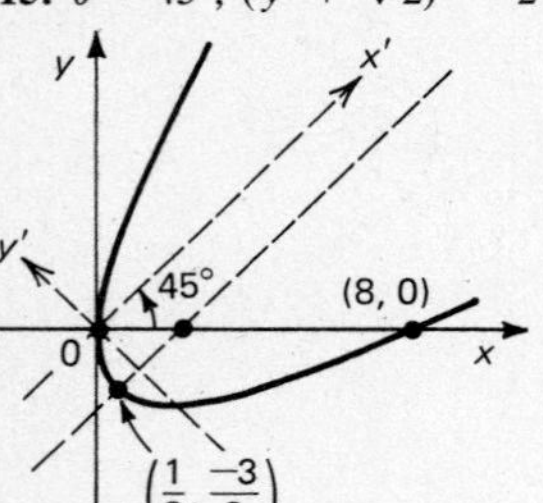

17. $\theta = \arctan 2$, $y'^2 = 9$

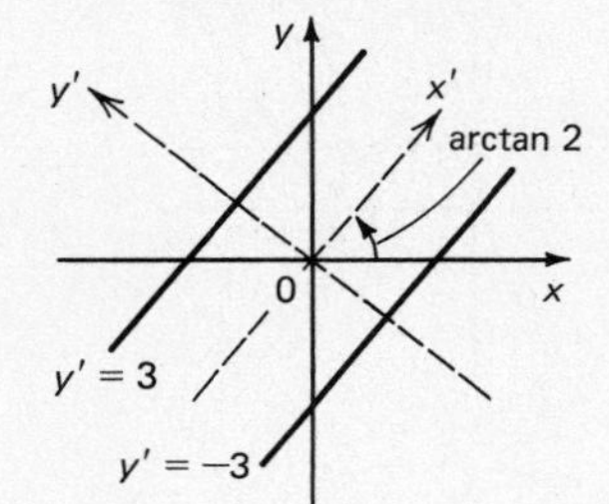

19. $\theta = \arctan 1/2$, $4x'^2 + y'^2 = 25$

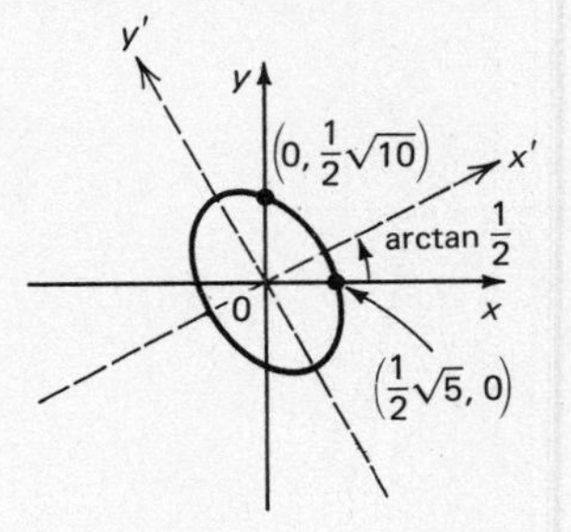

Exercises 7.7

1. $r \cos \theta = a$

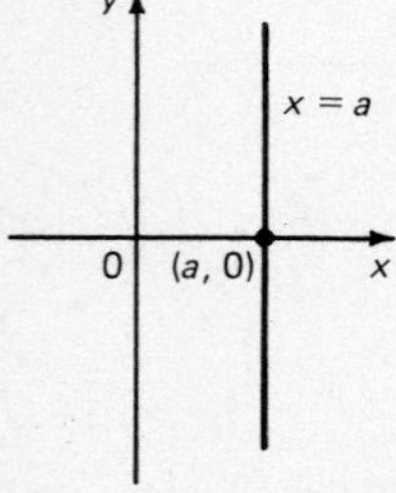

3. $\theta = 45°$

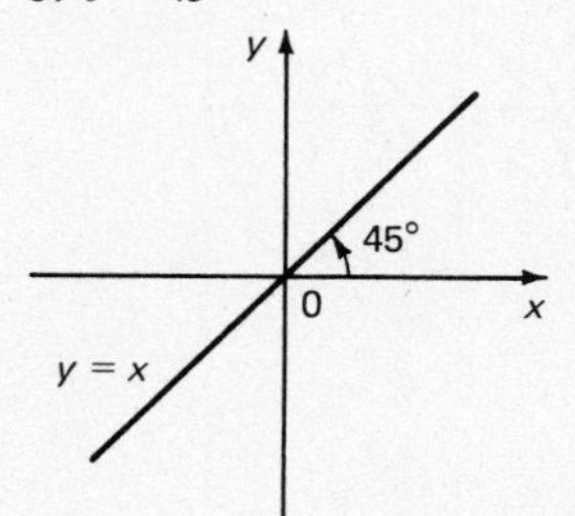

5. $r \tan \theta \sin \theta = 1$

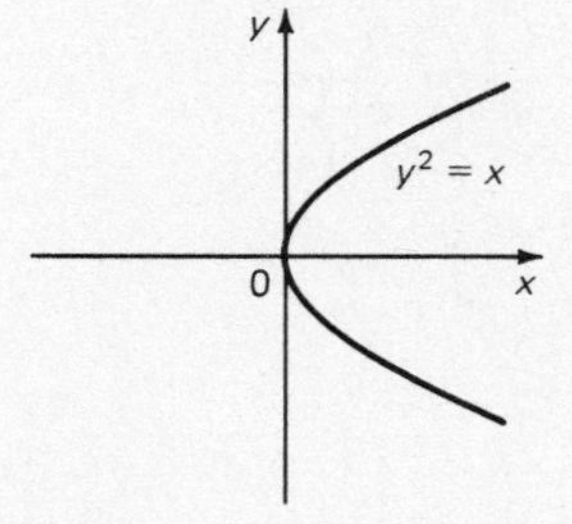

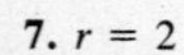
7. $r = 2$

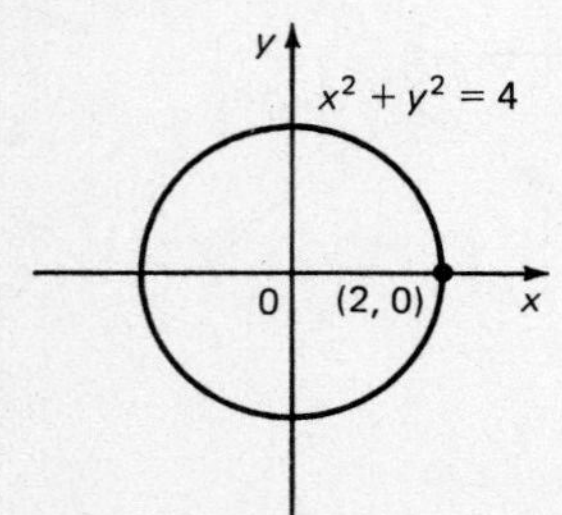

9. $r = \cos \theta$

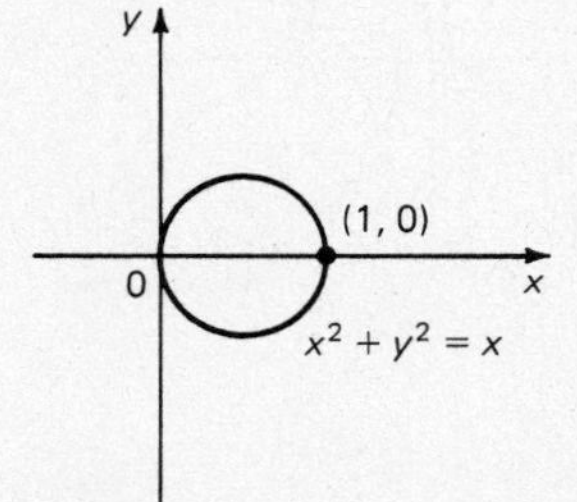

11. $r = -2 \sin \theta$

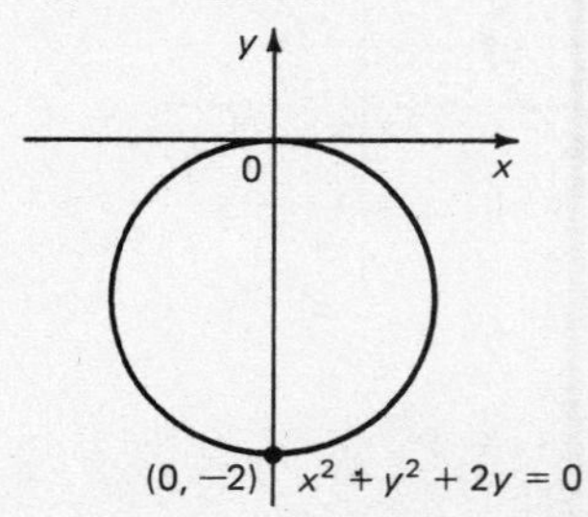

13. $x^2 + y^2 = 9$

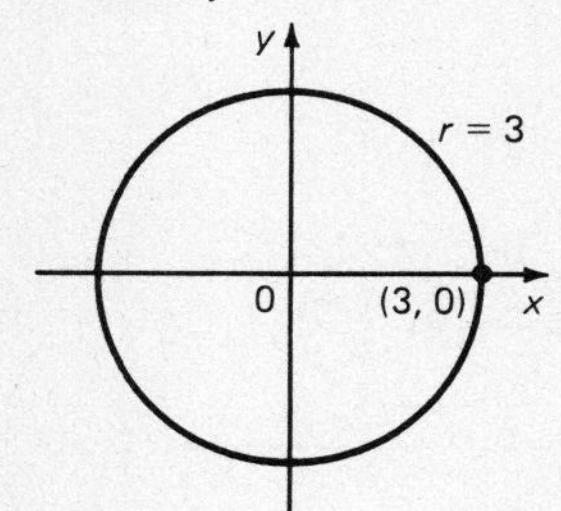

15. $x^2 - 3y^2 = 0$

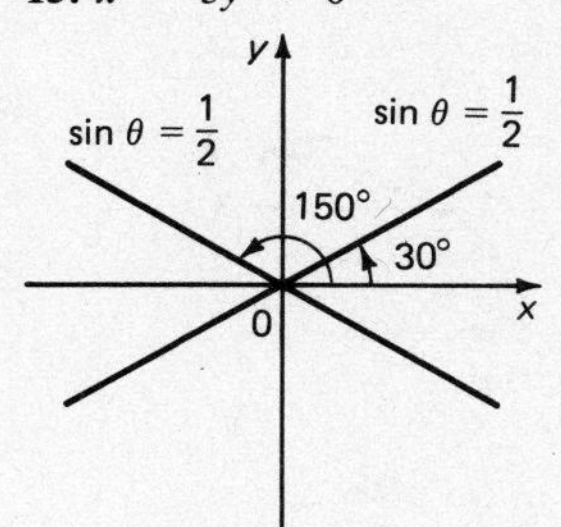

17. $x = 3$

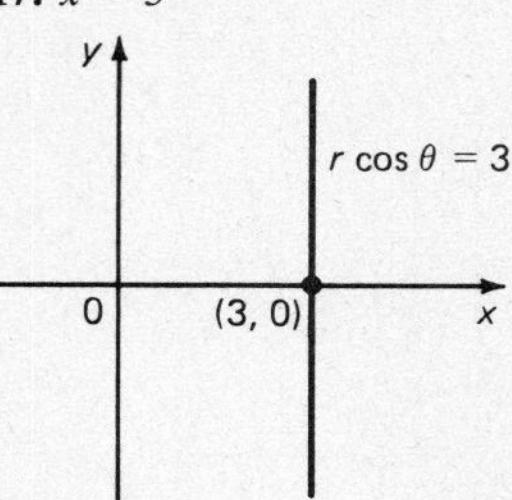

19. $x^2 + y^2 + 2x = 0$

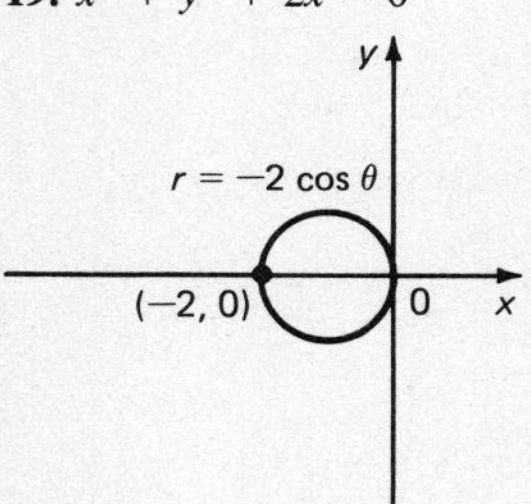

21. $y^2 = 2x + 1$

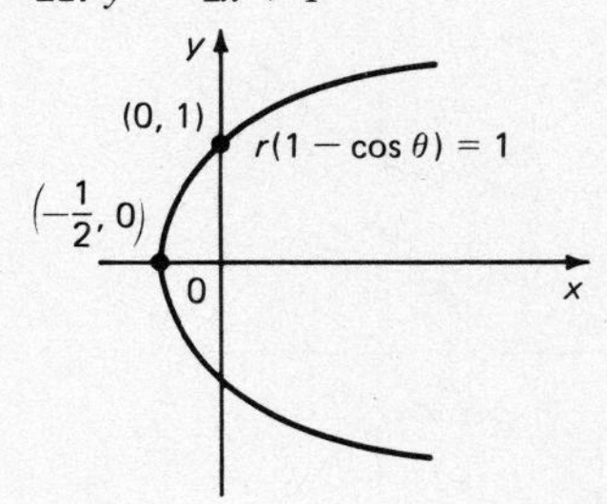

23. $x^2 - y^2 = 1$

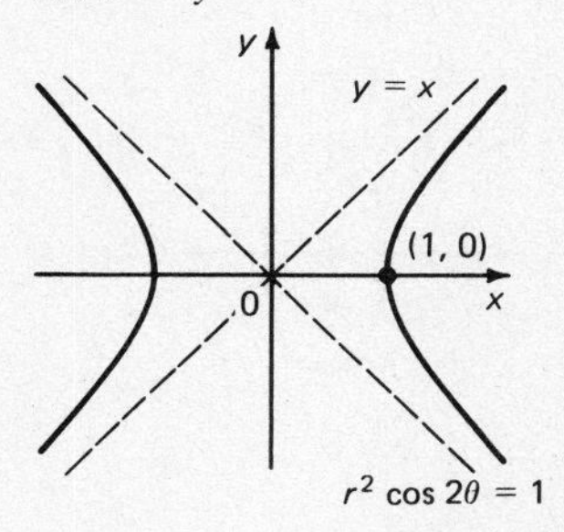

Exercises 7.8

1. Parabola

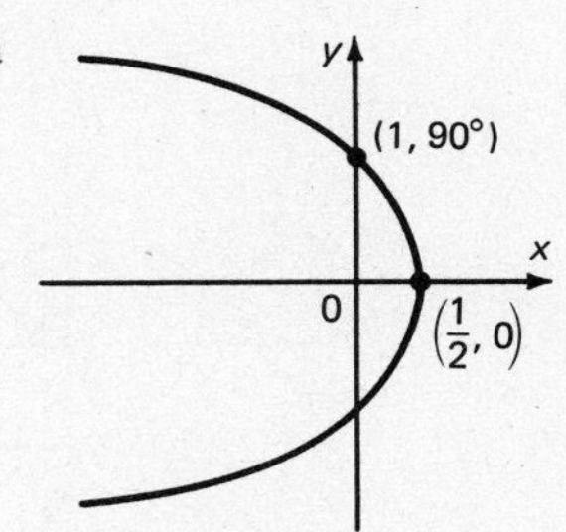

3. Ellipse

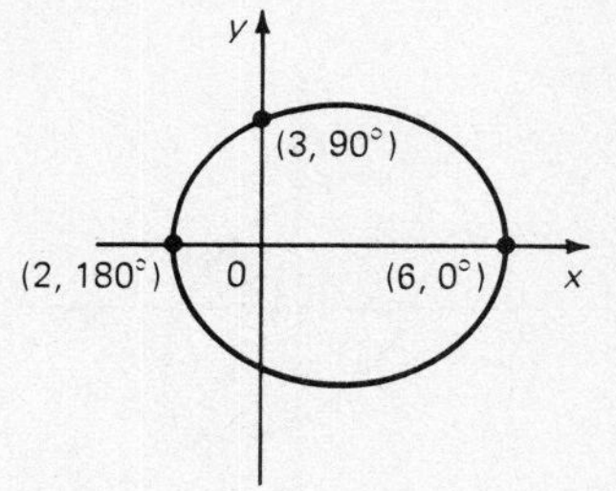

5. Ellipse

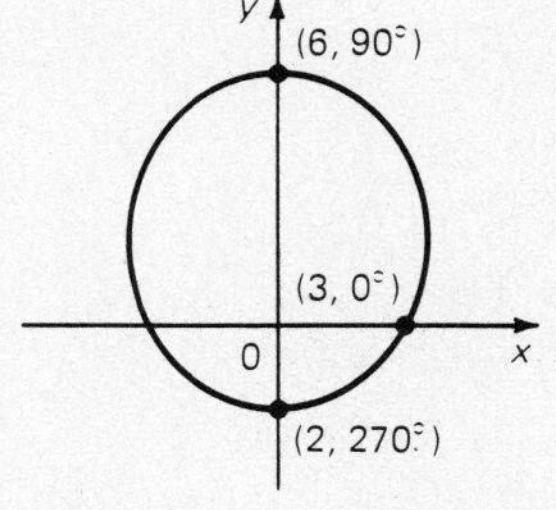

7. Parabola

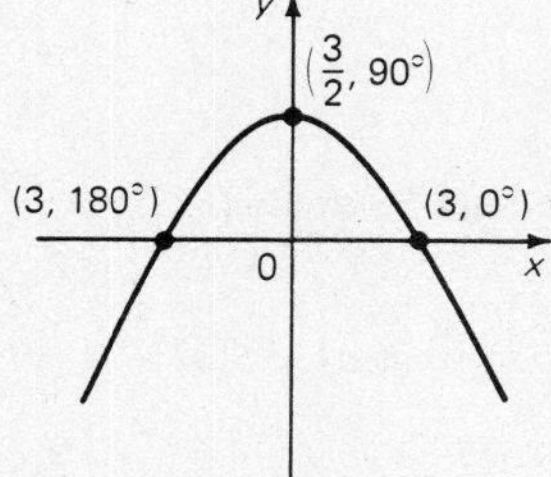

11. $r(1 - \sin\theta) = 4$ **13.** $r(3 + 4\sin\theta) = 36$ **15.** $r(1 - \cos\theta) = 4(10^6)$, 2 million miles; or $r(1 - \cos\theta) = 2(10^7)$, 10 million miles **17.** $r(1 + \sqrt{2}\cos\theta) = \sqrt{2}/2$, $e = \sqrt{2}$

Chapter 7 Review Exercises

1. $(y - 4)/(x + 3) = (1 - 4)/(2 + 3)$, or $(y - 1)/(x - 2) = (4 - 1)/(-3 - 2)$ **3.** $y = -3x/5 + 11/5$
5. $m = -3/5$, $\theta = 180° - \arctan 3/5 \approx 149°$, 11/3 and 11/5 **7.** $x + y + 1 = 0$ **9.** (a) $3x - 2y - 1 = 0$

(b) $2x + 3y - 5 = 0$ **11.** $x^2 + (y - 1/2)^2 = 13/4$ **13.** $(x + 2)^2 + (y - 3)^2 = 25$; center, $(-2, 3)$; radius, 5
15. $(x + 3/2)^2 + (y + 1)^2 = 5/2$; center, $(-3/2, -1)$; radius, $\sqrt{10}/2$

17.

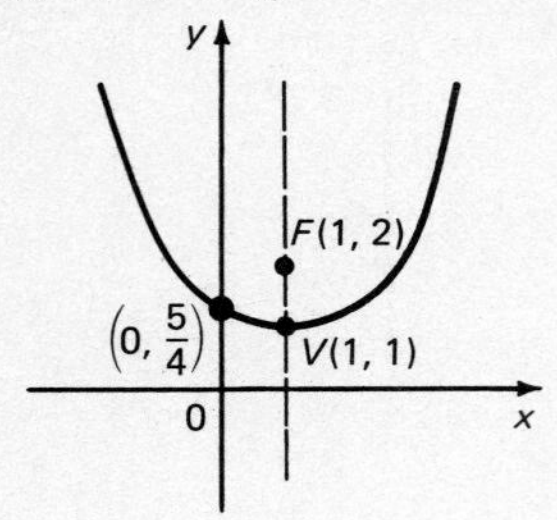

19.

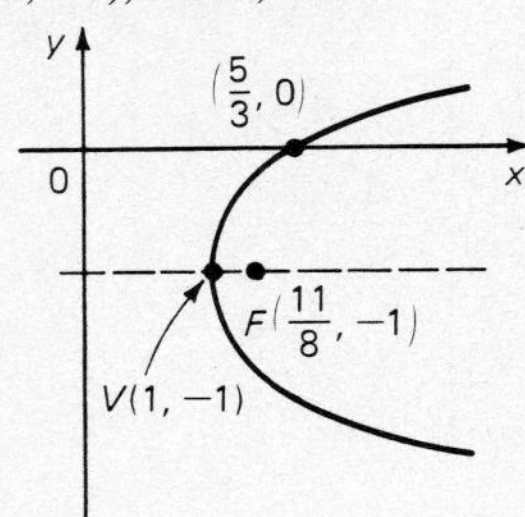

21.

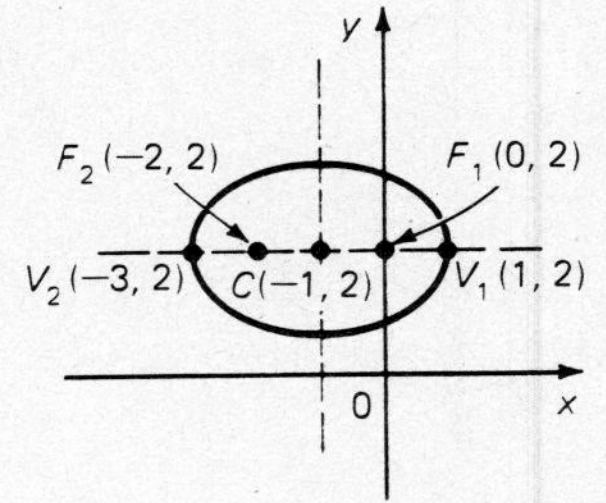

23.

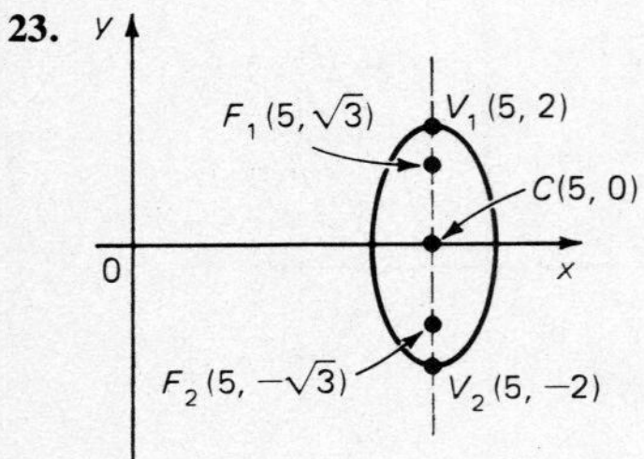

25.

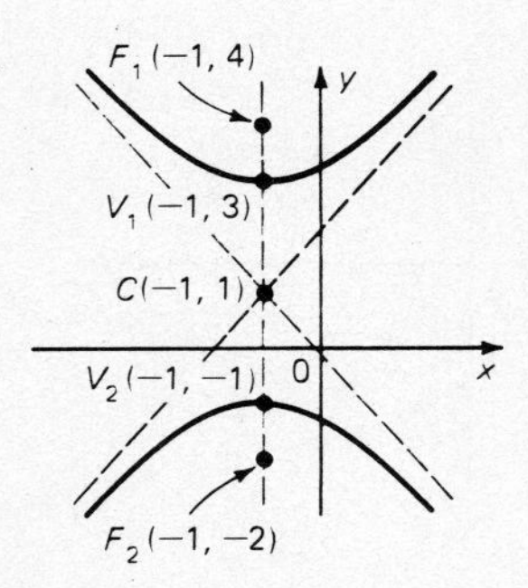

27. (a) $(x - 5/4)^2 = -(y - 25)/16$

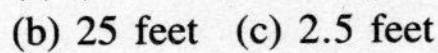
(b) 25 feet (c) 2.5 feet

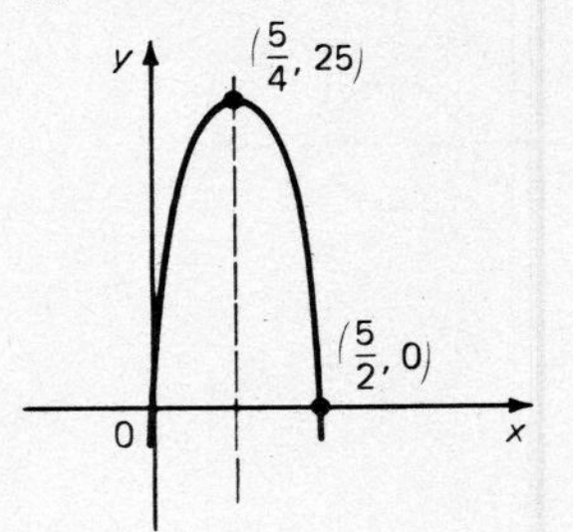

29. (4, 3) **31.** $x' = x + 1$, $y' = y - 1$, $4x'^2 + 5y'^2 = 20$

33. $9x'^2 + y'^2 = 90$

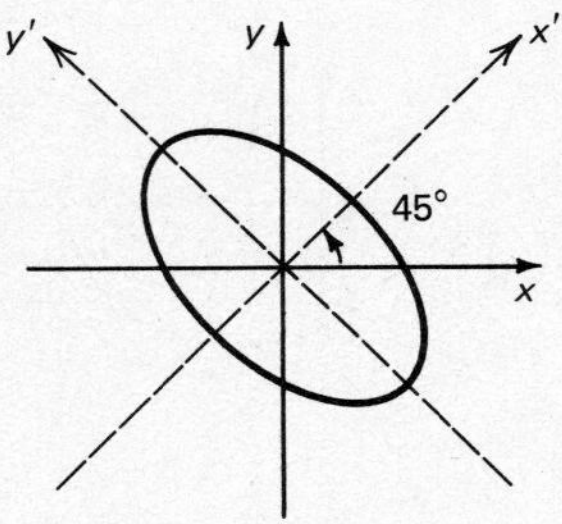

35. $\theta = \arctan 4$, $2x'^2 + y'^2 = 2$

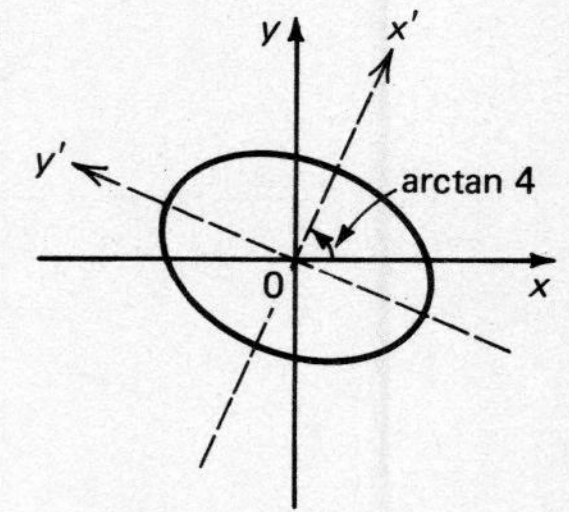

37. $\tan \theta = 2$ **39.** $r(1 - \cos \theta) = 2$ **41.** $x + 2 = 0$ **43.** $x^2 + 2y = 1$

Chapter 7 Miscellaneous Exercises

1. $6x - 5y + 7 = 0$, $y = 3$, $6x - y - 5 = 0$ **3.** $3x + 5y - 16 = 0$, $6x - y - 21 = 0$, $9x - 7y - 26 = 0$
5. (a) $a_1b_2 - a_2b_1 = 0$, (b) $a_1a_2 + b_1b_2 = 0$ **7.** $x/4 + y/3 = 1$ **9.** $x/5 + y/(-5/2) = 1$

11. $x/4 + y/(-3) = 1$

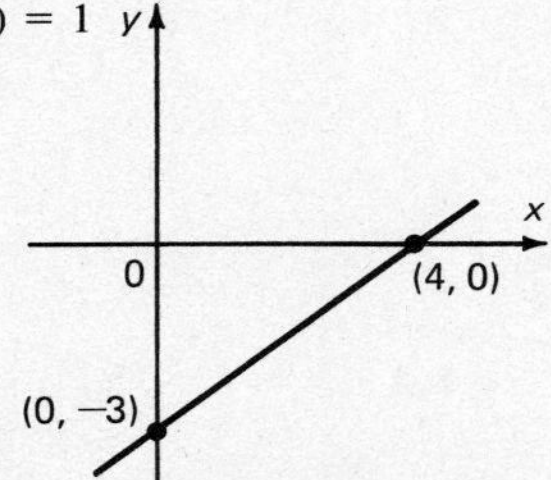

13. $x/4 + y/(-2) = 1$

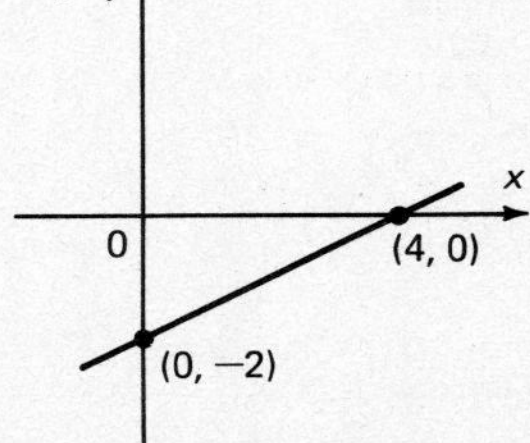

15. $x^2 - 4y^2 = 4$

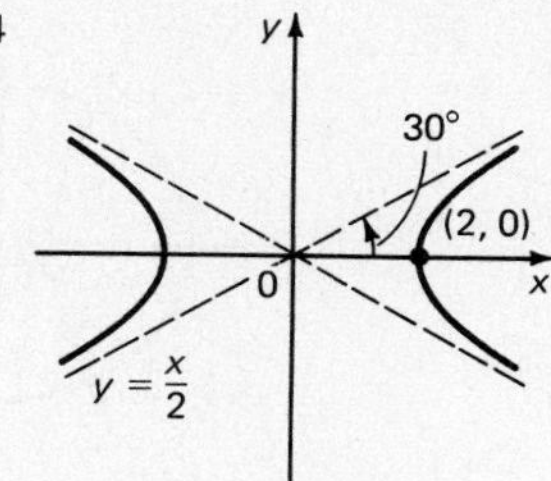

19. $r(1 + e \sin \theta) = 2ep$

CHAPTER 8

Exercises 8.1

1. $x = 2$, $y = 2$

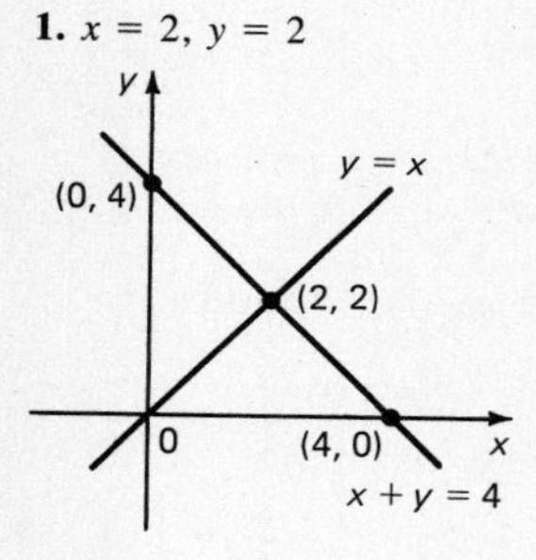

3. $x = 4$, $y = -3$

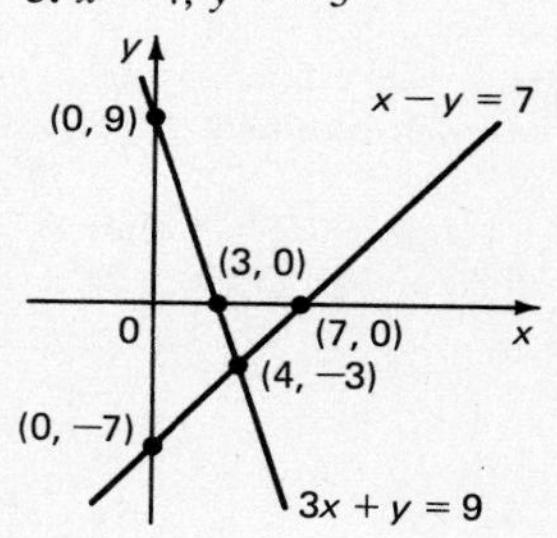

5. $x = -2$, $y = -1$, $x + y + 3 = 0$

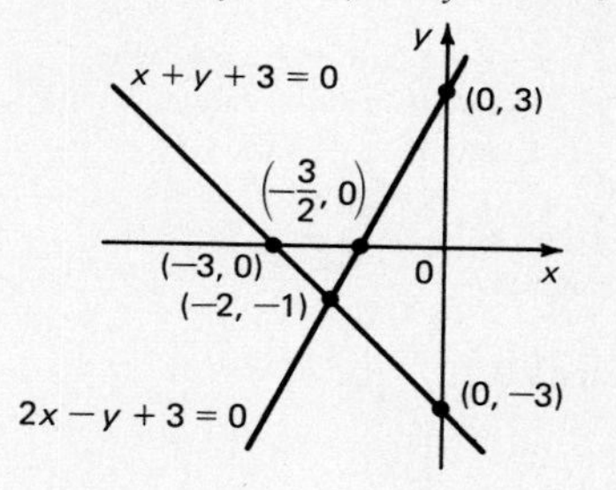

7. No solution. Lines parallel

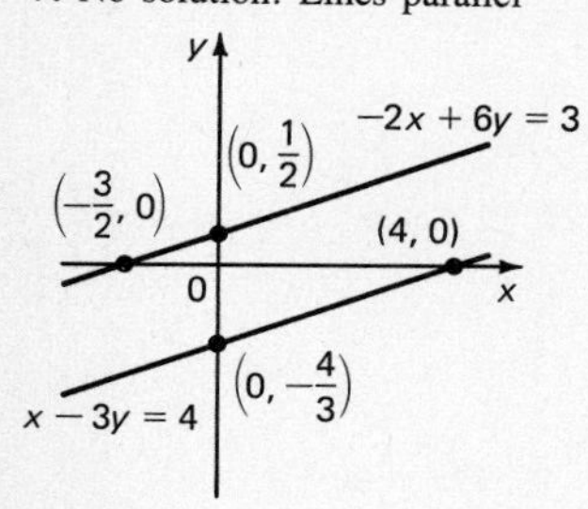

9. No solution. Lines parallel

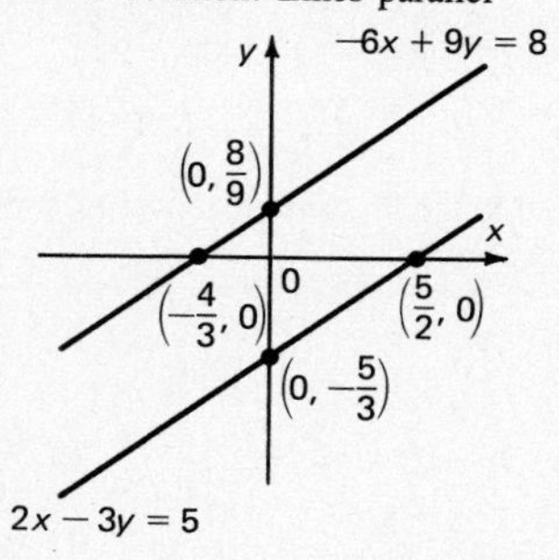

11. Infinte number of solutions. Dependent system.

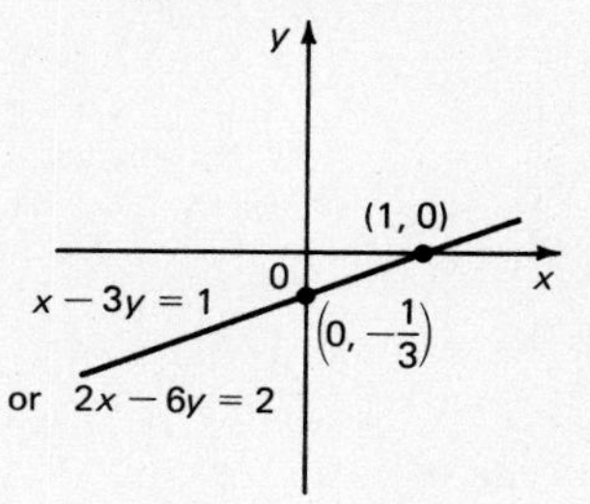

13. $x = 10/27$, $y = 11/9$ **15.** $x = -3$, $y = 2$ **17.** $x = 1/2$, $y = 1$

19.

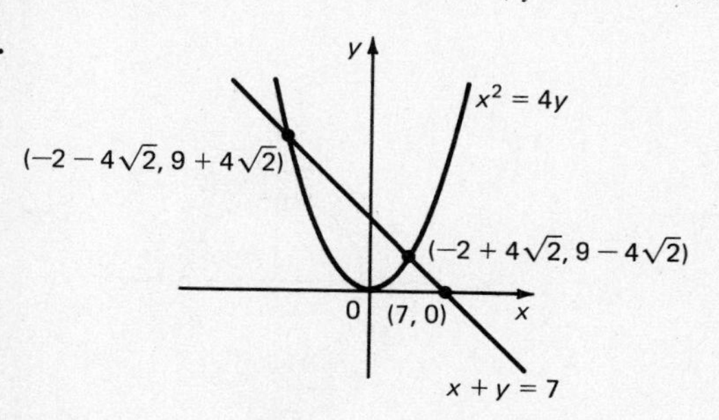

21.

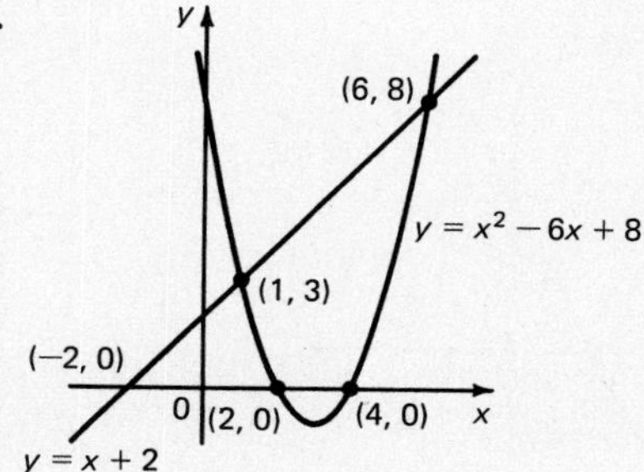

23.

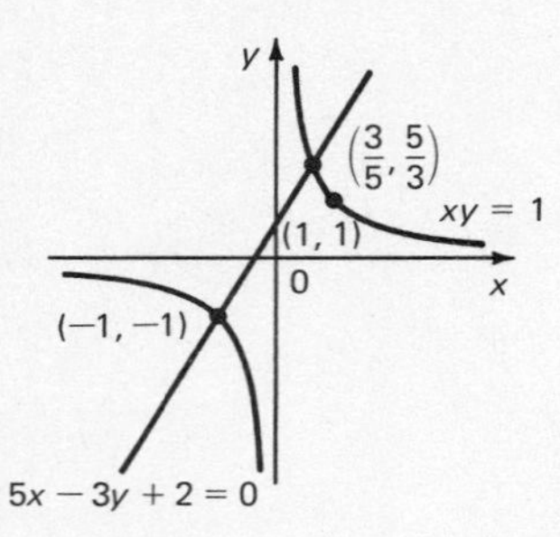

25.

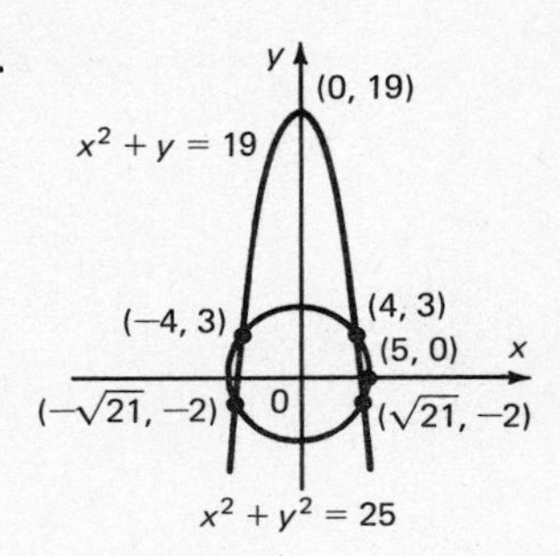

27.

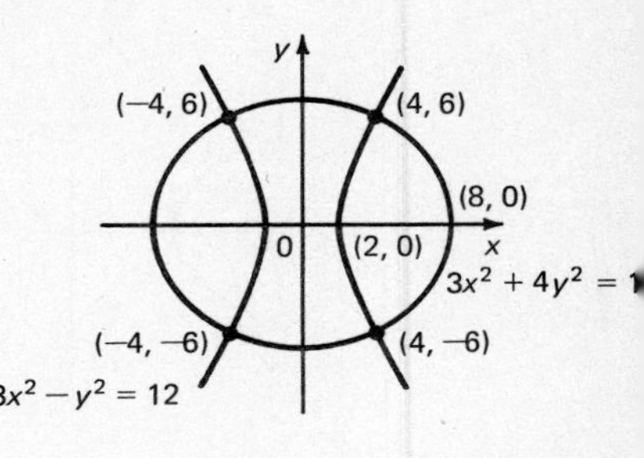

29. 21 dimes, 13 quarters **31.** \$400, \$600 **33.** 9 and 12, or −9 and −12 **35.** 12 meters by 35 meters **37.** $x_1 = 3$, $x_2 = 4$

Exercises 8.2

1. $x = 1$, $y = -1/2$, $z = 2$; consistent **3.** $x = -3$, $y = 1$, $z = 6$; consistent **5.** $x = 2/5$, $y = 1/5$, $z = 2/5$; consistent **7.** Inconsistent, no solution **9.** $x = 3$, $y = 2$, $z = 1$; consistent **11.** $x = -z$, $y = 3z$, $z = z$; dependent **13.** 2, −11, and 12 **15.** $y = x^2 - x$ **17.** $2x = -3y^2 + 11y - 6$ **19.** $x^2 + y^2 - x + y - 2 = 0$ **21.** $x^2 + y^2 + 4x - 4y - 17 = 0$ **23.** $-1/2(x - 1) + 1/3(x - 2) + 1/6(x + 1)$ **25.** $-1/x + 3/2(x - 1) - 1/2(x + 1)$ **27.** $-1/4x + 5/8(x - 2) + 5/8(x + 2)$ **29.** $i_1 = 2$ amp., $i_2 = 2$ amp., $i_3 = 0$ amp.

Exercises 8.3

1. $\begin{bmatrix} 1 & -2 & 0 \\ 0 & 7 & -1 \\ 0 & 0 & 1 \end{bmatrix}$ **3.** $\begin{bmatrix} -6 & -1 & 2 \\ 0 & 19/6 & -4/3 \\ 0 & 0 & 198/57 \end{bmatrix}$ **5.** $\begin{bmatrix} 1 & 1 & 1 & 1 \\ 0 & -2 & 0 & 0 \\ 0 & 0 & -2 & 0 \\ 0 & 0 & 0 & -2 \end{bmatrix}$ **7.** $\begin{bmatrix} 2 & -1 & 1 \\ 0 & 0 & 4 \end{bmatrix}$ **9.** $\begin{bmatrix} 1 & -1 & 1 \\ 0 & -2 & 0 \\ 0 & 0 & 2 \\ 0 & 0 & 0 \end{bmatrix}$

11. $\begin{bmatrix} 2 & -1 \\ 0 & 9/2 \\ 0 & 0 \\ 0 & 0 \end{bmatrix}$ **13.** $x = 5/2$, $y = -9$, $z = -3/2$ **15.** $y = (4 - 4x)/5$, $z = (1 - x)/10$ **17.** $x = 2$, $y = 1$, $z = 0$ **19.** $x = 1$, $y = 2$, $z = 1$, $w = 0$ **21.** $x = (9 - 4z)/3$, $y = (6 + z)/3$ **23.** No solution **25.** 10 lb., 6 lb., and 4 lb., respectively **27.** 12 nickels, 5 dimes, and 6 quarters **29.** $8y = 5x^2 + 8x - 5$

Exercises 8.4

1. 8 **3.** 5 **5.** $x = 6$ **7.** $x = 2, 3$ **9.** −10 **11.** 3 **13.** −20 **15.** $12x - 36$ **17.** $-2x^2 + 16x - 24$ **19.** −92 **21.** −10 **23.** $x = 4$, $y = 1$ **25.** $x = 0$, $y = -2$ **27.** Dependent, $y = 2x - 2$ **29.** $x = 2$, $y = -1$, $z = 3$ **31.** $x = 4$, $y = 3$, $z = 2$ **33.** $x = 3/13$, $y = -4/13$, $z = 0$ **35.** $x = -3/7$, $y = -1/2$, $z = 11/14$ **37.** $x = 10/3$, $y = -3$, $z = -4/3$ **39.** $x = 3$, $y = -1$, $z = 2$, $w = 1$ **41.** 14 **43.** 27/2

Exercises 8.5

1.

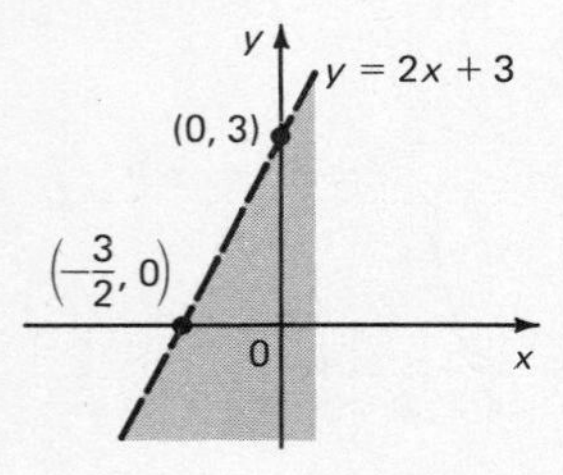

3.

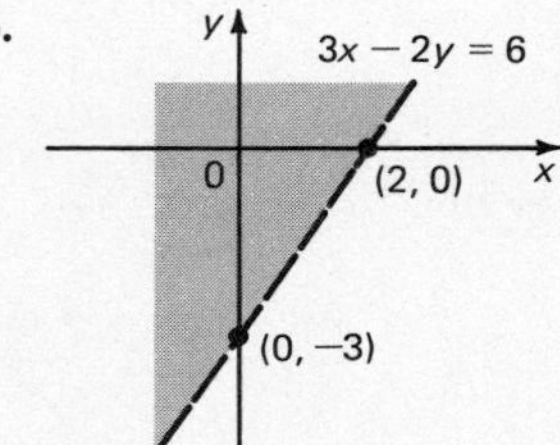

5.

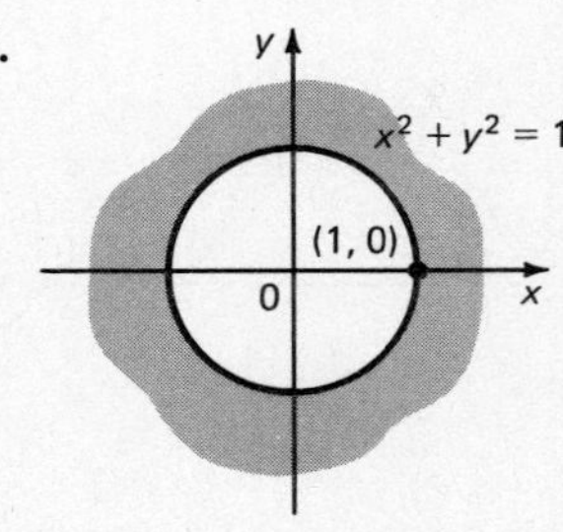

7.

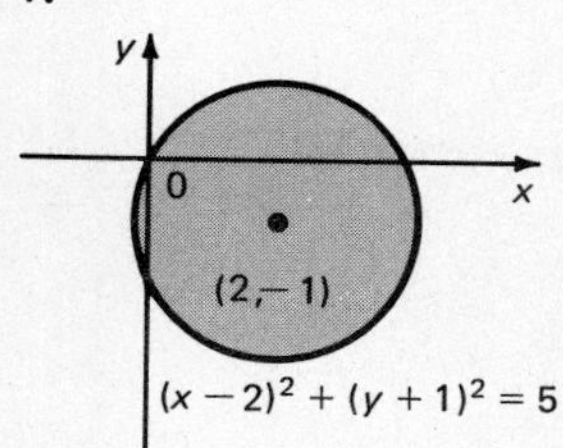

9.

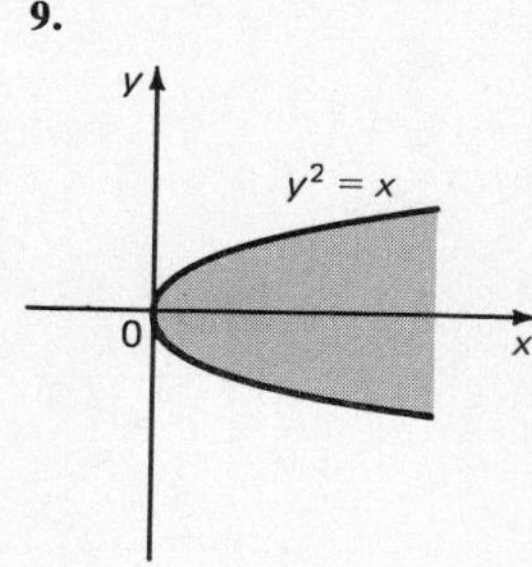

11.

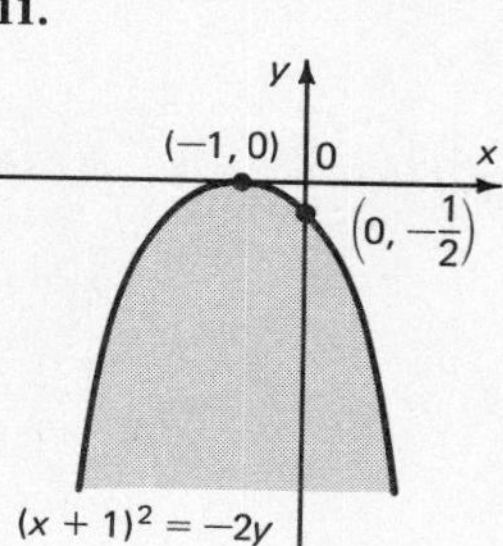

13.

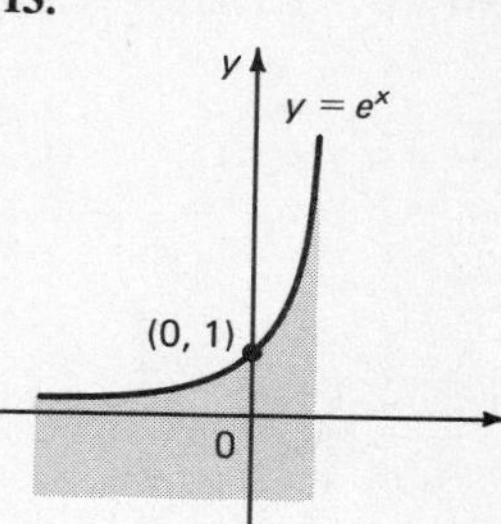

15.

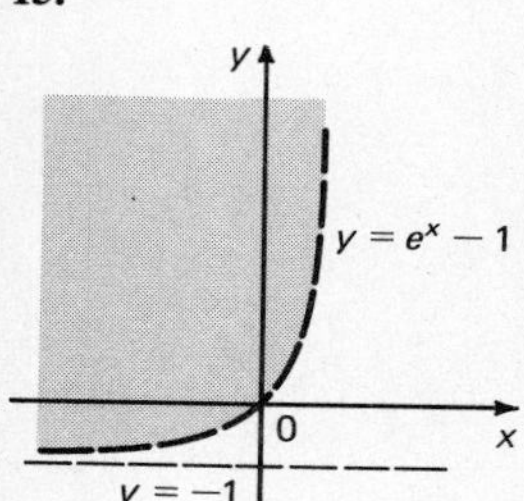

17.

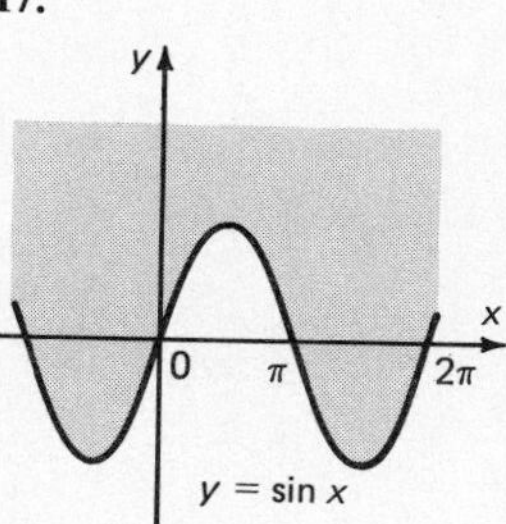

19.

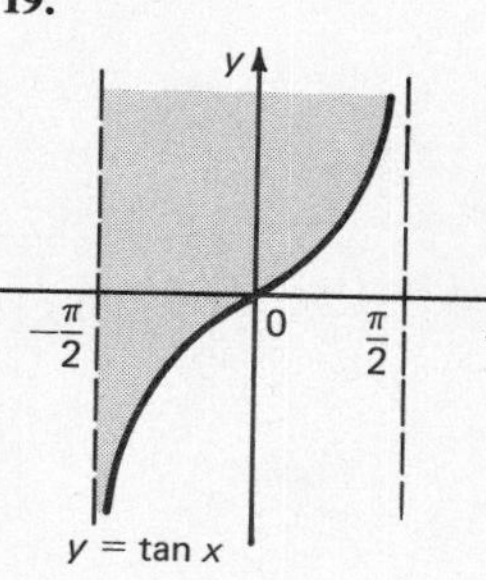

21.

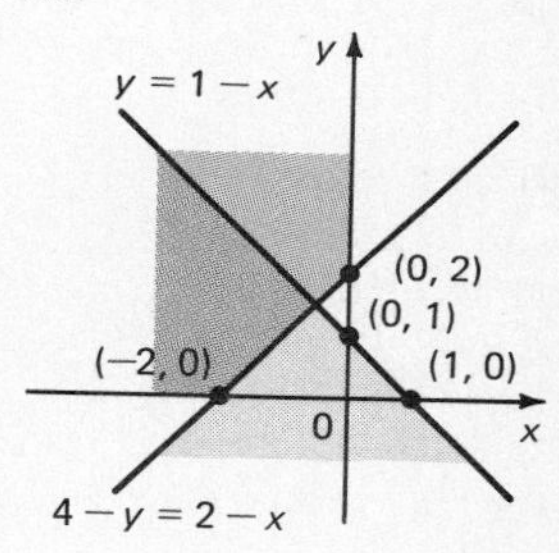

23.

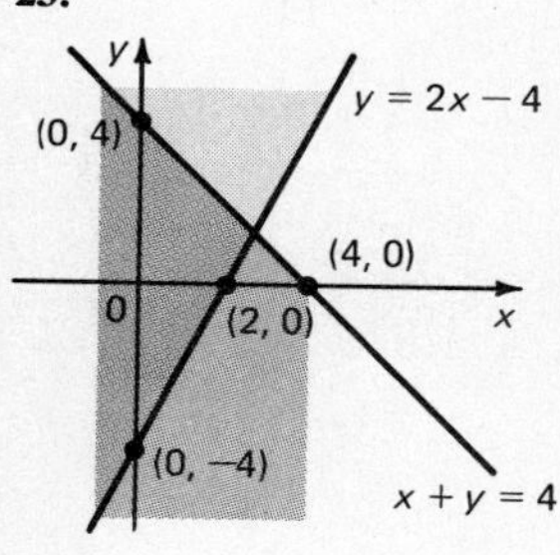

25.

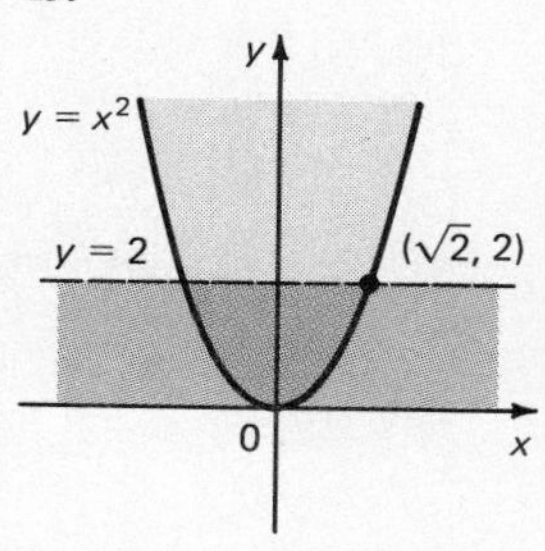

27.

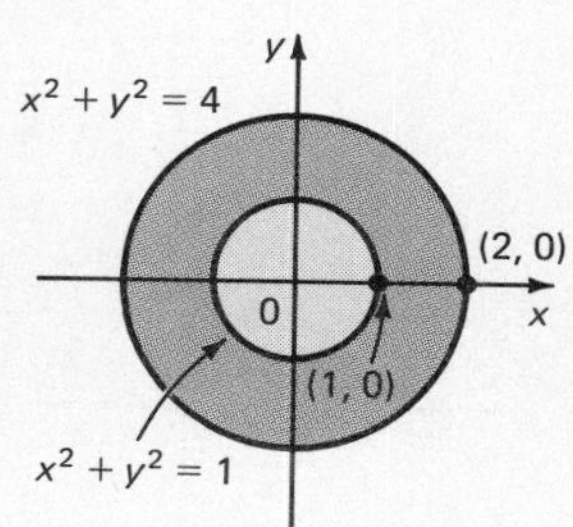

29.

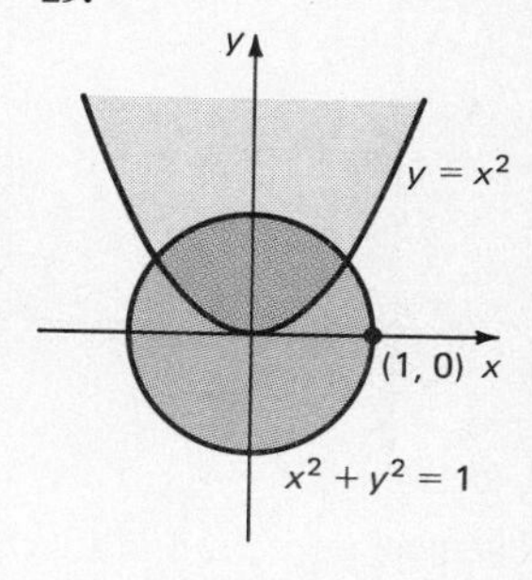

31.

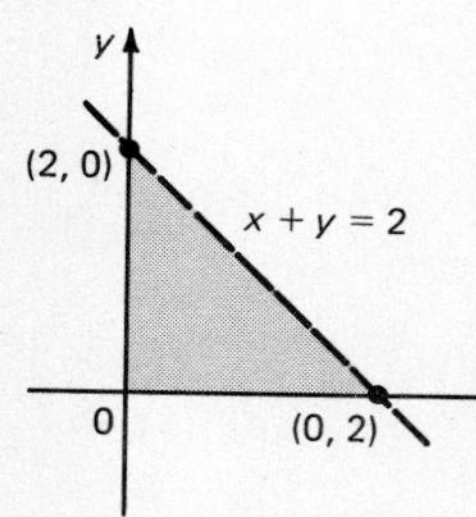

33.

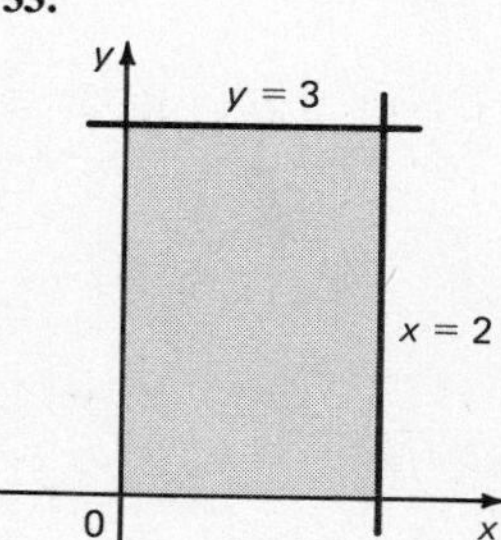

35.

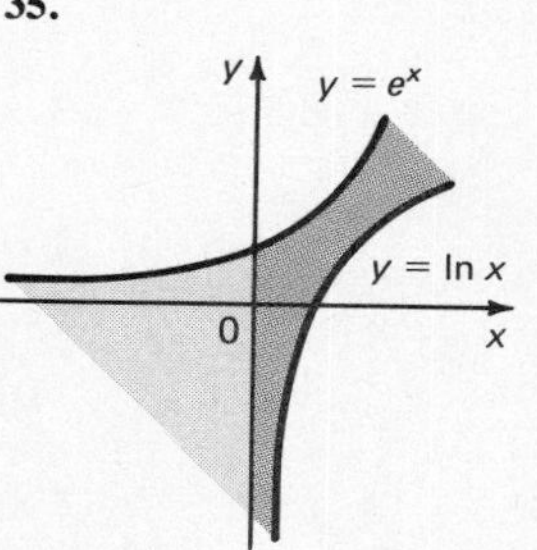

37.

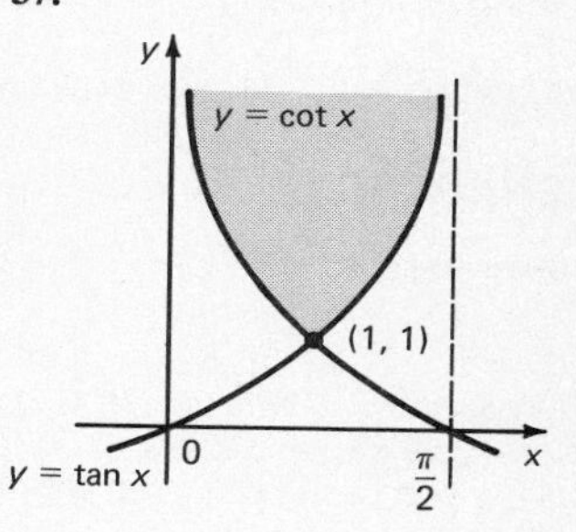

Exercises 8.6

1. Maximum, 65; minimum, 8 **3.** Maximum, 900; minimum, 360 **5.** 42/5 **7.** 87/4 **9.** 235/4 **11.** (a) 204 (b) 588/5 (c) 97 **13.** \$112 **15.** 8 of #1, 3 of #2

Chapter 8 Review Exercises

1. $x = 3, y = -2$ **3.** $x = 12, y = 12$ **5.** Dependent, $x = 3y + 4$ **7.** $x = 4, y = -2$ **9.** $x = 2, y = -5, z = -1$ **11.** $x = -3, y = 1, z = 6$ **13.** (0, 0, 0) **15.** $x = -1, y = -2, z = 3, w = -2$

17.

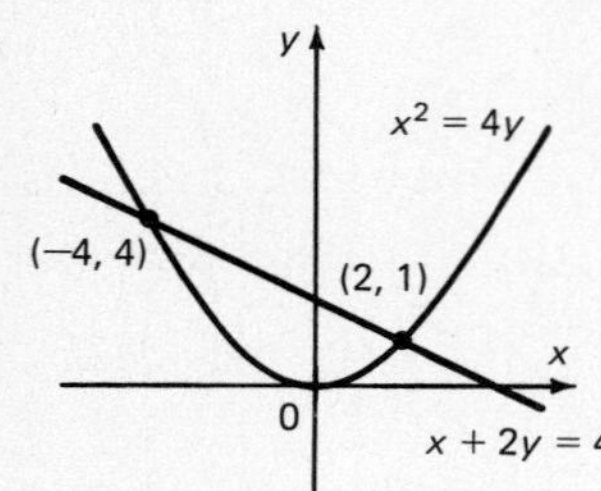

19.

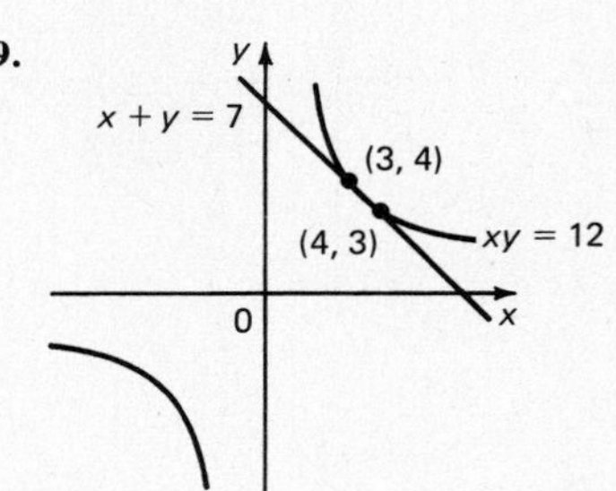

21. $\begin{bmatrix} 1 & -2 & 3 \\ 0 & 5 & -10 \\ 0 & 0 & 0 \end{bmatrix}$ **23.** $\begin{bmatrix} 1 & -1 & 4 \\ 0 & 3 & -9 \\ 0 & 0 & 0 \end{bmatrix}$ **25.** $\begin{bmatrix} 1 & 2 & -1 & 3 \\ 0 & -1 & 4 & 1 \\ 0 & 0 & -19 & 3 \\ 0 & 0 & 0 & -142 \end{bmatrix}$

27. -20 **29.** 0 **31.** 142 **33.** $x = -1, y = 3/2$ **35.** $x = 3, y = 12, z = -6$ **37.** $x = 1/3, y = 1, z = 8/3$ **39.** $x = 1, y = 2, z = -3, w = -1$

41.

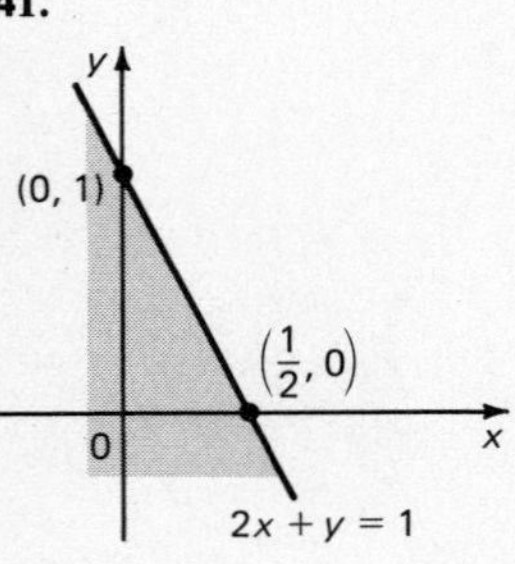

43.

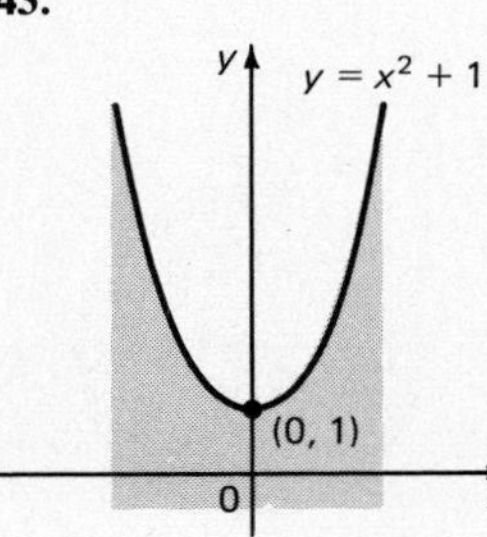

45.

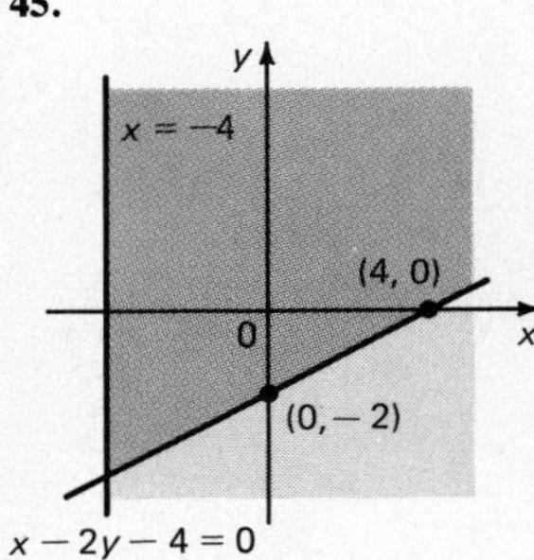

47.

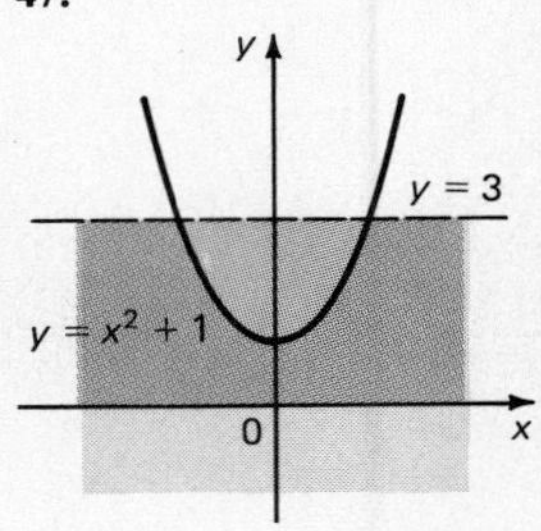

49. 33/2 **51.** 64/5 **53.** 1600 boxes to Chicago, 800 boxes to Dallas

Chapter 8 Miscellaneous Exercises

5. Collinear **7.** Non-collinear **9.** Collinear **11.** $x - y + 1 = 0$ **13.** $x = 2$

CHAPTER 9

Exercises 9.1

1. $-i$ **3.** -1 **5.** 1 **7.** -1 **9.** $-i$ **11.** 1 **13.** (a) $1 - 2i$ (b) $1 + 2i$ **15.** (a) $5 + i\sqrt{3}$ (b) $5 - i\sqrt{3}$ **17.** (a) $-1 + 2i\sqrt{2}$ (b) $-1 - 2i\sqrt{2}$ **19.** (a) $2 + 3i$ (b) $2 - 3i$ **21.** (a) $2 - 3i$ (b) $2 + 3i$ **23.** (a) $2i\sqrt{6}$ (b) $-2i\sqrt{6}$ **25.** $x = -2, y = 5$ **27.** $x = 4, y = 0$ **29.** $x = 3, y = -2$ **31.** $x = 3, y = 2$ **33.** $x = 2, y = 3$ **35.** $9 + 2i$ **37.** $5 + 2i$ **39.** $9 + i$ **41.** $-7 + 8i$ **43.** $-3 - 4i$ **45.** $7 - i$ **47.** $9 - 17i$ **49.** 5 **51.** $-2 + 3i$ **53.** $7/10 + (1/10)i$ **55.** $23/10 + (13/10)i$ **57.** $2/5 - (1/5)i$ **59.** $-i$ **61.** $x = 2i, -2i$ **63.** $x = 1/2 + (\sqrt{3}/2)i, -1/2 -(\sqrt{3}/2)i$ **65.** $x = -1 + i\sqrt{3}, -1 - i\sqrt{3}$ **67.** $x = 1/2 + (3/2)i, 1/2 - (3/2)i$ **69.** $x = -1/2 + (\sqrt{5}/2)i, -1/2 - (\sqrt{5}/2)i$ **71.** $x = 2, i, -i$ **73.** $x = 1/2, 1/2, -1/2 + (\sqrt{3}/2)i, -1/2 - (\sqrt{3}/2)i$

Exercises 9.2

1. $|z| = 3$

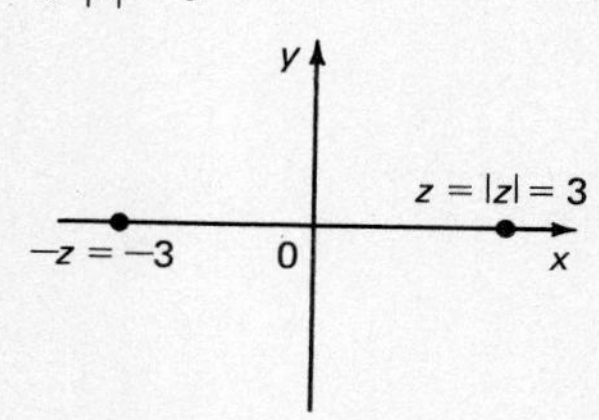

3.

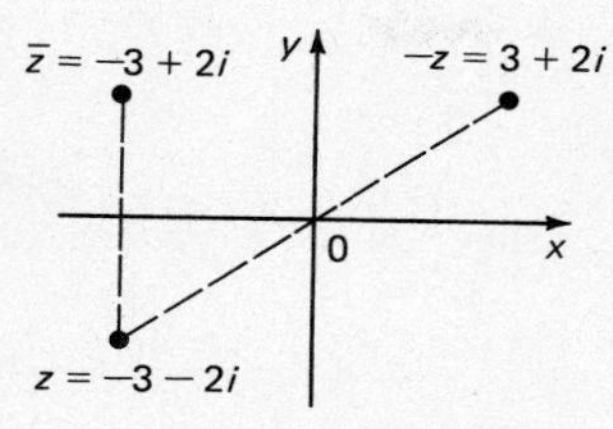

5.

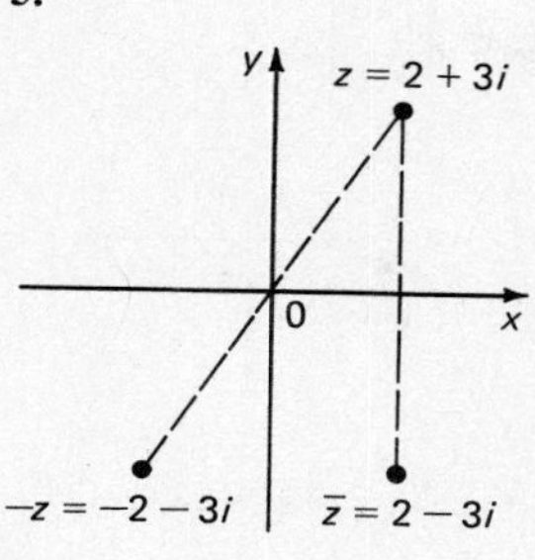

7.

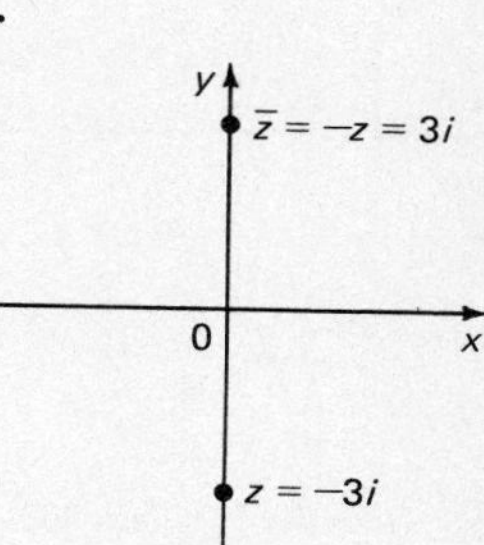

9.

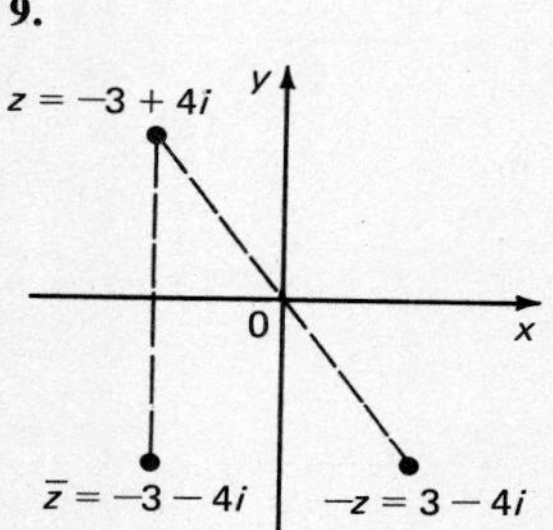

11.

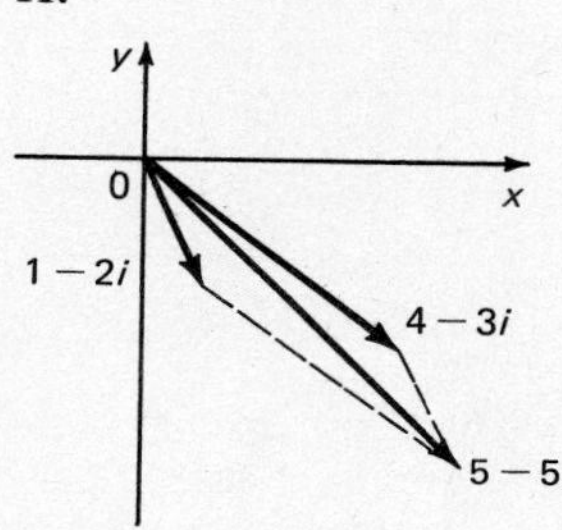

13.

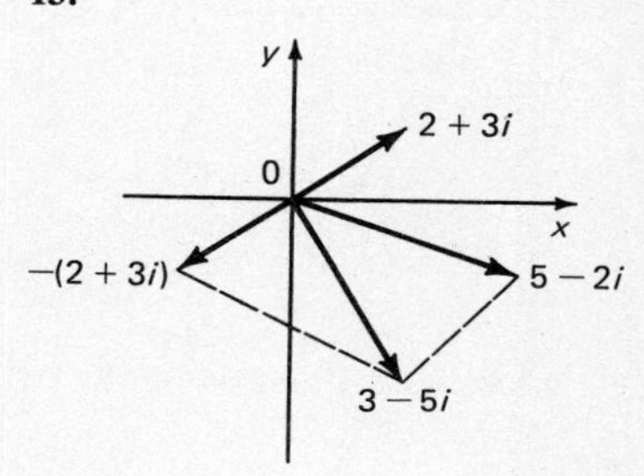

15.

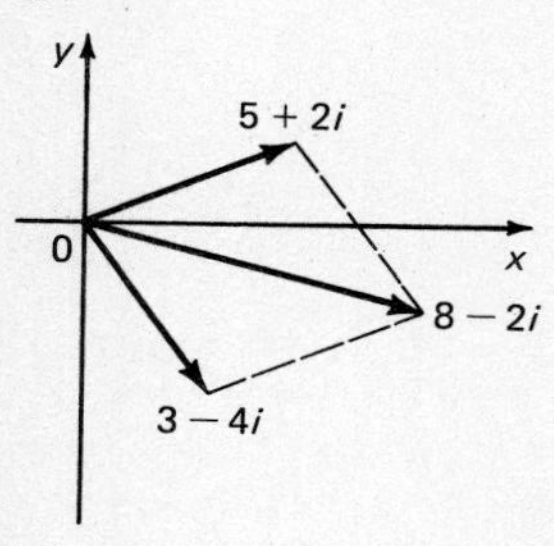

17.

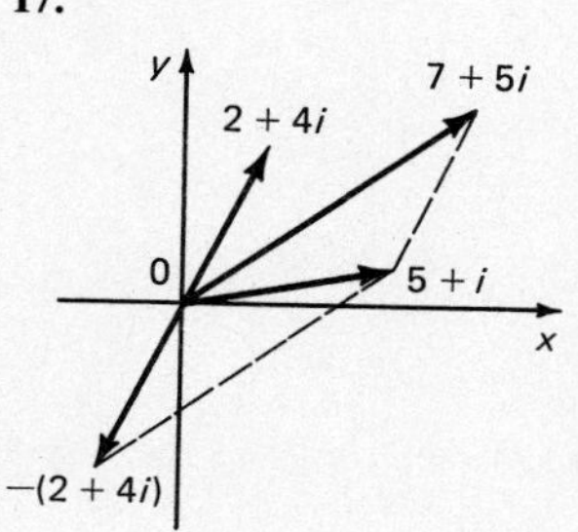

19. (b) $3\sqrt{2}/2 + (3\sqrt{2}/2)i$

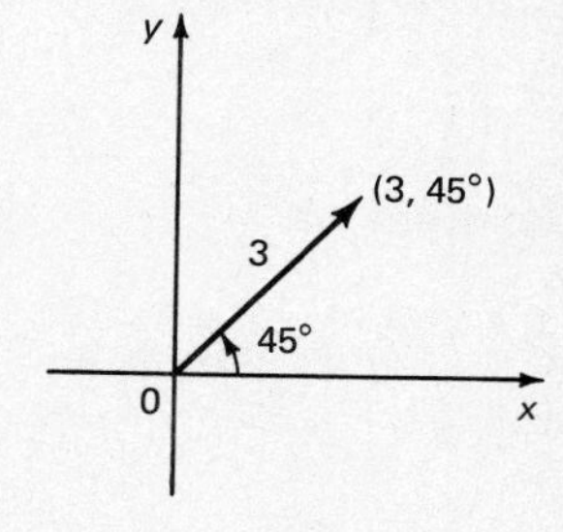

21. (b) -1

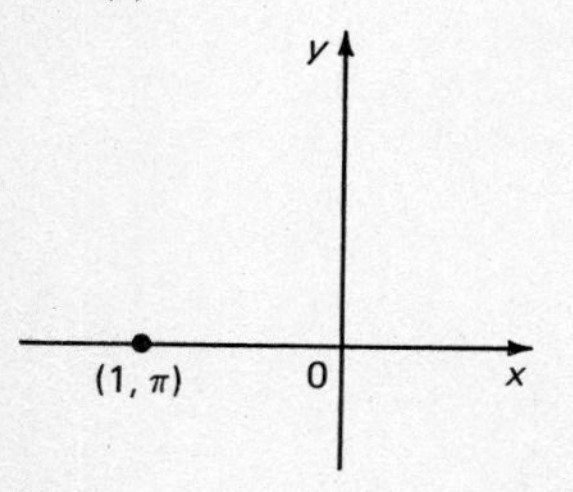

23. $3/2 + (3\sqrt{3}/2)i$

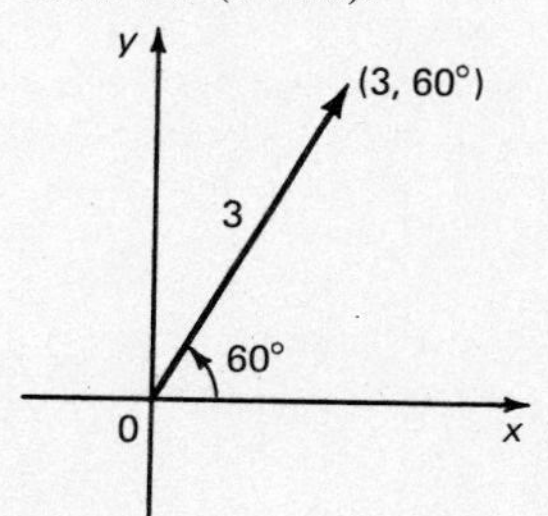

25. $(-5\sqrt{3}/2) + (5/2)i$

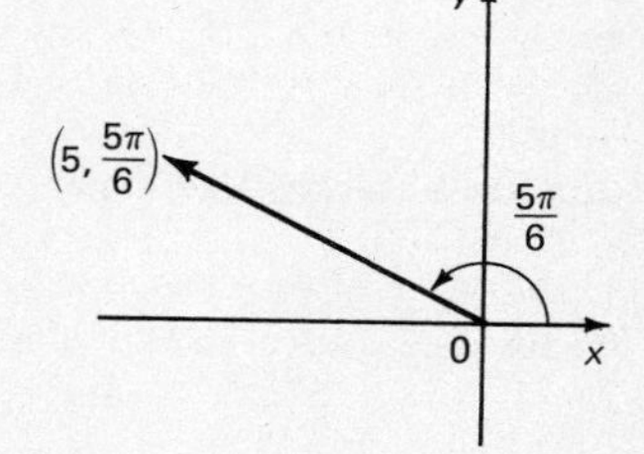

27. $-i$

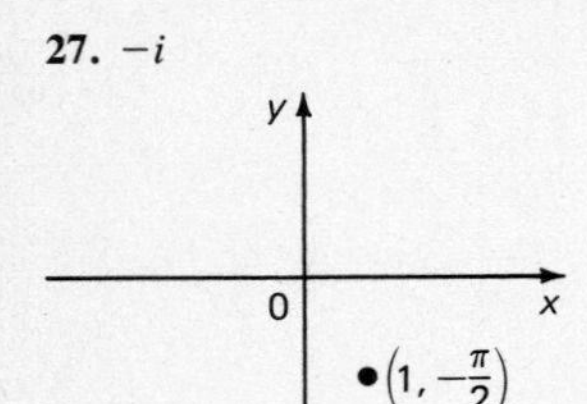

29. $2\sqrt{2}\,(\cos 45° + i \sin 45°)$

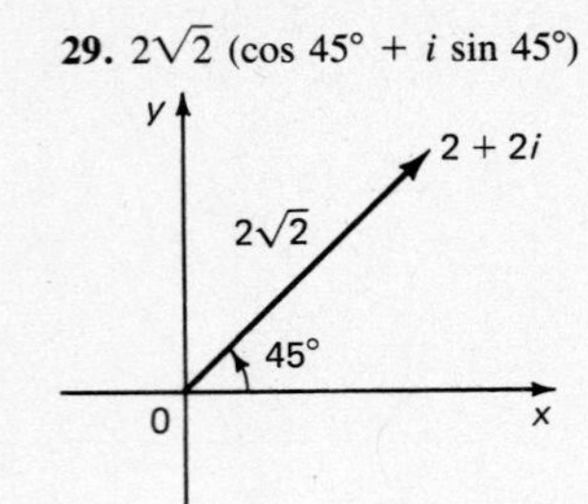

31. $3(\cos 180° + i \sin 180°)$

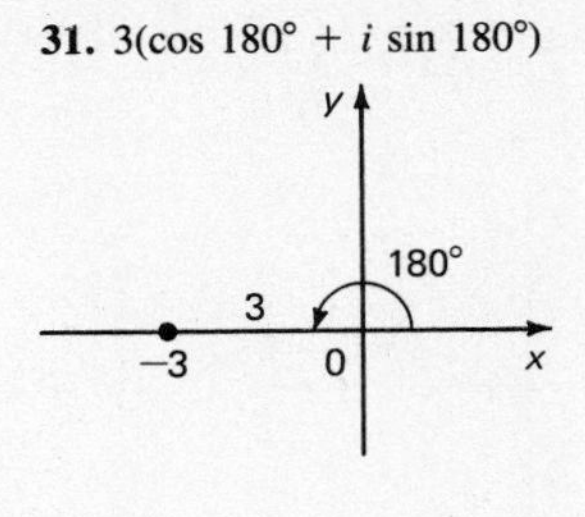

33. $1(\cos 300° + i \sin 300°)$

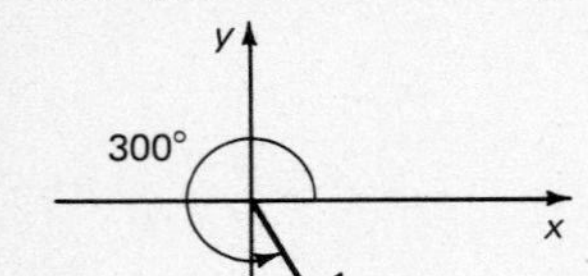

35. $\sqrt{2}(\cos 135° + i \sin 135°)$

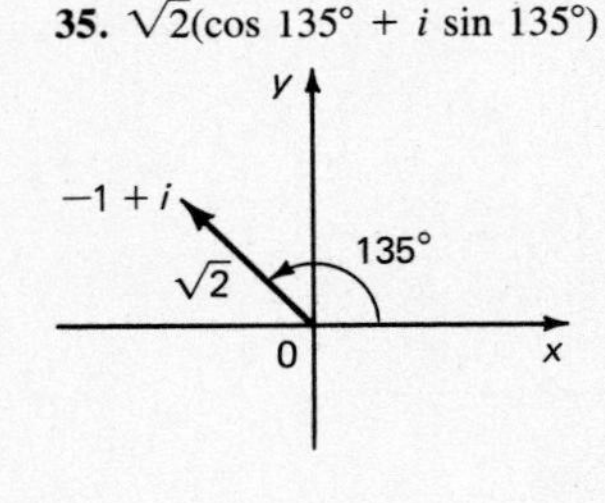

37. $2(\cos 270° + i \sin 270°)$

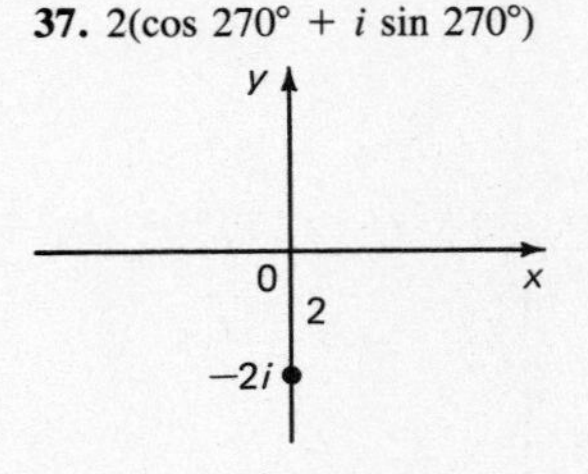

39. $\sqrt{10}(\cos 108.4° + i \sin 108.4°)$.

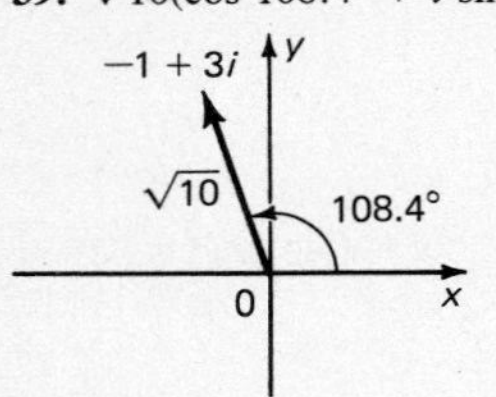

41. $0.423 + 0.906i$ **43.** $2.82 + 1.03i$ **45.** $0.383 + 0.924i$ **47.** $r = 2,\ \theta = 210°$ **49.** $r = 2,\ \theta = 5\pi/3$ **51.** $0°,\ -\infty < r < \infty$; $90°,\ -\infty < r < \infty$ **53.** $6(\cos 90° + i \sin 90°)$ **55.** $8(\cos \pi/2 + i \sin \pi/2)$ **57.** $1(\cos \pi/2 + i \sin \pi/2)$ **59.** $(1/4)(\cos 45° + i \sin 45°)$ **61.** $3(\cos \pi/2 + i \sin \pi/2)$ **63.** $5(\cos \pi/12 + i \sin \pi/12)$

Exercises 9.3

1. $4(\cos 4\pi/3 + i \sin 4\pi/3)$ **3.** $(1/16)(\cos 120° + i \sin 120°)$ **5.** $\cos 3\pi/2 + i \sin 3\pi/2$ **7.** $\cos 0° + i \sin 0°$ **9.** $-2 - 2i$ **11.** 64 **13.** $\sqrt{2}/2 - (\sqrt{2}/2)i$ **15.** -4 **17.** -16 **19.** $-1/2 - (\sqrt{3}/2)i$ **21.** $\sin 3\theta = 3 \sin \theta - 4 \sin^3\theta$ **23.** $\sin 4\theta = 4 \sin \theta \cos \theta - 8 \sin^3\theta \cos \theta$ **33.** $-\sqrt{3} + i,\ \sqrt{3} - i$ **35.** $\sqrt{2} - i\sqrt{2},\ -\sqrt{2} + i\sqrt{2}$ **37.** $\sqrt{3} + i,\ -\sqrt{3} + i,\ -2i$ **39.** $4i,\ -2\sqrt{3} - 2i,\ 2\sqrt{3} - 2i$ **41.** $3i,\ (3\sqrt{3}/2) - (3/2)i,\ (-3\sqrt{3}/2) - (3/2)i$ **43.** $1.53 + 1.29i,\ -1.88 + 0.68i,\ 0.35 - 1.97i$ **45.** $1.93 + 0.52i,\ -1.93 - 0.52i,\ -0.52 + 1.93i,\ 0.52 - 1.93i$ **47.** $\sqrt{2} + i\sqrt{2},\ -1.93 + 0.52i,\ 0.52 - 1.93i$ **49.** $0.77 + 0.64i,\ -0.77 - 0.64i,\ -0.17 + 0.98i,\ 0.17 - 0.98i,\ -0.94 + 0.34i,\ 0.94 - 0.34i$ **51.** $2,\ -2,\ 2i,\ -2i$ **53.** $-2 + 2i\sqrt{3},\ 2 - 2i\sqrt{3}$ **55.** $\sqrt{2} + i\sqrt{2},\ -\sqrt{2} - i\sqrt{2},\ -\sqrt{2} + i\sqrt{2},\ \sqrt{2} - i\sqrt{2}$ **57.** $0.35 + 1.97i,\ -1.88 - 0.68i,\ 1.53 - 1.29i$ **59.** $2,\ -1 + i\sqrt{3},\ -1 - i\sqrt{3}$ **61.** $1/2 + (\sqrt{2}/2)i,\ -1/2 - (\sqrt{2}/2)i,\ -1/2 + (\sqrt{2}/2)i,\ 1/2 - (\sqrt{2}/2)i$ **63.** $3,\ -3,\ 3i,\ -3i$ **65.** $\cos \pi/5 + i \sin \pi/5,\ \cos 3\pi/5 + i \sin 3\pi/5,\ -1,\ \cos 7\pi/5 + i \sin 7\pi/5,\ \cos 9\pi/5 + i \sin 9\pi/5$ **67.** $-2,\ -2(\cos 2\pi/5 + i \sin 2\pi/5),\ -2(\cos 4\pi/5 + i \sin 4\pi/5),\ -2(\cos 6\pi/5 + i \sin 6\pi/5),\ -2(\cos 8\pi/5 + i \sin 8\pi/5)$

Chapter 9 Review Exercises

1. $x = -1,\ y = 4$ **3.** $x = 3,\ y = 4$ **5.** (a) $5 - i$ (b) $-3 + 3i$ (c) $6 + 2i$ (d) $1/10 + (3/10)i$ **7.** (a) 4 (b) $10i$ (c) 29 (d) $-21/29 + (20/29)i$ **9.** $12(\cos 120° + i \sin 120°)$ **11.** $5(\cos \pi/2 + i \sin \pi/2)$

13.

$z = \left(1, \frac{2\pi}{3}\right)$ $\frac{2\pi}{3}$ 1 0 x y $\bar{z} = \left(1, \frac{4\pi}{3}\right)$ $-z = \left(1, \frac{5\pi}{3}\right)$

15.

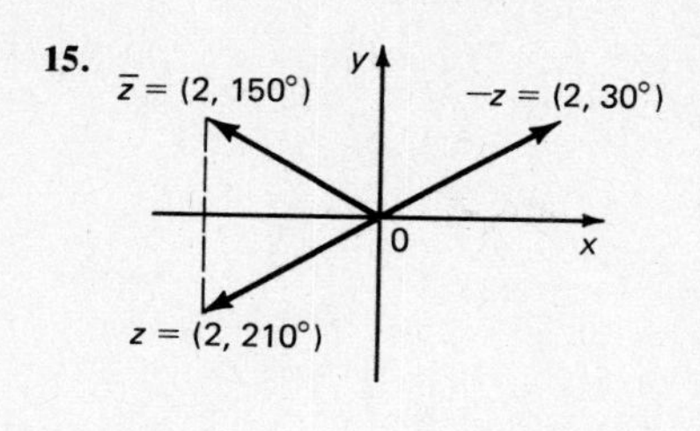

17. $\sqrt{3} + i$; argument, 30°; modulus, 2

19. $2(\cos 60° + i \sin 60°)$; argument, 60°; modulus, 2

21.

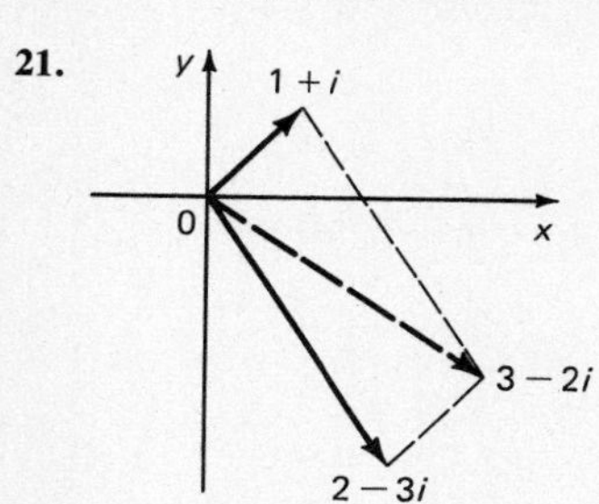

23. -1
25. $16 + 16i\sqrt{3}$
27. $\sqrt{3} + i,\ -\sqrt{3} - i,\ -\sqrt{3} + i,\ \sqrt{3} - i,\ 2i,\ -2i$
29. $\sqrt{6} + i\sqrt{2},\ -\sqrt{6} - i\sqrt{2}$

Chapter 9 Miscellaneous Exercises

13. $1, -1, \sqrt{2}/2 + i\sqrt{2}/2, -\sqrt{2}/2 - i\sqrt{2}/2, -\sqrt{2}/2 + i\sqrt{2}/2, \sqrt{2}/2 - i\sqrt{2}/2, i, -i$. 1 and -1 are square roots and 1, -1, i, and $-i$ are fourth roots. If $z^2 = 1$, then $z^8 = (z^2)^4 = 1^4 = 1$, so that square roots are also eighth roots. If $z^4 = 1$, then $z^8 = (z^4)^2 = 1^2 = 1$, so that fourth roots are also eighth roots.

CHAPTER 10

Exercises 10.1

1. 5, 8, 11, 14, 17 **3.** 1/2, 1/3, 1/4, 1/5, 1/6 **5.** $-1, 1, -1, 1, -1$ **7.** 1, 1/2, 1/6, 1/24, 1/120 **9.** -1, 1/2, $-1/6$, 1/24, $-1/120$ **11.** $5n - 2$ **13.** $2n + 3$ **15.** $n/(n + 1)$ **17.** $(-n)^n$ **19.** x^{2n} **21.** $1/nx$ **23.** $1 + 2 + 3 + 4 + 5$ **25.** $1(3) + 2(4) + 3(5) + 4(6)$ **27.** $2 + 3/2 + 2/3$ **29.** $1 + \sqrt[3]{2} + \sqrt[3]{3} + \sqrt[3]{4}$ **31.** $1 + 1/8 + 1/27$ **33.** $x + 2x^2 + 3x^3 + 4x^4 + 5x^5$ **35.** $x + 2x^2 + 6x^3 + 24x^4 + 120x^5 + 720x^6$ **37.** $\Sigma_{n=1}^{n=7} 2n$ **39.** $\Sigma_{n=1}^{n=5} 2^n$ **41.** $\Sigma_{n=1}^{n=9} (3n + 1)$ **43.** $\Sigma_{n=1}^{n=5} (3 - 2n)$ **45.** $\Sigma_{n=1}^{n=8} (-1)^n 11n$ **47.** $\Sigma_{n=1}^{n=12} x^n$ **49.** 40, 80, 120, 160, 200 **51.** 32°, 41°, 50°, 59°, 68°

Exercises 10.2

1. $d = 2, a_n = 2n + 1$ **3.** $d = 3, a_n = 3n - 14$ **5.** $d = 3/2, a_n = 3n/2 + 1/2$ **7.** $d = y, a_n = x + (n - 1)y$ **9.** $d = -2, a_n = 2x - 2n + 1$ **11.** $n = 11, S_{11} = 165$ **13.** $d = 2, S_{16} = 368$ **15.** $a_1 = 3, a_{12} = 47$ **17.** $a_1 = 3, S_{13} = 429$ **19.** $a_1 = 20, S_{13} = 26$ **21.** $d = 3, a_n = 3n - 1, S_{13} = 260$ **23.** $d = -6$, $a_n = 17 - 6n, S_7 = -49$ **25.** $d = -0.4, a_n = 3.4 - 0.4n, S_7 = 12.6$ **27.** 110 **29.** (a) 304 feet (b) 1600 feet

Exercises 10.3

1. $r = 2$; 24, 48, . . . , $3(2^{n-1})$ **3.** $r = 3$; 54, 162, . . . , $2(3^{n-1})$ **5.** $r = 3/4$; 27/8, 51/32, . . . , $8(3/4)^{n-1}$ **7.** $r = \sqrt{2}$; $2\sqrt{2}$, 4, . . . , $(\sqrt{2})^{n-1}$ **9.** $r = y$; xy^3; xy^4, . . . , xy^{n-1} **11.** $r = 3$; $a_{10} = 2(3^9)$ **13.** $r = -1/2$; $a_{10} = 3(-1/2)^9$ **15.** $r = z$; $a_{10} = xz^9/y$ **17.** (a) $r = -1/2$, (b) 11/16, (c) 2/3 **19.** (a) $r = 2/3$, (b) 211/27, (c) 9 **21.** (a) $r = -3/5$, (b) 421/125, (c) 25/8 **23.** (a) $r = 0.1$, (b) 4.4444, (c) $40/9 \approx 4.444 \ldots$

25. (a) $r = -1/10$, (b) 100001/550, (c) 2000/11 **27.** $n = 5$, $S_5 = 484/9$ **29.** $a_1 = 4$, $S_5 = 11/4$ **31.** 7/9 **33.** 1/11 **35.** 73/99 **37.** 1/7 **39.** 203/99 **41.** $x^{n-1}/(n-1)!$ **43.** $(-1)^{n-1}x^n/n$ **45.** $(-1)^{n-1}x^{2n-1}/(2n-1)!$ **47.** $(-1)^{n-1}x^{2n-1}/(2n-1)$ **51.** (a) 372 feet (b) 384 feet **53.** $64/729 \approx 8.8\%$

Exercises 10.5

1. $x^6 + 6x^5y + 15x^4y^2 + 20x^3y^3 + 15x^2y^4 + 6xy^5 + y^6$
3. $x^8 + 8x^7y + 28x^6y^2 + 56x^5y^3 + 70x^4y^4 + 56x^3y^5 + 28x^2y^6 + 8xy^7 + y^8$
5. $x^{10} - 10x^9y + 45x^8y^2 - 120x^7y^3 + 210x^6y^4 - 252x^5y^5 + 210x^4y^6 - 120x^3y^7 + 45x^2y^8 - 10xy^9 + y^{10}$
7. $x^5 + 10x^4y + 40x^3y^2 + 80x^2y^3 + 80xy^4 + 32y^5$ **9.** $8x^3 - 36x^2y + 54xy^2 - 27y^3$ **11.** $x^3 + 3x^2y + 3xy^2 + y^3$
13. $x^6 + 6x^5y + 15x^4y^2 + 20x^3y^3 + 15x^2y^4 + 6xy^5 + y^6$
15. $x^8 - 8x^7y + 28x^6y^2 - 56x^5y^3 + 70x^4y^4 - 56x^3y^5 + 28x^2y^6 - 8xy^7 + y^8$
17. $x^{10} - 10x^9y + 45x^8y^2 - 120x^7y^3 + 210x^6y^4 - 252x^5y^5 + 210x^4y^6 - 120x^3y^7 + 45x^2y^8 - 10xy^9 + y^{10}$
19. $x^3 + 3x^2y + 3xy^2 + y^3$ **21.** $x^6 + 6x^5y + 15x^4y^2 + 20x^3y^3 + 15x^2y^4 + 6xy^5 + y^6$
23. $128x^7 + 448x^6y + 672x^5y^2 + 560x^4y^3 + 280x^3y^4 + 84x^2y^5 + 14xy^6 + y^7$
25. $4096x^{12} + 73728x^{11}y + 608256x^{10}y^2 + 304280x^9y^3 + \ldots$ **27.** $\ldots + 2880x^2y^8 - 2560xy^9 + 1024y^{10}$
29. $3003x^{10}y^5$ **31.** $11440x^7y^9$ **33.** 21 **35.** 70 **37.** $-117 - 44i$ **39.** $316 - 12i$
41. $65 + 142i$ **43.** $23 - (10\sqrt{2})i$ **45.** 1030301 **47.** 96,059,601 **49.** $1 + x/4 - 3x^2/32 + \ldots$
51. $1 - 3x + 6x^2 - \cdots$ **53.** $1 - x^2 + x^4 - \ldots$ **55.** 3.16 **57.** 1.005
59. $x^3 + 3x^2y + 3xy^2 + y^3 + 3x^2z + 6xyz + 3y^2z + 3xz^2 + 3yz^2 + z^3$
65. $7! = 5040$; $10! = 3{,}628{,}800$; $12! = 479{,}001{,}600$

Chapter 10 Review Exercises

1. 11, 14, 17 . . . , $3n - 4$ **3.** $1/5\sqrt{5}, -1/6\sqrt{6}, 1/7\sqrt{7}, \ldots, (-1)^{n-1}/n\sqrt{n}$
5. $x^4/5^4, -x^5/5^5, x^6/5^6, \ldots, (-1)^{n-1}x^{n-1}/5^{n-1}$ **7.** $d = 3$, $a_8 = 23$, $S_8 = 100$ **9.** $d = 3$, $a_{10} = 16$, $S_{10} = 25$
11. $d = 3/4$, $a_5 = x + 3$, $S_5 = 5(2x + 3)/2$
13. $r = 2$, $a_5 = 80$, $S_5 = 155$; $\sum_{n=1}^{5} 5(2)^{n-1}$ **15.** $r = 3$, $a_6 = 1215$, $S_6 = 1820$; $\sum_{n=1}^{6} 5(3)^{n-1}$ **17.** $r = 2/3$, $a_4 = 8/81$, $S_4 = 19/27$; $\sum_{n=1}^{4} \frac{1}{3}\left(\frac{2}{3}\right)^{n-1}$ **19.** $r = -1/2$, $a_7 = 1/64$, $S_7 = 43/64$; $\sum_{n=1}^{7} (-1)^{n-1}\left(\frac{1}{2}\right)^{n-1}$
21. $r = 0.1$, $a_5 = 0.0002$, $S = 2.2222$; $\sum_{n=1}^{5} 2(0.1)^{n-1}$ **23.** 4/3 **25.** 4 **27.** 28/99 **29.** (a) \$11,000 (b) \$47,500 **31.** \$4095 **33.** 1364
39. $x^9 + 9x^8y + 36x^7y^2 + 84x^6y^3 + 126x^5y^4 + 126x^4y^5 + 84x^3y^6 + 36x^2y^7 + 9xy^8 + y^9$
41. $x^6 - 18x^5y + 135x^4y^2 - 540x^3y^3 + 1215x^2y^4 - 1458xy^5 + 729y^6$ **43.** $122 - 597i$ **45.** -4
47. $1 - x + x^2 - x^3 + \cdots$ **49.** 6.86

Chapter 10 Miscellaneous Exercises

3. 15 **5.** 45 **7.** 1 **9.** n **11.** 35 **13.** 6

Index